Handbook of Fuel Cells

Fundamentals Technology and Applications

EDITORIAL BOARD

EDITORS

Wolf Vielstich
IQSC
São Carlos
Universidade de São Paulo
Brazil

Arnold Lamm
DaimlerChrysler Research and Technology
Ulm
Germany

Hubert A. Gasteiger
Fuel Cell Activities
General Motors Corporation
Honeoye Falls
NY, USA

INTERNATIONAL ADVISORY BOARD

Masayuki Dokiya
Yokohama National University
Japan

Thomas F. Fuller
UTC Fuel Cells
South Windsor
CT, USA

Andrew Hamnett
University of Strathclyde
Glasgow, Scotland

Teresa Iwasita
Universidade de São Paulo
Brazil

Karl V. Kordesch
Technical University Graz
Austria

Anthony B. LaConti
Giner Inc, Newton
MA, USA

Gerd Sandstede
Consultant, Physikalischer Verein
Frankfurt-am-Main, Germany

Günther G. Scherer
Paul Scherrer Institut
Villigen, Switzerland

J. Robert Selman
Illinois Institute of Technology
Chicago, IL, USA

Harumi Yokokawa
National Institute of Advanced
Industrial Science and Technology
Ibaraki, Japan

Handbook of Fuel Cells

Fundamentals Technology and Applications

VOLUME 1
Fundamentals and Survey of Systems

Editors

Wolf Vielstich
IQSC, São Carlos, Universidade de São Paulo, Brazil

Arnold Lamm
DaimlerChrysler Research and Technology, Ulm, Germany

Hubert A. Gasteiger
Fuel Cell Activities, General Motors Corporation, Honeoye Falls, NY, USA

Wiley

Copyright © 2003 John Wiley & Sons Ltd,
The Atrium,
Southern Gate,
Chichester,
West Sussex
PO19 8SQ, England

Telephone   (+44) 1243 779777

Email (for orders and customer service enquiries): cs-books@wiley.co.uk
Visit our Home Page on www.wileyeurope.com or www.wiley.com

Reprinted in 2004, August 2005

All Rights Reserved. No part of this publication may be reproduced, stored in a retrieval system or transmitted in any form or by any means, electronic, mechanical, photocopying, recording, scanning or otherwise, except under the terms of the Copyright, Designs and Patents Act 1988 or under the terms of a licence issued by the Copyright Licensing Agency Ltd, 90 Tottenham Court Road, London W1T 4LP, UK, without the permission in writing of the Publisher. Requests to the Publisher should be addressed to the Permissions Department, John Wiley & Sons Ltd, The Atrium, Southern Gate, Chichester, West Sussex PO19 8SQ, England, or emailed to permreq@wiley.co.uk, or faxed to (+44) 1243 770620.

This publication is designed to provide accurate and authoritative information in regard to the subject matter covered. It is sold on the understanding that the Publisher is not engaged in rendering professional services. If professional advice or other expert assistance is required, the services of a competent professional should be sought.

Where articles in the Handbook of Fuel Cells have been written by government employees in the United States of America, please contact the publisher for information on the copyright status of such works, if required. Works written by US government employees and classified as US Government Works are in the public domain in the United States of America.

The Editor(s)-in-Chief, Advisory Board and Contributors have asserted their right under the Copyright, Designs and Patents Act, 1988, to be identified as the Editor(s)-in-Chief and Advisory Board of and Contributors to this work.

Cover Images
Foreground: Cutaway view of General Electric PEM fuel cell system for Gemini spacecraft, 3 stacks of 32 cells in parallel. (Reproduced by permission of Schenectady Museum.)
Background: Scheme of a H_2/O_2 fuel cell.

Other Wiley Editorial Offices

John Wiley & Sons Inc., 111 River Street,
Hoboken, NJ 07030, USA

Jossey-Bass, 989 Market Street,
San Francisco, CA 94103-1741, USA

Wiley-VCH Verlag GmbH, Boschstr. 12,
D-69469 Weinheim, Germany

John Wiley & Sons Australia Ltd, 33 Park Road,
Milton, Queensland 4064, Australia

John Wiley & Sons (Asia) Pte Ltd, 2 Clementi Loop #02-01,
Jin Xing Distripark, Singapore 129809

John Wiley & Sons Canada Ltd, 22 Worcester Road,
Etobicoke, Ontario, Canada M9W 1L1

Wiley also publishes its books in a variety of electronic formats. Some content that appears in print may not be available in electronic books.

British Library Cataloguing in Publication Data

A catalogue record for this book is available from the British Library

ISBN-10: 0-471-49926-9 (H/B)
ISBN-13: 978-0-471-49926-8 (H/B)

Typeset in 10/12.5pt Times by Laserwords Private Limited, Chennai, India
Printed and bound in Great Britain by Antony Rowe, Chippenham, Wiltshire.
This book is printed on acid-free paper responsibly manufactured from sustainable forestry
in which at least two trees are planted for each one used for paper production.

Contents

VOLUME 1: Fundamentals and Survey of Systems

Contributors to Volume 1	vii
Foreword	ix
Preface	xiii
Abbreviations and Acronyms	xv

Part 1: Thermodynamics and kinetics of fuel cell reactions — 1

1. The components of an electrochemical cell — 3
   A. Hamnett
2. The electrode–electrolyte interface — 13
   A. Hamnett
3. Thermodynamics of electrodes and cells — 21
   A. Hamnett
4. Ideal and effective efficiencies of cell reactions and comparison to carnot cycles — 26
   W. Vielstich
5. Kinetics of electrochemical reactions — 31
   A. Hamnett
6. Introduction to fuel-cell types — 36
   A. Hamnett

Part 2: Mass transfer in fuel cells — 45

7. Mass transfer at two-phase and three-phase interfaces — 47
   A. Weber/R. Darling/J. Meyers/J. Newman
8. Mass transfer in flow fields — 70
   K. Scott

Part 3: Heat transfer in fuel cells — 97

9. Low temperature fuel cells — 99
   J. Divisek
10. High temperature fuel cells — 115
    J. Divisek
11. Air-cooled PEM fuel cells — 134
    R. von Helmolt/W. Lehnert

Part 4: Fuel cell principles, systems and applications — 143

12. History of low temperature fuel cells — 145
    G. Sandstede/E. J. Cairns/V. S. Bagotsky/K. Wiesener
13. History of high temperature fuel cell development — 219
    H. Yokokawa/N. Sakai
14. Hydrogen/oxygen (Air) fuel cells with alkaline electrolytes — 267
    M. Cifrain/K. Kordesch
15. Hydrazine fuel cells — 281
    H. Kohnke
16. Phosphoric acid electrolyte fuel cells — 287
    J. M. King/H. R. Kunz
17. Aqueous carbonate electrolyte fuel cells — 301
    E. J. Cairns
18. Direct methanol fuel cells (DMFC) — 305
    A. Hamnett
19. Other direct-alcohol fuel cells — 323
    C. Lamy/E. M. Belgsir
20. Solid oxide fuel cells (SOFC) — 335
    P. Holtappels/U. Stimming
21. Biochemical fuel cells — 355
    E. Katz/A. N. Shipway/I. Willner
22. Metal/air batteries: The zinc/air case — 382
    O. Haas/F. Holzer/K. Müller/S. Müller
23. Seawater aluminum/air cells — 409
    J. P. Iudice de Souza/W. Vielstich
24. Energy storage via electrolysis/fuel cells — 416
    J. Divisek/B. Emonts

Contents for Volumes 2, 3 and 4	433
Subject Index	439

Contributors to Volume 1

V. S. Bagotsky
Mountain View, CA, USA

E. M. Belgsir
Université de Poitiers, France

E. J. Cairns
Lawrence Berkeley National Laboratory, and University of California, Berkeley, CA, USA

M. Cifrain
Graz University of Technology, Graz, Austria

R. Darling
UTC Fuel Cells, South Windsor, CT, USA

J. Divisek
Institute for Materials and Processes in Energy Systems, Forschungszentrum Jülich GmbH, Jülich, Germany

B. Emonts
Institute for Materials and Processes in Energy Systems, Forschungszentrum Jülich GmbH, Jülich, Germany

O. Haas
Paul Scherrer Institute, Villigen, Switzerland

A. Hamnett
University of Strathclyde, Glasgow, UK

P. Holtappels
Swiss Federal Institute for Materials Testing and Research (EMPA), Duebendorf, Switzerland

F. Holzer
Paul Scherrer Institute, Villigen, Switzerland

J. P. Iudice de Souza
Universidade Federale do Pará, Belém, Brazil

E. Katz
Institute of Chemistry and The Farkas Center for Light-Induced Processes, The Hebrew University of Jerusalem, Jerusalem, Israel

J. M. King
Manchester, CT, USA

H. Kohnke
i.H. Gaskatel, Kassel, Germany

K. Kordesch
Graz University of Technology, Graz, Austria

H. R. Kunz
Department of Chemical Engineering, University of Connecticut, Storrs, CT, USA

C. Lamy
Université de Poitiers, France

W. Lehnert
Center for Solar Energy and Hydrogen Research Baden-Wurttemberg, Ulm, Germany

J. Meyers
UTC Fuel Cells, South Windsor, CT, USA

K. Müller
Paul Scherrer Institute, Villigen, Switzerland

S. Müller
Euresearch, Bern, Switzerland

J. Newman
University of California, Berkeley, CA, USA

G. Sandstede
Physikalischer Verein von 1824 and Sandstede-Technologie-Consulting, Frankfurt am Main, Germany

N. Sakai
National Institute of Advanced Industrial Science and Technology, Ibaraki, Japan

K. Scott
University of Newcastle upon Tyne, Newcastle upon Tyne, UK

A. N. Shipway
Institute of Chemistry and The Farkas Center for Light-Induced Processes, The Hebrew University of Jerusalem, Jerusalem, Israel

U. Stimming
Technische Universität München, Garching, Germany

W. Vielstich
IQSC, São Carlos, Universidade de São Paulo, Brazil

R. von Helmolt
Adam Opel AG – Global Alternative Propulsion Center, Rüsselsheim, Germany

A. Weber
University of California, Berkeley, CA, USA

K. Wiesener
Kurt Schwabe Institut für Sensortechnik Meinsberg and Elektrochemische Energie, Dresden, Germany

I. Willner
Institute of Chemistry and The Farkas Center for Light-Induced Processes, The Hebrew University of Jerusalem, Jerusalem, Israel

H. Yokokawa
National Institute of Advanced Industrial Science and Technology, Ibaraki, Japan

Foreword

From the perspective of a newly formed fuel cell company (MTI Microfuel Cells, started early in 2001), the need at this point in time of an updated, comprehensive Handbook of Fuel Cells, is very clear. With the extensive developments in fuel cell science and technology during the last 10–15 years, most presently available sources of technical and scientific information in this area have become significantly outdated. The need for an updated source book, providing in-depth coverage of science and engineering aspects of all fuel cell technologies developed to date, has been strongly felt. One immediately noticeable feature of this updated collection of fuel cell technical material, is the significant component devoted to aspects of polymer electrolyte fuel cell (PEFC) technology. Strong recent development of PEFC science and engineering has brought it to "center stage" by the time this Handbook is published (2003) and this is in clear distinction from earlier reference books on fuel cells. These recent developments in PEFC technology must have also been behind the choice of the Editors to include a volume exclusively devoted to Electrocatalysis, as called for by the central role of interfacial catalysis in PEFC, as well as other low temperature fuel cell technologies. The Editors chose to provide the user of this Handbook with an extensive collection of individually authored chapters, resulting in a spectrum of insights and approaches. The list of contributors also reflects the significant geographical distribution of frontier R&D work performed in this technical area, a clear testimony for worldwide interest.

It is the highly interdisciplinary nature of this field of technology, that calls for such a unified, extensive source, covering the wide spectrum of all relevant scientific and technical aspects. Mechanical and electronics engineers in the fuel cell R&D community are provided here with relevant chapters in electrochemical thermodynamics, kinetics and catalysis, while, at the same time, electrochemists in the fuel cell community are provided with authoritative chapters on stack, fuel processor and power system engineering. Last but not least, effective introduction of newcomers into the community of fuel cell technologists will be particularly well assisted by the four volumes of this Handbook.

The interest in fuel cells has, for quite some time, spread well beyond the community of expert scientists and engineers. The technology has quickly become the subject of many discussions in the press and the electronic media, focusing primarily on fuel cells as the possible key for clean and efficient vehicles, and/or clean and efficient generation of electric power. Having received significant endorsement and having been so much in the public eye, it takes the very comprehensive collection of fuel cell technical material contained in the four volumes of this Handbook, to appreciate the science base and the magnitude of R&D effort that has provided the present foundation for fuel cell technology. Although not always fully appreciated, this science and technology base has clearly been the prerequisite for the development of the significant industrial interest and public awareness and endorsement.

The fundamental technical basis for the "aura" of the fuel cell, is a combination of high energy conversion efficiency and a potential for strong reduction of power source emissions. As an electrochemical device that converts the chemical energy of a fuel directly to electric energy, the fuel cell has some intrinsically highly attractive features, the most important of which is high efficiency achievable even at temperatures under $100\,^\circ C$. The theoretical conversion efficiency exceeds 80% for a fuel cell directly converting hydrogen, or methanol fuel to electric power at operation temperatures typical for a polymer electrolyte fuel cell, i.e., $30\,^\circ C$–$100\,^\circ C$, whereas the practical conversion efficiency ranges in different systems between 35% and 70%. Given that the practical energy conversion efficiency of heat engines is typically well below 30% and approaches 50% only for large gas turbines operating at much higher temperatures, a fuel cell operating near ambient temperature at higher efficiency independent on cell size, is obviously a highly attractive alternative. The argument for the low temperature of operation is particularly strong in transportation and portable power applications. In such applications, fast "cold start-up" is of high value and so is the elimination of excessive losses and materials challenges associated with thermal cycles of large amplitude.

Significant environmental benefits are expected from the introduction of fuel cells as new power source technology for transportation and/or electric power generation. These include lowering, or eliminating altogether hazardous tail pipe emissions and, in addition, reducing greenhouse gas emissions. No nitrogen oxide (NOx), a major pollutant emitted by internal combustion engines, will be generated by a fuel cell operating at a much lower temperature than that of an internal combustion engine. Other hazardous emissions, e.g., carbon monoxide, are also drastically lowered in a fuel cell powered vehicle and are, in fact, brought down to zero ("zero emission vehicle") when hydrogen is used as fuel for the PEFC, as water vapor becomes the only exhaust from the power system. Finally, CO_2 emissions per mile driven could be lowered by the ratio of efficiencies of combustion and electrochemical conversions of fuel energy. This could mean a potential drop in CO_2 emissions by as much as 50% per mile driven, if all cars were converted from the internal combustion engine to fuel cell power.

In addition to the PEFC, the other fuel cell technology which has received increasing attention most recently, is the solid oxide fuel cell (SOFC). This high temperature technology has the important advantage of operation on common hydrocarbon fuels, such as natural gas, and, at the same time, the potential to reach energy conversion efficiencies, fuel-to-electric power + heat, well exceeding 60%. The latter feature is enabled by the high value of heat generated at temperatures near 1000 °C. Very recent developments in SOFC technology, based on thinning down of oxide electrolyte layers and optimizing of electrode compositions and structures, are opening the door to SOFC's operating at lower temperatures than possible hitherto (as low as 600 °C), thereby enabling use of metal hardware components and, consequently, significant simplification of stack materials and fabrication. SOFC technology appears in 2003 as the lead among future high temperature fuel cell technologies, playing a role similar to that played recently by the PEFC among low temperature fuel cell technologies. Having made that statement, care must be exercised not to neglect the potential of other fuel cell technologies, particularly when some specific applications are considered. This may be true, for example, for the remaining interest in alkaline fuel cells for applications based on pure hydrogen fuel, such interest being driven primarily by the much smaller dependence of this low temperature technology on precious metal catalysts. Accordingly, the alkaline fuel cell, as well as the phosphoric acid and molten carbonate fuel cells, are all significantly covered in the four volumes of this Handbook.

From a year 2003 perspective, some further comments are in order on the potential application of smaller scale fuel cells for portable power. While transportation and off-grid power generation have been highlighted to date as the significant potential applications of PEFCs and/or SOFCs, the application of PEFCs and, particularly, of direct methanol fuel cells (DMFCs) as small scale, portable power sources, may actually provide the first market entry for low temperature fuel cells. The latter potential application enjoyed a significant increase in interest and activity starting around the year 2000, driven by a clear need in the consumer electronics industry for power sources of energy density exceeding that of secondary Li batteries. Such advanced power sources are required to power the present and, particularly, the next generation multifunctional hand held devices, targeting longer use times per recharge (or, rather, refuel). Portable power systems of such superior energy density and immediate refuelability, would serve particularly well in uses of prolonged duration away from re-charging sources. The projections of faster market entry in this area of fuel cell technology, are based on a real potential to significantly exceed the energy density of present portable power sources and, at the same time, the ability to compete favorably with the cost of present battery technology.

And with all this great potential of fuel cell technology having been revealed and confirmed in recent years, when trying to capture the actual "state of fuel cell technology in the year 2003", two facets of almost opposite nature stand out. On the one hand, we are looking at a very impressive list of technical advances and a significant number of recent industrial demonstrations of prototype fuel cell power systems. On the other hand, some significant barriers clearly remain at this point in time on the road to commercialization of any version of fuel cell technology. In the context of such remaining barriers and the probability of overcoming them, it is valuable to remember that PEFC technology – the leading fuel cell technology worldwide in terms of industrial investment at present – was considered fundamentally impractical for any terrestrial application as late as the mid-1980's. First introduced as part of the Gemini space program in the US, PEFCs developed for space applications used highly pure, cryogenic hydrogen and oxygen and very high precious metal loadings, as fully acceptable for power sources installed in a space vehicle. Terrestrial PEFC applications dictated, however, strong transition to a technology based on much lower precious metal loadings and using diluted and impure hydrogen as fuel and air, rather than oxygen. It also demanded the effective addressing of water redistribution in ionomeric membranes of PEFCs operating at high current densities. This list of challenges was considered by many in the relevant technical communities too daunting to deserve much effort. It was the strong belief on the part of a small number of engineers and scientists, in the real and important potential of polymer electrolyte fuel cells to provide "green and lean" power sources for terrestrial applications, that started in the mid-1980's some small scale but critical R&D and engineering efforts. The advancements in core technology that followed, were based on a series of technical breakthroughs between 1985 and 1995. These included development of effective PEFC membrane/electrode (MEA) technology based on precious metal loadings lower by more than an order of magnitude, effective approaches to PEFC operation with air and with impure hydrogen fuel, resolution to large degree of water management issues in the PEFC by use of thinner poly-perfluorocarbon sulfonic acid (PFSA) membranes, and, last but not least, the first demonstrations of power densities approaching 1 kW/liter in prototype PEFC stacks. All of the above, together with increasing public and industrial interest in "green" power sources, resulted in increase of the industrial investment in PEFC technology by three orders of magnitude between the mid-1980's and the mid-1990's. It is perhaps with this chapter of history of fuel cell technology as background, that barriers to fuel cell commercialization remaining in 2003, should be examined.

The prime barrier to market entry remaining at present, is the non-competitive cost. This is particularly true for PEFC technology applications in transportation, the largest potential market but the most demanding in terms of cost reduction. But the problem of a remaining cost barrier also applies to PEFC and SOFC technologies developed for distributed electric power generation. Can the record of success in advancing fuel cell technology during the last 20 years be further extended, to overcome remaining cost as well as any other identified remaining barriers? The answer must be "yes, at least to a large degree, as previous, higher cost and reliability barriers were overcome to a large degree through innovation originating within well targeted R&D projects. This latter resource is, in principle, not exhaustible. Admittedly, compared with the state of the technology 20 years ago, the remaining part of the "maturation curve" is much closer to the target asymptote and, consequently, the remaining road is more demanding and the pace may be relatively slower. Nevertheless, the recent history of developments in both PEFC and SOFC technologies, has been rife with innovative solutions which resolved highly daunting technical barriers and this record should provide the incentive for a successful remaining climb to the successful commercialization of fuel cell technology.

Reading through the chapters of this comprehensive Handbook, a researcher, or engineer working in this area should be able to find the updated, detailed background material describing the nature of the recent key developments in the relevant area of fuel cell science and engineering. He, or she, can then use these developments as the basis for further building of new, innovative technology tools and routes. Such an approach of inventing "within (or somewhat beyond) the box" proves, in many cases, more effective than inventing "completely out of the box". Along these lines, knowledge of past developments in a specific fuel cell technology of interest (e.g. PEFC) *and* in related fuel cell technologies (e.g. AFC and PAFC in that case), is essential for scientists and engineers new to the field to be effective; and the same might be said about the educational value of parallels between fuel processing catalysis and fuel cell electrocatalysis. The effective access consequently required to a wide scope of background material in this area of science and technology, is uniquely provided by this Handbook of Fuel Cells.

Having suggested an optimistic outlook for the resolution of remaining barriers to commercialization, one may still legitimately argue that "time to market" is becoming an

Figure 1. A UTC, PC 25™ 200 KW unit installed at Central Park, New York City, to power the local police station. (Reproduced by permission of UTC Fuel Cells.)

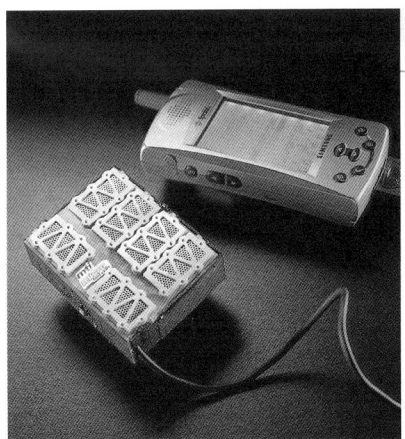

Figure 2. A developmental prototype of a direct methanol fuel cell (DMFC) power pack, shown here powering a "smart" cell phone. (Reproduced by permission of MTI Microfuel Cells.)

increasing challenge to fuel cell technology. Recent recognition of the probability of earlier entry of fuel cells into the market of portable power sources in the power range of 1 W-1 kW, makes this area likely to become a stronger focus of technology and product development. Considering this possibility, it is perhaps fit to end these introductory comments by presenting side-by-side the high and low ends of the power spectrum addressed by fuel cell technology: a 200 kW phosphoric acid power source fabricated by UTC Fuel Cells, (South Windsor, Connecticut; Figure 1), fueled by natural gas reformed upstream from the stack, and, next to it, a developmental prototype of a direct methanol fuel cell (DMFC) of output power near 1 W for consumer electronics applications, fabricated at MTI Microfuel Cells (Figure 2). The latter "micro" fuel-cell, is fueled from a methanol reservoir contained within the power pack shown in the photograph. The placing of these two photographs side by side, highlights the flexibility of fuel cell technology in accommodating a wide range of power demands and, at the same time, highlights an example of the last generation of fuel cell technology brought all the way to (niche) market next to an example of "something completely new and different" in this field of technology, signaling exciting potential future products.

Shimshon Gottesfeld
Chief Technology Officer and VP for R&D
MTI Microfuel Cells
Albany, New York, USA

Preface

Over the past few years, the direct conversion of chemical into electrical energy via fuel cells has been at the center of attention of electrochemical research and technology development. This is due not only to the scientifically fascinating complexity of fuel cell reactions and the general awareness of the technological potential of fuel cells, but is also a result of society's strive towards developing environmentally-friendly power generation. In the reaction between hydrogen and oxygen, for instance, the only chemical product is water. The fuel cell, initially developed in the sixties as an on-board power supply unit for spacecraft, has now found new applications in powering submarines, in decentralized power supply systems, in portable products, and in sensor technology. In the quest for a highly efficient, emission-free drive system, the development of mobile automotive fuel cell units is proving to be quite promising. The first bus series using gaseous hydrogen fuel will commence operations in 2003, initially in ten cities in eight European countries. This market introduction will be flanked by thirty European companies and universities, along with leading European enterprises from the petroleum and gas sectors. Activities on a similar scale are also being planned for Japan and the USA.

Fuel cell technology constitutes a highly varied, exacting interdisciplinary field which extends from the available fuels and their processing, through the fundamentals of electrochemical processes, especially electrocatalysis, right up to the numerous new concepts in systems technology for complete fuel cell aggregates including the control of gas, water and heat management.

Thousands of publications now appear annually in this field. The number of patent applications, especially from the industrial sector, has also risen sharply over the past five years. However, the monographs and symposium volumes recently published cannot provide an adequate overview of the entire field. This handbook, therefore, sets out to close this gap with its breadth and depth of content. The breadth is reflected in the wide range of topics which are intimately related to the design and application of fuel cell systems, as can be seen from the table of contents, while the depth is guaranteed by the choice of world-renowned experts on this subject.

Volume 1 conveys the fundamentals of electrochemistry, the underlying principles of mass and heat transfer in fuel cells, and, following a historical introduction, briefly presents the numerous fuel cell systems developed so far. Volume 2 is concerned with electrocatalysis, ranging from the theoretical fundamentals to practically all processes that take place within all different types of fuel cells, including the current understanding of their reaction mechanisms. This volume is the very first detailed presentation of such a crucial scientific field ever to be published and thus should be of keen interest well beyond the circle of scientists and engineers directly involved with fuel cells and their application.

Finally, in Volumes 3 and 4, the current state of development of materials and systems is presented in minute detail along with their practical applications. In addition, these volumes discuss current predictions of potential fuel cell markets and possible further technological and economical developments. Of key significance here is the interplay between technological progress and possible market penetrations of fuel cell systems based on economic considerations. The advantages and disadvantages of the various fuel cell systems are discussed in this context.

Materials research will remain a topic of key interest in the future; the objective in this regard is to enhance the efficiency of fuel cell stacks, which, by simplifying the overall system, will ultimately help bring about the necessary cost reductions. The search for improved catalysts is only at the beginning, as is the development of suitable membranes in the field of PEM technology. Even the matter of alternative automotive fuels for the future is still to be resolved. Although it is quite likely that hydrogen will become the automotive fuel of choice in the long term, the question still remains as to how the transformation to a hydrogen-based economy should be best effected. In the petroleum industry, endeavors are also intensifying world-wide to develop the production of synthetic gasoline. With a view to the direct electrochemical utilization of a synthetic fuel, ethylene glycol has proved to be on a par with methanol – even in acid solution – while the hope of directly converting ethanol has not yet abated.

Cost reduction in materials and components is still the greatest challenge facing the broad-based application of fuel cell systems. Considerable progress has been made here, notably in the automotive industry, e.g. with regard to noble metal content. However, especially in the proton exchange membrane (PEM) systems, the platinum loading must still be significantly reduced before market introduction can be effected on a broad basis. Of the low-temperature fuel cells which are considered for automotive applications, only the alkaline fuel cell (AFC) is currently showing any promise for the utilization of platinum-free catalysts. However, more efforts in development are required to increase the power density of AFCs to the level which has been achieved with PEMFCs. Regarding the required lifetime for stationary fuel cell systems, only the phosphoric acid fuel cell (PAFC) is currently capable of satisfying the required 40 000 hours. Considerable progress has been made with the molten carbonate fuel cell (MCFC), specifically with regard to longevity. Nevertheless, further improvements in service life are called for, not only in the case of the molten carbonate fuel cell, but also for the solid oxide fuel cell (SOFC) and the PEM fuel cell. Although a service life of only around 5 000 hours is now being aimed for in automotive applications, this target is yet to be reached as a result of the higher requirements on PEMFC systems regarding power density, dynamic behavior and operating temperature. Moreover, the inherent complexity of fuel cell systems must be reduced in order to minimize peripheral costs. Future units are therefore expected to undergo considerable simplification and improvement, particularly with regard to the power density of the entire fuel cell system.

Last but not least, we would like to express our sincere appreciation to the authors and to the Advisory Board for their invaluable contributions, as well as the publishing house for its interest and its willing assistance.

Wolf Vielstich
IQSC, São Carlos, Universidade de São Paulo, Brazil

Arnold Lamm
DaimlerChrysler Research and Technology, Ulm, Germany

Hubert A. Gasteiger
Fuel Cell Activities, General Motors Corporation, Honeoye Falls, NY, USA

Abbreviations and Acronyms

a.c.	Alternating Current	EC-NMR	Electrochemical Nuclear Magnetic Resonance
a.v.	Alternating Voltage	ECTDMS	Electrochemical Thermal Desorption Mass Spectroscopy
AES	Auger Electron Spectroscopy		
AFC	Alkaline Fuel Cell	EDX	Energy Dispersive X-Ray Spectroscopy
AFM	Atomic Force Microscopy	EIS	Electrochemical Impedance Spectroscopy
APU	Auxiliary Power Unit	EMF	Electromotive Force
ATR	Autothermal Reforming	EMIRS	Electrochemically Modulated Infrared Spectroscopy
ATS	Advanced Turbine System		
		EPMA	Electron Probe Microanalysis
BET	Brunner-Emmet-Teller Adsorption	ePTFE	Expanded Polytetrafluoroethylene
BIOS	Basic Input/Output System	EQMB	Electrochemical Quartz Crystal Microbalance
BOP	Balance-of-Plant	ESCA	Electron Spectroscopy for Chemical Analysis
		ESR	Electron Spin Resonance
CCM	Catalyst-Coated-Membrane	EVD	Electrochemical Vapor Deposition
CCS	Catalyst-Coated-Substrate	EW	Equivalent Weight
CE	Counter Electrode	EXAFS	Extended X-Ray Absorption Fine Structure Spectroscopy
CEM	Combined Compressor Expander Machine		
CGH	Compressed Gaseous Hydrogen		
CHP	Combined Heat and Power	FC	Fuel Cell
CNG	Compressed Natural Gas	FCPS	Fuel Cell Power System
CoTMPP	Cobalt Tetramethoxyphenylporphyrin	FCV	Fuel Cell Vehicle
CSTR	Continuously Stirred Tank Reactor	FTA	Fault Tree Analysis
CV	Cyclic Voltammogram	FTIR	Fourier-Transform Infrared Spectroscopy
CVD	Chemical Vapor Deposition		
		GDE	Gas-Diffusion Electrode
d.c.	Direct Current	GDL	Gas Diffusion Layer
d.c./a.c.	Direct Current/Alternating Current	GTL	Gas-to-Liquids
DEMS	Differential Electrochemical Mass Spectrometry		
DFT	Density Functional Theory	HDN	Hydrodenitrogenation
DG	Distributed Generation	HDS	Hydrodesulfurization
DHE	Dynamic Hydrogen Electrode	HER	Hydrogen Evolution Reaction
DL	Double-layer	HEVs	Hybrid Electrical Vehicles
DM	Diffusion Media	HHV	Higher Heating Value
DMAc	Dimethylacetamide	HMRDE	Hanging Meniscus RDE
DME	Dimethyl Ether	HOMO	Highest Occupied Molecular Orbital
DMFC	Direct Methanol Fuel Cell	HOR	Hydrogen Oxidation Reaction
DMSO	Dimethyl Sulfoxide	HPLC	Liquid Chromatography
DRIFT	Diffuse Reflectance Infrared Fourier Transform Spectroscopy	HRTEM	High-Resolution Transmission Electron Microscopy
DSA	Dimensionally Stable Anodes	HTS	High Temperature Shift
DSC	Differential Scanning Calorimetry	HTS/LTS	High and/or Low Temperature Shift Reaction
DSK	Double Skeleton	HVDC	High-Voltage Direct-Current

ICE	Internal Combustion Engine	PEMFC	PEM Fuel Cell
ICP	Inductively Coupled Plasma	PES	Poly-Arylene Ether Sulfone
IEC	Ion-Exchange Capacity	PFSA	Perfluorosulfonic Acid
IGBT	Insulated Gate Bipolar Transistor	PGM	Platinum Group Metal
IRRAS	Infrared Reflection-Absorption Spectroscopy	PHR	Power To Heat Ratio
ISS	Ion Scattering Spectroscopy	PM-IRRAS	Polarization Modulation Infrared Reflection-Absorption Spectroscopy
ITD	Initial Temperature Difference	POX	Partial Oxidation
LCA	Life-Cycle Analysis	PP	Polypropylene
LDOS	Local Density-of-States	PPBI	Phosphonated Poly(benzimidazole)
LEED	Low-Energy Electron Diffraction	PPS	Polyphenylenesulfide
LEIS(S)	Low Energy Ion Scattering (Spectroscopy)	PROX	Preferential, or Selective Oxidation
LFL	Lower Flammability Limit	PSA	Pressure Swing Adsorption
LH_2	Liquid Hydrogen	PSSA	Polystyrene Sulfonic Acid
LHV	Lower Heating Value	PSU	Poly Arylene Sulfone
LNG	Liquefied Natural Gas	PSU	Polysulfone
LPG	Liquefied Petroleum Gas	PTFE	Polytetrafluoroethylene
LSV	Linear Sweep Voltammetry	PV	Photovoltaic
LTS	Low Temperature Shift	PVD	Physical Vapor Deposition
		PVDF	Polyvinylidene Fluoride
MCFC	Molten Carbonate Fuel Cell	pzc	Potential of Zero Charge
MEA	Membrane-Electrode Assembly		
MH	Metal Hydride	QCM	Quartz Crystal Microbalance
MIEC	Mixed Oxide-Ion/Electronic Conductor		
ML	Monolayer	RDE	Rotating Disk Electrode
MS	Mass Spectrometry	rds	Rate-Determining Step
MTBE	Methyl Tertiary-Butyl Ether	RE	Reference Electrode
MTTF	Mean Time to Failure	REMPI	Resonant Multiphoton Ionization
		RFD	Reformulated Diesel
NEDC	New European Drive Cycle	RGA	Residual Gas Analysis
NEMCA	Non-Faradaic Electrochemical Modification of Chemical Activity	RH	Relative Humidity
		RHE	Reversible Hydrogen Electrode
NG	Natural Gas	RHEED	Reflection High Energy Electron Diffraction
NHE	Normal Hydrogen Electrode	RRDE	Rotating Ring-Disk Electrode
NMHC	Non-Methane Hydrocarbons		
NMR	Nuclear Magnetic Resonance	S/C	Steam/Carbon Ratio
		SANS	Small-Angle Neutron Scattering
OCV	Open-Circuit Voltage	SAXS	Small-Angle X-Ray Scattering
OEM	Original Equipment Manufacturer	SCE	Saturated Calomel Electrode
OER	Oxygen Evolution Reaction	ScSZ	Scandia-Stabilized Zirconia
ORR	Oxygen Reduction Reaction	SDC	Samaria-Doped Ceria
		SECM	Scanning Electrochemical Microscope
PAFC	Phosphoric Acid Fuel Cell	SECM	Scanning Electrochemical Microscopy
PAN	Polyacrylonitrile	SEIRAS	Surface Enhanced Infrared Absorption Spectroscopy
PBI	Polybenzimidazole		
PC	Propylene Carbonate	SEM	Scanning Electron Microscopy
PCS	Power Conditioning System	SFG	Sum Frequency Generation
PE	Polyethylene	SHE	Standard Hydrogen Electrode
PEEK	Poly-Arylene-Etheretherketone	SHG	Second Harmonic Generation
PEEKK	Poly-Arylene-Etheretherketoneketone	SIMS	Secondary Ion Mass Spectrometry
PEFC	Polymer Electrolyte Fuel Cell	SOC	State of Battery Charge
PEK	Polyarylene Ether Ketone	SOFC	Solid Oxide Fuel Cell
PEM	Polymer Electrolyte Membrane		

SPE	Solid Polymer Electrolyte	TPD/TPR	Temperature-Programmed Desorption/Temperature-Programmed Reduction
sPEEK	Sulfonated Polyetheretherketone		
SPEFC	Solid Polymer Electrolyte Fuel Cell		
sPEK	Sulfonated Poly(Etherketone)	UHV	Ultra-High Vacuum
sPS	Sulfonated Polystyrene	UPD	Under-Potential Deposition
SR	Steam Reforming	UPS	Ultraviolet Photoemission Spectroscopy
SSZ	Scandia-stabilized Zirconia	UV	Ultraviolet
STM	Scanning Tunneling Microscopy		
STP	Standard Temperature and Pressure	VPS	Vacuum Plasma Spraying
SULEV	Super Ultra Low Emission Gasoline Vehicles	VSC	Vacuum Slip Casting
SWV	Square-Wave Voltammetry		
SXS	Surface X-Ray Scattering	WE	Working Electrode
		WGS	Water-Gas Shift
TBA	Tetrabutylammonium		
TDS	Thermal Desorption Spectroscopy	XANES	X-Ray Absorption Near Edge Structure
TEM	Transmission Electron Microscopy	XPS	X-Ray Photoelectron Spectroscopy
TFA	Trifluoroacetic Acid	XRD	X-Ray Diffraction
TFE	Tetrafluoroethylene		
TGA	Thermogravimetric Analysis	YSZ	Yttria-Stabilized Zirconia
THF	Tetrahydrofuran		
TOF	Turn-Over Frequency	ZEV	Zero Emission Vehicle
ToF-SIMS	Time-of-Flight Secondary-Ion Mass Spectrometry		

Part 1

Thermodynamics and kinetics of fuel cell reactions

Chapter 1

The components of an electrochemical cell

A. Hamnett
University of Strathclyde, Glasgow, UK

1 INTRODUCTION TO ELECTROCHEMICAL CELLS

Electrochemistry is concerned with the study of the interface between an electronic and an ionic conductor, and traditionally has concentrated on: (i) the nature of the ionic conductor, which is usually an aqueous or (more rarely) a non-aqueous solution, polymer or superionic solid containing mobile ions; (ii) the structure of the electrified interface that forms on immersion of an electronic conductor into an ionic conductor; (iii) the electron-transfer processes that can take place at this interface, and the limitations on the rates of such processes.

Ionic conductors arise whenever there are mobile ions present. In electrolyte solutions, such ions are normally formed by dissolution of an ionic solid. Provided the dissolution leads to the complete separation of the ionic components to form essentially independent anions and cations, the electrolyte is termed *strong*. By contrast, *weak* electrolytes, such as organic carboxylic acids, are present mainly in the undissociated form in solution, with the total *ionic* concentration orders of magnitude lower than the formal concentration of the solute.

If the ions in such an electrolyte solution are subjected to an electric field, **E**, they will experience a force

$$\mathbf{F} = ze_o\mathbf{E} \qquad (1)$$

which will induce motion in or against the direction of the field depending on the sign of the charge on the ion i.e., whether z is positive or negative. This ion motion leads to the transport of charge and hence to the flow of electrical current through the electrolyte solution.

An electric field can be applied across an electrolyte solution quite straightforwardly by introducing two electronic conductors (solids or liquids containing free electrons, such as metals, carbon, semiconductors etc.), and applying a fixed potential difference. These electronic conductors are termed electrodes.

The actual arrangement of the electrodes is shown in Figure 1.[1] The electrical circuit between the electrodes is completed by a resistor, an ammeter and a constant voltage source connected by external wiring from one electrode to the other. The electrolyte solution in Figure 1 is formed from the dissolution of $CuCl_2$ in water, which leads to one Cu^{2+} and two Cl^- ions per formula unit. The electrodes themselves are formed from a suitable inert metal such as platinum.

2 CHEMICAL REACTIONS AND ELECTRON TRANSFER

When current flows in the cell, the negatively charged chloride ions migrate to the positive electrode and the positively charged ions to the negative electrode. At the phase boundary between ionic and electronic conductors, the ions arriving are transformed by capture or release of electrons. At the negative electrode, the Cu^{2+} ions are plated out as copper metal:

$$Cu^{2+} + 2e^- \longrightarrow Cu^0 \qquad (2)$$

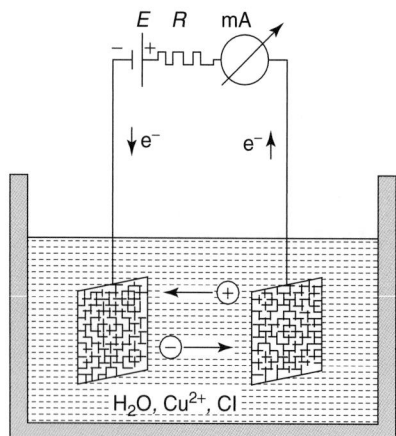

Figure 1. Electrochemical cell for the electrolysis of aqueous $CuCl_2$ solution. E: d.c. voltage; R: resistance; mA: galvanometer for current measurement.

and at the positive electrode the chloride ions release electrons to form chlorine gas:

$$2Cl^- \longrightarrow Cl_2 + 2e^-$$

It can be seen that the fundamental difference between charge transport through electrolyte solution by ion migration and through electronic conductors by electron migration is that the latter leaves the conductor essentially unaltered whereas the former leads to changes in the electrolyte. In the case above, the flow of current leads to the appearance of concentration differences, since the Cu^{2+} ions move from right to left and the chloride ions from left to right, and it also leads to changes in the total electrolyte concentration as both copper and chlorine are lost from the solution. From the addition of the two electrode reactions above, the overall cell reaction is:

$$\begin{array}{r} Cu^{2+} + 2e^- \longrightarrow Cu^0 \\ 2Cl^- \longrightarrow Cl_2 + 2e^- \\ \hline Cu^{2+} + 2Cl^- \longrightarrow Cu^0 + Cl_2 \end{array} \quad (3)$$

It should be emphasized at this point that a constant direct current through an ionic conductor is only possible if electrode reactions take place at the phase boundaries between the electronic and ionic components of the circuit. These reactions must permit electrons to be exchanged between the two phases, and it is clear that the nature of the interface between the electrodes and the electrolyte is critical to the electron-transfer process. The structure of this interface, or electrochemical double layer, is dealt with in detail in **The electrode–electrolyte interface**, Volume 1; suffice to say that in more concentrated electrolytes, the width of the interface i.e., the extent to which the solution structure is modified by the presence of the electrode, is very

small, usually below a nanometer, a distance comparable to the distance an electron can tunnel between electrode and acceptor ion in solution. It is this comparability of scale that is critical in electrochemical phenomena.

If the aqueous $CuCl_2$ solution in the electrochemical cell of Figure 1 is replaced by an aqueous solution of HCl, which dissociates into aquated hydrogen ions (protons) of approximate formula H_3O^+ and chloride ions, a d.c. current flow will again lead to the loss of Cl^- at the positive electrode. However, at the negative electrode, the H_3O^+ ions are reduced to hydrogen. Thus, the flow of charge is accompanied by the electrochemical decomposition of HCl to its component parts

$$\begin{array}{r} 2H_3O^+ + 2e^- \longrightarrow H_2 + 2H_2O \\ 2Cl^- \longrightarrow Cl_2 + 2e^- \\ \hline 2HCl_{aq} \longrightarrow H_2 + Cl_2 \end{array} \quad (4)$$

The electrochemical decomposition of a substance through the passage of an electrical current is termed electrolysis, and corresponds to the conversion of electrical to chemical energy.

For a significant rise in the current through an electrolysis cell, the potential difference between the electrodes, or cell voltage, E, must exceed a certain value, the decomposition voltage, E_D, as shown in Figure 2. The decomposition voltage for HCl at a concentration of 1.2 M has the value 1.37 V at 25 °C, and the value usually lies between one and four volts for other electrolytes.

If the electrolysis process is suddenly brought to a halt by removing the voltage source in the external circuit of Figure 1, and reconnecting the electrodes through a voltmeter such that no current is allowed to flow, a voltage of about 1 V is observed. If the electrodes are reconnected

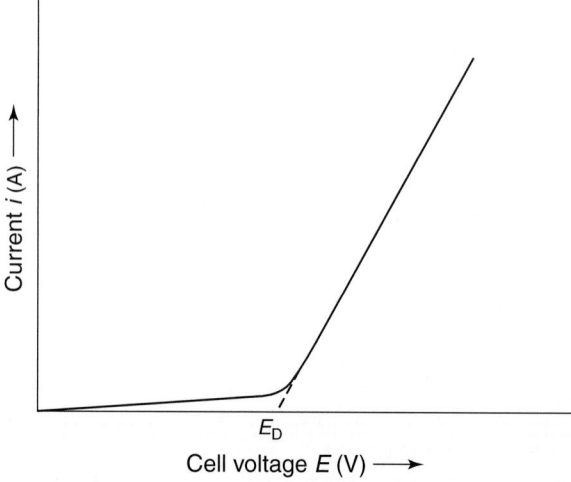

Figure 2. Electrolysis current as a function of the cell voltage E. E_D is the decomposition voltage.

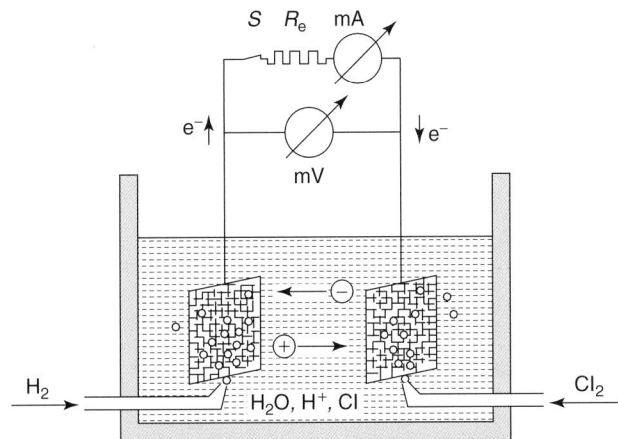

Figure 3. Galvanic cell based on the H_2/Cl_2 reaction. MV: voltage measurement; R_e: external resistance; S: switch.

by a resistor and ammeter, a current can also be observed. This current has its origin in the reversal of the two electrode reactions of equation (4); hydrogen is oxidized to hydrated protons, the electrons released travel round the external circuit and reduce chlorine gas to chloride ions. Actually, the current in this case would decay rapidly since the gases hydrogen and chlorine are only slightly soluble in water, and will, for the most part, have escaped from solution. However, if their concentrations can be maintained by bubbling the two gases over the electrodes, as shown in Figure 3, then the current will remain fairly constant with time; the cell can continually provide electrical energy from chemical energy.

3 ELECTROCHEMICAL GALVANIC CELLS

Electrochemical cells in which the electrode reactions take place spontaneously, giving rise to an electrical current, are termed galvanic cells, and are capable of the direct conversion of chemical to electrical energy. If the resistance between the electrodes is made very high, so that very little current flows, the observed potential difference between the electrodes of the galvanic cell becomes the rest voltage, E_r, which, under true equilibrium conditions, is related to the free energy of the overall cell reaction, as discussed in more detail below.

The most important types of galvanic cell are batteries and fuel cells. The main difference between batteries and fuel cells is that in the former, the chemical reactants are an inherent part of the device; a battery, in other words, carries its fuel around with it, whereas fuel must be supplied from an *external* source to a fuel cell. Unlike a battery, a fuel cell cannot 'go flat'; so long as fuel and combustant gases are supplied, the fuel cell will generate electricity. The problem of recharging is, therefore, peculiar to batteries, though the fuel management systems characteristic of most advanced fuel cells are not normally needed in battery operation. The commonest fuel cells use hydrogen as the fuel and oxygen or air as the combustant; such a cell could be realized in Figure 3 if the chlorine gas bubbling over the cathode is replaced by oxygen, giving the cell reactions:

$$2H_2O + H_2 \longrightarrow 2H_3O^+ + 2e^-$$
$$\tfrac{1}{2}O_2 + 2H_3O^+ + 2e^- \longrightarrow 3H_2O$$
$$\overline{H_2 + \tfrac{1}{2}O_2 \longrightarrow H_2O} \qquad (5)$$

Unlike the chlorine/hydrogen cell of Figure 3, however, the rest voltage of the hydrogen/oxygen cell is not normally well-defined, at least in the set-up shown. As will be seen below, this arises because the electron-transfer rate between the platinum cathode and the oxygen molecule is very slow at cell voltages near the decomposition voltage, with the result that other electrochemical processes can interfere.

Irrespective of whether electrolysis is taking place or the cell is behaving in a galvanic mode, the electrode at which negative charge enters the electrolyte solution is termed the cathode. Equivalently, it is at this electrode that positive charge may be said to leave the solution. Thus, typical cathode reactions are:

$$Cl_2 + 2e^- \longrightarrow 2Cl^-$$
$$Cu^{2+} + 2e^- \longrightarrow Cu^0$$
$$\tfrac{1}{2}O_2 + 2H_3O^+ + 2e^- \longrightarrow 3H_2O \qquad (6)$$

In these cases, the reactant, Cl_2, O_2 or Cu^{2+} is said to be reduced. In a similar fashion, at the anode negative charge leaves the electrolyte solution, or, equivalently, positive charge enters the solution, and typical anode reactions include:

$$2H_2O + H_2 \longrightarrow 2H_3O^+ + 2e^-$$
$$2Cl^- \longrightarrow Cl_2 + 2e^- \qquad (7)$$

In these cases, the reactant is said to be oxidized.

Clearly, for the electrolysis cell above, the cathode is that electrode at which hydrogen is evolved, whereas in the case of the galvanic cell, the cathode is that electrode at which chlorine gas is reduced to chloride. Since, in electrolysis, the positively charged ions migrate towards the cathode, they are termed cations, and the negative ions are termed anions. That part of the solution near the cathode is termed the catholyte, and that part near the anode is termed the anolyte.

If a galvanic cell is set up as shown in Figure 3, and both current and voltage are monitored simultaneously, it is found that as the current increases, the measured external voltage between the electrodes decreases, as shown schematically in Figure 4. In fact the total cell voltage is partitioned between internal (R_i) and external (R_e) resistances and the *measured* external voltage between the galvanic-cell electrodes is:

$$E = E_r - iR_i \tag{8}$$

where E is the measured cell voltage, and E_r the rest voltage for the cell as defined above. The power output, P, of the cell is the product of voltage and current:

$$P = iE = i(E_r - iR_i) \tag{9}$$

and this is a maximum for $i = E_r/2R_i$, and $E = E_r/2$ if R_i is independent of the current i, an approximation only valid if electron-transfer rates at both electrodes are fast at all voltages. A more normal situation for fuel cells is shown in Figure 5, in which there is an initial exponential fall in current with voltage, followed by a linear decrease. The initial fall is associated with poor electrode kinetics at voltages close to the rest voltage. As will be seen below, the rate of electron transfer increases as the voltage moves away from the rest voltage, and eventually ceases to be a limiting factor. At very high current densities, particularly with gas-feed electrodes, mass transport of reagents to the electrode/electrolyte interface becomes limiting, and the cell performance drops off catastrophically.

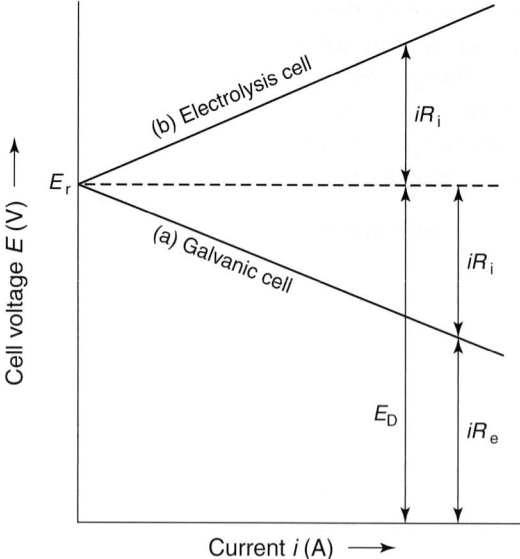

Figure 4. Schematic variation of cell voltage E, against load current i for (a) a galvanic cell; (b) an electrolysis cell. E_r is the voltage without current passing, also termed the electro-motive force, or EMF. E_D is the decomposition voltage, R_i the internal resistance of the cell and R_e the external resistance.

4 THE ELECTRODE–SOLUTION INTERFACE AND IONICS

As indicated above, critical to the understanding of modern electrochemistry is the understanding of the interface between solution and electrode, and we turn first to consideration of electrolyte solutions. Modern-day approaches to ionic solutions need to be able to contend with the following problems:

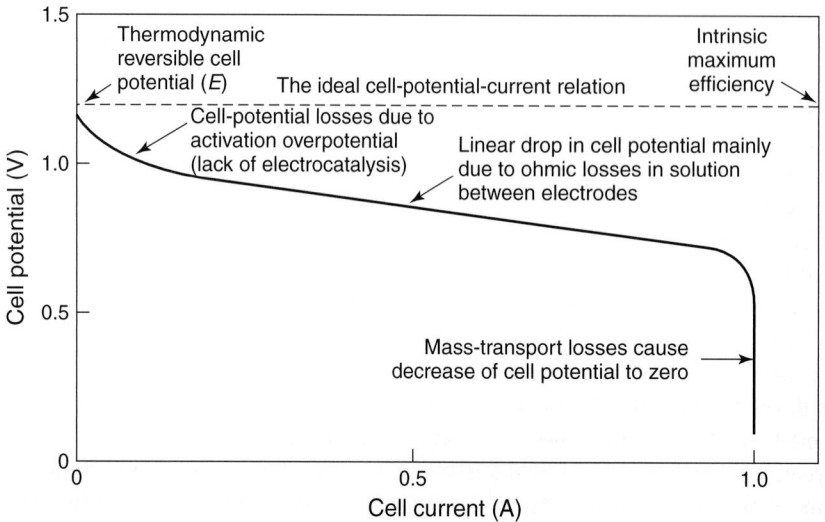

Figure 5. A typical plot of the cell voltage versus current for fuel cells, illustrating regions of control by various types of overpotential.

1. The nature of the solvent itself, and the interactions taking place in that solvent.
2. The changes taking place on dissolution of ionic electrolyte in the solvent.
3. Macroscopic and microscopic studies of the properties of electrolyte solutions.

Even the description of the solvent itself presents major theoretical problems. Unlike the gas phase, where densities are relatively low and intermolecular interactions can be reasonably treated as perturbations, or the solid state, where dynamic motion can be treated in first order in terms of collective oscillations of the entire solid, any theory of liquids must be able to account both for molecular motion and for substantial intermolecular forces. Direct calculation of the partition function for liquids has proved, as a result, quite impossible, and modern theories have primarily sought to make an indirect attack the problem by defining and calculating a distribution function $g(r)$. Assuming that the liquid is isotropic, we find that the number of molecules in a spherical shell at a distance r from a central molecule can be written:

$$N(r)\,dr = 4\pi r^2 \rho g(r)\,dr \qquad (10)$$

The function $g(r)$ is central to the modern theory of liquids, since it can be measured experimentally using neutron or X-ray diffraction and can be related to the inter-particle potential energy. Experimental data for water shows that $g(r)$ has a first strong maximum at 2.82 Å, very close to the intermolecular distance in the normal tetrahedrally bonded form of ice; moreover, the second peak in $g(R)$ is found at ca. 4.5 Å, which corresponds closely to the second-nearest-neighbor distance in ice, strongly supporting a model for the structure of water that is ice-like over short distances. This strongly structured model for water in fact dictates many of its anomalous properties. Further theoretical and experimental studies of water show the average coordination number of four decreases with increasing temperature. The picture should be seen as highly dynamic, with these hydrogen bonds forming and breaking continuously, with the result that the clusters of water molecules characterizing this picture are themselves in a continuous state of flux.

The complexity of the water structure in particular has defeated attempts to derive an analytical theory for liquid water, and has led, instead, to attempts to understand the behavior of water through simulation methods for calculating $g(r)$; these methods have come into their own in the past twenty years as the cost of computing has fallen. The two main approaches are the Monte-Carlo method, which depends essentially on identifying a Monte-Carlo approach to the integral defining the distribution function. The alternative simulation approaches are based on molecular dynamics calculations. This is conceptually simpler than the Monte Carlo method: the equations of motion are solved for a system of N molecules, and periodic boundary conditions are again imposed. This method permits both the equilibrium and transport properties of the system to be evaluated, essentially by solving the equations of motion numerically,

$$m\frac{d^2\mathbf{R}_k}{dt^2} = \sum_{j=1, j\neq k}^{N} \mathbf{F}(\mathbf{R}_{kj}) = -\sum_{j=1, j\neq k}^{N} \nabla_k U(\mathbf{R}_{kj}) \qquad (11)$$

where $\mathbf{R}_k$ is the coordinate of species k, and $\mathbf{F}(\mathbf{R}_{kj}) \equiv -\nabla_k U(\mathbf{R}_{kj})$ is the force on species k exerted by species j, and integration is over discrete time intervals δt. Details are given in Lee.[2]

There is, in essence, no limitation other than computing time to the accuracy and predictive capacity of molecular dynamic and Monte Carlo methods, and although the derivation of realistic potentials for water is a formidable task in its own right, we can anticipate that within a short period, accurate simulations of water will have been made. However, there remain major theoretical problems in deriving any analytical theory for water, and indeed any other highly polar solvent of the sort encountered in normal electrochemistry. It might be felt, therefore, that extension of the theory to analytical descriptions of ionic solutions was a well-nigh hopeless task. However, a major simplification of our problem is allowed by the possibility, at least in more dilute solutions, of smoothing out the influence of the solvent molecules, reducing their influence to such average quantities as the dielectric permittivity, ε_m, of the medium. Such a viewpoint is developed within the McMillan–Mayer theory of solutions, which essentially seeks to partition the interaction potential into three parts: that which is due to interaction between the solvent molecules themselves, that due to interaction between solvent and solute, and that due to interaction between the solute molecules dispersed within the solvent.

McMillan–Mayer theory allows us to develop a formalism similar to that of a dilute interacting fluid for solute dispersed in the solvent provided that a sensible description of the interaction potential between ions, $W(i|i)$, and between ion and solvent $W(i|s)$ can be given. This interaction potential will, of course, depend on the solvent, but the essence of the theory is that approximations can be made which usually correspond to the solvent being treated as a continuum.

The solute–solvent interaction $W(i|s)$ is a measure of the solvation energy of solute species at infinite dilution. The basic model for ionic hydration is shown in Figure 6;[3] there is an inner hydration sheath of water molecules

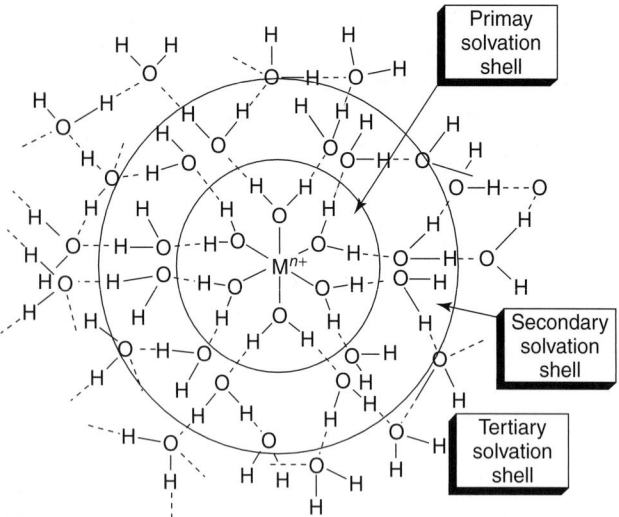

Figure 6. The localized structure of a hydrated metal cation in aqueous solution, the metal ion being assumed to have a primary hydration number of six.

whose orientation is essentially determined entirely by the field due to the central ion. The number of water molecules in this inner sheath depends on the size and chemistry of the central ion, being, for example, four for Be^{2+} but six for Mg^{2+}, Al^{3+} and most of the first-row transition ions. Outside this primary shell, there is a secondary sheath of more loosely bonded water molecules oriented essentially by hydrogen bonding, the evidence for which was initially indirect and derived from ion mobility measurements. More recent evidence for this secondary shell has now come from X-ray diffraction and scattering studies and IR measurements. A further highly diffuse region, the tertiary region, is probably present, marking a transition to the hydrogen-bonded structure of water described above. The ion, as it moves, will drag at least part of this solvation sheath with it, but the picture should be seen as essentially dynamic, with the well-defined inner sheath structure of Figure 6 being mainly found in highly charged ions of high electronic stability, such as Cr^{3+}. The enthalpy of solvation of *cations* primarily depends on the charge on the central ion and the effective ionic radius, the latter begin the sum of the normal Pauling ionic radius and the radius of the oxygen atom in water (0.85 Å). A reasonable approximate formula has

$$\Delta H_{hyd}^0 = \frac{-695 Z^2}{(r_+ + 0.85)} \text{ kJ mol}^{-1} \qquad (12)$$

In general, anions are less strongly hydrated than cations, but recent neutron diffraction data have indicated that even around the halide ions there is a well-defined primary hydration shell of water molecules, which, in the case of Cl^- varies from four to six in constitution, the exact number being a sensitive function of concentration and the nature of the accompanying cation.

5 IONIC CONDUCTIVITY[4, 5]

In the presence of an electric field, as indicated earlier, ions experience a force that will induce motion. Consider ions of charge $z_\alpha e_o$, accelerated by the electric field strength, $\mathbf{E}$, but being subject to a frictional force, $\mathbf{F}_R$, that increases with velocity, $\mathbf{v}$, and is given, for simple spherical ions of radius r_α, by the Stokes formula, $\mathbf{F}_R = 6\pi\eta r_\alpha \mathbf{v}$, where η is the viscosity of the medium. After a short induction period, the velocity attains a limiting value, $\mathbf{v}_{max}$, corresponding to the exact balance between the electrical and frictional forces:

$$z_\alpha e_o \mathbf{E} = 6\pi\eta r_\alpha \mathbf{v}_{max} \qquad (13)$$

and the terminal velocity is given by

$$\mathbf{v}_{max} = \frac{z_\alpha e_o \mathbf{E}}{(6\pi\eta r_\alpha)} \equiv u_\alpha \mathbf{E} \qquad (14)$$

where u_α is the mobility of ion α. It follows that for given values of η and $\mathbf{E}$, each type of ion will have a transport velocity dependent on the charge and the radius of the solvated ion, and a direction of migration dependent on the sign of the charge.

For an electrolyte solution containing both anions and cations, with the terminal velocity of the cations being v_{max}^+, and the number of ions of charge $z^+ e_o$ per unit volume being N^+, the product $AN^+ v_{max}^+$ corresponds just to that quantity of positive ions that passes per unit time through a surface of area A normal to the direction of flow. The product $AN^- v_{max}^-$ can be defined analogously, and the amount of charge carried through this surface per unit time, or the current per area A, is given by

$$I = I^+ + I^- = Ae_0(N^+ z^+ v_{max}^+ + N^- z^- v_{max}^-)$$
$$= Ae_0(N^+ z^+ u^+ + N^- z^- u^-) \times |\mathbf{E}| \qquad (15)$$

where the u are the mobilities defined above. If the potential difference experienced within the body of the electrolyte is ΔV, and the distance apart of the electrodes is l, then the magnitude of the electric field $|\mathbf{E}| = \Delta V / l$. Since $I = G\Delta V$, where G is the conductance, G is given by

$$G = \left(\frac{A}{l}\right) e_0 (N^+ z^+ u^+ + N^- z^- u^-) \qquad (16)$$

The conductivity is obtained from this by division by the geometric factor (A/l), giving

$$\kappa = e_0(N^+z^+u^+ + N^-z^-u^-) \quad (17)$$

It is important to recognize the approximations made here: the electric field is supposed sufficiently small that the equilibrium distribution of velocities of the ions is essentially undisturbed. We are also assuming that we can use the relaxation approximation, and that the relaxation time τ is independent of ionic concentration and velocity. These approximations break down at higher ionic concentrations: a primary reason for this is that ion–ion interactions begin to affect both τ and $\mathbf{F}_\alpha$. However, in very dilute solutions, the ion scattering will be dominated by solvent molecules, and in this limiting region equation (17) will be an adequate description.

Measurement of the conductivity can be carried out to high precision with specially designed cells. In practice, these cells are calibrated by first measuring the conductance of an accurately known standard, and then introducing the sample under study. Conductances are usually measured at about 1 kHz alternating current (a.c.) rather than with direct current (d.c.) voltages to avoid complications arising from electrolysis at anode and cathode.[4]

5.1 Conductivity of solutions

The conductivity of solutions depends, from equation (17), on both the concentration of ions and their mobility. Typically, for 1 M NaCl in water at 18 °C, a value of $0.0744\,\Omega^{-1}\,\text{cm}^{-1}$ is found; by contrast, 1 M H_2SO_4 has a conductivity of $0.366\,\Omega^{-1}\,\text{cm}^{-1}$ at the same temperature, but acetic acid, a *weak* electrolyte, has a conductivity of only $0.0013\,\Omega^{-1}\,\text{cm}^{-1}$.

In principle, the effects of concentration of ions can be removed by dividing equation (17) by the concentration. Taking Avogadro's constant as L, and assuming a concentration of solute $c\,\text{mol}\,\text{m}^{-3}$, then from the electroneutrality principle we have $N^+z^+ = N^-z^- = \nu_\pm z_\pm Lc$, and clearly

$$\Lambda \equiv \frac{\kappa}{c} = \nu_\pm z_\pm Le_0(u^+ + u^-) \equiv \nu_\pm z_\pm F(u^+ + u^-) \quad (18)$$

where Λ is termed the *molar conductivity* and F is the Faraday constant, which has the numerical value $96\,485\,\text{C}\,\text{mol}^{-1}$. In practice, even Λ shows some dependence on concentration, but extrapolation of Λ to Λ_0 as $c \to 0$ does give a well-defined quantity.

Λ_0 plays an important part in the theory of ionic conductivity since at high dilution, the ions should be able to move completely independently, and as a result, equation (18), expressed in the form

$$\Lambda_0 = \nu_\pm z_\pm F(u_0^+ + u_0^-) \equiv \nu_+\lambda_0^+ + \nu_-\lambda_0^- \quad (19)$$

is exactly true. Some individual limiting ionic conductivities in aqueous solution at 25 °C are given in Table 1.

At first sight, we would expect that the mobilities of more highly charged ions would be larger, but it is apparent from Table 1 that this is not the case; the *mobilities* of Na^+ and Ca^{2+} are comparable, even though equation (14) would imply that the latter should be about a factor of two larger. The explanation lies in the fact that the effective ionic radius, r, also increases with charge, which, in turn, can be traced to the increased size of the hydration sheath in the doubly charged species, since there is an increased attraction of the water dipoles to more highly charged cations.

One anomaly immediately obvious from Table 1 is the much higher mobilities of proton and hydroxide ions than expected from even the most approximate estimates of

Table 1.

Ion	$\lambda_0^+, \lambda_0^-/\Omega^{-1}\,\text{mol}^{-1}\,\text{cm}^2$	Ion	$\lambda_0^+, \lambda_0^-/\Omega^{-1}\,\text{mol}^{-1}\,\text{cm}^2$
		Ag^+	62.2
H^+	349.8	Na^+	50.11
OH^-	197	Li^+	38.68
K^+	73.5	$[Fe(CN)_6]^{4-}$	440
NH_4^+	73.7	$[Fe(CN)_6]^{3-}$	303
Rb^+	77.5	$[CrO_4]^{2-}$	166
Cs^+	77	$[SO_4]^{2-}$	161.6
		I^-	76.5
Ba^{2+}	126.4	Cl^-	76.4
Ca^{2+}	119.6	NO_3^-	71.5
Mg^{2+}	106	CH_3COO^-	40.9
		$C_6H_5COO^-$	32.4

their ionic radii. The origin of this behavior lies in the way in which these ions can accommodate to the water structure described above. Free protons cannot exist as such in aqueous solution: the very small radius of the proton would lead to an enormous electric field that would polarize any molecule, and in aqueous solution, the proton immediately attaches itself to the oxygen atom of a water molecule giving rise to an H_3O^+ ion. In this ion, however, the positive charge does not simply reside on a single hydrogen atom; NMR spectra show that all three hydrogen atoms are equivalent, giving a structure similar to that of the NH_3 molecule. The formation of a water cluster around the H_3O^+ ion and its subsequent fragmentation may then lead to the positive charge being transmitted across the cluster without physical migration of the proton, and the limiting factor in proton motion becomes H-bonded cluster formation and not conventional migration. It is clear that this model can be applied to the anomalous conductivity of the hydroxide ion without any further modification. Hydrogen-atom tunneling from a water molecule to a OH^- ion will leave behind a OH^- ion, and the migration of OH^- ions is, in fact, traceable to the migration of H^+ in the opposite direction.

This type of mechanism is supported by the observation of the effect of temperature. It is found that the mobility of the proton goes through a maximum at a temperature of 150 °C (where, of course, the measurements are carried out under pressure). This arises because as the temperature in increased from ambient, the main initial effect is to loosen the hydrogen-bonded local structure that inhibits reorientation. However, at higher temperatures, the thermal motion of the water molecules becomes so marked that cluster formation becomes inhibited. The complete hydration shell of the proton consists of both the central H_3O^+ unit and further associated water molecules; mass spectrometric evidence would suggest that a total of four water molecules form the actual $H_9O_4^+$ unit, giving a hydration number of four for the proton.

5.2 Conductivity of ionomeric polymers

Whilst aqueous solutions are the most carefully investigated, recent years have also seen the development of solid-polymer electrolytes. These have the immense advantage of castability, allowing the fabrication of thin, highly conducting but impervious structures that have found applications in chlor-alkali cells and fuel cells. By far the most important class of solid polymer electrolytes are the polymeric perfluorosulfonic acid derivatives, commonly referred to by the trade name Nafion®, the first to be marketed, which are singly-charged cation conductors that can be used both as separators and electrolytes. Other polymeric cation conductors include polyvinyl alcohols, though these have lower intrinsic resistance to aggressive chemicals. Anionically conducting polymers have been little used save in sensors.

The basic chemical form of the polymer chains in the perfluorosulfonic acid systems varies between manufacturers to some extent, but is shown schematically in Figure 7. It can be seen to consist of a perfluoro-CF_2-CF_2-backbone with pendant perfluoro side-chains ending in $-SO_3^-$ groups. Charge transport in these materials is usually by Na^+ or H^+ ions that wander from one sulfonic acid group to the next. These ions are normally solvated, and the membranes themselves are strongly hydrated. The internal structure of the membranes has been investigated by a variety of techniques, and is now known to consist of essentially inverse micelles joined through canals, as shown in Figure 8. The inner surfaces of these micelles contains $-SO_3^-$ groups, and the structure is apparently adopted to minimize repulsive interactions between water and the fluorocarbon backbone whilst maximizing solvation of the $-SO_3^-$ groups. An examination of the structure reveals that cations can easily transfer from one micelle to the next, whereas anions cannot traverse the canals due to repulsion from the $-SO_3^-$ groups in those canals.

The main advantages of these SPE membranes is their very high conductivity and the fact that they can be cast as thin pore-free films. Since the important parameter is the conductance per unit area, κ_A, it is usually this that is quoted, rather than the conductivity, σ. The relationship is simply

$$\kappa_A = \frac{\sigma}{d} \quad (20)$$

where d is the film thickness, and its units are $\Omega^{-1}\,\text{cm}^{-2}$ or $\Omega^{-1}\,\text{m}^{-2}$. If the current density is j A cm^{-2}, then the potential drop across the membrane is simply

$$V = \frac{j}{\kappa_A} \quad (21)$$

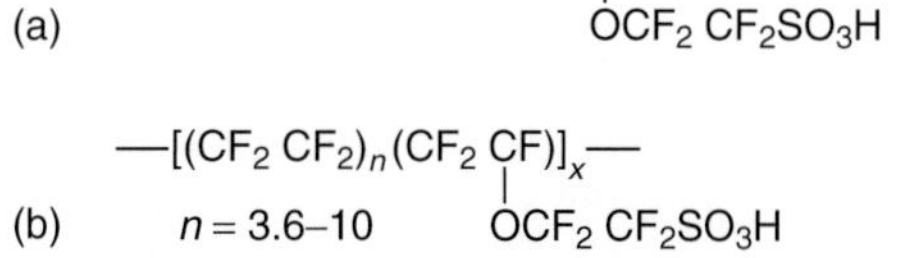

(a) —[(CF$_2$ CF$_2$)$_n$(CF$_2$ CF)]$_x$—
$n = 6.6$ OCF$_2$CFCF$_3$
                 |
                 OCF$_2$ CF$_2$SO$_3$H

(b) —[(CF$_2$ CF$_2$)$_n$(CF$_2$ CF)]$_x$—
$n = 3.6–10$ OCF$_2$ CF$_2$SO$_3$H

Figure 7. Molecular structures of (a) Nafion and (b) the Dow perfluorinated membranes.

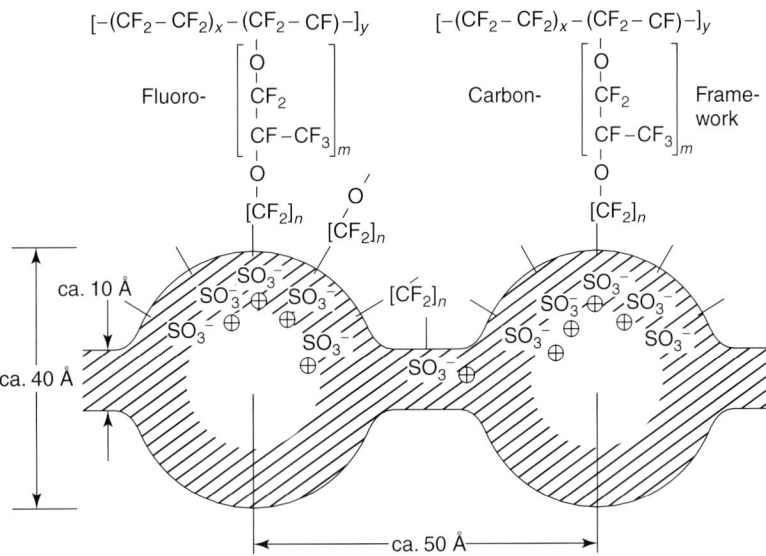

Figure 8. Details of the charge transport mechanism through the Nafion membrane: $x = 5-13$, $y \approx 1000$; $m = 0-3$; $n = 2-6$.

For Nafion used in chlor-alkali cells to separate saturated NaCl solutions, the thickness is normally ca. 0.5 mm and the value of $\kappa_A \sim 1\,\Omega^{-1}\,cm^{-2}$, corresponding to a σ value of $0.05\,\Omega^{-1}\,cm^{-1}$, comparable to concentrated aqueous NaCl. The corresponding potential drop across the membrane at a current density of $0.1\,A\,cm^{-2}$ is 0.1 V. For these films used in the chlor-alkali cells, the transport number of the cation is very close to unity; the contribution of Cl^- conduction amounts at most to a few percent.

5.3 Conductivity of superionic conductors

Of importance in high-temperature fuel cells are *solid* ionic conductors, which have also found application in certain modern batteries. In principle, a perfect lattice should not be capable of sustaining ionic conductivity, but no lattice is perfect above 0 K, with vacancies forming on lattice sites and interstitial sites becoming occupied, and there is a number of ways in which an ion can move under the influence of an electric field in a real crystal:

- from vacancy to vacancy;
- from interstitial to interstitial;
- in a concerted fashion in which an ion moves from an interstitial site to a lattice site expelling the ion on that site to an interstitial site.

An extreme form of the last case is found in so-called superionic conductors, such as AgI above 147 °C, in which one of the sublattices, in this case the silver-ion sublattice, in effect melts, whilst the iodide lattice remains solid. In this case the distinction between lattice site and interstitial breaks down, and the ion motion is much more reminiscent of a liquid. Such first-order transitions are easily observed as discontinuities in conductivity, often of many orders of magnitude, accompanied by strong maxima in the specific heat.

High ionic conductivity is also found in other materials. The ceramic $ZrO_2.(15\%)\,Y_2O_3$ is termed stabilized zirconia, and was developed initially to inhibit a structural transformation that takes place in pure ZrO_2 at 1100 °C. This transformation leads to a significant volume change and causes the ceramic to shatter, but admixture of ca. 15% Y_2O_3 stabilizes the cubic phase down to ambient temperatures. It was subsequently discovered that this material has good oxide-ion conductivity, with $\sigma > 0.01\,\Omega^{-1}\,cm^{-1}$. Subsequently, high conductivity was found in $Ce_{0.9}Gd_{0.1}O_{1.95}$ and related materials, all of which have large numbers of oxygen vacancies built into the lattice. Such materials are the basis of high-temperature 'solid-oxide' fuel cells; for $ZrO_2.(15\%)\,Y_2O_3$, conductivities again reach $\sim 10^{-1}\,\Omega^{-1}\,cm^{-1}$ but only at $\sim 1000\,°C$.

5.4 Conductivity of ionic melts

Finally, molten salts confer electrolytic conductivity in a similar manner to aqueous and non-aqueous electrolyte solutions. The analogy can be continued by distinguishing between strong and weak electrolytes, the former being molten salts derived from ionic lattices, such as those exhibited by alkali and alkaline-earth halides, hydroxides, nitrates, carbonates, sulfates etc., which are wholly dissociated within the melt; whereas the latter, usually arising from the molecular or semi-ionic lattices, such as $AlCl_3$, contain, in the melt, both ions and undissociated molecules. It is also possible to dissolve a salt into a melt to give a molten electrolyte; such melts are of considerable interest in modern battery research.

For strong electrolyte melts, the migration of ions can be considered in first order through a medium of viscosity η in the same way as for equation (13). If we neglect inter-ionic interactions, then the conductivity will be simply proportional to the concentration of charge carriers, and given that the conductivity of 0.1 M aqueous NaCl is ca. $0.01\, \Omega^{-1}\, cm^{-1}$, that the viscosity of the melt is comparable to that of an aqueous solution, and that the concentration of Na^+ ions in the melt is ca. 30 M, we expect a conductivity of ca. $3\, \Omega^{-1}\, cm^{-1}$. However, this estimate takes no account of the difference in temperature between the melt (850 °C) and the aqueous solution, and also neglects the obvious powerful inter-ionic forces, which might be expected strongly to inhibit ionic motion. In fact, the conductivity of molten NaCl is $3.75\, \Omega^{-1}\, cm^{-1}$ at 850 °C and $4.17\, \Omega^{-1}\, cm^{-1}$ at 1000 °C, suggesting that temperature plays little role, and that ionic interactions must be strongly shielded.

In the arena of fuel cells, the most significant application of ionic melts is found in molten carbonate systems, for which the cathode reaction is

$$\tfrac{1}{2}O_2 + CO_2 + 2e^- \longrightarrow CO_3^{2-} \qquad (22)$$

and the anode reaction is:

$$H_2 + CO_3^{2-} \longrightarrow H_2O + CO_2 + 2e^- \qquad (23)$$

Although inter-ionic interactions appear to be strongly shielded for ionic melts, in general such interactions do play a significant role, particularly in ionic solutions, where they have a very substantial impact on ionic mobilities; indeed, such interactions become dominant at high concentrations (in excess of 1 M). The quantitative treatment of this effect is complex, and outside the scope of this introductory essay, but the details of ionic motion in solutions and polymers does have a very direct impact on the engineering of fuel cells.

Inter-ionic interactions also have a major impact on the molar free energy or chemical potential of ions in solution. Again, the theory of this effect is complex, but substantial progress has been made in recent years, with the McMillan–Mayer theory offering the most useful starting point for an elementary theory of ionic interactions. At high dilution, we can incorporate all ion–solvent interactions into a limiting chemical potential, and deviations from solution ideality can then be explicitly connected with ion–ion interactions only. Detailed accounts are given elsewhere.[5]

REFERENCES

1. C. H. Hamann, A. Hamnett and W. Vielstich, 'Electrochemistry', Wiley-VCH, Weinheim, Germany (1998).
2. L. L. Lee, 'Molecular Thermodynamics of Non-ideal Fluids', Butterworths, Stoneham, MA (1988).
3. D. T. Richens, 'The Chemistry of Aqua-ions', Wiley, Chichester (1997).
4. R. A. Robinson and R. H. Stokes, 'Electrolyte Solutions', Butterworth, London (1959).
5. J. O'M. Bockris and A. K. N. Reddy, 'Modern Electrochemistry 1: Ionics', 2nd edition, Plenum Press, New York (1998).

Chapter 2
The electrode–electrolyte interface

A. Hamnett
University of Strathclyde, Glasgow, UK

1 THE ELECTRIFIED DOUBLE LAYER

Once an electrode, which for our purposes may initially be treated as a conducting plane, is introduced into an electrolyte solution, several things change. There is a substantial loss of symmetry, the potential experienced by an ion will now be not only the screened potential of the other ions but will contain a term arising from the field due to the electrode and a term due to the image charge in the electrode. The structure of the solvent is also perturbed: next to the electrode, the orientation of the molecules of solvent will be affected by the electric field at the electrode surface, and the net orientation will derive from both the interaction with the electrode and with neighboring molecules and ions. Finally, there may be a sufficiently strong interaction between ions and the electrode surface that the ions lose at least some of their inner solvation sheath and adsorb on the electrode surface.

The classical model of the electrified interface is shown in Figure 1,[1] and the following features are apparent:

1. There is an ordered layer of solvent dipoles next to the electrode surface, the extent of whose orientation is expected to depend on the charge on the electrode.
2. There is, or may be, an inner layer of specifically adsorbed *anions* on the surface; these anions have displaced one or more solvent molecules and have lost part of their inner solvation sheath. An imaginary plane can be drawn through the centers of these anions to form the inner Helmholtz plane (IHP).
3. The layer of solvent molecules not directly adjacent to the metal is the closest distance of approach of solvated *cations*. Since the enthalpy of solvation of cations is usually substantially larger than that of anions, it is normally expected that there will be insufficient energy to strip the cations of their inner solvation sheaths, and a second imaginary plane can be drawn through the centers of the solvated cations: this second plane is termed the outer Helmholtz plane (OHP).
4. Outside the OHP, there may still be an electric field and hence an imbalance of anions and cations extending in the form of a diffuse layer into the solution.
5. The potential distribution in this model obviously consists of two parts: a quasi-linear potential drop between the metal electrode and the IHP or OHP depending on the charge on the electrode surface and the corresponding planar ionic density, and a second part corresponding to the diffuse layer; as we shall see below, in this part, the potential decays roughly exponentially through screening. However, there are subtleties about what can actually be measured that need some attention.

2 THE ELECTRODE POTENTIAL

Any measurement of potential must describe a reference point, and we will take as this point the potential of an electron well separated from the metal and at rest *in vacuo*. By reference to Figure 2,[2] we can define the following quantities:

Handbook of Fuel Cells – Fundamentals, Technology and Applications, Edited by Wolf Vielstich, Hubert A. Gasteiger, Arnold Lamm.
Volume 1: *Fundamentals and Survey of Systems*. © 2003 John Wiley & Sons, Ltd. ISBN: 0-471-49926-9.

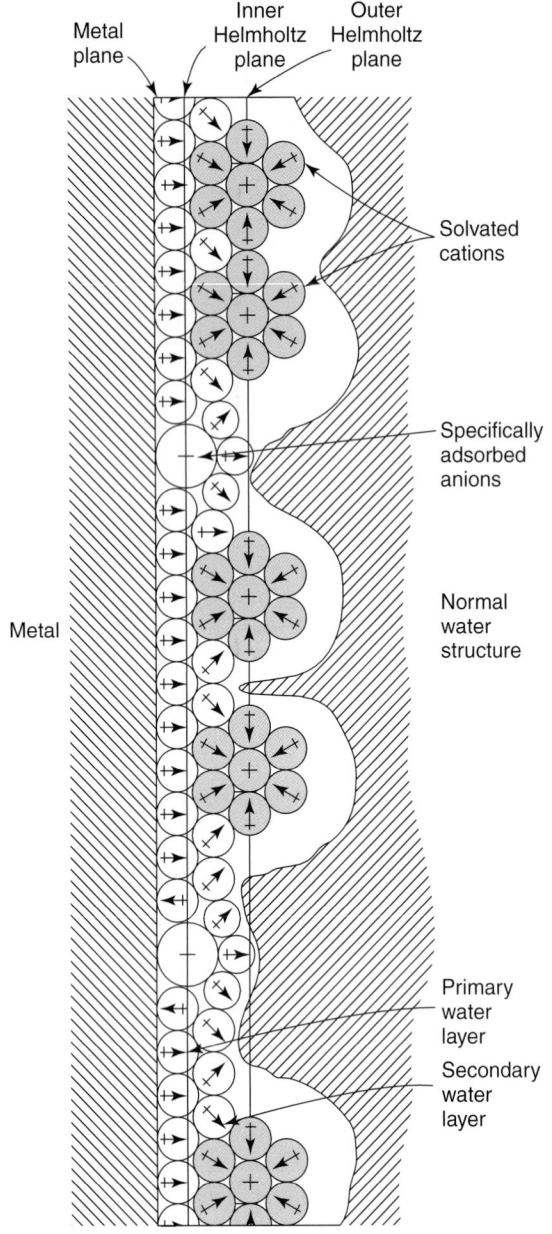

Figure 1. Hypothetical structure of the electrolyte double layer.

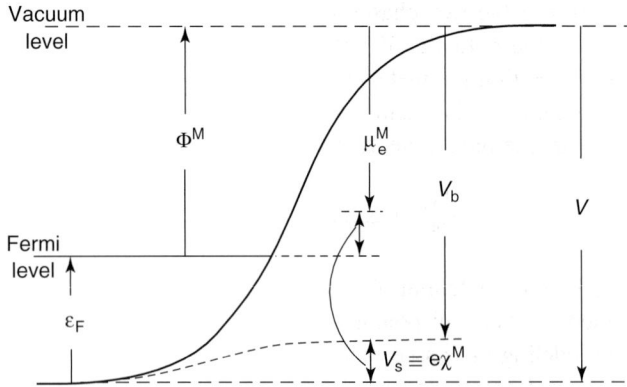

Figure 2. Potential energy profile at the metal–vacuum boundary. Bulk and surface contributions to V are separately shown. (Reproduced from Trassatti (1980)[2] with permission from Kluwer Academic/Plenum Publishers.)

1. The Fermi energy ε_F which is the difference in energy between the bottom of the conduction band and the Fermi level; it is positive and in the simple Sommerfeld theory of metals,[3] $\varepsilon_F = h^2 k_F^2/2m = h^2(3\pi n_e)^{2/3}/2m$, where n_e is the number density of electrons.
2. The work function Φ^M which is the energy required to remove an electron from the inner Fermi level to vacuum.
3. The surface potential of the electrode, χ^M, due to the presence of surface dipoles. At the metal–vacuum interface, these dipoles arise from the fact that the electrons in the metal can relax at the surface to some degree, extending outwards by a distance of the order of 1 Å, and giving rise to a spatial imbalance of charge at the surface.
4. The chemical potential of the electrons in the metal, μ_e^M, a negative quantity.
5. The electrochemical potential $\tilde{\mu}_e^M$ of the electrons in metal M is defined as $\mu_e^M - e_0\psi^M - e_0\chi^M$, where ψ^M is the mean electrostatic potential just outside the metal, which will tend to zero as the free-charge density, σ, on the metal tends to zero; as $\sigma \to 0$, then $\tilde{\mu}_e^M \to \mu_e^M - e_0\chi^M \equiv -\Phi^M$ from Figure 2. Hence, for $\sigma \neq 0$, $\tilde{\mu}_e^M = -\Phi^M - e_0\psi^M$.
6. The potential energy of the electrons, V, which is a negative quantity that can be partitioned into bulk and surface contributions as shown. Clearly, from Figure 2, $\mu_e^M = \varepsilon_F + V_b$.

Of the quantities shown in Figure 2, Φ^M is measurable, as is ε_F, but the remainder are not, and must be calculated. Values of 1–2 eV have been obtained for χ^M, though smaller values are found for the alkali metals.

If two metals with different work functions are placed in contact, there will be a flow of electrons from the metal with the lower work function to one with the higher work function; this will continue until the electrochemical potentials of the electrons in the two phases are equal. This change gives rise to a measurable potential difference between the two metals termed the contact potential or Volta potential difference. Clearly $\Delta_{M_2}^{M_1}\Phi = e_0\Delta_{M_1}^{M_2}\psi$, where $\Delta_{M_1}^{M_2}\psi$ is the Volta potential difference between a point close to the surface of M_1 and one close to the surface of M_2, both points being in the vacuum phase; this is an experimentally measurable quantity. The actual number of electrons transferred is very small, so that the electron densities of the two phases will be unaltered, and only the value of the potential V will have changed. If we assume that, on putting the metals together, the χ^M vanish, and we define the potential inside the metal as ϕ, then the equality of electrochemical potentials also leads to

$$-\mu_e^{M_1} + e_0 \Delta_{M_2}^{M_1} \phi + \mu_e^{M_2} = 0 \quad (1)$$

This internal potential, ϕ, is not directly measurable; it is termed the Galvani potential, and is the target of most of the modeling discussed below. Clearly, if the electrons are transferred across the free surfaces and vacuum between the two metals, we have $\Delta_{M_2}^{M_1}\Phi = \Delta_{M_2}^{M_1}\phi + \Delta_{M_1}^{M_2}\psi$.

Once a metal is immersed in a solvent, a second dipolar layer will form at the metal surface due to the alignment of the solvent dipoles. Again, this contribution to the potential is not directly measurable, and, in addition, the metal dipole contribution itself will change since the distribution of the electron cloud will be modified by the presence of the solvent. Finally, there will be a contribution from free charges both on the metal and in the electrolyte. The overall contribution to the Galvani potential difference between metal and solution then consists of these four quantities, as shown in Figure 3.[2] If the potential due to dipoles at the metal–vacuum interface for the metal is χ^M and for the solvent–vacuum interface is χ^S, then the Galvani potential difference between metal and solvent can be written either as

$$\Delta_S^M \phi = (\chi_M + \delta\chi_M) + (\chi_S + \delta\chi_S) \quad (2)$$

or as

$$\Delta_S^M \phi = \Delta_S^M \chi + \Delta_S^M \psi \quad (3)$$

where $\delta\chi^M$, $\delta\chi^S$, are the changes in surface dipole for metal and solvent on forming the interface. In equation (2) we pass across the interface, and in equation (3) we pass into the vacuum from both metal and solvent. As before, the value of $\Delta_S^M \psi$, the Volta potential difference, is measurable experimentally, but it is evident that we cannot associate this potential difference with that due to free charges at the interface, since there are changes in dipole contribution on both sides as well. Even if there are no free charges at the interface (at the point of zero charge (PZC)), the Volta potential difference is not zero unless $\delta\chi_M = \delta\chi_S$ i.e., the free surfaces of the two phases will still be charged unless the changes in surface dipole of solvent and metal balance exactly. In practice, this is not the case: careful measurements[4] show that $\Delta_{H_2O}^{Hg}\psi = -0.26 V$ at the PZC, showing that the dipole changes do not, in fact, compensate. Historically, this discussion is of considerable interest, since

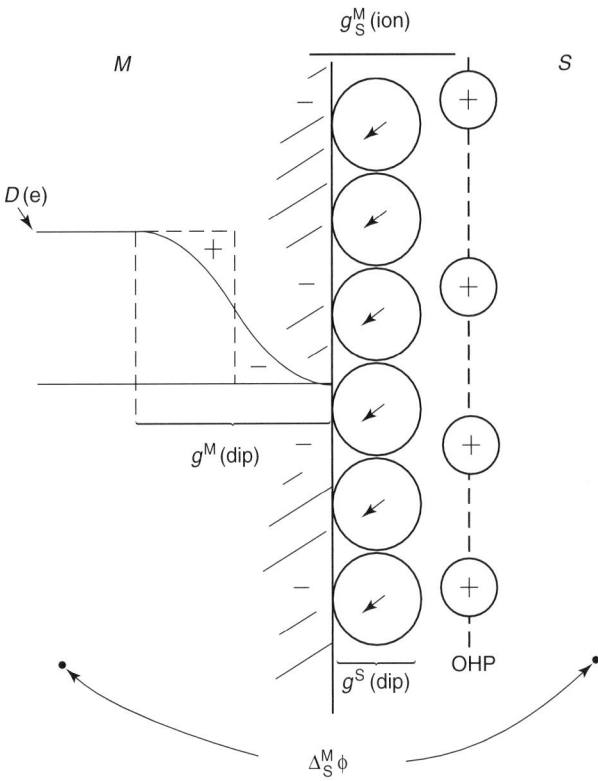

Figure 3. Components of the Galvani potential difference at a metal–solution interface. (Reproduced from Trassatti (1980)[2] with permission from Kluwer Academic/Plenum Publishers.)

a bitter dispute between Galvani and Volta over the origin of the electromotive force (EMF) when two different metals are immersed in the same solution could, in principle, be due just to the Volta potential difference between the metals. In fact, it is easy to see that if conditions are such that there are no free charges on either metal, the difference in potential between them, again a measurable quantity, is given by

$$\Delta E_{\sigma=0} = \Delta \Phi + (\Delta_S^M \psi)_{\sigma=0} \quad (4)$$

showing that the difference in work functions would only account for the difference in electrode potentials if the two Volta terms were actually zero.

3 INTERFACIAL STATISTICAL THERMODYNAMICS OF THE DIFFUSE LAYER

Development of a self-consistent theory for the double layer has proven extremely difficult, since the presence of an electrode introduces an essentially non-isotropic element into the equations. This manifests itself in the need for a term $w(x_k)$ in the potential energy of the ion deriving from the electrode itself, where x_k is the distance between the kth particle and the electrode surface. It is possible to work within the McMillan–Mayer theory of solutions, and a complete account of w must include the following contributions:

1. A short-range contribution, $w^s(x_k)$, which takes into account the nearest distance of approach of the ion to the electrode surface. For ions that do not specifically adsorb, this will be the OHP, distance h from the electrode. For ions that do specifically adsorb, $w^s(x_k)$ will be more complex, having contributions both from short-range attractive forces and from the energy of de-solvation.
2. A contribution from the charge on the surface, $w^{Q_e}(x_k)$. If this charge density is written Q_e, then elementary electrostatic theory shows that $w^{Q_e}(x_k)$ will have the unscreened form

$$w^{(Q_e)}(x_k) = \text{const.} + \frac{z_k e_0 Q_e}{\varepsilon \varepsilon_0} x_k \qquad (5)$$

3. An energy of attraction of the ion to its intrinsic image, $w^{im}(x_k)$, of unscreened form

$$w^{(im)}(x_k) = \frac{z_k^2 e_0^2}{16\pi\varepsilon\varepsilon_0 x_k} \qquad (6)$$

In addition, the energy of interaction between any two ions will contain a contribution from the mirror potential of the second ion; $u(r_{ii})$ is now given by a short-range term and a term of the form

$$u^{(el)}(r_{ij}) = \frac{z_i z_j e_0^2}{4\pi\varepsilon\varepsilon_0}\left(\frac{1}{r_{ij}} - \frac{1}{r_{ij}^*}\right) \qquad (7)$$

where r_{ij}^* is the distance between ion i and the image of ion j.

Note that there are several implicit approximations made in this model: the most important is that we have neglected the effects of the electrode on orienting the solvent molecules at the surface. This is highly significant: image forces arise whenever there is a discontinuity in dielectric function, and the simple model above would suggest that at the least, the layer of solvent next to the electrode should have a dielectric function rather different, especially at low frequencies, from the bulk dielectric function. Implicit in equations (5–7) is also the fact that the dielectric constant of water, ε_W, is assumed independent of x, an assumption again at variance with the simple model presented in Figure 2. In principle, these deficiencies could be overcome by modifying the form of the short-range potentials, but it is not obvious that this will be satisfactory for the description of the image forces, which are intrinsically long-range.

The zeroth-order solution is the Goüy–Chapman theory dating from the early part of the 20th century.[5, 6] In this solution, the ionic atmosphere is ignored, as is the mirror image potential for the ion. The variation of potential with distance from the electrode surface reduces to

$$\frac{d^2\phi}{dx^2} = -\frac{\sum_\alpha z_\alpha e_0 n_\alpha^0 \exp\{-\beta[z_\alpha e_0 \phi(x)]\}}{\varepsilon\varepsilon_0} \qquad (8)$$

where we have built in the further assumption that $w^s(x) = 0$ for $x > h$ and $w^s(x) = \infty$ for $x < h$. This corresponds to the hard sphere model introduced above. Whilst equation (8) can be solved for general electrolyte solutions, a solution in closed form can be most easily obtained for a 1–1 electrolyte with ionic charges $\pm z$. Under these circumstances, equation (8) reduces to

$$\frac{d^2\phi}{dx^2} = \frac{2ze_0 n^0}{\varepsilon\varepsilon_0} \sinh\left(\frac{ze_0}{kT}\phi\right) \qquad (9)$$

where we have assumed that $n_+^0 = n_-^0 = n^0$. Integration under the boundary conditions above gives:

$$Q_e = \sqrt{(8kT\varepsilon\varepsilon_0 n^0)} \sinh\left(\frac{ze_0\phi(h)}{2kT}\right) \qquad (10)$$

$$\phi(h) = \frac{2kT}{ze_0} \sinh^{-1}\left(\frac{Q_e}{\sqrt{(8kT\varepsilon\varepsilon_0 n^0)}}\right) \qquad (11)$$

$$\phi(x) = \frac{4kT}{ze_0} \tanh^{-1}\left\{e^{-\kappa(x-h)} \tanh\left(\frac{ze_0\phi(h)}{4kT}\right)\right\} \qquad (12)$$

and κ is given by $\kappa^2 = 2z^2 e_0^2 n^0/\varepsilon\varepsilon_0 kT$. These are the central results of the Goüy–Chapman theory, and clearly if $ze_0\phi(h)/4kT$ is small, then $\phi(x) \sim \phi(h)e^{-\kappa(x-h)}$, and the potential decays exponentially in the bulk of the electrolyte. The basic physics is similar to the Debye–Hückel theory of inter-ionic interactions, in that the actual field due to the electrode becomes screened by the ionic charges in the electrolyte. Clearly κ is a measure of the screening length; for a 1 M electrolyte of unit 1 : 1 charge, the value of κ is 3.2×10^9 m^{-1}, corresponding to a screening distance of just 3 Å, as indicated above. Developments of the theory have been the subject of considerable analytical investigation,[1] but there has been relatively little progress in devising more accurate theories, since the approximations made even in these simple derivations are very difficult to correct accurately.

3.1 Specific ionic adsorption and the inner layer

Interaction of the water molecules with the electrode surface can be developed through simple statistical models. Clearly, for water molecules close to the electrode surface, there will be several opposing effects: the hydrogen bonding tending to align the water molecules with those in solution, the electric field tending to align the water molecules with dipole moments perpendicular to the electrode surface, and dipole–dipole interactions tending to orient the nearest neighbor dipoles in opposite directions. Simple estimates[7] based on 20 kJ mol^{-1} for each H bond, suggest that the orientation energy pE, where p is the dipole moment of water, becomes comparable to this for $E \sim 5 \times 10^9$ V m^{-1}; such field strengths will be associated with surface charges of the order of 0.2–0.3 C m^{-2} or 20–30 µC cm^{-2} assuming $p = 6.17 \times 10^{-30}$ C m and the dielectric function for water at the electrode surface is taken here as approximately 6, corresponding to all molecules being strongly oriented, and only the electronic polarizability gives rise to the dielectric function. These are comparable to fields expected at quite moderate electrode potentials. Similarly, the energy of interaction of two dipoles lying anti-parallel to each other is $-p^2/(4\pi\varepsilon_0 R^3)$; for R $\sim$ 4 Å, the orientational field needs to be in excess of 10^9 V m^{-1}, a comparable number.

The simplest model for water at the electrode surface has just two possible orientations of the water molecules at the surface, and was initially described by Watts-Tobin.[8] The associated potential drop is given by

$$g(\text{dip}) = -\frac{(N_+ - N_-)p}{\varepsilon\varepsilon_0} \quad (13)$$

and if the total potential drop across the inner region of dimension h_i is $\Delta\phi$:

$$\frac{N_+}{N_-} = \exp\left[-\frac{\left(\frac{U_0 - 2p\Delta\phi}{h_i}\right)}{kT}\right] \quad (14)$$

where U_0 is the energy of interaction between neighboring dipoles. A somewhat more sophisticated model is to assume that water is present in the form of both monomers and dimers, with the dimers so oriented as not to give any net contribution to the value of dipolar potential drop.

A further refinement has come with the work of Parsons,[9, 10] building on an analysis by Frumkin. Parsons suggested that the solvent molecules at the interface could be thought of as being either free or associated to form clusters; in the second case, the net dipole moment would be reduced from the value found for perpendicular alignment since the clusters would impose their own alignment. The difficulty with such models is that the structure of the clusters themselves is likely to be a function of the electric field, and simulation methods show that this is indeed the case, as shown below.

Experimental data and arguments by Trassatti[11] show that at the PZC, the water dipole contribution to the potential drop across the interface is relatively small, varying from $\sim$0 V for Au to $\sim$0.2 V for In and Cd. For transition metals, values as high as 0.4 V are suggested. The basic idea of water clusters on the electrode surface dissociating as the electric field is increased has also been supported by in situ Fourier Transform infrared (FTIR) studies, as has the idea that re-orientation of water molecules takes place from oxygen up at negatively charged electrode to oxygen down as the electrode becomes positively charged,[12–14] and this model also underlies more recent statistical mechanical studies.[15] X-ray diffraction studies have also been used to probe the interface: these studies indicate that the surface produces a layering of the water molecules that extends up to three layers into the bulk. Interestingly, such studies also show that there are strong differences in this layered structure as a function of potential: for positively charged surfaces there is a large increase in the density of the first layer, which becomes compressed in towards the electrode surface.[16]

The model of the inner layer suggests that the interaction energy of water molecules with the metal will be at a minimum somewhere close to the PZC, a result strongly supported by the fact that adsorption of less polar organic molecules often shows a maximum at this same point.[4] However, particularly at anodic potentials, there is now strong evidence that simple anions may lose part of their hydration or solvation sheath and migrate from the OHP to the IHP. There is also evidence that some larger cations, such as [R$_4$N]$^+$, Tl$^+$, and Cs$^+$ also undergo specific adsorption at sufficiently negative potentials. The evidence for specific adsorption comes not only from classical experiments in which the surface tension of mercury is studied as a function of potential (electrocapillarity), and the coverage derived from rather indirect reasoning,[17] but from more direct methods such as measurement of the amount of material removed from solution, the use of radioactive tracers and ellipsometry. A critical problem is much of this work, particularly in those data derived from electrocapillarity, is that the validity of the Goüy–Chapman model must be assumed, an assumption that has been queried. The calculation of the free energy change associated with this process is not simple, and the following effects need to be considered:

1. The energy gained on moving from the OHP to the IHP. The electrostatic part of this will have the form

$z_k e_0 Q_e / \varepsilon \varepsilon_0$ ($x_{OHP} - x_{IHP}$), but the de-solvation part is much more difficult to estimate.
2. The fact that more than one molecule of water may be displaced for each anion adsorbed, and that the adsorption energy of these water molecules will show a complex dependence on the electrode potential.
3. The fact that a chemical bond may form between metal and anion, leading to at least partial discharge of the ion.
4. The necessity to calculate the electrostatic contribution to both the ion-electrode attraction and the ion-ion repulsion energies bearing in mind that there are at least two dielectric function discontinuities in the simple double-layer model above.
5. That short-range contributions to both ion–ion and ion–electrode interactions must be included.

These calculations have, as their aim, the generation of an *adsorption isotherm*, relating the concentration of ions in the solution to the coverage in the IHP and the potential (or more usually the charge) on the electrode. No complete calculations have been carried out incorporating all the above terms; in general, the analytical form for the isotherm is

$$\ln(f(\theta)) = \text{const.} + \ln a_\pm + AQ_e + g(\theta) \quad (15)$$

where $f(\theta)$ and $g(\theta)$ are functions of the coverage. For models where lateral interactions are dominant, $g(\theta)$ will have a $\sqrt{\theta}$ dependence; if multiple electrostatic imaging is important, a term linear in θ will be found, whereas if dispersion interactions between ions on the surface is important, then a term in θ^3 becomes significant. The form of $f(\theta)$ is normally taken as $\theta/(1-\theta)^p$ where p is the number of water molecules displaced by one adsorbed ion. Details of the various isotherms are given elsewhere[17] but modern simulation methods as reviewed below are needed to make further progress.

4 SIMULATION TECHNIQUES

The theoretical complexity of the models discussed above and the relative difficulty of establishing unequivocal models from the experimental data available has led to an increasing interest in Monte Carlo and particularly molecular dynamics approaches. Such studies have proved extremely valuable in establishing quite independent models of the interface against which the theories described above can be tested. In particular, these simulation techniques allow a more realistic explicit treatment of the solvent and ions on an equal footing. Typically, the solvent is treated within a rigid multipole model, in which the electrical distribution is modeled by a rigid distribution of charges on the various atoms in the solvent molecule. Dispersion and short-range interactions are modeled using Lennard–Jones or similar model potentials, and the interaction of water with the metal surface is generally modeled with a corrugated potential term to take account of the atomic structure of the metal. Such potentials are particularly marked for the metal–oxygen interaction, in the absence of an electrical charge on the electrode, the Pt–O interaction energy is usually given by an expression of the form:

$$U_{Pt-O} = [Ae^{-\alpha r} - Be^{-\beta r}]f(x,y) + Ce^{-\gamma r}[1 - f(x,y)] \quad (16)$$

where $f(x,y) = e^{-\lambda(x^2+y^2)}$ and the Pt–H interaction is weakly repulsive, of the form

$$U_{Pt-H} = De^{-\mu r} \quad (17)$$

This potential will lead to a single water molecule adsorbing at the PZC on Pt with the dipole pointing away from the surface and the oxygen atom pointing directly at a Pt-atom site (on-top configuration). More advanced ab initio calculations have also been attempted using either direct first principles calculation of the metal and a pseudopotential for the metal–water interactions, or fully first principles calculations using pseudopotential plane-wave density functional methods.[18, 19] These methods remain extremely costly in computing time and are only yet in their infancy.

The main difficulty in these simulations is the long-range nature of the Coulomb interactions, since both mirror plane images and real charges must be included, and the finite nature of the simulated volume must also be included. A more detailed discussion is given elsewhere,[20, 21] and the following conclusions have been reached:

1. Only at extremely high electric fields are the water molecules fully aligned at the electrode surface. For electric fields of the size normally encountered, a distribution of dipole directions is found, whose half-width is strongly dependent on whether specific adsorption of ions takes place. In the absence of such adsorption, the distribution function steadily narrows, but in the presence of adsorption, the distribution may show little change from that found at the PZC; an example is shown in Figure 4.[22]
2. The pair correlation functions g_{OO}, g_{OH} and g_{HH} have been obtained for water on an uncharged electrode surface. For Pt(100), the results are shown in Figure 5, and compared to the correlation functions for the second, much more liquid-like layer. It is clear that the first solvation peak is enhanced by comparison to the liquid, but is in the same position, emphasizing the importance

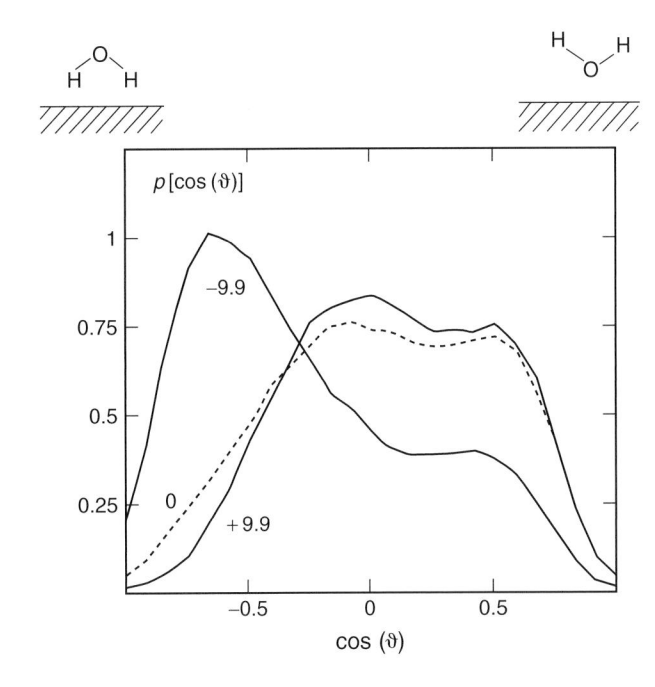

Figure 4. Orientational distribution of the water dipole moment in the adsorbate layer for three simulations with different surface charge densities (in units of $\mu C\, cm^{-2}$ as indicated). In the figure, $\cos\theta$ is the cosine of the angle between the water dipole vector and the surface normal that points into the aqueous phase. (Reproduced from Spohr (1999)[22] with permission from Elsevier Science.)

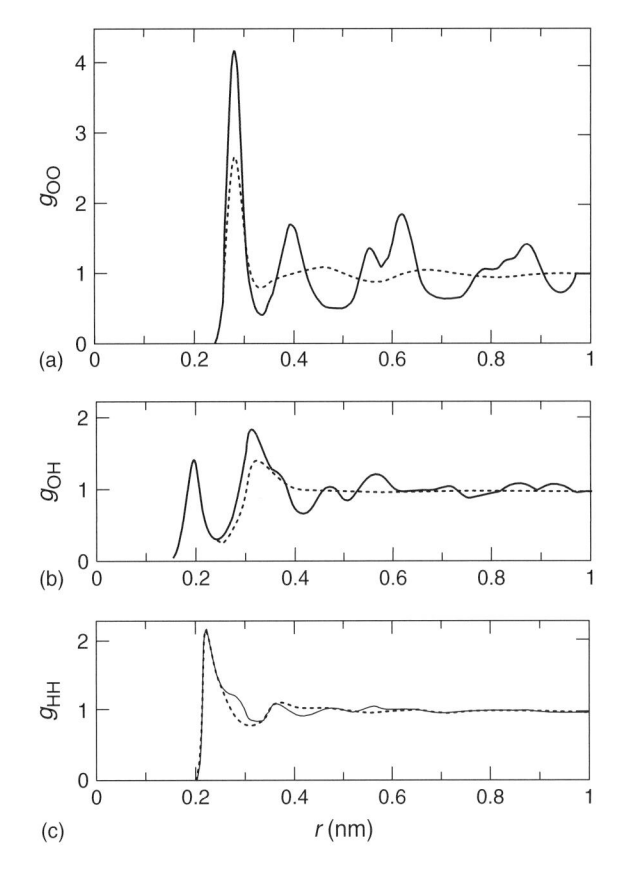

Figure 5. Water pair correlation functions near the Pt(100) surface. In each panel, the full line is for water molecules in the first layer, and the broken line is for water molecules in the second layer. (Reproduced from Benjamin (1997)[20] with permission from Kluwer Academic/Plenum Publishers.)

of H bonding in determining nearest O–O distances; however beyond the first peak, there are new peaks in the pair correlation function for the water layer immediately adjacent to the electrode that are absent in the liquid, and result from the periodicity of the Pt surface. By contrast, these peaks have disappeared in the second layer, which is very similar to normal liquid water.

3. Simulation results for turning on the electric field at the interface in a system consisting of a water layer between two Pt electrodes 3 nm apart show that the dipole density initially increases fairly slowly, but that between $10\,V\,nm^{-1}$ and $20\,V\,nm^{-1}$ there is an apparent phase transition from a moderately ordered structure, in which the ordering is confined close to the electrodes only, to a substantially ordered layer over the entire 3 nm thickness. Effectively at this field, which corresponds to the energy of ~ 4 H bonds, the system loses all ordering imposed by the H bonds and reverts to a purely linear array of dipoles.

4. For higher concentrations of aqueous electrolyte, the simulations suggest that the ionic densities do not change monotonically near the electrode surface, as might be expected from the Goüy–Chapman analysis above, but oscillate in the region $x < 10\,\text{Å}$. This oscillation is, in part, associated with the oscillation in the oxygen atom density caused by the layering effect occurring in liquids near a surface.

5. At finite positive and negative charge densities on the electrode, the counterion density profiles often exhibit significantly higher maxima, i.e., there is an overshoot, and the derived potential actually shows oscillations itself close to the electrode surface at very high concentrations.

6. Whether the potentials are derived from quantum mechanical calculations or classical image forces, it is quite generally found that there is a stronger barrier to adsorption of cations at the surface than anions, in agreement with that generally found experimentally.

We can summarize at this stage: the potential drop near the electrode surface primarily depends on the orientation of water dipoles and the specific adsorption of ions on the electrode surface or at the outer Helmholtz layer. In more concentrated (~ 1 M) electrolyte solutions, almost all the potential is, in any case, accommodated within a few angstroms of the electrode, and this region is, therefore, crucially important. For non-adsorbing ions, such as F^-, at potentials both positive and negative of the potential of zero charge (i.e., that potential at which there is no net surface charge density on the electrode), the region next to the electrode surface behaves rather like a parallel-plate capacitor, with one plate being the metal electrode surface, and the other plate the outer Helmholtz layer. If, as is quite common, the anion does specifically adsorb but the cation does not, then there will appear to be a sudden change in the double-layer capacitance for concentrated electrolytes at or near the potential of zero charge, as the second plate moves from the IHP to the OHP. At more dilute electrolyte concentrations, the behavior of the interface becomes more affected by the diffuse Goüy–Chapman layer, and the distance over which the potential change between electrode and electrolyte is accommodated increases, eventually, at dilute solutions ($\sim 10^{-2}$ M) showing a characteristic rather symmetrical hyperbolic behavior.

REFERENCES

1. C. A. Barlow and J. R. MacDonald, *Adv. Electrochem. Electrochem. Eng.*, **6**, 1 (1967).
2. S. Trassatti, in 'Comprehensive Treatise of Electrochemistry', J. O'M. Bockris, B. E. Conway and E. Yeager (Eds), Plenum Press, New York (1980).
3. N. W. Ashcroft and N. D. Mermin, 'Solid-state Physics', Holt, Rinehart and Winston, New York (1976).
4. A. N. Frumkin, O. A. Petrii and B. B. Damaskin, in 'Comprehensive Treatise of Electrochemistry', J. O'M. Bockris, B. E. Conway and E. Yeager (Eds), Plenum Press, New York, p. 221 (1980).
5. G. Goüy, *J. Phys.*, **9**, 457 (1910).
6. D. L. Chapman, *Phil. Mag.*, **25**, 475 (1913).
7. J. O'M. Bockris and S. U. Khan, 'Surface Electrochemistry', Plenum Press, New York (1993).
8. R. J. Watts-Tobin, *Phil. Mag.*, **6**, 133 (1961).
9. R. Parsons, *J. Electroanal. Chem.*, **53**, 229 (1975).
10. B. B. Damaskhin and A. N. Frumkin, *Electrochim. Acta*, **19**, 173 (1974).
11. S. Trassatti, in 'Trends in Interfacial Electrochemistry', A. Fernando Silva (Ed), NATO ASI series 179, Reidel, Dordrecht, (1986).
12. A. Bewick, K. Kunimatsu, J. Robinson and J. W. Russell, *J. Electroanal. Chem.*, **276**, 175 (1981).
13. M. A. Habib and J. O'M. Bockris, *Langmuir*, **2**, 388 (1986).
14. T. Iwasita and X. Xia, *J. Electroanal. Chem.*, **411**, 95 (1996).
15. R. Guidelli, in 'Trends in Interfacial Electrochemistry', A. Fernando Silva (Ed), NATO ASI series 179, Reidel, Dordrecht, p. 387 (1986).
16. J. G. Gordon, O. R. Melroy and M. F. Toney, *Electrochim. Acta*, **40**, 3 (1995).
17. M. A. Habib and J. O'M. Bockris, 'Comprehensive Treatise of Electrochemistry', J. O'M. Bockris, B. E. Conway and E. Yeager (Eds), Plenum Press, New York, Vol. 1, p. 135 (1980).
18. D. L. Price, *J. Chem. Phys.*, **112**, 2973 (2000).
19. S. Izvekov and G. A. Voth, *J. Chem. Phys.*, **115**, 7196 (2001).
20. I. Benjamin, *Modern Asp. Electrochem.*, **31**, 115 (1997).
21. W. Schmickler, *Chem. Rev.*, **96**, 3177 (1996).
22. E. Spohr, *Electrochim. Acta*, **44**, 1697 (1999).

Chapter 3
Thermodynamics of electrodes and cells

A. Hamnett

University of Strathclyde, Glasgow, UK

1 THERMODYNAMICS OF ELECTRIFIED INTERFACES[1-4]

If a metal such as copper is placed in contact with a solution containing the ions of that metal, such as aqueous copper sulfate, then we expect an equilibrium to be set up of the following form:

$$Cu^0 \longleftrightarrow Cu^{2+} + 2e_m^- \qquad (1)$$

where the subscript m refers to the metal. As indicated above, there will be a potential difference across the interface, between the Galvani potential in the interior of the copper and that in the interior of the electrolyte. The effects of this potential difference must be incorporated into the normal thermodynamic equations describing the interface, which is done, as above, by defining the *electrochemical potential* of a species with charge $z_i e_o$ per ion or $z_i F$ per mole. If one mole of z-valent ions is brought from a remote position to the interior of the solution, in which there exists a potential ϕ, then the work done will be $zF\phi$; this work term must be added to or subtracted from the free energy per mole μ depending on the relative signs of the charge and ϕ, and the condition for equilibrium for component i partitioned between two phases with potentials $\phi(I)$ and $\phi(II)$ is

$$\mu_i(I) + z_i F\phi(I) = \mu_i(II) + z_i F\phi(II) \qquad (2)$$

where $\phi(I)$ and $\phi(II)$ are the Galvani or inner potentials in the interior of phases (I) and (II). The expression $\mu_i + z_i F\phi$ is referred to as the electrochemical potential, $\tilde{\mu}_i$. We have

$$\tilde{\mu}_i = \mu_i + z_i F\phi = \mu_i^0 + RT \ln a_i + z_i F\phi \qquad (3)$$

and the condition for electrochemical equilibrium can be written for our copper system

$$\tilde{\mu}_{Cu}(M) = \tilde{\mu}_{Cu^{2+}}(S) + 2\tilde{\mu}_{e^-}(M) \qquad (4)$$

where the labels M and S refer to the metal and to the solution respectively. Assuming the copper atoms in the metal to be neutral, so that $\tilde{\mu}_{Cu^0} = \mu_{Cu^0}$, then we have

$$\mu_{Cu^0}^0(M) + RT \ln(a_{Cu}(M)) = \mu_{Cu^{2+}}^0(S) + RT \ln(a_{Cu^{2+}}) \\ + 2F\phi_S + \mu_{e^-}^0(M) \\ + 2RT \ln(a_{e^-}) - 2F\phi_M \qquad (5)$$

and given that the concentration of both copper atoms and electrons in copper metal will be effectively constant, so two of the activity terms can be neglected, we finally have, on rearranging equation (5),

$$\Delta\phi \equiv \phi_M - \phi_S = \frac{\mu_{Cu^{2+}}^0(S) + \mu_{e^-}^0(M) - \mu_{Cu^0}^0(M)}{2F} \\ + \frac{RT}{2F} \ln(a_{Cu^{2+}}) \equiv \Delta\phi_0 + \left(\frac{RT}{2F}\right) \ln(a_{Cu^{2+}}) \qquad (6)$$

where $\Delta\phi_o$ is the Galvani potential difference at equilibrium between electrode and solution in the case where $a_{Cu^{2+}}(aq.) = 1$, and is referred to as the standard Galvani potential difference. It can be seen, in general, that the Galvani potential difference will alter, in general, by a factor of $(RT/zF)\ln 10 \equiv 0.059/z$ V at 298 K, for every order of magnitude change in activity of the metal ion, where z is the valence of the metal ion in solution. Equation (6) is, in essence, a form of the celebrated Nernst equation.

2 THE NERNST EQUATION FOR REDOX ELECTRODES

In addition to the case of a metal in contact with its ions in solution, there are other cases in which a Galvani potential difference between two phases may be found. One case is the immersion of an inert electrode, such as platinum metal, into an electrolyte solution containing a substance S that can exist in either an oxidized or reduced form through the loss or gain of electrons from the electrode. In the simplest case, we have:

$$S_{ox} + e^- \longleftrightarrow S_{red} \tag{7}$$

an example being

$$Fe^{3+} + e^- \longleftrightarrow Fe^{2+} \tag{8}$$

where the physical process described is the exchange of *electrons* (not ions) between the electrolyte and the electrode: at no point is the electron conceived as being free in the solution. The equilibrium properties of the redox reaction (7) can, in principle, be treated in the same way as above. At equilibrium, once a double layer has formed and a Galvani potential difference set up, we can write:

$$\tilde{\mu}_{S_{ox}} + n\tilde{\mu}_{e^-}^M = \mu_{S_{red}} \tag{9}$$

and, bearing in mind that the positive charge on "ox" must exceed "red" by $|ne^-|$ if we are to have electroneutrality, then (9) becomes

$$\mu_{S_{ox}}^0 + RT\ln(a_{S_{ox}}) + nF\phi_S + n\mu_{e^-}^0$$
$$- nF\phi_M = \mu_{S_{red}}^0 + RT\ln(a_{S_{red}}) \tag{10}$$

whence

$$\Delta\phi = \phi_M - \phi_S = \frac{\mu_{S_{ox}}^0 + n\mu_{e^-}^0 - \mu_{S_{red}}^0}{nF} + \frac{RT}{nF}\ln\left(\frac{a_{S_{ox}}}{a_{S_{red}}}\right)$$

$$\equiv \Delta\phi^0 + \frac{RT}{nF}\ln\left(\frac{a_{S_{ox}}}{a_{S_{red}}}\right) \tag{11}$$

where the standard Galvani potential difference is now defined as that for which the activities of S_{ox} and S_{red} are equal. As can be seen, an alteration of this ratio by a factor of ten leads to a change of $0.059/n$ V in $\Delta\phi$ at equilibrium. It can also be seen that $\Delta\phi$ will be independent of the magnitudes of the activities provided their ratio is a constant. Equation (11) is the appropriate form of the Nernst equation for a redox couple in solution.

For more complicated redox reactions, a general form of the Nernst equation may be derived by analogy with (11). If we consider a stoichiometric reaction of the following type:

$$\nu_1 S_1 + \nu_2 S_2 + \cdots + \nu_i S_i + ne^- \longleftrightarrow \nu_j S_j + \cdots \nu_p S_p \tag{12}$$

which can be written in the abbreviated form

$$\sum_{ox} \nu_{ox} S_{ox} + ne^- \longleftrightarrow \sum_{red} S_{red} \tag{13}$$

then straightforward manipulation leads to the generalized Nernst equation

$$\Delta\phi = \Delta\phi^0 + \frac{RT}{nF}\ln\left(\frac{\prod_{ox} a_{ox}^{\nu_{ox}}}{\prod_{red} a_{red}^{\nu_{red}}}\right) \tag{14}$$

where the notation

$$\prod_{ox} a_{ox}^{\nu_{ox}} = a_{S_{ox_1}}^{\nu_{ox_1}} a_{S_{ox_2}}^{\nu_{ox_2}} \cdots a_{S_{ox_i}}^{\nu_{ox_i}} \tag{15}$$

As an example, the reduction of permanganate in acid solution follows the equation

$$MnO_4^- + 8H_3O^+ + 5e^- \longleftrightarrow Mn^{2+} + 12H_2O \tag{16}$$

and the Galvani potential difference between a platinum electrode immersed in a solution containing both permanganate and Mn^{2+} and the electrolyte is given by

$$\Delta\phi = \Delta\phi^0 + \frac{RT}{5F}\ln\left(\frac{a_{MnO_4^-} a_{H_3O^+}^8}{a_{Mn^{2+}}}\right) \tag{17}$$

assuming that the activity of neutral H_2O can be put equal to unity.

3 THE NERNST EQUATION FOR GAS ELECTRODES

The Nernst equation above for the dependence of the equilibrium potential of redox electrodes on the activity of solution species is also valid for uncharged species in the gas phase that take part in electron exchange reactions at the electrode–electrolyte interface. For the specific equilibrium process involved in the reduction of chlorine:

$$\text{Cl}_2 + 2e^- \longleftrightarrow 2\text{Cl}^- \quad (18)$$

the corresponding Nernst equation for the Galvani potential difference between the interior of the electrode and the interior of the electrolyte can easily be shown to be:

$$\Delta\phi = \Delta\phi^0 + \frac{RT}{2F}\ln\left(\frac{a_{\text{Cl}_2}(\text{aq.})}{a_{\text{Cl}^-}^2}\right) \quad (19)$$

where $a_{\text{Cl}_2}(\text{aq.})$ is the activity of the chlorine gas dissolved in water. If the Cl_2 solution is in equilibrium with chlorine at pressure p_{Cl_2} in the gas phase, then

$$\mu_{\text{Cl}_2}(\text{gas}) = \mu_{\text{Cl}_2}(\text{aq.}) \quad (20)$$

Given that

$$\mu_{\text{Cl}_2}(\text{gas}) = \mu_{\text{Cl}_2}^0(\text{gas}) + RT\ln\left(\frac{p_{\text{Cl}_2}}{p^0}\right)$$

and $\mu_{\text{Cl}_2}(\text{aq.}) = \mu_{\text{Cl}_2}^0(\text{aq.}) + RT\ln(a_{\text{Cl}_2}(\text{aq.}))$, where p^0 is the standard pressure of 1 atm ($\equiv 101\,325\,\text{Pa}$), then it is clear that

$$a_{\text{Cl}_2}(\text{aq.}) = \left(\frac{p_{\text{Cl}_2}}{p^0}\right)\cdot\exp\left(\frac{\mu_{\text{Cl}_2}^0(\text{gas}) - \mu_{\text{Cl}_2}^0(\text{aq.})}{RT}\right) \quad (21)$$

and we can write

$$\Delta\phi = \Delta\phi^{0'} + \left(\frac{RT}{2F}\right)\ln\left(\frac{p_{\text{Cl}_2}}{p^0 a_{\text{Cl}^-}^2}\right) \quad (22)$$

where $\Delta\phi^{0'}$ is the Galvani potential difference under the standard conditions of $p_{\text{Cl}_2} = p^0$ and $a_{\text{Cl}^-} = 1$.

4 THE MEASUREMENT OF ELECTRODE POTENTIALS AND CELL VOLTAGES

Although the results quoted above are given in terms of the Galvani potential difference between metal electrode and electrolyte solution, direct measurement of this Galvani potential difference between electrode and electrolyte is not possible, since any voltmeter or similar device will incorporate unknowable surface potentials into the measurement. In particular, any contact of a measurement probe with the solution phase will have to involve a second phase boundary between metal and electrolyte somewhere; at this boundary an electrochemical equilibrium will be set up and with it a second equilibrium Galvani potential difference, and the overall potential difference measured by this instrument will in fact be the difference of two Galvani voltages at the two interfaces. In other words, even at zero current, the actual voltage or electro motive force (EMF) measured for a galvanic cell will be the difference between the two Galvani potential differences $\Delta\phi(\text{I})$ and $\Delta\phi(\text{II})$ for the two interfaces, as shown in Figure 1 ([1], Figure 3-2).

Figure 1 shows the two possibilities that can exist, in which the Galvani potential of the solution, ϕ_S, lies between $\phi(\text{I})$ and $\phi(\text{II})$ and in which it lies below (or, equivalently, above) the Galvani potentials of the metals. It should be emphasized that Figure 1 is highly schematic: in reality the potential near the phase boundary in the solution changes initially linearly and then exponentially with distance away from the electrode surface, as we have seen above. The other point is that we have assumed that ϕ_S is a constant in the region between the two electrodes: this will only be true provided the two electrodes are immersed in the same solution and that no current is passing.

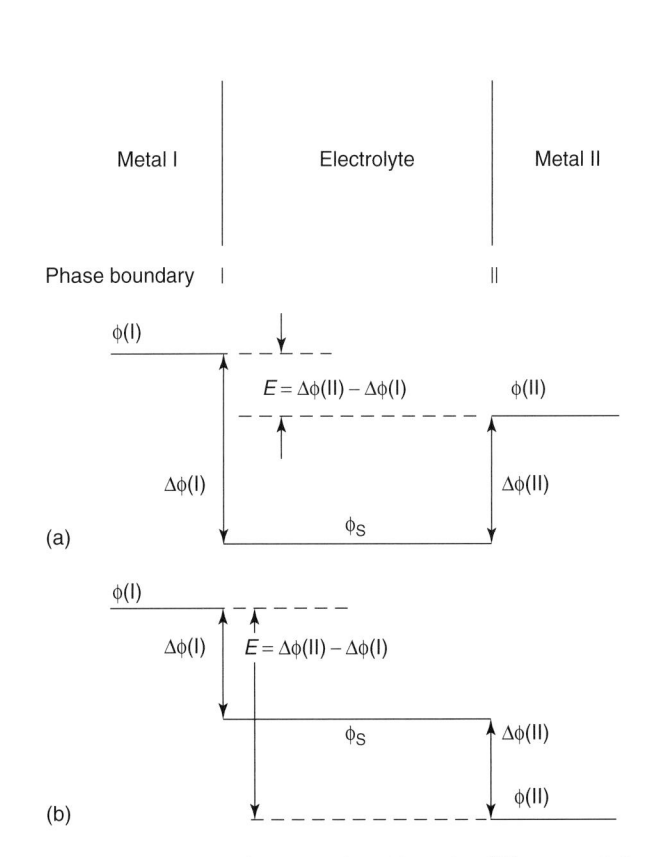

Figure 1. The EMF of a galvanic cell as the difference of the equilibrium Galvani potentials at the two electrodes: (a) $\Delta\phi(\text{I}) > 0$; $\Delta\phi(\text{II}) > 0$; (b) $\Delta\phi(\text{I}) > 0$; $\Delta\phi(\text{II}) < 0$.

It is clear from Figure 1 that the EMF or voltage, E, between the two metals is given by

$$E = \Delta\phi(\text{II}) - \Delta\phi(\text{I}) = \phi(\text{II}) - \phi(\text{I}) \quad (23)$$

where we adopt the normal electrochemical convention that the EMF is always equal to the Galvani potential difference for the electrode on the right of the figure minus the corresponding potential difference for the electrode on the left. It follows that once the Galvani potential difference of any one electrode is known, it should be possible, at least in principle, to determine the potential differences for all other electrodes. In practice, since the Galvani potential of no single electrode is known, the method adopted is to arbitrarily choose one *reference* electrode and assign a value for its Galvani potential difference. The choice actually made is that of the hydrogen electrode, in which hydrogen gas at one atmosphere pressure is bubbled over a platinized platinum electrode immersed in a solution of unit H_3O^+ activity. From the discussion in Section 3, it is clear that provided equilibrium can be established rapidly for such an electrode, its Galvani potential difference will be a constant, and changes in the measured EMF of the complete cell as conditions are altered at the other electrode will actually reflect the changes in the Galvani potential difference *at that electrode*.

Cells need not necessarily contain a reference electrode to obtain meaningful results; as an example, if the two electrodes in Figure 1 are made from the same metal, M, but these are now in contact with two solutions of the same metal ions, M^{z+} but with differing ionic activities, which are separated from each other by a glass frit that permits contact but impedes diffusion, then the EMF of such a cell, termed a concentration cell, is given by

$$E = \Delta\phi(\text{II}) - \Delta\phi(\text{I}) = \frac{RT}{z\text{F}} \ln\left(\frac{a_{M^{z+}}(\text{II})}{a_{M^{z+}}(\text{I})}\right) \quad (24)$$

Equation (24) shows that the EMF increases by $0.059/z$ V per decade change in the activity ratio in the two solutions.

5 CONVENTIONS IN THE DESCRIPTION OF CELLS

In order to describe any electrochemical cell, a convention is needed for writing down cells, such as the concentration cell described above. This convention should establish clearly where the boundaries between different phases exist and also what the overall cell reaction is. It is now standard to use vertical lines to delineate phase boundaries, such as those between a solid and a liquid or between two immiscible liquids. The junction between two miscible liquids, which might be maintained by the use of a porous glass frit, is represented by a single vertical dashed line, |, and two dashed lines, ||, are used to indicate two liquid phases joined by an appropriate electrolyte bridge adjusted to minimize potential differences arising from the different diffusion coefficients of anion and cation (so-called "junction potentials").

The cell is written such that the cathode is to the right when the cell is acting in galvanic mode, and electrical energy is being generated from electrochemical reactions at the two electrodes. From the point of view of external connections, the cathode will appear to be the positive terminal, since electrons will travel in the external circuit from the anode, where they pass from electrolyte to metal, to the cathode, where they pass back into the electrolyte. The EMF of such a cell will then be the difference in Galvani potentials of the metal electrodes on the right-hand side and the left-hand side. Thus, the concentration cell of Section 3 would be represented by $M|M^{z+}(\text{I})|M^{z+}(\text{II})|M$.

In fact, some care is needed with regard to this type of concentration cell since the assumption implicit in the derivation of equation (24), that the potential in the solution is constant between the two electrodes, cannot be entirely correct. At the phase boundary between the two solutions, which is here a semi-permeable membrane permitting the passage of water molecules but not ions between the two solutions, there will be a potential jump. This so-called liquid-junction potential difference will increase or decrease the measured EMF of the cell depending on its sign. Potential jumps at liquid–liquid junctions are in general rather small compared to normal cell voltages, and can be minimized further by suitable experimental modifications to the cell.

If two redox electrodes both use an inert electrode material such as platinum, the cell EMF can be written down immediately. Thus, for the hydrogen/chlorine fuel cell, which we represent by the cell $H_2(g)|Pt|HCl(m)|Pt|Cl_2(g)$, for which it is clear that the cathodic reaction is reduction of Cl_2 as considered in Section 3

$$\begin{aligned}E &= \Delta\phi\left(\frac{Cl_2}{Cl^-}\right) - \Delta\phi\left(\frac{H_2}{H_3O^+[\text{aq.}]}\right) \\ &= E^0 - \left(\frac{RT}{2\text{F}}\right)\ln(a_{H_3O}^2 a_{Cl^-}^2) \\ &\quad + \left(\frac{RT}{2\text{F}}\right)\ln\left\{\left(\frac{p_{H_2}}{p^0}\right)\left(\frac{p_{Cl_2}}{p^0}\right)\right\}\end{aligned} \quad (25)$$

where E^0 is the standard EMF of the fuel cell, or the EMF at which the activities of H_3O^+ and Cl^- are unity and the pressures of H_2 and Cl_2 are both equal to the standard pressure, p^0. Note that E^0 is temperature dependent, and to avoid confusion, the value under standard conditions of temperature (298 K), the symbol E^{00} can be used. Care should be taken in this area as symbol usage is not entirely standardized.

REFERENCES

1. C. H. Hamann, A. Hamnett and W. Vielstich, 'Electrochemistry', Wiley-VCH, Weinheim (1998).
2. V. S. Bagotzky, 'Fundamentals of Electrochemistry', Plenum Press, New York (1993).
3. A. J. Bard, R. Parsons and J. Jordan, 'Standard Potentials in Aqueous Solution', Marcel Dekker, New York (1985).
4. C. M. A. Brett and A. M. Oliveira Brett, 'Electrochemistry, Principles, Methods and Applications', Oxford University Press, Oxford (1993).

Chapter 4
Ideal and effective efficiencies of cell reactions and comparison to carnot cycles

W. Vielstich

IQSC, São Carlos, Universidade de São Paulo, Brazil

1 MAXIMUM OBTAINABLE WORK OF A CHEMICAL REACTION

The reaction in a fuel cell always corresponds to a chemical process, divided into two separated electrochemical reactions at anode and cathode. In a hydrogen/oxygen cell the overall chemical process is the formation of water

$$H_2 + \tfrac{1}{2}O_2 \longrightarrow H_2O \quad (1)$$

divided into the two electrochemical reactions

$$H_2 \longrightarrow 2H^+ + 2e^- \quad (2)$$

$$\tfrac{1}{2}O_2 + 2H^+ + 2e^- \longrightarrow H_2O \quad (3)$$

It is obvious that in order for the overall reaction to be balanced, the same number of electrons, n, must be exchanged in both reactions. If we allow one formula unit of reactants to be transformed, the electric work at our load will be nFE, E being the cell voltage during discharge. At a very slow discharge, i $\to$ 0, assuming for the moment that true electrochemical equilibrium can be established, we obtain the maximum work nFE^0 with E^0 as thermodynamic cell voltage.

The maximum electric work must correspond to the maximum useful work obtainable from a chemical reaction, in our example the work obtainable when gaseous H_2 and O_2 combine, resulting in water formation. For the transformation of one mole of reactants, the maximum work obtainable is designated as the difference in molar free energy, ΔG. The electric work obtained is a positive number, but it is conventional in thermodynamics to designate work done by the system on the surroundings as negative. From this follows:

$$\Delta G = -nFE^0 \quad (4)$$

If the reactants and products of the cell reaction are in their standard states, i.e., the gases are at unit pressure and temperature at 25 °C, then the free energy change (at constant pressure P and constant temperature T, $\Delta G_{PT} = G_f - G_i$, f = final and i = initial) is denoted by the standard value ΔG^0:

$$\Delta G^0 = -nFE^{00} \quad (5)$$

with E^{00} being the cell voltage under standard conditions.

If the fuel cell is in the discharge mode, then is $E < E^0$, and it follows

$$\Delta G + nFE < 0 \quad (6)$$

In the case that the current is flowing in the reverse direction and electrolysis takes place, one has

$$\Delta G + nFE > 0 \quad (7)$$

In general words, ΔG is a measure for the affinity and direction of a chemical process. The maximum obtainable useful energy is independent of the reaction routes taken.

2 FUEL CELL REACTIONS AND THERMODYNAMIC EFFICIENCIES

In order to discuss thermodynamic efficiencies, we have to introduce two other thermodynamic functions, the reaction enthalpy $\Delta Hr = Hf - Hi$ and the reaction entropy $\Delta Sr = Sf - Si$. Over the Gibbs–Helmholtz relation these functions are connected to the above free energy of reaction ΔGr:

$$\Delta Gr\,(P,\,T\,\text{constant}) = \Delta Hr + T\frac{\Delta Hr + T}{\{\delta/\delta T(\Delta Gr)\}}$$
$$= \Delta Hr - T\,\Delta Sr \quad (8)$$

The reaction enthalpy ΔHr is the heat delivered by the chemical reaction, the reaction entropy ΔSr is a measure of the change in the order of the system during the reaction. ΔHr also has a negative sign when heat is given out. Therefore, with $\Delta Sr < 0$, one has the case that $[\Delta G] < [\Delta H]$. One example is the H_2/O_2 Fuel Cell. A part of the chemical energy is transferred to heat in the case of electrochemical energy conversion, and this heat is in addition to the heat developed via the potential drop along the resistances inside the electrochemical cell (see Figure 1).

The thermodynamic or ideal efficiency of energy conversion is always related to the reaction enthalpy of the according chemical process

$$\xi_{th} = \frac{\Delta G}{\Delta H} = 1 - \frac{T\Delta S}{\Delta H} \quad (9)$$

With $\Delta S < 0$, we have efficiencies lower than 1.0. For $\Delta S > 0$ the thermodynamic efficiency is above 100%, heat from the surrounding is transferred into useful work.

If the thermal cell voltage is defined as $E_H^0 = -\Delta H/nF$, then the ideal efficiency can be expressed as

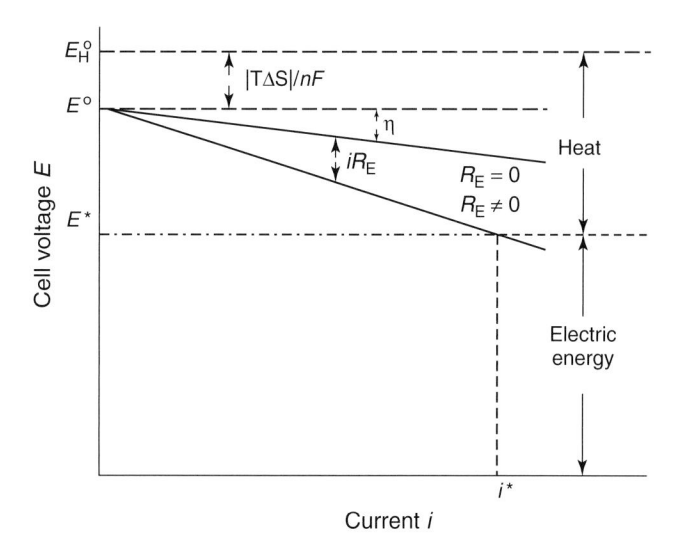

$$\xi_{th} = \frac{E^0}{E_H^0} \quad (10)$$

Figure 1. Dependence of the cell voltage of a galvanic cell (with reaction entropy $\Delta S < 0$) on the current load, i. R_E electrolyte resistance, η overvoltages at the electrodes, E^0 thermodynamic cell voltage, E_H^0 thermal cell voltage, E cell voltage under load. (Reproduced from Hamann et al. (1998)[1] with permission from Wiley-VHC.)

As an example, for the fuel cell reaction $H_2 + 1/2O_2 = H_2O$, for standard conditions we have $E^{00} = 1.23$ V and $E_H^0 = 1.48$ V, giving $\xi = 0.83$. In Figure 1 the relationships between the parameters discussed above are shown for $\Delta S < 0$. The respective data for several fuel cells are given in Table 1. For almost all fuel cell reactions one obtains $\Delta S < 0$, i.e., the amount of $T\Delta S$ is given as heat to the surrounding. Only for formic acid oxidation the free energy change ΔG is larger than the reaction enthalpy, i.e., with the conversion of chemical energy into electricity additional energy is gained via absorption of heat from the surroundings. The ideal efficiency becomes more than 100%. Another example for $\Delta S < 0$ is given in Table 2, the reaction $C + 1/2O_2 \rightarrow CO$.

Table 1. Thermodynamic Data for fuel cell reactions under standard conditions at 25 °C, ideal efficiency ξ_{th} (%).

Fuel cell	Reaction	n	$-\Delta H^0$ (kJ mol^{-1})	$-\Delta G^0$ (kJ mol^{-1})	E^{00} (V)	ξ_{th} (%)
Hydrogen	$H_2 + 1/2O_2 \rightarrow H_2O$ equation (1)	2	286.0	237.3	1.229	83.0
CO	$CO + 1/2O_2 \rightarrow CO_2$	2	283.1	257.2	1.066	90.9
Formic acid	$HCOOH + 1/2O_2 \rightarrow CO_2 + H_2O_l$	2	270.3	285.5	1.480	105.6
Formaldehyde	$CH_2O_g + O_2 \rightarrow CO_2 + H_2O_l$	4	561.3	522.0	1.350	93.0
Methanol	$CH_3OH + 3/2O_2 \rightarrow CO_2 + 2H_2O_l$	6	726.6	702.5	1.214	96.7
Methane	$CH_4 + 2O_2 \rightarrow CO_2 + 2H_2O_l$	8	890.8	818.4	1.060	91.9
Ammonia	$NH_3 + 3/4O_2 \rightarrow 1/2N_2 + 3/2H_2O_l$	3	382.8	338.2	1.170	88.4
Hydrazine	$N_2H_4 + O_2 \rightarrow N_2 + H_2O_l$	4	622.4	602.4	1.560	96.8
Zinc	$Zn + 1/2O_2 \rightarrow ZnO$	2	348.1	318.3	1.650	91.4

Table 2. Thermodynamic cell voltage E^0 and ideal efficiency ξ at different temperatures.

Fuel cell reaction	298 K		600 K		1000 K	
	E^0 [V]	ξ [%]	E^0 [V]	ξ [%]	E^0 [V]	ξ [%]
$H_2 + 1/2 O_2 \rightarrow H_2O_g$	1.18	94	1.11	88	1.00	78
$CH_4 + 2O_2 \rightarrow CO_2 + 2H_2O_g$	1.04	100	1.04	100	1.04	100
$CO + 1/2 O_2 \rightarrow CO_2$	1.34	91	1.18	81	1.01	69
$C + 1/2 O_2 \rightarrow CO$	0.74	124	0.86	148	1.02	178

3 EFFICIENCIES UNDER LOAD — CONVERSION OF CHEMICAL ENERGY INTO HEAT AND ELECTRIC ENERGY

The efficiencies of a fuel cell under load can be obtained, in the first approximation from plots as in Figure 1. Once the cell is under load, the efficiency will fall. The internal resistance loss in the electrolyte results in a voltage drop of iR_E, R_E being the electrolyte resistance. In addition, we have losses on both electrodes in the form of overvoltages η due to slow charge transfer and diffusion (see **Kinetics of electrochemical reactions**, Volume 1). The respective load efficiency ξ_{load} is defined with respect to the actual cell voltage $E(i)$ at current i as:

$$\xi_{load} = \frac{E(i)}{E_H^0} = \frac{1}{E_H^0 \{E^0 - iR_E - \sum |\eta(i)|\}} \quad (11)$$

For a hydrogen/oxygen fuel cell, current densities of technical interest can be obtained already at a cell voltage of 0.8 V. Divided by $E_H^0 = 1.48$ V, one has as load efficiency $\xi_{load} = 0.8/1.48 = 54\%$. All data will depend, of course on temperature. Examples are given in Table 2.

It should be noted that for hydrocarbon fuel cells, like the methane/oxygen cell in Table 1, at temperatures below 100 °C the reaction product water is in liquid form. This causes the entropy change to be negative and we have thermodynamic efficiencies below 100%. However at temperatures > 100 °C water is liberated as gas and we have no change in entropy, the efficiency being independent to temperature and equal to 100% (see Table 2).

The net balance for the production of heat per unit time can be written according to Figure 1:

$$W = i|T \Delta S|/nF + i \sum |\eta(i)| + i^2 R_E$$
$$= i(E_H^0 - E^0) + i \sum |\eta(i)| + i^2 R_E \quad (12)$$

The units of W are watts, and W is always positive. Its value is of great importance for engineering design.

For the effective efficiency, ξ_{eff} one has finally to consider losses in the active mass, as e.g., the loss of hydrogen during fuel cell module operation. If only 95% of the fuel is used, we have $\xi_{eff} = 0.54 \times 0.95 = 51.3$. In addition, it has to be included in the effective efficiency the energy necessary for pumping of fuel, heating or cooling, compression of gases as well as for auxiliary installations.

4 COMPARISON TO CARNOT CYCLES

The data in Tables 1 and 2 show that fuel cell reactions, as used in practice to produce electric energy in a direct way, do have ΔG values smaller than the respective ΔH values. Therefore the efficiency of a fuel cell should be compared to the theoretical efficiency of a heat engine as e.g., the internal combustion engine of our cars.

The maximum obtainable work A from a heat engine is given by the expression

$$A = \xi_{HE} |\Delta H| \quad (13)$$

with ΔH the reaction enthalpy or heat of reaction and the ξ_{HE} the well-known Carnot factor:

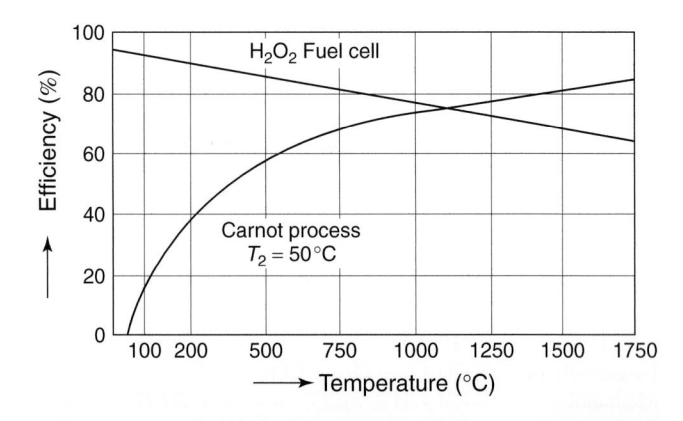

Figure 2. Comparison between the thermodynamic efficiency of a heat engine (Carnot efficiency) and the ideal efficiency of a H_2/O_2 fuel cell $\Delta G/\Delta H$ (calculated for the case that gaseous water is formed, see Table 2).

$$\xi_{HE} = \frac{(T_h - T_c)}{T_c} \quad (14)$$

T_h being the absolute temperature (°K) of the hot sink, and T_c the one of the cold sink. In Figure 2 the theoretical efficiency ξ_{HE} of the heat engine, the so-called *Carnot factor* is plotted versus the temperature (°C) and compared with the thermodynamic efficiency of a fuel cell. With increasing temperature T_h at constant temperature T_c the expression (14) is finally approaching 100%. In contrast to equation (9), for $\Delta S < 0$ the efficiency of the fuel cell falls only with increasing temperature (compare Table 2). While most thermodynamic efficiencies of fuel cells are near 90%, the thermodynamic efficiencies of heat engines, working ideally between temperatures T_h and T_c, are mostly less than 40%. Of course, as in the case of fuel cells anticipating losses by overvoltages, the practical data of heat engines are also less than expected, due to an increase of the effective T_c above the surrounding with T_h. Nevertheless, the total efficiencies of fuel cells ξ_{eff} show good performance, especially at low nominal load, as demonstrated in Figure 3. Due to the fact that with larger units the energy for auxiliary installations amounts to smaller percentages, efficiencies are increasing with power demand.

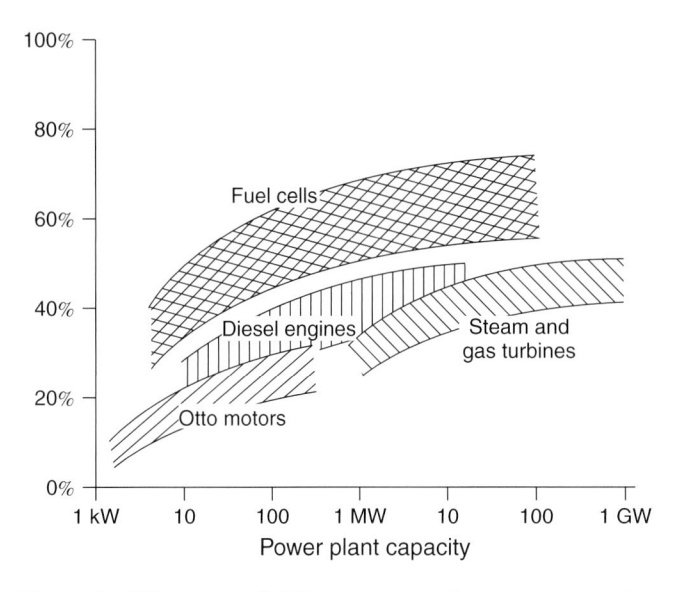

Figure 3. Efficiencies of different systems for power generation as function of nominal load.

5 EFFECTIVE EFFICIENCIES OF VARIOUS TRACTION SOURCES WITH ENERGY SUPPLY FROM COAL

The efficiency of energy conversion has to be considered anew if petrol and natural gas are no longer available in sufficient quantities and the main energy source is coal. Then

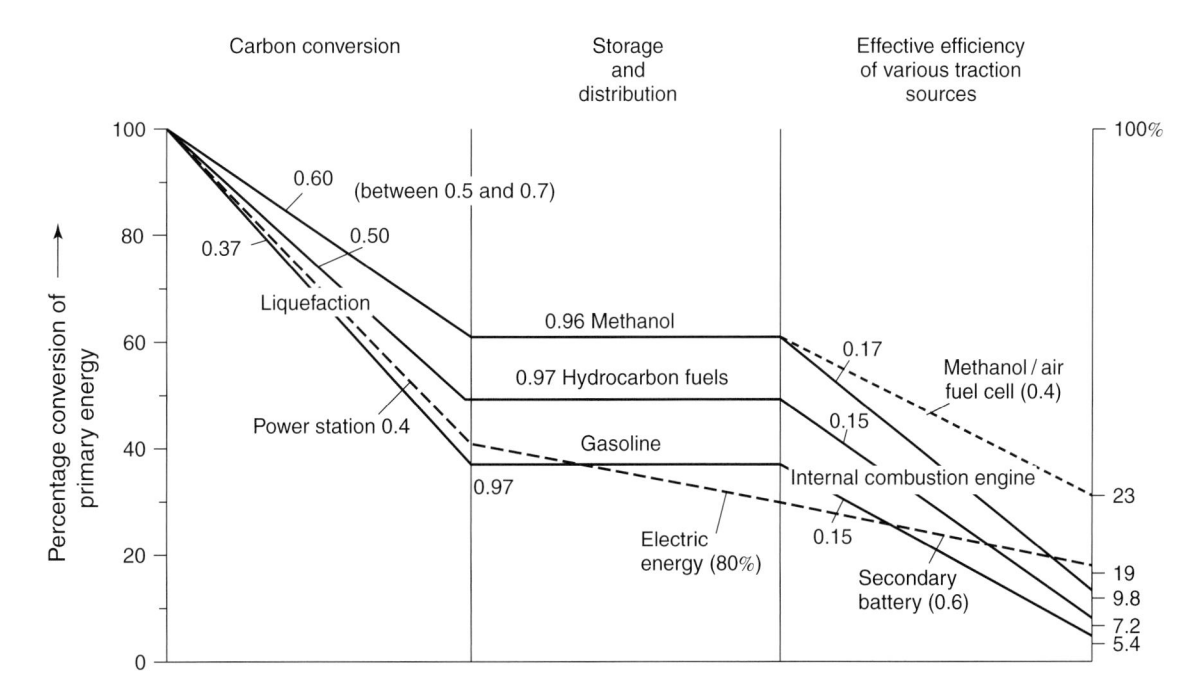

Figure 4. Simplified energy efficiency diagram with coal as the primary fuel source, for traction through liquefaction and electricity generation. The data on the right side of the diagram give the percentage of the primary energy (coal) available at the gear. The figures on each line are the approximate efficiencies for each process. (Reproduced from Hamann *et al.* (1998)[1] with permission from Wiley-VHC.)

we have to compare two energy conversion cycles: coal → liquid fuel → internal combustion engine, and coal → electrical energy → secondary battery → electric motor. This comparison is done in Figure 4, taking into account as a third possibility the cycle coal → methanol → direct methanol fuel cell → electric motor. In this simplified diagram the figures on each line are the approximate efficiencies of the respective steps of energy conversion. Via coal liquefaction (Fischer–Tropsch process) gasoline, hydrocarbons and also methanol can be obtained. The value of 0.8 for storage and distribution of electric energy includes the efficiency of the stationary charging equipment. The value of 0.6 for secondary batteries contains the efficiency of storage (0.67) and electric drive (0.9). The figure shows that clearly the electric vehicle would be ahead of the present combustion engines. This is clear the case for the battery-driven-vehicle, but very pronounced in case where a direct methanol fuel cell can be used.

REFERENCE

1. C. H. Hamann, A. Hamnett and W. Vielstich, 'Electrochemistry', Wiley-VCH, Weinheim, Germany (1998).

Chapter 5
Kinetics of electrochemical reactions

A. Hamnett
University of Strathclyde, Glasgow, UK

1 ELECTRICAL POTENTIALS AND ELECTRICAL CURRENT

The discussion in earlier sections has focused, by and large, on understanding the equilibrium structures in solution and at the electrode/electrolyte interface. In this last section, some introductory discussion will be given of the situation in which we depart from equilibrium by permitting the flow of electrical current through the cell. Such current flow leads not only to a potential drop across the electrolyte, which affects the cell voltage by virtue of an Ohmic drop $I \times R_i$, where R_i is the internal resistance of the electrolyte between the electrodes, but each electrode exhibits a characteristic current–voltage behavior, and the overall cell voltage will, in general, reflect both these effects.

1.1 The concept of overpotential[1]

Once current passes through the interface, the Galvani potential difference will differ from that expected from the Nernst equation above; the magnitude of the difference is termed the *overpotential*, which is defined heuristically as

$$\eta = \Delta\phi - \Delta\phi_r = E - E_r \qquad (1)$$

where the subscript r refers to the *rest* situation, i.e., to the potential measured in the absence of any current passing.

Provided equilibrium can be established, this rest potential will correspond to that predicted by the Nernst equation. Obviously, the sign of η is determined by whether E is greater than or less than E_r.

At low currents, the rate of change of electrode potential with current is associated with the limiting rate of electron transfer across the phase boundary between the electronically conducting electrode and the ionically conducting solution, and is termed the electron transfer overpotential. The electron transfer rate at a given overpotential has been found to depend on the nature of the species participating in the reaction, and the properties of the electrolyte and the electrode itself (such as, for example, the chemical nature of the metal). At higher current densities, the primary electron transfer rate is usually no longer limiting; instead, limitations arise through slow transport of reactants from solution to the electrode surface or conversely the slow transport of product away from the electrode (diffusion overpotential) or through the inability of chemical reactions coupled to the electron transfer step to keep pace (reaction overpotential).

Examples of the latter include adsorption or desorption of species participating in the reaction or the participation of chemical reactions before or after the electron transfer step itself. One such process occurs in the evolution of hydrogen from a solution of a weak acid, HA: in this case, the electron transfer from electrode to proton in solution must be preceded by the acid dissociation reaction taking place in solution.

2 THE THEORY OF ELECTRON TRANSFER[1–5]

The rate of simple chemical reactions can now be calculated with some confidence either within the framework of activated-complex theory or directly from quantum mechanical first principles, and theories that might lead to analogous predictions for simple electron-transfer reactions at the electrode–electrolyte interface have been the subject of much recent investigation. Such theories have hitherto been concerned primarily with greatly simplified models for the interaction of an ion in solution with an inert electrode surface. Specific adsorption of electroactive species has been excluded, and electron transfer is envisaged only as taking place between the strongly solvated ion in the outer Helmholtz layer and the metal electrode. The electron transfer process itself can only be understood through the formalism of quantum mechanics, since the transfer itself is a tunneling phenomenon that has no simple analogue in classical mechanics.

Within this framework, by considering the physical situation of the electrode double layer, the free energy of activation of an electron transfer reaction can be identified with the reorganization energy of the solvation sheath around the ion. This idea will be carried through in detail for the simple case of the strongly solvated Fe^{3+}/Fe^{2+} couple, following the change in ligand–ion distance as the critical reaction variable during the transfer process.

In aqueous solution, the reduction of Fe^{3+} can be conceived as a reaction of two aquo-complexes of the form

$$[Fe(H_2O)_6]^{3+} + e^- \longleftrightarrow [Fe(H_2O)_6]^{2+} \qquad (2)$$

The H_2O molecules of these aquo-complexes constitute the inner solvation shell of the ions, which are, in turn, surrounded by an external solvation shell of more or less uncoordinated water molecules forming part of the water continuum, as described in **The components of an electrochemical cell**, Volume 1. Owing to the difference in the solvation energies, the radius of the Fe^{3+} aquo-complex is smaller than that of Fe^{2+}, which implies that the mean distance of the vibrating water molecules at their normal equilibrium point must change during the electron transfer. Similarly, changes must take place in the outer solvation shell during electron transfer, all of which implies that the solvation shells themselves inhibit electron transfer. This inhibition by the surrounding solvent molecules in the inner and outer solvation shells can be characterized by an activation free energy $\Delta G^{\ddagger}$.

Given that the tunneling process itself requires no activation energy, and that tunneling will take place at some particular configuration of solvent molecules around the ion, the entire activation energy referred to above must be associated with ligand/solvent movement. Furthermore, from the Franck–Condon principle, the electron tunneling process will take place on a time-scale rapid compared to nuclear motion, so that the ligand and solvent molecules will be essentially stationary during the actual process of electron transfer.

Consider now the aquo-complexes above, and let x be the distance of the center of mass of each water molecule constituting the inner solvation shell from the central ion. The binding interaction of these molecules leads to vibrations of frequency $f = \omega/2\pi$ taking place about an equilibrium point x_0, and if the harmonic approximation is valid, the potential energy change U_{pot} associated with the ligand vibration can be written in parabolic form as

$$U_{pot} = \tfrac{1}{2}M\omega^2(x - x_0)^2 + B + U_{el} \qquad (3)$$

where M is the mass of the ligands, B is the binding energy of the ligands and U_{el} is the electrical energy of the ion–electrode system. The total energy of the system will also contain the kinetic energy of the ligands, written in the form $p^2/2M$, where p is the momentum of the molecules during vibrations:

$$U_{tot} = \frac{p^2}{2M} + \tfrac{1}{2}M\omega^2(x - x_0)^2 + B + U_{el} \qquad (4)$$

It is possible to write two such equations: for the initial state, i, (corresponding to the reduced aquo-complex $[Fe(H_2O)_6]^{2+}$) and the final state, f, corresponding to the oxidized aquo-complex and the electron now present in the electrode. Clearly

$$U_{tot}^i = \frac{p^2}{2M} + \tfrac{1}{2}M\omega^2(x - x_0^i)^2 + B^i + U_{el}^i \qquad (5)$$

with a corresponding equation for state f, and with the assumption that the frequency of vibration does not alter between initial and final states of the aquo-complex. During electron transfer, the system moves, as shown in Figure 1, from an equilibrium situation centered at x_0, along the parabolic curve labeled U_{pot}^i to the point x_s where electron transfer takes place; following this, the system will move along the curve labeled U_{pot}^f to the new equilibrium situation centered on x_0^f.

The point at which electron transfer takes place clearly corresponds to the condition $U_{pot}^i = U_{pot}^f$; equating equations (5) for the states i and f we find that

$$x_s = \frac{B^f + U_{el}^f - B^i - U_{el}^i + (M\omega^2/2)([x_0^f]^2 - [x_0^i]^2)}{M\omega^2(x_0^f - x_0^i)} \qquad (6)$$

The activation energy, U_{act}, is defined as the minimum additional energy above the zero-point energy that is needed

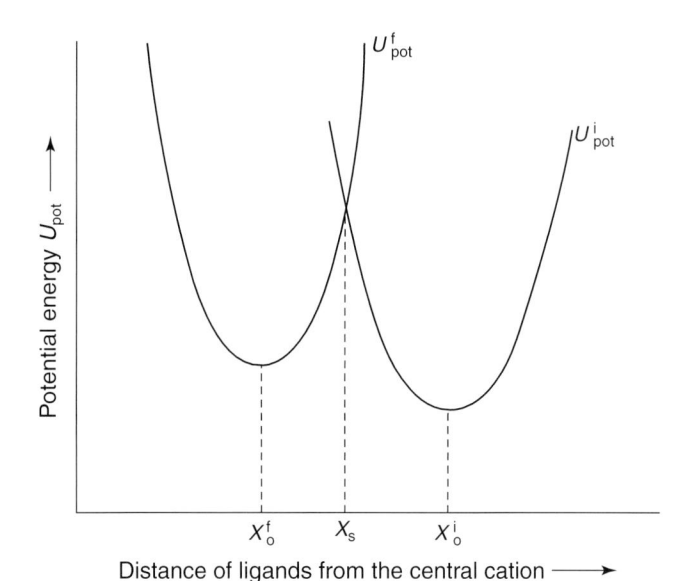

Figure 1. Potential energy of a redox system as a function of ligand–metal separation.

for a system to pass from initial to final state in a chemical reaction. In terms of equation (5), the energy of the initial reactants at $x = x_s$ is given by

$$U^i = \frac{p^2}{2M} + \frac{1}{2}M\omega^2(x_s - x_0^i)^2 + B^i + U_{el}^i \quad (7)$$

where $B^i + U_{el}^i$ is the zero-point energy of the initial state. The minimum energy required to reach the point x_s is clearly that corresponding to the momentum $p = 0$. By substituting for x_s from equation (6), we find

$$U_{act} = \frac{M\omega^2}{2}(x_s - x_0^i)^2 = \frac{(U_s + U_{el}^f - U_{el}^i + B^f - B^i)^2}{4U_s} \quad (8)$$

where U_s has the value $(M\omega^2/2)(x_0^f - x_0^i)^2$, and is termed the *reorganization energy*, since it is the additional energy required to deform the complex from initial to final value of x. It is common to find the symbol λ for U_s, and model calculations suggest that U_s normally has values in the neighborhood of 1 eV (10^5 J mol^{-1}) for the simplest redox processes.

In our simple model, the expression in equation (8) corresponds to the activation energy for a redox process in which only the interaction between the central ion and the ligands in the primary solvation shell is considered, and this only in the form of the totally-symmetrical vibration. In reality, the rate of the electron-transfer reaction is also influenced by the motion of molecules in the outer solvation shell, as well as by other vibrational modes of the inner shell. These can be incorporated into the model provided that each type of motion can be treated within the simple harmonic approximation. The total energy of the system will then consist of the kinetic energy of all the atoms in motion together with the potential energy arising from each vibrational degree of freedom. It is no longer possible to picture the motion as in Figure 1, as a one-dimensional translation over an energy barrier, since the total energy is a function of a large number of normal coordinates describing the motion of the entire system. Instead, we have two potential energy surfaces for the initial and final states of the redox system, whose intersection described the reaction hypersurface. The reaction pathway will proceed now via the saddle point, which is the minimum of the total potential energy subject to the condition $U_{pot}^i = U_{pot}^j$ as above.

This is a standard problem[2] and essentially the same result is found as in equation (8), save that the B^i and B^f now become the sum over all binding energies of the central ion in the initial and final states and U_s is now given by

$$U_s = \sum_j \frac{M_j \omega_j^2}{2}(x_{j,0}^f - x_{j,0}^i)^2 \quad (9)$$

where M_j is the effective mass of the jth mode and ω_j the corresponding frequency, and we still retain the approximation that these frequencies are all the same for the initial and final states.

With the help of U_s, an expression for the rate constant for the reaction

$$[Fe(H_2O)_6]^{2+} \longleftrightarrow [Fe(H_2O)_6]^{3+} + e_{M^-} \quad (10)$$

can be written

$$k_f = k_f^0 \exp\left(-\frac{U_{act}}{kT}\right)$$
$$= A \exp\left(-\frac{(U_s + U_{el}^f - U_{el}^i + B^f - B^i)^2}{4U_s kT}\right) \quad (11)$$

where A is the so-called frequency factor, and e_{M^-} refers to an electron in the metal electrode. The rate constant for the back reaction is obtained by interchanging the indices i and f in equation (11); it will be observed that under these circumstances, U_s remains the same, and we obtain

$$k_b = A \exp\left(-\frac{(U_s + U_{el}^i - U_{el}^f + B^i - B^f)^2}{4U_s kT}\right) \quad (12)$$

3 THE EXCHANGE CURRENT

It is now possible to derive an expression for the actual current density from equations (11) and (12), assuming reaction (10), and, for simplicity, assuming that the concentrations, c, of Fe^{2+} and Fe^{3+} are equal. The potential difference between the electrode and the outer Helmholtz layer, $\Delta\phi$, is incorporated into the electronic energy of the $Fe^{3+} + e_{M^-}$ system through a potential-dependent term of the form

$$U_{el}^{f} = U_{el,0}^{f} - e_0 \Delta\phi \tag{13}$$

where the minus sign in equation (13) arises through the negative charge on the electron. Inserting this into (11) and (12) and multiplying by concentration and the Faraday to convert from rate constants to current densities, we have

$$j^{+} = \mathbf{F}Ac \exp\left(-\frac{(U_s + U_{el,0}^{f} - U_{el}^{i} + B^{f} - B^{i} - e_0\Delta\phi)^2}{4U_s kT}\right) \tag{14}$$

$$j^{-} = -\mathbf{F}Ac$$
$$\times \exp\left(-\frac{(U_s + U_{el}^{i} - U_{el,0}^{f} + B^{i} - B^{f} + e_0\Delta\phi)^2}{4U_s kT}\right) \tag{15}$$

where we adopt the convention that positive current involves *oxidation*. At the rest potential, $\Delta\phi_r$, which is actually the same as the standard Nernst potential $\Delta\phi_0$ assuming that the activity coefficients of the ions are also equal, the rates of these two reactions are equal, which implies that the terms in brackets in the two equations must also be equal when $\Delta\phi = \Delta\phi_0$. From this, it is clear that

$$e_0 \Delta\phi_0 = U_{el,0}^{f} - U_{el}^{i} + B^{f} - B^{i} \tag{16}$$

and if we introduce the overpotential, $\eta = \Delta\phi - \Delta\phi_0$, evidently

$$j^{+} = \mathbf{F}Ac \exp\left[-\frac{(U_s - e_0\eta)^2}{4U_s kT}\right] \tag{17}$$

$$j^{-} = -\mathbf{F}Ac \exp\left[-\frac{(U_s + e_0\eta)^2}{4U_s kT}\right] \tag{18}$$

from which we obtain the exchange-current density as the current at $\eta = 0$:

$$j_0 = \mathbf{F}Ac \exp\left[-\frac{U_s}{4kT}\right] \tag{19}$$

and the activation energy of the exchange-current density can be seen to be $U_s/4$. If the overpotential is small, such that $e_0\eta \ll U_s$ (and recalling that U_s lies in the region of $\sim 1\,\mathrm{eV}$), the quadratic form of equations (17) and (18) can be expanded with neglect of terms in η^2; recalling also that $e_0/k_B = F/R$, then we obtain, finally

$$j = j^{+} + j^{-} = \mathbf{F}Ac \exp\left(-\frac{U_s}{4kT}\right)$$
$$\times \left\{\exp\left(\frac{F\eta}{2RT}\right) - \exp\left(-\frac{F\eta}{2RT}\right)\right\} \tag{20}$$

which is the simplest form of the familiar Butler–Volmer equation with a symmetry factor $\beta = 1/2$. This result arises from the strongly simplified molecular model that we have used above, and in particular the assumption that the values of ω_j are the same for all normal modes. Relaxation of this assumption leads to a more general equation:

$$j = j^{+} + j^{-} = \mathbf{F}Ac \exp\left(-\frac{U_s}{4kT}\right)$$
$$\times \left\{\exp\left(\frac{\beta F\eta}{RT}\right) - \exp\left(-\frac{(1-\beta)F\eta}{RT}\right)\right\} \tag{21}$$

$$= j^0 \left\{\exp\left(\frac{\beta F\eta}{RT}\right) - \exp\left(-\frac{(1-\beta)F\eta}{RT}\right)\right\} \tag{22}$$

Table 1. Standard exchange current densities for some simple electrode reactions.

System	Electrolyte	Temperature (°C)	Electrode	j_0 (A cm^{-2})	β
Fe^{3+}/Fe^{2+} (0.005 M)	1 M H_2SO_4	25	Pt	2×10^{-3}	0.58
$K_3Fe(CN)_6/K_4Fe(CN)_6$ (0.02 M)	0.5 M K_2SO_4	25	Pt	5×10^{-2}	0.49
$Ag/10^{-3}$ M Ag^+	1 M $HClO_4$	25	Ag	1.5×10^{-1}	0.65
$Cd/10^{-2}$ M Cd^{2+}	0.4 M K_2SO_4	25	Cd	1.5×10^{-3}	0.55
$Cd(Hg)/1.4 \times 10^{-3}$ M Cd^{2+}	0.5 M Na_2SO_4	25	Cd(Hg)	2.5×10^{-2}	0.8
$Zn(Hg)/2 \times 10^{-2}$ M Zn^{2+}	1 M $HClO_4$	0	Zn(Hg)	5.5×10^{-3}	0.75
Ti^{4+}/Ti^{3+} (10^{-3} M)	1 M acetic acid	25	Pt	9×10^{-4}	0.55
H_2/OH^-	1 M KOH	25	Pt	10^{-3}	0.5
H_2/H^+	1 M H_2SO_4	25	Hg	10^{-12}	0.5
H_2/H^+	1 M H_2SO_4	25	Pt	10^{-3}	0.5
O_2/H^+	1 M H_2SO_4	25	Pt	10^{-6}	0.25
O_2/OH^-	1 M KOH	25	Pt	10^{-6}	0.3

For a more general reaction of the form Ox + ne⁻ ↔ Red, with differing concentrations of Ox and Red, the exchange-current density is given by

$$j^0 = n\mathbf{F}A(c_{Ox}^{\beta} \cdot c_{Red}^{1-\beta}) \exp\left(-\frac{U_{act}}{kT}\right) \quad (23)$$

Some values for j^0 and β for electrochemical reactions of importance are given in Table 1, and it can be seen that the exchange currents can be extremely dependent on electrode material, particularly for more complex processes such as hydrogen oxidation. Much modern electrochemical study is spent understanding the origin of these differences in electrode performance.

REFERENCES

1. C. H. Hamann, A. Hamnett and W. Vielstich, 'Electrochemistry', Wiley-VCH, Weinheim (1998).
2. W. Schmickler, 'Interfacial Electrochemistry', Oxford University Press, Oxford (1996).
3. V. S. Bagotzky, 'Fundamentals of Electrochemistry', Plenum Press, New York (1993).
4. C. M. A. Brett and A. M. Oliveira Brett, 'Electrochemistry, Principles, Methods and Applications', Oxford University Press, Oxford (1993).
5. J. Ulstrup, 'Charge Transfer Processes in Condensed Media', Springer-Verlag, Berlin (1979).

Chapter 6
Introduction to fuel-cell types

A. Hamnett
University of Strathclyde, Glasgow, UK

1 FUEL CELLS AND ELECTROCATALYSTS[1-5]

The galvanic cell of **The components of an electrochemical cell**, Volume 1, Section 3, operating with hydrogen and oxygen as fuel and combustant, represents the archetypal fuel cell, and the operation of such fuel cells represents one of the most attractive and elusive goals of modern day electrochemistry. One immediate problem is that the fuel used is hydrogen, and therein lies one of the major problems for fuel cells. All attempts to drive fuel cells *directly* using primary fuels, such as coal oil, natural gas, etc., have failed, at least with low-temperature devices; either high temperatures or fuel reforming or both have proved necessary, leading to high costs associated with the use of expensive catalysts or complex engineering. These interconnected problems have bedeviled efforts to build commercially attractive fuel cell systems. Since the work of Grove, more than 150 years ago, which led to the realization of the first simple fuel cell (based on reactions (5) of **The components of an electrochemical cell**, Volume 1, Section 3), hydrogen/oxygen fuel cells of steadily increasing efficiencies have been fabricated, but this enhanced performance has not been sufficient to justify the costs of isolation of H_2 from the primary fuels available. The dominance of coal in the 19th century led to innumerable attempts to explore the latter's electrochemistry, attempts only abandoned when it was realized that not only would the ash severely contaminate the molten KOH favored as the electrolyte, but that this electrolyte actually reacted directly with carbon when moisture was present.

The end of the 19th century saw, initially in America, the rise of oil as a primary fuel, and this led to renewed efforts to realize fuel cells based on direct hydrocarbon oxidation at the anode. Unfortunately, facile splitting of the non-polar C–C bond has never been achieved at low temperatures by electrochemical means, and the activation of the C–H bond to allow controlled electrochemical oxidation also remains a very difficult reaction. As we shall see below, internal reforming of natural gas (i.e., its conversion with steam to CO_2 and H_2) remains a real possibility for at least one type of modern fuel cell, based on molten carbonate, but for both coal and heavy hydrocarbon fractions, reforming externally with steam remains the only option, resulting in fuel cell systems that are both technically complex and prone to the production of sulfur-containing poisons.

The emergence of natural gas as a major energy vector has altered the equation in a number of ways: methane can be reformed with high efficiency, and considerable ingenuity has been expended on the design of reformers that are significantly less complex than those required for heavier fuels. The innate purity of natural gas has also led to considerable advantages in terms of poisoning, allowing internal reforming as a possibility, as well as the coupling of external reformers to low-temperature fuel cells that are particularly prone to poisoning. Methane can even be used directly in very high temperature fuel cells of the solid-oxide type as discussed below. Even more attractive is the possible use of methanol as a major fuel, since this is a liquid that can be transported easily and would be ideal for traction. Methanol is appreciably more reactive than methane. It can be reformed at temperatures of ca. 300 °C and is even sufficiently electroactive for

---

Handbook of Fuel Cells – Fundamentals, Technology and Applications, Edited by Wolf Vielstich, Hubert A. Gasteiger, Arnold Lamm.
Volume 1: *Fundamentals and Survey of Systems*. © 2003 John Wiley & Sons, Ltd. ISBN: 0-471-49926-9.

direct oxidation at the anode at temperatures in excess of ca. 60 °C. A direct methanol fuel cell, as implied by the latter possibility, would indeed be a serious competitor for traction applications, since the costs of reforming need no longer be considered. The downside of natural gas is, inevitably, the limited lifetime of the current supplies, though new discoveries are likely to prolong its availability well into the 21st century.

In rehearsing the problems of the anode, it would not be appropriate to lose sight of the fact that even for a hydrogen based fuel cell there are other problems that reduce both efficiency and practicability. Perhaps the most serious is the relatively poor performance of electrocatalysts on the cathode side. We have already seen the effects of electrode irreversibility in our discussion in **The components of an electrochemical cell**, Volume 1, Section 3. In practice, at least at lower temperatures, electro-reduction of oxygen does not take place at an appreciable rate until relatively high overpotentials (~0.3 V), providing a further major limitation on realizable efficiencies.

2 THE ALKALINE FUEL CELL[3, 4]

The alkaline fuel cell using H_2 as a fuel is the simplest low-temperature fuel cell in concept and operation, and schematically operates in a similar manner to the cell of Figure 3 in **The components of an electrochemical cell**, Volume 1, Section 3, but with an electrolyte of concentrated KOH. It operates at temperatures of ca. 70 °C, and even at room temperature has power levels of ca. 50% of those at the nominal operating temperature. Within Europe, stacks have been constructed and tested by Elenco and Siemens particularly, and considerable experience gained in operation of fuel cells in general.

The electrode construction is critically important in this, as in all low-temperature systems, with the primary requirements being: (a) good electronic conductivity to reduce ohmic losses; (b) adequate mechanical stability and suitable porosity; (c) chemical stability in the rather aggressive alkaline electrolyte; (d) stable electroactivity of the catalyst with time. These are quite demanding considerations, especially as the cell must operate over a number of years. Normally, the electrodes are fabricated from a mixture of carbon and polytetrafluoroethylene (PTFE), the latter controlling both the macro-porosity of the electrode and its hydrophobicity. By using a multi-layer structure, with an outer uncatalyzed but highly hydrophobic layer to allow ingress of gas without egress of electrolyte, freestanding electrodes can be made. Furthermore, the cost of the current collector is relatively low in these systems, since nickel mesh can be used.

The primary cost of these electrodes lies in the catalyst, most particularly for the cathode. The Elenco electrodes used platinum, though with a relatively low loading (0.6 mg cm^{-2}), but an immense advantage of alkaline electrolytes is that the oxygen reduction reaction is relatively facile, and non-noble metal catalysts can be used: the Siemens electrodes, for example, use Ti-doped Raney Nickel at the anode and silver at the cathode, the latter with a rather high loading of 60 mg cm^{-2}.

The primary problems associated with the alkaline fuel cell are: (a) operation in air is problematic owing to the presence of CO_2, which is absorbed into the alkaline electrolyte, generating the relatively insoluble K_2CO_3 which, in turn, can deposit on and foul the cathode. CO_2 must, therefore, be scrubbed from the air if it is used as the oxidant; (b) the poisoning of the electrodes, which is particularly severe for platinum, since the anode can be poisoned by traces of CO in the hydrogen arising from sub-optimal efficiency of the reformer, or by sulfur-containing compounds derived from the primary fuel; (c) removal of the main combustion product, water, which would otherwise dilute the KOH and reduce performance. This is achieved in the Siemens cell by a remarkable piece of engineering, in which the electrolyte is pumped through the cell, carrying away waste heat, and is then passed through an evaporator, in which a hydrophobic diffusion membrane eliminates excess water from the electrolyte, passing it back at its normal strength into the stack. It should however, be emphasized that the evaporator is comparable in size to the stack, giving a rather large overall system.

3 THE PHOSPHORIC ACID CELL[3, 5]

The difficulties associated with the use of air as an oxidant in alkaline fuel cells sparked considerable interest in the exploration of acidic media, where CO_2 would present fewer problems. The common mineral acids, however, present quite serious problems in this regard: adequate conductivities can only be attained at temperatures close to boiling, at which the thermal instability of the commoner strong oxy-acids, particularly in contact with powerful noble-metal catalysts, can lead to decomposition products that poison the electrodes. Perchloric acid was found to be explosively unstable in contact with the fuel, the hydrohalic acids are extremely corrosive, and perfluorosulfonic acids, though possessing considerable advantages such as high ionic conductivity, thermal stability, high oxygen solubility and low adsorption on the platinum catalysts, have led to electrode flooding through their high wettability on PTFE and have presented considerable concentration management problems, as well as being much more expensive. The restrictions on temperature associated with these acids is particularly troublesome as the kinetics of the oxygen reduction reaction are poor in acid at low temperature, a drawback that can most easily be overcome by increasing the temperature to take advantage of the positive enthalpy of activation.

Phosphoric acid at room temperature is only slightly dissociated, showing very low conductivity even in concentrated solution, and oxygen reduction at low temperatures on platinum shows very poor kinetics owing to strong competitive adsorption of phosphate ions. However, at temperatures above ca. 150 °C, the pure acid is found predominantly in the polymeric state, as pyrophosphoric acid, which is a strong acid of high conductivity and the large size and low charge density of the polymeric anions leads to low chemisorption on platinum, facilitating the oxygen reduction reaction. Other advantages of phosphoric acid include: good fluidity at high temperatures; tolerance to CO_2 and to moderate CO concentrations at higher temperatures; low vapor pressure; high oxygen solubility; low corrosion rate and large contact angle. The result is that since its introduction in 1967, the phosphoric acid fuel cell has undoubtedly come to dominate the low-temperature fuel cell market, and is the only commercially available fuel cell.

The basic principles of the cell are shown in Figure 1. The electrode reactions are given above, and the basic design of the stack is bipolar, with separator plates used to delineate the individual cells. Within each cell, impervious graphitic carbon sheets are used for the bipolar plates, and each electrode is formed of a ribbed porous carbon-paper substrate onto which is affixed a catalyst layer composed of a high-surface-area carbon powder compacted with PTFE, which acts both as a binder and to control the hydrophobicity of the electrode. Between anode and cathode is a porous matrix formed from PTFE-bonded silicon carbide impregnated with phosphoric acid that acts as the electrolyte and also as a separator to prevent the oxidant and fuel gases mixing; since there is some loss of electrolyte, means of replenishment are incorporated into the cell. The operating temperature lies between 190 and 210 °C, and active cooling systems, using air, water or dielectric liquid, are essential.

The development of the phosphoric acid fuel cell (PAFC) has taken two directions. The most important is the construction of relatively large stationary power units, such as the 11 MW system at Goi, near Tokyo in Japan, with smaller systems in the US and Europe, which have allowed technologists to study the operation of these systems under real conditions. Smaller scale local combined heat and power (CHP) systems are also under active development, with the key advantage being the lack of any efficiency penalty for partial-load operation, even under 50% load; this balance is particularly suitable for applications such as small hospitals. There has also been considerable investigation of vehicular applications, using reformed methanol as the fuel, which has been directed towards buses as the primary commercial goal. Again, it would seem that niche markets may dictate the success of such systems in the next few years.

4 MOLTEN CARBONATE FUEL CELL[2, 4, 5]

The necessity of employing noble metal catalysts and the sensitivity of the phosphoric acid fuel cell to significant

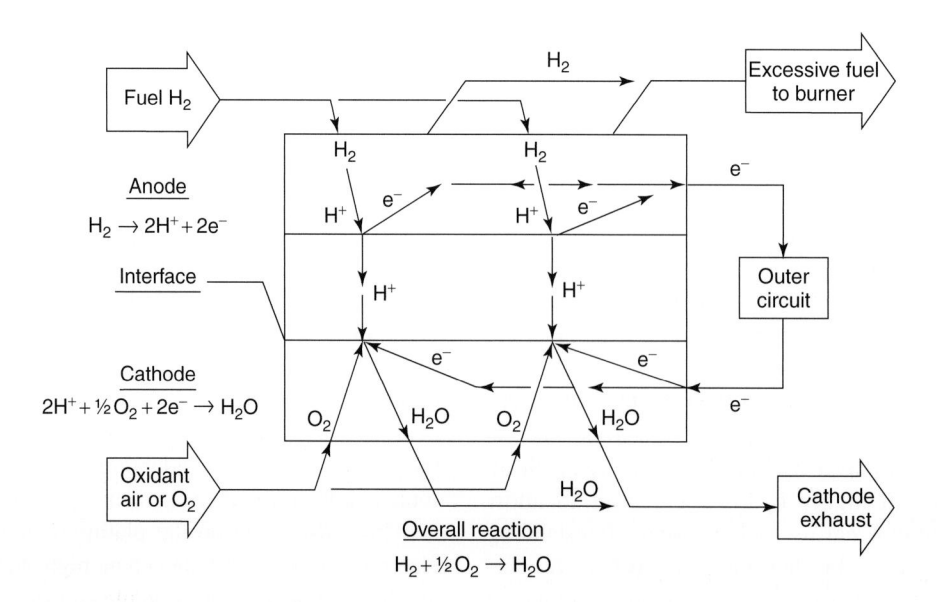

Figure 1. The principles of operation of a phosphoric acid fuel cell.

quantities of CO are serious drawbacks; only by working at higher temperatures can the kinetics of the oxygen reduction reaction be sufficiently accelerated for the use of cheaper catalysts, and as indicated above, high temperatures also favor the desorption of adsorbed CO. In addition, heat from higher temperature fuel cells can be used in combined cycle operation, allowing us to increase the system efficiency quite substantially, albeit at the cost of flexibility of operation. An additional advantage of high-temperature operation is improved compatibility with the operating temperature of the reformer, allowing us, at least in principle, the attractive prospect of combining the reformer with the fuel cell stack.

The basic principle of operation of molten carbonate fuel cells is shown in Figure 2. At the cathode, made from porous lithiated NiO ($Li_xNi_{1-x}O$: $0.022 \leq x \leq 0.04$), the reaction is

$$\tfrac{1}{2}O_2 + CO_2 + 2e^- \longrightarrow CO_3^{2-} \qquad (1)$$

The electrolyte is a eutectic mixture of 68% Li_2CO_3/32% K_2CO_3, at a temperature of 650 °C, retained in a porous γ-$LiAlO_2$ tile. At the anode, made from a porous Ni/10 wt% Cr alloy (the Cr preventing sintering), the reaction is

$$H_2 + CO_3^{2-} \longrightarrow H_2O + CO_2 + 2e^- \qquad (2)$$

In a practical cell, the CO_2 produced at the anode must be transferred to the cathode, which can be carried out either by burning the spent anode stream with excess air and mixing the result with the cathode inlet stream after removal of water vapor, or by directly separating CO_2 at the anode exhaust point. A major advantage of this cell is that CO, far from being a problem, can actually serve as the fuel, probably through the water–gas shift reaction

$$CO + H_2O \longrightarrow CO_2 + H_2 \qquad (3)$$

which at 650 °C equilibrates very rapidly on Ni.

We have seen that at least in principle, the reformer and stack temperatures are sufficiently comparable for the two to be combined, and three configurations have been suggested: (a) indirect reforming, in which a conventional reformer is provided operating separately from the stack; (b) direct internal reforming, in which the reformer catalyst pellets are incorporated into each anode gas inlet channel, and therefore operate at 650 °C, a temperature maintained by direct heating of the anode compartment; (c) indirect internal reforming, in which the reformation takes place in separate catalyst chambers installed in a stack between a set of cells so that waste heat from the fuel cell stack is supplied to help maintain the reformer temperature during the endothermic reformation process. The most elegant solution is (b), which is also the most cost-effective solution, but contamination from electrolyte vapor in the anode space leads to rapid degradation of the catalyst performance.

The overall performance figures for the molten carbonate fuel cell (MCFC) reflect the fact that the E^0 value for the hydrogen–oxygen reaction is reduced from the room temperature value of 1.23 V to a value of 1.02 V at 650 °C. The cell voltage drops almost linearly with increasing current; at ambient pressures, state-of-the-art cells show a voltage of 0.70–0.75 V at 150–160 mA cm^{-2}, though this can be appreciably enhanced with pressure.

Problems with the MCFC are: (a) the oxygen reduction reaction remains relatively slow, and its mechanism depends critically on the electrolyte, with the initial reduction product being either peroxide in Li-rich mixtures or superoxide in K-rich electrolytes, with both present at the eutectic composition; (b) dissolution of NiO leads to serious problems since the Ni(II) can migrate to the anode where, at the low potentials, it deposits as the metal. The deposition of metal grains in the pores of the tile can lead eventually to the formation of electronically conducting pathways through the electrolyte, effectively short-circuiting the cell. The dissolution of Ni is related to the partial pressure of CO_2, and may involve the reaction

$$NiO + CO_2 \longrightarrow Ni^{2+} + CO_3^{2-} \qquad (4)$$

(c) water must be present in the feed gas to avoid carbon deposition through the Boudouard reaction

$$2CO \longrightarrow CO_2 + C \qquad (5)$$

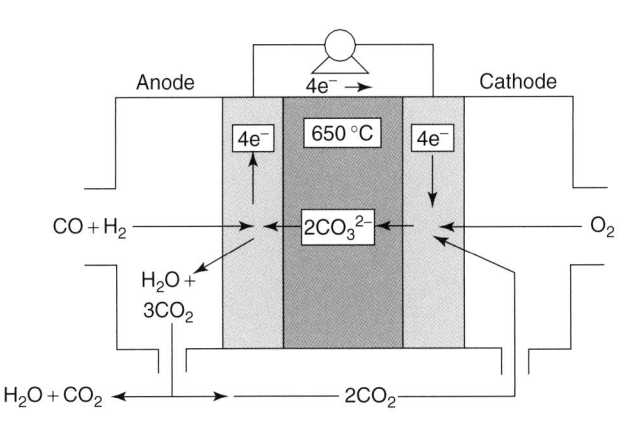

Figure 2. A schematic diagram of a molten-carbonate fuel cell. (Reproduced from Hall (2000)[2] with permission from Cambridge University Press.)

(d) Electrolyte management is a key problem in the MCFC; the control of the three-phase boundary relies on controlling the extent to which the molten carbonate is drawn out of the tile at the cell operating temperature by capillary action. By careful choice of pore size in electrode and tile, and by partially pre-filling the porous electrodes themselves, an initially optimum distribution of electrolyte can be assured, but throughout the lifetime of the cell, a slow but steady loss of electrolyte has been found, leading to gradual performance decay. This decay may, however, become catastrophic if loss of electrolyte from the tile leads to the tile becoming permeable to the fuel gases; gas crossover is usually accompanied by intense local hotspots and rapid loss of performance, and it is therefore essential that a liquid layer is maintained at all times.

5 SOLID-OXIDE FUEL CELL[2, 3, 6]

At a sufficiently high temperature, all kinetic limitations at the cathode will disappear, and it also becomes possible to utilize solid ceramic oxide-ion conductors that show very high conductivities above ca. 900 °C.

The principle of operation of these cells is shown in Figure 3. The electrolyte is typically ZrO_2 with 8–10 mol% Y_2O_3; the latter not only stabilizes the fluorite structure, preventing transition at lower temperatures to the baddeleyite phase with resultant shattering of the ceramic, but confers a substantial ionic conductivity through the presence of mobile O_2^- ions. Research has been carried out into improved electrolytes, which must: (a) show high oxygen ion conductivity and minimum electronic conductivity; (b) have good chemical stability with respect to the electrodes and inlet gases; (c) have a high density to inhibit fuel crossover; (d) have a thermal expansion compatible with other components. Materials such as Bi_2O_3 and doped-CeO_2 do show higher oxide-ion conductivity at lower temperatures than stabilized zirconia, but are more easily reduced to electronic conductors at the anode.

The anode requirements are: effective oxidation catalyst; high electronic conductivity; stability in the reducing environment; thermal expansion compatible with the electrolyte and other fuel cell components; a physical structure offering low fuel transport resistance; chemical and mechanical stability; and tolerance to sulfur contaminants. The anode most closely satisfying these requirements is a porous ca. 35% $Ni/ZrO_2/Y_2O_3$ cermet (i.e., an intimate mixture of ceramic and metal) with good electronic conductivity. Unlike conventional electrodes, there is no necessity for a well defined three-phase boundary, since oxidation of the fuel gas can take place over the entire electrode surface in this mixed electronic/ionic conductor. The anode reaction is:

$$H_2 + O^{2-} \longrightarrow H_2O + 2e^- \quad (6)$$

The cathode must: (i) exhibit good electrocatalytic activity for O_2 reduction; (ii) show good electronic conductivity since it must serve as the current collector. The combination of high temperature and an oxidizing atmosphere leads to severe materials problems at the cathode, which must be stable over very wide ranges of oxygen pressure, involatile, not show any destructive phase changes, should adhere strongly to the electrolyte over a very wide temperature range, possessing, therefore, a coefficient of expansion essentially identical to ZrO_2, and should form a junction with the zirconia of very low resistance. The cathode material most closely satisfying these requirements is porous perovskite manganite of the form $La_{1-x}Sr_xMnO_3$ ($0.10 < x < 0.15$), which shows a transition from small polaron to metallic conduction near 1000 °C. It also exhibits mixed ionic/electronic conduction, again allowing reduction of the oxygen to take place over the entire surface. The cathode reaction is

$$\tfrac{1}{2}O_2 + 2e^- \longrightarrow O^{2-} \quad (7)$$

Realization of the cell in practical terms is through three competing designs, a planar geometry, similar to conventional designs, a tubular design shown in Figure 4, and a monolithic design. All these designs, however, still need to overcome quite serious interconnect and fabrication technology problems.

Although the solid oxide fuel cell (SOFC) clearly possesses a number of advantages, there remain some severe problems:

(a) The thermodynamic cell voltage is only 0.9 V, at 1000 °C. However, because there are essentially no kinetic limitations on the cathode or anode reactions,

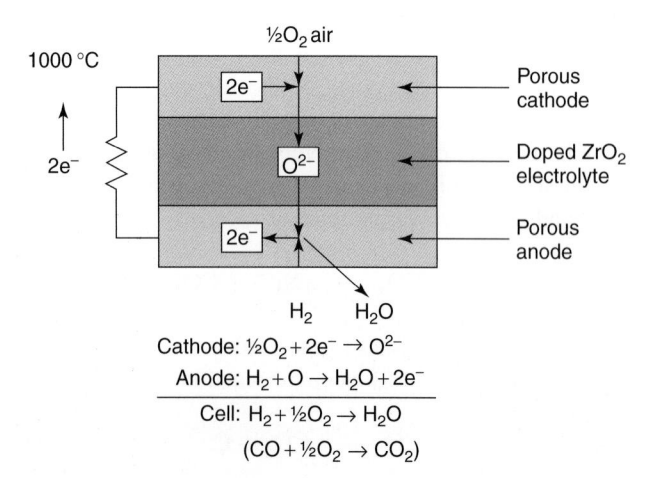

Figure 3. The principles of operation of a high-temperature solid-oxide-electrolyte fuel cell.

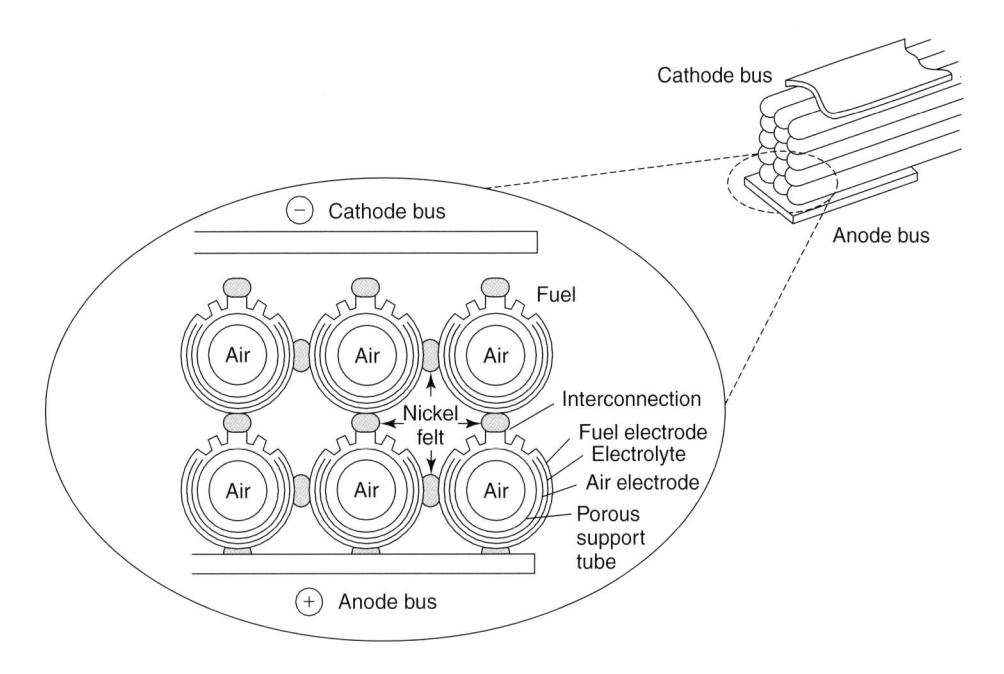

Figure 4. A schematic diagram showing the Westinghouse tubular solid-oxide fuel cell system in its bundle configuration.

reasonable current densities can be obtained at voltages of 0.75 V.

(b) The materials problems, particularly those related to thermal expansion and stability to mechanical damage, are proving extremely difficult to solve.

The intolerance of SOFCs to repeated shutdown–startup cycles, which is related to the problem of compatibility of thermal expansion, suggests that the most likely type of application envisaged for SOFCs is that of stationary power generation under steady-state conditions. Westinghouse, in particular, have constructed 25 kW and 100 kW systems, and there is strong interest in both the US and Australia in developing this technology to commercialization.

6 SOLID-POLYMER FUEL CELLS[1–3, 6]

Related to the solid-oxide-based fuel cells above, which are dependent on oxide-ion conducting electrolytes, it should be possible to find analogous fuel cells based on solid *proton* conductors. In fact, high-temperature solid proton conductors appear not to exist; at these elevated temperatures, all hydrated oxides tend to lose water and if conduction takes place at all, it will be through the metal or the oxide ion. However, at low temperatures, (below ca. 150 °C), there are several types of solid-proton conductor, both inorganic and organic, that can be used in so-called solid-polymer-electrolyte (SPE) fuel cells.

Cathode
$½O_2 + 2H^+ + 2e^- \rightarrow H_2O$

Anode
$H_2 \rightarrow 2H^+ + 2e^-$

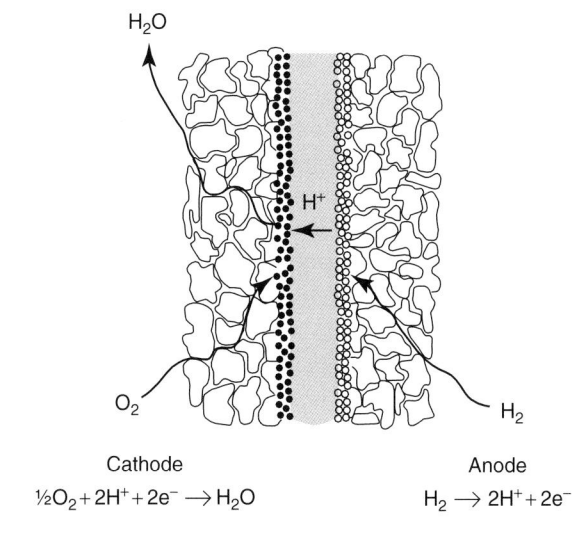

Figure 5. A schematic diagram of the operation of a solid-polymer-electrolyte H_2/O_2 fuel cell. (Reproduced from Hall (2000)[2] with permission from Cambridge University Press.)

A schematic diagram of this type of cell is shown in Figure 5: the electrolyte itself is a polymeric membrane proton conductor, as described in **The components of an electrochemical cell**, Volume 1, Section 5.2, and anode and cathode are, in the simplest designs, formed either directly from metal particles or from catalyzed carbon particles bound to the membrane. The current collectors are porous carbon or graphite plates, and the cell reactions are, at the anode

$$H_2 \longrightarrow 2H^+ + 2e^- \qquad (8)$$

and at the cathode

$$\tfrac{1}{2}O_2 + 2H^+ + 2e^- \longrightarrow H_2O \qquad (9)$$

The low operating temperatures dictate the use of noble metal catalysts, with particular problems being experienced at the cathode.

The performance of such cells, particularly under a few bar pressure, is quite remarkable, as shown in Figure 6, where power densities in excess of $1\,W\,cm^{-2}$ are seen. These high performances should not, however, disguise the fact that if application of the SPE fuel cell to transport is considered, particularly if methanol is used as the fuel and a reformer employed, there are several additional sources of loss. Operation of the fuel cell at 5 bar leads to an air compressor loss of $\sim 12\%$ of stack output, and reformer losses of 10–15% based on the heating value of the input methanol fuel are expected. Further losses to ancillary units, and in power converters etc., can lead to overall power efficiencies from methanol chemical energy to wheels of ca. 15%.

In contrast, moreover, to the high temperature cells above, the SPE fuel cell shows very considerable sensitivity to contaminants in the fuel gas, and in particular CO must be reduced to well below 10 ppm to avoid deterioration in the anode performance. This type of purity can only be achieved by using multi-stage reformers, and considerable effort has recently been expended on developing promoted platinum catalysts that are less sensitive to residual CO.

The high power density of SPE fuel cells makes them extremely attractive for traction applications, where the size and weight of the fuel cell should be as small as possible. Ballard have recently designed and constructed a bus that uses compressed hydrogen in cylinders as the fuel, and have been testing this in city driving in Vancouver.

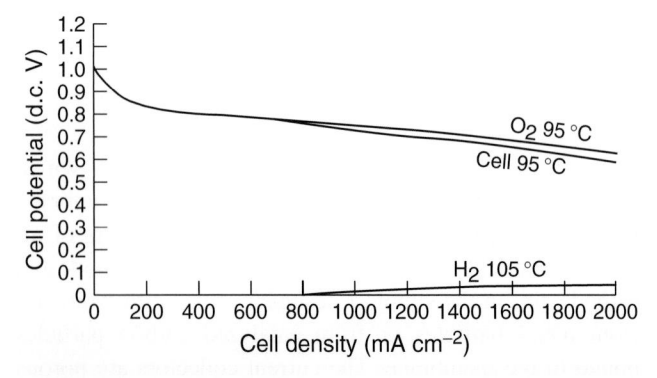

Figure 6. Cell and half-cell potentials vs. current density for a single cell with Dow membrane (thickness $125\,\mu m$) and Pt-sputtered Prototech electrodes (Pt loading $0.45\,mg\,cm^{-2}$) operating at $95\,^\circ C$ with H_2/O_2 at 4–5 atm.

7 THE DIRECT METHANOL FUEL CELL

All the cells previously described have been based on hydrogen as a fuel, but as the Ballard bus shows, transport of this fuel clearly requires a considerable weight penalty. The alternative, which is to use methanol as a fuel and use a reformer, is both costly and fraught with engineering difficulties. Highly desirable would be a fuel cell that could *directly* oxidize methanol at the anode, but retain the high power/weight ratio of the SPE fuel cell described above. Such a fuel cell is now under active development in Europe and the USA, and the principles are shown in Figure 7. The anode reaction is

$$CH_3OH + H_2O \longrightarrow CO_2 + 6H^+ + 6e^- \qquad (10)$$

and the cathode reaction is

$$\tfrac{3}{2}O_2 + 6H^+ + 6e^- \longrightarrow 3H_2O \qquad (11)$$

Methanol possesses a number of advantages as a fuel: it is a liquid, and therefore easily transported and stored and dispensed within the current fuel network; it is cheap and plentiful, and the only products of combustion are CO_2 and H_2O. The advantages of a direct methanol fuel cell are: changes in power demand can be accommodated simply by alteration in supply of the methanol feed; the fuel cell operates at temperatures below ca. $150\,^\circ C$ so there is no production of NO_x, methanol is stable in contact with the acidic membrane, and is easy to manufacture.

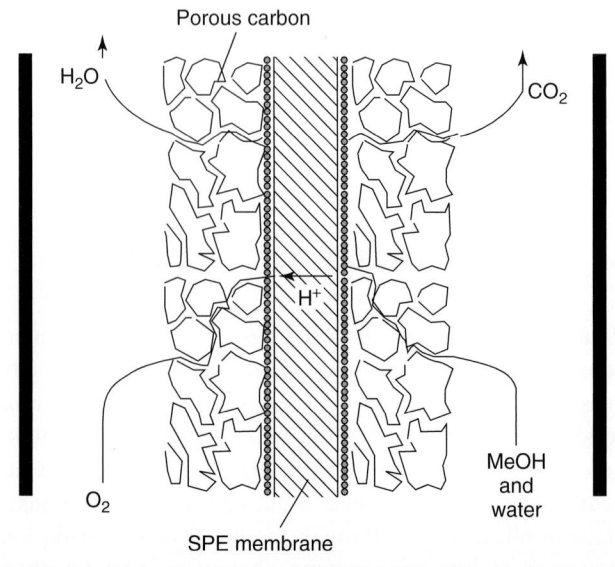

$3/2O_2 + 6H^+ + 6e^- = 3H_2O \quad CH_3OH + H_2O = CO_2 + 6H^+ + 6e^-$

Figure 7. The SPE fuel cell configuration for the direct methanol fuel cell.

The basic problems currently faced by the direct methanol fuel cell are: (a) the anode reaction has poor electrode kinetics, particularly at lower temperatures, making it highly desirable to identify improved catalysts and to work at as high a temperature as possible; (b) the cathode reaction, the reduction of oxygen, is also slow, though the problems are not so serious as with aqueous mineral acid electrolytes. Nevertheless, the overall power density of the direct methanol fuel cells is much lower than the $600+$ mW cm^{-2} envisaged for the hydrogen-fuelled SPE fuel cells; (c) perhaps of greatest concern at the moment is the permeability of the current perfluorosulfonic acid membranes to methanol, allowing considerable fuel crossover. This leads both to degradation of performance, since a mixed potential develops at the cathode, and to deterioration of fuel utilization. Methanol vapor also appears in the cathode exhaust, from which it would have to be removed.

In spite of these difficulties, the direct methanol fuel cell (DMFC)-SPE does have the capability of being very cheap and potentially very competitive with the internal combustion engine, particularly in niche city driving applications, where the low pollution and relatively high efficiency at low load are attractive features. Performances from modern single cells are highly encouraging, with power densities in excess of 300 mW cm^{-2} in oxygen; in air power densities of 200 mW cm^{-2} have been attained under pressure.

REFERENCES

1. C. H. Hamann, A. Hamnett and W. Vielstich, 'Electrochemistry', Wiley-VCH, Weinheim, Germany (1998).
2. N. Hall, 'The New Chemistry', Cambridge University Press, Cambridge (2000).
3. L. J. M. J. Blomen and M. N. Mugerwa, 'Fuel Cell Systems', Plenum Press, New York (1993).
4. K. Kordesch and G. Simader, 'Fuel Cells and their Applications', VCH, Weinheim, Germany (1996).
5. D. G. Lovering, 'Fuel Cells', Elsevier, Barking, UK (1990).
6. P. G. Bruce, 'Solid-State Electrochemistry', Cambridge University Press, Cambridge (1995).

Part 2

Mass transfer in fuel cells

Chapter 7
Mass transfer at two-phase and three-phase interfaces

A. Weber[1], R. Darling[2], J. Meyers[2] and J. Newman[1]

[1] *University of California, Berkeley, CA, USA*
[2] *UTC Fuel Cells, South Windsor, CT, USA*

1 INTRODUCTION

Mass transport losses represent a significant inefficiency in fuel cell systems. In fuel cells, the major mass transport losses are associated with the transport of charged and uncharged species in the electrolytic phase and the transport of gas phase reactants and products. The causes for these losses and their mathematical descriptions are the principal issues discussed in this chapter. Other mass transport related processes, such as catalyst sintering and membrane poisoning, may also affect fuel cell performance. In this review, we consider only the effects of mass transport that pertain to the movement of critical reactants and products. This chapter begins with a simplified model of a fuel cell that introduces many of the key components. This is followed by brief descriptions of porous electrode theory, gas phase mass transport, convective transport of liquids, transport in electrolytic solutions, and numerical simulations of the cathode of a proton exchange membrane (PEM) fuel cell.

Figure 1 shows a cross-sectional view of a PEM fuel cell operating on reformed hydrocarbon fuel and air. Table 1 gives typical through-plane dimensions for the various components. The membrane area may range from 1 cm² for a lab-scale unit to 1 m² for an automotive or stationary power plant. All types of fuel cells discussed in this chapter contain similar components, the major difference being the nature of the electrolyte. Commonly encountered fuel cell electrolytes include phosphoric acid, potassium hydroxide, cation exchange membranes, molten carbonate, and solid oxides.

The fuel is fed into the anode flow field, moves through the diffusion medium, and reacts electrochemically at the anode catalyst layer. The diffusion medium is typically a carbon cloth or carbon paper, possibly treated with Teflon®. The catalyst layer usually contains a platinum alloy supported on carbon and an ionomeric membrane material such as Nafion®. For an acid fuel cell operating with hydrogen as the fuel, the hydrogen oxidizes according to the reaction

$$H_2 \longrightarrow 2H^+ + 2e^- \qquad (1)$$

The oxidant, usually oxygen in air, is fed into the cathode flow field, moves through the diffusion medium, and is reduced at the cathode according to the reaction

$$4H^+ + 4e^- + O_2 \longrightarrow 2H_2O \qquad (2)$$

The water, either liquid or vapor, produced by the reduction of oxygen at the cathode exits the fuel cell through either the cathode or the anode flow field. This movement must be accomplished without hindering the transport of reactants to the catalyst layers in order for the fuel cell to operate efficiently. Adding equations (1) and (2) yields the overall reaction

$$2H_2 + O_2 \longrightarrow 2H_2O \qquad (3)$$

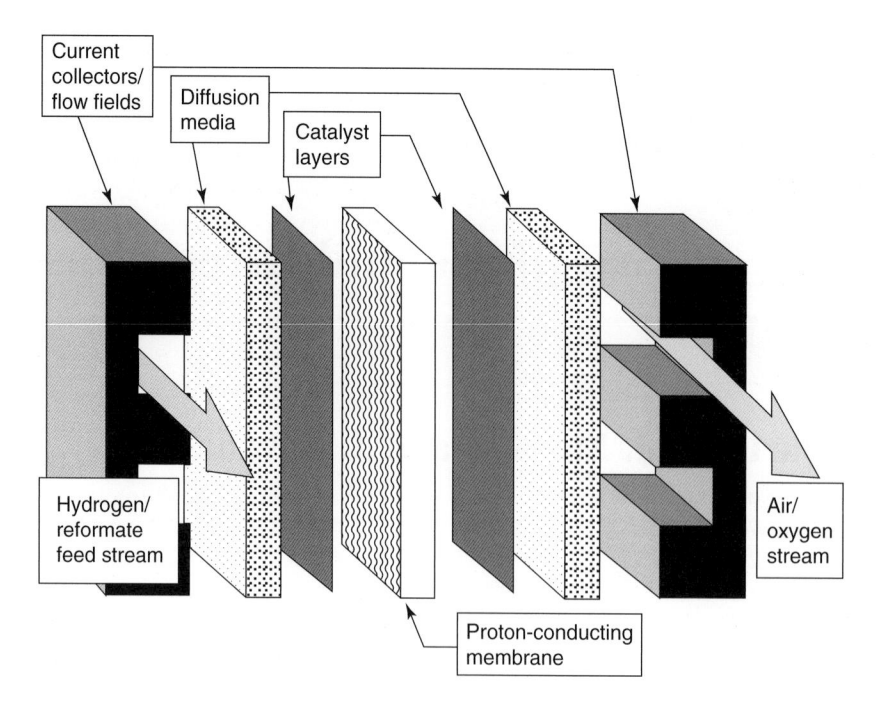

Figure 1. Schematic diagram of a hydrogen polymer electrolyte membrane fuel cell.

Table 1. Thickness of fuel cell components.

Component	Thickness (μm)	Reference
Flow channel	3000	40
Diffusion medium	100–300	43
Catalyst layer	5–25	10
Membrane	10–200	40

The electrons generated at the anode pass through an external circuit and may be used to perform work before they are consumed at the cathode. The maximum work that the fuel cell can deliver may be found by considering the theoretical open-circuit potential of the cell. A typical PEM cell can be represented as

$$
\begin{array}{c|cc|c|cc|c}
\alpha & \multicolumn{2}{c|}{\beta} & \gamma & \multicolumn{2}{c|}{\beta'} & \alpha' \\
\text{Graphite} & Pt_{(s)}, H_{2(g)} & & \text{Membrane} & & Pt_{(s)}, O_{2(g)} & \text{Graphite} \\
& H_2O_{(g)} & & H^+_{(m)}, H_2O_{(m)} & & H_2O_{(g)} &
\end{array} \quad (4)
$$

where each Greek letter identifies a distinct phase and the wavy lines imply that the membrane phase boundary is not sharp; rather, the membrane extends into adjacent regions and may include water activity gradients. The potential of this cell is

$$FU = -F(\Phi^\alpha - \Phi^{\alpha'}) = \mu_{e^-}^\alpha - \mu_{e^-}^{\alpha'} \quad (5)$$

where F is Faraday's constant, U is the thermodynamically defined reversible cell potential, Φ^α is the electrical potential of phase α, and $\mu_{e^-}^\alpha$ is the electrochemical potential of electrons in phase α. After introducing expressions for the activities of the various components, this becomes

$$FU = FU^\theta + \frac{RT}{2}\ln a_{H_2}^\beta + \frac{RT}{4}\ln a_{O_2}^{\beta'}$$
$$- \frac{RT}{2}\ln a_{H_2O}^{\beta'} + (\mu_{H^+}^\beta - \mu_{H^+}^{\beta'}) \quad (6)$$

where a_i^β is the activity of species i in phase β, R is the ideal gas constant, T is the absolute temperature, and U^θ is the standard cell potential, a combination of appropriately chosen reference states. This equation reduces to the familiar Nernst equation when the gases are assumed to be ideal and activity gradients in the electrolyte are neglected.

1.1 Introductory model

A good way to introduce the subject is to begin with a simple steady state model that captures the gross behavior of a fuel cell. For our example, we consider a phosphoric acid fuel cell operating on a reformed hydrocarbon fuel and air at atmospheric pressure and a temperature of 190 °C. If the fuel and air streams flow cocurrently, and all of the product water leaves with the air stream, then it is possible to relate the gas compositions in the two flow channels to the fuel utilization by means of material balances and Faraday's law. The hydrogen utilization at any point along the channel can be defined as

$$u = \frac{X_{H_2}^0 - X_{H_2}}{X_{H_2}^0} \quad (7)$$

where X_{H_2} is the molar flow rate of hydrogen divided by the molar flow rate of inerts, principally carbon dioxide, in the fuel stream. The superscript 0 denotes the value at the inlet of the flow field. Faraday's law can then be used to relate the amount of hydrogen reacted to the current

$$\frac{d}{dy}(X_{H_2} F_a) = -\frac{iW}{2F} \quad (8)$$

where i is the local current density, W is the width perpendicular to the flow, y is the distance down the flow field channel, and F_a is the molar flow rate of inerts at the anode. Similar expressions can be written for water and oxygen at the cathode. Using the reaction stoichiometry and Faraday's law, we get

$$X_{H_2}^0 u = (X_{H_2O} - X_{H_2O}^0)f = 2(X_{O_2}^0 - X_{O_2})f \quad (9)$$

where X_i is the ratio of moles of component i to moles of inerts in the same stream and f is the ratio of moles of inerts in the air stream to moles of inerts in the fuel stream. The first equality arises directly from our assumption that all of the product water exits the fuel cell in the air stream. While this assumption is not generally true, it does serve as a useful starting point. Thus, the gas compositions in the flow channel are entirely determined by u, f, and the inlet gas composition; such a simple relationship is impossible when the gases are not fed cocurrently. However, this analysis should provide a reasonable approximation of the true behavior when one of the electrodes is limiting. Figure 2 shows the mole fractions when a dry, equimolar mixture of hydrogen and carbon dioxide is fed to the anode, and dry air, 20% in excess of the stoichiometric amount, is fed to the cathode. In practice, performance gains due to adding more air must be balanced against the costs and parasitic power losses of pumping the air.

Figure 3 shows a typical fuel cell polarization curve. The curve includes a sharp drop in potential at low current densities due to the sluggish kinetics of the oxygen reduction reaction (ORR). This part of the polarization curve is commonly called the kinetic regime. At moderate current densities, the cell enters an ohmic regime where the cell potential varies nearly linearly with current density. At high current densities, mass transport resistance dominates, and the potential of the cell declines rapidly as the concentration

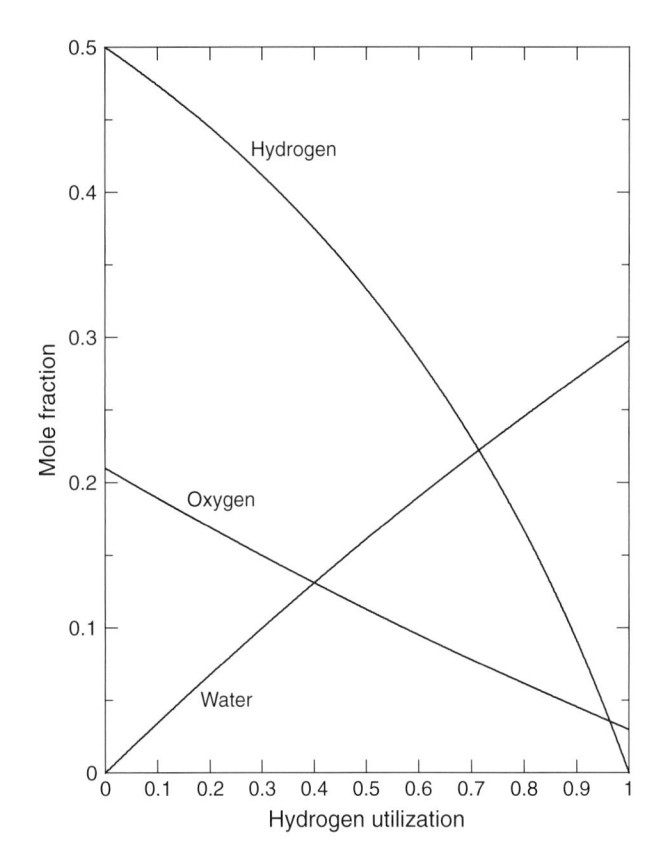

Figure 2. Composition of gas streams. The feed is 50% hydrogen in carbon dioxide, 20% excess air, and $f = 2$.

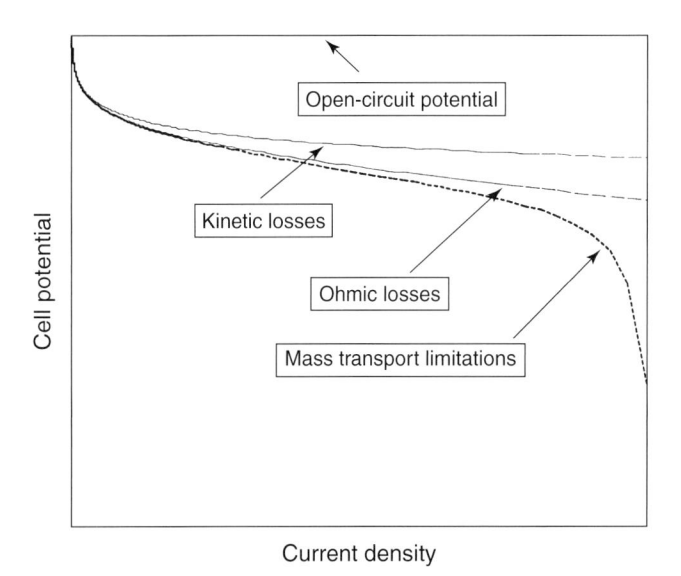

Figure 3. Example of a polarization curve showing the losses associated with irreversibilities in a fuel cell.

of one of the reactants approaches zero at the corresponding catalyst layer. This defines the limiting reactant. In a typical PEM cell operating at temperatures below 80 °C, much of the water produced by the ORR is liquid, and this liquid water may flood parts of the fuel cell, dramatically increasing the resistance to mass transfer.

A simplified equation[1] describing the major features of a typical polarization curve is

$$V = U^\theta + \frac{RT}{\alpha F}\ln(ai_0 L) - \frac{RT}{\alpha F}\ln\left(\frac{i}{p_{O_2}}\right) + \frac{RT}{2F}\ln(p_{H_2}) - R'i \quad (10)$$

where V is the cell potential and i is the superficial current density. The first term in the equation, U^θ, is the standard cell potential, 1.144 V at 190 °C. The second and third terms arise from the assumption that the ORR follows Tafel kinetics, with a first-order dependence on the partial pressure of oxygen. α is the cathodic transfer coefficient, which normally has a value of 1, a is the interfacial area of the catalyst per unit volume of electrode, i_0 is the exchange current density of the ORR, and L is the thickness of the cathode catalyst layer. Thus, the quantity aL is a roughness factor, a ratio of catalyst area to superficial electrode area. The first and second terms may be combined to form a potential intercept, U'; this quantity is a convenient way to group terms pertaining to (possibly unknown) thermodynamic and kinetic constants. The third term describes the potential loss at the cathode at the specified current density i, and subject to the oxygen partial pressure present at the electrode interface. The fourth term is an equilibrium expression for the hydrogen oxidation reaction (HOR). This is usually a good approximation unless the anode catalyst is poisoned. Finally, R' is the effective ohmic resistance, which includes the resistance of the separator as well as the residual contact resistances between cell components. The detailed reaction rate distributions within the porous electrodes are neglected in this analysis, and the electrodes are treated as planar with enhanced surface area.

The use of equation (10) requires that interfacial partial pressures of oxygen and hydrogen be known. These partial pressures will be lower than those in the channels because of mass transport losses in the backing layers. Conversely, the partial pressure of water will be higher at the cathode catalyst layer than in the cathode gas channel.

At sufficiently high utilizations near the exit of the anode flow channel, hydrogen will be a minor component in the anode fuel stream and, under these conditions, diffusion of hydrogen through the stagnant inert gases in the diffusion layer can be modeled with Fick's law. If we define the limiting current as the current density at which the hydrogen partial pressure goes to zero at the anode catalyst layer, then rearrangement of Fick's law yields

$$x_{H_2}^i = x_{H_2}^b \left(1 - \frac{i}{i_{\lim,H_2}}\right) \quad (11)$$

where x_{H_2} is the mole fraction of hydrogen, and the superscripts i and b refer to the catalyst layer interface and bulk, respectively. The hydrogen limiting current varies with utilization according to

$$i_{\lim,H_2} = \frac{2FD_{H_2,CO_2}X_{H_2}^0 p}{RT\delta}\frac{1-u}{1+X_{H_2}^0(1-u)} \quad (12)$$

where $D_{i,j}$ is the diffusion coefficient for species i moving through species j, p is the total gas pressure, and δ is the diffusion length. As a reference, using values given by Newman,[1] the limiting current is 43.2 A cm^{-2} at zero utilization and 1.69 A cm^{-2} at 98% utilization with an equimolar feed of hydrogen and carbon dioxide.

In the cathode diffusion medium, oxygen diffuses through stagnant nitrogen and counter diffusing water vapor. The Stefan–Maxwell equations are appropriate for describing this process. An analytic solution is possible for a three-component system like that described here; see, for example, Bird et al.[2]

Figure 4 shows polarization curves at various hydrogen utilizations, using the parameters $U' = 0.694$ V and $R' = 0.278\,\Omega\,cm^2$. As the hydrogen utilization increases, the partial pressure of hydrogen in the gas channel drops and the effects of mass transfer resistance increase. As specified in equation (9), and under the conditions of cocurrent flow, oxygen utilization and, hence, oxygen mass transport limitations, increase along with hydrogen utilization. The limiting currents in figure 4 are caused by the oxygen partial pressure going to zero at the cathode catalyst layer. Thus, oxygen is the limiting reactant and the introduction of additional oxygen should be considered.

Figure 5 shows the current density as a function of hydrogen utilization at three different cell potentials. Again we see that the performance of the cell decreases as the partial pressures of the reactants drop. Since, during normal operation, the cell potential remains uniform along the flow channel, even as hydrogen and oxygen are being consumed, Figure 5 also indicates how the local current density decreases with position as the air and fuel flow through the cell.

In the remainder of this chapter, we focus on transport processes in the direction normal to the face of the electrodes. Many of the key drivers of cell performance are controlled by the motion of gases and ions across the thickness of the cell, and a great deal of information about factors limiting cell performance might be gleaned by modeling

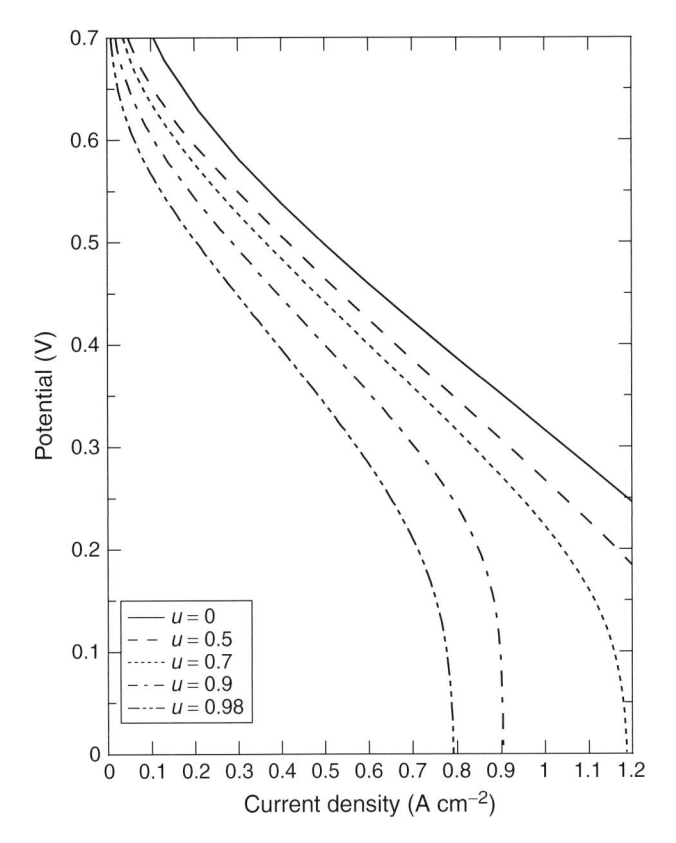

Figure 4. Polarization curves at hydrogen utilizations of 0, 0.5, 0.7, 0.9, and 0.98; calculated using the introductory model.

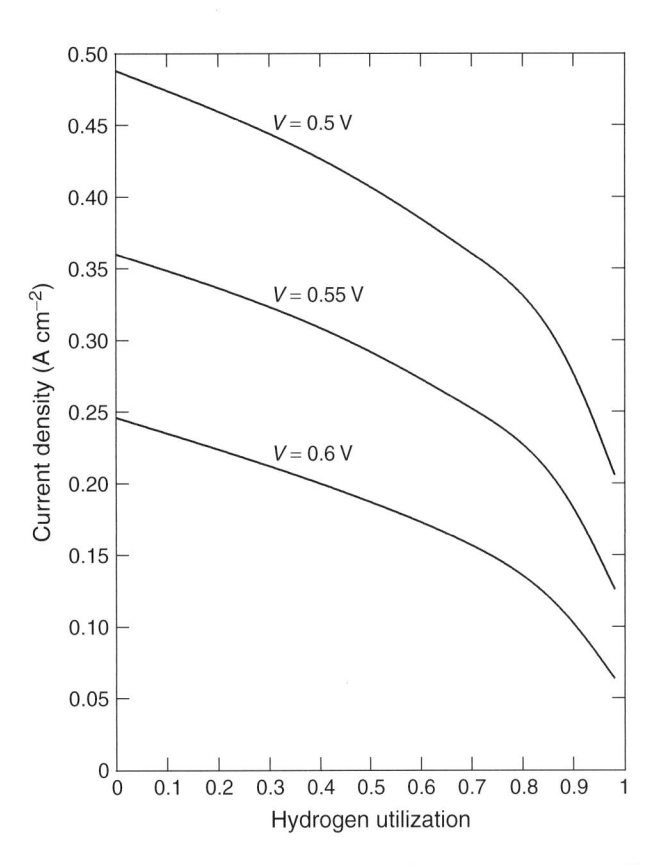

Figure 5. Current density across a cell at three constant cell potentials as a function of hydrogen utilization; calculated using the introductory model.

one-dimensional transport through the various layers. When the resistance to transport across the cell is much greater than the mass transport resistance along the channel, one can decouple the two spatial dimensions and integrate down the channel to determine the cell performance in a manner similar to that set forth above.

2 POROUS ELECTRODES

Porous electrodes are used in fuel cells in order to maximize the interfacial area of the catalyst per unit geometric area. In many fuel cells, an electrolytic species and a dissolved gas react on a supported catalyst. Thus, the electrode must be designed to maximize the available catalytic area while minimizing the resistances to mass transport in the electrolytic and gas phases, and the electronic resistance in the solid phase. Clearly, this is a stringent set of requirements. We want to construct a three-dimensional structure with continuous transport paths in multiple phases. Porous electrode theory provides a mathematical framework for modeling these complex electrode structures in terms of well defined macroscopic variables.

The behavior of porous electrodes is inherently more complicated than that of planar electrodes because of the intimate contact between the solid and fluid phases. Reaction rates can vary widely through the depth of the electrode due to the interplay between the ohmic drop in the solid phase, kinetic resistances, and concentration variations in the fluid phases. The number and complexity of interactions occurring make it difficult to develop analytic expressions describing the behavior of porous electrodes except under limiting conditions. Thus, the governing equations must usually be solved numerically.

Porous electrode theory has been used to describe a variety of electrochemical devices including fuel cells, batteries, separation devices, and electrochemical capacitors. In many of these systems, the electrode contains a single solid phase and a single fluid phase. Newman and Tiedemann reviewed the behavior of these flooded porous electrodes.[3] Many fuel cell electrodes, however, contain more than one fluid phase, which introduces additional complications. The classical gas diffusion electrode, for example, contains both an electrolytic phase and a gas phase in addition to the solid, electronically conducting phase. Earlier reviews of gas diffusion electrodes for fuel cells include those of Chizmadzhev et al.[4] and Bockris and Srinivasan.[5] This section deals with general aspects of macroscopic porous electrode theory and does not delve into detailed reaction mechanisms and electrode morphology. These issues should, of course, be considered when constructing a model of a particular system.

2.1 Macroscopic approach

We follow the macroscopic approach for modeling porous electrodes as described by Newman and Tiedemann.[3] In the macroscopic approach, the exact geometric details of the electrode are neglected. Instead, the electrode is treated as a randomly arranged porous structure that can be described by a small number of variables such as porosity and surface area per unit volume. Furthermore, transport properties within the porous structure are averaged over the electrode volume. Averaging is performed over a region that is small compared to the size of the electrode, but large compared to the pore structure. A detailed description of the averaging can be found in Dunning's dissertation.[6] The macroscopic approach to modeling can be contrasted to models based on a geometric description of the pore structure. Many early models of flooded porous electrodes treated the pores as straight cylinders arranged perpendicular to the external face of the electrode. Further examples of this type of approach are the flooded agglomerate models of Giner and Hunter[7] and Iczkowski and Cutlip.[8] These models are frequently used to describe fuel cells, and treat the electrode as a collection of flooded catalyst-containing agglomerates, which are small compared to the size of the electrode and are connected by hydrophobic gas pores.

2.2 Declaration of variables

A useful place to begin the formulation of the model is to determine the number of independent variables. The first consideration is the number of phases present. For example, consider the catalyst layer in a state of the art PEM fuel cell containing a supported platinum-on-carbon (or platinum-alloy-on-carbon) catalyst, a polymeric membrane material, and a void volume. For reference, the primary carbon particles are approximately 40 nm in diameter,[9] and the platinum crystallites are approximately 2 nm in diameter.[10] If we lump the solid phases comprising the supported catalyst together, and treat the polymeric membrane as a single phase, then we end up with four phases: solid, membrane, gas, and liquid. This last phase corresponds to liquid water infiltrating the gas pores. This model differs, conceptually, from those of Bernardi and Verbrugge[11] and Springer et al.[12] by explicitly accounting for a void volume containing gas and liquid water. In Bernardi and Verbrugge's model, oxygen and hydrogen within the catalyst layer travel as dissolved species within the ionomer. Springer et al. propose a similar picture, but fit the permeability of the catalyst layer to experimental data. The basic mathematics of the different models are fundamentally similar, a strength of the macroscopic approach. Each phase is assumed to be electrically neutral, an idea that we will return to later. We neglect double layer processes in this chapter, although these may be important in the simulation of transient phenomena. If desired, it is possible to include the interfacial regions between the different macroscopic phases as additional phases with negligible volume, but having the ability to store mass.[13]

Now we turn our attention to the number, M, of degrees of freedom that need to be specified. The Gibbs phase rule allows the reduction of an arbitrarily large number of species to a set of independent components of size M

$$M = C - R - P + 2 \qquad (13)$$

where C is the number of species, R is the number of equilibrated homogeneous reactions, P is the number of phases, and the 2 indicates the selection of temperature and pressure. If we have unequilibrated reactions, they would not contribute to R, and if we do not assume phase equilibrium (treating mass transfer separately), we would treat each phase separately, taking $P = 1$. If we also treat pressure and temperature separately, say, by fluid mechanics and an energy balance, we would leave off the 2, yielding $C - R - 1$. For a gas phase of three components and no equilibrated reactions, $C - R - 1 = 2$; specify two mole fractions. In phases without charged species, like the gas phase in our PEM example, we do not need an electrical state variable.

As a second example, consider a sulfuric acid electrolyte containing H^+, HSO_4^-, SO_4^{2-}, H_2SO_4, and H_2O. In this case, $C = 5$ and $R = 2$, which yields $N = 2$, interpreted to mean the electric potential and the concentration of sulfuric acid. $R = 2$ implies that we have two equilibrium relationships among the 5 species, one for bisulfate equilibrium and one for sulfuric acid equilibrium. For Nafion®, we have water, membrane, and protons, $C = 3$. Now, $C - R - 1 = 3 - 0 - 1 = 2$; specify water concentration and potential. This approach allows the electrical systems to be handled without exception, by turning that last degree of freedom into an electric state variable, namely, the potential.

In all of these systems, the corresponding number of transport properties required is

$$\frac{N(N-1)}{2} \qquad (14)$$

where N is the number of independent species. Newman[14] gives a more complete counting of transport properties, with simple examples, including thermal and electrical variables.

2.3 Average quantities

In the macroscopic approach, the electrode is treated as a superposition of all phases present. Thus, all variables are defined at all positions within the electrode. The concentration of species i in phase k, $c_{i,k}$, is averaged over the pore volume of phase k, ensuring that the concentration profile is continuous entering or leaving the electrode. $\mathbf{N}_{i,k}$ is the superficial flux density of species i in the pores of phase k averaged over the cross-sectional area of the electrode. These averages apply to regions that are large compared to the pore structure, but small compared to regions over which macroscopic variations occur. The interstitial flux density of species i in phase k is

$$\mathbf{N}_{i,k}^{\text{interstitial}} = \frac{\mathbf{N}_{i,k}}{\varepsilon_k} \qquad (15)$$

where ε_k is the volume fraction of phase k. This equation assumes that the medium is isotropic. The superficial current density in phase k is

$$\mathbf{i}_k = F \sum_i z_i \mathbf{N}_{i,k} \qquad (16)$$

where F is Faraday's constant and z_i is the valence of species i.

2.4 Ohm's law in the solid phase

In many fuel cell systems, the conductivity of the solid phase is much greater than the conductivity of the electrolytic phase. In this case, it is permissible to treat the solid phase potential as a constant. More generally, the transport of current in the solid phase may be treated with Ohm's law:

$$\mathbf{i}_1 = -\sigma \nabla \Phi_1 \qquad (17)$$

where σ is the electrical conductivity of the solid. This conductivity may be adjusted for porosity and tortuosity with a Bruggeman[15] correction

$$\sigma = \sigma_0 \varepsilon_1^{1.5} \qquad (18)$$

where ε_1 and σ_0 are the volume fraction and the bulk conductivity of the solid phase, respectively. The conductivity will take on some effective value based on the conductivities of the different solid phase materials. For example, many PEM electrodes contain carbon, platinum, and Teflon® (polytetrafluoroethylene) with bulk conductivities of approximately 25, 10^5, and 10^{-4} S cm^{-1} at 25 °C, respectively.[16] The electronic current will flow through the platinum and carbon phases, and the carbon will effectively determine the conductivity. Bernardi[17] discusses the averaging of the conductivities of electronically conducting phases in a porous electrode. She advocates Maxwell's model, which assumes a continuous phase and several discrete phases. However, the prediction of conductivities of packed bed materials is not fully resolved; thus, experimental values are preferred when available.

2.5 Electrode kinetics

A single electrochemical reaction can be written schematically as

$$\sum_k \sum_i s_{i,k,h} M_i^{z_i} \longrightarrow n_h e^- \qquad (19)$$

where $s_{i,k,h}$ is the stoichiometric coefficient of species i, residing in phase k, and participating in electron transfer reaction h, n_h is the number of electrons transferred in reaction h, and $M_i^{z_i}$ represents the chemical formula of i having valence z_i. The electrons reside in the solid phase.

The rate of an electrochemical reaction depends upon the potential drop across the interface and the concentrations of the various species. It is not possible to write down completely general rate equations for electrochemical reactions. However, it is frequently possible to begin with a Butler–Volmer equation.

Two electrochemical reactions of particular interest in fuel cells are the HOR and the ORR. In the absence of catalyst poisons, the HOR is fast, and a detailed reaction mechanism may be unnecessary. Instead, we use a Butler–Volmer equation of the form

$$i = i_0 \left[\frac{p_{H_2}}{p_{H_2}^{\text{ref}}} \exp\left(\frac{(1-\beta)nF}{RT}(\Phi_1 - \Phi_2)\right) - \exp\left(\frac{\beta nF}{RT}(\Phi_1 - \Phi_2)\right) \right] \qquad (20)$$

where i_0 is the exchange current density per unit catalyst area evaluated at the reference conditions, $n = 2$ is the number of electrons transferred, β is a symmetry factor having a typical value of 0.5, p_{H_2} is the hydrogen partial pressure, $p_{H_2}^{\text{ref}}$ is the hydrogen partial pressure at which the exchange current density is specified, typically 1 bar, Φ_1 is the potential in the solid phase, and Φ_2 is the potential in the proton conducting phase. In this discussion, the potential in the proton conducting phase is measured with a normal hydrogen electrode (NHE), defined as a platinum metal electrode exposed to hydrogen at 1 bar and a solution of 1 N acid, measured at the same temperature as the solution of interest. This kinetic expression reduces to the Nernst relationship when the ratio i/i_0 becomes small. Springer et al.[18] present a treatment of the HOR applicable to a platinum catalyst in a PEM fuel cell when carbon monoxide poisoning of the electrode is significant. This treatment involves consideration of the Tafel and Volmer reaction steps.

The ORR, on the other hand, is slow and represents the principal inefficiency in many fuel cells. The ORR behaves irreversibly and may be modeled reasonably well with Tafel kinetics with a first-order dependence upon oxygen partial pressure. Appleby[19] suggests the following form for oxygen reduction in acid electrolytes

$$i_n = -i_0 \left(\frac{p_{O_2}}{p_{O_2}^{ref}}\right) \exp\left(\frac{-\alpha F}{RT}(\Phi_1 - \Phi_2 - U^\theta)\right) \quad (21)$$

where p_{O_2} is the oxygen partial pressure, i_0 is the exchange current density, α is the cathodic transfer coefficient, for which Appleby recommends a value of 1, and U^θ is the standard potential for oxygen reduction, 1.229 V at 25 °C. A linear fit on a Tafel plot of surface overpotential, $\eta_s = \Phi_1 - \Phi_2 - U^\theta$, versus the log of the current density yields the commonly reported Tafel slope, b

$$b = 2.303 \frac{RT}{\alpha F} \quad (22)$$

Equation (21) is first order in oxygen concentration and applies for acid electrolytes; another form is required for basic electrolytes. Additionally, the rate of the ORR may depend upon proton concentration; see, for example, Kinoshita.[20] Furthermore, the ORR is sensitive to the presence of adsorbed anions and surface oxides on platinum. Finally, one may prefer to use the dissolved oxygen concentration in equation (21) instead of the oxygen partial pressure. Typically, these can be related through Henry's law.

Further refinement of the kinetic expressions may be necessary to describe the effects of competitively adsorbed poisons or unwanted by-products. It is usually possible to incorporate such effects into a general numerical model.

2.6 Material balances

Reactions occurring within the porous electrode are mathematically treated as source or sink terms, rather than as boundary conditions, as reactions at a planar electrode might be treated. It is necessary to write a material balance for each independent component in each phase. In the following discussion, we assume that there is only one solid, electronically conducting phase, denoted by subscript 1. The differential form of the material balance for species i in phase k is

$$\frac{\partial \varepsilon_k c_{i,k}}{\partial t} = -\nabla \cdot \mathbf{N}_{i,k} - \sum_h a_{k,1} s_{i,k,h} \frac{i_{h,k}}{n_h F}$$
$$+ \sum_l s_{i,k,l} \sum_{p \neq k} a_{k,p} r_{l,k-p} + \sum_g s_{i,k,g} R_{g,k} \quad (23)$$

The term on the left side of the equation is the accumulation term, which accounts for the change in the total amount of species i held in phase k within a differential control volume. The first term on the right side of the equation keeps track of the material that enters or leaves the control volume by mass transport. The remaining three terms account for material that is gained or lost due to chemical reactions. The first summation includes all electron transfer reactions that occur at the interface between phase k and the solid, electronically conducting phase. The second summation accounts for all other interfacial reactions that do not include electron transfer, and the final term accounts for homogeneous reactions in phase k.

In the above expression, $c_{i,k}$ is the concentration of species i in phase k, and $s_{i,k,l}$ is the stoichiometric coefficient of species i in phase k participating in heterogeneous reaction l. When we specify $s_{i,k,l}$, we necessarily assume that species i exists in phase k immediately prior to reaction or upon formation. If i can exist in more than one phase, care must be taken to ensure that phase equilibrium is properly addressed. $a_{k,p}$ is the specific surface area (surface area per unit total volume) of the interface between phases k and p. We assume that this is a nonspecific surface area for the interface; for example, if only a particular crystalline surface participates in an interfacial reaction, the portion of the total surface area covered by that particular crystal face must be included in the kinetic rate constants, rather than grouped into the $a_{k,p}$ term. This simplifies the terminology of this discussion considerably. $i_{h,k}$ is the normal anodic interfacial current transferred per unit interfacial area across the interface between the solid, electronically conducting phase and phase k due to electron transfer reaction h. We note that a current $i_{h,k}$, written with two subscripts, implies an interfacial, or transfer, current density. Conversely, a current $\mathbf{i}_k$, written in boldface and with a single subscript, indicates the total current density carried within phase k. This current density is, strictly speaking, a vector quantity. $r_{l,k-p}$ is the rate of the heterogeneous reaction l per unit of interfacial area between phases k and p. $R_{g,k}$ is the rate of a strictly homogenous reaction g in phase k per unit volume of phase k.

The electron transfer reactions could also be treated as simple interfacial reactions taking place between a given phase and the solid. Faraday's law can be used to relate this reaction rate to the normal transfer current density for reaction h

$$i_{h,k} = n_h F r_{h,k-1} \quad (24)$$

where $r_{h,k-1}$ is the rate of reaction h occurring at the interface between phase k and the solid, electronically conducting phase and $i_{h,k}$ is, as described above, the normal interfacial current density flowing to phase k from the solid phase due to an anodic reaction h.

2.7 Electroneutrality and conservation of charge

Since a large electrical force is required to separate charge over an appreciable distance, a volume element in the electrode will, to a good approximation, be electrically neutral. We further assume that each phase within the electrode is electrically neutral. Thus, for each phase

$$\sum_i z_i c_{i,k} = 0 \quad (25)$$

The assumption of electroneutrality implies that the diffuse double layer, where there is significant charge separation, is small compared to the volume of the electrode, which is normally the case. The assumption of electroneutrality also leads us to the conclusion that the divergence of the total current is zero

$$\sum_k \nabla \cdot \mathbf{i}_k = 0 \quad (26)$$

Using the subscript 1 to again denote the solid, electronically conductive, phase, we may relate the divergence of the electronic current to the rates of the electrochemical reactions. In other words, $-\nabla \cdot \mathbf{i}_1$ represents the total anodic rate of electrochemical reactions per unit volume of electrode. This can be related to the average transfer current density, and when combined with equation (24), yields

$$-\nabla \cdot \mathbf{i}_1 = \sum_k \sum_h a_{k,1} i_{h,k} \quad (27)$$

The above charge balance assumes that faradaic reactions are the only electrode processes; double layer charging is neglected.

3 GAS PHASE TRANSPORT

The gaseous reactants enter the fuel cell through the gas channels and are transported through the diffusion media into the catalyst layers where they react. Some gases may dissolve into the membrane and leak across the cell to react at the opposite electrode, thereby reducing the electrical work delivered by the fuel cell. The transport of the gases across the membrane is discussed later; here, only true gas phase mass transport is considered. Thus, the regions of interest are the diffusion media and the catalyst layers. Gas phase mass transport limitations may become important at the anode, the cathode, or both. These limitations could have severe effects on the operation of the fuel cell, as was shown with the introductory model.

Diffusion describes the movement of a given species relative to the motion of other species in a mixture. Several important modes of diffusion are ordinary (molecular), Knudsen, configurational, and surface diffusion. Ordinary diffusion almost always occurs; Knudsen diffusion is important in small pores; configurational diffusion takes place when the characteristic pore size is on the molecular scale; and surface diffusion involves the movement of adsorbates on surfaces. In a typical fuel cell, the pores are large enough that configurational diffusion does not occur. Surface diffusion may play an important role in interfacial reactions, but it is not treated in this chapter.

3.1 Ordinary diffusion

Fick's law may be used to describe the molecular movement of a dilute component in a mixture

$$\frac{\mathbf{N}_i}{\varepsilon} = -D_i c_T \nabla x_i \quad (28)$$

where $\mathbf{N}_i$ is the superficial flux density of species i, ε is the porosity, D_i is the Fickian diffusion coefficient of species i in the mixture, which varies inversely with pressure and to approximately the 1.81 power with absolute temperature,[2] and c_T is the total concentration or molar density, which is assumed constant. Since we are discussing single phase phenomena, we have dropped the subscript k identifying the phase in the above and subsequent equations. In more concentrated and multicomponent systems, where it is necessary to take into account mutual interactions among different species, the generalized Stefan–Maxwell equations for constant temperature and pressure are appropriate

$$c_i \nabla \mu_i = \frac{RT}{c_T} \sum_{j \neq i} \frac{c_i c_j}{D_{i,j}} (\mathbf{v}_j - \mathbf{v}_i) \quad (29)$$

where c_i is the concentration of i, μ_i is the chemical potential of i, $D_{i,j}$ is the binary interaction parameter between i and j, and $\mathbf{v}_i$ is the interstitial velocity of i relative to some reference velocity. As long as the reference frame is applied consistently, it need not be specified. The reason for this is that the frictional interactions depend only the relative and not the absolute values of the velocities. These equations involve only binary interaction parameters, and yield the correct number of transport coefficients as determined by species analysis.

Using the relationship between flux density and velocity

$$\frac{\mathbf{N}_i}{\varepsilon} = c_i \mathbf{v}_i \quad (30)$$

equation (29) can be rewritten, and for ideal mixtures at constant pressure and temperature, the Stefan–Maxwell equations take the form

$$\nabla x_i = \sum_{j \neq i} \frac{x_i \mathbf{N}_j - x_j \mathbf{N}_i}{\varepsilon c_\mathrm{T} D_{i,j}} \quad (31)$$

where x_i is the mole fraction of i. By the Onsager reciprocal relationships or by Newton's third law of motion, $D_{i,j} = D_{j,i}$. Bird et al.[2] provide correlations of $D_{i,j}$ for various gas pairs. Since the transport is occurring in pores, the diffusion coefficients need to be corrected for tortuosity. Frequently, a Bruggeman[15] equation can be used for this correction

$$\tau = \varepsilon^{-0.5} \quad (32)$$

The diffusion coefficients do not need to be corrected for porosity since it was explicitly accounted for by using interstitial properties and superficial fluxes.

The above set of equations yields $N - 1$ independent equations. In addition to determining the velocities of species relative to one another, the reference velocity must be determined, either by fluid dynamics or by the fixing the movement of a particular species with a properly chosen reference frame.

3.2 Knudsen diffusion

As the pore size decreases, molecules collide more often with the pore walls than with each other. This movement, intermediated by these molecule-pore-wall interactions, is known as Knudsen diffusion, named after the first person to study this type of flow comprehensively.[21] Knudsen diffusion occurs when the mean free path of the molecule is on the same order as the diameter of the pore. Using the kinetic theory of gases to express the mean free path of the molecule, the Knudsen diffusion coefficient is given by

$$D_{K_i} = \frac{d}{3}\left(\frac{8RT}{\pi M_i}\right)^{1/2} \quad (33)$$

where d is the pore diameter and M_i is the molecular weight of i. This diffusion coefficient is independent of pressure whereas ordinary diffusion coefficients have an inverse dependence on pressure.

Knudsen diffusion and ordinary or Stefan–Maxwell diffusion may be treated as mass transport resistances in series, and combined to yield

$$\nabla x_i = -\frac{\mathbf{N}_i}{\varepsilon c_\mathrm{T} D^e_{K_i}} + \sum_{j \neq i} \frac{x_i \mathbf{N}_j - x_j \mathbf{N}_i}{\varepsilon c_\mathrm{T} D^e_{i,j}} \quad (34)$$

where the superscript e indicates that the diffusion coefficients have been corrected for tortuosity. In effect, the pore wall, with zero velocity, constitutes another species with which the diffusing species interact, and it determines the reference velocity used for diffusion. From an order of magnitude analysis, when the mean free path of a molecule is less than 0.01 times the pore diameter, bulk diffusion dominates, and when it is greater than 10 times the pore diameter, Knudsen diffusion dominates. This means that Knudsen diffusion should be considered when the pore diameter is less than 100 nm at atmospheric conditions. For reference, a typical carbon diffusion medium has pores between 100 nm and 20 μm[22] in diameter and a catalyst layer contains pores on the order of 50 nm in diameter.[9]

3.3 Pressure driven flow

In most porous systems, the full Navier–Stokes equations are not solved; instead, a simplified construct is used. The reason for this simplification is that the actual porous structure is often unknown and a volume average approach is taken. This averaging is not valid over a bimodal distribution.[23] The convective flow encountered in a fuel cell is mainly pressure driven. If we further assume that the flow is steady and laminar (i.e., creeping flow), then we may use Darcy's law

$$\mathbf{v} = -\frac{k}{\mu}\nabla p \quad (35)$$

where $\mathbf{v}$ is the superficial mass average velocity, μ is the viscosity, and k is the permeability coefficient or Darcy's constant. The permeability coefficient is best determined from experiment, but should be on the order of the reciprocal of the square of the pore diameter. The Carman–Kozeny equation may be used to estimate the permeability[23]

$$k = \frac{\varepsilon^3}{(1-\varepsilon)^2 k' S_o^2} \quad (36)$$

where S_o is a shape factor defined as the surface area to volume ratio of the solid phase and k' is the Kozeny constant, which has a value of approximately 5. If desired, equation (35) can be written as a flux in the absence of diffusion by using equation (30).

3.4 Combined transport

The inclusion of Knudsen diffusion coefficients is analogous to imposing a frictional interaction between each species in the gas mixture and the pore wall. When the characteristic pore diameter is small enough that the frictional interactions between gas species and pore wall are of the same order as, or larger than, the Stefan–Maxwell frictional interactions, then the reference velocity, which is necessary to specify the absolute velocities of all the species in the mixture, is set by fixing the velocity of the solid pore wall in the laboratory reference frame. In such a case, the inclusion of the Knudsen diffusion coefficients is mathematically similar to including a species $N + 1$ to represent the static pore wall in an N-component gas mixture. As mentioned above, the matrix of Stefan–Maxwell equations is not linearly independent, so one row must be removed and replaced with the specification of the reference velocity. In the case where Knudsen diffusion coefficients interact significantly with the remaining gas phase species, the reference velocity is that of the pore wall.

As the pore diameter becomes larger, however, the frictional interactions between the solid pore wall and the diffusing gas species become weaker and the coupling between them tenuous. In this case, the pore wall ceases to interact significantly with the diffusing gas species, and the interactions become too weak for the pore wall to specify the reference velocity. In such a case, one must resort to the original $N \times N$ matrix of Stefan–Maxwell relations, which becomes, in the absence of significant frictional interactions between the species and the pore wall, a singular matrix. In such a case, a single row of the matrix of equations is removed in order to specify a different reference velocity, namely, the mass average velocity of the gas mixture given by Darcy's law. The inclusion of Knudsen diffusion and bulk flow in the Stefan–Maxwell framework is discussed at great length by Mason and Malinauskas.[24]

4 LIQUID PHASE TRANSPORT

The transport of liquid water is an important consideration in the design of fuel cells intended to operate below 100 °C.[25] In a PEM fuel cell, for example, water may be transported as both a liquid and a gas. In general, the exchange between the two may be given by a kinetic expression for the evaporation rate. However, equilibrium between the pure liquid and the water vapor can often be assumed

$$x_{H_2O} = \frac{p^{vap}}{p} \quad (37)$$

where x_{H_2O} is the mole fraction of water in the vapor phase and p^{vap} is the vapor pressure of water, a strong function of temperature. Additionally, if the pores are small, it may be necessary to apply the Kelvin equation to correct the activity of water for curvature effects.[23]

Darcy's law, using the gradient of the liquid pressure as the driving force, may be used to describe the bulk transport of liquid water in the gas diffusion media and catalyst layers. These media are typically unsaturated, introducing an additional complication because the permeability is typically a strong function of the saturation level. Frequently, a power law relationship holds between permeability and the saturation level[23]

$$k = k_{sat} S^m \quad (38)$$

where k_{sat} is the permeability at complete saturation, S is the saturation level, and the exponent m usually has a value of approximately 3. The saturation is defined as the fraction of the pore volume filled with liquid. When a measured value for k_{sat} is unavailable, the Carman–Kozeny equation can be used to make an estimate. The saturation is often related through porosimetry data to the capillary pressure, p_c, which is defined as the liquid pressure minus the gas pressure. The capillary pressure is related to the pore diameter by[23]

$$p_c = -\frac{4\sigma \cos \theta}{d} \quad (39)$$

where σ is the surface tension, d is the pore diameter, and θ is the contact angle, which is related to the hydrophobicity of the pore.

5 ELECTROLYTIC AND MEMBRANE PHASE TRANSPORT

To model the behavior of a PEM fuel cell system, the transport of protons across the membrane must be described. It is the movement of protons that carries current and thus permits fuel cell operation. In order to relate current density to potential drop across the membrane, the flux of charged species resulting from a potential gradient in an electrolytic solution has to be expressed. In this section, we examine the frameworks of dilute solution theory and concentrated solution theory, and examine the limitations of each. The complications of water transport due to electroosmotic drag are specifically considered, and the challenges of describing pressure driven flow in a solid polymer electrolyte are discussed.

5.1 Dilute solution theory

The simplest way to describe the movement of charged species in an electrolytic medium is by dilute solution theory. In dilute solution theory, one considers an uncharged solvent, charged solute species, and (perhaps) uncharged minor components. The superficial flux density of each dissolved species in terms of its interstitial concentration is given by:

$$\frac{\mathbf{N}_i}{\varepsilon} = -z_i u_i^e F c_i \nabla \Phi - (D_i^e + D_a) \nabla c_i + \frac{c_i \mathbf{v}}{\varepsilon} \quad (40)$$

where $\mathbf{N}_i$ is the superficial flux density of species i. The first term in the expression is a migration term, representing the motion of charged species that results from a potential gradient. The migration flux is related to the potential gradient $(-\nabla \Phi)$ by the charge number of the species z_i, its concentration c_i, and the tortuosity corrected mobility of the species u_i^e. The second term relates the diffusive flux to the concentration gradient and the Fickian diffusion coefficient D_i^e, which has been corrected for tortuosity. D_a is a dispersion coefficient representing the effect of axial dispersion. Strictly speaking, it is not a transport property and depends on fluid flow parameters; it disappears when convection is absent. Dispersion is usually ignored due to the relatively low velocities found in fuel cells. The final term is a convective term and represents the motion of the species as the bulk motion of the solvent carries it along. In general, the velocity of the solvent, $\mathbf{v}$, is determined by mass and momentum balances, e.g., by the Navier–Stokes equation. But as discussed above, in a porous network it is often easier to use Darcy's law.

Dilute solution theory considers only the interactions between each dissolved species and the solvent. The motion of each species is described by its transport properties, namely, the mobility and the diffusion coefficient. These transport properties can be related to one another at infinite dilution via the Nernst–Einstein equation:

$$D_i = RT u_i \quad (41)$$

So long as the solute species are sufficiently dilute that the interactions among them can be neglected, material balances can be written based upon the above expression for the flux, and the concentration and potential profiles in the electrolyte can be determined. Neglecting dispersion, and using equation (41) and the definition of the current

$$\mathbf{i} = F \sum_i z_i \mathbf{N}_i \quad (16)$$

equation (40) can be rewritten as a modified version of Ohm's law

$$\mathbf{i} = -\kappa \nabla \Phi - F \sum_i z_i D_i^e \nabla c_i \quad (42)$$

where κ is the conductivity, defined as

$$\kappa = F^2 \sum_i z_i^2 u_i^e c_i = \frac{F^2}{RT} \sum_i z_i^2 D_i^e c_i \quad (43)$$

The velocity does not appear in equation (42) due to electroneutrality, or, in other words, the bulk motion of a fluid with no charge density can contribute nothing to the current density.

5.2 Electroneutrality in the membrane

In PEM systems, in general, there is only one cation and one anion, the dissociated proton and the sulfonic acid site. The acid site is bound to the membrane, as seen in Figure 6 for the case of Nafion®. The sulfonic acid sites are distributed more or less evenly throughout the membrane, and, because the dissociation of the sulfonic acid groups is nearly complete in the presence of water, it is safe to assume that the charged acid sites are also distributed evenly throughout the membrane in the presence of water. Electroneutrality holds across the membrane

$$\sum_i z_i c_i = 0 \quad (25)$$

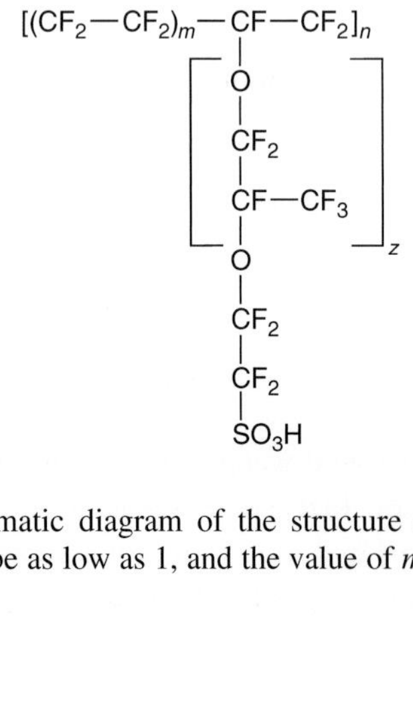

$$[(CF_2-CF_2)_m-CF-CF_2]_n$$
$$\begin{bmatrix} O \\ | \\ CF_2 \\ | \\ CF-CF_3 \\ | \\ O \\ | \\ CF_2 \\ | \\ CF_2 \\ | \\ SO_3H \end{bmatrix}_z$$

Figure 6. Schematic diagram of the structure of Nafion®. The value of z may be as low as 1, and the value of m ranges between 6 and 13.

except for a very thin double layer region near electrodes and other interfaces. Combining this equation with the requirement that the concentration of the anions be uniform across the membrane, it can be shown that the concentration of mobile protons is also fixed. In such a system, then, there will only be a migration and a convection term to describe the flux of other species, as is the case in a PEM based fuel cell system. If the details of the water concentration profiles and fluxes can be neglected in a well hydrated PEM fuel cell, Ohm's law can be used to relate the current density to the potential gradient in the membrane. Amphlett et al.[26] describe the performance of a fuel cell in this manner.

5.3 Water transport in the membrane

The framework of dilute solution theory allows for just two transport properties in a system composed of acid sites, protons, and water. If a model is to describe migration of protons, convective transport of water across the membrane, and electroosmotic drag, there must be a third parameter to quantify the fluxes. Furthermore, in most electrolytic systems, anions and cations move relative to a solvent whose velocity is determined by fluid dynamics. In the PEM fuel cell system, the anion sites are fixed, and the cations and water move relative to those sites. This peculiarity of having fixed acid sites changes the reference frame for transport in a PEM. Protons, water, and other minor species move relative to a fixed polymer framework, rather than all species moving relative to the convection of an aqueous solvent. The requirement that the anion species remain fixed in space is a key issue in describing transport in the membrane, and it results in the inclusion of the electrokinetic driving force.

Bernardi and Verbrugge[11, 27] use dilute solution theory to describe transport in the PEM system, for a system composed of a membrane, catalyst layer, and gas diffusion medium. The motion of the protons is related to the potential gradient and to the motion of the water in the membrane; the movement of the water is specified by pressure drop and electroosmotic drag:

$$\mathbf{v}_{H_2O} = -\left(\frac{k_p}{\mu}\right)\nabla p_L - \left(\frac{k_\Phi}{\mu}\right) z_f c_f F \nabla \Phi \quad (44)$$

where k_p and k_Φ are the hydraulic permeability and the electrokinetic permeability, respectively, p_L is the hydraulic or liquid pressure, and z_f and c_f refer to the charge and concentration of fixed ionic sites, respectively.

The movement of water with the passing of current, even in the absence of a gradient of pressure or water activity, is known as electroosmotic drag and is an important factor in the design of fuel cell systems. Proper water management is critical for fuel cell operation (too much water leads to flooding at the cathode; too little water dries out the membrane and increases ohmic losses), so understanding the details of water transport is critical. Measurements of electroosmotic drag coefficients indicate that, near saturation, anywhere from 1 to 2.5 water molecules are dragged across the membrane with each proton.[28, 29] Electroosmotic drag delivers considerably more water to the cathode than stoichiometry dictates would be generated through the reduction of oxygen alone.

It is interesting to calculate the hydraulic pressure difference that, according to the above model, must be applied across the membrane to balance the electroosmotic drag with convection, and yield zero net water flux. When the net water flux is zero, the current density obeys Ohm's law. Combining this fact with equation (44) and assuming constant physical properties leads to the following expression:

$$\Delta p_L = \frac{k_\Phi z_f c_f F L i}{k_p \kappa_m} \quad (45)$$

Thus, the required pressure difference is proportional to the current density and the thickness of the membrane, while it is inversely proportional to the conductivity of the membrane. In order to improve water management, it is common to operate with a higher pressure at the cathode than the anode.

5.4 Electrokinetic phenomena

There have been two principal approaches to describing the motion of water relative to the fixed polymer; the first is by the consideration of electrokinetic phenomena, and the second is the treatment of water as a dissolved species in the membrane. The treatment of electrokinetic phenomena, as Pintauro, Verbrugge, and others have adopted,[30, 31] is to describe the membrane as a series of pores with charged walls, through which liquid water filled with positively charged protons is allowed to move. The movement of the protons with the passing of current tends to drag water along with it, via the frictional interactions between the protons and the associated water. This approach presupposes that all the water in the membrane behaves as a fluid, and that there is a contiguous pore network that allows water to be transported from one side of the membrane to the other. In such a system, the movement of protons can be attributed to a potential gradient and a pressure gradient. The movement of water is determined primarily by a permeability of water subject to a pressure gradient, moving through the pore network. This approach is quite useful for describing fuel cell systems where the membrane is very well hydrated, but it requires that the water content be uniform across the membrane, with only a pressure gradient as a driving force for water movement. Such a treatment does not necessarily lend itself to describing the flux of water resulting when there is a water activity gradient across the membrane.

The second approach, which treats all of the species as solute species absorbed in a single membrane phase, allows for a concentration gradient of water across the membrane. Thus, it can deal with issues of incomplete humidification of the membrane. The difficulty with this approach, however, is related to the details of pressure driven flow, which will be discussed later. The intricacies of Schroeder's paradox, namely, that the membrane takes up different equilibrium amounts of water when exposed to liquid water and to steam, complicate the picture even further.

5.5 Concentrated solution theory

Under the rules of concentrated solution theory, the framework for describing the transport of species is altered slightly from dilute solution theory. Instead of describing only the interactions between each solute species and the prescribed solvent, concentrated solution theory also describes the interactions of the species with each other; hence, it is more generally valid, especially for nonideal systems. In concentrated solution theory, a force balance is written that equates a thermodynamic driving force to a sum of frictional interactions. At constant temperature and pressure, the driving force is the gradient of electrochemical potential, and the frictional interactions are specified by the motion of the species relative to one another:

$$c_i \nabla \mu_i = \frac{RT}{c_T} \sum_{j \neq i} \frac{c_i c_j}{D_{i,j}^e} (\mathbf{v}_j - \mathbf{v}_i). \qquad (46)$$

Here the $D_{i,j}$ term is a binary diffusion coefficient specifying the frictional interaction between species i and j. This equation is just a restatement of the generalized Stefan–Maxwell diffusion equation except now μ_i represents the electrochemical instead of chemical potential. Furthermore, as with Stefan–Maxwell diffusion, it may be necessary to choose explicitly a reference velocity, whether it is the velocity of a particular species, the mass average velocity, or the volume average velocity. For mass transport in the membrane, the membrane velocity is often used. The presence of a high molecular weight polymer is allowable under the Stefan–Maxwell framework. The diffusion coefficients are determined from experimental measurements. An appropriate choice of concentration scale is necessary to interpret the measurements and assign values to the diffusion coefficients; as long as the assumptions about the molecular weight of the polymer are applied consistently to both data analysis and system modeling, the framework will hold.

In a multicomponent system composed of N species, there are $N(N-1)/2$ independent transport properties. For a system composed only of protons, acid sites, and water, the framework of concentrated solution theory specifies three transport properties: the membrane-proton binary diffusion coefficient, the water-proton binary diffusion coefficient, and the water-membrane binary diffusion coefficient. These three coefficients can be arranged to yield the ionic conductivity of the membrane, the electroosmotic drag coefficient of water, and the diffusion coefficient of water in the membrane.

Concentrated solution theory has been used to describe transport in fuel cell membranes.[12, 32] By careful inversion of the Stefan–Maxwell equations in special cases (e.g., in the absence of current or in the absence of a water activity gradient) $D_{i,j}$ can be related to measurable transport properties; namely, D_{H_2O}, the diffusion coefficient of water; κ, the ionic conductivity; and ξ, the electroosmotic drag coefficient, which is the number of water molecules carried across the membrane with each hydrogen ion in the absence of a water activity gradient. Re-casting the Stefan–Maxwell equations in terms of these properties, we find[33]

$$\mathbf{i} = -\frac{\kappa \xi}{F} \nabla \mu_{H_2O} - \kappa \nabla \Phi \qquad (47)$$

$$\mathbf{N}_{H_2O} = -\left(D_{H_2O} + \frac{\kappa \xi^2}{F^2}\right) \nabla \mu_{H_2O} - \frac{\kappa \xi}{F} \nabla \Phi \qquad (48)$$

The multicomponent diffusion equation may be combined with the material balance equations and the condition of electroneutrality to provide a consistent description of transport processes in concentrated electrolytic systems.

5.6 Pressure driven flow in the membrane

In order to describe the effect of a pressure gradient on the transport of species in a membrane, it has been proposed that the electrochemical potential gradient be replaced with a more generalized driving force, suggested by Hirschfelder, Curtiss, and Bird[34]

$$d_i = c_i \left[\nabla \mu_i + \overline{S_i} \nabla T - \frac{M_i}{\rho} \nabla p - X_i + \frac{M_i}{\rho} \sum_j X_j c_j \right] \qquad (49)$$

where d_i is the driving force per unit volume acting on species i, M_i is the molecular weight of species i, $\overline{S_i}$ is the molar entropy of species i, ρ is the density of the solution, and the X_i terms refer to body forces per mole acting on species i. Bennion[35] applies this treatment to a membrane by requiring that the stresses acting on the membrane constitute the external body force X_i, a force that is absent for the solid species. If there is only a single body force acting, then a mechanical force balance requires that

$$c_m X_m = \nabla p \qquad (50)$$

This relationship allows one to simplify the above driving force for a membrane.

In general, for a species containing N solute species, there are $N + 3$ variables that must be determined: N compositional variables, the potential (Φ), the temperature, and the pressure. Either by the energy balance or by the requirement of isothermal conditions, the temperature profile can be determine, and N material balances, then, will allow for the concentration profiles to be solved. The requirement of electroneutrality provides an additional equation for the potential. Thus, there is required one additional equation to determine the pressure drop.

In Pintauro and Bennion's[36] development, they back off from this framework and, in a treatment of desalinization, consider a membrane equilibrated with liquid water and exposed to NaCl solutions. They write a matrix that relates fluxes of water, current, and NaCl salt to driving forces in pressure, electrolyte concentration, and potential. Assuming uniform water content, they solve for the relevant variables by material balances on water and salt concentration, and by electroneutrality. The implicit assumption of uniform water content also forces them to consider the membrane as a separate phase, thereby allowing the Gibbs-Duhem equation to specify the chemical potential of the water, rather than allowing for an additional degree of freedom.

Meyers[37] addresses the problem from a slightly different perspective. He assumes that the water and ion content of the membrane is all part of a single, homogeneous phase. He argues that in a single phase system, a pressure gradient cannot be supported because there is nothing to withstand the force; a pressure gradient should impose net movement of the membrane itself. He treats pressure driven flow by allowing for a discontinuity in pressure at the membrane/solution interface, and argues that additional mechanical stresses compressing the membrane should be indistinguishable from the thermodynamic pressure. In his treatment, the thermodynamic pressure might be discontinuous, as the total force acting on the membrane consists of a liquid solute pressure combined with the mechanical stress of the solid structures (gas diffusion media, flow fields, etc.) that support the membrane. Equilibrium is imposed on all soluble species by requiring equality of electrochemical potential of all soluble species at the membrane/solution interfaces.

Since the PEM in a fuel cell is not always saturated with water, the uptake of water at lower activities needs to be considered. If, as seems likely, acid sites do not induce condensation of bulk-like water in the pores over the entire range of water activities that a fuel cell is likely to experience, then the behavior of water that is associated strongly with the membrane must be treated and is not bulk-like. While some of the water in a well hydrated membrane might behave similarly to liquid water collected in open pores, at lower water content, the water behaves more like a solute species dissolved in the membrane.

For a fuel cell system that is not completely hydrated, the strong dependence of ionic conductivity on water content[38] demands that careful attention is paid in determining the concentration profile of water in the membrane; this profile, in turn, has a strong effect on the ohmic losses in the membrane. Perhaps the true nature of PEMs is one that contains elements of both models: a phase that specifically absorbs some water and imparts low levels of ionic conductivity and a second phase that behaves primarily like bulk water. These two approaches have not, as of yet, been reconciled into an overarching, complete theory of membrane transport that addresses all of the relevant issues.

5.7 Transport properties

Regardless of the details of model construction, reliable data on transport properties are necessary to quantify transport in PEMs. Perfluorosulfonate membranes with structures like that shown in Figure 6, including Nafion®, have been characterized fairly extensively. The ionic conductivity is generally measured by ac impedance techniques,[39] and all of the models described above rely on conductivity data to predict ohmic losses. The saturation of the membrane can affect the conductivity and transport properties greatly, and thus these effects must be taken into account. Furthermore, the thickness can also appreciably affect the transport properties.[40]

Fuller and Newman[28] devised a technique to measure the electroosmotic drag coefficient of water in a PEM by relating the drag coefficient to the open-circuit potential difference measured across a membrane with different water concentrations in the membrane sections adjacent to the two electrodes. This technique was repeated by researchers at Los Alamos,[29] who also reported on volumetric measurements performed on a membrane exposed to liquid water. Water diffusion coefficients have proved difficult to measure in the PEM system, although they have been estimated by NMR techniques.[41] Also of interest are molecular simulations to calculate the transport properties of perfluorosulfonate membranes from *ab initio* electronic structure calculations and dielectric continuum modeling.[42]

5.8 Transport of uncharged species in the membrane

In general, the transport of additional uncharged species is neglected in the models presently available in the open literature. There are exceptions; notably, Bernardi and Verbrugge[11] discuss the crossover of hydrogen and oxygen in their paper, treating the transport of the gases by convection and diffusion in the steady state. Springer et al.[43] discuss the importance of the diffusion of hydrogen and oxygen in the membrane and note that transport of these species appears to be much higher in the catalyst layers of the membrane electrode assembly (MEA) than would be expected from measurements of crossover through a membrane separator.

One system in which the movement of uncharged species is particularly important is the direct methanol fuel cell (DMFC), where methanol crossover occurs and compromises fuel cell efficiency. Several models exist that describe methanol crossover.[44, 45] These models generally treat methanol as a minor species in the solution, transported by diffusion and convection along with the water that is present in much higher concentrations.

6 NUMERICAL SIMULATIONS

The complexity of the governing differential equations that we have outlined for modeling porous electrodes generally precludes analytic solution, and numerical simulations must be used. Numerical models lend themselves to the incremental inclusion of ever greater levels of detail, which cannot usually be said of analytic solutions. In fact, while it may be possible to reduce the number of variables through mathematical manipulation, this is often not the best course of action because it reduces the flexibility of the program. Furthermore, the increase in computation time is usually modest, at least for a one-dimensional model. When constructing a numerical model, it is best to include a complete picture of the cell sandwich in order to capture interactions among the various components. Thus, for example, a one-dimensional model of a PEM fuel cell could include the regions outlined in Figure 1.

A useful approach to numerical modeling of porous electrodes is as outlined below. Conservation equations are cast in control volume form, as depicted schematically in Figure 7. Patankar[46] describes this approach in detail. The integer j is the spatial mesh point, the integer n is the discretized time. Vectors are defined at half mesh points, while scalars are defined at full mesh points. Reaction terms are evaluated at quarter mesh points for each half mesh box. In the case of material balances, this approach

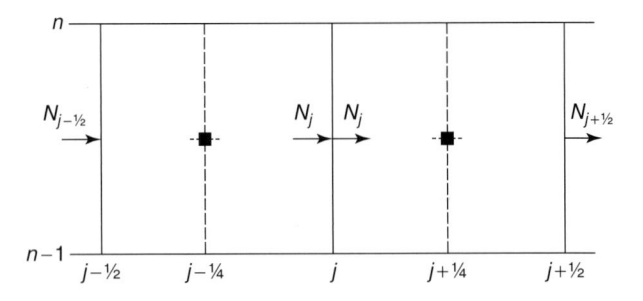

Figure 7. Schematic of the control volume approach for numerical simulations.

rigorously conserves mass, which cannot be said of finite difference methods. The coupled differential equations in the spatial domain may be solved with a banded solver, such as that described in Appendix C of Newman.[47] Crank-Nicholson time stepping, which involves averaging the equations symmetrically in time, is recommended in order to achieve stability and second order accuracy in time.

6.1 Cathode simulation

In this section, we present steady state results from a numerical model of the cathode of a PEM fuel cell. The model presented in this work is similar to that of Springer and co-workers[39, 43] and Perry et al.[48] Springer and coworkers demonstrated how to use a model like this to fit experimental data from lab-scale fuel cells. Perry et al. examined limiting cases due to either oxygen or ionic mass transport limitations within the catalyst layer. We have selected the cathode for our example because it is the most important electrode, in terms of polarization losses, in PEM fuel cells operating on either hydrogen or a reformed hydrocarbon fuel. A number of publications have dealt with simulations of the cathode and limiting cases that may arise under certain conditions. We discuss both of these topics in this section. Furthermore, we mention some of the diagnostic techniques that may be used to understand what is limiting the performance of a fuel cell electrode.

The model treats the diffusion of oxygen through the diffusion medium and the catalyst layer, ohmic drop in the ionomeric and solid phases within the catalyst layer, and oxygen reduction kinetics. Oxygen diffusion within the catalyst layer is treated with Fick's law, Ohm's law is used to describe the potential drop within the ionomeric and solid phases, and the ORR is assumed to follow Tafel kinetics with a first order dependence on the oxygen partial pressure and a cathodic transfer coefficient of 1. The total gas pressure and temperature are assumed constant. The model does not treat the water generated by the reduction of oxygen. Thus, the flow of water through the cathode catalyst layer and the diffusion medium does not affect the oxygen transport. This is an important simplification that limits the applicability of a model of this type, since water management is a key to the proper design and operation of a PEM fuel cell. The conductivity of the ionomer within the catalyst layer is assumed to be constant, implying that the hydration level of the ionomer is uniform.

The model contains the following five variables: Φ_1, the potential in the solid phase; Φ_2, the potential in the ionomeric phase; i_1, the current density in the solid phase; i_2, the current density in the ionomeric phase; and x_{O_2}, the oxygen mole fraction. The interface between the membrane and the catalyst layer is located at $z = 0$, while the interface between the catalyst layer and the diffusion medium is located at $z = L$. The boundary conditions are

$$\Phi_2|_{z=0} = 0$$
$$\Phi_1|_{z=L} = V$$
$$i_1|_{z=0} = 0$$
$$i_2|_{z=L} = 0$$
$$\nabla x_{O_2}|_{z=0} = 0$$

and

$$x_{O_2}|_{z=L} = x_{O_2}^b \left(1 - \frac{i_1|_{z=L}}{i_{\lim}}\right) \qquad (51)$$

Thus, the ionomeric potential is arbitrarily set to 0 at $z = 0$, the solid phase potential is set at the interface between the catalyst layer and the diffusion medium, all of the current is carried in the ionomeric phase at the membrane interface, and all of the current is carried by the solid phase at the diffusion medium interface. The oxygen mole fraction at $z = L$ is set in accordance with a mass transfer resistance in the diffusion medium consistent with the limiting currents in Table 2. In this model, it is assumed that no oxygen diffuses past the interface between the catalyst layer and the membrane.

The simulations were run using a program written in FORTRAN using Newman's BAND(J), MATINV, and AUTOBAND subroutines.[47] The governing equations were cast in finite difference form.

Table 2 gives the parameters used in the simulations. These values are representative of state of the art PEM fuel cells. The interfacial area per unit volume was selected to give a catalyst surface area of $50 \, m^2 \, g^{-1}$. The exchange current density was selected to give a mass specific current density of $200 \, mA \, mg_{Pt}^{-1}$ at a cathode potential of $0.9 \, V$ and an oxygen partial pressure of 1 bar. The value of the limiting current on air is typical for a PEM system and is assumed to be proportional to the oxygen mole fraction in the gas channel.

Table 2. Input and varied parameters used for the numerical simulations of a cathode under a base case, an ohmically limited case, and a mass transfer limited case.

Input parameters	
Temperature	$65 \,°C$
Total gas pressure	1 bar
Gas relative humidity	100%
Standard potential	1.195 V
Electrode thickness	$10 \, \mu m$
Platinum loading	$0.4 \, mg \, cm^{-2}$
Cathodic transfer coefficient	1
Catalyst surface area	$50 \, m^2 \, g^{-1}$
Interfacial area per unit volume	$2.00 \times 10^5 \, cm^2 \, cm^{-3}$
Exchange current density	$1.51 \times 10^{-8} \, A \, cm^{-2}$
Limiting current (air)	$2.13 \, A \, cm^{-2}$
Limiting current (oxygen)	$10.1 \, A \, cm^{-2}$
Channel oxygen mole fraction (air)	0.16
Channel oxygen mole fraction (oxygen)	0.75
Variable parameters	
Base ohmic resistance	$1.00 \times 10^{-5} \, \Omega \, cm^2$
High ohmic resistance	$0.2 \, \Omega \, cm^2$
Base oxygen diffusion coefficient	$2.30 \times 10^{-2} \, cm^2 \, s^{-1}$
Low oxygen diffusion coefficient	$2.00 \times 10^{-4} \, cm^2 \, s^{-1}$
Calculated parameters[a]	
Roughness factor	200
$i \, (0.9 \, V, 1 \, bar \, O_2)$	$200 \, mA \, mg_{Pt}^{-1}$
$i \, (0.9 \, V, 1 \, bar \, O_2)$	$400 \, \mu A \, cm_{Pt}^{-2}$

[a]Derived from Ref. [40].

Three cases are considered. Case 1, the base case, simulates a cathode with negligible ohmic drop and negligible oxygen mass transfer limitations within the catalyst layer. Case 2 describes a cathode with significant ohmic limitations, but negligible mass transfer limitations. Finally, case 3 simulates an electrode with negligible ohmic limitations, but significant mass transfer limitations within the catalyst layer. The parameters for cases 2 and 3 were selected to give similar polarization losses on air. The ohmically limited case may correspond to an electrode with a low ionomer content, while the mass transfer limited case may correspond to an electrode with a high ionomer content. In practice, there is probably a natural trade-off between ohmically limited electrodes and mass transfer limited electrodes.

Figure 8 shows polarization curves for the three cases on fully humidified air and oxygen. At low current densities, all of the electrodes are kinetically limited and perform similarly. As the current density increases, the polarization associated with the ohmic and mass transfer effects becomes significant, and the performance of the three electrodes on air begins to deviate. At a current density of $1 \, A \, cm^{-2}$, the ohmic and mass transfer cases are approximately $50 \, mV$ below the kinetic case. All three cases have the same limiting current, corresponding to zero oxygen

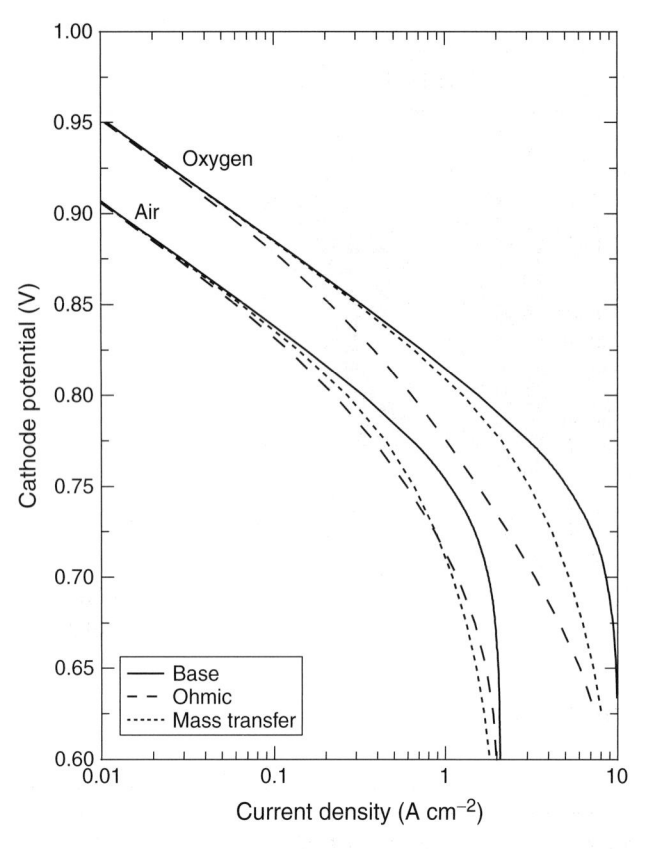

Figure 8. Polarization curves for the base, ohmically limited, and mass transfer limited cases of the cathode simulation for feeds of humidified air and oxygen. The base case is as given in Table 2, the ohmically limited case increases the ohmic resistance of the base case, and the mass transfer limited case decreases the oxygen diffusion coefficient of the base case.

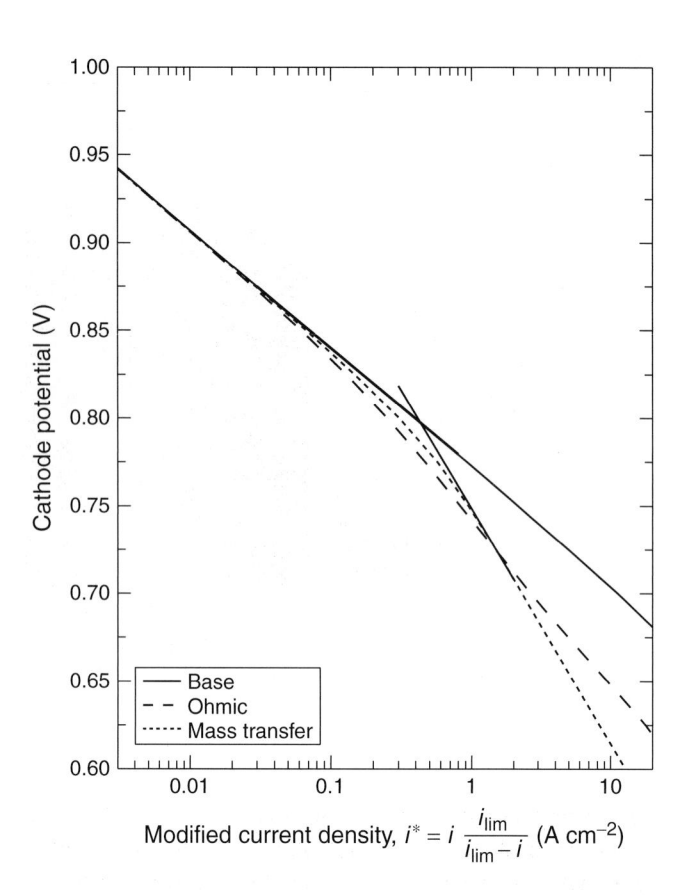

Figure 9. Polarization curves modified to remove external mass transfer effects for the three cases of the cathode simulation on air. The bold lines represent lines with single and double Tafel slopes.

partial pressure at the interface between the electrode and the diffusion medium.

The oxygen polarization curves are instructive. The ohmically limited electrode begins to deviate from the base electrode at the same current density on oxygen as it does on air. In the absence of external mass transfer limitations, the differences in potential between the base electrode and the ohmic electrode at a fixed current density would be identical for oxygen and air. The mass transport limited electrode, on the other hand, begins to deviate from the base electrode at the same cathode potential on oxygen as it does on air. In this case, the differences in current density between the base and mass transfer electrodes at a fixed cathode potential are the same for oxygen and air. Both of these results were reported by Perry et al.[48]

Figure 9 shows the cathode potential as a function of a mass transfer corrected current for the air simulations on all three electrodes. This type of plot corrects for external mass transport limitations, and allows for changes in the Tafel slope due to phenomena occurring within the catalyst layer to be seen more clearly. All three electrodes show a single Tafel slope at high potentials, as expected, since there are only kinetic limitations. This single slope extends through the entire potential range in the case of the kinetically limited electrode. The mass transfer electrode shows a distinct double Tafel slope at lower cathode potentials due to a non uniform reaction rate in the electrode caused by the strong mass transfer limitations. The ohmic electrode shows an increased slope, but not a double Tafel slope on this scale. Mass transport limitations internal to the catalyst layer are, in the absence of ohmic limitations, a function of potential, while ohmic limitations, in the absence of mass transfer limitations, are a function of current density. This explains why the double Tafel slope due to mass transfer limitations is visible on this plot while the double Tafel slope due to ohmic limitations is not. Thus, it may be possible to discern between mass transfer limited and ohmically limited electrodes by plotting data this way.

The difference in potential between the oxygen and air curves at a given current density, the oxygen gain, is simply

$$V_{O_2} - V_{air} = \frac{RT}{\alpha F} \ln\left(\frac{x_{O_2,ox}}{x_{O_2,air}}\right) \quad (52)$$

The oxygen gain increases with increasing current density as mass transport losses become more important, and it is larger for the mass transfer limited cathode than it is for either the ohmic electrode or the kinetic electrode. It can be shown that, in the absence of external mass transfer limitations, the oxygen gain for an ohmically limited system retains its value, while the oxygen gain for the mass transfer limited electrode doubles. Figure 10 elaborates on this point by plotting the ratio of current density measured on oxygen to the current density measured on air at the same cathode potential. This ratio is equal to $x_{O_2,ox}/x_{O_2,air} = 4.76$ for both the kinetic and mass transfer limited electrodes at all potentials. Thus, these electrodes show a first order dependence on oxygen concentration at all potentials. The ratio is 4.76 for the ohmically limited electrode at high potentials, where the electrode kinetics dominate. As the electrode potential drops, and the current density increases, the ratio falls, approaching the square root of 4.76 (line A in Figure 10). This behavior was explained by Perry et al.[48] At higher current densities, mass transport resistance in the diffusion medium, which is first order in oxygen, dominates, and the ratio again approaches 4.76. Thus, ohmic limitations within the electrode can lead to an apparent reaction order of 1/2, while mass transfer limitations may lead to an apparent reaction order of 1, with respect to oxygen.

Figures 11 and 12 show the current, potential, and oxygen distributions at 0.7 V on air for the ohmically limited and mass transfer limited electrodes, respectively. The current densities for these cases are approximately 1.1 and 1.2 A cm^{-2}, respectively. In the ohmically limited case, the reaction is shifted towards the interface between the membrane and the catalyst layer. The potential drop in the ionomer is about 70 mV, or approximately $L/(3\kappa)$. For reference, a value of $L/(2\kappa)$ is expected for a uniform current distribution. In the mass transfer limited case, the current is shifted towards the interface between the catalyst

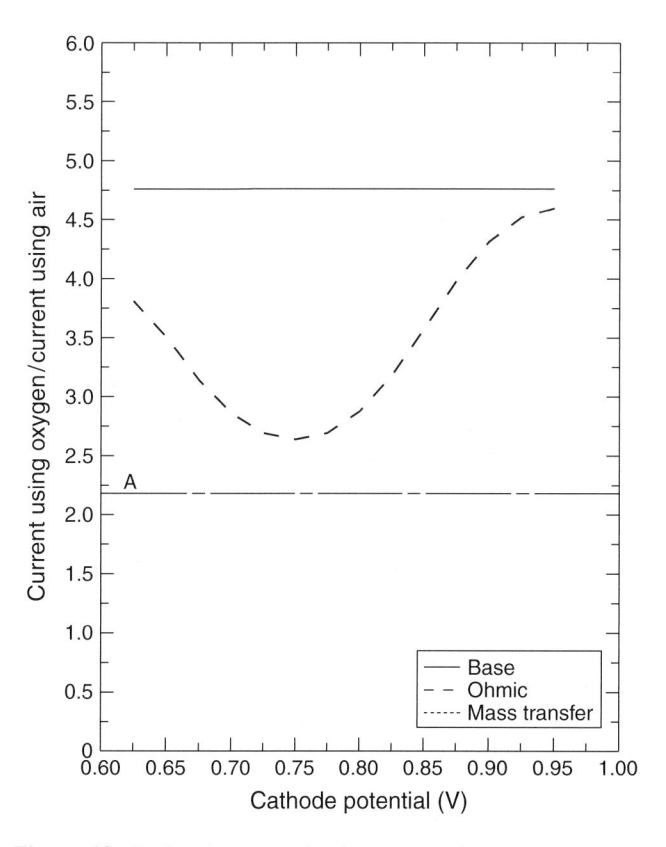

Figure 10. Ratio of current density measured on oxygen to the current density measured on air as a function of the potential of the cathode for all three cases of the cathode simulation. The mass transfer limited and base cases are the same, and equal the theoretical ratio for a kinetically or mass transfer limited electrode. Line A is the theoretical ratio for a completely ohmically limited electrode.

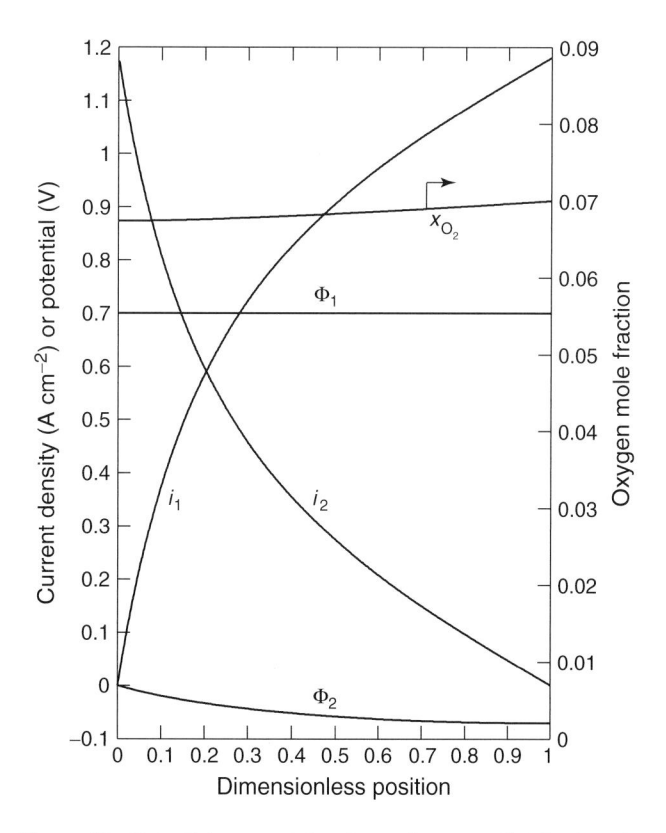

Figure 11. Potential, current density, and oxygen distributions in the solid and electrolyte phases of the cathode for the ohmically limited case. The cathode is at a potential of 0.7 V, and the feed is humidified air.

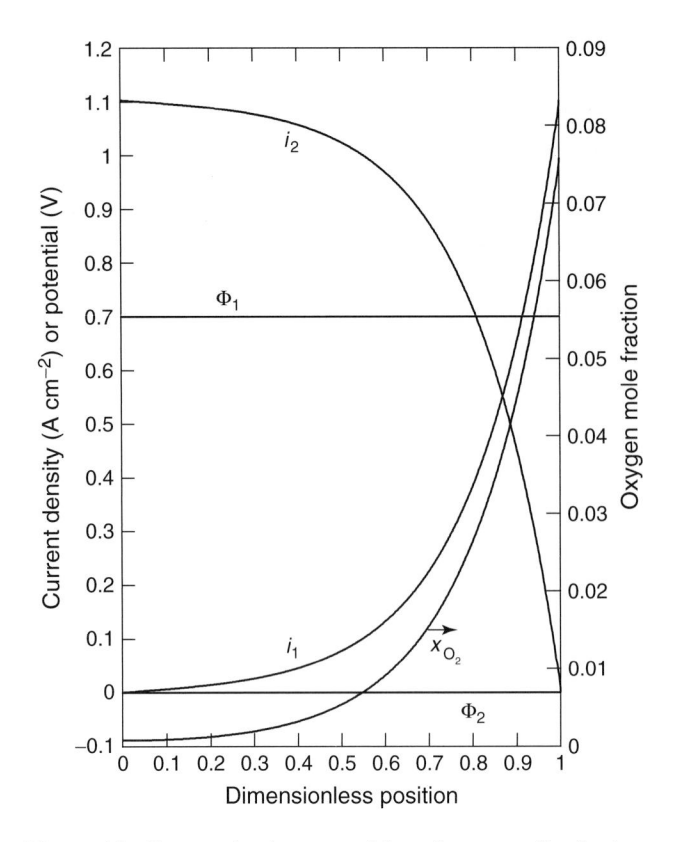

Figure 12. Current density, potential, and oxygen distributions in the solid and electrolyte phases of the cathode for the mass transfer limited case. The cathode is at a potential of 0.7 V, and the feed is humidified air.

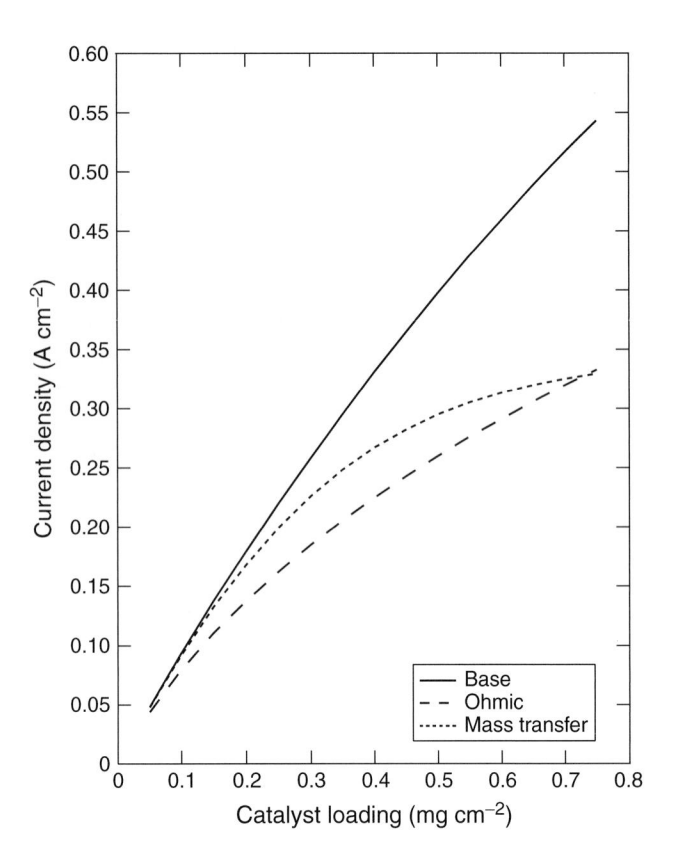

Figure 13. Effect of catalyst loading on current density at a cathode potential of 0.8 V and a feed of humidified air for all three cases of the cathode simulation.

layer and the diffusion medium; nearly all of the current is transferred between $z = 0.5$ and $z = 1$. These profiles agree with the double Tafel slope seen in Figure 9.

Figure 13 shows the effect of catalyst loading on air performance at a cathode potential of 0.8 V. The thickness of the catalyst layer was assumed to be proportional to the catalyst loading. If the catalyst were uniformly accessible, the result would be a straight line, but even the base case curves at high current densities because of the mass transfer resistance in the diffusion medium. The ohmic and mass transfer cases lose a significant amount of performance at high catalyst loadings. At very high loadings, the current in the mass transport limited electrode becomes independent of loading as the oxygen only penetrates a small region near the interface with the diffusion medium. Similarly, in the ohmically limited electrode, there is no advantage to increasing the loading beyond a certain point, as the ohmic limitations confine the reaction to a thin layer near the interface between the membrane separator and the catalyst layer. These limitations are clearly visible in Figures 11 and 12. Perry *et al.*[48] and Newman[47] present dimensionless groups that may be used to assess the importance of kinetic, ohmic, and mass transfer effects.

7 FUEL CELL MODELS

The cathode simulations described in the preceding section provide an introduction to the mathematical modeling of fuel cells, emphasizing the relationships between kinetic, mass transfer, and ohmic effects. The particular system was selected in order to yield interesting, important, and realistic results. Simple models like the one described in this work are useful as they can guide our thinking about the behavior of fuel cells. However, it is worthwhile to consider important additional effects that could be incorporated into a fuel cell model.

Consideration of an entire fuel cell cross-section represents an important advance since it allows one to examine interactions among the different layers. In the PEM case, communication of water between adjacent layers is of particular interest since proper water management is a key aspect of the design of PEM fuel cells. Bernardi and Verbrugge[11] treat the water balance in a fuel cell when the membrane is fully saturated at all times. The transport of liquid water through the various components is treated with Darcy's law, assuming constant permeabilities. Interestingly, in their simulations, the diffusion media are more resistive to water flow than the Nafion® membrane. Since the porosities of the two components are reasonably similar, and the pores are probably significantly larger in the substrate than in the ionomer, this suggests that the saturation level of the substrate is quite low. A low saturation level is desired in the diffusion media in order to minimize the resistance to gas phase mass transfer.

Fuller and Newman[33] treat the water balance when the gas streams, and therefore, the membrane, are subsaturated. They use concentrated solution theory and their model solves explicitly for water and potential profiles in the membrane. Fuller and Newman integrated in the channel direction, following a procedure similar to the one described in this chapter. This allowed them to look at reactant depletion, water generation, and thermal effects. Nguyen and White[32] constructed a model in the channel direction applicable to cocurrent and countercurrent flow arrangements and examined various humidification schemes.

The models that we have focused on thus far involve a one-dimensional description of transport in the direction normal to the membrane. Two-dimensional effects may be important because the ribs in a fuel cell usually occlude part of the membrane. Kulikovsky et al.,[49] among others, have modeled this behavior. Finally, if uncoupling the different length scales is considered undesirable, full three-dimensional models can be pursued.[50]

8 CONCLUSIONS

In this chapter, we reviewed some of the models that have been developed to describe fuel cell performance. Even the simplest models provide some insight into the selection of proper operating conditions and, as such, can be very instructive tools. In designing a fuel cell, the details of mass transport within the layers of the cell should be examined. It is the interaction of several simultaneous processes, namely, ionic resistance, gas phase mass transport, kinetic losses, and liquid water removal, which make fuel cell operation possible. A detailed knowledge of these interrelated processes is necessary to develop a system that behaves optimally. It is perhaps worth noting that, were it possible to neglect completely any of these effects, one might conclude that the PEM fuel cell system is far from optimized. The fact that all of these effects must be considered in concert implies that the PEM fuel cell is nearly optimized for the present class of materials, as the improvement of one process often comes at the detriment of another. Modeling and analysis based upon the techniques reviewed in this chapter will assist fuel cell scientists and engineers to optimize fuel cells based upon the materials that will be the focus of future developmental efforts.

LIST OF SYMBOLS

Roman

a_i^α	activity of species i in phase α
$a_{k,p}$	interfacial surface area between phases k and p per unit volume, cm^{-1}
b	Tafel slope, V
c_i	concentration of species i per unit pore volume, mol cm^{-3}
$c_{i,k}$	concentration of species i in phase k, mol cm^{-3}
c_T	total solution concentration or molar density, mol cm^{-3}
C	number of species
d	pore diameter, cm
d_i	driving force per unit volume acting on species i, J cm^{-4}
D_i	Fickian diffusion coefficient of species i in a mixture, cm^2 s^{-1}
D_a	dispersion coefficient, cm^2 s^{-1}
$D_{i,j}$	diffusion coefficient of i in j, cm^2 s^{-1}
D_{K_i}	Knudsen diffusion coefficient of species i, cm^2 s^{-1}
f	the ratio of moles of inerts in the air stream to moles of inerts in the fuel stream
F_a	molar flow rate of carbon dioxide at the anode, mol s^{-1}
F	Faraday's constant, 96487 C eq^{-1}
$\mathbf{i}_k$	current density in phase k, A cm^{-2}
i_0	exchange current density, A cm^{-2}
$i_{h,k}$	transfer current density per unit interfacial area between phase k and the solid phase due to reaction h, A cm^{-2}
$i_{\lim}$	limiting current density, A cm^{-2}
k	permeability, cm^2
k_{sat}	permeability measured at complete saturation, cm^2
k_p	hydraulic permeability, cm^2
k_Φ	electrokinetic permeability, cm^2
k'	Kozeny constant
L	catalyst layer thickness, cm
m	exponent in equation (38), usually equal to 3
M	number of degrees of freedom in equation (13)
M_i	molecular weight of species i, g mol^{-1}
$M_i^{z_i}$	symbol for the chemical formula of species i in phase k having charge z_i
n_h	number of electrons transferred in electrode reaction h
N	number of species
$\mathbf{N}_i$	superficial flux density of species i, mol cm^{-2} s^{-1}
$\mathbf{N}_{i,k}$	flux density of species i in phase k, mol cm^{-2} s^{-1}

p	total gas pressure, bar	μ_i	chemical potential of species i in equation (29), J mol^{-1}
p_i	partial pressure of species i, bar	μ_i	electrochemical potential of species i, J mol^{-1}
p_c	capillary pressure, bar	μ_i^α	electrochemical potential of species i in phase α, J mol^{-1}
p_L	hydraulic or liquid pressure, bar		
p^{vap}	vapor pressure of water, bar	θ	contact angle, degrees
P	number of phases in equation (13)	ρ	solution density, g cm^{-3}
$r_{l,k-p}$	rate of reaction l per unit of interfacial area between phases k and p, mol s^{-1} cm^{-2}	σ	surface tension in equation (39), N cm^{-1}
		σ	conductivity in the solid phase, S cm^{-1}
$R_{g,k}$	rate of strictly homogenous reaction g in phase k, mol s^{-1} cm^{-3}	τ	tortuosity
		ξ	electroosmotic drag coefficient
R'	ohmic resistance, Ω cm^2		
R	universal gas constant, 8.3143 J mol^{-1} K^{-1}		

R number of equilibrated reactions in equation (13)

Subscripts

0	solvent or bulk value
1	solid, electronically conducting phase
2	ionic conducting phase
f	fixed ionic site in the polymer membrane electrolyte
g	homogeneous reaction number
h	electron transfer reaction number
i	generic species
j	generic species
k	generic phase
l	heterogeneous reaction number
m	membrane
p	generic phase

$s_{i,k,l}$	the stoichiometric coefficient of species i in phase k participating in reaction l
S	saturation
S_o	surface area to volume of solid phase, cm^{-1}
$\bar{S}_i$	molar entropy of species i, J mol^{-1} K^{-1}
t	time, s
T	absolute temperature, K
u	hydrogen utilization
u_i	mobility of species i, cm^2 mol J^{-1} s^{-1}
U	reversible cell potential, V
U'	potential intercept for the polarization equation, V
U^θ	standard potential for oxygen reduction, 1.229 V at 25 °C
$\mathbf{v}_i$	interstitial velocity of species i, cm s^{-1}
V	cell potential, V
W	width perpendicular to flow in gas channel, cm
x_i	mole fraction of species i
X_i	extensive term referring to body forces per mole acting on species i
X_i	ratio of moles of species i to moles of inerts in the same stream
y	distance down the flow field channel, cm
z	distance across the cell sandwich, cm
z_i	valence or charge number of species i

Superscripts

0	inlet or initial value
b	bulk value
e	effective, corrected for tortuosity
i	electrode interface value

REFERENCES

1. J. Newman, *Electrochim. Acta*, **24**, 223 (1979).
2. R. B. Bird, W. E. Stewart and E. N. Lightfoot, 'Transport Phenomena', 2nd edition, John Wiley & Sons, New York (2001).
3. J. Newman and W. Tiedemann, *AIChE J.*, **21**, 41 (1975).
4. Y. A. Chizmadzhev, V. S. Markin, M. R. Tarasevich and Y. G. Chirkov, 'Makrokinetika protsessov v poristykh sredakh (Toplivnye elementy)', Izdatel'stvo "Nauka", Moscow (1971).
5. J. Bockris and S. Srinivasan, 'Fuel Cells: Their Electrochemistry', McGraw-Hill, New York (1969).
6. J. S. Dunning, 'Analysis of Porous Electrodes with Sparingly Soluble Reactants', Unpublished Ph.D. Dissertation, University of California, Los Angeles, CA (1971).

Greek

α	transfer coefficient for ORR
β	symmetry factor
δ	diffusion length, cm
ε	porosity
ε_k	volume fraction of phase k
Φ_1	potential in the solid phase, V
Φ_2	potential in the ionically conducting phase, V
η_s	surface overpotential, V
κ	conductivity of the electrolytic phase, S cm^{-1}
μ	viscosity, g cm^{-1} s^{-1} or Pa s

7. J. Giner and C. Hunter, *J. Electrochem. Soc.*, **116**, 1124 (1969).

8. R. P. Iczkowski and M. B. Cutlip, *J. Electrochem. Soc.*, **127**, 1433 (1980).

9. M. Uchida, Y. Fukuoka, Y. Sugawara, N. Eda and A. Ohta, *J. Electrochem. Soc.*, **143**, 2245 (1996).

10. T. R. Ralph, G. A. Hards, J. E. Keating, S. A. Campbell, D. P. Wilkinson, M. Davis, J. St-Pierre and M. C. Johnson, *J. Electrochem. Soc.*, **144**, 3845 (1997).

11. D. M. Bernardi and M. W. Verbrugge, *J. Electrochem. Soc.*, **139**, 2477 (1992).

12. T. E. Springer, T. A. Zawodzinski and S. Gottesfeld, *J. Electrochem. Soc.*, **138**, 2334 (1991).

13. B. Pillay, 'Design of Electrochemical Capacitors for Energy Storage', Unpublished Ph.D. Dissertation, University of California, Berkeley, CA (1996).

14. J. Newman, *Ind. Eng. Chem. Res.*, **34**, 3208 (1995).

15. D. A. G. Bruggeman, *Ann. Phys.*, **24**, pp. 636 (1935).

16. K. Kinoshita, 'Carbon Electrochemical and Physiochemical Properties', John Wiley & Sons, New York (1988).

17. D. Bernardi, 'Mathematical Modeling of Lithium (Alloy) Iron Sulfide Cells and the Electrochemical Precipitation of Nickel Hydroxide', Unpublished Ph.D. Dissertation, University of California, Berkeley, CA (1986).

18. T. Springer, T. Zawodzinski and S. Gottesfeld, 'Modeling of Polymer Electrolyte Fuel Cell Performance with Reformate Feed Streams: Effects of Low Levels of CO in Hydrogen', Electrochemical Society, PV 97–13 (1997).

19. A. J. Appleby, *J. Electrochem. Soc.*, **117**, 328 (1970).

20. K. Kinoshita, 'Electrochemical Oxygen Technology', John Wiley & Sons, New York (1992).

21. M. Knudsen, 'The Kinetic Theory of Gases', Methuen, London (1934).

22. L. R. Jordan, A. K. Shukla, T. Behrsing, N. R. Avery, B. C. Muddle and M. Forsyth, *J. Appl. Electochem.*, **30**, 641 (2000).

23. F. A. L. Dullien, 'Porous Media: Fluid Transport and Pore Structure', Academic Press, New York (1979).

24. E. A. Mason and A. P. Malinauskas, 'Gas Transport in Porous Media: The Dusty-Gas Model', Elsevier, Amsterdam (1983).

25. D. M. Bernardi, *J. Electrochem Soc.*, **137**, 3350 (1990).

26. J. C. Amphlett, R. M. Baumert, R. F. Mann, B. A. Peppley, P. R. Roberge and T. J. Harris, *J. Electrochem. Soc.*, **142**, 1 (1995).

27. D. M. Bernardi and M. W. Verbrugge, *AIChE J.*, **37**, 1151 (1991).

28. T. F. Fuller and J. Newman, *J. Electrochem. Soc.*, **139**, 1332 (1992).

29. T. A. Zawodzinski, Jr, J. Davey, J. Valerio and S. Gottesfeld, *Electrochim. Acta*, **40**, 297 (1995).

30. P. N. Pintauro and M. W. Verbrugge, *J. Membr. Sci.*, **44**, 197 (1989).

31. A. G. Guzmán-Garcia, P. N. Pintauro, M. W. Verbrugge and R. F. Hill, *AIChE J.*, **36**, 1061 (1990).

32. T. V. Nguyen and R. E. White, *J. Electrochem. Soc.*, **140**, 2178 (1993).

33. T. F. Fuller and J. S. Newman, *J. Electrochem. Soc.*, **140**, 1218 (1993).

34. J. O. Hirshfelder, C. F. Curtiss and R. B. Bird, 'Molecular Theory of Gases and Liquids', John Wiley & Sons, New York (1954).

35. D. N. Benion, 'Mass Transport of Binary Electrolyte Solutions in Membranes, Water Resources Center Desalination Report No. 4', Tech. Rep. 66–17, Department of Engineering, University of California, Los Angeles, CA (1966).

36. P. N. Pintauro and D. N. Bennion, *Ind. Eng. Chem. Fundam.*, **23**, 230 (1984).

37. J. Meyers, 'Simulation and Analysis of the Direct Methanol Fuel Cell', Unpublished Ph.D. Dissertation, University of California, Berkeley, CA (1998).

38. T. A. Zawodzinski, Jr, T. E. Springer, J. Davey, R. Jestel, C. Lopez, J. Valerio and S. Gottesfeld, *J. Electrochem. Soc.*, **140**, 1981 (1993).

39. T. E. Springer, T. A. Zawodzinski, M. S. Wilson and S. Gottesfeld, *J. Electrochem. Soc.*, **143**, 587 (1996).

40. S. Srinivasan, D. J. Manko, H. Kock, M. A. Enayetullah and A. J. Appleby, *J. Power Sources*, **29**, 367 (1990).

41. T. A. Zawodzinski, Jr, M. Neeman, L. O. Sillerud and S. Gottesfeld, *J. Phys. Chem.*, **95**, 6040 (1991).

42. S. J. Paddison, R. Paul and T. A. Zawodzinski, Jr, *J. Electrochem. Soc.*, **147**, 617 (2000).

43. T. E. Springer, M. S. Wilson and S. Gottesfeld, *J. Electrochem. Soc.*, **140**, 3513 (1993).

44. K. Scott, W. Taama and J. Cruickshank, *J. Power Sources*, **65**, 159 (1997).

45. S. F. Baxter, V. S. Battaglia and R. E. White, *J. Electrochem. Soc.*, **146**, 437 (1999).

46. S. Patankar, 'Numerical Heat Transfer and Fluid Flow', McGraw-Hill, New York (1999).

47. J. S. Newman, 'Electrochemical Systems', Prentice Hall, Englewood Cliffs, NJ (1991).

48. M. L. Perry, J. Newman and E. J. Cairns, *J. Electrochem. Soc.*, **145**, 5 (1998).

49. A. A. Kulikovsky, J. Divisek and A. A. Kornyshev, *J. Electrochem. Soc.*, **146**, 3981 (1999).

50. A. Dutta, S. Shimpalee and J. W. Van Zee, *J. Appl. Electrochem.*, **30**, 135 (2000).

Chapter 8
Mass transfer in flow fields

K. Scott
University of Newcastle upon Tyne, Newcastle upon Tyne, UK

1 INTRODUCTION TO FLOW FIELDS

A fuel cell consists of a thin composite structure of anode, cathode and electrolyte. Good electrochemical performance of the cell requires effective electrocatalysts. The electrocatalysts in a fuel cell are positioned on either side of an electrolyte, typically in the form of a polymer, ceramic or immobilized acid or alkali, to form the cell assembly. The reaction gases are, in practical operation, fed to the back faces of the electrodes. Flow fields are used to supply and distribute the fuel and the oxidant to the anode and the cathode electrocatalyst respectively. The distribution of flow over the electrodes should ideally be uniform to try to ensure a uniform performance of each electrode across its surface. The flow field allows gas to flow along the length of the electrode whilst permitting mass transport to the electrocatalyst normal to its surface.

In most practical systems, fuel cells are connected in series to produce useful overall voltages. In principle, this is achieved by simply connecting the edges of the electrodes. However because the electrode structures are thin, and of relatively low electrical conductivity, they would introduce a small but significant loss of voltage, especially at high current loads. Thus, electrical connection in stacks is usually achieved using bipolar plates, which make electrical connection over the surface of the electrode. A second function of the bipolar plates is to separate the anode and cathode gases. Hence, not surprisingly, the functions of the bipolar plate and flow field are incorporated into one unit, sometimes referred to as the "flow field plate". This function is depicted in Figure 1. Plates have to be thin, electrical conducting, nonpervious to gases, noncorrosive, low weight and low cost. These factors introduce several challenges in plate selection and design. For example, whilst the flow field enables access of gas to the electrode structure in its open spaces, it prevents electrical contact at these points. Electrical contact should be as frequent and as large as possible to mitigate against long current flow path lengths. However large areas of electrical contact could lead to problems of access of reactant gases to regions under the electrical contact. Thus overall the flow field design and the flow therein are a critical factor in fuel cell operation.

2 FLOW FIELD PLATE DESIGNS

One of the simplest flow field designs consists of a series of narrow parallel rectangular channels (Figure 2) where fuel or oxidant is fed at one end and removed from the opposite end. Such "ribbed" designs are commonly used in phosphoric acid cells and proton exchange membrane (PEM) cells. The plates can be machined graphite or metal, or produced by compression or injection molding of carbon polymer composites. The direction of anode and cathode gas flows can be parallel, either counter-current or co-current, or at an angle of 90°.

In phosphoric acid fuel cells the ribbed flow channels (Figure 2) are either formed on opposite sides of the bipolar plate connector (referred to as ribbed separator type) or in a porous substrate (referred to as the ribbed substrate type). The latter has the advantage of presenting a flat porous surface to the electrode assembly, which maximizes electrical contact whilst achieving a more uniform diffusion of gases to the electrodes.

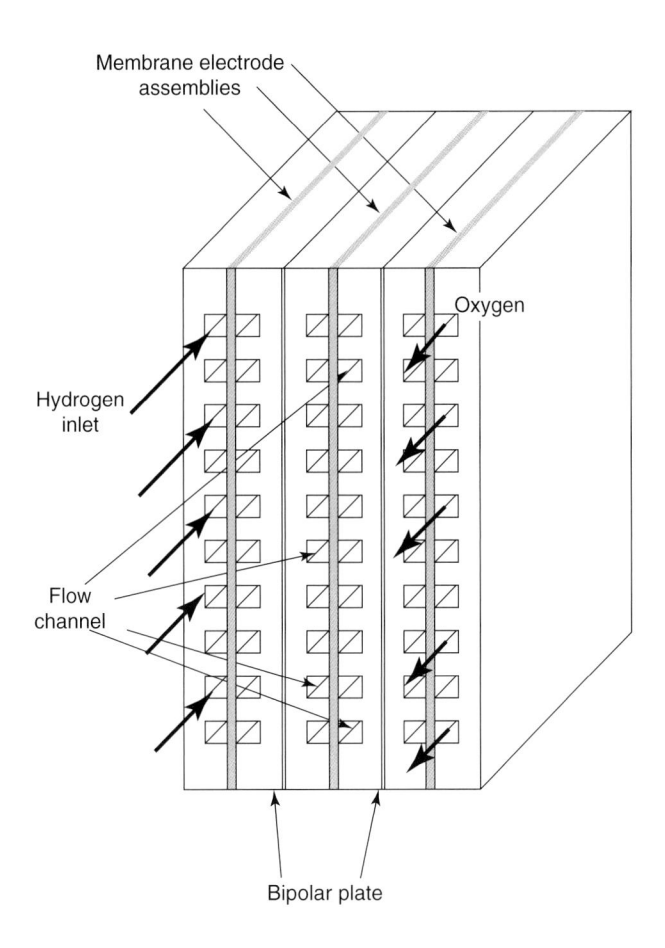

Figure 1. Function of flow field plate in fuel cells.

In molten carbonate fuel cells the parallel channels can be formed by using corrugated plates (stainless steel with Ni coated anode side) divided by a nonporous separator. An alternative flow field design, borrowed from a plate heat exchanger concept, uses internal manifolds to feed reactants at the corners of the flow fields (see Figure 3). Flow of fuel and oxidant in the end triangular sections is thus in cross flow and will influence mass transport in the cell.

The design of solid oxide fuel cells is somewhat unusual compared to other systems, in that stack designs are based on flat sheet and tubular electrode assemblies. In tubular assemblies oxygen is fed through the center of the tube and fuel is fed over the periphery. In stacks rows of cells are connected to one another by individual contacts. In principle the flow in the tubular channel is uniform and makes for uniform conditions of mass transport. In solid oxide fuel cells, an alternative, monolith design has also been proposed by the Argonne National Laboratory in which ceramic cell components are mutually supporting. A tape calendering process is used to form the anode/electrolyte/cathode tape that is then corrugated to produce flow channels for fuel and oxidant (Figure 4).

In polymer electrolyte (PEM) cells, flow field designs are based on a number of different concepts; varying from simple parallel channels to serpentine flow to rather complex designs in which flow can be in a zig-zag manner (Figure 5). In the serpentine design the flow snakes backwards and forwards, from one edge of the cell to the other, in a small number of channels grouped together. This creates a long flow path for reactants in the cell.

An alternative to the use of open channels is to use a thin porous structure (1–2 mm thick) such as metal foam or sintered metal or fiber mat which directly contacts the membrane electrode assembly (MEA) (Figure 5). In practice two porous structures are separated by a thin conducting metal, or carbon, nonporous sheet.

An alternative flow arrangement is the so-called interdigitated flow field (Figure 5). This design has inter-linked

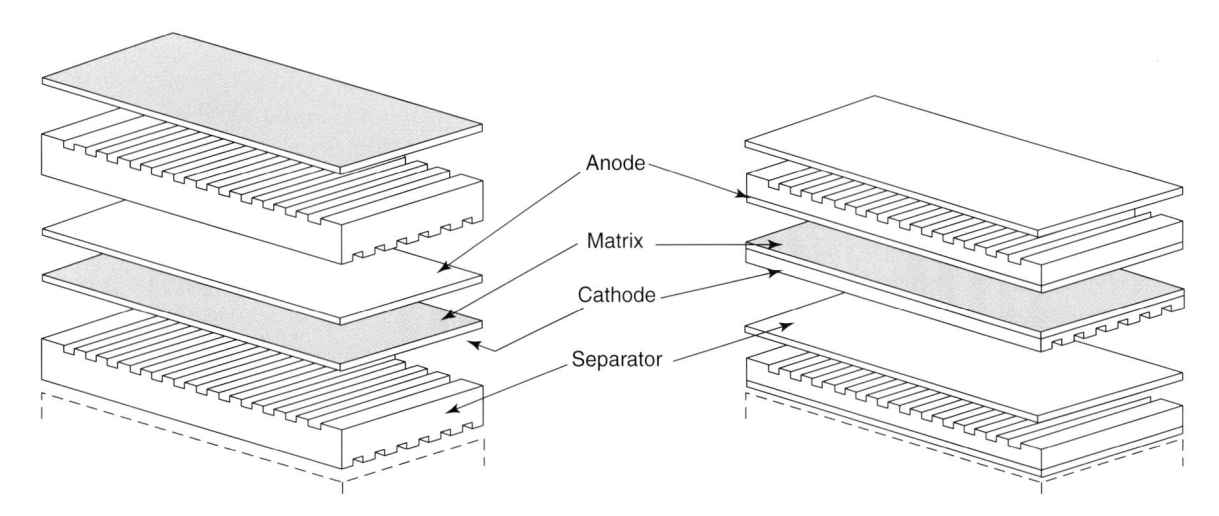

Figure 2. Types of parallel channel ribbed flow field plates. (Reproduced from Blomen and Mugerwa (1993)[35] by permission of Plenum Press.)

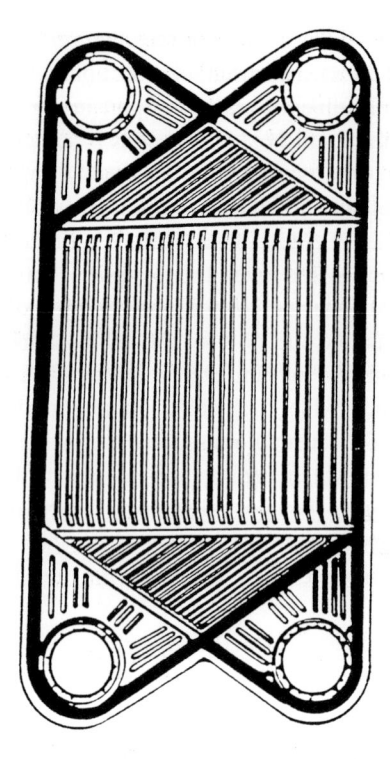

Figure 3. Flow field plate design based on a heat exchanger concept. (Reproduced from Blomen and Mugerwa (1993)[35] by permission of Plenum Press.)

finger like channels with dead ends. Interdigitated flow fields with the dead end flow design have been reported to improve performance in PEM and direct methanol fuel cells (DMFC). The improvement is attributed to the dead end channels changing the transport mechanism is the porous layers to a forced convection transport rather than predominantly diffusion. In addition the sheer force of the gas flow helps remove a large amount of liquid water that may become trapped in the electrocatalyst layers and thus alleviates any potential problems of flooding.

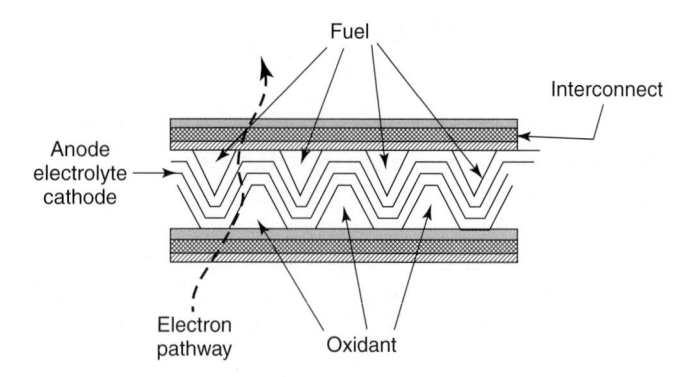

Figure 4. Cross-section through the monolith solid oxide fuel cell.

3 FLUID DYNAMICS AND MASS TRANSPORT IN FLOW FIELDS

As an example of the treatment of mass transport and hydrodynamics in a fuel cell the case of a proton conducting membrane (PEM) fuel cell using hydrogen as the fuel and air (or oxygen) as oxidant is considered. As depicted in Figure 6 the system can be represented by a number of regions

1. flow channels, for fuel and oxidant
2. porous diffusion layers, for anode and cathode
3. porous electrocatalyst layers
4. membrane.

For a PEM fuel cell it is known that the reduction of oxygen is kinetically much slower than the oxidation of hydrogen. Thus typically hydrodynamic and mass transport limitations affect the cathode side of the cell very much more than the anode side. This is especially the case when air is used as a source of oxygen. However, it should not be overlooked that, if cells are run with near stoichiometric

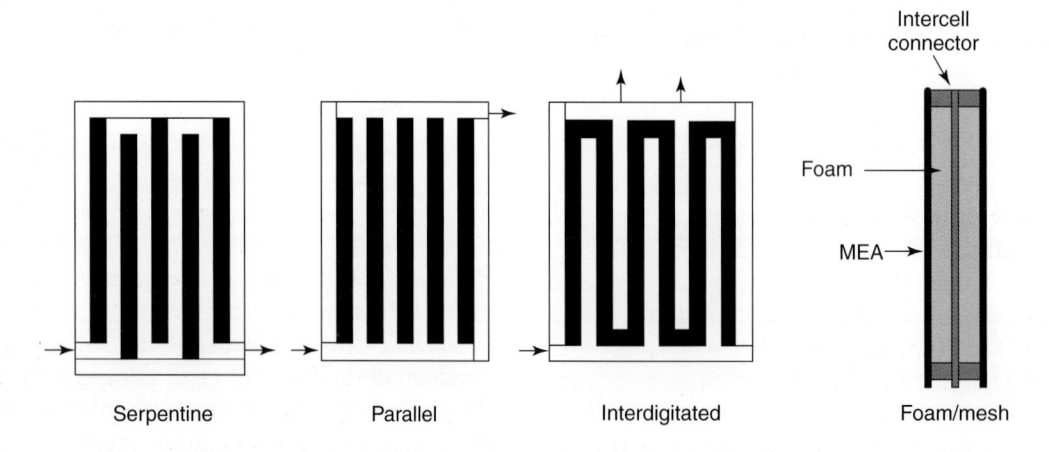

Figure 5. Schematic diagram of serpentine, interdigitated, parallel and porous (or foam) flow fields.

Water is produced from the electrochemical reaction and also can be supplied externally in fuel or oxidant. The purpose of external water supply is to ensure that the polymer membrane is sufficiently hydrated to maintain its high ionic conductivity. This overall requirement is influenced by the various transport modes of water in the PEM cell:

1. water is produced at the cathode;
2. water can diffuse back to the anode, through the membrane, by diffusion if the cathode has a greater activity of water;
3. water will be dragged by electroosmosis with protons produced by hydrogen oxidation;
4. water will be removed by the oxygen-depleted air in the cathode;
5. excess water present at the anode may cause diffusion through the membrane to the cathode.

In PEM cells, which use Nafion®, or equivalent membranes, problems of membrane dehydration and thus loss of membrane conductivity are often experienced at temperatures above 60 °C. This is due to the fact that with typical dry air flows, with SR values greater than 2, the generation of water is not sufficient to balance the water removal by the air stream. Operation of cells below temperatures of 60 °C can lead to significantly reduced power losses due to slower kinetics and thus higher temperatures are desired. Overall the PEM cell system design should address the issue and requirements for external humidification. External humidification adds an additional cost to the system but is frequently desirable to maximize power output.

Flow field design and flow direction can have an influence on water transport in PEM cells and thereby affect the overall mass transport behavior. For example consider a cell that is not externally humidified in which the flow of air and hydrogen are arranged counter-currently. Water produced at the cathode near the hydrogen inlet will diffuse from cathode to anode at a rate greater than electroosmotic drag, as the air at that point will be fully humidified. On the anode side, as the hydrogen flows along the cell flow bed, the rate of diffusion falls as the amount of water on the cathode side falls. At the hydrogen exit the predominant transport of water is by electroosmosis as the air at inlet is dry. In this way a better water balance is maintained in the cell than if the flow of hydrogen and air were co-current, where typically the membrane may become dry at the cell inlet.[1]

In addition to water transport the major species which is of interest in oxygen. Cell performance is influenced by the diffusion of oxygen through a mixture of water vapor and nitrogen. Diffusion occurs in both the flow channel and in the wet proofed pores of the carbon backing or diffusion

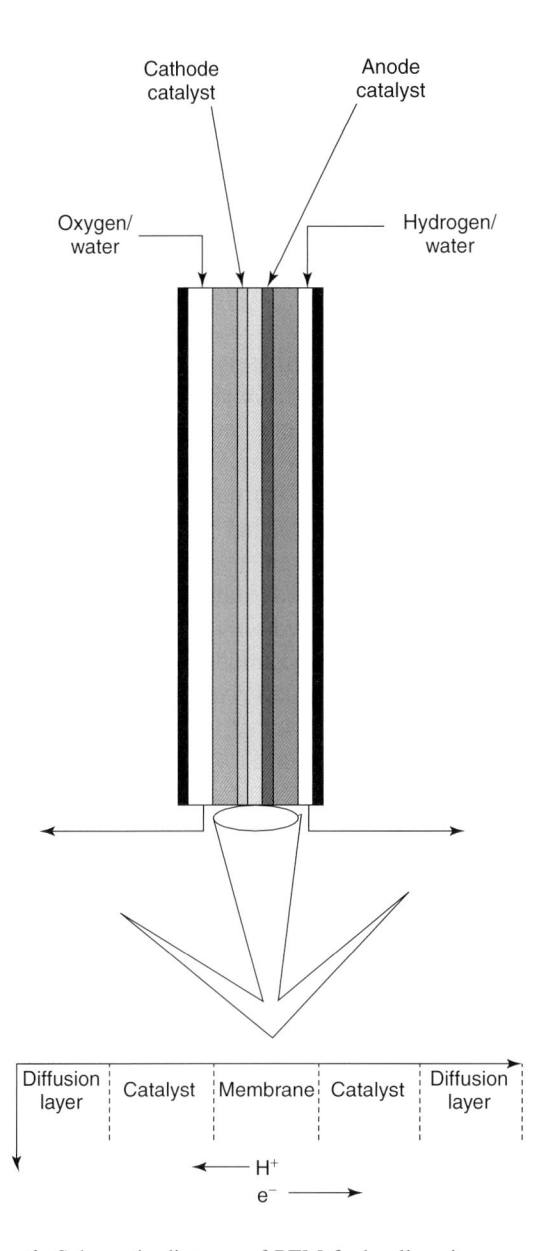

Figure 6. Schematic diagram of PEM fuel cell regions.

quantities of hydrogen, then, at some point when hydrogen consumption is high, hydrodynamics at the anode may become an issue, for example, if water produced by the cell reactions becomes a significant species present in the anode channel. Regardless of this, it is often assumed that, because the typical stoichiometric ratio (excess) (SR) of hydrogen is between 1.15 and 1.3, the hydrodynamic limitations to cell performance are restricted to the cathode.

3.1 Species transport in PEM cells

In a fuel cell, the species present (ignoring fuel impurities) are predominantly hydrogen, oxygen, water and nitrogen.

layer. Macro-pores (5–50 nm) are typically coated with polytetrafluoroethylene (PTFE) and are thus hydrophobic and facilitate the transport of gas. Micro-pores, which are not coated with PTFE, are hydrophilic and thus facilitate water transport. Movement of water in the micro-pores can be viewed as capillary and surface tension induced flow of micro-droplets or streams. It is suggested that formation of the water droplet phase is through ejection of the water formed at the hydrophilic cathode catalyst. A large accumulation of water in the catalyst and diffusion layer leads to water flooding which restricts access of oxygen to the catalyst particles in the electrode. Oxygen has to diffuse through a relatively thick water layer at the catalyst surface and thus mass transport limitations start to be seen in cell performance.

3.2 Fundamental equations

In general the treatment of flow and mass transport in flow fields and porous backing layers will require the coupling of, and solution of, equations of motion or momentum transfer, equations of continuity, equations of multi-component mass transfer, source equations for material transfer generation or consumption and boundary conditions. This will enable the prediction of the effect of variation in fluid velocity on material transfer to and from the MEA in the cell and thus its effect on local electrode potentials and thus current distribution.

3.2.1 Gas diffusion

For a multi-component mixture of gases, the diffusion coefficient of a species in the gas mixture depends on the nature, concentration and fluxes of the other species present. For an isothermal gas in the absence of a pressure gradient the diffusion flux of a component k is given by the Stefan–Maxwell equation.[2]

$$\nabla x_i = \sum_{j=1}^{n} \frac{RT}{PD_{ij}} (x_i N_{j,g} - x_j N_{i,g}), \quad i = 1, 2, \ldots, n \quad (1)$$

where, ∇ is the gradient (or spatial derivative), N is the molar flux, n is the number of components, x is the mole fraction, P is the pressure, T is temperature, R is the gas constant and D_{ij}^{eff} is a binary diffusion coefficient for the i–j pair in the porous medium.

In the absence of pressure drop and with no variation in velocity, for example as associated with friction at surfaces of a flow channel, this equation becomes redundant, as there is no variation in concentration in a plane normal to the electrode surface. In practice the flow in the flow fields of a fuel cell can be laminar and thus this two-dimensional velocity distribution will influence the molecular diffusion of species. This aspect will be considered later. The influence of diffusion in porous media is first considered because of the interaction there will exist between the flow of gases in the flow fields and in the diffusion layers of PEM cells.

3.2.2 Diffusion in porous media

The treatment of gas diffusion here is relevant to the case of porous types of flow fields, interdigitated flow fields and to gas diffusion layers in cells. The latter case is treated in terms of relevant mass transport phenomena.[3, 4]

Adjacent to the flow field in a PEM cell is the gas diffusion layer where gas can flow and diffuse. For diffusion an effective diffusion coefficient can be defined which allows for the porous structure of the layer, i.e., its porosity, and tortuosity. The effective diffusion coefficient can typically be related to the diffusion coefficient of the species in free space, D_{ij}, using a Bruggeman-type relationship

$$D_{ij}^{\text{eff}} = D_{ij} \varepsilon^{1.5} \quad (2)$$

where, ε, is the porosity of the gas-diffusion electrode. The free diffusion coefficients D_{ij} can be estimated from an expression developed from a combination of kinetic theory and corresponding-states arguments:

$$D_{ij} = \frac{C_1 T^{C_2}}{p} \quad (3)$$

where C_2 is a tabulated, empirically determined constant and C_1 is a function of the critical properties of i and j. This diffusion coefficient is virtually independent of composition.

3.2.3 PEM fuel cell model

The model considers the case of one-dimensional diffusion in the porous media, normal to the MEA plane. Pressure changes in the flow channel and porous media are assumed to be small. In the fuel cell the reactant partial pressures in the inlet flow channels will vary with both the humidification level of the inlet streams and the consumption rates of oxygen and hydrogen. In the model it is assumed that the inlet gases are fully saturated with water at inlet conditions.[4]

Air feed
In the case when the inlet flows contain an inert diluent, i.e., a mixture of oxygen and nitrogen, saturated with water vapor at the fuel cell operating temperature, we know the

mole fractions, x, and partial pressures, p, are related to the total pressure, P, by

$$x_{O_2} + x_{N_2} + x_{H_2O,vap} = 1$$

or

$$p_{O_2} + p_{N_2} + p_{H_2O,vap} = P \quad (4)$$

The derivative of equation (4) with respect to z, the gas diffusion path through the electrode, gives

$$\frac{dx_{O_2}}{dz} + \frac{dx_{N_2}}{dz} + \frac{dx_{H_2O,vap}}{dz} = 0 \quad (5)$$

With constant pressure and temperature in the flow channels, i.e., the partial pressure of water vapor is at the saturation vapor pressure, the gas-phase flux of water normal to the cathode face is zero. As nitrogen is not removed from the diffusion channel, the net nitrogen gas-phase flux normal to the cathode face is also zero:

$$N_{H_2O,g} = 0$$
$$N_{N_2,g} = 0 \quad (6)$$

Thus by combining the above equations the one-dimensional Stefan–Maxwell equation for nitrogen is

$$\frac{dx_{N_2}}{dz} = \frac{RT}{PD^{eff}_{N_2,O_2}}(x_{N_2} N_{O_2,g}) \quad (7)$$

The gas-phase flux of oxygen is given by

$$N_{O_2,g} = \frac{j}{4F} \quad (8)$$

where j is the current density.

Thus equation (7) yields the effective mole fraction of nitrogen at the gas/liquid film interface of the cathode catalyst layer, $x^{interface}_{N_2}$, as

$$x^{interface}_{N_2} = x^{channel}_{N_2} \exp\left[\frac{RTj\ell_d}{4FPD^{eff}_{N_2,O_2}}\right] \quad (9)$$

where ℓ_d is the thickness of the gas diffusion layer and $x^{channel}_{N_2}$ is the average mole fraction of nitrogen in the humidified gas at the interface of the cathode and the cathode gas flow channel.

The effective partial pressure of oxygen at the gas/liquid interface of the catalyst layer can now be calculated using equation (4)

$$p^{interface}_{O_2} = P\left[1 - x^{sat}_{H_2O} - x^{interface}_{N_2}\right] \quad (10)$$

Combining equations (9) and (10) gives

$$p^{interface}_{O_2} = P\left[1 - x^{sat}_{H_2O} - x^{channel}_{N_2} \exp\left(\frac{RTj\ell_d}{4FPe^{1.5}D}\right)\right] \quad (11)$$

where e is the voidage which represents a correction factor for the diffusivity to account for the void fraction, tortuosity and D is the diffusivity of nitrogen in oxygen given by the Slattery and Bird correlation[5]

$$PD = 0.0002745 \left(\frac{T}{\sqrt{T^{crit}_{O_2} T^{crit}_{N_2}}}\right)^{1.832} \left(P^{crit}_{O_2} P^{crit}_{N_2}\right)^{1/3}$$
$$\times (T^{crit}_{N_2} T^{crit}_{O_2})^{5/12} \left(\frac{1}{M_{N_2}} + \frac{1}{M_{O_2}}\right)^{1/2} \quad (12)$$

With appropriate values for the known constants equations (11) and (12) yield

$$p^{interface}_{O_2} = P\left[1 - x^{sat}_{H_2O} - x^{channel}_{N_2} \exp\left(\frac{0.1j}{e^{1.5}T^{0.832}}\right)\right] \quad (13)$$

A similar development for the interfacial hydrogen partial pressure on the anode side yields

$$p^{interface}_{H_2} = P\left[1 - x_{H_2O,anode} - x^{channel}_{CO_2} \exp\left(\frac{0.061j}{e^{1.5}T^{0.832}}\right)\right] \quad (14)$$

where $x_{H_2O,anode}$ is the average water mole fraction in the anode channel.

Oxygen feed

In the case when the cathode feed stream contains only water vapor and oxygen, equation (4) indicates that the drop in oxygen partial pressure along the diffusion path will result in a corresponding drop in the total pressure. There will thus be a molecular flux of water vapor "out" just sufficient to balance the bulk motion of the gas due to the diffusion of oxygen "in" and the net flux of water vapor through the cathode is zero.

In this case the Stefan–Maxwell equation leads to

$$\frac{dx_{O_2}}{dz} = \frac{RT}{PD^{eff}_{H_2O,O_2}}\left(x_{H_2O}\frac{j}{4F}\right) \quad (15)$$

Since $x_{O_2} + x_{H_2O} = 1$, integration of this expression from the flow channel to the catalyst surface, assuming the product $PD^{eff}_{i,j}$ is constant, gives

$$x^{interface}_{H_2O} = x^{channel}_{H_2O} \exp\frac{RTj\ell_d}{4FPD^{eff}_{H_2O,O_2}} \quad (16)$$

As the partial pressure of water vapor in the cathode is fixed at the saturation level, the partial pressure of oxygen at the interface is

$$p_{O_2}^{\text{interface}} = (1 - x_{H_2O}^{\text{interface}}) \left(\frac{p_{H_2O}^{\text{sat}}}{x_{H_2O}^{\text{interface}}} \right) \quad (17)$$

Combining equations (16) and (17) yields

$$p_{O_2}^{\text{interface}} = (p_{H_2O}^{\text{sat}}) \left[\frac{1}{\exp\left(\frac{RTj\ell_d}{4FPD_{H_2O,O_2}^{\text{eff}}}\right) x_{H_2O}^{\text{channel}}} - 1 \right] \quad (18)$$

Evaluation of the Slattery–Bird equation for $PD_{H_2O,O_2}^{\text{eff}}$ gives

$$p_{O_2}^{\text{interface}} = (p_{H_2O}^{\text{sat}}) \left[\frac{1}{\exp\left(\frac{(1.397j)}{e^{1.5}T^{1.334}}\right) x_{H_2O}^{\text{channel}}} - 1 \right] \quad (19)$$

A similar development for hydrogen on the anode side yields the equation

$$p_{H_2}^{\text{interface}} = (0.5 p_{H_2O}^{\text{sat}}) \left[\frac{1}{\exp\left(\frac{(0.551j)}{e^{1.5}T^{1.334}}\right) x_{H_2O}^{\text{channel}}} - 1 \right] \quad (20)$$

The above equations determine the partial pressures of hydrogen and oxygen necessary to calculate the anode and cathode overpotentials and thus, in conjunction with estimates of potential drops in the membrane, determine the cell voltage. To assist these calculations it is, in principle, necessary to consider transport of dissolved gases through any water films which may cover catalyst sites in the anode or cathode layer. It is often assumed that any water layer on the anode side will be extremely thin; no water is being produced at the anode, and the membrane retains water. Thus the film will not present a significant barrier to hydrogen diffusion to the catalyst, and the dissolved hydrogen gas concentration at the anode will be in equilibrium with the bulk gas and can be estimated from Ref. [4].

$$C_{H_2}^* = \frac{p_{H_2}^{\text{interface}}}{1.09 \times 10^6 \exp\left(\frac{77}{T}\right)} \quad (21)$$

Although the production of water may give rise to a relatively thick water layer it can be assumed that the concentration of dissolved oxygen at the gas/liquid interface is defined by a Henry's law expression of the form

$$c_{O_2}^{\text{interface}} = \frac{p_{O_2}^{\text{interface}}}{5.08 \times 10^6 \exp\left(\frac{-498}{T}\right)} \quad (22)$$

To use the above model requires approximations to the mean concentrations of gases in the flow channels. The partial pressures of oxygen and hydrogen in the humidified inlet gas streams can be determined from the dry gas compositions and the water partial pressure.

$$p_{H_2,\text{inlet}}^{\text{hum}} = \left(\frac{P - (p_{H_2O,\text{anode}})}{P} \right) p_{H_2,\text{inlet}}^{\text{dry}} \quad (23)$$

$$p_{O_2,\text{inlet}}^{\text{hum}} = \left(\frac{P - (p_{H_2O,\text{cathode}})}{P} \right) p_{O_2,\text{inlet}}^{\text{dry}} \quad (24)$$

The outlet dry gas consumption at the cathode can be calculated as a function of the dry inlet mole fractions, the oxygen stoichiometric ratio (SR) (the ratio of the inlet flow rate of oxygen to the rate at which it is consumed), and the total pressure

$$p_{O_2}^{\text{out dry}} = \left[\frac{x_{O_2}^{\text{in dry}} - \frac{x_{O_2}^{\text{in dry}}}{SR}}{\left(x_{O_2}^{\text{in dry}} - \frac{x_{O_2}^{\text{in dry}}}{SR} \right) + x_{N_2}^{\text{in dry}}} \right] P \quad (25)$$

As approximations, the effective oxygen partial pressure is given by a log-mean average of the inlet and outlet oxygen partial pressures.[5]

$$p_{O_2}^{\text{avg}} = \frac{p_{O_2,\text{in}}^{\text{hum}} - p_{O_2,\text{out}}^{\text{hum}}}{\ln\left(\frac{p_{O_2,\text{in}}^{\text{hum}}}{p_{O_2,\text{out}}^{\text{hum}}} \right)} \quad (26)$$

and the effective hydrogen partial pressure as an arithmetic mean

$$p_{H_2}^{\text{avg}} = \frac{p_{H_2,\text{in}}^{\text{hum}} + p_{H_2,\text{out}}^{\text{hum}}}{2} \quad (27)$$

3.2.4 Partially humidified gas streams

A PEM cell may be operated with any degree of humidification of anode and cathode gases which will affect the transport, by diffusion, of water and other species in cathode and anode diffusion layers. The amount of water produced by the electrochemical reaction is balanced by the amount of water evaporated through the pores of the gas-diffusion electrode. We can therefore write

$$\frac{j}{2F} = 2N_{O_2} = N_{H_2} = -(N^a_{H_2O} + N^c_{H_2O}) \quad (28)$$

where $N^a_{H_2O}$ and $N^c_{H_2O}$ are the water flux in the anode and cathode side of the cell. For convenience the quantity f_a is defined as the fraction of the water produced that evaporates through the gas-diffusion anode;[3]

$$f_a = \frac{N^a_{H_2O}}{N^a_{H_2O} + N^c_{H_2O}} = 1 - \frac{N^c_{H_2O}}{N^a_{H_2O} N^c_{H_2O}} \quad (29)$$

Estimation of the partial pressures of hydrogen and oxygen requires solution of the Stefan–Maxwell equation as illustrated earlier.

Gas-diffusion anode

Hydrogen gas diffuses through water vapor to the membrane-electrode interface where it reacts. The anode water flux is related to the hydrogen flux

$$N^a_{H_2O} = -f_a N_{H_2} \quad (30)$$

The analysis of the Stefan–Maxwell diffusion problem with the boundary conditions at $z = 0$, $x_{H_2O} = x^a_{H_2O}$, and at $z = l_c$, $x_{H_2O} = x^{m,a}_{H_2O}$ gives

$$v = \frac{1}{2} \frac{\delta_a p_a D^{eff}_{H_2O-H_2}}{\delta_c p_c D^{eff}_{H_2O-N_2}} \ln\left(\frac{f_a + (1-f_a)x^{m,a}_{H_2O}}{f_a + (1-f_a)x^a_{H_2O}}\right) \quad (31)$$

The concentration distribution $x_{H_2O}(z)$ in the pores of the gas diffusion anode is

$$x_{H_2O}(z) = \frac{f_a + (1-f_a)x^a_{H_2O}}{1 - f_a}$$
$$\times \left[\frac{f_a + (1-f_a)x^{m,a}_{H_2O}}{f_a + (1-f_a)x^a_{H_2O}}\right]^{z/\delta_a} - \frac{f_a}{1-f_a}, \quad f_a \neq 1 \quad (32)$$

where

$$v \equiv \frac{N_{O_2} RT l_c}{P_c D^{eff}_{H_2O,N_2}} \quad (33)$$

Cathode

Oxygen diffuses through both water vapor and nitrogen gas at the cathode. From the solution of the appropriate Stefan–Maxwell equations for this system with the boundary conditions

$$z = 0, \quad x_{O_2} = x^c_{O_2}$$

and

$$x_{N_2} = x^c_{N_2}$$

and at $z = \delta_c$

$$x_{O_2} = x^m_{O_2} = 1 - x^{m,c}_{H_2O} - x^m_{N_2}$$

we obtain

$$x^m_{N_2} = x^c_{N_2} \exp(v[r_{N_2} - 2(1-f_a)]) \quad (34)$$

$$1 - x^{m,c}_{H_2O} - x^c_{N_2}\left(1 + \frac{1}{K}\right)\exp(v[r_{N_2} - 2(1-f_a)])$$
$$- \frac{1}{2f_a - 1} - \left(x^c_{O_2} - \frac{1}{2f_a - 1} + \frac{x^c_{N_2}}{K}\right)$$
$$\times \exp[v(2f_a - 1)r_{H_2O}] = 0 \quad (35)$$

where

$$K = 1 + 2(1-f_a)\left(\frac{1 - r_{H_2O}}{r_{N_2} - r_{H_2O}}\right)$$

$$r_{H_2O} = \frac{D^{eff}_{H_2O-N_2}}{D^{eff}_{H_2O-O_2}}, \quad r_{N_2} = \frac{D^{eff}_{H_2O-N_2}}{D^{eff}_{N_2-O_2}}$$

The expressions for the concentration profiles are

$$x_{O_2(Z)} = -\frac{x_{N_2(Z)}}{K} + \frac{1}{2f_a - 1} + \left(x^c_{O_2} - \frac{1}{2f_a - 1} + \frac{x^c_{N_2}}{K}\right)$$
$$\times \exp\left[v(2f_a - 1)r_{H_2O}\frac{z}{\delta_c}\right] \quad (36)$$

where

$$x_{N_2(Z)} = x^c_{N_2} \exp\left(\frac{z}{\delta_c}v[r_{N_2} - 2(1-f_a)]\right) \quad (37)$$

To use the above equations requires estimations of the anode and cathode water compositions as used in the boundary conditions.

An expression for $x^a_{H_2O}$ in terms of the stoichiometric flow rate ratio (SR) ζ_a can be obtained from a material balance on hydrogen

$$\zeta\frac{jA}{2F} = x^a_{H_2}C_a + \frac{jA}{2F} \quad (38)$$

where

$$\zeta_a = \frac{x^o_{H_2}Cv^o_a}{N_{H_2}A}$$

and v^o_a is the inlet flow rate of hydrogen.

The material balance on water vapor

$$x_{H_2O}^{o,a} C v_a^o + f_a \frac{jA}{2F} = x_{H_2O}^a C v_a \quad (39)$$

yields

$$x_{H_2O}^a = \frac{\zeta_a \dfrac{x_{H_2O}^{o,a}}{x_{H_2}^o} + f_a}{f_a - 1 + \zeta_a \left(1 + \dfrac{x_{H_2O}^{o,a}}{x_{H_2}^o}\right)} \quad (40)$$

Similarly for the cathode chamber

$$x_{H_2O}^c = \frac{\zeta_c \dfrac{x_{H_2O}^{o,c}}{x_{O_2}^o} + 2(1 - f_a)}{\zeta_c \dfrac{x_{H_2O}^{o,c}}{x_{O_2}^o} + 2(1 - f_a) + \zeta_c \left(1 + \dfrac{x_{N_2}^o}{x_{O_2}^o}\right)} - 1 \quad (41)$$

where

$$\zeta_c = \frac{x_{O_2}^o c v_c^o}{N_{O_2} A}$$

The expression for $x_{H_2O}^a$ is now used to determine the quantity f_a. Once f_a is determined, the current density, j, to maintain a water balance is then determined from equations (27) and (33).

The above analysis has been used by Bernardi[3] to identify operating conditions that result in a balance of water produced and removed in a PEM cell. This factor is important in operation to ensure that the membrane is sufficiently hydrated but on the other hand that too much water does not lead to flooding of the catalyst and introduce mass transport limitations to the catalyst sites. The water balance is particularly sensitive to changes in air humidity and flow rate. The model can be used to assess the level of humidity for either gas streams. In use of the model, due consideration as to the mode of water transport through the membrane should be made and in particular with respect to transport by electroosmosis and diffusion.

From the relevant equation above the influence of diffusion in a porous backing layer of flow field can be determined and the effective concentration of reactant gas, e.g., oxygen, at the edge of the catalyst layer calculated to enable estimation of electrode polarization phenomena.

Amphlett et al.[4] have used the above model as part of a performance model of a PEM fuel cell stack supplied by Ballard, that combines gas diffusion with semi-empirical expressions for electrode kinetics and membrane ionic transport. Correlation of the empirical model with experimental data, as shown in Figure 7, is good.

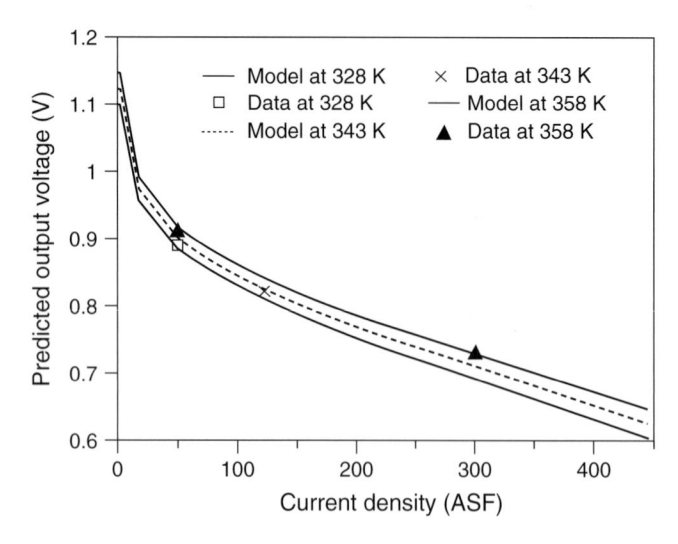

Figure 7. Comparison of model prediction and experimental response of a Ballard Mark IV PEM fuel cell. ASF (amp per square foot $\approx$ mA cm^{-2}). (Reproduced from Amphlett et al. (1995)[4] by permission of the Electrochemical Society, Inc.)

3.2.5 Influence of temperature and friction

In a PEM cell, oxygen reduction at the cathode leads to water formation which, if the gas phase is fully saturated, is present as liquid. This liquid phase can change the effective porosity of the diffusion layer by partly filling the pores in which gas flows. This therefore influences the gas mass transport rate in the porous structure. The extent to which the porous structure is filled with liquid depends upon the governing mechanism for water transport through the structure. In addition the transport of reactant gases and thus local partial pressures are generally affected by both Stefan–Maxwell diffusion, Knudsen diffusion and friction pressure losses. Knudsen diffusion arises when the mean free path of the gas molecules is of a similar magnitude to the pore dimensions, i.e., molecular and pore wall interactions. Pressure changes are due to friction and can be simply modeled on the basis of laminar flow in a capillary or Poiseulle flow.

For gas transport in a fuel cell, we therefore in general must consider, for a one-dimensional system, several interactive phenomena which predict both dynamic and steady state behavior:[6]

1. Material balances of gases which take into account changes in gas voidage simultaneously with gas partial pressure and allow for a change in gas volume associated with a change in temperature, i.e., $dV/dt (= -p/RT^2 \, dT/dt$ for an ideal gas);
2. Material balance of water vapor; which includes the influence of condensation or evaporation of water depending upon the saturation partial pressure and the content of water in the vapor;

3. Material balance of liquid water which includes an appropriate description for the mechanism of water transport;
4. Energy balance of gas and solid phase which predict the local values of temperature;
5. Gas transport.

In the development of a model the following assumptions are assumed to be good approximations of the system:

1. Temperature of the gas phase is identical to the solid phase at every position;
2. No heat transport in the gas phase;
3. No flooding of the cathode, i.e., sufficiently rapid water transport.

The governing equations for heat and material transfer in the diffusion and reaction layer are described as follows. The diffusion layer is described by a set of differential equations for gas and water transport:

Material balance for a gas describes the change in the number of moles with time in terms of the change in molar flux

$$-\frac{\varepsilon_g p_j}{RT^2}\frac{dT}{dt} + \frac{p_j}{RT}\frac{d\varepsilon_g}{dt} + \frac{\varepsilon_g}{RT}\frac{dp_j}{dt} = -\frac{dN_j}{dz} \quad (42)$$

Similarly the material balance of water vapor is written in terms of the change in molar flux of water vapor and the pressure difference (right side of the equation) which defines the condensation and evaporation of water.

$$-\frac{\varepsilon_g p_{H_2O}}{RT^2}\frac{dT}{dt} + \frac{p_{H_2O}}{RT}\frac{d\varepsilon_g}{dt} + \frac{\varepsilon_g}{RT}\frac{dp_{H_2O}}{dt}$$
$$= -\frac{dN_{H_2O,V}}{dz} - \frac{\alpha_V k_1}{RT}(p_{H_2O} - p_{H_2O}^{sat}) \quad (43)$$

where k_1 is the mass transfer coefficient for transfer of water vapor and α_V is the specific surface of condensation.

The saturation pressure of water vapor is calculated according to the Antoine equation. If the water vapor pressure exceeds the saturation partial pressure, water will condense until the saturation partial pressure is reached. Conversely water will evaporate (if the water loading is greater than zero) until the saturation pressure is reached. Typically, however, it can be assumed that water saturates the gas at all times, due to the large evaporation surface, i.e., the large surface of the carbon.

The liquid water content in the layer is defined in terms of the mass ratio of water to catalyst or the liquid loading, X_s. The material balance of liquid water is

$$\frac{\rho_s}{M_{H_2O}}\frac{dX_s}{dt} = -\frac{dN_{H_2O,L}}{dz} + \frac{\alpha_V k_1}{RT}(p_{H_2O} - p_{H_2O}^{sat}) \quad (44)$$

The equivalent energy balance for non-isothermal conditions considers energy transfer due to the flow of fluids, heat conduction and evaporation/condensation:

$$\rho_s c_s \frac{dT}{dt} = -\sum_{j=1}^{N}\left(N_j M_j c_{p,j}\frac{dT}{dz}\right) + \lambda_s \frac{d^2 T}{dz^2}$$
$$+ \frac{\alpha_V k_1 M_{H_2O}}{RT}\Delta H_c (p_{H_2O} - p_{H_2O}^{sat}) \quad (45)$$

In the energy balance, gas and solid phases are considered simultaneously, i.e., gas and solid are isothermal at every point.

Gas transport considers the change in partial pressure due to Stefan–Maxwell diffusion (intermolecular) and Knudsen diffusion (molecule/pore-wall interaction) and Poiseuille flow.

$$-\frac{dp_j}{dz} - p_j \frac{r_p^2}{\eta_g D_j 8}\frac{dp}{dz} = RT\left(\frac{N_j}{e_g^{1.5} D_j}\right.$$
$$\left.+ \sum_{i=1, i\neq j}^{k}\frac{p_i N_j - p_j N_i}{e_g^{1.5} D_{ij}}\right) \quad (46)$$

where $e_g^{1.5}$, allows for the porosity and tortuosity of the porous region and

$$D_j = \frac{4}{3}r_p \sqrt{\frac{2RT}{\pi M_j}}$$

Liquid water transport is defined in terms of a liquid surface diffusion mechanism

$$N_{H_2O,L} = -\frac{\rho_s}{M_{H_2O}} D_{H_2O,L}\frac{dX_s}{dz} \quad (47)$$

where the water fraction is related to the gas porosity according to

$$\varepsilon_g = \varepsilon_g^o\left[1 - \left(\frac{X_s}{X_s^{max}}\right)^m\right] \quad (48)$$

with a maximum water fraction given by

$$X_s^{max} = \frac{\rho_{H_2O,L}}{\rho_s}\frac{\varepsilon_g^o}{\varepsilon_s} \quad (49)$$

$$\varepsilon_g^o = 1 - \varepsilon_s - \varepsilon_{el} \quad (50)$$

The variation in the production of water and its diffusion effects the local value of the water fraction and thus the gas porosity. With greater water loading, pores are closed by the water for gas transport and vice versa. Water-filled pores therefore do not contribute to gas transport. This approach can be supported by the theory of capillary condensation. The effect of decreasing porosity is a diffusion hindrance of gas transport by pore flooding which will finally reduce the partial pressure of the gas. In many systems a large surface diffusion coefficient of water (which results in fixed porosity of the electrode) can often be assumed.

The model of the gas flow field or porous diffusion layer cannot be considered in isolation as there is clearly significant interaction with the catalyst reaction layer as well as the membrane in the fuel cell. In the case of the catalyst layer the equations used for the diffusion layer are valid and must in addition include chemical transformations associated with local reactions. The physico–chemical effect of local oxygen consumption and water generation affects the material balances of these two species and the energy balance as follows:

Material balance of oxygen

$$-\frac{\varepsilon_g p_{O_2}}{RT^2}\frac{dT}{dt} + \frac{p_{O_2}}{RT}\frac{d\varepsilon_g}{dt} + \frac{\varepsilon_g}{RT}\frac{dp_{O_2}}{dt} = -\frac{dN_{O_2}}{dz} - \frac{1}{nF}j \quad (51)$$

Material balance of liquid water

$$\frac{\rho_s}{M_{H_2O}}\frac{dX_s}{dt} = -\frac{dN_{H_2O,L}}{dz}$$
$$+ \frac{\alpha_V k_1}{RT}(p_{H_2O} - p_{H_2O}^{sat}) + \frac{2}{nF}j \quad (52)$$

Energy balance

$$\rho_s c_s \frac{dT}{dt} = -\left(\sum_{j=1}^{n} N_j M_j c_{pj}\frac{dT}{dz}\right) + \lambda_s \frac{d^2T}{dz^2} + \frac{\alpha_V k_1 M_{H_2O}}{RT}$$
$$\times \Delta H_C(p_{H_2O} - p_{H_2O}^{sat}) + \left(\frac{T(-\Delta^R S)}{nF} + \eta\right)j \quad (53)$$

The local values of current density are computed from an appropriate kinetic equation(s) for oxygen reduction reaction, variation in local potential and thus current density due to ionic (proton) conduction, electronic conduction in the catalyst and catalyst support solid phase and a volumetric current balance as follows:

Butler–Volmer equation for oxygen reduction

$$j = \alpha_{act} j_o \left[\frac{p_{O_2}}{p_{O_2}^o}\exp\left(\frac{-(1-\alpha)nF}{RT}\eta\right) - \exp\left(\frac{\alpha nF}{RT}\eta\right)\right] \quad (54)$$

where, $\alpha_{act} j_o$ is the apparent exchange current density at open circuit potential and the overpotential is given by

$$\eta_D = U_o + \phi_{el} - \phi_s \quad (55)$$

Ionic conduction in the electrode structure is given by

$$j_{H^+} = -\varepsilon^{1.5} K_{el}\frac{d\phi_{el}}{dz} \quad (56)$$

Electronic conduction in the carbon support matrix is given by

$$j_{H^+}(0) - j_{H^+} = -\varepsilon^{1.5} K_s \frac{d\phi_s}{dz} \quad (57)$$

The volumetric current balance is

$$-\frac{dj_{H^+}}{dz} = i \quad (58)$$

Boundary conditions for the problem are such that the gradient of the solid phase potential, ϕ_s, at the membrane interface is zero as there is no electron flow into the membrane, i.e., $z = 0$, values of parameters are specified as

$$T,\ N_j,\ N_{H_2O,L},\ \phi_{el},\ j_{H^+} = j_{appl},\quad \frac{d\phi_s}{dz} = 0$$

At the interface between the reaction and the diffusion layers, the gradient of the electrolyte phase potential, ϕ_{el}, is zero as there is no ionic current flow into the diffusion layer, i.e.,

$$\frac{d\phi_{el}}{dz} = 0,\quad j_{H^+} = 0$$

at $z = L$ (gas channel), values of parameters are specified

$$T,\ X_s,\ p_j$$

The above model for the PEM cell has been used by Bevers et al.[6] to simulate the polarization characteristics of the cell. The data, shown in Figure 8, clearly demonstrate the influence of mass transport phenomena on the polarization characteristics particularly at high current densities, where a rapid fall in cell potential, characteristic of PEM cells, is observed. Factors that can influence the cell polarization include:

1. Active catalyst area. A decrease in active catalyst area decreases the effective exchange current density that causes a more rapid loss in cell voltage with current density that also causes mass transport limitations to occur at lower current densities.
2. Electrolyte conductivity. In PEM cells the effective electrolyte conductivity is modified by incorporating more or less ionomer (Nafion®) in the catalyst structure. Increasing the effective electrolyte conductivity increases the utilization of the electrocatalyst area by enabling a greater penetration of current into the electrocatalyst region. At low electrolyte conductivity the

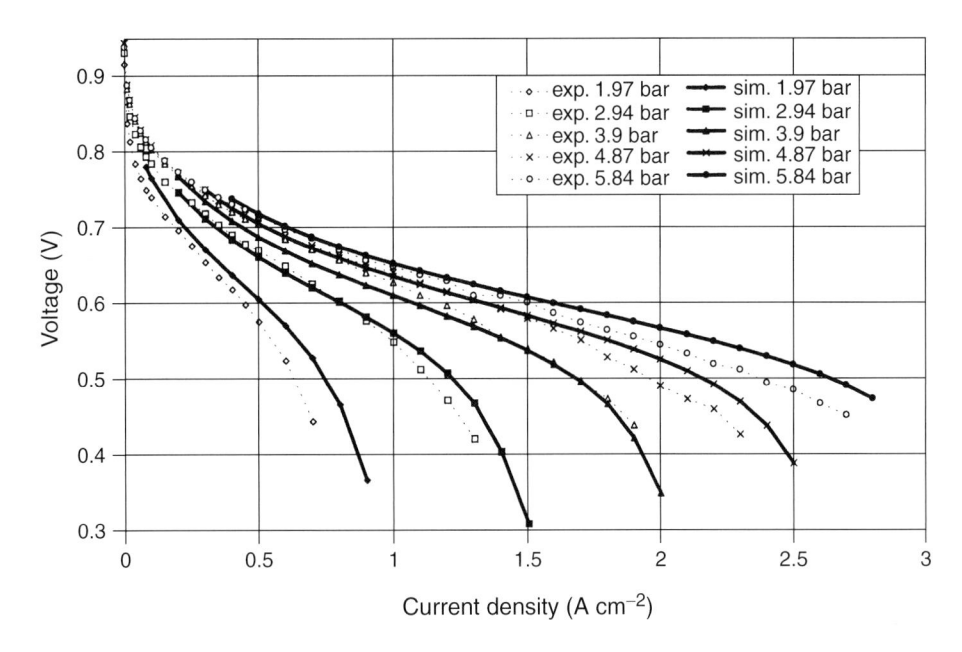

Figure 8. Model and experimental IR corrected polarization curves for PEM cells. (Reproduced from Bevers *et al.* (1997)[6] by permission of Kluwer Academic.)

active catalyst region is restricted to areas near the membrane. As a consequence of the effective reduced catalyst area, mass transport effects become more significant and thus lower limiting current densities for the cell are produced at lower conductivity.

3. Pore radius and porosity. These two parameters affect predominantly the mass transport characteristics. A decrease in either parameter results in a reduction in partial pressure of reactant and causes mass transport limiting current densities to occur at lower current density.

3.2.6 Two-dimensional models

Kulikovsky *et al.*[7] have considered a two-dimensional model of the cathode compartment of a PEM fuel cell. The model is based on the continuity equations for the gases, Poisson's equation for potentials of the carbon catalyst support phase and membrane phase, Butler–Volmer kinetics and Stefan–Maxwell and Knudsen diffusion. The influence of hydrodynamics was not considered. Regardless of this latter point, the model shows some interesting effects for a simple, parallel, flow field channel configuration.

The model showed that there were two-dimensional distributions in oxygen concentration, overpotential and current density (see Figure 9). The model identified potential dead zones in the active layers in front of the fuel channel where the reaction rate is small. Current density is also higher at positions of the electrocatalyst layer facing the edges of the current collectors. It has been shown that it is theoretically possible to remove up to 50% of the electrocatalyst from the areas facing the central parts of the flow channels without significant loss in power performance.

From an analysis of the problem of current distribution with standard parallel channel flow fields, a suggestion to prevent the shielding of catalyst layers by the flow field current collectors has been put forward.[8] The solution is to use embedded types of current collectors that are positioned inside the backing and catalyst layers. Thus current collection is normal to the flux of protons into the membrane and results in a uniform current distribution. Preparation of this type of structure may lead to practical problems and has yet to be realized.

It is possible to develop simpler expressions for the flow in fuel cells. An approach is to model the flow as an ideal gas in a straight channel of cross-section S and ignore the influence of viscosity and consider the momentum balance (Euler) equation

$$\rho(v\nabla)v = -\nabla p \tag{59}$$

Mass transfer through the bottom surface causes a momentum flux in this direction. For the cathode channel, oxygen and water transport influence the momentum balance. The bottom of the channel is in contact with backing layer and the change of flux along the channel due to the mass flux through the bottom surface is given by

$$\left[\frac{d(\rho v_z)}{\partial z}\right] = \frac{1}{h}(\rho v_x) \tag{60}$$

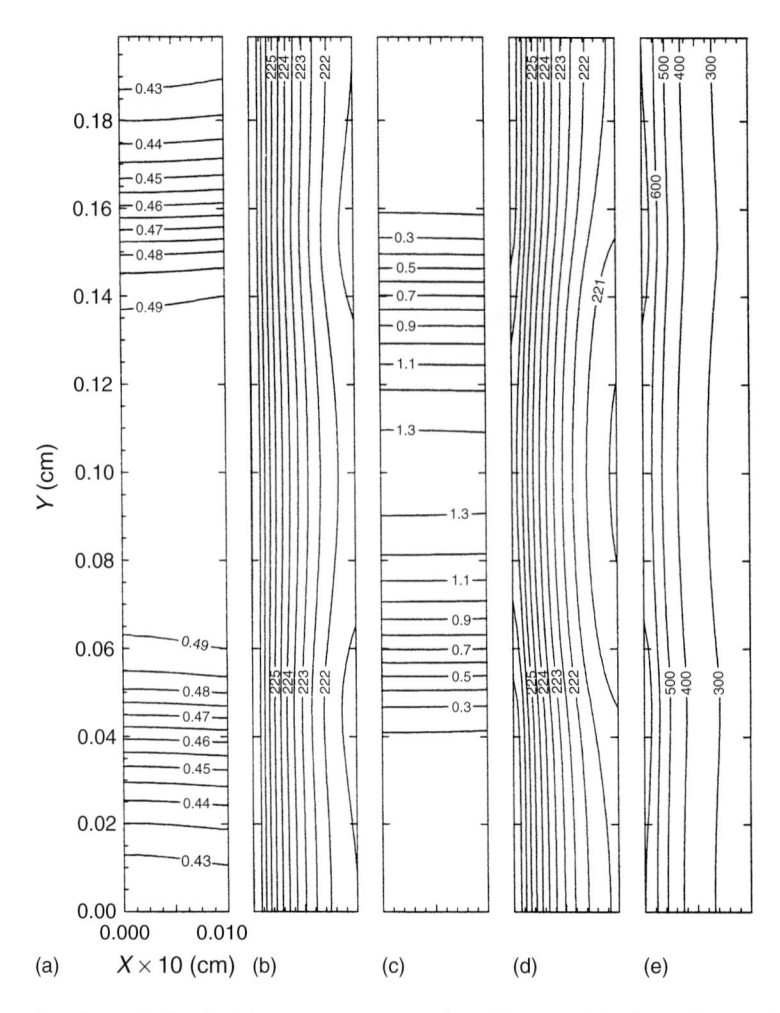

Figure 9. Contour lines showing the variation in (a) oxygen concentration, (b) potential of membrane phase, (c) potential of carbon phase, (d) overpotential in mV and (e) current density in mA cm^{-2}. X is the dimension across the channel width. (Reproduced from Kulikovsky *et al.* (1999)[7] by permission of the Electrochemical Society, Inc.)

where z is measured along the channel axis, a is the width and h is the height of the channel.

The mass flux is

$$\rho v_x = -j\left(\frac{2(1+\lambda_{H_2O})M_{H_2O} - M_{O_2}}{4F}\right) \quad (61)$$

where λ_{H_2O} is the drag coefficient for water transport across the membrane.

The momentum balance equation for the channel in one dimension is

$$\rho v_z \frac{\partial v_z}{\partial z} = -\frac{\partial p}{\partial z} - \frac{\sum(\rho_i v_x)v_z}{h} \quad (62)$$

The term $\sum(\rho_i v_x)v_z/h$ represents the change in momentum due to transfer of oxygen and water from the channel.

For an ideal gas we can write

$$\frac{\partial p}{\partial z} = \frac{RT}{M}\frac{\partial \rho}{\partial z} \quad (63)$$

where M is the mean molecular weight of the mixture M.

Equations (60) and (62) are two differential equations for two unknowns; v_z and ρ. Solving these coupled equations will enable the variation in density and thus partial pressure of oxygen to be calculated and thus computation of the local current density, using, for example, a Butler–Volmer relation for the cathode kinetics.

3.3 Three-dimensional modeling and momentum transfer

The flow in flow channels of fuel cells is usually laminar unless high stoichiometric excess of fuel or oxidant is used. Hence as an approximation the flow in porous flow fields or in porous backing layers can also be considered to be laminar. The influence of hydrodynamics in the flow fields is to change the local values of flow velocity, which has a direct influence on the mass transport or diffusion flux of species. In general this variation in velocity occurs in three dimensions which can result in a three-dimensional variation in diffusion. Consequently in fuel cells we can expect that there will be multidimensional variation in local reactant gas partial pressure and thus local current density.

The treatment of mass transport and hydrodynamic effects in flow fields requires the simultaneous solution of the coupled equations for momentum transfer and multicomponent mass transfer: equations for multicomponent mass transport are presented above.

The equations of momentum transfer or motion for isothermal systems are well documented and can be written, for a Newtonian fluid, as

$$\rho \frac{D\upsilon_x}{Dt} = -\frac{\partial p}{\partial x} + \frac{\partial}{\partial x}\left[2\mu\frac{\partial \upsilon_x}{\partial x} - \frac{2}{3}\mu(\nabla \cdot \upsilon)\right]$$
$$+ \frac{\partial}{\partial y}\left[\mu\left(\frac{\partial \upsilon_x}{\partial y} + \frac{\partial \upsilon_y}{\partial x}\right)\right]$$
$$+ \frac{\partial}{\partial z}\left[\mu\left(\frac{\partial \upsilon_z}{\partial x} + \frac{\partial \upsilon_x}{\partial z}\right)\right] + \rho g_x$$

$$\rho \frac{D\upsilon_y}{Dt} = -\frac{\partial p}{\partial y} + \frac{\partial}{\partial x}\left[\mu\left(\frac{\partial \upsilon_y}{\partial x} + \frac{\partial \upsilon_x}{\partial y}\right)\right]$$
$$+ \frac{\partial}{\partial y}\left[2\mu\frac{\partial \upsilon_y}{\partial y} - \frac{2}{3}\mu(\nabla \cdot \upsilon)\right]$$
$$+ \frac{\partial}{\partial z}\left[\mu\left(\frac{\partial \upsilon_z}{\partial y} + \frac{\partial \upsilon_y}{\partial z}\right)\right] + \rho g_y \quad (64)$$

$$\rho \frac{D\upsilon_z}{Dt} = -\frac{\partial p}{\partial z} + \frac{\partial}{\partial x}\left[\mu\left(\frac{\partial \upsilon_z}{\partial x} + \frac{\partial \upsilon_x}{\partial z}\right)\right]$$
$$+ \frac{\partial}{\partial y}\left[\mu\left(\frac{\partial \upsilon_z}{\partial y} + \frac{\partial \upsilon_y}{\partial z}\right)\right]$$
$$+ \frac{\partial}{\partial z}\left[2\mu\frac{\partial \upsilon_z}{\partial z} - \frac{2}{3}\mu(\nabla \cdot \upsilon)\right] + \rho g_z$$

where

$$\frac{D\upsilon}{Dt} = \frac{\partial \upsilon}{\partial t} + (\upsilon \cdot \nabla)\upsilon$$

These equations determine the pressure, density and velocity components in a flowing isothermal fluid. A complete description is obtained with appropriate equations of state $p = p(\rho)$, the density dependence of viscosity $\mu = \mu(\rho)$, and the boundary and initial conditions. Invariably restricted forms of the equations of motion are used for convenience;

For constant ρ and constant μ, by means of the equation of continuity [$(\nabla \cdot \upsilon) = 0$]

$$\rho \frac{D\upsilon}{Dt} = -\nabla \rho + \mu\nabla^2 \upsilon + \rho g \quad (65)$$

where $\nabla^2 = \partial^2/\partial x^2 + \partial^2/\partial y^2 + \partial^2/\partial z^2$ is called the Laplacian operator

For constant ρ and $\mu = 0$, equation (64) reduces to

$$\rho \frac{D\upsilon}{Dt} = \nabla P \quad (66)$$

which is the Euler equation, first derived in 1755, in which we have used the quantity P defined by $\nabla P = -\nabla p + \rho g$.

The above set of equations describes the three-dimensional variation in velocity which occurs in flow channels of known dimensions. The above equations must be solved with suitable boundary conditions for the geometry of the flow channel.

Several models of PEM fuel cells have been developed, and solved, which have coupled transport effects of species with the electrochemical characteristics of the MEA to estimate cell performance. Fuller and Newman[9] developed a heat and mass transport model that accommodated variations in temperature and membrane hydration along the flow channel. The model considered transport through a diffusion backing layer but assumed that the conditions and thus concentration was well mixed in the flow channel.

Nguyen and White[10] modeled two-dimensional heat and water transport in a PEM cell that enabled prediction of temperature and water hydration along the channels. The model neglected the effect of concentration variation along the channel and the influence of the porous backing layer. Extensions of these models have included the introduction of both liquid and water flow along the anode and cathode flow channels[11] and two-dimensional flow in the cathode flow channels.[12] These models have been shown to give adequate predictions of axial water distribution in flow channels in PEM cells. However the accuracy in predicting current distribution is not as good, due to several assumptions applied and the neglect of three-dimensional flow in the flow bed channels. This latter problem is complex and can be approached by using commercial computational fluid dynamic (CFD) software packages to solve the hydrodynamic flows of gases. When considering a fuel cell channel the mass transfer through the bottom surface, (in direction z, see Figure 10) which provides a momentum flux, due to species transport to, or from, the MEA to support electrochemical reaction must be included.

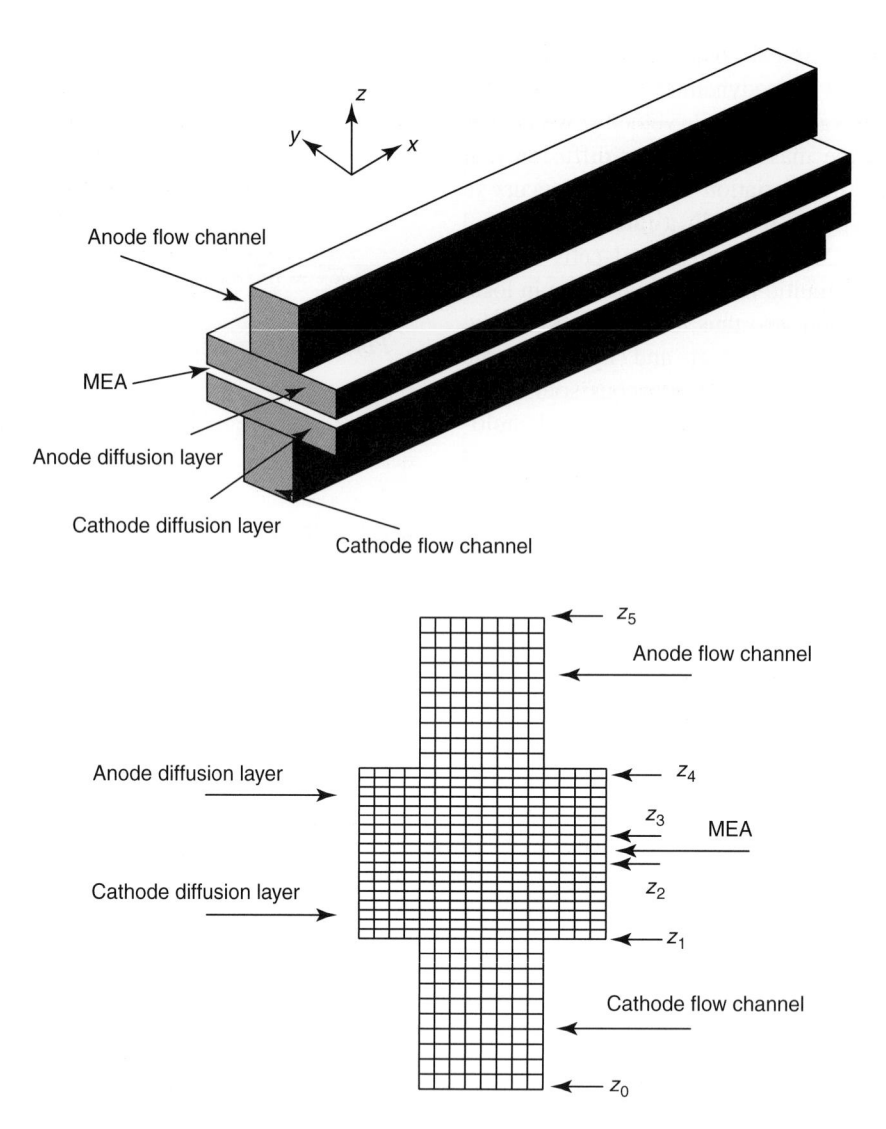

Figure 10. Schematic of the computational domain for a straight channel fuel cell with diffusion layers on the anode and cathode sides of the MEA. (Reproduced from Dutta *et al.* (2000)[13] by permission of Kluwer Academic.)

3.3.1 Three-dimensional model

Dutta *et al.*[13] considered a three-dimensional model of the flow in a PEM cell. They used the three-dimensional form of the equation of motion (Navier Stokes equation) and solved this with the aid of a CFD code (Fluent, Inc., Lebanon, NH) modified to account for the electrochemical reactions which affect the momentum transfer equations. The approach uses a control volume based discretisation of the computational domain for a straight channel fuel cell (Figure 10) to obtain the velocity and pressure distributions in the flow channels and the gas diffusion layers. The model equations for the fuel cell flow channel computation are reproduced in Table 1. The model equations include several volumetric source terms (S) defined on a basis of the control volumes. These terms correspond to the consumption of hydrogen and oxygen and production of water and its transport by electroosmosis and diffusion. The momentum transfer equations have source terms for the porous media used for flow through porous media based on a Darcy permeability law.

In the porous backing layer it is convenient to use an alternative equation of motion based on Darcy's Law in the form

$$\nabla p = -\frac{\mu v_o}{\kappa} + \mu \nabla^2 v_o \tag{67}$$

Figure 11 shows typical results of current distribution at selected axial positions along the flow channel obtained from the model. The data were calculated using parameter estimates from data correlations, shown in Table 2, for water transfer, electroosmotic drag etc. The distributions in current density are linked to the secondary flow patterns in the flow channel. The observed variations in current

Table 1. Governing equations and source terms.

Governing equations	Mathematical expressions		Nonzero volumetric source terms and location of application (Figure 2)	
Conservation of mass	$\dfrac{\partial(\rho u)}{\partial x} + \dfrac{\partial(\rho v)}{\partial y} + \dfrac{\partial(\rho w)}{\partial z} = S_m$	(1)	$S_m = S_{H_2} + S_{aw}$ at $z = z_3$ $\quad S_m = S_{O_2} + S_{cw}$ at $z = z_2$	(7)
Momentum transport	$u\dfrac{\partial(\rho u)}{\partial x} + v\dfrac{\partial(\rho u)}{\partial y} + w\dfrac{\partial(\rho u)}{\partial z}$ $= -\dfrac{\partial P}{\partial x} + \dfrac{\partial}{\partial x}\left(\mu\dfrac{\partial u}{\partial x}\right) + \dfrac{\partial}{\partial y}\left(\mu\dfrac{\partial u}{\partial y}\right)$ $+ \dfrac{\partial}{\partial x}\left(\mu\dfrac{\partial u}{\partial z}\right) + S_{px}$	(2)	$S_{px} = -\dfrac{\mu u}{\beta_x}, \quad S_{py} = -\dfrac{\mu v}{\beta_y},$ $S_{pz} = -\dfrac{\mu w}{\beta_z}$ at $z_1 \leq z \leq z_4$	(8)
	$u\dfrac{\partial(\rho v)}{\partial x} + v\dfrac{\partial(\rho v)}{\partial y} + w\dfrac{\partial(\rho v)}{\partial z}$ $= -\dfrac{\partial P}{\partial y} + \dfrac{\partial}{\partial x}\left(\mu\dfrac{\partial v}{\partial x}\right) + \dfrac{\partial}{\partial y}\left(\mu\dfrac{\partial v}{\partial y}\right)$ $+ \dfrac{\partial}{\partial z}\left(\mu\dfrac{\partial v}{\partial z}\right) + S_{py}$			
	$u\dfrac{\partial(\rho w)}{\partial x} + v\dfrac{\partial(\rho w)}{\partial y} + w\dfrac{\partial(\rho w)}{\partial z}$ $= -\dfrac{\partial P}{\partial z} + \dfrac{\partial}{\partial x}\left(\mu\dfrac{\partial w}{\partial x}\right) + \dfrac{\partial}{\partial y}\left(\mu\dfrac{\partial w}{\partial y}\right)$ $+ \dfrac{\partial}{\partial z}\left(\mu\dfrac{\partial w}{\partial z}\right) + S_{pz}$			
Hydrogen transport (anode side)	$u\dfrac{\partial(\rho m_{H_2})}{\partial x} + v\dfrac{\partial(\rho m_{H_2})}{\partial y} + w\dfrac{\partial(\rho m_{H_2})}{\partial x}$ $= \dfrac{\partial(N_{x,H_2})}{\partial x} + \dfrac{\partial(N_{y,H_2})}{\partial y} + \dfrac{\partial(N_{z,H_2})}{\partial x} + S_{H_2}$	(3)	$S_{H_2} = -\dfrac{j(x,y)}{2F}$ at $M_{H_2}A_{cv}$ at $z = z_3$	(9)
Water transport (anode side)	$u\dfrac{\partial(\rho m_{aw})}{\partial x} + v\dfrac{\partial(\rho m_{aw})}{\partial y} + w\dfrac{\partial(\rho m_{aw})}{\partial z}$ $= \dfrac{\partial(N_{x,aw})}{\partial x} + \dfrac{\partial(N_{y,aw})}{\partial y} + \dfrac{\partial(N_{z,aw})}{\partial z} + S_{aw}$	(4)	$S_{aw} = -\dfrac{\alpha(x,y)}{F}j(x,y)M_{H_2O}A_{cv}$ at $z = z_3$	(10)
Oxygen transport (cathode side)	$u\dfrac{\partial(\rho m_{O_2})}{\partial x} + v\dfrac{\partial(\rho m_{O_2})}{\partial y} + w\dfrac{\partial(\rho m_{O_2})}{\partial z}$ $= \dfrac{\partial(N_{x,O_2})}{\partial x} + \dfrac{\partial(N_{y,O_2})}{\partial y} + \dfrac{\partial(N_{z,O_2})}{\partial z} + S_{O_2}$	(5)	$S_O = -\dfrac{j(x,y)}{4F}M_{O_2}A_{cv}$ at $z = z_2$	(11)
Water transport (cathode side)	$u\dfrac{\partial(\rho m_{cw})}{\partial x} + v\dfrac{\partial(\rho m_{cw})}{\partial y} + w\dfrac{\partial(\rho m_{cw})}{\partial z}$ $= \dfrac{\partial(N_{x,cw})}{\partial x} + \dfrac{\partial(N_{y,cw})}{\partial y} + \dfrac{\partial(N_{z,cw})}{\partial z} + S_{cw}$	(6)	$S_{cw} = \dfrac{1 + 2\alpha(x,y)}{2F}j(x,y)M_{H_2O}A_{cv}$ at $z = z_2$	(12)

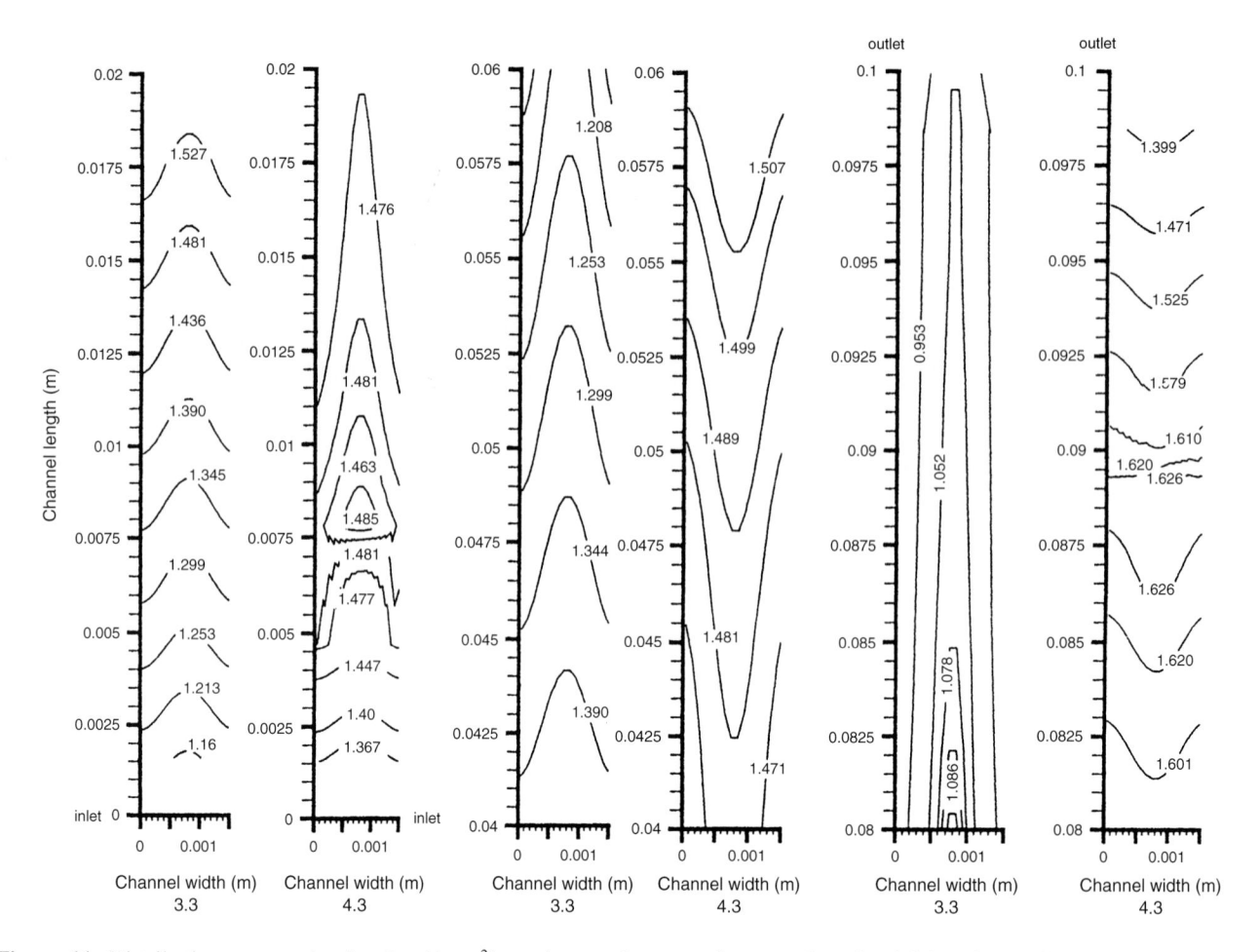

Figure 11. Distribution current density $I_{x,y}$ (A m^{-2}) on the membrane surface at selected axial locations. (Channel dimensions are 1 mm depth and 0.8 mm wide with 0.4 mm current collection width). Anode side inlet conditions: 0.43 m s^{-1} velocity, 0.48 mol fraction hydrogen, 0.16 mol fraction CO_2, 0.36 mol fraction water. Cathode side inlet conditions: 0.56 m s^{-1} velocity, 0.13 mol fraction oxygen, 0.51 mol fraction nitrogen, 0.36 mol fraction water. Diffusion layer permeability 2.3×10^{-11} m^2. Membrane thickness, 175 μm. Cell voltage, 1.08 V. 3.3 and 4.3 refer to co-flow and to counter-flow of anode and cathode streams respectively. (Reproduced from Dutta et al. (2000)[13] by permission of Kluwer Academic.)

density are due to variations in water activity and not H_2 or O_2 limitations. Counter-flow of gas streams gives a more uniform distribution of current density. The results of the model show that a diffusion layer in the MEA serves to create a more uniform current density and that membrane thickness and cell voltage have a significant effect on the axial distribution of current density and the net rate of water transport.

3.3.2 Two phase flow

The two-phase flow of liquid water in the reactant gas streams is likely to occur in PEM cells. This is because water is generated at the cathode of the cell and also that water is typically added with the cathode and/or the anode gas streams to maintain a wet polymer membrane. The treatment of two-phase to model low temperature PEM cells is thus, in principle, necessary, especially when excess liquid water is carried out with the exhaust gas from the cathode.

Bevers et al.[6] included the generation of liquid water in the one dimensional model of the PEM cell. This model considered the influence of liquid water on the voidage of the catalyst and diffusion layers but did not consider the implications of two-phase flow in the flow bed channels. Scott et al.[14, 15] have developed simple models of two phase flow in PEM cells (including direct methanol) based on capillary pressure theory and have used these to predict performance of operating PEM cells.

An initial assessment of the occurrence of two-phase flow in flow channels can be made from simple material balance calculations. In this, for single-phase conditions, the water generated by the cell reaction plus that transferred across the membrane is equal to that transferred by mass transport through the porous cathode and into the channel. As current

Table 2. Equations for modeling electrochemical effects.

Diffusion mass flux of species 1 in ζ direction	$N_{i,1} = \rho D_{t,1} \dfrac{\partial m_{K,1}}{\partial \zeta}$	(13)
Binary diffusion coefficient[10]	$\dfrac{PD_{1,j}(x,y)}{(P_{c-i}x\,P_{c-j})^{1/3} \times (T_{c-i}T_{c-j})^{5/12} \times \left(\dfrac{1}{M_t}+\dfrac{1}{M_t}\right)^{1/2}} = 3.64 \times 10^{-8} \left(\dfrac{T_{\text{cell}}}{\sqrt{T_{c-i}T_{c-j}}}\right)^{2.334}$	(14)
Net water transfer coefficient per proton	$\alpha(x,y) = \lambda_d(x,y) - \dfrac{F}{j(x,y)} Dw(x,y) \left(\dfrac{C_{wc}(x,y)-C_{wa}(x,y)}{t_m}\right)$	(15)
Electroosmotic drag coefficient	$\lambda_d(x,y) = 0.0049 + 2.02 a_a - 4.53 a_a^2 + 4.09 a_a^3 \quad (a_a \leq 1)$	(16)
	$= 1.59 + 0.159(a_a - 1) \quad (a_a > 1)$	
Water diffusion coefficient for cases similar to Yi and Nguyen	$D_w = n_d 5.5 \times 10^{-11} \exp\left[2416\left(\dfrac{1}{303}-\dfrac{1}{T_s}\right)\right]$	(17)
Water diffusion coefficient for cases similar to Fuller and Newman	$D_w = 3.5 \times 10^{-6} \lambda \exp\left(-\dfrac{2436}{T_s}\right)$	(18)
Water concentration for anode and cathode surfaces of the MEA	$C_{wK}(x,y) = \dfrac{\rho_{m,\text{dry}}}{M_{m,\text{dry}}} \left(0.043 + 17.8 a_K - \dfrac{39.8 a_K^2}{36.0 a_K^3}\right) \quad (a_K \leq 1)$	(19)
	$= \dfrac{\rho_{m,\text{dry}}}{M_{m,\text{dry}}} (14 + 1.4(a_K - 1)) \quad (a_K > 1), \quad \text{where K = a or c}$	
Water activity	$a_K = \dfrac{X_{w,K} P(x,y)}{P_{w,k}^{\text{sat}}}$	(20)
Local current density	$j(x,y) = \dfrac{\sigma_m(xy)}{t_m} \{V_{oc} - V_{\text{cell}} - n(xy)\}$	(21)
Local membrane conductivity	$\sigma_m(x,y) = \left(0.00514 \dfrac{M_{m,\text{dry}}}{\rho_{m,\text{dry}}} C_{wa}(x,y) - 0.00326\right) \exp\left(1268\left(\dfrac{1}{303}-\dfrac{1}{T_s}\right)\right) \times 10^2$	(22)
Local overpotential	$n(x,y) = \dfrac{RT_s}{0.5F} \ln\left(\dfrac{j(x,y)}{j_0 P_{O_2}(x,y)}\right)$	(23)

density increases, the amount of water transport increases and eventually the air reaches saturation. Above this current density liquid water appears and two-phase conditions prevail. The current density at which two phase flow occurs depends upon the inlet humidity of the air, the temperature, the flow velocity of air in the flow field and the mass transport conditions in porous media and at the surface of the cathode in the flow channel. Thus it is possible to give a first order estimation of the occurrence of two phase flow under given conditions as a function of air velocity (or Reynold Number, Re) and current density as shown in Figure 12. In this figure we see that at a given temperature of 60 °C and at a Reynolds number of 20, i.e., a SR of approximately 1.2, two-phase flow will occur at current densities above 1000 mA cm^{-2}. It should be stressed that this estimation is for a short electrode and assumes appropriate conditions of mass transport in the cathode porous media and flow channel. At longer electrodes two phase flow will occur at higher velocities under similar operating conditions.

The two-phase flow and transport in the air cathode of a PEM cell has been modeled by Wang et al.[16] The model is in two dimensions and considers multi-component and two-phase flow in the porous cathode, which is assumed to be a homogeneous porous media. The model uses the equations of conservation of momentum and mass and

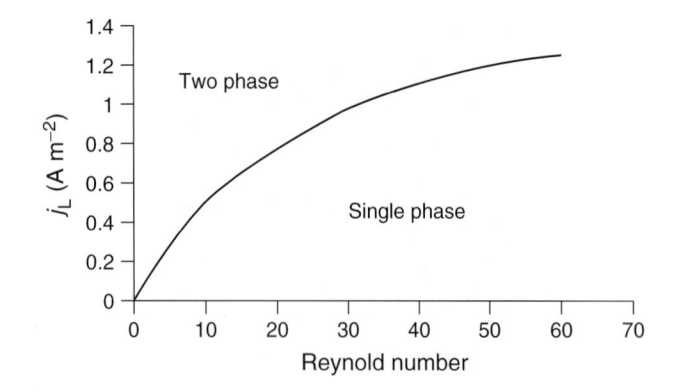

Figure 12. The influence of stoichiometric ratio and current density on two phase flow in PEM cells. 80 °C, channel height 1.0 mm, Pressure 1 bar, air as oxidant. j_l is the current density at which two phase flow occurs.

redefines parameters in terms of two-phase mixture average quantities. Mass flux in the porous media is determined by capillary flow, i.e., the capillary pressure due to meniscus curvature.[17] The model deals with the situation where both single-phase flow and two-phase flow can co-exist and predicts water and oxygen distribution as well as PEM cell polarization characteristics.

Nguyen[18] has also considered the implication of two-phase flow on the performance of an air cathode. A two-dimensional model, that considers multi-component mass transport and two-phase flow, was developed to assess the use of an interdigitated flow field on performance. The transport of gas was defined by Darcy's Law whilst that of water was by shear force of the gas and capillary diffusion. The model predicted two-dimensional variations in velocity, pressure, oxygen and water distribution and the variation of current density along the channel.

4 OTHER FUEL CELLS

The treatment of mass transport phenomena in other fuel cells can be based on the same fundamental models as described for PEM cells. Fundamental differences are that, with cells operating at higher temperatures, formation of liquid product in cells in not an issue and that in certain cases alternative fuels, to hydrogen, are used. These fuels include hydrocarbons and alcohols, e.g., methanol. In the case of methanol, new and significant advances have been made for low temperature operation in PEM cells. For such "direct methanol" fuel cells, mass transport in flow channels is particularly important in determining cell performance due to the two-phase flow that can develop.

4.1 High temperature fuel cells

The modeling of high temperature fuel cells such as solid oxide and molten carbonate has mainly focused on evaluating the influence of electrochemical parameters and heat transfer on the temperature distribution and the corresponding current density distribution in the cell and stacks. Achenbach[19] presented a three-dimensional and time dependent simulation of a planar solid oxide cell stack which included heat transfer between gases and cell and simple material balances for gas composition changes. Bessette et al.[20] developed a similar model for a tubular solid oxide cell and included a mass transport resistance term to allow for mass transport polarization at anode and cathode. Gas transport in solid oxide cells has been modeled by Lehnert et al.[21] using the Stefan Maxwell and Knudsen diffusion model and Darcy's law for combined diffusion and permeation through the porous electrode.

For modeling of molten carbonate fuel cells, He and Chen have used computational fluid dynamic methods for three-dimensional simulations.[22] Fujimura et al.[23] have modeled heat and mass transfer in the MCFC and determined performance and temperature distributions. Potyomkin et al.[24] have modeled the gas flow distribution in two dimensions along a channel in a MCFC using the Euler approximation for laminar flow and percolation model for diffusion through the electrode.

4.2 Direct methanol fuel cells

Conventional PEM fuel cells operate on hydrogen as a fuel and oxygen or air as an oxidant. The gas streams are typically humidified to keep the membrane (e.g., Nafion® 112, Du Pont de Nemours) fully hydrated to obtain sufficient proton conductivity. However hydrogen is inconvenient to store and transport, and a more satisfactory approach would be to electro-oxidize a liquid fuel at the anode. Methanol is a convenient liquid fuel that has substantial electroactivity and can be directly oxidized to carbon dioxide and water in a direct methanol fuel cell (DMFC) shown schematically in Figure 13. The DMFC can function with either a liquid or vaporized aqueous methanol solution as fuel. Due to the fact that water is also present in the fuel feed, external hydration of reactant streams is, in principle, not required. The vapor feed DMFC has similarities with the hydrogen PEM cell and models for this system are similar to those for the PEM cell with due allowance for appropriate differences in physical, chemical and transport properties. The liquid feed DMFC, based on a solid polymer electrolyte (SPE) offers inherently simple operation, e.g., conversion of a dilute aqueous based fuel into an gaseous carbon dioxide product.

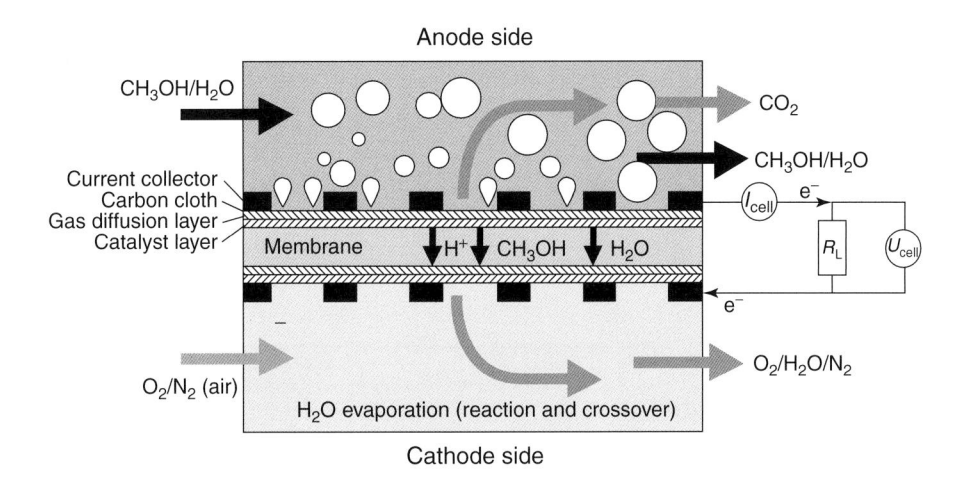

Figure 13. Schematic representation of direct methanol fuel cell. R_L is the load.

Currently significant power densities have been achieved with the use of catalysts based on Pt–Ru: typically higher than $0.18 \, W \, cm^{-2}$.[25, 26] For high performance of a DMFC relatively low concentrations of methanol are required. At concentrations higher than approximately $2.0 \, mol \, dm^{-3}$, the cell voltage declines significantly due to permeation of methanol through the membrane, i.e., methanol crossover. This permeation results in a mixed potential at the cathode with a significant loss in oxygen reduction performance and also poor fuel utilization. At lower methanol concentrations power density performance is lower, due to lower limiting currents.

Limiting current densities for methanol oxidation at SPE based electrodes have been reported by Ravikumar and Shukla[27] and Kauranen and Skou.[28] The latter attributed this limiting current behavior to saturation coverage of absorbed OH on the platinum surface. Ravikumar and Shukla[27] have reported data for the DMFC using platinum–ruthenium catalysts for methanol oxidation. The DMFC is a complex system based on porous electrocatalytic electrodes. Its operation as a liquid feed system is complicated by the evolution of carbon dioxide gas from the anode surface of the MEA. In this respect the DMFC has similarities with other gas evolving electrodes as has been observed in flow visualization of gas evolution by Scott et al.[15] Figure 14 shows a set of images captured during the flow visualization study. Figure 14(a,b) compares the amounts of gas present in the flow bed for the large and the small cell under similar operating conditions. In the large cell, gas tended to form small fast moving bubbles, while in the small cell large gas slugs were formed. A flow field design based on a plate heat exchanger concept has been developed from this work, which gives excellent performance with the formation of fine, carbon dioxide bubbles. Direct methanol fuel cell flow beds have also been fabricated from stainless steel mesh flow beds.[29]

Mass transport is a factor that generally limits the performance of solid polymer electrolyte fuel cells that operate at relatively high current densities. The mass transport behavior effects hydrogen and methanol fuel cells in different ways. For hydrogen fuel cells, these mass transport limitations are predominantly associated with oxygen transport in the gas diffusion layers and in the catalyst layers. For the vapor-fed DMFC mass transfer limitations occur at the anode, similar to those for the oxygen cathode, although the cell is fed with methanol/water vapor and produces CO_2, flowing counter current to the fuel. In the liquid-fed DMFC mass transport limitations arise due to counter current gas, liquid flow in the gas diffusion and catalyst layers.

4.2.1 Vapor feed DMFC

The fuel to the anode is a mixture of methanol in water vapor at a maximum molar ratio of approximately 1 : 27, methanol to water (2.0 M methanol solution). Thus, the water mol fraction will not change significantly and is assumed constant. Mass transfer of this system is described by the Stefan–Maxwell equation. With the approximations that the crossover of methanol is small in comparison to water transferred across the membrane and that the effective diffusivities for methanol in water and carbon dioxide are equal, the mol fraction of methanol in the anode side, x_1 is given by

$$\frac{x_{1i} + \dfrac{B}{A}}{x_{1c} + \dfrac{B}{A}} = \exp\left[\frac{k_{12} l_{ca}(1 + \alpha_W) j}{6F}\right] \quad (68)$$

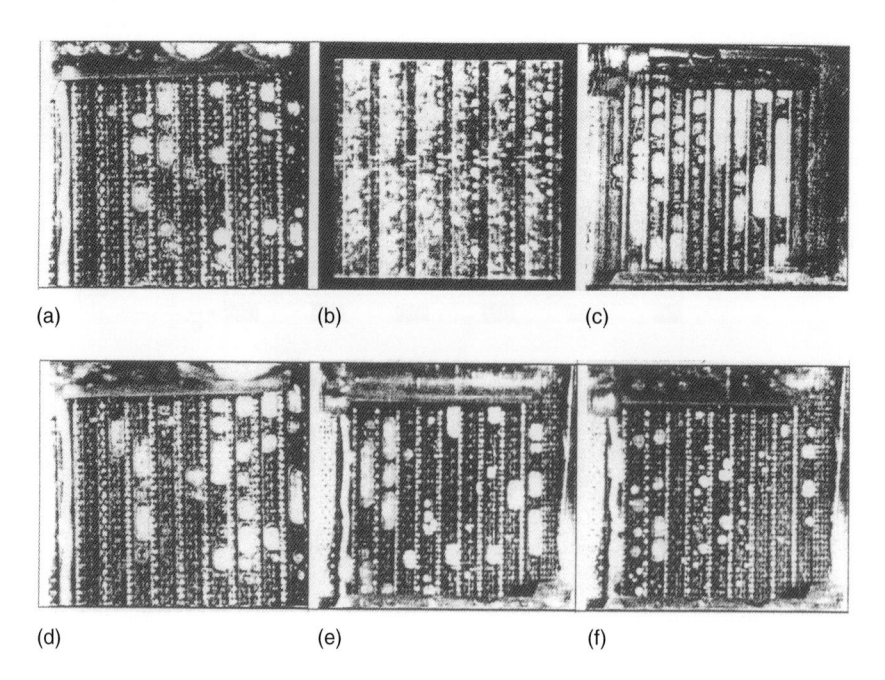

Figure 14. Flow visualization and bubble flow images in the DMFC. (a) Carbon cloth in the small cell (30 mA cm^{-2}, anode inlet flow rate 206 ml min^{-1}). (b) Carbon cloth in the large cell (30 mA cm^{-2}, anode inlet flow rate 206 ml min^{-1}). (c) Carbon paper in the small cell (30 mA cm^{-2}, anode inlet flow rate 206 ml min^{-1}). (d) Carbon cloth in the small cell (30 mA cm^{-2}, anode inlet flow rate 3.1 ml min^{-1} channel^{-1}). (e) Carbon cloth in the small cell (80 mA cm^{-2}, anode inlet flow rate 1137 ml min^{-1}). (f) Carbon cloth in the small cell (50 mA cm^{-2}, anode inlet flow rate 1137 ml min^{-1}).

where, i and c, refer to the catalyst interface and the flow channel interface, $B/A = -(1 + \alpha_W)^{-1}$, α_W is the ratio of total water flux to the molar water flux and $k_{12} = RT/PD_{12}^{\text{eff}}$.

Figure 15 shows the variation of mol fraction of methanol vapor with current density as a function of gas voidage (at 80 °C and with $l_{ca} = 0.3$ mm). At relatively high current densities the mole fraction of methanol can fall significantly

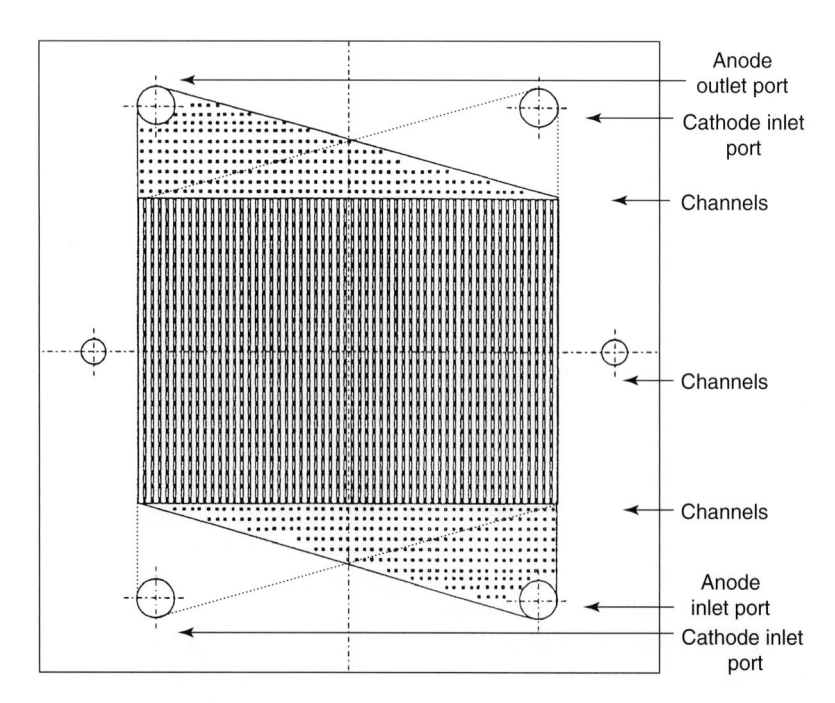

Figure 15. Variation in methanol mole fraction with current density. Ratio of total water flux to molar flux, $\alpha_W = 15$. Value of gas voidage, e_g; ♦, 0.3; □, 0.2; △, 0.15.

below the value and thus illustrates why mass transport limitations are observed in the DMFC. These limitations occur at lower current densities than usually experienced in hydrogen fuel cells. Figure 16 shows how this mass transport effect influences the DMFC cell voltage characteristics and compares it to experimental data.

Kulikovsky et al.[8] have developed a two-dimensional model of the DMFC based on the Stefan–Maxwell and Knudsen diffusion mechanism for gas transport in the backing layers (see Section 3.2.4).

4.2.2 Flow field mass transport

The analysis of mass transport in flow fields for vapor feed DMFC can be based on models described in Section 3.2.4. A simple model for gas flow in the direct methanol fuel cell can be based on the analysis using the Euler equation in which viscous effects are ignored. This model essentially balances a loss in momentum with a change in mass flow due to the anode methanol oxidation reactions.

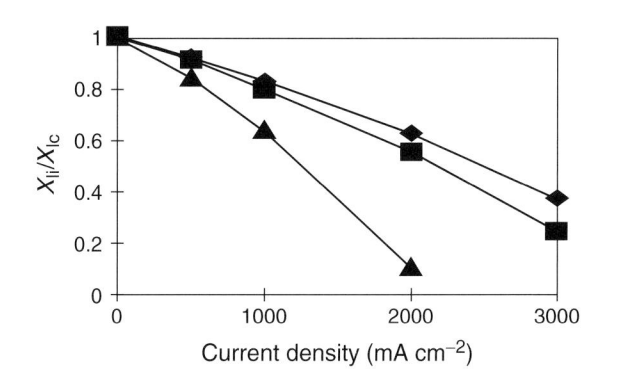

Figure 16. Effect of mass transport on DMFC performance with vapor feed. 80 °C. Experiment; 0 bar (▲) and 2 (■) bar oxygen. Model (○). Cell area 9 cm^2.

4.2.3 Liquid feed DMFC

Ravikumar and Shukla[27] have reported data for the DMFC using platinum–ruthenium catalysts for methanol oxidation. The electrode structure comprises a Nafion® membrane onto which are pressed Nafion® bonded carbon-supported catalysts. Limiting current densities are obtained, at 0.5–2.0 mol dm^{-3} methanol concentrations, which are approximately proportional to the methanol concentration. Noticeably the limiting current densities observed by Ravikumar and Shukla,[27] shown in Table 3, are very much lower, at equivalent methanol concentrations, than those of Kauranen and Skou.[30] In two sets of the data,[27, 31] the mass transfer coefficient is approximately constant whereas in a third set[28] this is not the case; values of mass transfer coefficient decrease with an increase in methanol concentration. However it should be noted that the electrode structures used however were significantly different.

The DMFC is a complex porous electrode system due to the thin three-dimensional structure of the electrocatalyst region, the porous diffusion layers and the generation of carbon dioxide gas. The limiting current behavior of the DMFC is due to mass transport limitations of methanol supply to the anode catalyst caused by:

1. Hydrodynamic mass transfer from the feed flow to the surface of the carbon cloth backing layer, where the gas bubbles are released into the flowing methanol solution;
2. Diffusion mass transfer in the carbon cloth and anode carbon diffusion layer;
3. Diffusion mass transfer in the catalyst layer.

The combined effect of the three, above mentioned, mass transfer effects is simply expressed as

$$\frac{1}{k_{\text{eff}}} = \frac{1}{k_l} + \frac{1}{k_{\text{cl}}} + \frac{\ell}{\zeta k^*} \quad (69)$$

Table 3. Effect of methanol concentration on limiting current densities and effective mass transfer coefficients for methanol oxidation.

Methanol concentration (mM)	Limiting current density (mA cm^{-2})			Mass transfer coefficient (k_L) (ms^{-1})		
	Kauranen[28]	Ravikumar[27]	Kauranen[31]	Kauranen[28]	Ravikumar[27]	Kauranen[31]
20			20/80 °C			1.7×10^{-5}
30	75/80 °C			4.0×10^{-5}		
50			50/80 °C			1.7×10^{-5}
100	150/80 °C		90/80 °C	2.5×10^{-5}		1.5×10^{-5}
200			190/80 °C			1.6×10^{-5}
300	290/80 °C			1.6×10^{-5}		
500		110/70 °C			3.7×10^{-6}	
		180/95 °C			6.0×10^{-6}	
1000		220/70 °C			3.7×10^{-6}	
		360/95 °C			6.0×10^{-6}	
1500		360/70 °C			4.0×10^{-6}	

where, k_l, is the hydrodynamic mass transport coefficient, k_{cl} is the carbon cloth mass transport coefficient and ζk^* is the effective electrochemical methanol oxidation rate constant in the catalyst layer, where ζ is an effectiveness factor for the catalyst.

The effective mass transfer coefficients, k_{eff}, determined from

$$k_{eff} = \frac{j}{nFC_{MeOH}} \quad (70)$$

are given in Table 3, as calculated from published data.[27, 28, 31] Equation (66) enables the overall methanol mass transfer behavior of the DMFC to be analyzed in terms of three separate, but interactive, components. The porous electrocatalyst layer is initially considered as a thin pseudo two-dimensional structure where, due to the potential distribution, highest activity is at a region next to the membrane surface. The value of k_l in equation (66) depends on the hydrodynamics in the channel as well as on gas bubble liberation at the surface. Diffusion mass transfer in the cloth depends upon the liquid void fraction and the cloth structure and can be represented by

$$j = nFk_{cl}^o e^m \Delta C_{cl} \quad (71)$$

where, ΔC_{cl} is the concentration change over the cloth thickness, e is the liquid voidage, m is frequently $= 1.5$, and $k_{cl}^o = D_{MeOH}/\ell_{cl}$, where ℓ_{cl} is the cloth thickness.

Mass transport due to gas evolution
In the case of the DMFC gas liberation is from the surface of a carbon cloth positioned away from the active electrode region. The gas evolution is not, as observed in flow visualization studies, uniformly distributed and has similarities to gas evolution observed at planar electrodes. Mass transfer at gas evolving electrodes has, for example, been measured for oxygen evolution[32] at vertical electrodes and is correlated, for carbon dioxide evolution by the expression

$$k_l = 1.87 \times 10^{-6} \left(\frac{j}{3}\right)^{0.32} \quad (72)$$

where k_l is the average mass transfer coefficient. The correlation for gas evolving electrodes gives mass transfer coefficients in the range of $0.4–1.8 \times 10^{-5}$ m s^{-1}, for current densities in the range $100–2000$ A m^{-2}. The experimental determined mass transfer coefficients (Table 3) are in the range of those predicted by correlation. The above model equation predicts that mass transfer coefficients for the DMFC will exhibit a broad maximum in value with increasing methanol concentration. The mass transfer coefficients as expected increase with temperature of operation.

The overall mass transfer behavior of the liquid feed DMFC is influenced by the gas liberation at the surface of the diffusion layer and diffusion in the carbon cloth which will decrease as the volume of gas in the diffusion layer increases as current density increases. The importance of the mass transfer resistance can be quickly assessed assuming mass transfer through a stationary methanol solution in the carbon cloth. For example $k_{cl}^o = D_{MeOH}/\ell_{cl} \approx 2.8 \times 10^{-9}/2.8 \times 10^{-4} \approx 10^{-5}$ m s^{-1}. For a 50% flooded cloth and a 1.0 M methanol solution, the limiting current density is

$$i = nFk_c^* \varepsilon^{1.5} C_{MeOH} = 6 \times 96\,485 \times 10^{-5}$$
$$\times 0.5^{1.5} \times 1000 = 2047\,\text{A m}^{-2}$$

Such a limiting current places a potentially large restriction on the performance of the DMFC and indicates the need for effective electrode design and suitable mass transport analysis. Overall the effect of increased current density on the mass transfer characteristics of carbon cloth based anodes in the DMFC is a combined effect of; enhancement due to increased gas evolution at the cloth surface and suppression due to decreased liquid volume in the cloth.

Figure 17 shows experimentally measured mass transport coefficients for the DMFC as a function of methanol concentration. The mass transport coefficients are in the range of approximately $2.5–6.0 \times 10^{-5}$ m s^{-1}. The data indicates that there is a combined effect of increased methanol concentration, i.e., limiting current density, whereby mass transport is increased due to bubble mass transport enhancement and suppressed due to gas accumulation in the carbon cloth.

A one-dimensional model for predicting the cell voltage of the DMFC, that includes mass transfer behavior

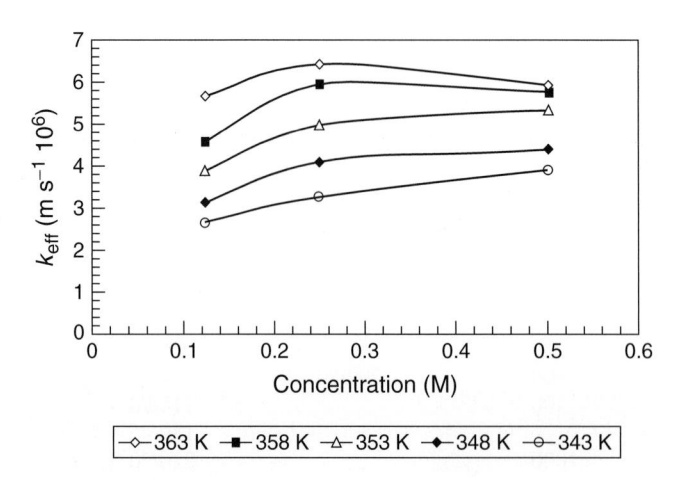

Figure 17. Experimental mass transfer coefficients, k_{eff}, for the DMFC. ■, 85 °C; △, 80 °C; ◇, 90 °C; ♦, 75 °C; ○, 70 °C.

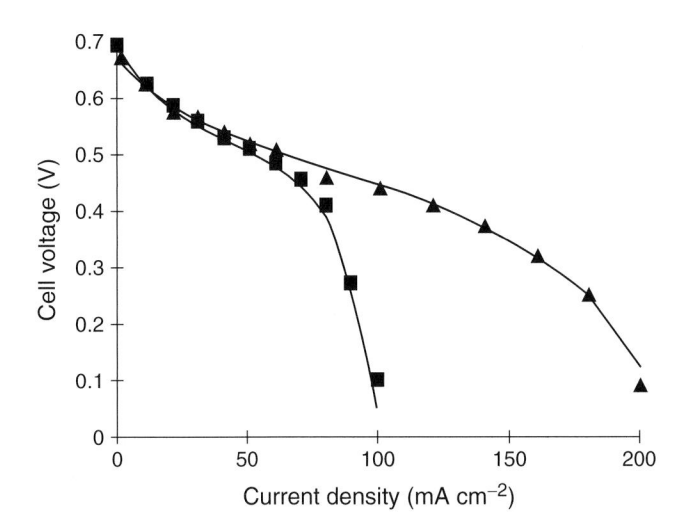

Figure 18. Comparison of experimental and model cell polarization curves for the DMFC, 80 °C, 1.5 bar air. ■, 0.125 mol dm^{-3}; ▲, 0.25 mol dm^{-3}.

associated with the two-phase flow in the MEA diffusion layer has been produced.[33, 34] The comparison between experiment data and the model prediction is shown in Figure 18 for low concentrations of methanol solution. Agreement between the model and the experiment is generally good over the full range of current densities. Clearly the model is applied to the particular type of MEA used in this study and it remains to be seen whether it can be used to predict behavior of other MEA structures and materials.

LIST OF SYMBOLS

p_j^o	partial pressure of component j at reference state
$p_{H_2O}^{sat}$	saturation partial pressure of water vapor
$D_{H_2O,L}$	surface diffusion coefficient of water
D_{ij}^{eff}	effective binary diffusivity coefficient of the ij gas pair
M_{H_2}	molecular weight of hydrogen
M_{O_2}	molecular weight of oxygen
ΔS_o^o	standard state entropy change
E_0^o	standard state reference potential
$P_{w,K}^{sat}$	vapor pressure of water in stream K
A_{CV}	specific surface area of the control volume (CV)
C_i	concentration of species i
c_{pj}	heat capacity of component j
c_s	heat capacity of the solid
C_{wK}	concentration of water at K interface of the membrane
D_{ij}	binary diffusion coefficient between components i and j
D_j	Knudsen diffusion coefficient of component j
$d_{rea/diff}$	thickness of the reaction layer/diffusion layer
D_w	diffusion coefficient of water
F	Faraday's constant
H_i	Henry's law constant for species i
I	local volumetric current
J	local current density
j_H^+	ionic current density
j_O	exchange current density
k_{cl}	mass transfer coefficient for cloth
K_{el}	conductivity of the electrolyte (bulk)
k_l	material transfer coefficient
K_s	conductivity of the solid
k_{eff}	effective mass transfer coefficient
L	thickness
l_d	thickness of the gas diffuser
M_j	molar weight of component j
$m_{K,l}$	mass fraction of the species l in stream K (dimensionless)
$M_{m,dry}$	equivalent weight of a dry membrane
n	number of electrons involved in reaction
N_j	molar flux of the component j
P	pressure
p_j	partial pressure of component j
R	universal gas constant
T	temperature of the system
T	time
T	temperature
t_m	membrane thickness
T_s	surface temperature at the anode
U_o	open circuit potential
V_{cell}	cell voltage
V_{oc}	cell open-circuit voltage
X	channel length measured from anode inlet
x_i	mole fraction of species i
X_s	water loading $kg_{H_2O}\, kg_{solid}^{-1}$
X_{wK}	mole fraction of water in stream K
Z	distance in the "horizontal direction" (direction of diffusion of gases)
z_i	charge on species i
ΔH_c	heat of water condensation/evaporation
ΔS	reaction entropy

Superscripts and subscripts

A	anode
Act	activation
c	cathode
channel	conditions at the interface of the cathode gas flow channel and the cathode
cl	cloth

CO_2	carbon dioxide
crit	critical values
dry	dry gas conditions
eff	effective (i.e., average) conditions
equil.	equilibrium
H_2	hydrogen
H_2O	water
hum	humidified gas conditions
in	gas conditions at the inlet to the fuel cell
interface	conditions at the interface of the gas diffusion cathode
N_2	nitrogen
O_2	oxygen
out	gas conditions at the outlet of the fuel cell
sat	saturated
sat	saturation conditions
W	water

Greek symbols

ϕ_s	potential of the electric conducting material
ϕ_{el}	potential of the electrolyte
ε_g^o	volume fraction of pore (gas) volume in the water free electrode
σ_m	membrane conductivity
α_{act}	specific active surface
u, v, w	velocities in x, y and z directions, respectively
ζ	effectiveness of electrocatalyst
λ	electroosmotic drag coefficient (number of water molecules carried per proton)
v	velocity
•	cathodic charge transfer coefficient
•	net water flux to molar water flux
α_K	activity of water in stream K (dimensionless)
α_V	specific surface of condensation
Δ	gradient or spatial derivative
ε	diffusivity correction factor
ε_{el}	volume fraction of electrolyte
ε_g	volume fraction of pore (gas) volume
ε_s	volume fraction of solid volume
ζ	dummy variable for direction x, y or z
η	overvoltage
η_{act}	overvoltage due to activation
η_{ohmic}	overvoltage due to ohmic resistance
λ_s	heat transfer coefficient of the solid
μ	dynamic viscosity
ρ	density of the mixture
$\rho_{m,dry}$	density of a dry membrane
ρ_s	density of the solid
σ_{ij}	Lennard–Jones force constant

REFERENCES

1. F. N. Buchi and S. Srinivasan, *J. Electrochem. Soc.*, **144**, 2767 (1997).
2. C. F. Curtis and J. O. Hirscfelder, *J. Chem. Phys.*, **17**, 550 (1949).
3. D. M. Bernardi, *J. Electrochem. Soc.*, **137**, 3344 (1990).
4. J. C. Amphlett, R. M. Baumert, R. F. Mann, B. A. Peppley and P. R. Roberge, *J. Electrochem. Soc.*, **142**, 1 (1995).
5. R. B. Bird, W. E. Stewart and E. N. Lightfoot, 'Transport Phenomena', John Wiley & Sons, New York (1960).
6. D. Bevers, M. Wohr, K. Yasuda and K. Oguro, *J. Appl. Electrochem.*, **27**, 1254 (1997).
7. A. A. Kulikovsky, J. Divisek and A. A. Kornyshev, *J. Electrochem. Soc.*, **146**, 3981 (1999).
8. A. A. Kulikovsky, J. Divisek and A. A. Kornyshev, *J. Electrochem. Soc.*, **147**, 953 (2000).
9. T. F. Fuller and J. Newman, *J. Electrochem. Soc.*, **140**, 1218 (1993).
10. T. V. Nguyen and R. E. White, *J. Electrochem. Soc.*, **140**, 2178 (1993).
11. J. S. Yi and T. V. Nguyen, *J. Electrochem. Soc.*, **145**, 1149 (1998).
12. V. Gurau, H. Liu and S. Kakac, *AIChE J.*, **44**, 2411 (1998).
13. S. Dutta, S. Shimpalee and J. W. Van Zee, *J. Appl. Electrochem. Soc.*, **30**, 135 (2000).
14. K. Scott, P. Argyropoulos and K. Sundmacher, *J. Electroanal. Chem.*, **477**, 97 (1999).
15. K. Scott, W. M. Taama and P. Argyropoulos, *Electrochim. Acta*, **44**, 3575 (1999).
16. Z. H. Wang, C. Y. Wang and K. S. Chen, *J. Power Sources*, **94**, 40 (2000).
17. A. E. Scheidegger, 'The Physics of Flow Through Porous Media', University of Toronto Press, Toronto (1958).
18. T. V. Nguyen, 'Tutorials in Electrochemical Engineering-Mathematical Modeling', "Electrochemical Society Proceedings", The Electrochemical Society, Pennington, NJ, PV99-14, pp. 223–241 (1999).
19. E. Achenbach, *J. Power Sources*, **49**, 333 (1994).
20. N. F. Bessette, W. J. Wepfer and J. Winnick, *J. Electrochem. Soc.*, **142**, 3792 (1995).
21. W. Lehnert, J. Meusinger and F. Thom, *J. Power Sources*, **87**, 57 (2000).
22. W. He and Q. Chen, *J. Power Sources*, **55**, 25 (1995).
23. H. Fujimura, N. Kobayashi and K. Ohtsuka, *ISME Int. J.*, Series II **35**, 125 (1992).
24. G. A. Potyomkin, N. G. Kozhuhar, A. V. Ahissin, N. N. Batalov, S. I. Malevanny, A. Y. Malishev and A. Y. Postnikov, 'Carbonate Fuel Cell Technology', "Proceedings of the 5th International Symposium", The Electrochemical Society, Pennington, NJ, PV99-20, 80 (1999).

25. K. Scott, W. M. Taama and J. Cruickshank, *J. Power Sources*, **65**, 40 (1998).
26. X. Ren, W. Henderson and S. Gottesfeld, *J. Electrochem. Soc.*, **144**, L267 (1997).
27. K. Ravikumar and A. K. Shukla, *J. Electrochem. Soc.*, **143**, 2601 (1996).
28. P. S. Kauranen and K. Skou, *J. Electroanal. Chem.*, **408**, 189 (1996).
29. K. Scott, P. Argyropoulos, P. Yiannopoulos and W. M. Taama, *J. Appl. Electrochem.*, **31**, 823 (2001).
30. P. S. Kauranen and K. Skou, *J. Appl. Electrochem.*, **26**, 909 (1996).
31. P. S. Kauranen and K. Skou, *J. Electrochem. Soc.*, **143**, 143 (1996).
32. M. G. Fouad and G. H. Sedahmed, *Electrochim. Acta*, **17**, 665 (1972).
33. K. Scott, S. Kraemer and K. Sundmacher, '5th European Symposium on Electrochemical Engineering', Exeter, I. Chem. E. Symp. Ser. No. 145 March 24–26 (1999).
34. K. Sundmacher, T. Schultz, S. Zhou, K. Scott, M. Ginkel and E. D. Gilles, *Chem. Eng. Sci.*, **56**, 333 (2001).
35. M. J. Blomen Leo and M. N. Mugerwa, 'Fuel Cell Systems', Plenum, New York (1993).

Part 3

Heat transfer in fuel cells

Chapter 9
Low temperature fuel cells

J. Divisek
Institute for Materials and Processes in Energy Systems, Forschungszentrum Jülich GmbH, Jülich, Germany

1 ENERGY BALANCE AND HEAT TRANSFER IN FUEL CELLS

The principle of energy conservation determines the energy balance equation, which can be generally formulated as the sum of single rates of

$$\text{energy input} - \text{energy ouput} + \text{energy production}$$
$$= \text{accumulation of energy} \quad (1)$$

This balance equation is true for a particular system or control volume. In a fuel cell, the balance takes into account the thermal and the electrical energy. Enthalpy change, ΔH, within a closed constant molar volume system, V, can be written as

$$\Delta H = Q_{\text{mol}} - W + V\Delta P \quad (2)$$

In equation (2), Q_{mol} denotes the molar heat supplied to the system, W is the work done and P is the system pressure. An electrochemical system, however, is an open one. For this kind of system, the thermal energy balance is based on the difference between the enthalpy flow into the cell and the flow out of the cell. An accumulation of the thermal energy generally occurs during the changes from one operating condition to another, e.g., during start-up or shut-down. The time variation of energy balances is usually manifested in temperature changes. The energy input into the cell is associated with the enthalpy content of the inflowing process fluids, the fuel and the oxidant. The major contribution of the input energy is associated with the reaction enthalpy of the fuel combusting reaction.

The rate of enthalpy outflow is given by the enthalpy content of the outgoing process fluids, the joule heat and heat losses from the reactor surfaces. An important issue of a fuel cell reaction is the fact that, in contrast to the usual chemical combustion, the complete fuel cell reaction occurs at two different electrodes, i.e., we have to deal with the delocalization of the chemical energy source. This may become a complication, especially when concerning the estimation of the local entropy sources, which are usually known only for the reaction as a whole. When we consider a fuel cell as a whole, it is possible to formulate the following general energy balance equation for the accumulation of energy in the cell (left-hand side of equation (3)) in the form

$$\sum m_i c_{pi} \frac{dT}{dt} = -I \left(\frac{\Delta H}{nF} + E_c \right)$$
$$+ \sum N_i A c_{pi}(T_{\text{in}} - T_{\text{out}}) - Q_{(-)}$$
$$E_c = E_0 - \sum |\eta| - IR_i, \quad E_0 = -\frac{\Delta G}{nF} \quad (3)$$

In equation (3), A and I are the total system area and the total current flow, respectively, ΔH is the enthalpy change due to the fuel cell reaction, $\sum N_i A c_{pi}(T_{\text{in}} - T_{\text{out}})$ is the enthalpy change between the inlet and outlet for all the components involved (T_{out} being generally the cell temperature, T), N_i is the mass flow of each component, m_i the mass of each component, E_c the cell voltage and $Q_{(-)}$ are the enthalpy heat losses in the cell. E_0 is the voltage without current passing, usually termed as the electromotive force (also the open circuit voltage).

The heat generated by a fuel cell, $I(\Delta H/nF + E_c)$, differs therefore from the reaction entropy by the addition of some irreversible components, such as the heat resulting from the anodic and cathodic overvoltages, $I\sum|\eta|$, as well as the internal joule heat production $\sum I^2 R_i$, R_i being the internal cell resistivity. A reduction in these energy losses can be achieved by using high performance materials for the electrodes, electrocatalysts and the membrane, thus minimizing polarization and ohmic voltage losses. Usually, these losses can be further reduced by enhancing the operating temperature, which undergoes, however, technological restrictions in the case of low-temperature fuel cells (LTFCs). Small inter-electrode gaps and a general reduction of the cell size are an essential requirement for loss reduction.

The heat losses, $Q_{(-)}$, in a fuel cell are caused by the heat transfer through the walls and the vaporization of the water. The corresponding expression for this is

$$Q_{(-)} = R_{vap}\Delta H_{vap} + K_\lambda A(T - T_f) \quad (4)$$

Since the heat transfer through the walls is caused mainly by heat convection, the second term on the right in equation (4) can be considered as an extended form of Newton's law of cooling. In this equation, A denotes the cell wall surface, K_λ the overall heat transfer coefficient, $(T - T_f)$ the temperature difference between the fuel cell and the fluid surrounding the cell, H_{vap} the enthalpy of vaporization and R_{vap} the rate of water vaporization. An analysis of both terms in equation (4) needs a closer consideration of the general heat transport phenomenon in a fuel cell.

1.1 Mechanism of heat transport in low temperature fuel cells

There are three modes of heat transfer process that occur in a fuel cell: conduction, convection and radiation.

Heat conduction is the mechanism of internal energy exchange from one body to another by the exchange of the kinetic energy of motion of the molecules by direct communication (electric insulators) or by the drift of free electrons (electric conductors). This energy flow passes from the higher energy molecules to the molecules with lower energy or from a higher temperature region to a lower temperature level. It takes place within the boundaries of a body or across the boundaries between two bodies placed in contact to each other without a displacement of the matter comprising the body. Heat convection applies to the heat transfer mechanism that occurs in a fluid by mixing one portion of the fluid with another portion due to movements of the mass of the fluid. This movement may be caused either by forced or free convection or natural convection.

Finally, thermal radiation should also be mentioned, even if this phenomenon takes place mainly in high temperature fuel cells. This term describes the electromagnetic radiation that can be observed by emission at the body surface when it is thermally excited. Good general references for this material are Refs. [1–3].

1.1.1 Conduction heat transfer

The basic law governing heat conduction may be illustrated as shown in Figure 1. When a temperature gradient exists in a body, experience shows that the rate of heat flow is directly proportional to the area, A and the temperature difference across the body. This can be expressed as

$$Q_x = -\lambda A \frac{\partial T}{\partial x} \quad (5)$$

where Q_x is the heat transfer rate in the x direction, $\partial T/\partial x$ denotes the temperature gradient in the direction of the heat flow, A is the total area and the constant λ is called thermal conductivity of the material. Strictly speaking $\partial T/\partial x$ should be written as the normal gradient that may vary over the surface. For the purpose of computing the heat transfer across the whole body of the fuel cell, a heat conduction problem consists of finding the temperature at any time and at any point within the cell.

To develop the necessary general heat conduction equation, we start with equation (5). The case in which the heat-conducting material may have internal sources of heat generation must also be considered. This is the usual situation in the fuel cell as the result of electrochemical reactions, energy dissipation and liquid-gas state changes.

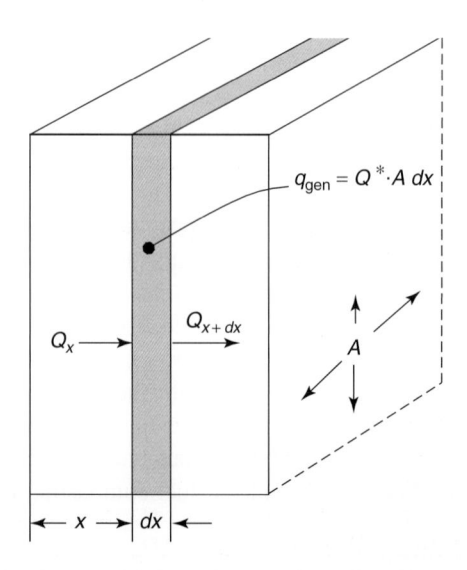

Figure 1. Elemental volume for one-dimensional heat-conduction analysis.

Considering Figure 1 then for the body element of thickness dx, the following energy balance can be made, if Q^* is the heat generated per unit volume

Heat conduction into left face(Q_x)

+ heat generation within element(Q^*)

= rate of heat storage $\left(\rho c_p A \dfrac{\partial T}{\partial t} dx\right)$

+ heat released by right face(Q_{x+dx})

$$-\lambda A \dfrac{\partial T}{\partial t}\bigg|_x + Q^* A\, dx = \rho c_p A \dfrac{\partial T}{\partial t}\, dx - \lambda A \dfrac{\partial T}{\partial t}\bigg|_{x+dx} \quad (6)$$

By the use of Taylor's expansion, these heat quantities give the following relations for one-dimension:

$$-\lambda A \dfrac{\partial T}{\partial x} + Q^* A\, dx = \rho c_p A \dfrac{\partial T}{\partial t}\, dx$$
$$- A\left[\lambda \dfrac{\partial T}{\partial x} + \dfrac{\partial}{\partial x}\left(\lambda \dfrac{\partial T}{\partial x}\right) dx\right]$$
$$\dfrac{\partial}{\partial x}\left(\lambda \dfrac{\partial T}{\partial x}\right) + Q^* = \rho c_p A \dfrac{\partial T}{\partial t} \quad (7)$$

In the space, one obtains analogously

$$\rho c_p \dfrac{\partial T}{\partial t} = \dfrac{\partial}{\partial x}\left(\lambda \dfrac{\partial T}{\partial x}\right) + \dfrac{\partial}{\partial y}\left(\lambda \dfrac{\partial T}{\partial y}\right) + \dfrac{\partial}{\partial z}\left(\lambda \dfrac{\partial T}{\partial z}\right) + Q^*$$
$$\rho c_p \dfrac{\partial T}{\partial t} = \nabla(\lambda \nabla T) + Q^* \quad (8)$$

For constant thermal conductivity equation (8) is written as

$$\dfrac{1}{\alpha}\dfrac{\partial T}{\partial t} = \dfrac{\partial^2 T}{\partial x^2} + \dfrac{\partial^2 T}{\partial y^2} + \dfrac{\partial^2 T}{\partial z^2} + \dfrac{Q^*}{\lambda}$$
$$\dfrac{1}{\alpha}\dfrac{\partial T}{\partial t} = \nabla^2 T + \dfrac{Q^*}{\lambda} \quad (9)$$

where the quantity $\alpha = \lambda/\rho c_p$ is called the thermal diffusivity of the material used and is seen to be a physical property of this material. A high value of α could result either from a high value of thermal conductivity or from a low value of thermal heat capacity ρc_p. The latter means that less of the energy moving through the cell is absorbed and used to raise the temperature of the cell material.

Sometimes, it is useful to write equation (9) in terms of cylindrical or spherical coordinates. The transformation reads:

cylindrical coordinates:

$$\dfrac{1}{\alpha}\dfrac{\partial T}{\partial t} = \dfrac{\partial^2 T}{\partial r^2} + \dfrac{1}{r}\dfrac{\partial T}{\partial r} + \dfrac{1}{r^2}\dfrac{\partial^2 T}{\partial \phi^2} + \dfrac{\partial^2 T}{\partial z^2} + \dfrac{Q^*}{\lambda}$$

spherical coordinates:

$$\dfrac{1}{\alpha}\dfrac{\partial T}{\partial t} = \dfrac{1}{r}\dfrac{\partial^2 rT}{\partial r^2} + \dfrac{1}{r^2 \sin\theta}\dfrac{\partial}{\partial \theta}\left(\sin\theta \dfrac{\partial T}{\partial \theta}\right)$$
$$+ \dfrac{1}{r^2 \sin^2\theta}\dfrac{\partial^2 T}{\partial \theta^2} + \dfrac{Q^*}{\lambda} \quad (10)$$

and the corresponding elemental volume is shown in Figure 2.

A very important quantity is the thermal conductivity λ. Since the thermal energy can be transported in a solid either by lattice vibrations or by transport of electron gas, there are two different ranges of the values of this constant. Since the first mode of heat transport is not as large as the electron transport, good electrical conductors are also good heat conductors. In contrast to it, electric insulators are usually good heat insulators, i.e., poor heat conductors. Examples of thermal conductivities of some relevant materials are given in Table 1. In general, thermal conductivity is a function of temperature.

More thermal conductivity constants are given in Refs. [1, 2]. The thermal conductivity of some important materials for fuel cell technology such as carbon or electrolyte membrane cannot be given properly. For carbon, it varies in a wide range between the values of 0.07 W m^{-1} K^{-1} (soot) and 0.5 W m^{-1} K^{-1} (charcoal) to the very high values of graphite. Concerning graphite, there is a strong dependency on the heat flow direction. At a heat flow parallel to the crystallographic c-axis, graphite has a thermal conductivity, λ, of 5.7 W m^{-1} K^{-1} (300 K) and at a heat flow perpendicular to the c-axis a value of 2000 W m^{-1} K^{-1} at 300 K.[4] For the membrane, depending on the cell type and the working conditions, it has values of water (liquid direct-methanol fuel cell (DMFC)] or of plastic materials in dry conditions.

1.1.2 Convection heat transfer

The term convection heat transfer is used in a fluid because of a combination of conduction within the fluid and heat transport due to the fluid motion itself. This can be expressed by extending equation (9) in the form of equation (11), $u_{x,y,z}$ being

$$\dfrac{\partial T}{\partial t} + u_x \dfrac{\partial T}{\partial x} + u_y \dfrac{\partial T}{\partial y} + u_z \dfrac{\partial T}{\partial z}$$
$$= \alpha\left(\dfrac{\partial^2 T}{\partial x^2} + \dfrac{\partial^2 T}{\partial y^2} + \dfrac{\partial^2 T}{\partial z^2}\right) + \dfrac{Q^*}{\rho c_p} \quad (11)$$

the fluid velocity components (cf. also Figure 3).

The consequence is that this kind of heat transfer is more complicated and makes at least an understanding of

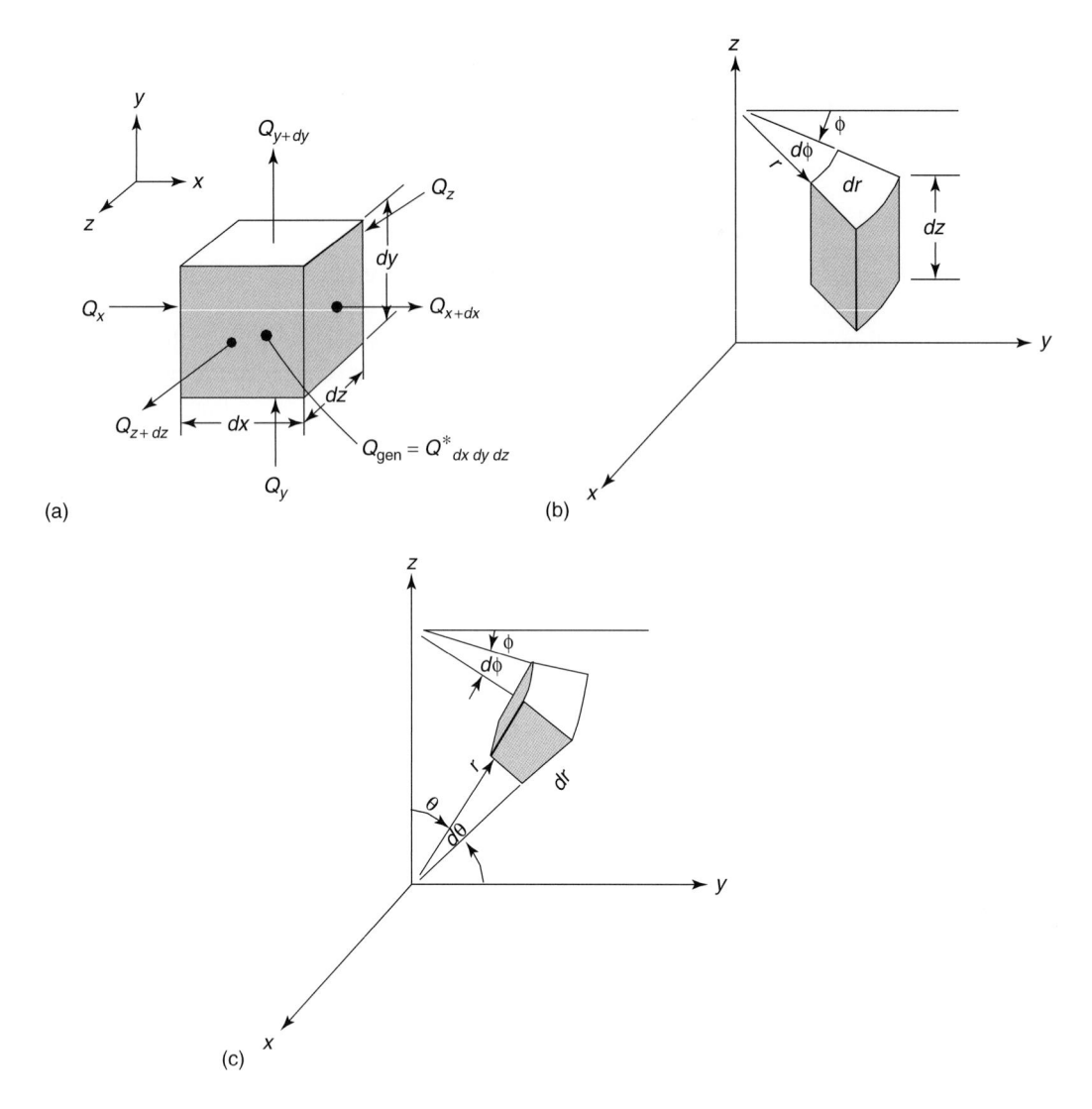

Figure 2. Elemental volume for three-dimensional heat-conduction analysis; coordinates. (a) Cartesian, (b) cylindrical and (c) spherical.

Table 1. Thermal conductivity of various materials at 0 °C.

Material	Thermal conductivity (λ, W m^{-1} K^{-1})
Iron (pure)	73
Carbon steel (1% C)	43
Chrome-nickel steel (18% Cr, 8% Ni)	16
Copper (pure)	385
Graphite	170
Ceramic	2
Glass	0.80
Water	0.56
Hydrogen gas	0.175
Air	0.024
Water vapor (saturated)	0.021
Carbon dioxide	0.015

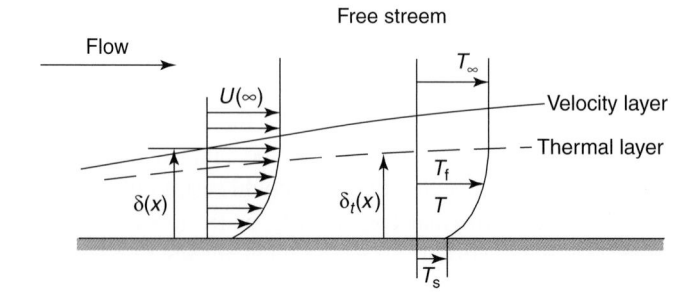

Figure 3. Velocity and thermal boundary layers.

some of the principles of fluid dynamics necessary. The most fundamental principle in this respect is the boundary layer concept. The thermal boundary layer can be defined similarly to the velocity boundary layer. It concerns the region where the effect of the wall on the motion of the fluid is significant (cf. Figure 3). Outside the boundary layer, it is assumed that the effect of the wall may be neglected. The limit of the layer is usually taken to be at the distance from the wall at which the fluid velocity is equal to the main stream value, being approximately between 95 and 99%. The flow conditions can be both laminar or turbulent. If the solid is maintained at a temperature T_s, which is different from the fluid temperature T_f, a variation in the fluid temperature is observed, i.e., the fluid temperature changes from T_s to T_f far away from the wall, designated as $T_{f\infty}$, the temperature of the free stream as illustrated in Figure 3. The thermal boundary layer is generally not coincident with the velocity boundary layer. Even if the physical mechanism of heat transfer at the wall itself is a conduction process, the temperature gradient on the wall is dependent on the rate at which the fluid carries the heat away and is a function of the flow field. The whole process is therefore very complex. All the complexities involved in equation (11) may be lumped together in terms of single parameters by the introduction of Newton's law of cooling in the simple form

$$Q = hA(T_s - T_{f\infty}) \quad (12)$$

The quantity h is the convection heat transfer coefficient or unit thermal conductance. This means that this coefficient is not a material property like the thermal conductivity λ. It is a complex function of the composition of the fluid, the geometry of the solid surface and the hydrodynamic conditions in the surroundings of the surface. Usually, it must be determined experimentally. The typical range of convection heat transfer coefficient values is indicated in Table 2.

If we compare equations (5) and (12), we recognize that the temperature gradient $\partial T/\partial x$ is at least partially included in the heat transfer coefficient, h. From Table 2, it can be anticipated that, in contrast to the thermal conductivity, the heat transfer coefficient can only be estimated approximately within some range, which in fact reveals the whole complexity included in this quantity.

1.2 Heat transfer without change of state

Newton's law of convective heat transfer, e.g., equation (12), postulates a linear relationship between the heat flux and the temperature difference between the wall and the bulk of the fluid. It may be assumed that both phases have the same temperature at their boundary, which varies from the value at the wall to that in the bulk across the thermal boundary layer. In the immediate vicinity of the boundary, heat is transferred through the fluid flow perpendicularly to the wall by conduction. For laminar flow, only the molecular conductivity contributes to this process. In turbulent flows, a very thin viscous laminar sublayer exists in which the molecular conduction of heat occurs. Beyond this layer, velocity fluctuations normal to the main stream direction predominantly affect the heat exchange, which leads to an increase of the effective conductivity by orders of magnitude. Thus, the main heat resistance in the case of turbulent flow exists in the viscous sublayer. Therefore, the temperature gradient is very high near the wall and flat in the bulk. For laminar flow the temperature profile is well rounded and the heat conduction is also significant inside the fluid, but if the main flow is turbulent, a practically constant temperature $T_{f\infty}$ can be calculated within the fluid. In both cases the driving temperature differences, ΔT, in equation (12) results from the difference between the wall temperature and the bulk temperature averaged across the cross section. In turbulent flow the bulk temperature is close to the temperature determined at some distance from the wall. For this case, Figure 4 schematically shows the temperature profile.

The form of the temperature profile in Figure 4 can, in principle, be derived from the relation

$$\lambda \left(\frac{\partial T}{\partial x}\right)_{wall} \approx \lambda_{fl} \left(\frac{\partial T}{\partial x}\right)_{inner} \quad (13)$$

The effect of the high value of the coefficient λ_{fl} in the inner of a turbulent flow is that in the steady-state case no high temperature gradients can exist there, so that $(\partial T/\partial x)_{wall}$ will be higher than $(\partial T/\partial x)_{inner}$ by some decades.

Equation (12) can also be used for laminar core flow in which there is no uniform internal temperature $T_{f\infty}$. $T_{f\infty}$ can then be defined, for example, as the fluid temperature averaged over the entire flow cross-section, but other definitions are also possible. If formulas for the convection heat transfer coefficient h are used, the respective definition of $T_{f\infty}$ must be taken into consideration because it influences the

Table 2. Approximate values of convection heat transfer coefficient.

Mode	Heat transfer coefficient (h, W m^{-2} K^{-1})
Free convection in air	5–25
Free convection in water	500–1000
Forced convection in air	10–500
Forced convection in water	100–15 000
Boiling water in a container	2500–35 000
Boiling water flowing in a tube	5000–100 000
Condensation of water vapor	5000–25 000

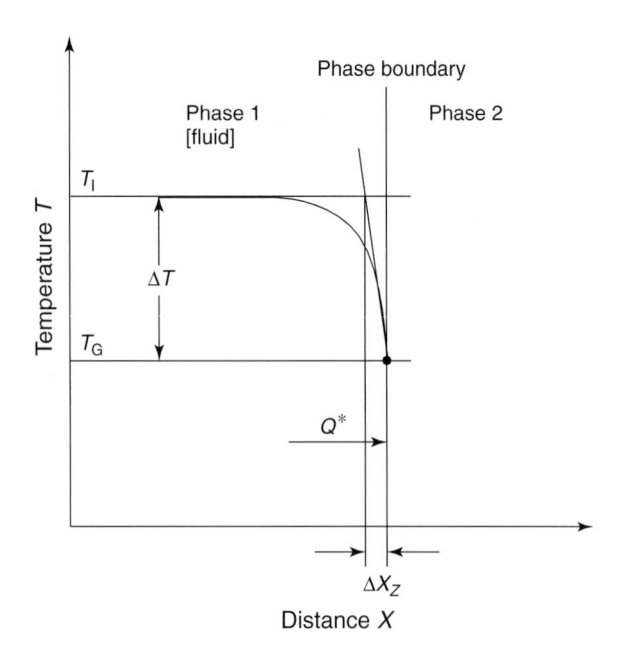

Figure 4. Temperature profiles for heat transfer at turbulent flow conditions; T_I: inner fluid temperature, T_G: boundary wall temperature.

value of h. In the contrary, an estimation can be made for h for a turbulent core flow, i.e., at practically constant $T_{f\infty}$ on the basis of a simple model assuming pure heat conduction in a wall zone of the fluid, which – provided that the wall is not curved too much – results in a nearly linear temperature profile in the laminar sublayer which changes into the zone of constant temperature at $T = T_I$, as indicated for one-dimension in Figure 4. The fictitious thickness, Δx, of the wall zone results from approximation (12) in the form of

$$Q = hA\Delta T = \lambda A \frac{\Delta T}{\Delta x}, \quad h = \frac{\lambda}{\Delta x} \qquad (14)$$

From equation (14), it is possible to estimate the order of magnitude of Δx and to compare it with that of the boundary layer thickness δ as known from fluid dynamics. The same also applies to influencing the layer thickness and thus h by fluid mechanics measures. Thus, for example, the quantity of h for free convection in air in Table 2 results from the following estimation: assuming a boundary layer for air flows of the order of $\Delta x \approx \delta \approx 1$ mm, a value of $h = 0.024/0.001 = 24$ W m^{-2} K^{-1} is obtained from equation (14) with $\lambda_{air} = 0.024$ W m^{-1} K^{-1} (cf. Table 1). However, this value should only be considered as a guide value. Generally, in order to calculate the heat rate by means of equation (12), the heat transfer coefficient h must be known. Therefore, for the representation of the dependence of the heat transfer coefficient, h, on the decisive parameters of influence in the practice the dimensionless physical parameters are used.

For a large variety of technical configurations, such as tubes, channels, heat exchangers, plates, etc., h has been determined experimentally or theoretically. To reduce the number of empirical equations, the results are usually presented in a dimensionless form. Dimensionless groups can be deduced, which contain the characteristic quantities governing the heat transfer process. These are thermal and fluid properties, characteristic velocities and characteristic length scales. Six dimensionless groups important for the fuel cell technology can be found in the literature (see Refs. [1, 2]) and are mentioned here: the Prandtl number (Pr), the Reynolds number (Re), the Grashof number (Gr), the Nusselt number (Nu), the Schmidt number (Sc), and the Sherwood number (Sh). The definitions are:

$$\mathrm{Re} = \frac{u_\infty l}{\nu}, \quad \mathrm{Gr} = \frac{gl^3\beta(T_s - T_{f\infty})}{\nu^2}, \quad \mathrm{Pr} = \frac{\nu}{\alpha}$$

$$\mathrm{Nu} = \frac{hl}{\lambda}, \quad \mathrm{Sc} = \frac{\nu}{D}, \quad \mathrm{Sh} = \frac{\gamma l}{D} \qquad (15)$$

In equations (15), β is the volume coefficient of thermal expansion, u_∞ the free-stream velocity, ν the kinematic viscosity, l the characteristic distance, g the acceleration of gravity and γ the material transfer coefficient.

The forced convective flow is described by the Reynolds number, Re. This number represents a measure of the magnitude of the inert forces in the fluid to the viscous forces. Its values indicate either the laminar or the turbulent flow. As an example, for a tube the flow is turbulent for Re > 2400. This value can, however, be different for other fuel cell components under consideration. The Nusselt number, Nu, represents the dimensionless heat transfer coefficient h. It is the ratio of the convective heat flux to heat conducted through a fluid layer of the thickness l. Knowledge of this quantity enables the determination of the heat flux through the wall. The convective mass transfer can be analogously treated as the convective heat transfer. In this case a Nusselt number of the second type, the Sherwood number, Sh, is defined and instead of h it describes the mass transfer coefficient γ.

Applying the three-fold analogy between momentum, heat and mass transfer, two further groups result. Both members are composed from physical properties only. For the convective heat transfer the Prandtl number, Pr, influences the heat rate. The Prandtl number can be understood as the ratio of momentum diffusivity to temperature diffusivity. For air at ambient conditions Pr = 0.7. With increasing Pr the convective heat transfer improves. The same meaning as the Prandtl number for the heat transfer, the Schmidt number, Sc, has for the mass transfer. Analogously, it is the ratio of momentum diffusivity mass diffusivity. With increasing Sc, the convective mass transfer improves. In a resting fluid or at low velocities buoyancy effects may occur, which initiate a free natural convective flow or conjugated natural and forced convection. In this case, the heat transfer depends on the Grashof number, Gr, which is the ratio of the buoyancy forces to inert forces and the square of the friction forces.

As mentioned above, the listed groups are suitable for providing the user with information about the heat and mass transfer coefficients. Empirical formulas for the different technical components can be found in the literature.[1–3]

$$\text{Forced convective heat transfer}: \quad \text{Nu} = f_1(\text{Re}, \text{Pr})$$
$$\text{Natural convection}: \quad \text{Nu} = f_2(\text{Gr}, \text{Pr})$$
$$\text{Forced convective mass transfer}: \quad \text{Sh} = f_1(\text{Re}, \text{Sc})$$
$$\text{Natural convective mass transfer}: \quad \text{Sh} = f_2(\text{Gr}, \text{Sc})$$

The functions f_1 and f_2 can frequently be expressed as simple relationships

$$\text{Nu} = a\text{Re}^n\text{Pr}^m + b$$
$$\text{Nu} = c(\text{Gr}, \text{Pr})^p + d \quad (16)$$

with the dimensionless constants a, b, c, d, m, n and p.

For the heat transfer equations, the same functions apply as for the mass transfer. In the case of the established laminar flow in tubes and channels, the Nusselt number, Nu, and the Sherwood number, Sh, become independent of the Reynolds and Prandtl numbers. Thus, we have Nu = const. and Sh = const. To set this constant, a value of const. = 4 is a good approximation for the flow channels of a fuel cell.

1.3 Heat transfer with change of state

In the LTFCs, processes also take place which involve a change of state. This includes, in particular, the condensation and evaporation of the water contained in the cell. Of particular significance in this connection is the evaporation due to heat transferred by convection according to equation (12). The temperature difference ΔT, however, is determined by the two temperatures T_s and T_b, where T_s is the solid temperature and T_b is the boiling temperature. The convection heat transfer coefficient h is state-dependent and its value can vary over several decades. During evaporation of the water, its value changes depending on the state of boiling, whether subcooled nucleate boiling or film boiling occurs.

If water vapor or a vapor/gas mixture comes into contact with a surface, whose temperature T_s is below the temperature T_b of the vapor, it is condensed on this surface and the condensation heat must be removed through that surface. The velocity of this process and thus the convection heat transfer coefficient h is determined for pure vapors by the heat transport from the fluid surface of the condensed substance through this substance to the interface. In the case of a vapor/gas mixture, the vapor must first pass through a gas-rich layer to the fluid surface, which is a rate determining step significantly slowing down the overall process. Even in the case of a small inert gas content (about 1%), therefore, the h value significantly decreases.

2 GENERAL ENERGY BALANCE EQUATION

In the general energy balance equation for the accumulation of energy in the cell (3), the first term on the right represents the heat generated by the cell. Both the current, I, and the cell voltage, E_c, are considered to be positive. The cell voltage, E_c, may be written as

$$E_c = -\frac{\Delta G}{nF} - \sum |\eta| - \sum IR_i = E_0 - \sum |\eta| - \sum IR_i \quad (17)$$

The single components of equation (17) represent Gibb's free energy (reversible cell voltage), both anodic and cathodic overvoltages and ohmic losses. In the absence of the two last losses, which of course only have a hypothetical value, the expression $\Delta H/nF + E_c$ in equation (3) as the difference between the reaction enthalphy of the fuel cell reaction and the reversible cell voltage is representative of the reaction entropy. In contrast to the usual chemical combustion, both the anodic and cathodic parts of the full cell reaction are separated in space from each other. For this reason, the entropy of the cell reaction may also be divided into two parts, which should be considered separately inside the membrane electrode assembly (MEA) structure. Generally, however, due to the voltage losses, irreversible heat is generated internally and must be removed. This heat is a source of the reduction of process efficiency. The rate of the vaporization term R_{vap} in equation (4) depends firstly on the difference between the total system pressure, P, and the water partial pressure, P_{H_2O} and secondly on the difference between the actual water partial pressure P_{H_2O} and the fully saturated water atmosphere with the water partial pressure, $P_{H_2O}^{sat}$.

2.1 Energy balances for single phases

The following considerations concern the conservation of energy in the bipolar plates and distributor structures. For the general case of a fluid flowing in a channel, equation (11) is valid with Q^* as the heat due to chemical reactions and other volume-related sources/sinks. Furthermore, the influence of the heat conduction in the direction of flow can be estimated in comparison to heat transfer by convection. The characteristic variable here is the Peclet number, Pe:

$$\text{Pe} = \text{Re} \cdot \text{Pr} = \frac{u_\infty l}{\alpha} \quad (18)$$

With large Pe numbers, the heat conduction can be neglected. The following is obtained as an estimate for the polymer electrode membrane fuel cells (PEMFCs) and DMFCs: Pe_{gas} between 10 and 150; Pe_{fluid} between 10 and 450.

The heat transfer from the solid to the gas and the enthalpies given off or taken up by the educt and product streams can be taken into consideration as sources or sinks in equation (11). The heat transfer coefficient, h, can be determined from the Nusselt number (cf. also equation (15)). Empirical correlations exist for the Nusselt number, Nu, both for turbulent and laminar flow. In a laminar flow along tubes and plates, a value of $\text{Nu} = 3.66$ holds for a developed thermal layer and with constant heat flux $\text{Nu} = 4.36$. Neither value is given for the case of the fuel cell, so that a mean value of $\text{Nu} = 4$ is used for the calculations.

In addition to the mass flows leaving the channel which are spent and the mass flows which are produced in the catalyst layer before entering the channel, heat is exchanged at the interface of the solid and channel and water is evaporated or condensed up to saturation of the air. The water produced in the catalyst layer is located in the diffusion layer and the catalytic layer in accordance with the local temperature as a mixed gas/liquid phase and is transported into the channel in this form. Due to the temperature prevailing there, a phase transition may occur again in the channel. The following equations are, therefore, obtained for the convective fractions of the heat conduction in the boundary layers, e.g., for the fuel and air channels:

$$Q_{\text{conv, fuel}} = \frac{I}{n_{\text{fuel}}F} M_{\text{fuel}} c_{p,\text{fuel}} T_{\text{f, s}} + \sum N_j A c_{pj} T_{\text{s, f}}$$

$$Q_{\text{conv, ox}} = \frac{I}{n_{\text{ox}}F} M_{\text{ox}} c_{p,\text{ox}} T_{\text{f, s}} + \sum N_j A c_{pj} T_{\text{s, f}} \quad (19)$$

The following descriptive mass balance equations are obtained for the heat balance in the cell channels outside the MEA structure due to the described effects:

$$\rho c_p \left(u_x \frac{\partial T_f}{\partial x} + u_y \frac{\partial T_f}{\partial y} + u_z \frac{\partial T_f}{\partial z} \right) dV$$
$$= -Q_{\text{conv, fuel}} - K_\lambda (T_s - T_f) dA$$

$$\rho c_p \left(u_x \frac{\partial T_f}{\partial x} + u_y \frac{\partial T_f}{\partial y} + u_z \frac{\partial T_f}{\partial z} \right) dV$$
$$= -Q_{\text{conv, ox}} - \left[K_\lambda (T_s - T_f) + \frac{N_{H_2O}}{M_{H_2O}} H_{\text{vap}} \right] dA \quad (20)$$

The irreversible losses in the solid phase of the cell lead to heat production composed of the entropy losses, the losses due to overvoltages and heating as a result of the ohmic resistance. This corresponds to the difference between the theoretically achievable output, i.e., the reaction enthalpy and the actual output.

$$Q_{\text{loss}} = -\frac{I}{nF} \Delta H - E_c I + N_{i,\text{inner}} \frac{A}{M_i} \Delta H$$
$$= -\frac{T_s \Delta S I}{nF} + I \sum |\eta| + I^2 R \quad (21)$$

The term $N_{i,\text{inner}} \Delta H$ is an additional term characterized by the internal chemical secondary reaction of the fuel involved as a conversion factor, such as the chemical combustion of the permeated methanol at the DMFC cathode.

The solid energy balance takes account of the three-dimensional heat exchange by conduction between the cells and on the cell level as well as convective heat transport by the liquid water flowing through the balance space. With regard to equations (19) and (21), as a whole the mass balance equation is obtained

$$\rho c_{p,s} \frac{\partial T_s}{\partial t} dV = \lambda_{\text{eff}} \nabla^2 T_s \, dV - \frac{N_{H_2O}}{M_{H_2O}} \Delta H_{\text{vap}} dA$$
$$+ \sum K_\lambda (T_f - T_s) dA + \sum Q_{\text{conv}} + Q_{\text{loss}}$$
$$(22)$$

Heat conduction and radiation occurs within the solid structure. However, with low-temperature fuel cells the radiation can be neglected due to these low temperatures. λ_{eff} is an effective value composed of the individual MEA components. The anode and cathode essentially consist of graphite mesh and flowing medium. The catalyst and polymer can be neglected in the heat conduction. In calculating the effective value, it is assumed that the two components have the same temperature. Weighting is performed according to the mass fractions. It is evident that the influence of the thermal conductivity of graphite predominates in the anodic and cathodic diffusion and catalyst layer. The membrane consists of polymer and water. Since water has much better thermal conductivity, the conductivity of water is used for the membrane. The effective conductivity of the MEA is obtained in accordance with:

$$\frac{\text{thickness}_{\text{tot}}}{\lambda_{\text{eff}}} = \frac{\text{thickness}_{\text{anode}}}{\lambda_{\text{anode}}} + \frac{\text{thickness}_{\text{membrane}}}{\lambda_{\text{membrane}}} + \frac{\text{thickness}_{\text{cathode}}}{\lambda_{\text{cathode}}} \quad (23)$$

If real dimensions are used, a value of about $1.3\,\text{W}\,\text{m}^{-1}\,\text{K}^{-1}$ results.

The heat capacities of the anode and cathode are also obtained from the mass weighting of graphite and water. In the membrane, the heat capacity of water is used. The effective value is obtained from a mass weighting of the anode, cathode and membrane.

2.2 Delocalization of reaction entropy in LTFCs

Due to the splitting of the whole fuel combusting reaction into the anodic and cathodic partial reactions, one has to take into consideration that the reaction entropy also must be divided into two parts. The entropic local heat is different from the total heat effect, which is well known and consists of the sum of both the anodic and cathodic heat evolution. However, their asymmetry may be substantial. Based on the theory developed by Ratkje et al.,[5, 6] we can perform a calculation for one of the interesting cases for PEMFCs, namely the DMFC.

The DMFC is schematically given by

$$(\text{Pt/Ru})\text{C}|\text{CH}_3\text{OH}(c_{\text{CH}_3\text{OH}}, T_a)|\text{Nafion}117$$
$$|\text{O}_2(P_{\text{O}_2}, T_c)|\text{C(Pt)} \quad (24)$$

where T_a and T_c are the anode and cathode temperatures, respectively. To find the local entropy heat effects, we write the single electrode reactions

Anode: $\text{CH}_3\text{OH}_{(c)} + \text{H}_2\text{O}_{(l)} \longrightarrow \text{CO}_{2(g)} + 6\text{H}^+_{(l)} + 6\text{e}^-$

Cathode: $3/2\text{O}_{2(g)} + 6\text{H}^+_{(l)} + 6\text{e}^- \longrightarrow 3\text{H}_2\text{O}_{(l)} \quad (25)$

Usually, the methanol concentration in the analyte is about 1 M, i.e., the concentration ratio of water/methanol is approximately 55 : 1. For this reason, even if the methanol crossover across the cell has a very big influence on the cathodic overvoltage, its transported entropy can be considered to be negligible in comparison to that of water. Therefore, across the membrane we only consider a reversible co-transport of protons and water from the left (l) to the right (r), depending on the membrane transference coefficient of water t_w, better known as the drag factor or drag coefficient in the literature about fuel cell membrane technology:

$$\text{H}^+_{(l)} + t_w\text{H}_2\text{O}_{(l)} \longrightarrow \text{H}^+_{(r)} + t_w\text{H}_2\text{O}_{(r)} \quad (26)$$

The total cell reaction is thus

$$\frac{1}{6}\text{CH}_3\text{OH} + \left(\frac{1}{6} + t_w\right)\text{H}_2\text{O}_{(l)} + \frac{1}{4}\text{O}_2$$
$$\longrightarrow \frac{1}{6}\text{CO}_2 + \left(\frac{1}{2} + t_w\right)\text{H}_2\text{O}_{(r)} \quad (27)$$

The reversible heat balance of an electrode defines the Peltier heat, which is the difference between the reaction enthalpy, ΔH, and the free Gibb's reaction enthalpy, ΔG. This is the heat that must be supplied to the electrode compartment to maintain constant temperature when current is passing. Using the same procedure as in Ref. [5], according to equation (27), the Peltier heat at the anode π_a is given as

$$\frac{d\pi_a}{dt} = T_a\left[S^*_{\text{HM}} + \frac{1}{6}S_{\text{CO}_2,\text{g}} - \left(\frac{1}{6} + t_w\right)S_w - \frac{1}{6}S_{\text{CH}_3\text{OH}}\right]\frac{I}{F} \quad (28)$$

Accordingly, the Peltier heat at the cathode π_c is then

$$\frac{d\pi_c}{dt} = T_c\left[-\frac{1}{4}S_{\text{O}_2,\text{g}} - S^*_{\text{HM}} + \left(\frac{1}{2} + t_w\right)S_w\right]\frac{I}{F} \quad (29)$$

In equations (28) and (29), the symbol S defines the corresponding entropy values: S_w = partial molar entropy of water in bulk solution, S_{O_2} = oxygen, $S_{\text{CH}_3\text{OH}}$ = methanol and S^*_{HM} is the transported entropy of protons in the membrane. The entropies are functions of temperature and pressure, the membrane transference coefficient of water is a function of temperature only (full membrane saturation with water is assumed in liquid DMFC). Whereas, the entropy pressure dependence can be simply expressed as the known function

$$\Delta S = -R\ln\frac{p_2}{p_1} \quad (30)$$

its temperature dependence may be calculated on the basis of published thermodynamic data[7, 8] as expressions of the form $S_i(T) = \alpha l_iT + \beta_i$. All these values are given in Table 3.

Table 3. Reaction entropies and water transference coefficient: temperature dependencies.

$t_w(T) = 1.337 \times 10^{-4}(T - 273)^2 + 9.116 \times 10^{-3}(T - 273) + 1.791$
$S_{\text{H}^+} = 195\,\text{J}\,\text{mol}^{-1}\,\text{K}^{-1}$
$S_{\text{O}_2}(T) = 8.55 \times 10^{-2} \cdot T + 180\,\text{J}\,\text{mol}^{-1}\,\text{K}^{-1}$
$S_{\text{H}_2\text{O}}(T) = 0.218 \cdot T + 4.974\,\text{J}\,\text{mol}^{-1}\,\text{K}^{-1}$
$S_{\text{CO}_2}(T) = 0.113 \cdot T + 180\,\text{J}\,\text{mol}^{-1}\,\text{K}^{-1}$
$S_{\text{MeOH}}(T) = 0.235 \cdot T + 56.71\,\text{J}\,\text{mol}^{-1}\,\text{K}^{-1}$

3 WATER AND HEAT MANAGEMENT FOR PEMFCs (H_2 AND DMFCs)

Proper water and heat management is essential for obtaining high power density performance at high energy efficiency for PEMFCs. There are three important aspects concerning the thermal management of the PEMFC: drying out the polymer membrane at the hydrogen cells,[9, 10] engineering aspects for the design of the cell cooling system[11] and the water crossover at the DMFC, connected with the methanol crossover.[12] It was shown that at high current densities ohmic loss in the membrane accounts for a large fraction of the voltage losses in the cell and back diffusion of water from the cathode side of the membrane is insufficient to keep the membrane hydrated, i.e., conductive.[10] For this reason the anode stream must be humidified. Similarly, heat management modeling allows the influence of some geometric and working parameters to be studied (size, shape, number of channels, inlet temperature, flow rates of cooling water and air, gas pressures) on stack performance.[11] In the following, some examples of the thermal modeling of the cell and process analysis will be given.

A fuel cell stack consists of a number of single cells connected electrically in series. Typically the stack is made up of different components that must be considered during the calculation procedure.

3.1 Heat transfer in PEMFCs

Thermal cell modeling can be performed on three different levels: single cell modeling, stack modeling and finally, process analysis.

The most important part of the single cell is the MEA, the structure of which is given in Figure 5. As an example, the structure of a DMFC is shown here. For the same structure the principle of the corresponding MEA thermal mode is given in Figure 5(b).

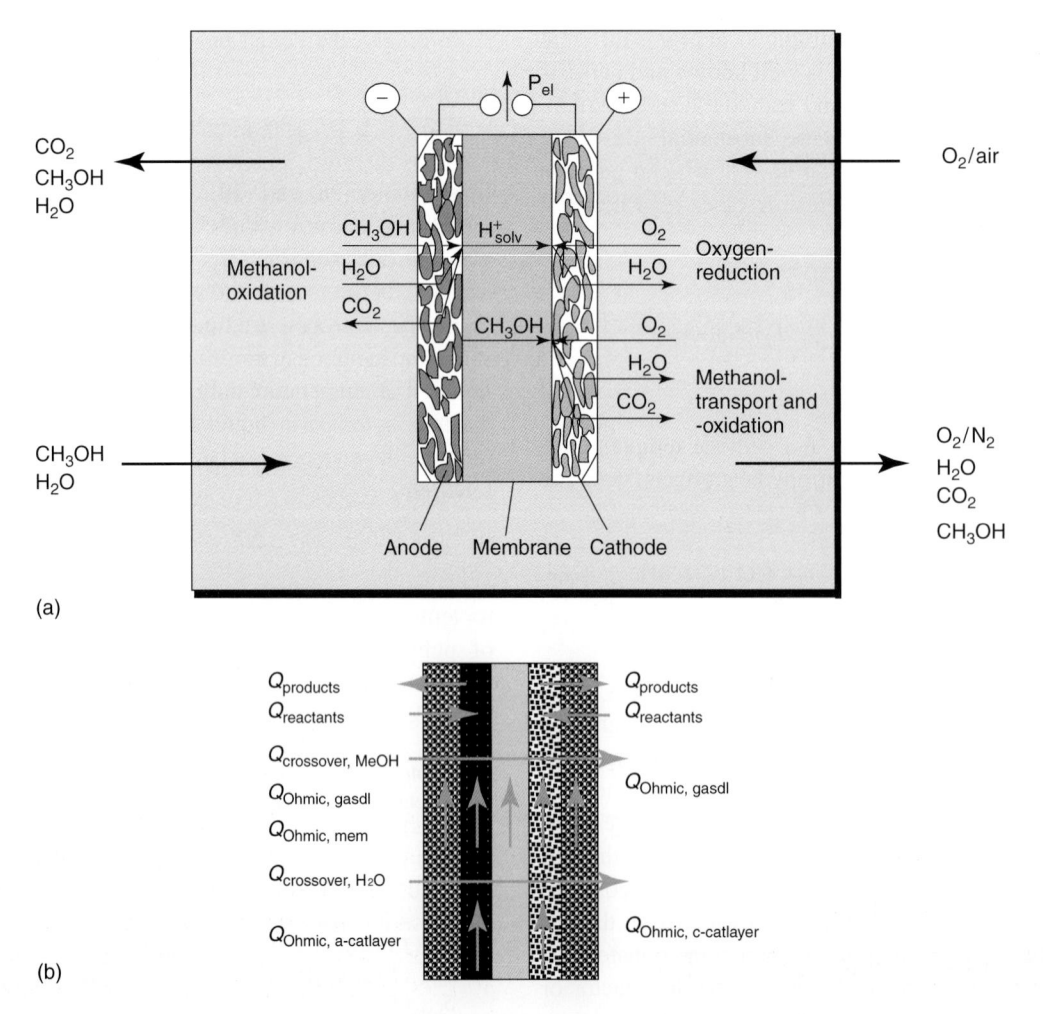

Figure 5. Schematic of the DMFC cell. (a) MEA functional scheme, (b) thermal model. (Reproduced from Argyropoulos *et al.* (1999)[16] with permission from Elsevier Science.)

The MEA consists of two catalyst electrode layers, one as cathode and other as anode, attached to the PEMFC, which serves as the cell separator and electrolyte. Each electrode is divided into two sublayers: the catalyst layer, directly contacting the membrane and the outer diffusion layer (backing layer), which serves both as the current collector and the mass distributor. The MEA is sandwiched between the bipolar plates, which have a ribbed structure, forming the mass and electricity distributing system. The schematic shown in Figure 5 represents the MEA between the channel area of the bipolar plate. The heat generated in the MEA structure consists of the following parts (cf. Figure 5b). Ohmic heat is produced in both the anodic and cathodic diffusion and catalyst layers as well as in the membrane. Joule heating is the only heat source inside the membrane. In the remaining MEA sublayers conditions are more complex. The anode and cathode mass streams of the products and reactants act as heat transfer fluids and determine the amount of heat carried away from or into the cell. Further heat is transferred by the electroosmotic flow of water and methanol across the membrane from anode to cathode. All the governing energy balance equations for the MEA layers (diffusion, catalytic, membrane) for one-dimension are given by equation (31). In these equations the following assumptions are made:

The pressure difference between the actual water vapor pressure and the saturation pressure is the driving force for the condensation/evaporation of water. In the catalyst layers, both the electronic and ionic currents are present. In the anodic catalyst layer only the fuel (methanol) current is operative, whereas in the cathodic catalyst layer both the oxygen reduction and the parasitic methanol oxidation currents are heat sources. The local entropy is given by equations in Section 2.2.

Diffusion layer:

$$\bar{\rho}c_p \frac{dT}{dt} = -\sum_{j=1}^{N}\left(N_j c_{pj}\frac{dT}{dx}\right) + \bar{\lambda}\frac{d^2T}{dx^2} + \frac{\alpha_0 \gamma}{RT}\Delta H_{vap}$$
$$\times (P_{H_2O} - P_{H_2O}^{sat}) + \sigma\left(\frac{dE}{dt}\right)^2$$

Catalyst layer cathode:

$$\bar{\rho}c_p \frac{dT}{dt} = -\sum_{j=1}^{N}\left(N_j c_{pj}\frac{dT}{dx}\right) + \bar{\lambda}\frac{d^2T}{dx^2} + \frac{\alpha_0 \gamma}{RT}\Delta H_{vap}$$
$$\times (P_{H_2O} - P_{H_2O}^{sat}) + \frac{T\Delta S_{O_2}}{4F}(j_{O_2} + j_{Me})$$
$$- \eta_{O_2}j_{O_2} + \eta_{Me}j_{Me} - \frac{\Delta H_{Me}}{6F}j_{Me} + \sum \sigma\left(\frac{dE}{dx}\right)^2$$

Catalyst layer anode:

$$\bar{\rho}c_p \frac{dT}{dt} = -\sum_{j=1}^{N}\left(N_j c_{pj}\frac{dT}{dx}\right) + \bar{\lambda}\frac{d^2T}{dx^2}$$
$$+ \frac{\alpha_0 \gamma}{RT}\Delta H_{vap}(P_{H_2O} - P_{H_2O}^{sat}) + \frac{T\Delta S_{Me}}{6F}j_{Me}$$
$$+ \eta_{Me}j_{Me} + \sum \sigma\left(\frac{dE}{dx}\right)^2$$

Membrane:

$$\bar{\rho}c_p \frac{dT}{dt} = -\sum_{j=1}^{N}\left(N_j c_{pj}\frac{dT}{dx}\right) + \bar{\lambda}\frac{d^2T}{dx^2}$$
$$+ \sigma_{el}\left(\frac{dE_{el}}{dx}\right)^2 \tag{31}$$

The system of equation (31) can only serve as the definition of the heat sources for the heat management calculation of the cell or the cell stack. Usually, the mass flows N_i of the corresponding components i in the gas phase can be calculated by a system of Stefan–Maxwell equations (not shown here), which describe the multicomponent mass transport in porous medium. Analogously, the mass flow N_i in the liquid phase is usually defined by the Darcy law equations (diffusion layers) or by a combination of the Darcy law with ionic mass flow caused by electrophoresis (catalyst layers and membrane). The value of the material transfer coefficient γ in porous systems is approximately of about $0.001\,m\,s^{-1}$.

As an example of the modeling of a liquid DMFC single cell, a three-dimensional cell model is presented here enabling the local distribution of mass, energy and current density to be calculated.[13] The membrane, the two catalyst layers and the two diffusion layers are combined in this model into an integral MEA element. In this element, the electrochemically required volumes of methanol, water and oxygen enter and the produced volumes of water and CO_2 are discharged. Furthermore, a crossover flow of methanol leaves the fuel channel ($N_{MeOH,\,cross}$) and at the cathode is converted with oxygen into CO_2 and water. It is furthermore assumed that CO_2 diffuses through the membrane ($N_{CO_2,\,cross}$). Water is transported through the membrane by electro-osmotic drag and diffusion ($N_{H_2O,\,cross}$). The performance data and cross-over flows are taken locally from an experimental performance chart in a temperature-, concentration- and pressure-dependent manner. The mass and flow distribution is calculated by solving the mass balances and the three-dimensional Navier–Stokes equations. The presentation of the heat balance corresponds to that in Section 2.1. Applied to the DMFC cell, with complete MeOH conversion at the cathode from equation (19) follows

$$Q_{conv,\,fuel} = \frac{j}{6F}(M_{MeOH}c_{p,MeOH}T_f + M_{H_2O}c_{p,H_2O}T_f)$$
$$+ N_{MeOH,\,cross}Ac_{p,MeOH}T_f + N_{H_2O,\,cross}Ac_{p,H_2O}T_f$$
$$- N_{CO_2}Ac_{p,CO_2}T_s \qquad (32)$$

The complete CO_2 term is composed of the fraction diffusing through the membrane (crossover) and that entering the fuel channel:

$$N_{CO_2,\,tot} = \frac{jM_{CO_2}}{6F} = N_{CO_2} - N_{CO_2,\,cross} \qquad (33)$$

Water evaporates at the cathode up to the level of saturation predetermined by the local temperature ($N_{H_2O,\,gas}$). The water transported through the membrane and that produced in the cathodic catalyst layer (reaction and conversion of permeated methanol) enters into the air channel as a mixture of liquid and gaseous water.

$$N_{H_2O} = \frac{jM_{H_2O}}{2F} + N_{H_2O,\,cross} + 2N_{MeOH,\,cross}$$

$$= N_{H_2O,\,gas} + N_{H_2O,\,fl} \qquad (34)$$

For the convection flow at the interface with the air channel, equation (35) is obtained from equation (19)

$$Q_{conv,\,ox} = \frac{j}{4F}M_{O_2}c_{p,O_2}T_f - N_{H_2O,\,gas}Ac_{p,H_2O,\,gas}T_s$$
$$- N_{H_2O,\,l}Ac_{p,H_2O}T_s - N_{CO_2,\,cross}Ac_{p,CO_2}T_s$$
$$+ N_{MeOH,\,cross}A(c_{p,CO_2}T_f - c_{p,CO_2}T_s) \qquad (35)$$

The irreversible losses in the solid phase of the cell can be determined according to equation (21) using the relation $N_{i,\,inter} = N_{MeOH,\,cross}$.

The physical properties of the mixture of substances in the fuel and air channel are obtained from the weighting of the local mass fractions. Since the fluid in the channels is composed from the gas and liquid, the corresponding main value for the density is estimated from the volume fraction. The local thermal conductivity of the integral MEA element is obtained from equation (23).

Solving the coupled equations (19)–(23) yields the heat distribution in a cell. As an example, Figure 6 shows the local temperature in the MEA, fuel and air channel of a

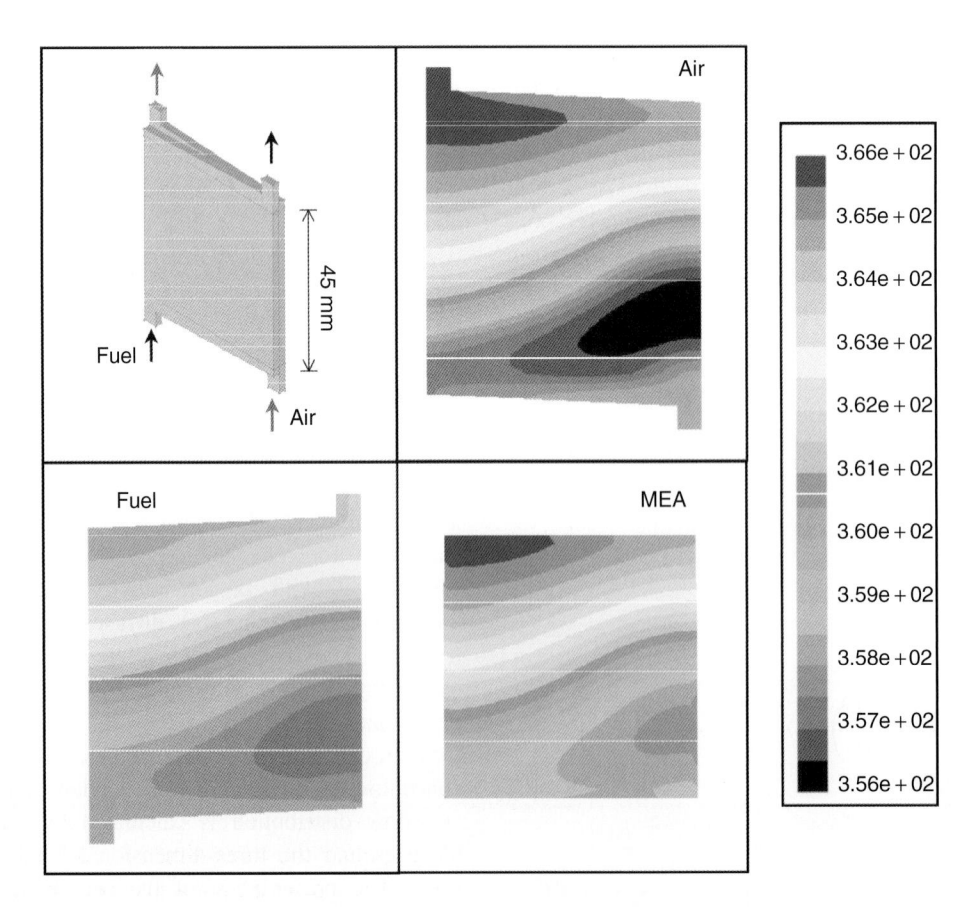

Figure 6. Schematic of local temperature distributions in DMFC: air channel, fuel channel, MEA.[13]

Table 4. DMFC single cell parameters.[a]

Power data	Air	Fuel
$E_c = 0.25\,V$	$N_{air}A = 4 \times 10^{-6}\,kg\,s^{-1}$	$N_{fuel}A = 3 \times 10^{-5}\,kg\,s^{-1}$
$j_{average} = 180\,mA\,cm^{-2}$	$T_{air,\,in} = 358\,K$	$T_{fuel,\,in} = 358\,K$
	$v_{air,\,el} = 3$	$v_{MeOH,\,el} = 5$
	dry air ($P_{O_2} = 0.2$)	MeOH concentration $1 \times 10^3\,mol\,m^{-3}$

[a]Reproduced from Bewer[13].

20 cm² cell. The cell parameters are given in Table 4. Periodic boundary conditions were applied in the calculation, i.e., the fuel and air channel are connected to each other by a wall.

The cooling in the inlet region of the channel by the evaporation of water can be clearly seen. This cooling reduces the temperature in the air inlet region by 2 °C. After air-saturation, the temperature rises by produced heat up to 366 K. This gives a total temperature difference in the air channel of 10 °C. Due to the higher mass fluxes and the higher heat capacity of the liquid methanol/water mixture, the temperature differences in the fuel channel are at about 6 °C. The cooling effect at the inlet is reduced. The temperature distribution in all three regimes is dominated by the velocity distribution.

A fuel cell stack consists of a number of single cells connected electrically in series. The typical components are the end plates, MEA and the bipolar plates. They have machined flow channels on both surfaces and are in electrical contact with the MEA backing layer. The membrane is electrically insulated to avoid a cell short circuit. A typical cell stack construction is shown in Figure 7.

There are several thermal models for H_2/PEMFCs.[14, 15] However, the most interesting case for the thermal management of PEMFCs is that of the DMFC, since this has the highest thermal sensitivity of cell performance as a temperature function. A one-dimensional thermal model for the DMFC stack was presented by Scott et al.[16]

The thermal model developed in Ref. [16] is based on the assumption that the total inlet flow rate is equally divided between all the cells present in the stack. To simplify the mathematical treatment, some assumptions were made, which, nevertheless allow a correct model description. In this model, both the reversible and irreversible heat losses are given by equations (2) and (21) and the enthalpy balances for both half cell reactions are given by equations that correspond to equations (28) and (32) for the anodic reaction and to equations (29) and (35) for the cathodic reaction. Similarly, the equations for both the anode and cathode gas diffusion layers and both catalyst layers including the membrane phase are defined analogously, as has been shown in the system equation (31). A schematic representation of the model is presented in Figure 5(b). In addition to the above equations, the conditions regarding the bipolar and the end plates have also been taken into account. Bipolar plates provide the electrical connection between the cells and also act as the main heat transfer area between the cells. The end plates create the boundaries of the stack structure. Usually they are made of the same material as the bipolar plates. If we neglect the radiation heating which is negligible in the low temperature fuel cells and take into consideration the fact that inside the plates neither the entropy changes nor the overvoltage phenomena take place the equations describing the conditions in both the plate elements are given in the reduced form of equation (22) as equation (36).

$$\rho\,dV c_{p,s}\frac{\partial T_s}{\partial t} = \lambda_{eff}\nabla^2 T_s\,dV + \sum K_\lambda (T_f - T_s)\,dA + Q_{loss} \quad (36)$$

By using the above equations, it was possible to predict the temperature profile for a two-cell stack.[16] The calculated temperature profile for a two-cell stack is shown in Figure 8.

The process conditions are as follows: *anode*: 1.0 dm³ min⁻¹ aqueous methanol solution flow rate per cell, 2 M methanol concentration, 80 °C solution inlet temperature, current density 100 mA cm⁻²; *cathode*: 2 bar

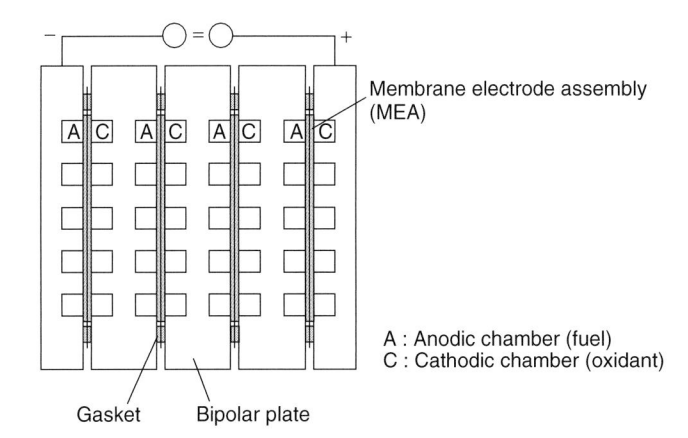

Figure 7. PEFC stack scheme.

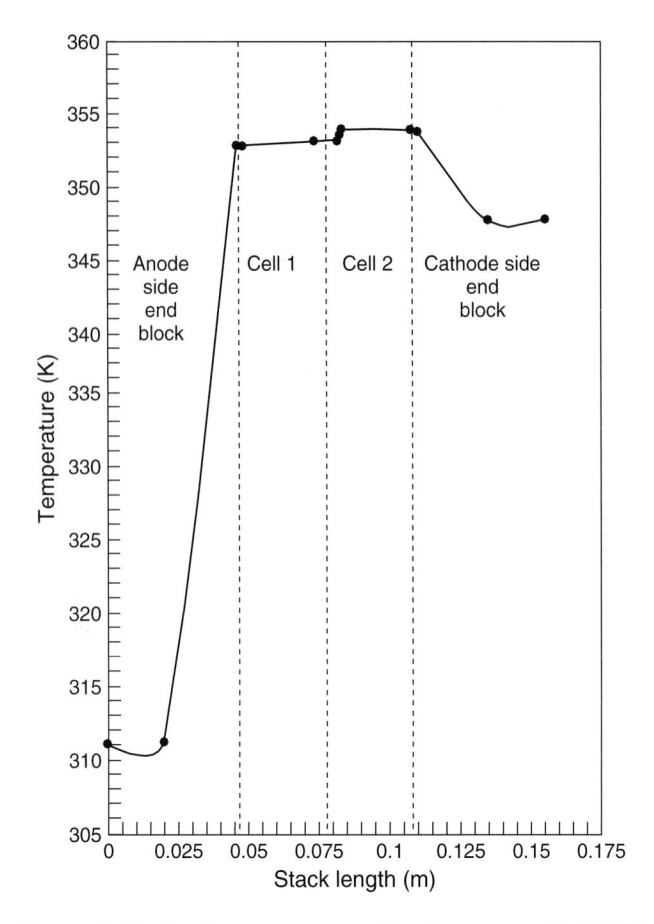

Figure 8. Predicted temperature profile for a two-cell DMFC stack. (Reproduced from Argyropoulos et al. (1999)[16] with permission from Elsevier Science.)

cathode gauge pressure, 22 °C air inlet temperature, 1.0 dm³ min⁻¹ air flow rate per cell.

The model describes the thermal behaviour of the cell stack using the material and energy balance in one dimension, along the length of the stack. Using this model, the temperature profiles along the stack length can be predicted.[17]

3.2 Performance analysis of PEMFCs considering heat management

Generally, high power densities of the fuel cell do not guarantee high system efficiency. For this reason, an analysis of both the mass and energy flows in the cell system is necessary. The direct methanol fuel cell can be considered as an example of such an analysis. The reason is that this cell is much more sensitive with respect to heat management than the usual hydrogen cells. In the case of liquid fuel supply, heat management is strongly influenced by the water vaporization rate at the cathode and the water crossover through the cell membrane. Further operation parameters influencing the heat effects are pressure, temperature, air flow and methanol permeation rate. The most important system parameter characterizing the cell energy efficiency is the net system efficiency, defined as the ratio of the net power and fuel input corresponding to the lower heating value of methanol (638.5 kJ mol⁻¹). In the electrical net power output, the power requirements of the system auxiliaries also have to

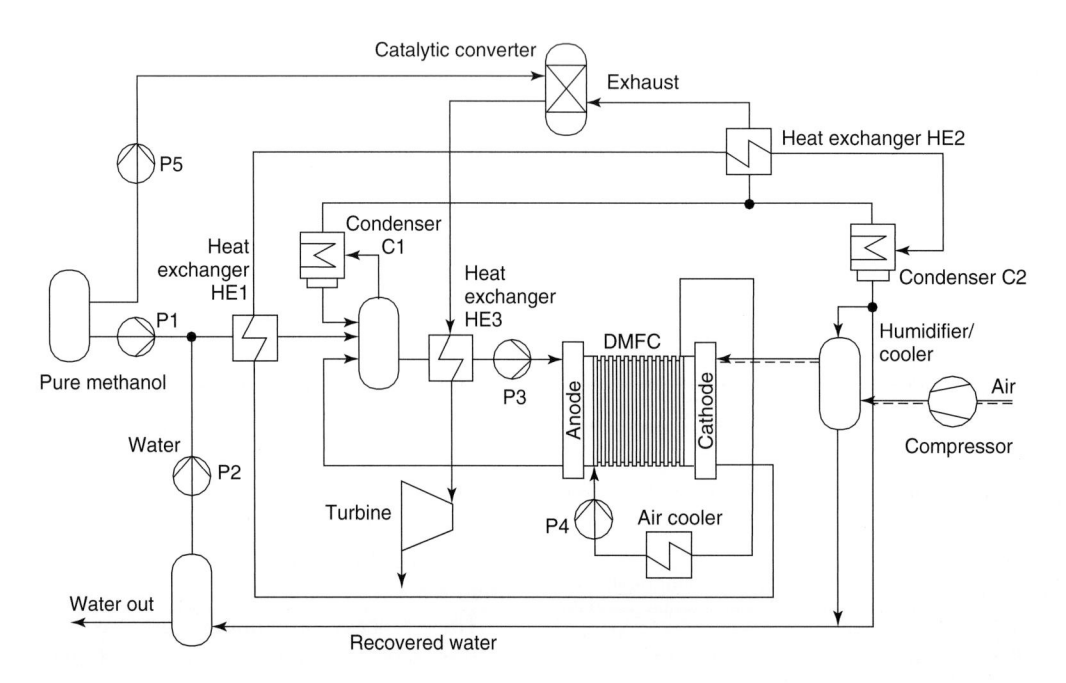

Figure 9. Flowsheet of a liquid-feed DMFC system. (Reproduced from Andrian and Meusinger (2000)[12] with permission from Elsevier Science.)

be taken into account. Considering the thermal balance of the DMFC stack, the heat duty as a measure of the waste heat output (waste heat negative) or the heat input (waste heat positive) is a further important parameter.

An extensive process analysis of the liquid-feed DMFC system was recently performed by von Andrian and Meusinger.[12] For this analysis, a liquid-feed fuel cell, 1 M MeOH solution as a working liquid in the anodic compartment and a constant water transference coefficient (drag factor) of an average value of four were considered (electrolyte Nafion® 117). According to the data of Table 3, this corresponds to an average cell working temperature of 100 °C.

The flow sheet of this fuel cell system is shown in Figure 9. The operating conditions used for the calculation of the system efficiency are given in Table 5.

Both the system efficiency and the heat duty depend very strongly on the water and methanol permeation rate. As an example, the values in the cell voltage/current

Table 5. Fixed system design parameters.[a]

Design parameter	Value
Fuel utilization	80%
Pump efficiencies	80%
Adiabatic compressor efficiency	60%
Adiabatic expander efficiency	50%
Pressure loss in heat exchanger	20 mbar
Pressure loss in fuel cell	50 mbar
Catalytic converter temperature	125–175 °C

[a]Reproduced from Andrian and Meusinger (2000)[12].

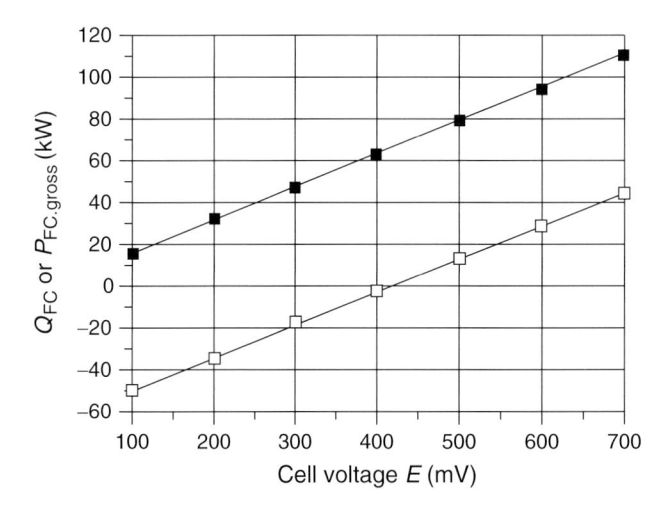

Figure 10. Dependence of cell voltage, E, on heat duty and electric power of a DMFC stack (without MeOH permeation $T = 85\,°C$, $P_{system} = 1.5\,bar$, 1 M MeOH, $v_{air,el} = 2.5$, $t_w = 4$). Heat duty, Q_{FC} (■); gross electric power, $P_{FC,gross}$ (□). (Reproduced from von Andrian and Meusinger (2000)[12] with permission from Elsevier Science.)

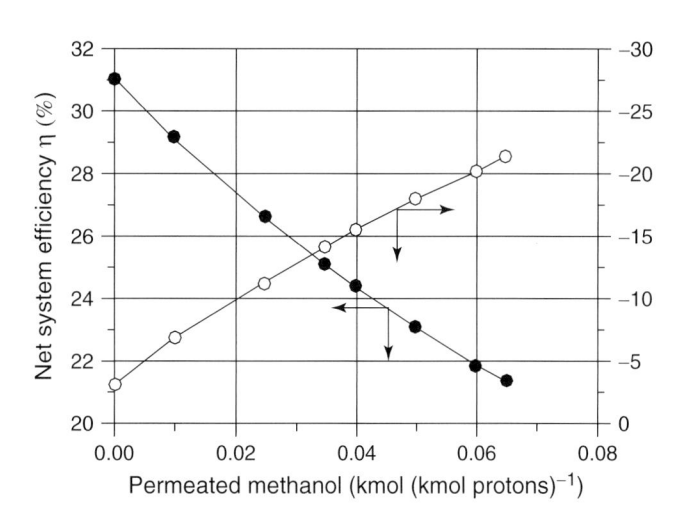

Figure 11. Influence of methanol permeation on system efficiency (η left) and heat duty (kW, right) ($T = 85\,°C$, $P_{system} = 1.5\,bar$, $E = 0.4\,V$, 1 M MeOH, $v_{air,el} = 2.5$, $t_w = 4$). (Reproduced from von Andrian and Meusinger (2000)[12] with permission from Elsevier Science.)

density working point of $(400\,mV/400\,mA\,cm^{-2})$ should be discussed. Then the net system efficiency, without considering the methanol permeation, would be in the range of 26–29% as a function of the working temperature and system pressure. Figure 10 shows the energy balance of the cell stack. At a 400 mV cell voltage there is a heat balance of $-2.5\,kW$. The zero net energy balance is achieved at a cell voltage of 420 mV. Increasing cell voltage necessitates a heat input since the water permeation needs additional heat on the cathodic side of the cell stack for water vaporization. This additional need of heat can be partially compensated by the methanol permeation, which produces excess heat on the cathode side as shown in Figure 11.

This additional heating results in a drastic reduction of system efficiency, of course, so that efforts must be made to achieve a reasonable balance between the heat management and the net system efficiency.

REFERENCES

1. J. P. Holman, 'Heat Transfer', McGraw-Hill, New York (1990).

2. A. J. Chapman, 'Heat Transfer', Macmillan, New York (1984).

3. B. Elves, S. Hawkins and G. Schulz (Eds), 'Ullmann's Encyclopedia of Industrial Chemistry', VCH, Weinheim, Vol. B1, Chapter 4 (1990).

4. I. S. Grigoriev and E. Z. Meilikhov (Eds), 'Handbook of Physical Quantities', CRC Press, Boca Raton, FL, p. 421 (1997).

5. S. K. Ratkje, M. Ottøy, R. Halseid and M. Strømgård, *J. Membr. Sci.*, **107**, 219 (1995).

6. S. K. Ratkje and S. Møller-Holst, *Electrochim. Acta*, **38**, 447 (1993).
7. 'JANAF Thermochemical Tables', 3rd Edition, American Chemical Society and the American Institute of Physics for NBS, New York (1985).
8. I. Barin, 'Thermochemical Data of Pure Substances', VCH Verlagsgesellschaft, Weinheim (1995).
9. T. F. Fuller and J. Newman, *J. Electrochem. Soc.*, **140**, 1218 (1993).
10. T. V. Nguyen and R. E. White, *J. Electrochem. Soc.*, **140**, 2178 (1993).
11. G. Maggio, V. Recupero and C. Mantegazza, *J. Power Sources*, **62**, 167 (1996).
12. S. von Andrian and J. Meusinger, *J. Power Sources*, **91**, 193 (2000).
13. T. Bewer, 'Mass Fluxes and Current Density Distribution in Liquid DMFC', Ph.D. Dissertation, RWTH Aachen (2002).
14. J. C. Amphlett, R. F. Mann, B. A. Pepley, P. R. Roberge and A. Rodrigues, *J. Power Sources*, **61**, 183 (1996).
15. J. H. Lee and T. R. Lalk, *J. Power Sources*, **73**, 229 (1998).
16. P. Argyropoulos, K. Scott and W. M. Taama, *J. Power Sources*, **79**, 169 (1999).
17. P. Argyropoulos, K. Scott and W. M. Taama, *J. Power Sources*, **79**, 184 (1999).

Chapter 10
High temperature fuel cells

J. Divisek
Institute for Materials and Processes in Energy Systems, Forschungszentrum Jülich GmbH, Jülich, Germany

1 RADIATION HEAT TRANSFER

All three modes of heat transfer can occur in high temperature fuel cells: conduction, convection and radiation. For the first two, see **Low temperature fuel cells**, Volume 1. Because of the higher working temperature of high temperature fuel cells, between 600 and 1000 °C, in addition to the heat transfer by conduction and convection radiation heat transfer also exists. The mechanism in this case is electromagnetic radiation propagated as a result of temperature difference, called thermal radiation. The rate at which thermal energy is emitted is given by the Stefan–Boltzmann law.

$$Q_{em} = \sigma_{SB} A T^4 \qquad (1)$$

In equation (1) σ_{SB} is the Stefan–Boltzman constant (5.669×10^{-8} W m^{-2} K^{-4}). The net radiation exchange between two surfaces is proportional to the difference in absolute temperature to the fourth power, corrected by a complex function of emissivity ε, which takes account of the deviation of the black-body character and geometric shape factor F_s. To take into account both situations then we have the radiation exchange as

$$Q_{em} = \varepsilon F_s \sigma_{SB} A (T_1^4 - T_2^4) \qquad (2)$$

The shape factor is a purely geometric property of the system; it is nevertheless an important quantity when considering the heat transfer in the tubular solid oxide fuel cell (SOFC). For this reason, the standard procedure for its formulation should be given here.

Consider the two surfaces in question denoted A_1 and A_2 as shown in Figure 1. Elements of each surface are denoted by dA_1 and dA_2 with corresponding normals N_1 and N_2. The line connection has length r and makes polar angles ψ_1 and ψ_2 with the corresponding normals.

The radiation shape factors $F_{s,ij}$ are defined as follows: F_{12} is the fraction of energy leaving surface 1 which strikes surface 2; F_{21} is the fraction of energy leaving surface 2 which strikes surface 1; F_{mn} is the fraction of energy leaving surface m which strikes surface n. The factors $F_{s,ij}$ are calculated[1] as the expressions

$$F_{ij} = \int_{A_i} \int_{A_j} \frac{\cos \psi_i \cos \psi_j}{\pi r^2} dA_j dA_i \qquad (3)$$

A simple radiation problem of radiation in an enclosure when we have a heat transfer surface at temperature T_1 completely enclosed by a much larger surface maintained at temperature T_2 can be solved by the equation[1]

$$Q_{em} = \varepsilon \sigma_{SB} A (T_1^4 - T_2^4) \qquad (4)$$

2 HEAT TRANSFER IN HTFCs

The general energy balance equation in the high temperature fuel cells has to consider three special circumstances which are different from the corresponding equations in the low temperature fuel cells. Because of pure gas phase reactions the liquid state is missing, i.e., the evaporation/condensation equilibrium is not present, the heat transfer in the porous system due to the fast diffusion

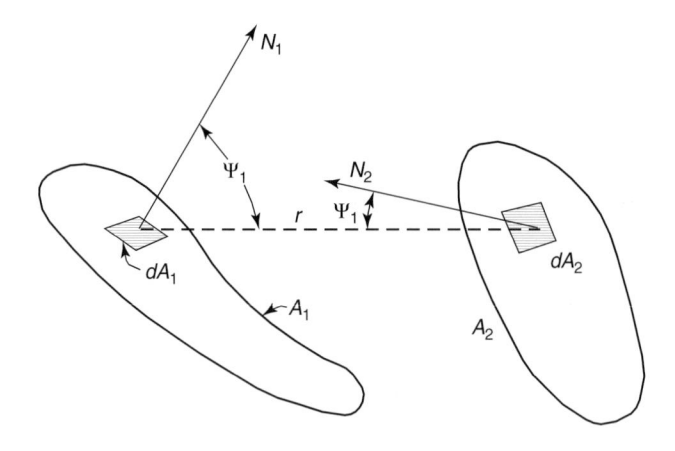

Figure 1. Area elements used in deriving radiation shape factor.

mass transfer inside the porous systems is more effective and, finally, additional chemical heat sources as the fuel-reforming reactions are present. Generally, the temperature differences across the cell are much bigger in high temperature fuel cells (HTFCs) than in the low temperature fuel cells (LTFCs).

The energy equation is then given as

$$\rho c_p \frac{\partial T}{\partial t} = -u_{x_i} \rho c_p \frac{\partial T}{\partial x_i}$$

$$+ \frac{\partial}{\partial x_i}\left(\bar{\lambda}\frac{\partial T_s}{\partial x_i} - \sum_j D_j \frac{\partial(\rho_j/M_j)}{\partial x_i} H_{T,j}\right) + Q^*$$

$$H_{T,j} = \int_{T_1}^{T_2} c_{p,j} \, dT \tag{5}$$

Since there is noticeable energy transport through the porous system in HTFCs (diffusion layers, or substrate in planar SOFC), the general energy equation (5) in porous media is used with some modifications as equation (6):

$$\frac{\partial}{\partial t}(\phi \rho_f c_{p,f} T_f + (1-\phi)\rho_s c_{p,s} T_s) = -u_{x_i} \rho_f c_{p,f} \frac{\partial T_f}{\partial x_i}$$

$$+ \frac{\partial}{\partial x_i}\left(\bar{\lambda}\frac{\partial T_s}{\partial x_i}\right) - \phi \frac{\partial}{\partial x_i}\left(\sum_j D_j \frac{\partial(\rho_j/M_j)}{\partial x_i} H_{T_f,j}\right)$$

$$+ \phi Q_f^* + (1-\phi)Q_s^* \tag{6}$$

The heat generation within the solid skeleton matrix Q_s^*, the complementary porosity of which is characterized by the expression $(1-\phi)$, is caused in the HTFC usually by the irreversible heat losses Q_{loss} (see **Low temperature fuel cells**, Volume 1, equation (21)). The heat generation within pore element Q_f^*, the porosity of which is characterized by the porosity factor ϕ in the HTFC cells is usually a chemical reforming reaction.

3 HEAT SOURCE TERMS FOR HIGH TEMPERATURE FUEL CELLS

As a typical heat generation source Q_f^* we can consider methane-water reforming accompanied by a fast parallel water shift reaction according to the reaction scheme:

$$CH_4 + H_2O \Longleftrightarrow 3H_2 + CO,$$
$$\Delta H_{298}^0 = +206.1 \, \text{kJ mol}^{-1}$$

$$CO + H_2O \Longleftrightarrow H_2 + CO_2,$$
$$\Delta H_{298}^0 = -41.2 \, \text{kJ mol}^{-1} \tag{7}$$

At low temperatures ($T < 500\,°C$) or slight water vapor fractions ($H_2O/CH_4 < 1.5$), carbon formation has been observed so that two further reactions must be considered to describe carbon deposition:

$$CH_4 \Longleftrightarrow 2H_2 + C_{\text{solid}}, \quad \Delta H_{298}^0 = +74.8 \, \text{kJ mol}^{-1}$$

$$2CO \Longleftrightarrow CO_2 + C_{\text{solid}}, \quad \Delta H_{298}^0 = -173.3 \, \text{kJ mol}^{-1} \tag{8}$$

The first of the two reactions (8) is preferred at high temperatures and the second (Boudouard equilibrium) at low temperatures. The equilibrium constants K_p at $T = 650\,°C$ for reactions (7) and (8) are in sequence $K_p = 2.0$, 1.9, 0.4, 3.0.[2] The temperature dependence of the constant K_p of the water shift reaction from equation (7) is given by the relation

$$K_p = \exp\left(-\frac{\Delta G^0}{RT}\right)$$

$$\Delta G^0(T) = -39.25 + 40.56 \times 10^{-3} T$$
$$- 4.37 \times 10^{-6} T^2 \, \text{kJ mol}^{-1} \tag{9}$$

The temperature dependence of the reaction constants for the two reactions (7) must be represented in different ways. Whereas the water shift reaction proceeds very rapidly and can be regarded as a reaction in permanent equilibrium, this is not the case for the reforming reaction. Since the latter is a reaction that proceeds heterogeneously, on the one hand reaction formulations according to homogeneous reaction kinetics are not admissible and on the other hand the expected dependencies are fairly complicated. As an example of a description of the reaction rate k_{CH_4} of the reforming reaction (7), a formal kinetic equation after Langmuir–Hinselwood will be given:[2]

$$k_{CH_4} = k_{CH_4}^0 \Theta_{CH_4} \Theta_{H_2O}$$

$$= k_{CH_4}^0 \frac{K_{CH_4} K_{H_2O} \cdot P_{CH_4} P_{H_2O}}{(1 + K_{CH_4} P_{CH_4} + K_{H_2O} P_{H_2O})^2} \tag{10}$$

When the reaction proceeds on Ni particles embedded in YSZ, this is of particular relevance for the internal reforming reaction in SOFCs. The following relations are given:[2]

$$K_{H_2O} = K^0_{H_2O} \exp\left(\frac{\Delta H^{ads}_{H_2O}}{RT}\right), \quad K^0_{H_2O} = 1.3 \times 10^{-3} \text{ kPa}^{-1},$$

$$\Delta H^{ads}_{H_2O} = 39 \text{ kJ mol}^{-1}$$

$$K_{CH_4} = \text{const.} = 0.16 \text{ kPa}^{-1}$$

$$k^0_{CH_4} = k^{00}_{CH_4} \exp\left(\frac{-E_a}{RT}\right), \quad k^{00}_{CH_4} = 660$$

$$\times 10^6 \text{ mol}_{CH_4} (\text{mol}_{Ni} \text{ s})^{-1}, \quad E_a = 50 \text{ kJ mol}^{-1} \quad (11)$$

The expressions are reaction kinetically rigorous and correct. They are also used to calculate the reforming kinetics in a separate reactor connected in front of the fuel cell. However, if the reforming reaction (7) is present in the form of a reaction proceeding internally in the actual fuel cell, other simpler expressions are used to calculate the reaction constant k_{CH_4} which are not dependent on the partial pressure of water P_{H_2O}.[3] Accordingly, this simpler reforming reaction constant $\bar{k}_{CH_4}$ is defined by equation (12).

$$\bar{k}_{CH_4} = \bar{k}_{refCH_4} \cdot \frac{P_{CH_4}}{RT}, \quad \frac{1}{\bar{k}_{refCH_4}} = \frac{1}{\bar{k}_{DCH_4}} + \frac{1}{\bar{k}_{rCH_4}}$$

$$\bar{k}_{rCH_4} = RT\bar{k}^0_{rCH_4} \exp\left(\frac{-E_a}{RT}\right), \quad \bar{k}^0_{rCH_4} = 4274$$

$$\times 10^{-5} \text{ mol } (\text{m}^2 \text{ Pa s})^{-1}, \quad E_a = 82 \text{ kJ mol}^{-1}$$

$$\bar{k}_{DCH_4} = 4\frac{D_{CH_4}}{d_{hyd}} Re \cdot Sc \cdot \frac{P}{P - P_{CH_4}} \quad (12)$$

d_{hyd} is the hydraulic diameter (4 × cross section area/circumference).

4 PERFORMANCE ANALYSIS OF MCFC CONSIDERING CURRENT DENSITY AND TEMPERATURE DISTRIBUTIONS

There are two types of HTFC to be taken into consideration, the molten carbonate fuel cell (MCFC) and the ceramic SOFCs. The main difference between the two of them is the electrolyte. Whereas the MCFC operates with molten salt as an electrolyte, the SOFC needs a ceramic matrix, based on yttrium-stabilized zirconium oxide (YSZ).

The electrochemical reactions considered in the MCFC electrodes are the following:

Anode:

$$H_{2(g)} + CO_{3(l)}^{2-} \longrightarrow H_2O_{(g)} + CO_{2(g)} + 2e^-$$

Cathode:

$$0.5O_{2(g)} + CO_{2(g)} + 2e^- \longrightarrow CO_{3(l)}^{2-} \quad (13)$$

Overall cell reaction:

$$H_{2(g)} + 0.5O_{2(g)} + CO_{2(g)cathode} \longrightarrow H_2O_{(g)} + CO_{2(g)anode}$$

If the inlet fuel gas has been externally preformed, then besides the cell reaction there is an additional heat source caused by the anodic gas water shift reaction (7). The temperature dependence of the enthalpy change of this reaction is given by the relation.[4]

$$\Delta H_{shift}(T) = -43729 + 9.4657T \text{ J mol}^{-1} \quad (14)$$

The operating temperature of the MCFC lies between 600 and 700 °C. At a temperature lower than 600 °C the cell performance drops significantly. At a temperature higher than 700 °C material corrosion is a problem. To make stable performance during the targeted operation period of 40 000 h possible, the stack temperature must be controlled within a specified range. This temperature control can be achieved in two ways. The first is construction of a stack with direct or indirect internal reforming. In this case, the endothermic reforming reactions (7) absorb the excess heat generated from the cell stacks and lead to a state of thermal balance. This problem and the method of solving it is common to both the MCFC and SOFC (see Section 5.1 for more details). The other way of temperature control is cooling with an excess amount of process gas which lowers the stack temperature by convective heat transfer. Both equations (5) and (6) can be used as the conservation equation for heat transfer. In contrast to the SOFC, a CO_2 gas flow circulation must also be considered in this process in the MCFC. Accordingly, for the calculation of irreversible heat losses Q^*_s in equation (6) by using equation (21) from **Low temperature fuel cells**, Volume 1, the reversible cell voltage term $-\Delta G/nF$ in equation (17) of **Low temperature fuel cells**, Volume 1, must be written in the case of the hydrogen/oxygen reaction as

$$-\frac{\Delta G}{nF} = E^0 + \frac{RT}{2F}$$

$$\times \ln\left(\frac{P_{H_2(g)anode}\sqrt{P_{O_2(g)cathode}}}{P_{H_2O(g)anode}} \frac{P_{CO_2(g)cathode}}{P_{CO_2(g)anode}}\right)$$

$$E^0 = 1.272 - 2.7645 \times 10^{-4}T \quad (15)$$

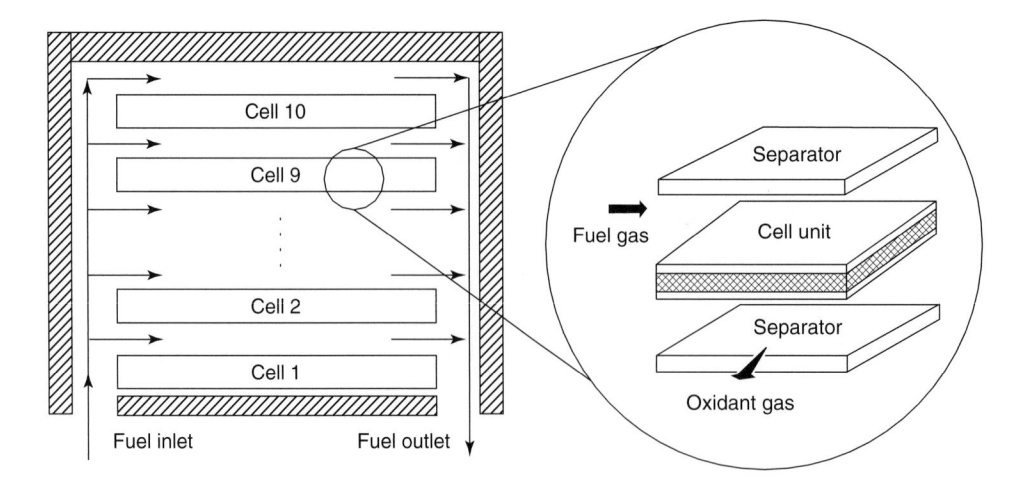

Figure 2. Schematic representation of MCFC fuel cell stack. (Reproduced from He and Chen (1998)[6] with permission from Elsevier Science.)

Generally, the temperature and current density profiles in the MCFC can today be calculated reliably and well by means of numerical analysis.[4–8] The MCFC cells are constructed in a planar way (cf. Figure 2).

There is a difference between the co-flow and cross-flow stack design.[7] It can be shown that the current density distribution is a function of the temperature distribution.[6] Such analysis is presented in reference.[6] The computational grids are shown in Figure 3.

The grid definition for the analysis is a unit of $20 \times 20\,\text{mm}^2$. A real cell unit with dimensions of $1 \times 1\,\text{m}^2$ will consists of 50×50 such basic units. As an example the correlation between the current density and the temperature distribution in the case of the cross-flow geometry is given in Figure 4.

Apparently, the current density increases in a manner similar to the cell temperature, following the direction of the oxidant gas flow. The model has been applied to calculate a stack which had 10 cells with dimensions of $0.36\,\text{m}^2$. From the experimental point of view it is easier to compare the calculated temperature distribution with the theoretical prediction than the measurements of current density distribution. For this reason the experimental control is mostly only given for the temperature distribution fields. Figure 5 shows the predicted temperature profiles which agree reasonably well with the simulation results.[6]

The heat transfer calculations and the temperature regime of the MCFC are known mainly for the planar stack design. In case of the SOFC the tubular geometry is also of importance as shown in Sections 5.2 and 5.3.

5 SOFC CELLS

The operation principle of the solid oxide fuel cell is different from the MCFC and is determined by the ceramic

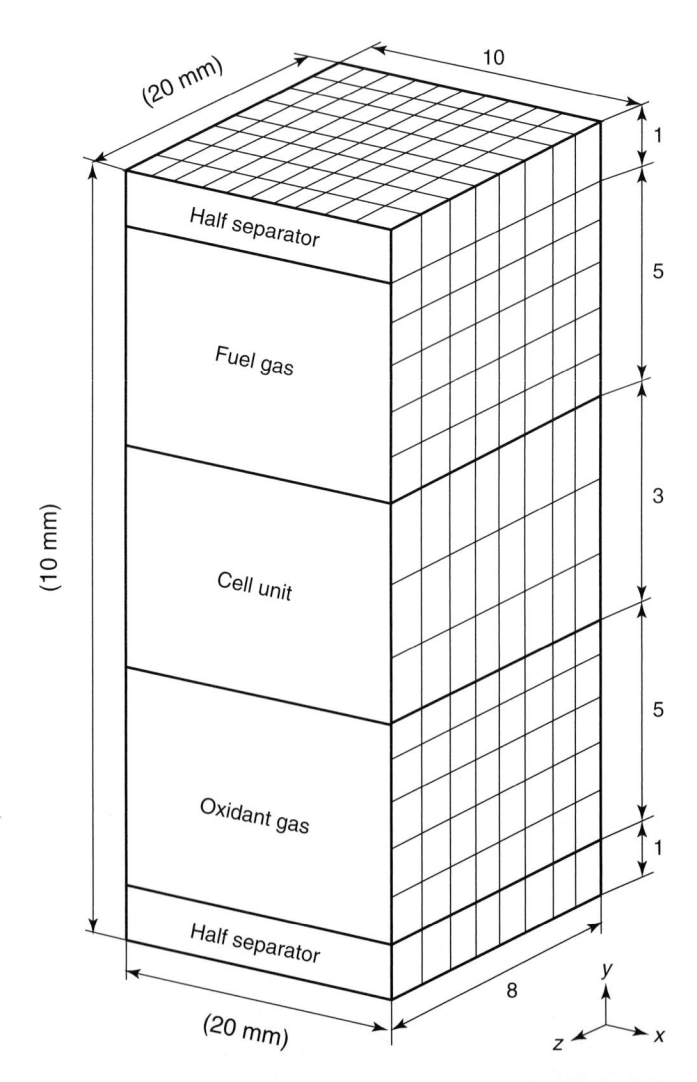

Figure 3. Dimensions and computational grids of an analysis unit. (Reproduced from He and Chen (1998)[6] with permission from Elsevier Science.)

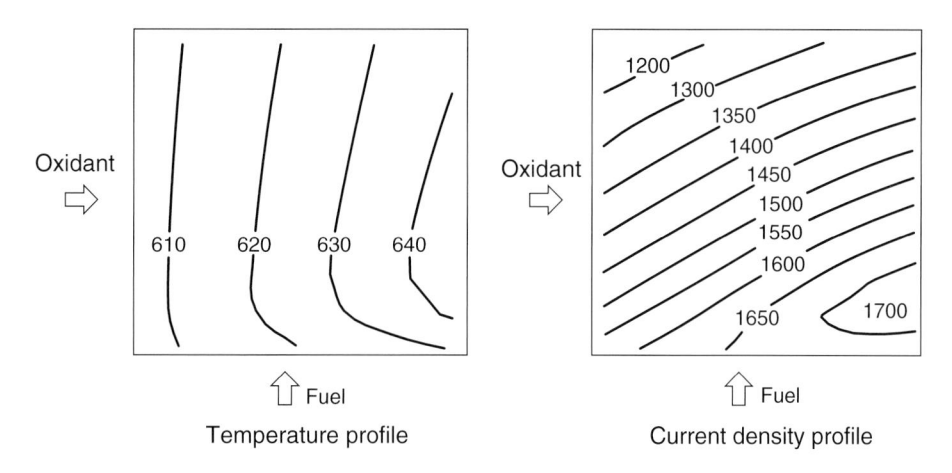

Figure 4. Temperature profile (°C) and current density (A m^{-2}) responses at a transverse cell section. (Reproduced from He and Chen (1998)[6] with permission from Elsevier Science.)

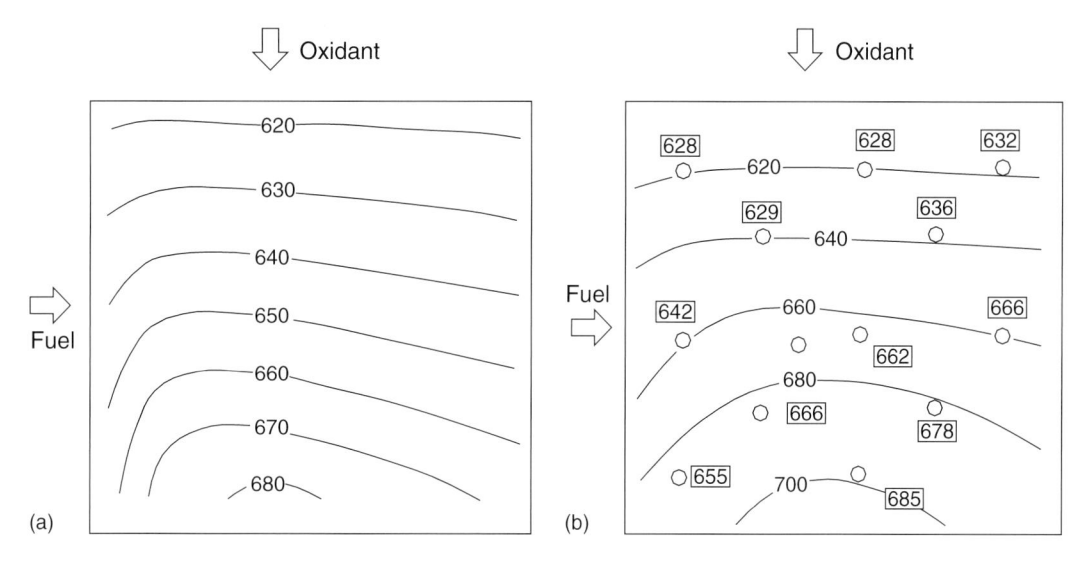

Figure 5. Comparison of temperature distributions in MCFC fuel cell (°C): (a) computed results; (b) analytical profiles and experimental data. (Reproduced from He and Chen (1998)[6] with permission from Elsevier Science.)

character of the membrane electrolyte. The electrolyte of choice is yttria-stabilized ZrO_2. The solid state of the components does not impose any restrictions on the cell configuration, but does cause sealing problems.

The electrochemical reactions occurring in the SOFC are in principle the same as with LTFCs, but due to the high working temperature combustion reactions of some carbon containing compounds are also possible. The utilization reaction of hydrogen is given by equation (16):

Anode:

$$H_{2(g)} + O_{(s)}^{2-} \longrightarrow H_2O_{(g)} + 2e^-$$

Cathode:

$$0.5O_{2(g)} + 2e^- \longrightarrow O_{(s)}^{2-} \quad (16)$$

Overall cell reaction:

$$H_{2(g)} + 0.5O_{2(g)} \longrightarrow H_2O_{(g)}$$

In the SOFC both the CO and CH_4 can also be used as fuel. In this case the electrochemical reactions are

Anode:

$$CO_{(g)} + O_{(s)}^{2-} \longrightarrow CO_{2(g)} + 2e^-$$

$$CH_{4(g)} + 4O_{(s)}^{2-} \longrightarrow CO_{2(g)} + 2H_2O + 8e^-$$

Cathode:

$$0.5O_{2(g)} + 2e^- \longrightarrow O_{(s)}^{2-} \quad (17)$$

Overall cell reaction:

$$CH_{4(g)} + CO_{(g)} + 2.5O_{2(g)} \longrightarrow 2CO_{2(g)} + 2H_2O$$

Instead of the direct electrochemical methane-combusting reaction, which is part of the system equations (17), the methane-reforming reaction (7) is normally used for hydrogen production which can react electrochemically according to equations (16). Because of the endothermic reaction (7) heat management and temperature distribution are strongly influenced as a function of the place where the methane reforming takes place.

5.1 Heat transfer in planar-type SOFC

For the sake of simplicity, most work on SOFC heat management has been performed with the planar-type geometry shown in Figure 6.

There are two SOFC cell types based on this configuration: the monolith cell and the electrode-(anode or cathode) supported cell. The energy balance of the SOFC monolith cell was described theoretically for the first time by Vayenas et al.[8] A similar model was used for the simulation of the Allied Signal monolithic cell.[9] Also the fundamental transport equations were combined in a differential form and electrochemical processes were treated theoretically.[10] The consumption rate of the methane-reforming reaction (7) was given by a general kinetic expression in this work.

It has become evident that the internal methane reforming is a crucial point to be considered since the cooling effect of this reaction strongly affects the operating regime of the SOFC cell stack. The first heat transfer calculation implementing the methane-water reforming reaction was performed by Achenbach.[3, 11] In the SOFC we can generally assume that not hydrogen but methane or the products of methane reforming are used as a fuel. In this case the mass balance for the fuel channels is made with respect to methane, hydrogen, water vapor, CO and CO_2, that for the air channel with respect to oxygen. The potential dependences and the conversion rates of the gas components can be expressed by the combination of the Butler–Volmer equations taking into account the corresponding stoichiometry.[3] The general mass balance equations are of the form

$$\frac{1}{M_i}\frac{\partial N_i}{\partial x} = v_i \bar{k}_{CH_4}\frac{1}{h_f} + v_i \Delta N_{SHIFT} + v_{i,el}\frac{j}{2Fh_f} \quad (18)$$

The computation of the molar conversion rate of the water shift reaction (7) is given by equation (19).

$$\Delta N_{SHIFT} = X_1 \pm \sqrt{X_1^2 - X_2}$$

$$X_1 = \frac{K_p(N_{CO}/M_{CO} + N_{H_2O}/M_{H_2O}) + N_{CO_2}/M_{CO_2} + N_{H_2}/M_{H_2}}{2(K_p - 1)}$$

$$X_2 = \frac{K_p \cdot N_{CO}N_{H_2O}/M_{CO}M_{H_2O} - N_{CO_2}N_{H_2}/M_{CO_2}M_{H_2}}{(K_p - 1)} \quad (19)$$

The constant K_p is determined by equation (9), the reforming constant $\bar{k}_{CH_4}$ by equation (12).

The energy balance for the gas flows can be obtained for both the fuel and the air by equation (20).

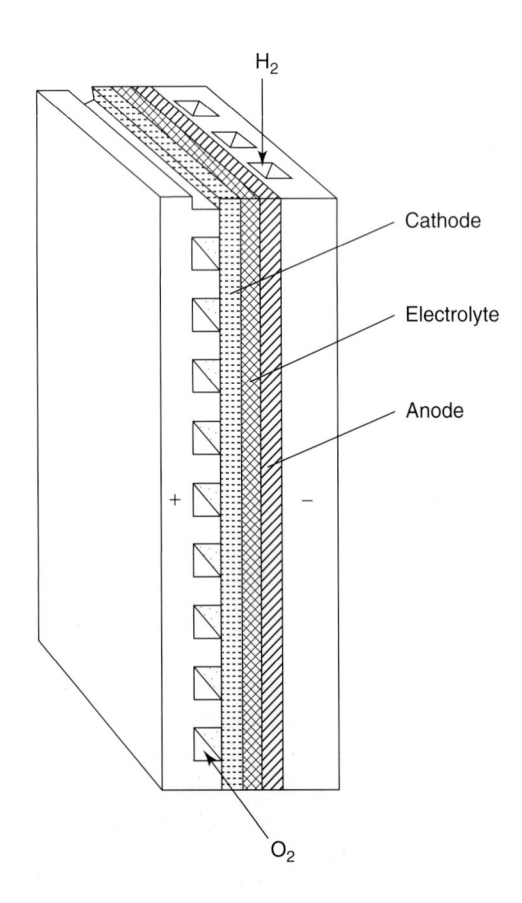

$$\frac{1}{M_{fuel}}\frac{\partial (N_{fuel}c_{p,fuel}T_f)}{\partial x}$$
$$= \frac{h(T_s - T_f) + \bar{k}_{prod}c_{p,prod}T_s - \bar{k}_{educt}c_{p,educt}T_f}{h_f}$$

$$\frac{1}{M_{ox}}\frac{\partial (N_{ox}c_{p,ox}T_f)}{\partial x}$$
$$= \frac{h(T_s - T_f) + \bar{k}_{prod}c_{p,prod}T_s - \frac{j}{4F}c_{p,O_2}T_f}{h_a} \quad (20)$$

Figure 6. Schematic of a planar SOFC, cross-flow geometry.

The convection heat transfer coefficient h is determined by the Nusselt number Nu. The value of Nu for laminar flow conditions in the channels is

$$Nu = \frac{h d_{\text{hyd}}}{\lambda} = 4 \qquad (21)$$

The energy balance of the solid phase is given by equation (8) in **Low temperature fuel cells**, Volume 1. The solid is considered to be a quasi-homogeneous material with respect to heat conductivity. In this context an effective heat conductivity coefficient is defined accounting for the heat paths through the solid structure and the thermal radiation as expressed in equation (22). The source term Q^* in equation (3) in **Low temperature fuel cells**, Volume 1, contains the sum of the reaction enthalpies and the irreversible heat of the electrochemical reactions. The convective term appears as the boundary condition of the third order according to equation (4) or equation (12) in **Low temperature fuel cells**, Volume 1. The radiation for the planar SOFC according to equation (2) can be approximated by[3]

$$Q_{\text{em}} = 4\sigma_{\text{SB}} \left(\frac{T_1 + T_2}{2}\right)^3 A(T_1 - T_2) \qquad (22)$$

The solid temperature distribution of the SOFC depends on the relative flow direction of the fuel with respect to the oxidant: cross-flow, counter-flow or co-flow. As an example, all three possibilities of flow directions in a SOFC are presented in Figure 7.

Defining a reference cell, the parameters of which are given in Table 1, a temperature optimization of the planar SOFC can be made.[3, 11] The geometry of the reference cell is as follows: $130 \times 130\,\text{mm}^2$ in size, $100 \times 100\,\text{mm}^2$ electrochemically active area surrounded by a frame of 15 mm.

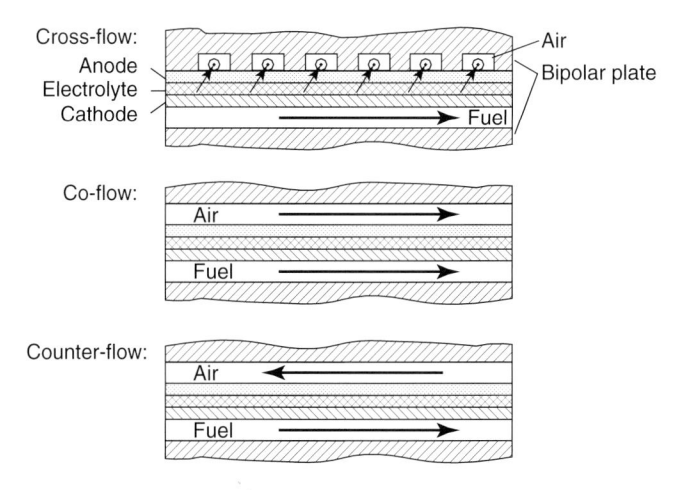

Figure 7. Geometric alternatives of the gas flow in planar SOFC.

Table 1. Gas composition and temperatures of the reference SOFC cell.[a]

Conditions	Inlet	Outlet
Flow rate fuel	$0.67\,\text{mol}\,\text{h}^{-1}$	$0.90\,\text{mol}\,\text{h}^{-1}$
Flow rate air	$10.97\,\text{mol}\,\text{h}^{-1}$	$10.64\,\text{mol}\,\text{h}^{-1}$
Gas composition (%):		
H_2	26.26	7.85
CH_4	17.10	0.00
H_2O	49.34	73.92
CO	2.94	2.58
CO_2	4.36	15.65
Mean air temperature (K)	1173	1276
Mean fuel temperature (K)	1173	1260

[a]Data from Achenbach (1994).[11]

A comparison of the solid temperature distribution for all three configurations above in an internally reforming planar SOFC is shown in Figure 8.[11]

Due to the fast endothermal reforming reaction, for cross-flow the temperature drops at the fuel inlet. The temperature drop is not so dramatic for the co-flow since the cooling air with an inlet temperature of about 900 °C supplies the reforming reaction with the necessary heat. The best results concerning the temperature distribution and performance are achieved with counter-flow because the air has the highest temperature at the fuel inlet, i.e., at that location where the heat for the reforming reaction is required. For this reason both the mean cell temperature and the electrochemical reaction rate are highest and the internal ohmic resistances lowest. This results in a better cell efficiency compared with the cross-flow and co-flow cells. The cell comparison overview at the mean current density of $3000\,\text{A}\,\text{m}^{-2}$ is given in Table 2.

Also the external boundary conditions in the cell stack have an important influence on cell operation. At stack surface temperatures of $T = 1000\,°\text{C}$ the equivalent heat transfer coefficient $h_{\text{er}} = 4\sigma_{\text{SB}}[(T_1 + T_2)/2]^3$ for radiant heat transfer has the value of approx. $470\,\text{W}\,\text{m}^{-2}\,\text{K}$ (cf. equation (22)). This means that if there is a temperature difference of only 10 °C between the stack surface and the ambient surroundings walls, about $5\,\text{kW}\,\text{m}^{-2}$ is radiated from the stack. This demonstrates that an extremely effective thermal insulation of the vessels must be provided to protect the stack from cooling down. As an example Figure 9 shows the heat-loss effect on the temperature distribution in a stack radiating heat to the surroundings at 900 °C from the top plate only. Figure 9(a) refers to a stack of 50 cells with ceramic bipolar plates, and Figure 9(b) to a stack with metallic bipolar plates. It is evident that the heat losses affect the top region of the stack. The

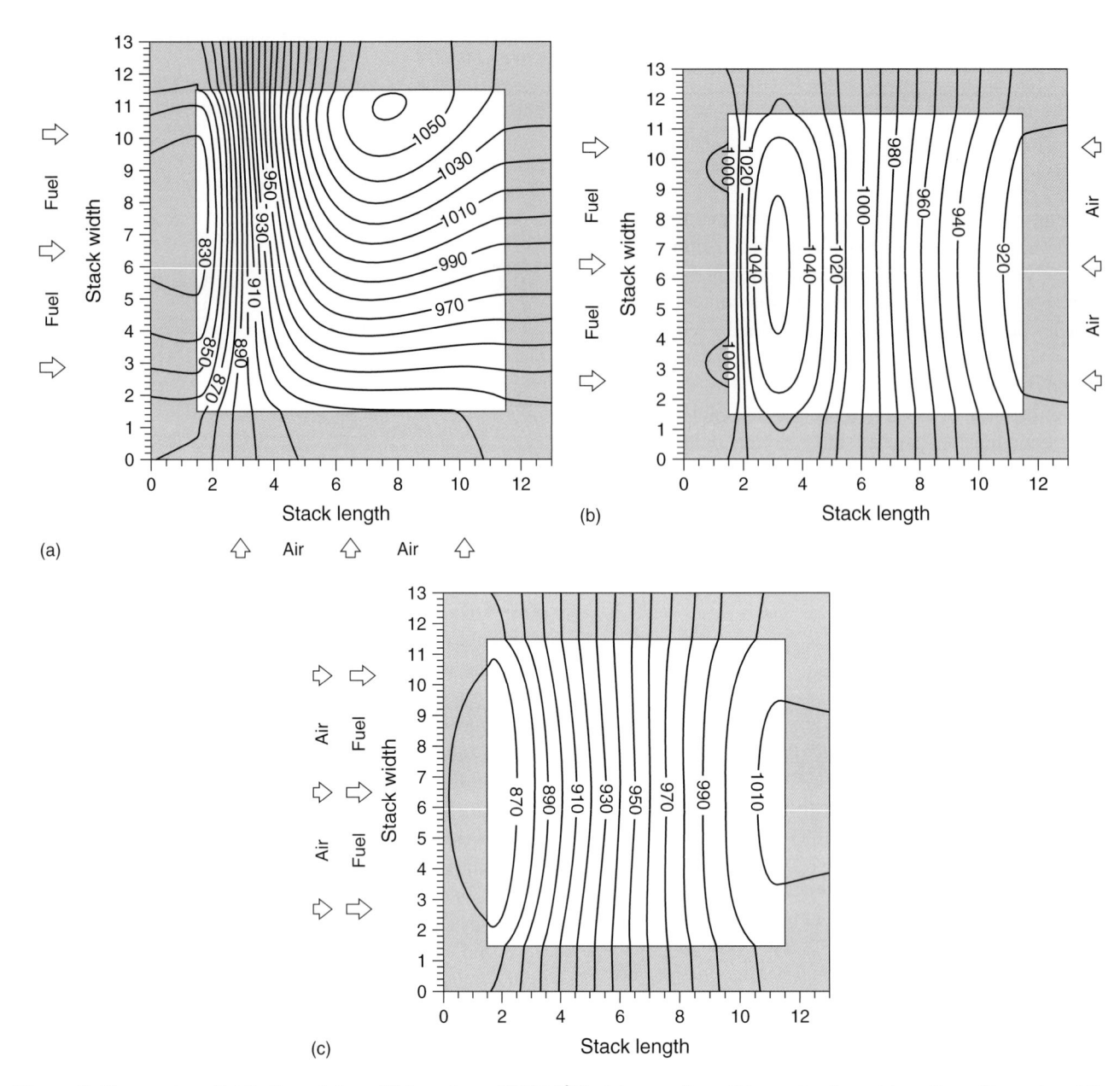

Figure 8. Temperature distribution of the solid in a planar SOFC (°C): (a) cross-flow; (b) counter-flow; (c) co-flow. (Reproduced from Achenbach (1994)[11] with permission from Elsevier Science.)

Table 2. Performance data of planar SOFC cells in cross-, co- and counter-flow working mode.[a]

	E_c (V)	P (W)	$T_{s,max}$ (K)	$T_{s,min}$ (K)	$T_{air,out}$ (K)	dT/dl (K mm^{-1})
Cross-flow	0.682	20.46	1334	1096	1276	7.09
Co-flow	0.674	20.21	1286	1134	1281	5.08
Counter-flow	0.708	21.23	1325	1183	1282	5.13

[a] Data from Achenbach (1994).[11]

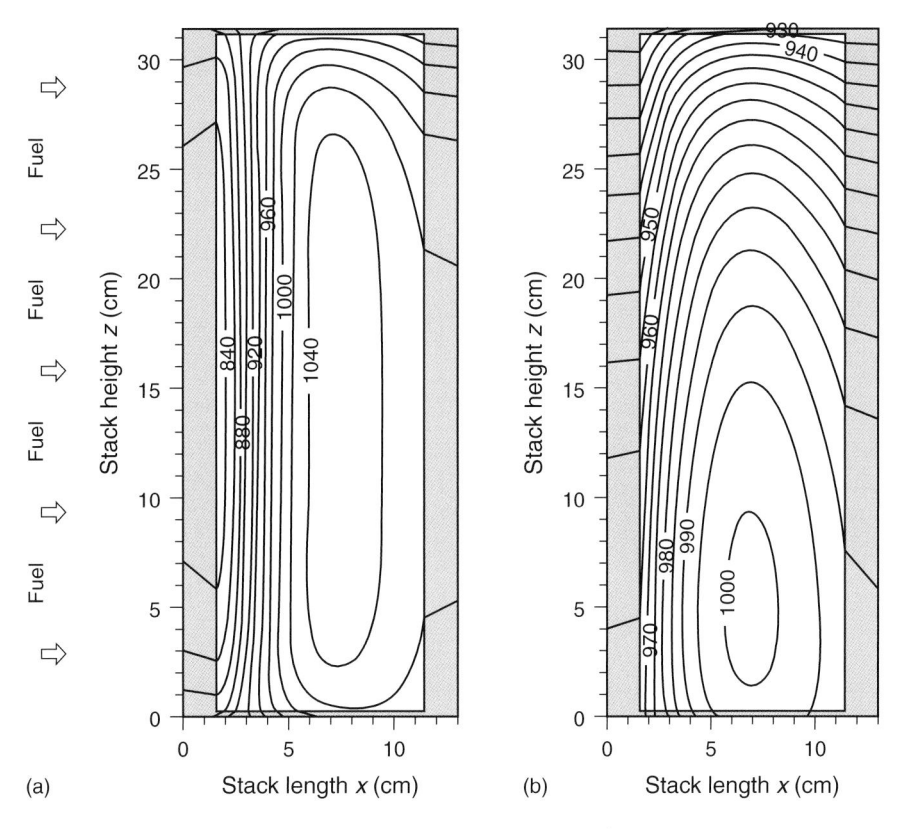

Figure 9. Temperature distribution of the solid in a 50-cell planar SOFC stack (°C): (a) ceramic bipolar plates; (b) metallic bipolar plates. (Reproduced from Achenbach (1994)[11] with permission from Elsevier Science.)

radiation heat rate is 7% for the ceramic and 22% for the metallic stack with respect to the total amount of waste heat.[11]

Similar performance analysis of a planar SOFC unit has recently been made by Iwata et al.,[12] where the authors considered pre-reformed and perfectly shifted methane. In contrast to Achenbach,[11] the gas concentration changes through the porous electrode were obtained by solving the Stefan–Maxwell equations. The calculated temperature profiles in cross-flow, counter-flow or co-flow cell geometry are similar to those presented in Figure 8.

5.2 SOFCs with integrated heat exchanger

SOFCs with integrated heat exchanger, which are already being projected for large-scale applications, usually have rotational symmetry, either as the tubular cells or planar circular cell. The tubular design, easy to manufacture with a sealless design, is particularly difficult for modeling, however. The nonsymmetrical cell geometry, relatively long gas flow paths and three-dimensional distribution of the temperature gradients all contribute to the problem. The heat management in the tubular SOFC was calculated either for the purpose of the SOFC as an energy source,[13, 14] both of them dealing with the 3 kW Westinghouse cell at Osaka Gas, or as a high temperature electrochemical reactor for the production of chemicals.[15]

The Nernst equation for the calculation of electrode potential is based on the cell reactions given by equations (16) and (17), excluding, however, the direct methane oxidation in equation (17). This is the usual approach when the water/methane reforming reaction is present or possible. Since the reforming reaction is highly endothermic, the methane is partially pre-reformed in an external reformer (cf. Figure 10) with only a residual amount entering the cell.

The principal structure of the tubular cell is given in Figure 11. The relative dimensions of this cell are given in Table 3. On the inside of the cell is an injection tube made of aluminum oxide. This tube allows for the injection of oxidant and forms the oxidant flow annulus in the active region of the cell. The next layer is the porous support tube made of calcia-stabilized zirconia. This tube is used to provide the shape and structure of the cell. The air electrode (cathode) is positioned on the outer side of the support tube. This electrode is covered with a layer of electrolyte made of YSZ. The fuel electrode (anode) is sintered onto the electrolyte. On top of the cell is the interconnection, which provides the electron path contacting the next cell.

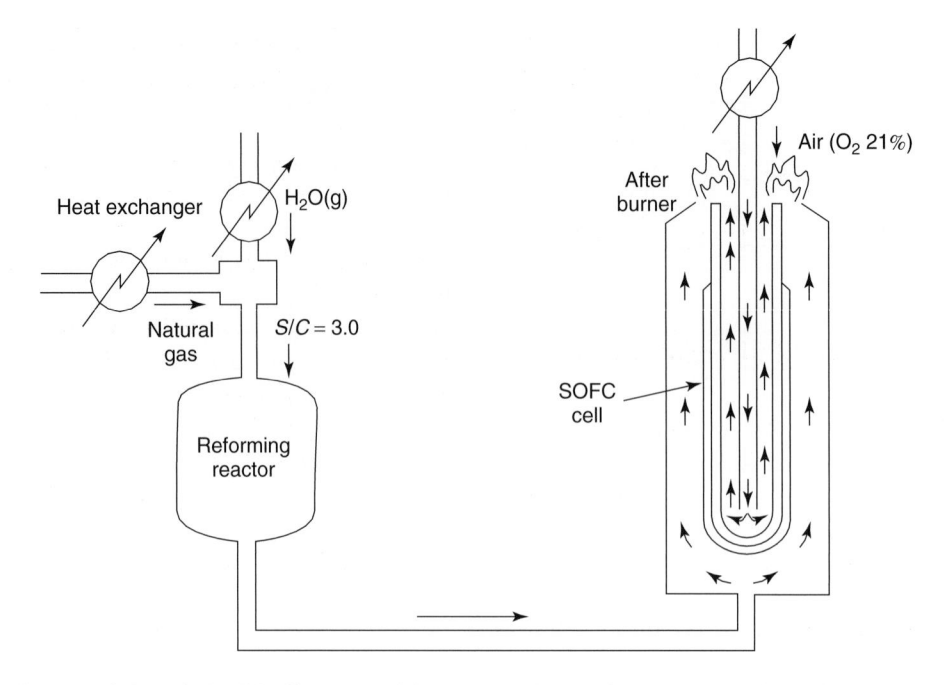

Figure 10. Block diagram of the tubular SOFC system with an external pre-reformer. (Reproduced from Hirano et al. (1992)[13] by permission of The Electrochemical Society, Inc.)

Table 3. Dimensions of a tubular SOFC.[a]

Thickness	
Support tube	1400–1600 μm
Cathode	600–800 μm
Electrolyte	40 μm
Anode	100 μm
Diameter	
Aluminia tube inside	approx. 4 mm
Aluminia tube outside	approx. 6 mm
Support tube: inside	10 mm
Support tube: outside	13 mm
Tube axial length:	300 mm

[a]Data according to Hirano et al.[13]

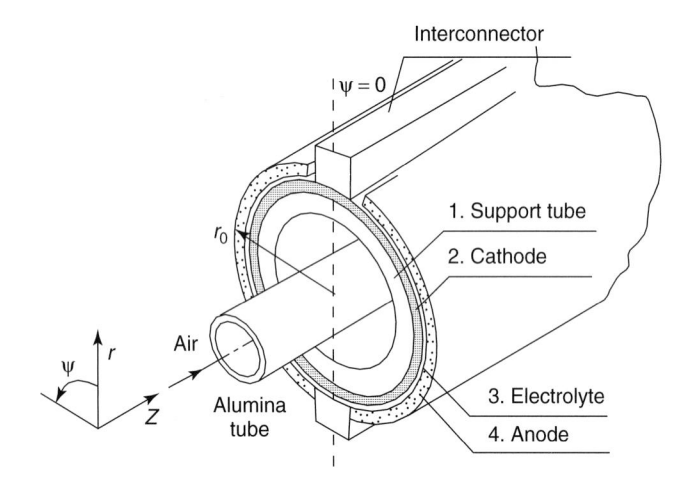

Figure 11. Structure of the tubular SOFC used for calculations. (Reproduced from Hirano et al. (1992)[13] by permission of The Electrochemical Society, Inc.)

5.3 Performance analysis of circular SOFC considering current density and temperature distributions

To make a mathematical calculation of the tubular SOFC system, we restrict calculations here to a single cell only. Because of the long gas flow path shown in Figure 10, both the oxidant and the fuel must diffuse through the hydrodynamic boundary layer. This leads to diffusion polarization η_D, characterized by the diffusion limiting current I_L which can be given by the known expression

$$\eta_D = \frac{RT}{nF} \ln\left(1 - \frac{I}{I_L}\right) \quad (23)$$

where I is the current through the fuel cell. The data of the methane reforming reaction are given by equation (11). If the concentration of component i in free stream is C_i^0, the limiting current calculation is given by equation (24).[14]

$$I_L = \frac{C_i^0 n F}{\left(\dfrac{1}{\gamma A}\right) + \left(\dfrac{\dfrac{r}{|r|}\ln(r_{\text{outer}}/r_{\text{inner}})}{D_i A}\right)} \quad (24)$$

For the anode, the second term in the denominator is zero. For the cathode, the outer radius r_{out} of the interspace is the inner wall radius of the support tube, the inner radius r_{inner} of the interspace is the outer wall radius of the injection tube (cf. Figure 11). The value of the convective mass-transfer coefficient γ for the cathode can be calculated using the empirical equation for the Sherwood number Sh at laminar flow conditions as the expression (25).[14]

$$Sh = \frac{\gamma d_{hyd}}{D_i} = 0.83 \left(\frac{r_{inner}}{r_{outer}}\right) + 4.02 \qquad (25)$$

For the fuel stream, the Sherwood number Sh can be assumed to be constant at 4.02. As in the case of the planar SOFC, the activation polarization and the conversion rates of the gas components can be expressed by the combination of the Butler–Volmer equations taking into account the corresponding stoichiometry. For the calculations, all the heat transport modes – conduction, convection and radiation – are used in a similar way as for the planar SOFC. A schematic overview of the used modes of heat transfer is given in Figure 12.

For the tubular cell geometry, the energy balance of the solid phase expressed in cylindrical coordinates equation (10) in **Low temperature fuel cells**, Volume 1, is applied. The source term Q^* in equation (10) in **Low temperature fuel cells**, Volume 1, contains the sum of the reaction enthalpies, irreversible heat rates of the electrochemical reaction and the ohmic heating. The boundary conditions involve both the heat convection according to equation (4) or equation (12) in **Low temperature fuel cells**, Volume 1, and the radiation according to equation (2). In this latter equation the shape factor between all the elements must be determined using equation (3).

The SOFC tubular model developed in Refs. [13, 14] was tested with the experimental data obtained from the Westinghouse 3 kW test module operated by Osaka Gas in Japan. The calculated curves matched the performance data well. A comparison between the calculated and measured cell temperature as a function of the cell length is given in Figure 13.

This figure shows an increase in the cell temperature as a function of axial position at the fuel electrode/electrolyte interface. The position zero point is defined at the bottom of the support tube. The cell temperature increases following the temperature profile of the fuel.[13, 14] Since the system uses air as a coolant, the air temperature inside the alumina injection tube decreases, whereas the temperature of the reverse air stream outside the injection tube increases, analogously to the fuel stream as shown in Figure 14.[13, 14]

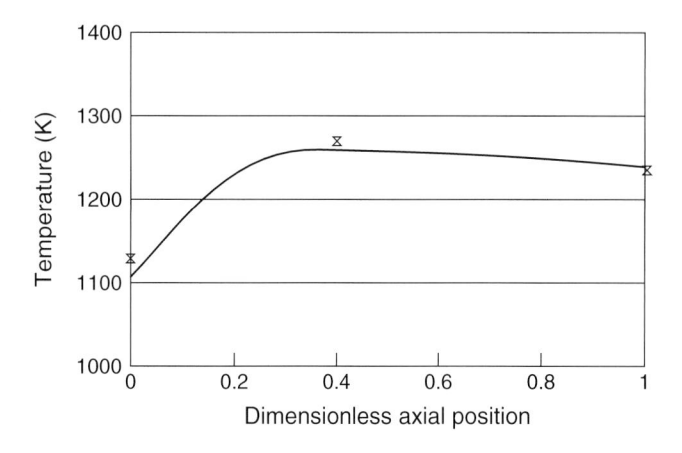

Figure 13. Comparison of calculated and measured cell temperatures. (Reproduced from Bessette et al. (1995)[14] by permission of The Electrochemical Society, Inc.)

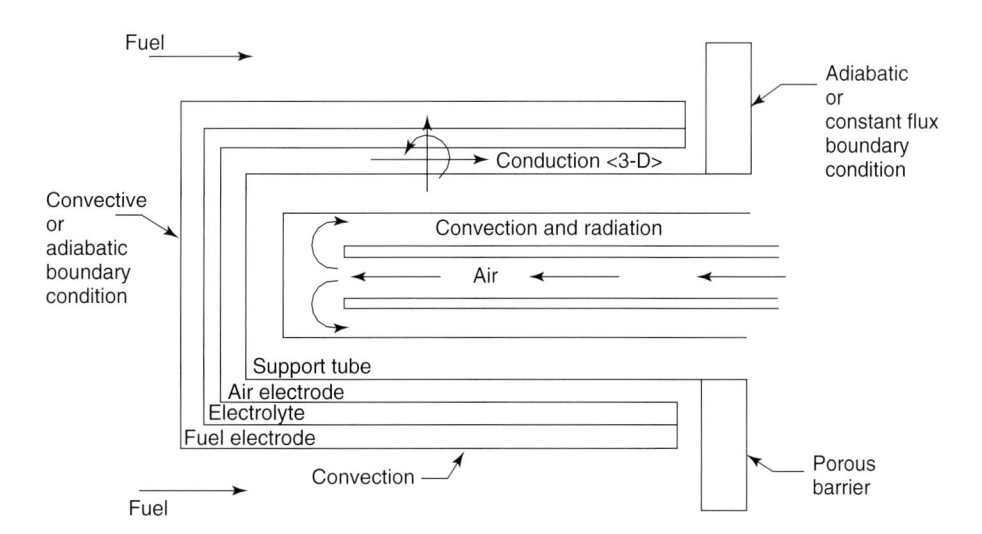

Figure 12. Modes of heat transfer in a tubular SOFC. (Reproduced from Bessette et al. (1995)[14] by permission of The Electrochemical Society, Inc.)

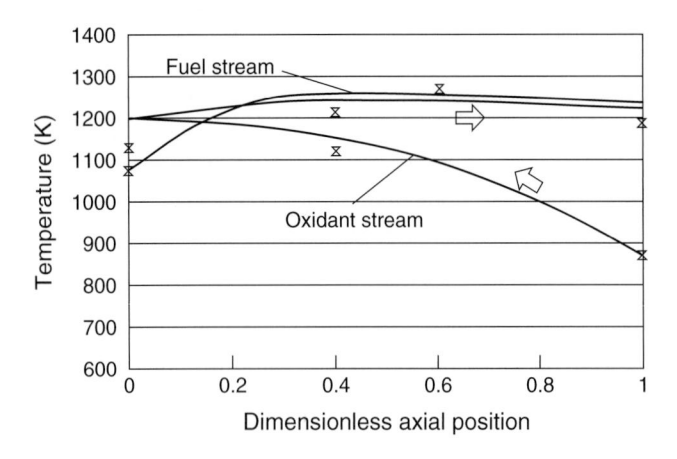

Figure 14. Comparison of calculated and measured gas stream temperatures. (Reproduced from Bessette *et al.* (1995)[14] by permission of The Electrochemical Society, Inc.)

An important and interesting thermal feature of SOFC cells (both the planar and tubular) is their response concerning internal reforming. Figure 15 shows the effect that the internal reforming of methane has on the fuel stream temperature for three increasing inlet mole fractions of methane. Due to the high rate of the reforming kinetics the main amount of methane is reformed within the first 10 cm of the cell length. One can estimate that even low percentages of methane of up to 20% lead to a severe decrease in temperature at the inlet of the fuel stream.

Analogously, the air stream temperatures show a drop of 200 K for the same increase of three methane inlet compositions, which lowers the average cell temperature and causes a degradation in performance.[14]

A solid oxide fuel cell with integrated air preheater having a planar circular shape has the benefit that the temperature gradients in the solid structure can be reduced.[16] A scheme of such a stack is shown in Figure 16.

Fuel is fed in at the center of each circular cell and then flows to the outer rim of the anode compartment. Air is fed in at the outer rim of each cell. It flows into the preheater chamber towards the cell center and enters the cathode compartment. Then it flows in the opposite direction to the outer rim so that the flow geometry is co-flow within the electrochemically active part of the cell. Due to the cylindrical symmetry and the co-flow geometry the composition and flow rate of the reacting species depend only on the radial coordinate. The geometric parameters of the SOFC are given in Table 4.

Heat management of the hydrogen/oxygen cell following the reaction equations (16) is calculated as a two-dimensional model which is a cylindrical function of the radial and axial coordinates (r and z).[16] The electrode polarizations are described by the Butler–Volmer equation, in which the concentration polarization was corrected in the usual way. As a criterion for the concentration polarization,

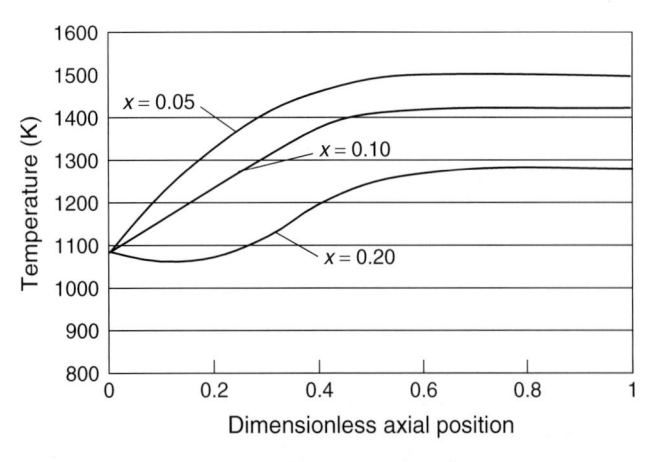

Figure 15. Effect of endothermic reforming on fuel stream temperature (x, molar fractions of methane). (Reproduced from Bessette *et al.* (1995)[14] by permission of The Electrochemical Society, Inc.)

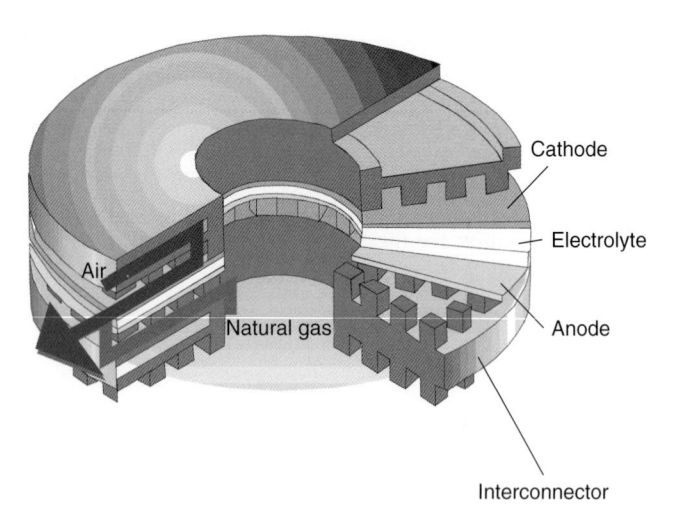

Figure 16. Scheme of a planar circular SOFC stack. (Reproduced from Costamagna and Honegger (1998)[16] by permission of The Electrochemical Society, Inc.)

Table 4. Some design parameters of the single planar circular cell.[a]

Inner cell radius	0.01 m
Inner radius of the electrochemically active area	0.02 m
Outer cell radius	0.06 m
Anode thickness	1×10^{-4} m
Cathode thickness	0.5×10^{-4} m
Channel height	1×10^{-3} m
Rib dimension	1.5×10^{-3} m

[a]Data according to Costamagna and Honegger (1998).[16]

the Thiele modulus for the spherical catalyst shape was used. The macroscopic mass balance is analogously to equation (18) given by equation (26).

$$\frac{1}{r}\frac{\partial \left(N_i \frac{1}{M_i} r\right)}{\partial r} = v_i \frac{j}{h_f v_{i,el} F} \quad (26)$$

Using the geometric factor B, which describes the ratio gas-solid heat exchange area/cell area, the energy balance for the gas flows can be obtained similarly to equation (20) for both the fuel and the air by equation (27):

$$\frac{1}{M_{\text{fuel}}}\frac{\partial (N_{\text{fuel}} c_{p,\text{fuel}} T_f)}{\partial r} = \frac{hB(T_f - T_s)}{h_f}$$

$$+ c_{p,\text{fuel}} \frac{1}{r} \frac{\partial \left(N_i \frac{1}{M_i} r\right)}{\partial r}(T_f - T_s)$$

$$\frac{1}{M_{\text{ox}}}\frac{\partial (N_{\text{ox}} c_{p,\text{ox}} T_f)}{\partial r} = \frac{hB(T_{\text{ox}} - T_s)}{h_f}$$

$$+ c_{p,\text{ox}} \frac{1}{r} \frac{\partial \left(N_i \frac{1}{M_i} r\right)}{\partial r}(T_{\text{ox}} - T_s)$$

(27)

The heat transfer coefficient h is calculated from equation (21). The energy balance of the solid is written as equation (28).

$$\frac{1}{w} hB(T_f - T_s) - \frac{1}{w}\left(\frac{\Delta H}{nF} + E\right) j$$

$$+ \bar{\lambda}\left[\frac{1}{r}\frac{\partial}{\partial r}\left(r\frac{\partial T_s}{\partial r}\right)\right] + \bar{\lambda}\frac{\partial^2 T_s}{\partial z^2} = 0 \quad (28)$$

A comparison between the experimental data and the calculated anode gas temperature is given in Figure 17.

Due to the small cross section of the channels, a sharp increase of the anode gas temperature from the input value to that of the solid temperature is visible in Figure 17. The slight temperature increase along the cell radius is due to the dissipation of the ohmic and electrode processes, whereas the temperature increase at the cell rim is given by the afterburning reaction. In the planar circular SOFC the temperature gradients in the solid are almost negligible and increase slightly with increasing the hydrogen fuel rate. Figure 18 shows the simulated results of temperature level in the solid along the radial coordinate at a high fuel utilization rate of 90%, which decreases in the radial direction. A comparison with Figure 13 suggests that the air cooling is more effective with the planar SOFC than with the tubular SOFC.

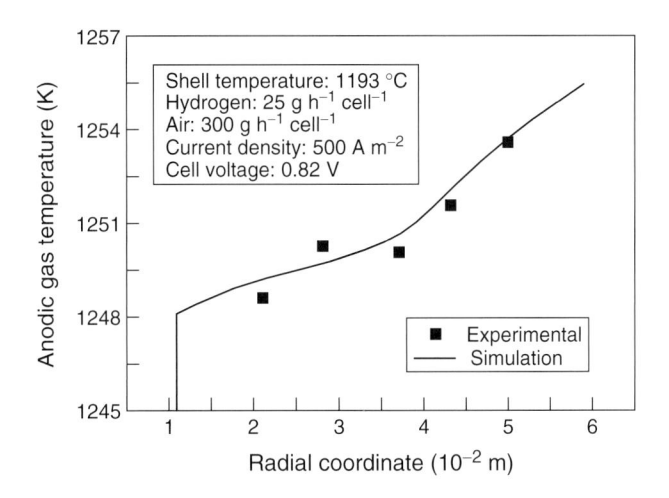

Figure 17. Anode gas temperature along the radial coordinate for planar circular stack. (Reproduced from Costamagna and Honegger (1998)[16] by permission of The Electrochemical Society, Inc.)

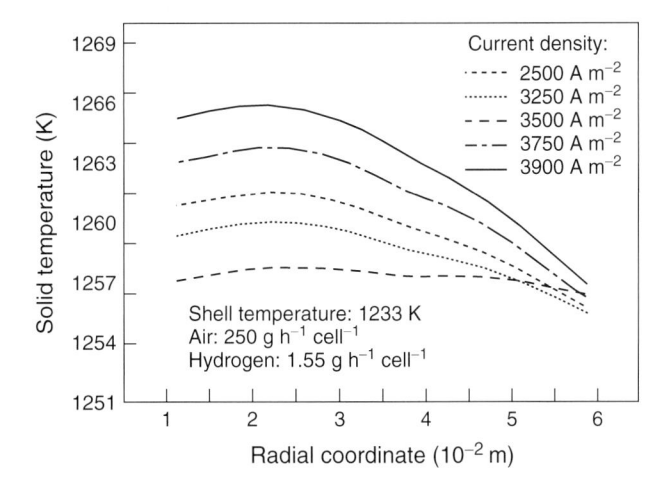

Figure 18. Temperature distribution along the radial coordinate at different current densities. (Reproduced from Costamagna and Honegger (1998)[16] by permission of The Electrochemical Society, Inc.)

Model results for the temperature distribution along the stack radius at high fuel utilization rates are shown in Figure 19. Even changing the shell temperature does not significantly change the temperature gradient in the solid along the cell radius as shown in this figure.

This means that the SOFC stack with integrated air preheater in the given planar configuration is expected to operate safely at high fuel utilization and no significant problems of mechanical stresses due to the temperature gradient in the ceramic material should occur.

5.4 Heat removal for cogeneration in SOFCs

SOFC systems offer the possibility of cogeneration of heat and electricity. Since the SOFC exhaust is available at

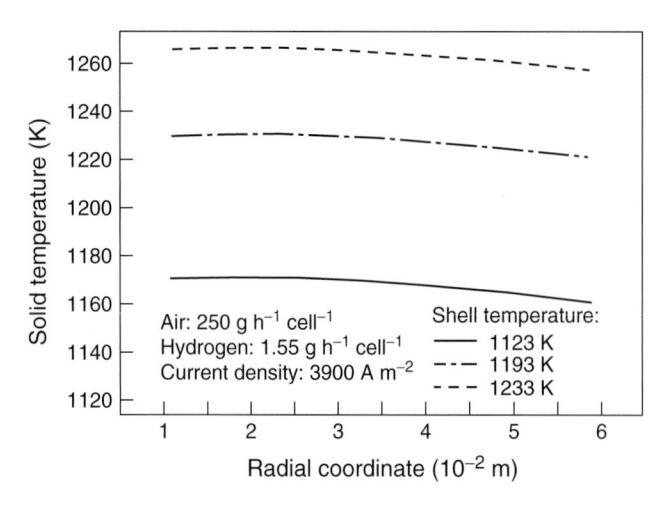

Figure 19. Temperature distribution along the radial coordinate at variation of shell Temperature. (Reproduced from Costamagna and Honegger (1998)[16] by permission of The Electrochemical Society, Inc.)

high temperatures, recovered heat can take the form of process steam or hot water. Waste heat produced in nonideal electrochemical processes can be supplied as useful heat at various temperature levels. The flowsheet of a 200 kW SOFC cogeneration power plant is shown in Figure 20.[17] The SOFC reference data are listed in Table 5.

Table 5. Some selected 200 kW SOFC reference data.[a]

Degree of methane reforming	50%
Fuel temperature at stack inlet	850 °C
Air temperature at stack inlet	850 °C
Air temperature increase in stack	100 K
Fuel utilization	80%
Cell voltage	0.75 V

[a]Data from Riensche et al.(1998).[17]

The natural gas stream (originally 400 kW LHV) is compressed to overcome the pressure losses in the components. Before entering the prereformer it is mixed with steam produced in a heat integrated boiler. The prereformer is heated recuperatively by the hot gas leaving the afterburner. The large air stream has an energy demand of 41 kW for compression. Then the air is preheated recuperatively up to 850 °C. In the SOFC stack a gross electric direct current (d.c.) power of 231 kW is produced. After subtracting the energy loss of the inverter and the energy demand for compression a net alternating current (AC) power output of 172 kW remains. Thus, the electrical plant efficiency is 43%. Taking into account the useful heat of 94 kW additionally produced, a total plant efficiency of 67% is obtained.

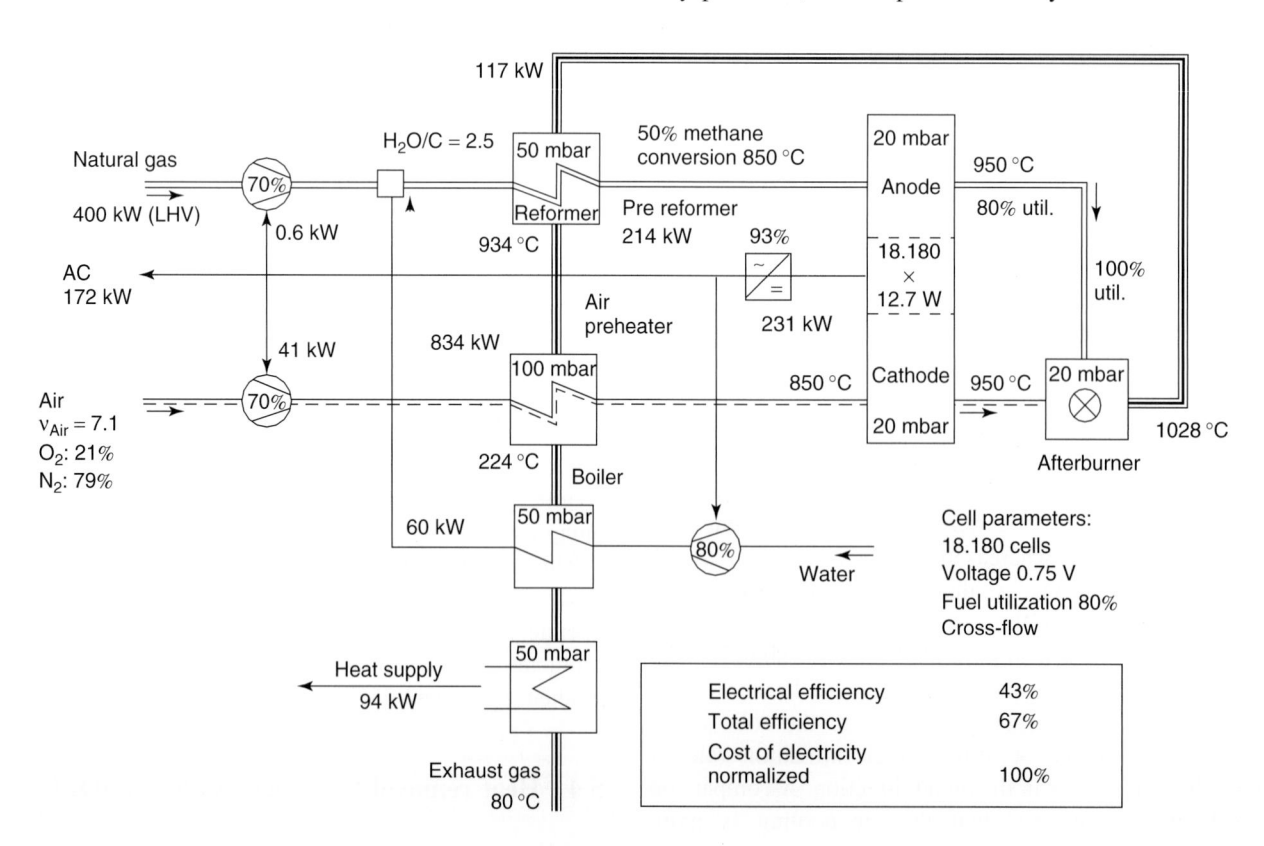

Figure 20. Basic concept of a SOFC combined heat and power plant. (Reproduced from Riensche et al. (1998)[17] with permission from Elsevier Science.)

Both these efficiencies can be further optimized by:

(i) variation of process parameters
(ii) by variation of the flowsheet.

The second method (flowsheet variation) is discussed here briefly.[17] In the plant concept described, the steam production is external and heat integration is realized by recuperative heat exchange for fuel and air preheating and steam production. Alternatively, gas recycling of the stack outlet gases leads to better process variants. The first possibility is given by cathode gas recycling. In this case, electrical efficiency of 46–49% the total efficiency of 77–78% can be achieved. In the next case, instead of external steam production the water produced in the electrochemical reaction (16) can be used when a part of the hot anode outlet gas is recycled to the prereformer inlet. Gas recycling can be realized either by gas blowers, hot gas fans, jet pumps, or after condensing the steam. Anode gas recycling provides an internal steam circuit. Therefore the steam concentration in the exhaust gas and the corresponding heat loss are reduced. For lower air ratios (internal reforming or cathode gas recycling), the temperature increase in the afterburner is higher so that the ratio of unused/used temperature range of the exhaust gas becomes smaller. All this results in a plant efficiency of 81%. If complete 100% internal reforming after condensing the steam is possible, the air ratio is essentially decreased, which leads to a high temperature increase in the afterburner so that the portion of unused heat decreases and the total efficiency reaches 93%.

Two cogeneration SOFC systems are possible: for planar cells and for tubular cells. Parametric studies show that the long current path in the tubular electrodes along the circumference of the tube results in an additional voltage drop of the order of 100 mV for usual operating conditions. This disadvantage of the tubular concept can, however, be compensated by pressurized operation. The simulated performance characteristics show that in comparison to atmospheric pressure the cell voltage can be about 60 mV higher for 5 bar and approximately 100 mV for 10 bar.[18] For this reason, besides the atmospheric tubular 100 kWe SOFC, Westinghouse also suggested pressurized systems for tubular stacks.[19] In the Westinghouse stack design, the cells are oriented with the axis vertical and the closed end down. The prototype module is capable of 200 kW d.c. using the 150 cm cells. The process schematic for atmospheric power generation is given in Figure 21.

In this design, the reforming process is to a large extent directly heat-integrated with the electrochemical and ohmic heat evolution in the cells. In a first reforming step, the higher hydrocarbons and a small amount of methane are converted in an adiabatic prereformer located in the anode recycle loop. In a second reforming step, methane is completely converted in an in-stack reformer. Its construction principle is shown in Figure 22.

The reformer is radiantly heated by the adjacent rows of cells. Therefore the corresponding fraction of the electrochemical excess heat is directly transferred to the reforming process and is not transported via the cooling air stream.

In elevated pressure systems, turbine work is extracted from the gas stream by an expander before the exhaust

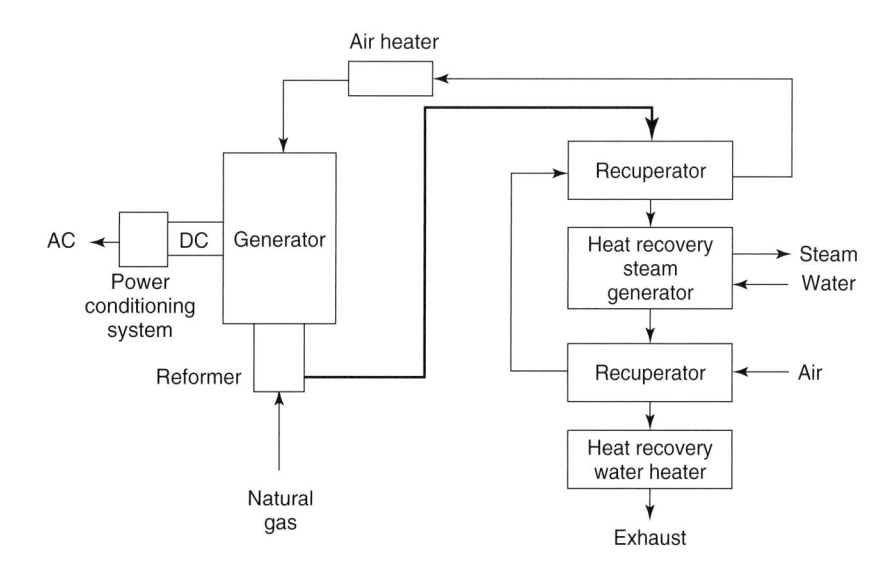

Figure 21. Process schematic for atmospheric pressure SOFC co-generation system. (Reproduced from Veho and Forbes (1998)[19] by permission of European Fuel Cell Forum, Dr. Ulf Bossel.)

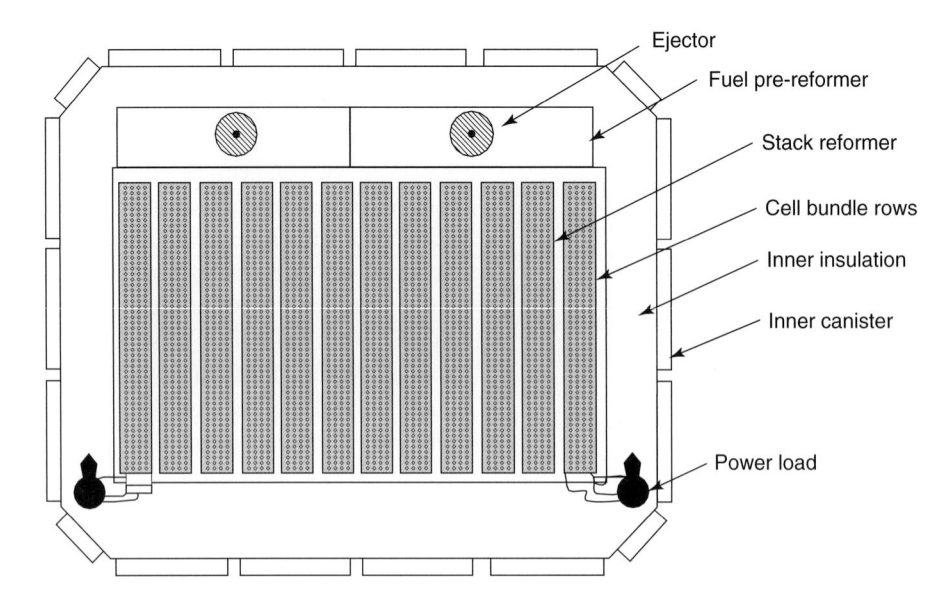

Figure 22. Cross-section of SOFC stack as electricity and heat generator. (Reproduced from Veho and Forbes (1998)[19] by permission of European Fuel Cell Forum, Dr. Ulf Bossel.)

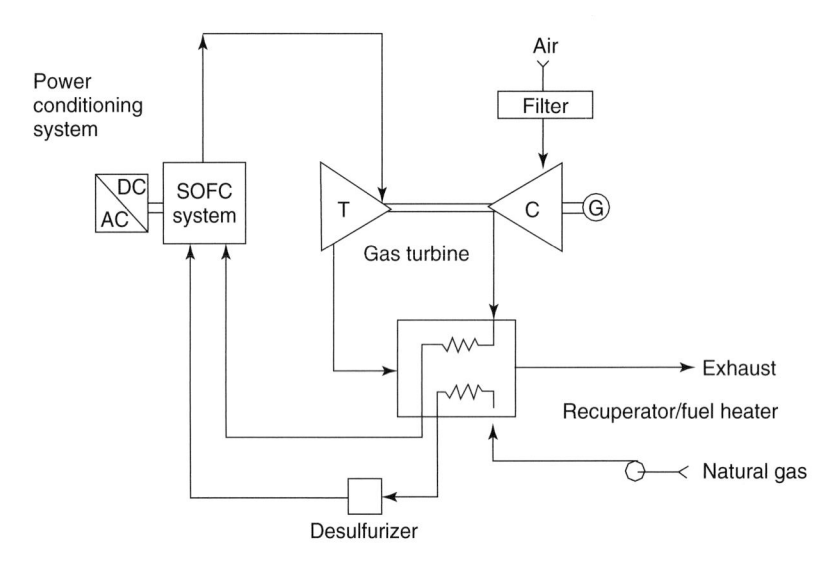

Figure 23. Process schematic for SOFC/gas turbine combined pressurized system. (Reproduced from Veho and Forbes (1998)[19] by permission of European Fuel Cell Forum, Dr. Ulf Bossel.)

enters the recuperator. Depending on the turbine under consideration and the process capacity, such systems can be configured in a number of ways, thus achieving efficiencies between 60 and 70%. Figure 23 shows a process schematic for such a pressurized SOFC combined cycle. In this case, the expander directly drives the compressor.

The cell stack shown schematically in Figure 23 operating at a pressure of 3.5 atm with a 50 kW micro-turbine generator is expected to achieve an electrical efficiency of 60%.[19] Recent performance calculations of this system[20] support this assumption. As an example, a comparison between the SOFC stack efficiency (SOFC group) and the hybrid system SOFC/micro-turbine generator (hybrid plant) is given in Figure 24.[20]

Depending on the proposed design point (DP) (A,B,..., G), overall efficiencies of between 56 and 62% can be achieved. Figure 24 shows the response of the overall plant and the SOFC group to a variation of the fuel flow rate from the design point DP to the part load conditions A–G. The efficiencies are expressed in terms of dimensionless plant power (P_N/P_{DP}) at a fixed turbine rotational speed of 85 000 rpm. T_{FC} is the temperature of the SOFC stack, U_0 the oxygen utilization factor and TIT the turbine inlet temperature.

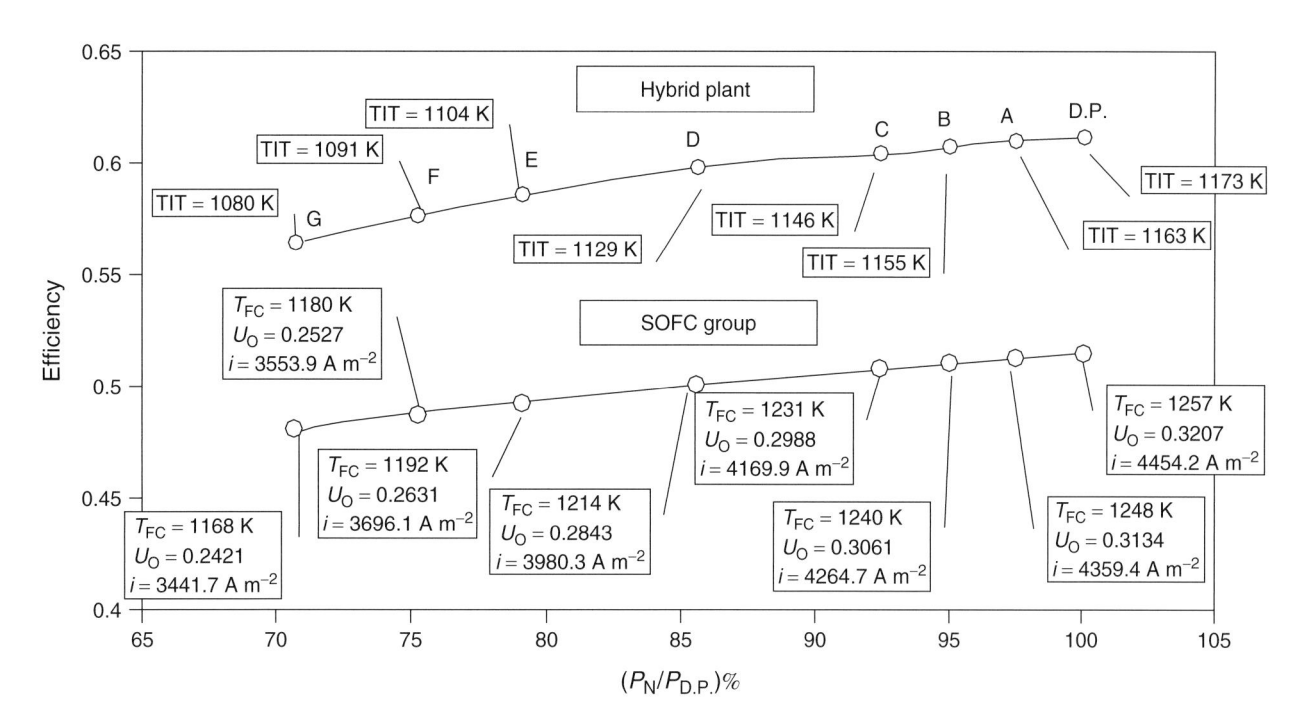

Figure 24. Hybrid plant and SOFC efficiencies vs. dimensionless plant power (P_N/P_{DP}). (Reproduced from Costamagna *et al.* (2001)[20] with permission from Elsevier Science.)

ACKNOWLEDGEMENTS

I would like to acknowledge the useful and fruitful discussions with Elmar Achenbach, who has read and commented on some parts of the manuscript.

LIST OF SYMBOLS

a	constant (–)
A	total area (m^2)
b	constant (–)
B	ratio gas-solid heat exchange area/cell area (–)
c	constant (–)
c_p	heat capacity (J (kg K)$^{-1}$)
C_i^0	concentration of component i in free stream (mol m^{-3})
d	constant (–)
D_i	diffusion coefficient of component i (m^2 s^{-1})
d_{hyd}	hydraulic diameter (m)
E_c	cell voltage (V)
E^0	standard cell voltage (V)
E_0	EMF (open circuit voltage (OCV)) (V)
F	Faraday constant (A s mol^{-1})
F_s	radiation shape factor (–)
g	gravity constant (m s^{-2})
ΔG	Gibb's free energy (J mol^{-1})
Gr	Grashof number (–)
h	convection heat transfer coefficient (W (m^2 K)$^{-1}$)
h_{er}	equivalent heat transfer coefficient for radiation (W (m^2 K)$^{-1}$)
h_a	air channel height (m)
h_f	fuel channel height (m)
ΔH_i	enthalpy change of component i (J mol^{-1})
H_{vap}	enthalpy of vaporization (J mol^{-1})
I	total current flow (A)
j	current density (A m^{-2})
j_i	local volumetric current of component i (A m^{-3})
k_{CH_4}	reaction velocity of reforming reaction (mol$_{CH_4}$ (mol$_{Ni}$)$^{-1}$ s^{-1})
$\bar{k}_{educt}$	surface reaction rate of educts (mol m^{-2} s^{-1})
$\bar{k}_{prod}$	surface reaction rate of products (mol m^{-2} s^{-1})
$\bar{k}_{CH_4}$	surface reaction rate of reforming reaction (mol$_{CH_4}$ m^{-2} s^{-1})
K_λ	overall heat transfer coefficient (W (m^2 K)$^{-1}$)
K_p	equilibrium constant at constant pressure (–)

K_i	adsorption equilibrium constant of component I (Pa^{-1})		β	volume coefficient of thermal expansion (K^{-1})
			β_I	constant (W s (mol K)$^{-1}$)
l	distance (m)		γ	material transfer coefficient (m s^{-1})
m	constant (–)		δ	boundary layer thickness (m)
m_i	mass of component i (kg)		ε	emissivity (–)
M_i	molar weight of component i (kg mol^{-1})		η	overvoltage (V)
N_i	mass flow of component i (kg (m^2 s)$^{-1}$)		Θ_i	degree of coverage with component i (–)
ΔN_{SHIFT}	molar conversion rate of the water shift reaction (mol m^{-3} s^{-1})		λ	thermal conductivity (W (m K)$^{-1}$)
			$\bar{\lambda}$	average thermal conductivity of the structure (W (m K)$^{-1}$)
Nu	Nusselt number (–)			
n	constant (also the number of transferred electrons) (–)		λ_{eff}	effective thermal conductivity of the structure (W (m K)$^{-1}$)
p	constant (–)		v	kinematic viscosity (m^2 s^{-1})
P	system pressure (Pa)		v_i	stoichiometry factor of component I (–)
P_i	partial pressure of component i (Pa)		$v_{i,\text{el}}$	electrochemical stoichiometry factor (–)
Pr	Prandtl number (–)		ρ	density (kg m^{-3})
$Q, Q_{(-)}, Q_{\text{em}}$	heat transfer (W)		$\bar{\rho}$	average medium density (kg m^{-3})
Q_x, Q_y, Q_z	heat transfer rate (W)		σ	specific conductivity (S m^{-1})
Q^*	heat generated per unit volume (W m^{-3})		σ_{SB}	Stefan–Boltzman constant (W (m^2 K^4)$^{-1}$)
r	radius (m)		ϕ	porosity (–)
R	molar gas constant (J (mol K)$^{-1}$)			
R_i	internal resistivity (Ω)			
R_{vap}	rate of water vaporization (mol s^{-1})			
Re	Reynolds number (–)			
S_i	entropy of component i (J (mol K)$^{-1}$)			
S_{HM}^*	transported entropy of protons in the membrane (J (mol K)$^{-1}$)			
Sc	Schmidt number (–)			
Sh	Sherwood number (–)			
t	time (s)			
t_{w}	transference coefficient of water (–)			
T	temperature (K)			
T_{b}	boiling temperature (K)			
T_{s}	solid temperature (K)			
T_{f}	fluid temperature (K)			
u	fluid velocity (m s^{-1})			
u_∞	free-stream velocity (m s^{-1})			
u_x, u_y, u_z	fluid velocity components (m s^{-1})			
U_0	oxygen utilization factor (–)			
V	volume (m^3)			
W	molar work (J mol^{-1})			
w	channel width (m)			
x	co-ordinate (m)			
y	co-ordinate (m)			
z	co-ordinate (m)			

Greek

α	thermal diffusivity (m^2 s^{-1})
α_0	specific condensation area (m^2 m^{-3})
α_I	constant (W s (mol K^2)$^{-1}$)

REFERENCES

1. J. P. Holman, 'Heat Transfer', McGraw-Hill, New York (1990).
2. I. Drescher, "Kinetics of the Methene Steam Reforming", PhD thesis RWTH Aachen, Forschungszentrum Jülich (1999).
3. Ch. Rechenauer and E. Achenbach, 'Three-dimensional Mathematical Modelling of Steady and Unsteady State Conditions for Solid Oxide Fuel Cells', PhD thesis RWTH Aachen, Jül-2752, Forschungszentrum Jülich GmbH (1993).
4. J.-H. Koh, B. S. Kang and H.Ch. Lim, *J. Power Sources*, **91**, 161 (2000).
5. W. He and Q. Chen, *J. Power Sources*, **55**, 25 (1995).
6. W. He and Q. Chen, *J. Power Sources*, **73**, 182 (1998).
7. F. Yoshiba, N. Ono, Y. Izaki, T. Watanabe and T. Abe, *J. Power Sources*, **71**, 328 (1998).
8. C. G. Vayenas, P. G. Debenedetti, I. Yentekakis and L. L. Hegedus, *Ind. Eng. Chem. Fundam.*, **24**, 316 (1985).
9. S. Ahmed, Ch. McPheeters and R. Kumar, *J. Electrochem. Soc.*, **138**, 2712 (1991).
10. R. Selman, R. Herbin, M. Fluck and R. Gruber, in Proc. Workshop on Modelling and Evaluation of Advanced SOFC, Hertenstein, Switzerland, June 24–29, Swiss Federal Office of Energy, Bern, pp. 17–45 (1990).
11. E. Achenbach, *J. Power Sources*, **49**, 333 (1994)
12. M. Iwata, T. Hikosaka, M. Morita, T. Iwanari, K. Ito, K. Onda, Y. Esaki, Y. Sakaki and S. Nagata, *J. Power Sources*, **132**, 297 (2000).
13. A. Hirano, M. Suzuki and M. Ippommatsu, *J. Electrochem. Soc.*, **139**, 2744 (1992).

14. N. F. Bessette II, W. J. Wepfer and J. Winnick, *J. Electrochem. Soc.*, **142**, 3792 (1995).
15. P. Costamagna, E. Arato, P. L. Antonucci and V. Antonucci, *Chem. Eng. Sci.*, **51**, 3013 (1996).
16. P. Costamagna and K. Honegger, *J. Electrochem. Soc.*, **145**, 3995 (1998).
17. E. Riensche, J. Meusinger, U. Stimming and G. Unverzagt, *J. Power Sources*, **71**, 306 (1998).
18. E. Riensche, E. Achenbach, D. Froning, M. R. Haines, W. K. Heidug, A. Lokurlu and S. von Andrian, *J. Power Sources*, **86**, 404 (2000).
19. S. E. Veyo and C. A. Forbes, in Proc. Third European Solid Oxide Fuel Cell Forum, 2–5 June 1998, Nantes, France, P. Stevens (Ed), Publisher Dr. Ulf Bossel, p. 79 (1998).
20. P. Costamagna, L. Magistri and A. F. Massardo, *J. Power Sources*, **96**, 352 (2001).

Chapter 11
Air-cooled PEM fuel cells

R. von Helmolt[1] and W. Lehnert[2]

[1] Adam Opel AG – Global Alternative Propulsion Center, Rüsselsheim, Germany
[2] Center for Solar Energy and Hydrogen Research Baden-Wurttemberg, Ulm, Germany

1 BASIC CONSIDERATIONS

1.1 Requirements, comparison to liquid-cooled PEM fuel cells

At peak power, proton exchange membrane (PEM) fuel cells operated with hydrogen are usually operated at a voltage of around 0.7 V. This means that power conversion in a PEM fuel cell usually has a thermal efficiency of about 50%. Nowadays high performance PEM fuel cells reach electrical power densities of up to $0.8\,\text{W}\,\text{cm}^{-2}$, referring to active membrane area. Correspondingly, a heat flux density in the order of $1\,\text{W}\,\text{cm}^{-2}$ has to be removed from the fuel cell. Heat conductivity of the bipolar plates is usually far from sufficient to remove this heat flux density without creating significant heat gradients. Therefore, a cooling medium has to flow through the bipolar plate and the heat has to be removed by convection of this medium. Most PEM fuel cells are cooled with de-ionized water, which is cheap, not toxic and has a high heat capacity.

The widespread liquid cooling has however some disadvantages. For example, the coolant is not only in thermal but also in electrical contact with the bipolar plates. In order to prevent electrolysis, which might affect the plates as well as the coolant, the coolant has to have very low conductivity, which is usually achieved by high water purity. Ion exchangers are necessary to maintain the water purity during operation, as ions increase its conductivity. Furthermore, the coolant can be subject to freezing. It is therefore worth thinking about other possibilities for the cooling of fuel cells. It is obvious that air cooling can be an option, especially for mobile applications such as automobiles. As the complete liquid cooling loop is no longer necessary, an air-cooled system might also be much simpler. This is illustrated by simple block diagrams in Figure 1.

Because of large openings and air channels the air-cooled stack might be larger and heavier than the liquid-cooled stack but the overall fuel cell system can be lighter, smaller and cheaper. This however depends very much on the details of the specific application; e.g., if the off-heat is used or if a liquid cooling loop is needed for other components. Small vehicles and power generators seem to be good applications for air-cooled fuel cells, whereas in combined heat and power stationary systems usually liquid coolants are easier to handle.

Before coming to some design rules and examples for air cooled fuel cells, it is worth considering the heat that has to be removed by the cooling air stream. Heat flux from the stack is carried by:

- (air) cooling;
- product water evaporation;
- reactant gases heat capacity.

Heat is produced due to the irreversible cell losses. The media mentioned above also carry heat to the stack, which can be significant, e.g., if humidified gases are supplied and therefore no product water can evaporate or even condensate (e.g., for fully humidified hydrogen), or if the supplied gases are hot, e.g., from reforming (fuel) or compression (air). The irreversible cell losses, which characterize the

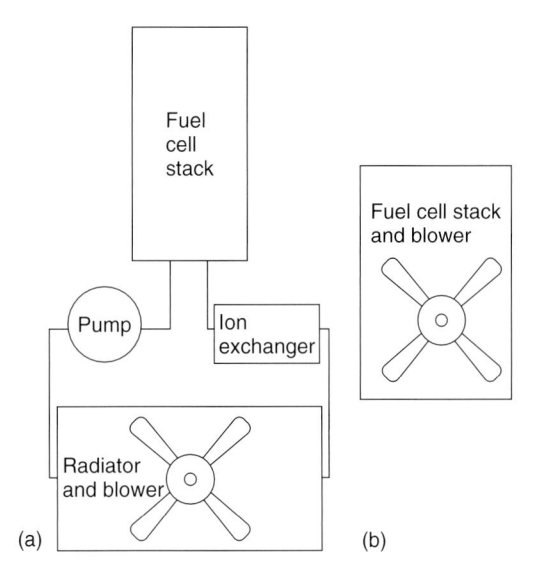

Figure 1. Block diagram for a liquid (a) and an air cooled (b) system.

polarization curve, consist in thermodynamic irreversibility, electrode kinetics, ohmic losses and mass transport irreversibility.

The different contributions for a fuel cell can be written as follows

$$\sum P_i = 0 \tag{1}$$

the sum is mainly taken over cell losses

$$P_{th} = I(U_{th} - U_{cell}) \tag{2}$$

(with $U_{th} = 1.48\,V$ for an H_2/air system under standard conditions), product water evaporation,

$$P_{vap} = Q_{SD}(\dot{m}_{H_2O-in} - \dot{m}_{H_2O-out}) \tag{3}$$

and reactant gases heat capacity

$$P_j = c_{Pj}(T_{j-in}\dot{m}_{j-in} - T_{j-out}\dot{m}_{j-out}) \tag{4}$$

$$P_{fuel} = c_{P(fuel)}(T_{fuel-in}\dot{m}_{fuel-in} - T_{cell}\dot{m}_{fuel-out}) \tag{5}$$

(assuming negligible pressure loss in the gas channel, and exiting gases at cell temperature).

The cooling air mass flow can now be calculated to balance the above equation

$$P_{cool} = c_{P(air)}\dot{m}_{c_{air}}\Delta T \tag{6}$$

As an example, we calculated the contributions for a hydrogen/air fuel cell; the results are shown in the Table 1 below. The operating conditions are assumed as follows: $T_{cell} = 80\,°C$, Air stoich = 2 at $80\,°C$ and relative humidity (RH) = 50%, H_2 stoich = 1 at $80\,°C$ and RH = 100%, pressures 200 kPa. The assumption was made that the product water is in liquid state and it can be evaporated up to the saturation value.

When decreasing (increasing) the relative humidity of the air at the entrance more (less) product water can be evaporated. As a consequence the need for external cooling decreases (increases). Table 2 shows the results of the calculations for Case 1. The results stress how important it is to decrease the relative humidity at the entrance of the stack.

Table 1. The contributions for a hydrogen/air fuel cell for the variables in equation (6). (Data from Ref. [1].)

	Case 1	Case 2	Case 3
U_{cell} (mV)	800	750	650
P_{el} (kW)	80	80	80
P_{th} (total) (kW)	68	77.8	102.2
P_{vap} (kW)	−14.30	−15.33	−17.69
$\sum P_j$ (kW)	5.40	5.83	6.73
P_{cool} (kW)	59.2	68.5	91.4

1.2 Design rules

For liquid-cooled fuel cells, especially water cooled fuel cells, coolant channels in the bipolar plate are sufficient for effective heat transfer from the plate to the coolant. For cooling air, however, an increased surface area has to be provided, in order to transfer reaction heat from the cell to the cooling air. Therefore air cooled fuel cells have increased volume, compared to liquid cooled cells. This can be shown easily if the air flow is considered which is necessary for carrying 0.1 kW heat out of the stack. If $\Delta T = 20\,K$ is allowed, we get a flow of $5\,g\,s^{-1}$, or $248\,l\,min^{-1}$. If we assume a pipe with a length of e.g., 0.25 m and the requirement for effective cooling, that the differential pressure must be lower than $\Delta p = 10\,hPa$, we can calculate a pipe diameter of 1 cm (turbulent flow). If, instead, water is used as a coolant, we get a flow of $1\,g\,s^{-1}$, which corresponds to $0.07\,l\,min^{-1}$. Water pumps are more effective and also a lower volume has to be pumped, so we assume $\Delta p = 10\,kPa$. The pipe now needs only to be 0.1 cm in diameter (laminar flow).

Table 2. The results of the calculations for Case 1 from Table 1 of increasing RH.

RH_{cath} (%)	0	20	40	60	80	100
P_{cool} (kW)	45.3	50.4	56.04	62.31	69.31	77.19

For a more realistic view, let us look at a single cell with an active area of 25 × 25 cm. The heat to be removed by the cooling is 370 W. The cooling flow field has 25 straight channels. The channel width is 5 mm. The pressure losses and temperature gradients are the same as above. If the heat is removed by air, the channel height is 0.4 mm. When using water for cooling, the channel height could be 0.2 mm. This very rough calculation shows that each cell of an air-cooled stack is about 0.4 mm thicker than of a liquid cooled stack. A comparison of a liquid-cooled and an air-cooled stack can be seen in Figure 2. Both stacks have five cells and the same gross power. In the case of the liquid-cooled stack a single cell has a thickness of 0.45 cm whereas a single cell of the air-cooled stack has a thickness of 0.95 cm.

For low cost and weight, thin bipolar plates are used, and therefore, lateral heat conductivity is low. Cooling has to be provided between cells, because otherwise lateral heat transport in the bipolar plate would lead to high temperature gradients. This is a design rule that also applies to liquid cooled cells. In air-cooled cells, however, the temperature spread between inlet and outlet of the coolant is much higher; usually, the air enters at ambient temperature, which is significantly lower than the cell operating temperature. Therefore, a temperature profile along the coolant channel is caused. This temperature profile has to be taken into account when designing the cell for optimized water management. The cell has to be balanced between dry (increased membrane resistance) and wet conditions (condensed water affects gas transport). This implies that the temperature gradient is along the cathode air channel, and increased water uptake of the cathode air goes along with increased temperature. The graph in Figure 3 shows the water uptake of the cathode air along the channel length, expressed as dewpoint, for dry cathode inlet and also for air humidified at 30 °C.

Ideally, one would assume that the optimum cell temperature profile would match the dewpoint curve. Due to kinetics, i.e. diffusion processes in the electrode and diffusion medium (and also due to hydrophilic parts in the electrode, which results in lowered water vapor pressure at the surface,) the membrane/electrode should have a higher temperature than the dewpoint temperature.

The temperature profile of the cell, T_{MEA}, is mainly defined by the temperature increase of the cooling air T_{air}. If heat production in the cell is homogeneous, the temperature T_{air} rises linearly in the channel. T_{MEA} is slightly higher, and the offset defined by the heat transfer, which is also assumed to be constant in this case (Figure 4).

This however is true only if lateral heat conductivity is completely neglected. The disadvantage of such a configuration is that the membrane electrode assembly (MEA) temperature gradient between inlet and outlet is the same as for the cooling air. Typically ΔT_{air} is about 20 K. As a consequence, parts of the cell might below the optimum operating temperature.

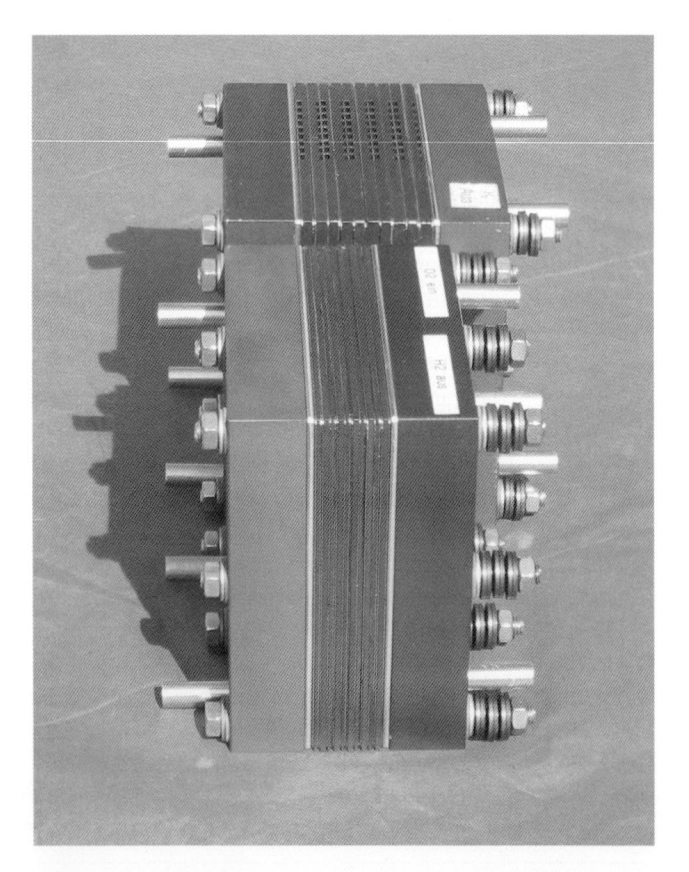

Figure 2. Comparison of liquid-cooled (front) and air-cooled stacks from the Zentrum für Sonnenenergie und Wasserstoff Forschung (ZSW).

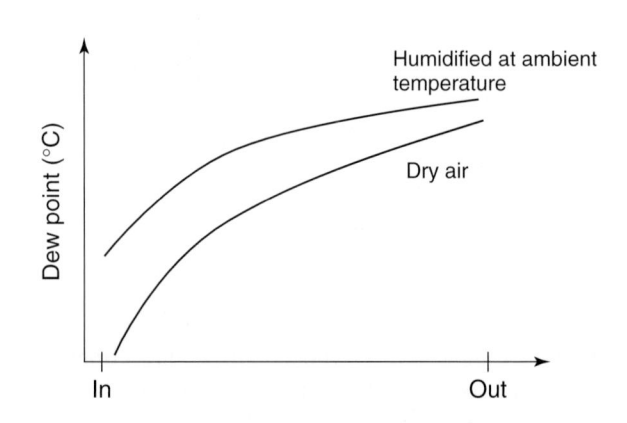

Figure 3. Water uptake of the cathode air.

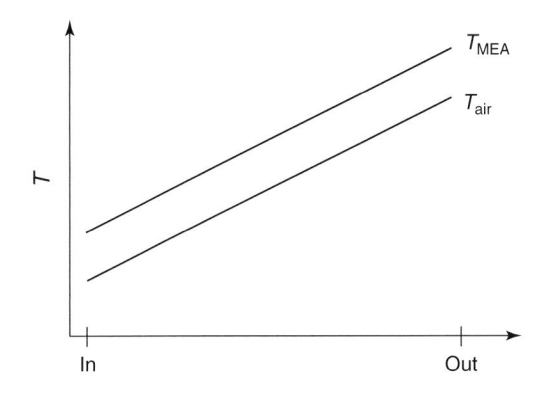

Figure 4. Temperature profile of cell and air (no lateral heat conductivity).

The heat flux emitted to the coolant can be written in the following form[2]

$$q = \int_A \alpha(T_{\text{carrier}} - T_{\text{fluid}})\frac{dA}{A} \quad (7)$$

where q = heat flux, α = local heat transfer coefficient, T_{fluid} = local temperature of coolant, T_{carrier} = local temperature of carrier material and A = total heat transfer surface.

The heat flux inside the (isotropic) heat-conducting cell-material is

$$q = k\nabla T \quad (8)$$

where q = heat flux, k = thermal conductivity and T = temperature of support material.

The equations show that there are possibilities for changing the design or the material in order to decrease the temperature gradient inside the cell.

If the heat conductivity in the plate along the channel is large, heat is transported from the outlet to the inlet and the difference between coolant temperature and cell temperature is highest at inlet. This leads to a decrease of the temperature gradient in the cell but also to high temperature gradients at the inlet. This is shown schematically in Figure 5.

If heat conductivity of the plates is low, a similar profile can be obtained by increasing the heat transferring area part $\Delta A/A$ or the local heat transfer coefficient α in the flow direction. This can be obtained e.g., by cooling fins with higher density or surface area at cell outlet than at the inlet. Figure 6 shows the profiles for a cell where we assume a linear increase of the heat transfer coefficient from inlet to outlet.

In order to keep the system simple there is no temperature adjustment of the cooling air before entering the cell. This is in contrast to liquid-cooled stacks where the cooling water is kept at a certain value independent of the ambient temperature. Therefore, in air-cooled fuel cells low ambient

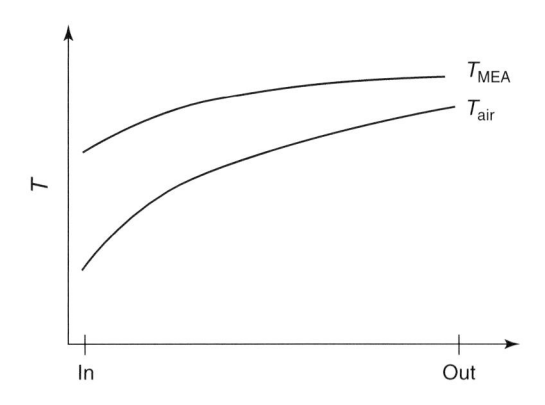

Figure 5. Temperature profile of cell and air. Heat conductivity in the bipolar plates taken into account.

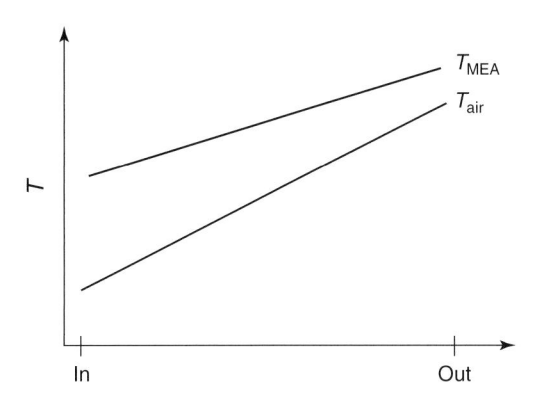

Figure 6. Temperature profile of cell and air. Linear increase of heat transfer coefficient from inlet to outlet.

temperatures might cause a high temperature spread in the cell which can misbalance the water management, and a large area of the cell is at a temperature well below optimum operating temperature. Control of air temperature and flow is especially critical at low ambient temperature, and the mean stack performance and efficiency may be significantly lower than at higher temperatures. Figure 7

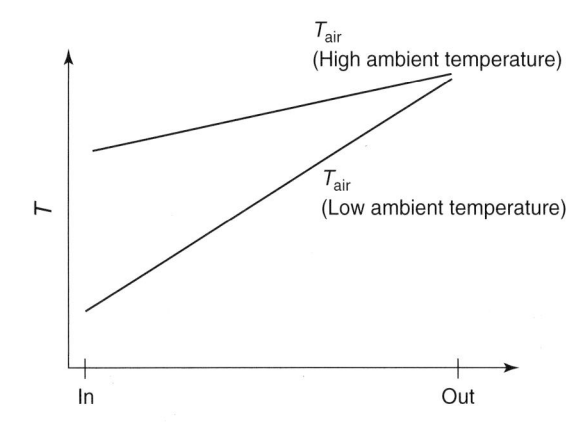

Figure 7. Air temperature rise inside the cooling channel for low and high ambient temperatures.

shows the temperature rise of the cooling air along the channel for different ambient temperatures. For low ambient temperatures the mass flow of the cooling air has to be reduced in order to keep the cell at operating temperature.

2 EXAMPLES FOR AIR-COOLED PEM FUEL CELLS

Two basic applications of air cooled PEM fuel cells can be distinguished. One is the automotive-type PEM fuel cell stack, which has high power densities of typically $>0.5\,W\,cm^{-2}$, $>0.5\,W\,g^{-1}$ and $>0.5\,W\,cm^{-3}$, and a gross power of 20–100 kW. The other application is a lower power PEM fuel cell with typically 1 kW gross power. In this section some examples will be presented.

2.1 Automotive-type air-cooled stacks

Besides power, the range of environmental parameters for the automotive PEM fuel cell is usually higher. A separate control of process air and cooling air is desirable, so stack temperature and process gas flow can be adjusted independently and according to stack temperature, outside temperature and power setting. Requirements for process air are more restrictive with respect to dust, humidity etc. If for example, cooling air would be fed to the fuel cell cathode, huge filters would be needed to prevent fuel cell contamination. Therefore, most fuel cells of this type currently have separate manifolds for coolant and cathode air. An exception, however, is the Proton Motor fuel cell.

In an European Community (EC) project, Siemens demonstrated with its project partners, ECN (Netherlands) and PSI (Switzerland) the adaptation of existing technology in solid polymer fuel cells to the demands of commercial applications. The focus was on highly efficient operation with air at low overpressure, and a simplified construction of the electrochemical cell suitable for mass production at low cost. Under assumption of the rated performance of $0.35\,W\,cm^{-2}$ and an automated production of 10^5 stacks per year, material costs of $36\,€\,kW^{-1}$ can be reached. (However, current prototypes based on this design will have material costs around $2800\,€\,kW^{-1}$. The corresponding labour costs will exceed even this value.) Design and construction were crucial issues in achieving the cost goals. Besides the material costs, the manufacturing has turned out to be an important factor. Therefore, feasibility of mass production of stack components and easy assembly of the fuel cell stack have to be considered.

A 35 cell demonstration stack based on embossed thin-sheet metal and a simplified stack design was assembled and operated in 1999. In the H_2/air mode it yielded 4.0 kW at 0.7 V. The peak power was 6.2 kW (all values at 1.5 bar_a). The power volume and weight was 1.8 l kW^{-1} and 2.0 kg kW^{-1} (extrapolated to a 100 cell stack with optimized end plates), including the inserted cooling gap which renders an external heat exchanger unnecessary.

The stack concept is based on thin-sheet metal as construction material. The sheet metal technology enables a simple manufacturing of the components by normal embossing and punching operations. These established processes are suitable for mass production. The construction also enables an easy assembly since it uses only clamps, thereby avoiding the use of any additional seals or glues. The disassembly of such a construction for repair or recycling purposes is also easy. Furthermore, the concept allows for a compact lightweight fuel cell stack, which still enables air cooling (Figure 8).[3]

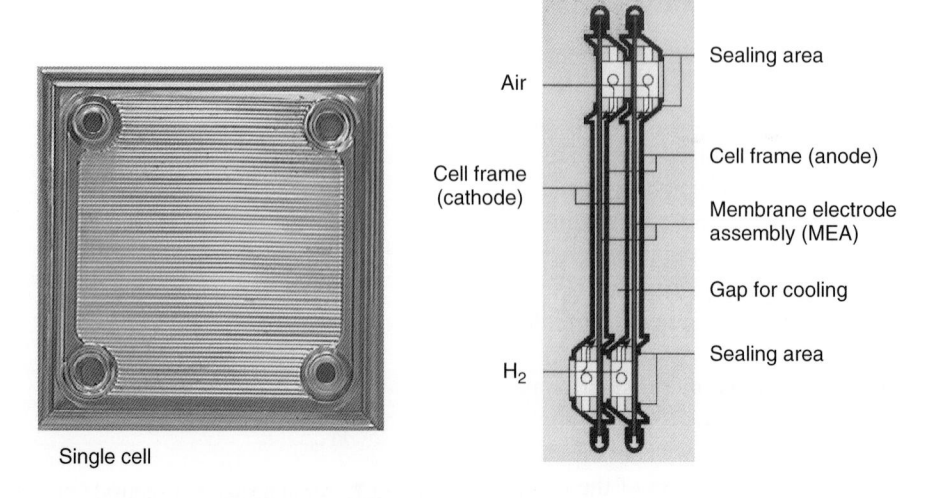

Figure 8. Schematic view of the Siemens low-cost cell design (patented in 1994).

The concept also includes the use of the electrolyte membrane itself as gasket at the cell edges. Therefore, the gas-tightness of the cells had to be verified using different membranes of different thicknesses. Membranes with a thickness higher than 100 μm can be used without any additional seals, but thinner membranes require an additional sealing, e.g., thin foil covering the edges.

The feasibility of air-cooling was demonstrated by operating short stacks of 100 and 310 cm² cell size. It was shown that efficient cooling could be provided at current densities up to more than 350 mW cm^{-2}. The power consumption of the blower was typically below 3% of the stack power. The experiments were done at ambient temperature (25 °C). An early laboratory set-up of an air-cooled six cell stack is shown in Figure 9.

A 35 cell stack of 310 cm² active area (Figure 10) was demonstrated with the operational parameters shown in Table 3. The pressure drop was measured in front of and behind a stack; the result is shown in Figure 11. For the 35 cell demonstrator and an air inlet temperature of 40 °C (death valley) the required volume flow is 277 m³ h^{-1} for an outlet temperature of 73 °C. This means that the

Figure 10. 35 cell stack with active cell area of 310 cm. during gas leakage test (Siemens).

Table 3. Operational parameters for a 35 cell stack of 310 cm² active area.

Air pressure	1.5 bar$_a$
Air stoichiometry λ	2–3
Air dewpoint	62 °C
Hydrogen pressure	1.5 bar$_a$
Hydrogen stoichiometry	1.2–2
Hydrogen dewpoint	68 °C
Inlet cooling air temperature	25 °C
Outlet cooling air temperature	50–80 °C

Figure 9. Set-up of a five cell air cooled battery with an active electrode size of 100 cm². The thermocouples, which can be seen in the picture, were used to examine and verify the temperature distribution more in detail (Siemens).

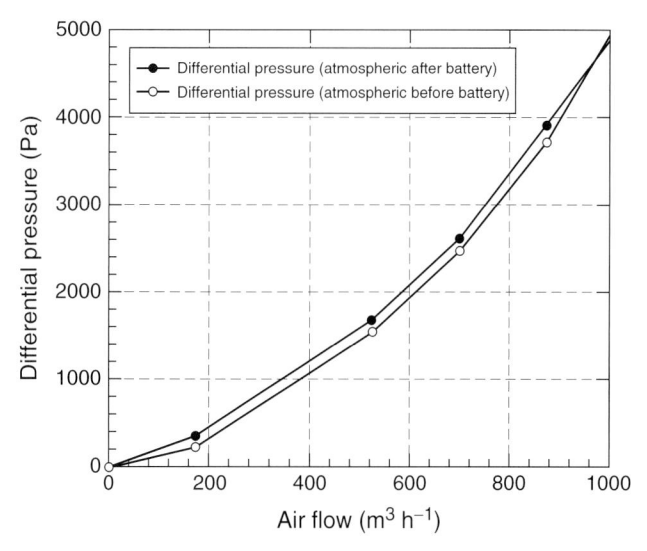

Figure 11. Cooling air characteristic: pressure drop as a function of volume flow for a 35 cell stack (Siemens).

pressure drop will be around 750 Pa (7.5 mbar). This can be generated with conventional low-cost fans.

2.2 Air-cooled PEM fuel cells for stationary and portable application

PEM fuel cells for low power application are often designed for operating temperatures of 50 °C or below. Durability and environmental factors are often not as critical as for automotive application. Because of the low operating temperature, external humidification might be not necessary, and the overall system could become very simple. The layout and a picture of such a system is shown in Figure 12.[4]

The stack integrated into the system was made from machined graphite composite bipolar plates having internal manifolding for hydrogen and external manifolding for air (compare Figure 2). A separate cooling layer is integrated into each bipolar plate. The hydrogen side can be operated in dead ended mode. Air is distributed by a simple ladder flow-field. The outlet air is partly recirculated in order to humidificate the cathode air. A separate cooling fan achieves airflow. Separation of reaction air and cooling air reduces problems associated with water management. The stack has 20 cells with active area of 100 cm² per cell. The maximum power is 350 W at a stack temperature of 55 °C.

The system is completely self-starting by admitting hydrogen to the anode. Figure 13 shows the characteristic

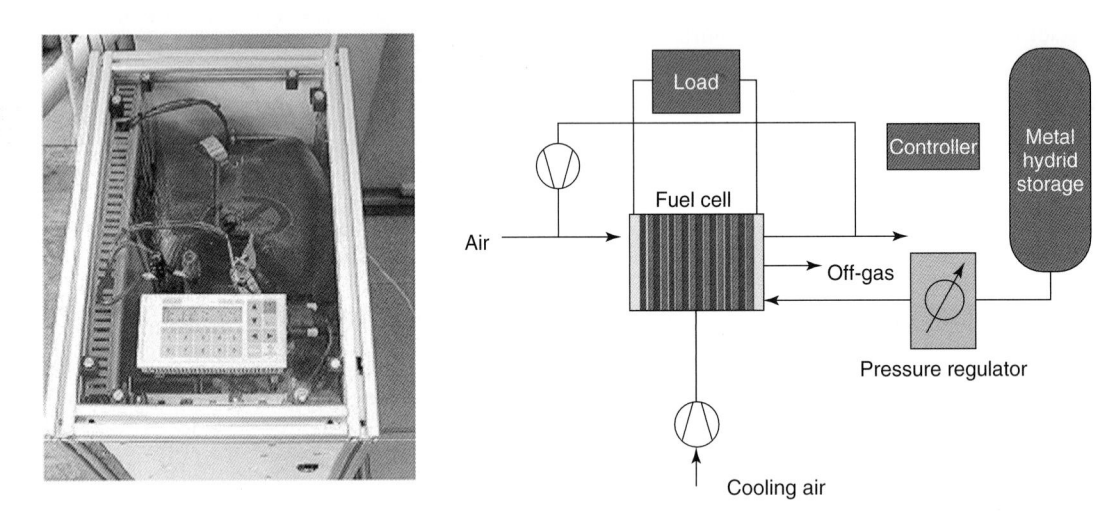

Figure 12. Photo and layout of a portable 250 W fuel cell system from ZSW.

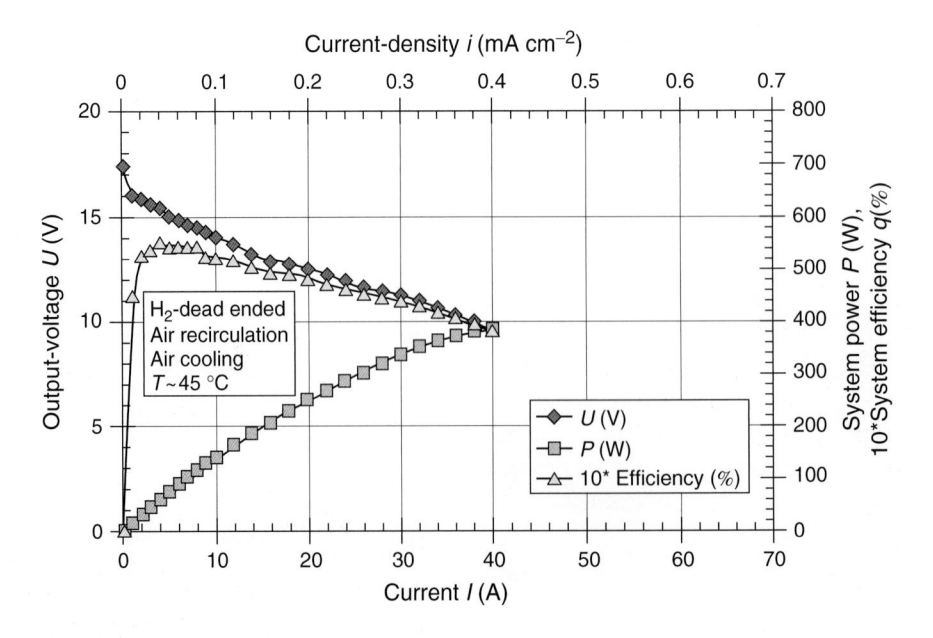

Figure 13. Current voltage curve of the portable 250 W fuel cell system.

2.3 Low power air-cooled PEM fuel cells

A lot of progress has been made in developing PEM fuel cells in the low power range. Figure 15 shows a 90 W fuel cell with a weight of 223 g including electronics, blower and control valves. The stack is operated in dead end mode with pure hydrogen, and has no separate cooling layer. The cooling air will be also used for feeding the cathode via an open cathode. Therefore only a single blower is necessary, for providing cooling and process air.[9]

The Fraunhofer Initiative on miniature fuel cells developed a miniaturized power supply for a camcorder based on a PEM fuel cell. The completely integrated fuel cell system with ventilators, valves, the fuel cell stack and metal

Figure 14. 1.2 kW fuel cell system Nexa™ from Ballard Power Systems.[5]

data for the system. The system exceeds 50% efficiency at about 200 W power output. Operation of the system can be sustained at a power of 250 W. The maximum power consumption of the auxiliary equipment never exceeds 15 W.

In 2002 Ballard Power Systems introduced the Nexa™ power module in the 1 kW range. At this time it was the first volume produced PEM fuel cell module designed for integration into stationary and portable power generation applications such as uninterruptible power supply systems, emergency power generators, and recreational and portable products. The system was designed for operation with dry hydrogen. The weight of the system shown in Figure 14 is 13 kg. The maximum power is 1200 W. At this operating point the voltage is 26 V and the current is 46 A.[6–8]

Figure 15. 90 W fuel cell stack from NovArs, Germany (www.novars.de).

Figure 16. Fuel cell system for a camcorder from the Fraunhofer Initiative on Miniature Fuel Cells. (Reproduced by permission of Christopher Hebling.)

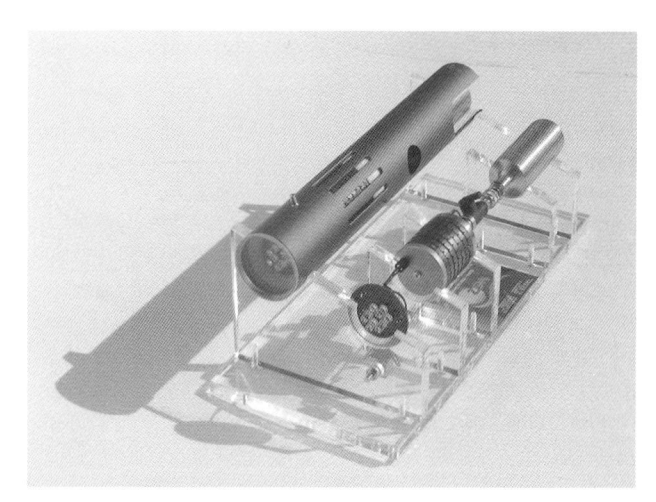

Figure 17. 2 W PEM fuel cell torch (ZSW).

hydride storage is shown in Figure 16. The system provides about 10 W of power.

The system becomes even simpler if convection and diffusion are sufficient to provide cooling and cathode oxygen; self-breathing and convection cooling however only works for very low power applications below 10 W. An example is shown in Figure 17. The stack works without any peripheral components. The peak power is 4 W. Typical applications are toys and torches.

3 OUTLOOK ON NEW TECHNOLOGIES

Air cooled PEM fuel cells are attractive because, in principle, the balance-of-plant can be very simple. This however is only true if a strategy can be found which allows for correct temperature and humidification, as current membranes can only be operated in a comparably narrow parameter range. The number of applications might significantly increase if new membranes widen this range. Higher operating temperatures and reduced need for humidification are therefore important goals for membrane development. Air-cooled fuel cells of the automotive type in particular will be enabled by a wider membrane operating parameter range.

ACKNOWLEDGEMENTS

The authors would like to thank the colleague who contributed data and pictures, and Laszlo Küppers for performing the calculations.

REFERENCES

1. VDI-Wärmeatlas, Springer-Verlag Berlin, 8. Auflage, (1997).
2. W. M. Kays and M. E. Crawford, "Convective Heat and Mass Transfer", McGraw-Hill (1993).
3. Siemens AG, "Fuel cells and Batteries made thereof", European Patent EP 0 795 205 B1, German Patent DE 44 42 285 C1.
4. J. Scholta, L. Jörissen and J. Garche, "Portable Fuel Cells and their Competitors", *Eur. Fuel Cell News*, **7** (2000).
5. K. Bonhoff, "In Serie gefertigtes 1,2 kW PEM-Brennstoffzellenmodul NEXA jetzt verfügbar," Proceedings of the Otti Energie Kolleg, neuntes Fachforum, Brennstoffzellen, 7–9 October (2002).
6. F. N. Büchi, C. A. Marmy, R. Panozzo, G. G. Scherer and O. Haas, "Development of a 1.6 kW PE Fuel Cell System," PSI Scientific Report, V, pp. 92f (1999).
7. F. N. Büchi, C. A. Marmy and G. G. Scherer, "Development of a 2 kW PE Fuel Cell System," Proceedings of the 3rd International Fuel Cell Conference, November 30–December 3, Nagoya, Japan (1999).
8. Avista Labs, "A Proton Exchange Membrane Fuel Cell Power System," PCT patent applied for, WO 99/27599; and http://www.avistalabs.com
9. H. Chang, P. Koschany, C. Lim and J. Kim, *J. New Mater. Electrochem. Systems*, **3**, 55–59 (2000).

Part 4

Fuel cell principles, systems and applications

Chapter 12
History of low temperature fuel cells

G. Sandstede[1], E. J. Cairns[2], V. S. Bagotsky[3] and K. Wiesener[4]

[1] *Physikalischer Verein von 1824 and Sandstede-Technologie-Consulting, Frankfurt am Main, Germany*
[2] *Lawrence Berkeley National Laboratory, and University of California, Berkeley, CA, USA*
[3] *Mountain View, CA, USA*
[4] *Kurt Schwabe Institut für Sensortechnik Meinsberg and Elektrochemische Energie, Dresden, Germany*

1 GENERAL SCIENTIFIC BACKGROUND BEFORE AND DURING THE BEGINNING OF FUEL CELL RESEARCH

1.1 Introduction: review of earlier papers on the history of fuel cells

The history of fuel cells is certainly a topic in its own right, but on the other hand, it may be considered to be a part of the history of science and technology of energy and energy production devices, especially of electrical energy generators. Therefore, this chapter may be started with some remarks about the development of the science of energy and energy technology and of the use of the various kinds of energy.

Beginning with a general description of the history of fuel cells given in the literature, the reader should first be pointed to the following: The first review paper was written by Baur and Tobler in 1933.[1] Ostwald's book "Elektrochemie, ihre Geschichte und Lehre"[2] (Electrochemistry: History and Theory),[3] contains interesting information about energy and fuel cells in the 19th century. In 1945, the US committee on the chemical utilization of coal also covered the subject of fuel cells.[4] There are two books by Euler, published in 1966[5] and 1974,[6] and Ketelaar provides an interesting discussion in Ref. [7] from 1983. In addition, we refer to another book by Liebhafsky and Cairns, from 1968,[8] which contains a prominent chapter of the history. And also, the introduction and closing remarks to the first "Grove Symposium" by Appleby in 1989[9, 10] can be recommended in this respect. Finally, there is the thesis of Schaeffer;[11] although it bears the title "Fuel Cells for the Future", it contains a number of thorough sections about the history of fuel cells. And last but not least, a lot of information about the history of certain subjects can be found on the internet, especially in the Encyclopaedia Britannica.[12, 13] As far as we know, there is only one recent publication about the history of fuel cells, which is by Sandstede and was published in 2000.[14, 15] But also an older comprehensive treatment of the literature about fuel cells up to 1937 by Davtyan may be mentioned here.[16] Furthermore, there were two lectures by Schnurnberger and Schlapbach, which have not yet been published.[17, 18] Apart from this, Moebius wrote a history of solid electrolyte fuel cells in 1997[19] and Bossel wrote a book about the birth of the fuel cell in 2000.[20] To conclude, the project of the Smithsonian Institution should be mentioned: "Fuel Cells: Collecting History with the World Wide Web".[21]

1.1.1 Development of the energy idea and technology

Energy conversion and conservation
Only since the end of the 18th century, has it been known that energy cannot be generated but only converted from one form into another. At that time, the term

energy was not yet known and chemistry had just become a science.[12, 13] The term 'physical chemistry' was used for the first time by Lomonossow[22, 23] (see below) and the terms 'energy' and 'work' were coined by Young in 1807 and by Poncelet in 1827, respectively.[5, 6] Young was an English physician and physicist, who established the principle of interference of light and, thus, resurrected the century-old wave theory of light. In addition, he was an Egyptologist who deciphered the Rosetta Stone,[13] while Poncelet was a French mathematician and engineer who contributed to the development of projective geometry.[13]

Although Sadi Carnot (1796–1832), the French scientist, defined his degree of efficiency in 1824, the term 'energy' has been broadly applied only since the middle of the 19th century. Up to the middle ages, the only 'energy' mankind used was their own muscle power and that of animals – apart from heat. Physical or technical connections between the various forms of forces and energies – heat, sun light, muscle power and electricity – had not been realized until the end of the 18th century.[24] The famous law of the conservation of energy was formulated by Julius Robert Mayer[25] as the conservation of a 'living force' in 1845. Interestingly, it is little known that William Grove had postulated this law of conservation of energy in his book about the correlation of physical forces a few years earlier.[26–28] von Helmholtz[13] wrote about the conservation of force in 1847. In addition, Gibbs[13] may be mentioned because of his thermodynamic relations and definition of enthalpy and, furthermore, Prescott Joule[6] in this respect, who discovered the equivalence of mechanical and electrical energy and heat in 1846. Last but not least Lord Kelvin, alias William Thomson, may be mentioned, who defined energy as the effect a system exercises outside of the system by changing its (physical or chemical) state in 1854.[6] William Thomson should not be confused with Sir Benjamin Thomson, alias Duke of Rumford, who discovered in 1798 that friction results in heat.[13]

First energy sources: wood, water and wind
The first energy source used by humans was firewood. Some 25 000 years ago, Stone Age men and women needed firewood to cook their food, as well as to heat and light their caves and huts. About 5000 years ago, people in the Middle East used the heat energy from wood for the winning and refining of tin and copper after they had taken the ore out of the ground. These were the first metals from which durable tools and weapons were made.[29, 30]

Water power and wind power have been used for several thousand years. At first only undershot water wheels were known, but in the beginning of the first millennium overshot water wheels could be detected. In the 11th century, William the Conqueror counted more than 5600 water wheels in England. In the mid-1800s, the construction of small dams to generate hydroelectric power began. Towards the end of the 19th century, experiments began to use windmills to generate electrical power. Up until then, windmills had been used for centuries to produce the mechanical power to grind grains and corn.

Magnetism and electrostatics
Also about 5000 years ago, most likely in China, it was discovered that a type of iron ore rock called magnetite possessed a mysterious force: The rock would attract pieces of iron and hold them fast because it develops a magnetic force, which could be transferred to other iron material. They magnetized a needle, which was fastened to the end of a string, and, thus, was able to turn. Such a needle would always point north: the magnetic compass was invented. A further 2000–3000 years passed before electric force was discovered. A Greek scientist and philosopher named Thales noticed – and reported in 600 BC – that when he rubbed fur against a piece of amber, the amber would attract things to itself, such as feathers and lint. It was not until 1600 AD that Gilbert first used the word electricity in describing such a static force.[31]

Coal
About a 1000 years ago, people living in China discovered that the hard, black rocks they found in the ground burned steadily and developed plenty of heat. China is a country rich in coal today. The explorer Marco Polo went to China in 1275, and when he returned to Italy 20 years later, he wrote a book about his experiences there.[29] In his book, Europeans read about these 'black stones', and they began to burn pieces of coal, too. People in the Netherlands were the first to find this fossil fuel underground. When the English discovered how useful coal was for heating, they also started mining it, and by 1660, England was doing booming business, providing most of the coal for the world. In 1712, Thomas Newcomen in England used coal to power the first practical steam engine to pump water out of deep mines. The steam engine made it possible to mine more coal and iron than ever before. James Watt improved the steam engine in 1765, making it able not only to pump, but also turn a wheel to run a machine: even to drive a 'vehicle'. In 1804, a Cornish engineer built the first steam locomotive, and in 1807, the first steam boat was invented by Robert Fulton in America. In 1820, the first American steamboat to cross the Atlantic left Savannah, Georgia, bound for Liverpool. It made the crossing in 25 days. In 1830 and 1831, the first American steam locomotives were run in York and Baltimore, where they carried 26 passengers over 13 miles in about 1 h.[30]

Coal gas
In the early years of the 19th century, coal gas was invented. One of the inventors, Walter Murdock, was Scottish. He developed lamps that burned 'coal gas', which was formed by heating coal in the absence of air. Soon others discovered that if coal was heated in the absence of air but to higher temperatures, a liquid called 'coal oil', which was later named kerosene, was created. In 1807, coal oil street lamps lighted the streets of London. On the other side of the Atlantic in 1813, David Melville received a patent for his apparatus for making coal gas, which was used for street lighting. Baltimore became the first city in the USA to establish a gas company to provide coal gas for lighting city streets. The first time natural gas was used was in Fredonia (New York State) in 1825, where the house of General Lafayette was brilliantly illuminated by natural gas with about 30 burners, which was regarded as a great curiosity.

Oil
In the middle of the 19th century, another big change was about to happen. In America in 1859, Edwin Laurentine Drake was trying to find a way to pump crude petroleum oil out of the ground in Titusville, Pennsylvania. This was the first of many oil wells, and Titusville became the first oil boom town as more and more wells were drilled. The first major oil pipeline to successfully transport crude petroleum was completed by Samuel Van Syckel of Titusville. It was about 5 miles long. Wrought iron pipes, 2-in. in diameter, were laid underground in 15 foot sections. Two pumping stations supplied the power. In 1861, petroleum was first exported to Europe; after 7 weeks the shipment arrived at the Victoria Docks in London, England, on January 9, 1862. The "Elizabeth W", a 224 ton ship captained by Charles Bryant, carried 1329 barrels of oil. The crew was shanghaied, since most sailors would not work above a cargo of oil for fear of fire and explosions. Seven years later, the first oil tanker, the "Charles" of Antwerp, Belgium, began carrying oil from the USA to Europe. The ship had 59 iron tanks arranged in rows in the ballast and below the deck, with a bulk capacity of 7000 barrels of oil.

Automobiles
In about 1860, with this new fuel available, French inventor Etienne Lenoir constructed the first practical internal combustion engine, in which gasoline exploded to drive a piston inside the engine.[32, 33] Earlier, between 1807 and 1813, a car with an internal combustion engine was designed by Issac de Rivaz (Switzerland). The engine had been gas-driven and used a mixture of hydrogen and oxygen to generate energy. Another gas engine was invented by Captain Samuel Morey of Oxford, New Hampshire, in 1826. The engine had two cylinders, 180° cranks, poppet valves, a carburetor, an electric spark and a water cooling device and was powered by spirits of turpentine vapor. It lasted quite some time, until 1876, which was when the German Nikolaus August Otto built his famous 4 stroke engine. Then in 1886, two German engineers, Gottlieb Daimler and Karl Friedrich Benz, hooked this engine up to the wheels of a carriage, and the first real automobile was on the road. Otto's small shop grew into Gasmotorenfabrik Deutz, where Gottlieb Daimler later worked as the chief engineer. The French claim a complete car had been first developed between 1860 and 1884 by Delamare-Deboutteville. Siegfried Marcus was a prolific inventor living in Vienna. He had a 4 stroke engine running in 1870, using some sort of petroleum fuel and a 2 foot high carburettor. Starting in 1893, America's first automobile was built by Charles and Frank Duryea. The two brothers designed and built the car together, working in a rented loft in Springfield, Massachusetts. They were the first in the country to manufacture cars for sale and by 1896 had built 13 cars.[33, 34]

Steam-driven cars
On the other hand the ICE-driven car was not the first automobile because there were already steam-powered cars in broad use and electric cars in development. A full-size road carriage powered by steam was built by Richard Trevithick in 1801. Much earlier, namely in 1769, Nicholas-Joseph Cugnot in France designed a steam truck with a speed of 2.5 m.p.h. Steam-powered stage coaches were in regular service between many towns in Britain from 1820 onwards. Also in America and other countries steam-powered cars were used for over a century. US president Theodore Roosevelt still used such a car officially in 1907. And, by the way, the name 'chauffeur' stems from the French word for heater.

There were nearly 14 000 automobiles on the road in the USA in 1900. A total of 40% were steam powered, 38% were electric and just 22% were powered by gasoline burnt in an internal combustion engine.

Electric vehicles
By the invention of the primary battery by an Italian physicist, named Alessandro Volta, giving the world its first steady supply of electrical energy, in 1800, a new power came into existence. In the course of time, electricity was used for chemistry, movement and lighting. But a working electric motor was not built until 1833. Thomas Davenport, a Vermont blacksmith, conceived it after observing a demonstration of an electromagnet. Davenport patented his motor in 1837. This invention set the stage for the incorporation of a battery into an electric vehicle. Davenport had in fact built a model electric locomotive as early as 1834, powered by primary cells.[31, 35, 36] In Frankfurt am

Main (Germany) a self-made physicist of the local Physical Society, named Johann Philipp Wagner, who developed electrical equipment for medical purposes, and invented, for instance, the interrupter for the electrical bell. Using this principle, he developed a motor, which he used for driving a model car by applying the new primary batteries of John F. Daniell from the UK. Wagner applied for a patent and got one from the city of Frankfurt in 1840, but his application for the area of the German federation was refused because the patent officers decided after a 3-year examination that the 'electric engine' was much too expensive compared with the coal-fired steam engine. Wagner needed for a 1 horsepower/40 lb model about 10 lbs of zinc per hour.[37, 38]

In 1847, Moses Farmer of Massachusetts designed an electric locomotive that, powered by 48 × 1 pint cells, could carry two people along an 18 inch wide track. About the same time, Professor Charles Page of Washington, DC, built a locomotive which, using 100 cells and a 16 horsepower motor, carried 12 people on the Washington Railroad at up to 19 m.p.h. In 1847, Lilly and Colton of Pittsburg built a locomotive that received its power, produced from a central station, through an electrified rail. Such rail-bound vehicles were a great leap forward, but it would be 30 years before another major advance. In 1888, electric cars suddenly began appearing on the scene both in the USA and abroad. The first really successful electric automobile was the carriage built by William Morrison of Des Moines, Iowa, in 1890. Morrison's car used high, spoked wagon wheels to negotiate the rutted roads of America, and an innovative guidance system, which included patented rack-and-pinion steering. Morrison's car was capable of running for 13 consecutive hours at 14 m.p.h. Other literature quotes that Thomas Edison's car, named the Electro-Runabout, was successfully manufactured starting in 1889.

Electrical energy and lighting
A usable capacitor was designed by Humphry Davy in 1802, after the Leidener Flasche and similar equipment had already been experimented with in the middle of the 18th century. Michael Faraday was one of the most brilliant men in the history of electricity. In this connection, he discovered the electromagnetic effect in 1831, and, thus, built the first electric generator. In 1866, Werner von Siemens was able to improve the generator into a technical device by introducing the dynamo-electric principle: the first alternating current (a.c.) generator. In 1871, a really practical direct current (d.c.) generator was built by Thomas Alva Edison in America. Edison's many inventions included the phonograph and an improved printing telegraph. In 1878, Joseph Swan, a British scientist, invented the incandescent filament lamp and within 1 year Edison had made a similar invention in America. He discovered carbonized cotton filaments and produced a light bulb that would burn for 40 h. Swan and Edison later set up a joint company in New York to produce the first practical filament lamp. Prior to this, electric lighting had been by crude arc lamps. At the same time when the Edison Electric Light Company was started at the East coast, at the West coast the California Electric Light Company began selling electricity in San Francisco for lighting Brush arc light lamps. Edison used his d.c. generator – the first one to be successful – to provide electricity to illuminate the first New York street to be lit by electric lamps, in September 1882.

Secondary batteries
The most useful electrochemical secondary cell, even up to now, has been the lead acid battery, which was invented by the German medical doctor Josef Sinsteden in 1854 and a little later – for practical use – by Gaston Planté in France in 1859. The first alkaline battery, the silver/zinc battery, was developed by Dun and Haßlacher in 1887, then in 1899, the nickel/zinc battery was invented by de Michalowski and in 1901, the nickel/cadmium battery by Jungner[39] and in the same year the nickel/iron battery by Thomas Edison, who also equipped an electric car with it.[40–43]

With this background, the discovery and development of the fuel cell shall be discussed in the following sections.

1.1.2 The roots of the fuel cell in physical chemistry

Before the historical interdisciplinary interactions to develop the fuel cell (see Figure 1) can be discussed, a few facts about the development of physical chemistry shall be mentioned. Physico–chemical methods have enabled mankind to investigate the influence of greenhouse gases as well as gases that are dangerous to human health – all of which are connected to the use of fossil energy. Physico–chemical methods and theories likewise had to be in existence in order to form a starting position for the discovery and development of fuel cells and, in addition,

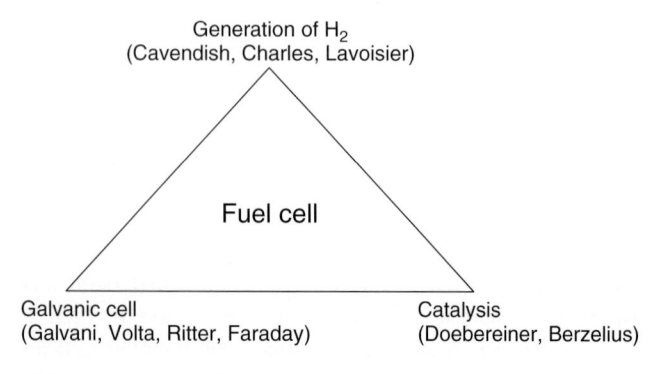

Figure 1. Interdisciplinary interaction of scientific fields for the birth of the fuel cell.

physico–chemical methods especially have been added to the electrochemistry of the fuel cell. Also, physical chemistry is the basis for technology. Thus, looking for the scientific roots of the fuel cell, they can certainly be found in several areas of physical chemistry about 200 years ago. One area lies in the production of hydrogen, another in the field of catalysis and a third in a way, even the basis, is the electrochemistry as well as the fundamentals of a galvanic cell.

The birth of physical chemistry, as one of the first interdisciplinary scientific fields in general, took place in the second half of the 18th century. The name had already been created by the Russian scientist Mikhail Vasilyevich Lomonossow[22] in 1741, who not only was a famous scientist but also a poet and the first great Russian linguistic reformer. He is known for the modernization of the Saint Petersburg's Imperial Academy of Sciences and the establishment of the university in Moscow, which was named after him. His philosophical considerations, discussed previously, dealt with a universal law of conservation of matter and energy. It was a time when the chemists and physicists still believed in alchemy and the phlogiston theory. In 1697, the German physician and chemist Georg Ernst Stahl set up the theory that all combustible substances contained a volatile gaseous substance, which escapes during the combustion of the substance. This product was called phlogiston by him. Even after the discovery of oxygen, the supporters of the phlogiston theory did not give up and called oxygen 'dephlogistonized air'. Nowadays, this fact may be surprising, but one can consider this theory to contain a true nucleus because the released energy was also called phlogiston and not distinguished from the likewise invisible gaseous combustion product.

Shortly before oxygen became known, hydrogen had been discovered. Interestingly enough, the first application of hydrogen was not based on its chemical properties, namely its fuel character, but on its physical property of outstandingly low density, see the description of the first hydrogen balloon later on. As a prerequisite of the understanding of the fuel character of a substance, the discovery of hydrogen was very important. Between 1775 and 1780, one of the most famous chemists of all time, Antoine-Laurent Lavoisier, investigated hydrogen and oxygen thoroughly and arrived at a new theory of combustion, which revolutionized chemistry (see Figure 2).

Another important result for the characterization of a fuel and the process of combustion was the discovery of the catalytic reaction. In 1823, the German chemist Johann Wolfgang Döbereiner detected the instantaneous combustion of hydrogen when coming into contact with powdered platinum (see Figure 3). The ignition took place at room temperature. Thus, the knowledge of catalysis

Figure 2. Antoine-Laurent Lavoisier. (Reproduced by permission of Académie des Sciences.)

Figure 3. Wolfgang Döbereiner. (Reproduced from Platinum Metals Review by permission of Johnson Matthey.)

started, which was supplemented by the work of Jöns Jacob Berzelius. He was a Swedish scientist and also called one of the founders of modern chemistry. Among many other scientific achievements, he recognized many catalytic reactions and introduced the term 'catalysis'.

So far the two fields of physical chemistry that were a prerequisite for the understanding and the development of the fuel cell have been dealt with. The third contribution to the background of the fuel cell is electrochemistry, which goes without saying is the basis also of a gaseous cell, being a galvanic cell too. Thus, we have three roots for the discovery of an electrochemical gas reaction.[14, 20] These fundamental fields are

- the generation of hydrogen,
- the electrochemical cell and
- the catalytical reaction.

It is interesting to note that in all three areas Johann Wolfgang von Goethe was involved with both the physical experimentation and implementation of the theory. At that time, he was a member of the government of the state of Weimar and he appointed professors at the university of Jena, some of whom became his friends.[44]

1.1.3 Discovery of oxygen and hydrogen and their first applications

Distribution
If you think about the fact that hydrogen and oxygen are the most commonly occurring elements in nature, then you could find it surprising that so little was known about them till the end of the 18th century, – including their productly of combination, water. Hydrogen is the most abundant element in the universe and its position within the row of the elements in the lithosphere is number 9, whereas oxygen is the most abundant element in the earth's crust (= lithosphere). This layer has a thickness of 10 miles and contains 49.4% of oxygen and 15.4% of hydrogen. The product of combustion of hydrogen, water, covers three quarters of the surface of the earth. It is the most important substance of at all, being fundamentally for life.

Scheele and Priestley
The first person who discovered oxygen in 1772 was Carl Wilhelm Scheele, the famous Swedish chemist,[45] but he did not publish it this fact before he had written (in German) his only book "Chemical Observations and Experiments on Air and Fire" in 1777. In this book, Scheele reported about his experiments that the atmosphere is composed of two gases, one that supports combustion and the other one that prevents it. He called the former 'fire-air' and prepared it from nitric acid, potassium nitrate, manganese dioxide or mercuric oxide. So he obtained the gas 2 years earlier than Priestley. Scheele was born on 9th December 1742 in Stralsund, Pomerania (Germany) and died 21st May 1786 in Köping, Sweden. He is also known for his discoveries of a great number of other substances, among them also the toxic gases chlorine, hydrogen sulfide, hydrogen cyanide and hydrogen fluoride.

Apparently the best known person with respect to the discovery of oxygen was Joseph Priestley, who was born 13th March 1733 in Birstall, UK, and died 6th February 1804 in Northumberland, PA, USA. He was a clergyman (Calvinist), political theorist and physical scientist. On the one hand, he contributed to advances in liberal political and religious thought and, on the other hand to experimental chemistry, especially to the chemistry of gases. But Priestley's interest was broader. In 1765, he met the American scientist and statesman Benjamin Franklin, who encouraged him to publish "The History and Present State of Electricity, with Original Experiments" (1767). In this book, Priestley communicated that he was not only a prominent experimentalist but also a philosopher by discussing scientific methodology of achieving progress. But the result was that he preferred facts instead of hypotheses.

The ten gases Priestley discovered were all called 'air': nitrous air (nitric oxide), red nitrous vapour (nitrogen dioxide), inflammable nitrous air (nitrous oxide, also called 'laughing gas'), marine acid air (hydrogen chloride), alkaline air (ammonia), vitriolic acid air (sulfur dioxide), fluor acid air (silicon tetrafluoride), phlogisticated air (nitrogen), a gas later identified as carbon monoxide and, most importantly, dephlogisticated air (oxygen). Furthermore, he propagated the drinking of carbonated water (containing carbon dioxide), which promoted the soda water industry.

Lavoisier
A few months after the discovery of oxygen in 1774, by heating red mercuric oxide, Priestley visited Lavoisier in Paris, where he told him about his discovery of dephlogisticated air. This meeting between the two scientists became very significant for the future of chemistry. Lavoisier repeated Priestley's experiments and carried out further investigations, which resulted in his theory of combustion and respiration. That was the end of the phlogiston theory. Unfortunately, Priestley remained a follower of the phlogiston theory, so that he no longer participated in the revolution of chemistry. Lavoisier was so proud of his success that he arranged a public spectacle in 1789: His wife – in a white mini-costume, representing oxygen – threw the books by Stahl about the phlogiston theory into a fire.

Hydrogen
Around 1780, Lavoisier also investigated the 'inflammable air' and arrived at the result that by the combustion of hydrogen water was formed. This result was also achieved by Henry Cavendish, English physicist and chemist, who discovered hydrogen in 1766. He was not the first to detect inflammable air, but he was the first to identify its nature by performing quantitative experiments. The first to be known to have handled hydrogen is Theophrastus Paracelsus (1493–1541) in the 16th century since he noted that an inflammable gas was generated when a metal was dissolved in acid. A French scientist, Turquet de Mayerne (1573–1655), had the same experience. Other reporters communicated that Paracelsus only noticed the light gas and not its inflammability, but this was found by de Mayerne. Also Robert Boyle made the experiment of dissolution of iron in hydrochloric acid and detected the inflammability of the evolved gas in 1671. In the middle of the next century,

Lomonossow in Moscow also experimented dissolving metals in acids and detected the combustible air but did not go further.[46]

Cavendish

Coming back to Cavendish, who was an enormously talented man but a shy personality. He was a real outsider of society and avoided the public wherever possible, although he did accept such honors as being made a Fellow of the Royal Society (1760) and being elected one of the eight foreign associates of the Institute de France (1803). After he inherited a fortune, he was even more whimsical and avoided speaking whenever possible. Not only did he not marry, he apparently never formed a sympathetic attachment with any person outside of his family. The costume he usually wore consisted of a faded, crumpled suit of an earlier time, with a high collar and frilled cuffs, and a three-cornered hat. Only a few fellow scientists came to visit him to discuss scientific results in various fields or just to visit his huge library, which he gladly opened to them. Cavendish was born 10th October, 1731, Nice, France and died 24th February, 1810, in London, UK.

In 1776, J. Warltire noticed that water was formed when hydrogen was burnt. Cavendish's experiments with hydrogen and air, described in 1784–1785, led to the discovery that water is not an element but a compound. Priestley had noted that when a mixture of hydrogen and air is exploded by means of an electric spark, the walls of the vessel are covered with moisture, a fact to which Priestley did not pay any attention. After a careful repetition of Priestley's experiment, Cavendish concluded that this moisture was mainly water. An analogous conclusion was reached at about the same time by James Watt, the Scottish engineer and who communicated this to Priestley and the Royal Society, which led to a dispute that also concerned Lavoisier. In 1785, Lavoisier achieved the splitting of water into hydrogen and oxygen by heating it in a copper tube.

Cavendish's electrical research was equally remarkable. He developed the basic law of electrostatics, Coulomb's law, a few years earlier than the French physicist, Charles-Augustin de Coulomb. He also anticipated a result obtained by Faraday in demonstrating that the capacity of a condenser depends on the substance inserted between its plates. Likewise, he anticipated Georg Simon Ohm, a German physicist, by establishing his law decades earlier. These and further results were the more remarkable since Cavendish had no exact means of measuring current and managed this for instance by turning his own body into a meter, estimating the strength of the current by grasping the ends of the electrodes. All these findings were subsequently repeated after their discovery in his notebooks and manuscripts over a century later. Publishing only part of his research was part of the strange behavior he showed, particularly in later years.

Balloons

Because of the lack of a current source (a prerequisite for electrolysis) so far, hydrogen was generated from dissolution of metal in acid, which is a reaction of an electrochemical nature and, therefore, a forerunner of electrolysis. The first application of hydrogen, however, was not of a chemical but physical nature, although the characteristic property of hydrogen, its combustibility, led to its discovery and first notation. After Cavendish (see Figure 4) had measured the density and in 1782 Tiberius Cavallo in Italy and afterwards Jacques Alexandre Cesar Charles in Paris had demonstrated the small density by means of soap bubbles, the Montgolfier brothers, Joseph and Étienne, had the idea that 'combustible air' might render possible the flight of a balloon. They failed, however, because of the leakage of the bags used made from paper and silk and, immediately, invented the 'hot air' balloon. On 21st November, 1783, the first manned flight took place when Jean-François Pilâtre de Rozier and François Laurent, Marquis d'Arlandes, sailed over Paris in a Montgolfier's hot air balloon. They burned wool and straw to keep the air in the balloon hot. Their flight covered 5.5 miles in about 23 min.

Figure 4. Henry Cavendish. (Reproduced by permission of AKG, London.)

Goethe
At this stage Johann Wolfgang von Goethe can be cited, whose 250th birthday was celebrated in 1999. This 'Goethe year' has made it known to the general public that he, besides being the greatest German poet, also was an ingenious scientist (though not necessarily an exact one). In the 1780s, Goethe experimented together with Georg Christoph Lichtenberg in Göttingen and with Samuel Thomas von Soemmerring in Frankfurt am Main with hydrogen by generating it by the dissolution of metal in acid. They watched the buoyancy of soap bubbles and dreamt of filling a balloon.[47, 48]

Charles
At the same time, in 1783, this was achieved by Charles in Paris. J.A.C. Charles was a French mathematician and physicist (born in Beaugency, France, on 12th November, 1746, and died in Paris on 7th April, 1823). In 1785, he was elected to the Academy of sciences and became a Professor of Physics. In about 1787, he anticipated Gay-Lussac's law of the expansion of gases with heat. Together with Faujas and the Robert brothers, Charles constructed a hydrogen balloon. It was filled with 40 m^3 of hydrogen, which was prepared from iron and sulfuric acid. On the 27th August, 1783, the first flight of the hydrogen balloon took place; it consisted of rubberized silk and was called Charliere. But the balloon failed because of the dirty hydrogen generated from the acid.[49] Therefore, Charles invented in a way the first chemical gas technology for the gathering (from many barrels), cleaning and drying of the gas before filling it into the balloon. Within less than 1 year Charles, as a kind of project manager, had carried out the whole project starting with the research stage over the development of the production plant for hydrogen as well as its investment and erection and, in addition, the construction of the first hydrogen balloon up to the second successful start on the 1st December, 1783. At the same time, he had conceived the essential elements of chemical technology and the gas industry, such as gas washing, cleaning and drying as well as heat exchanging and construction in technical dimensions (1000 lbs of iron). Charles, accompanied by Nicolas-Louis Robert, flew the hydrogen balloon 2000 feet up into the air 27 miles over Paris for 2 h. This first hydrogen balloon flight was a great success. Goethe only regretted that he had not succeeded earlier. He continued experimenting with hydrogen in Weimar, this time concerning himself with its chemical properties, which resulted occasionally in explosions over the course of demonstrations.

Samuel Thomas von Soemmerring
Another application of hydrogen may be mentioned in this respect. Samuel Thomas von Soemmerring was a medical professor in Frankfurt am Main, then in Munich and later again in Frankfurt. He invented the first electrical telegraph by using the electrolytic generation of hydrogen as a signal. Being a friend of Goethe, he demonstrated the complete equipment at a lecture in Munich in 1809 on the occasion of Goethe's 60th birthday.

Further work of Lavoisier
In 1784, Lavoisier produced hydrogen from water vapor and iron. In 1785, after quantitative experiments regarding the formation of water from hydrogen and oxygen, he established the name hydrogenium (from Greek, where hydor = water and genes = generating). In 1777, Lavoisier proposed the name oxygenium (acid producer) because he was of the opinion that all acids result from the union of oxygen with other substances. Lavoisier (born on 26th August, 1743, in Paris) was also an important public official and was executed under the Guillotine on 8th May, 1794.

Further applications of hydrogen
To finish this section, some more historic dates will be mentioned, which are closely related to hydrogen. In 1874, nearly a century later, Jules Verne, the great first science fiction author from France,[50] wrote the novel "The Mysterious Island", in which the engineer tried to answer a difficult question: What can mankind use for heating after natural fuels have been exhausted? To which he answered *"water, but split into its chemical elements and doubtlessly split by electrolysis"*.

In 1879, Augustin Muchot, a Frenchman and founder of solar technology, tried to create hydrogen by electrolysis of water with an electrical current, which he generated with a steam power station, for which the heat was obtained with the help of large mirrors collecting the sun's energy.

In 1898, James Dewar, a British chemist and physicist, invented the liquefaction of hydrogen.

On 6th May, 1937, the zeppelin 'Hindenburg' had an accident when landing in Lakehurst, New Jersey, USA. A total of 200 000 m^3 of hydrogen burnt, luckily 62 out of 97 passengers were saved. The cause of the ignition was not the hydrogen, but the lacquer by electrostatic charging.

1.1.4 Development of electrochemistry

In 1790, electrochemistry began with the observation of Aloysio Galvani (see Figure 5) that the legs of a dead frog twitched when exposed to an electrical spark. Further related experiments convinced Galvani that electricity was produced by organisms and was an essential basis of life. 'Animal electricity' became an important scientific topic. For some time, it was believed that animal matter was a source of electricity. During the 1790s, a number of scientists investigated this animal electricity and the behavior

Figure 5. Luigi Galvani. (Reproduced from Reference [3].)

of frogs' legs when exposed to various stimuli through the contact of various materials (including metals) to the nerves and muscles of the frogs' legs. Two opposing schools of thought developed: the animal electricity school, and the group whose experiments showed that electricity could be produced in electrochemical systems that contained no animal matter.

The Italian physicist Allesandro Volta (see Figure 6) initially accepted the idea of animal electricity, but his experiments soon led him to the conclusion that the frogs' legs were simply an indicator of electrical current rather than a source of it. He performed many experiments with a variety of metals, leading to the development of a precursor to the electromotive series.

During this active decade of the 1790s, Johann W. Ritter (see Figure 7) played a key role in clarifying the relationship between galvanic and chemical phenomena. By using

Figure 6. Alessandro Volta. (Reproduced from Reference [3].)

Figure 7. Johann Wilhelm Ritter. (Reproduced by permission of Deutsches Museum, München.)

various pairs of metals in electrochemical experiments, he shed light on the chemical transformations brought about by electrochemical cells. This work contributed to the formulation of the ideas that now take the form of the electromotive series.

The era of electrochemical energy conversion was opened by Volta's report in March, 1800, of Volta's Pile (see Figure 8). The pile consisted of alternating layers of zinc (or tin), silver and cardboard or leather soaked with electrolyte. The pile provided for the first time a reliable, continuous source of an electrical current at a significant voltage. This permitted a large variety of electrochemical experiments on an unprecedented scale. It is curious that Volta did not report any chemical effects or changes in his pile. These observations were made by others.

Very soon after Volta's report, both Nicholson and Carlysle carried out the electrolysis of water with a pile. Their work was important in pointing out the chemical action produced by the electricity, a point not discussed by Volta. During the following several years, Berzelius and Hisinger performed experiments on the decomposition of salts by electrolysis, leading in 1810 to the ideas relating electricity and chemical affinity.

Davy's former assistant, Michael Faraday (see Figure 9), continued electrochemical research with the very important development of Faraday's Laws in 1834:

1. The amount of electrochemical reaction produced by an electrical current is proportional to the amount of electrical charge passed.
2. The amounts of different substances produced or consumed by a given amount of charge are proportional to their equivalent weights.

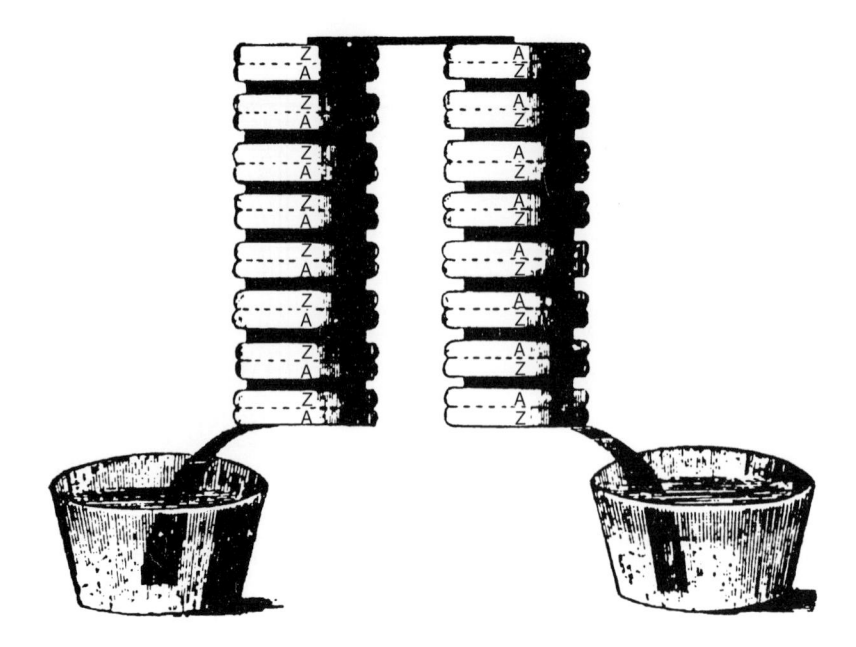

Figure 8. Volta's Pile. (Reproduced by permission of Deutsches Museum, München.)

Figure 9. Michael Faraday. (Reproduced from Reference [3].)

1.1.5 Discovery and development of the battery

Volta's discovery of the Volta Pile opened a new era: the development of electrochemical energy conversion. This field became very active, with many investigators developing electrochemical cells for producing and storing electrical energy. Relatively large currents at high voltages for significant periods of time were available for the first time in history. The various versions of the Pile were constructed and used widely, in spite of the lack of understanding of how the cells functioned. As we know now, the zinc was oxidized and the native oxide layer on the silver was reduced. The cells were 'recharged' by disassembling them and exposing the silver electrodes to air, which reoxidized them. Inevitably, other electrochemical cells and batteries were developed.

John F. Daniell reported his two-fluid cell in 1836. The negative electrode was amalgamated zinc, and the positive electrode was copper. The arrangement of the cell is shown in Figure 10. The copper electrodes were placed in (porous) porcelain jars, which were surrounded by cylindrical zinc electrodes and placed in a larger container. A copper sulfate solution was placed in the copper electrode's compartment and sulfuric acid was put in the zinc electrode compartment. Other forms of this cell include a version that eliminates the porcelain separator and relies on the density difference between the two solutions to keep them separated. This

Figure 10. Daniell Cell. (Reproduced from Reference [3].)

reduced the cell resistance, permitting higher currents to be drawn.

Sir William R. Grove, inventor of the fuel cell, developed a two-electrolyte cell related to the Daniell cell in 1839. Grove used fuming nitric acid at a platinum electrode (the positive) and zinc in sulfuric acid (the negative). Variants on this formulation were widely used for a number of years. Of course, all of these cells were primary cells, in that they could be discharged only once, and then had to be reconstructed with fresh materials.

In 1860 Gaston Planté reported the development of the first rechargeable battery (see Figure 11). It was comprised of lead sheet electrodes with a porous separator between them, spirally wound into a cylindrical configuration. The electrolyte was sulfuric acid. During the first charging cycles, the positive electrode became coated with a layer of PbO_2. The charging operation was carried out using primary batteries, a laborious process. Due to the low cost and the ruggedness of these batteries, they remain in widespread use today, with only evolutionary modifications.

1.1.6 Discovery of catalysis

At this point Professor Johann Wolfgang Döbereiner has to be mentioned again. He also belonged to the university professors in Jena, who were appointed by Johann Wolfgang von Goethe. Döbereiner (born on 13th December, 1780, in Hof an der Saale, Germany, and died on 24th March, 1849, in Jena, Germany) was fond of experimenting with platinum and also interested in its chemistry. So one day, he prepared powdered platinum by igniting ammonium hexachloroplatinate(IV). When he exposed this powder to hydrogen and admitted air, he observed that water had been formed at room temperature. Thus, he discovered the catalytic property of platinum on the 27th July, 1823.[51] Further, when a flow of hydrogen gas touched platinum black in air it ignited. As the effect (and also the term 'catalysis') was not known, he discussed with his friend Johann Wolfgang von Goethe. They explained it as a dynamic process, where electric phenomena might be involved. Döbereiner's lighter (catalytic ignition of hydrogen) has become world famous.

The importance of electrochemistry can likewise be learnt from Goethe.[44] One day in 1812, he wrote to his scientific friend Wolfgang Döbereiner in Jena: "*Verläßt man nie den herrlichen elektrochemischen Leitfaden, so kann uns das übrige auch nicht entgehen*", which translates as "*if you don't depart from the wonderful electrochemical guide, you will also understand the remaining*". In spite of this recognition of many scientists, this has been neglected over the last 200 years by mankind with respect to fuel cells because the heat engines were so powerful and cheap.

Amongst the other research that Döbereiner occupied himself with was the investigation of the repetition of certain properties of elements, thus anticipating the development of the periodic system of the chemical elements. Chlorine, bromine and iodine showed similarities, and the properties of the middle element were approximately equal to the average values of the other two. This held also for calcium, strontium and barium and sulfur, selenium and tellurium.[52]

The next step in the development of catalysis was the definition and discovery of a term for designation. Here comes the appearance of Jöns Jacob Berzelius, who was one of the founders of modern chemistry. He is especially known for his determination of atomic weights, the introduction of modern chemical symbols, his electrochemical theory, the development of classical analytical techniques, the discovery of many elements and the investigation of minerals, isomerism and catalysis. Berzelius was a Swedish scientist, born on 20th August, 1779, near Linköping, Sweden, and died on 7th August, 1848, Stockholm.[23] He invented the name catalysis and also the name isomerism.

Concerning the use of hydrocarbon fuels in a fuel cell system, in addition to the fuel cell stack and the fuel processing unit, the catalytic reactions of the fuel are of great importance and must be considered. Therefore, it would be interesting to learn more about the development of heterogeneous catalysts. However, this is too large a field to discuss fully and applies more so to the petrol industry.

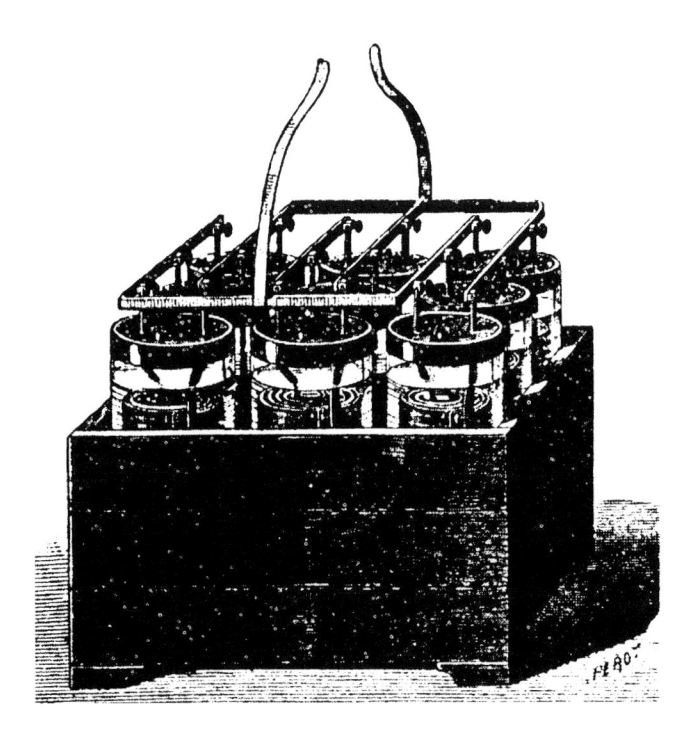

Figure 11. Planté Battery. (Reproduced from Reference [361].)

On the other hand, it can be seen that there is a large area of study into the improvement of present catalysts available.

1.2 How fuel cells were discovered

1.2.1 The investigation of the hydrogen/oxygen chain

Usually, Sir William Robert Grove is considered to be the first person to discover the fuel cell in 1839. At least, this is widely accepted by the scientific community and it includes the meaning that he knew the name and what a fuel cell is. But the exact details are more difficult because his publication came out a month later than that of Professor Christian Friedrich Schoenbein in the January issue of the Philosophical Magazine in 1839 that for the first time described an experiment, and showed the fuel cell effect, as Ulf Bossel called it.[20] Hydrogen and oxygen were filled in tubes, which were immersed in dilute sulfuric acid and which contained in the centre of them a platinum foil each. Incidentally, the hydrogen and oxygen could be created by electrolysis. After disconnection from a current/voltage source (a Volta Pile), a voltage between the platinum foils could still be measured: the voltage of the hydrogen/oxygen chain, a gaseous chain. William Grove found the same, which will be described later on. In the next section, it will also be explained why Grove, nevertheless, is the inventor of the fuel cell, regarding it as an electricity generator. This real fuel cell he invented in 1842 or, in a way, only in 1845.

Schoenbein, a Swiss scientist, was a Professor of Physics and Chemistry at the University of Basel in Switzerland. Schoenbein (born on 18th October, 1799, in Metzingen, Germany, and died on 29th August, 1868) is known for his discovery of ozone and gun cotton. (Gun cotton, by the way, was discovered by Rudolf Christian Boettger in Frankfurt am Main at the same time.) Schoenbein's paper in January 1839 was titled "On the Voltaic Polarization of certain Solid and fluid substances"[53] and it closed "...*we are entitled to assert that the current in question is caused by the combination of hydrogen with oxygen (contained dissolved in water) and not by contact*". This conclusion shows that Schoenbein understood the voltage was measured as the cell voltage of a gaseous cell (the fuel cell effect according to Bossel).

Previous to this, gaseous and liquid reactants had already been used to form a galvanic or electrolytic chain. In one of the preceding discussions, the discovery of the galvanic effect and the invention of a galvanic cell and a bipolar multi-cell battery by Volta in the year 1799 has been described. In addition, the application of this Volta Pile for carrying out electrolysis of water has already been mentioned. During the electrolysis, two Englishmen, Nicholson and Carlyle, in 1800 observed a white powder at one electrode (they were made from brass), whereas hydrogen developed at the other electrode. In a later publication, they reported about experiments where hydrogen and oxygen had been developed in the right proportions.[54] The same year Johann Wilhelm Ritter from Jena, who has already been mentioned as the discoverer of the secondary cell (secondary battery), repeated the electrolysis of water, but used gold electrodes, which were not oxidized in the acid solution.[55, 56] Thus, he observed the development of hydrogen and oxygen in a ratio of 2 : 1, probably independently and possibly earlier. Like Bossel,[20] it can be assumed that Ritter also saw a voltage after disruption of the electrolysis current because he was familiar with electrode potentials from the voltage row of metals he discovered at the same time as Volta did.

In 1802, Sir Humphry Davy, whom Goethe, incidentally, admired, built the first element with liquid reactants. His publication was titled "Carbonaceous Fuel Cell with Aqueous Electrolyte at low-temperature",[57] cited by Schaeffer.[346] So he was the first to use the designation 'fuel cell', whereas Grove did not. As a matter of fact, the term fuel cell was not used until 1889, when Mond and Langer coined the term again. Davy (born on 17th December, 1778, and died on 29th May, 1829) also developed a primary cell. It consisted of a carbon anode and nitric acid as the cathodic depolarizer. In addition, it should be mentioned that Davy is very well known for his melting electrolysis for the formation of sodium and potassium. Davy's picture is to be seen in Figure 12.

Also in 1802, Gautherot observed a voltage between two platinum wires, which had been used for the electrolysis of water.[58] According to Schönbein,[59] the Geneva

Figure 12. Humphry Davy. (Reproduced by permission of AKG, London.)

physicist, A.A. de la Rive had observed secondary currents (inverse current after switching off the electrolysis) in 1827. Schönbein also reported similar experiments by Edmond Becquerel,[59] as referred by Bossel.[20] In a way, this is already the discovery of the gas chain. Another researcher also carried out experiments with hydrogen and oxygen and published the results in 1838.[60] His name was Carlo Matteucci, Professor of Physics in 1832 in Bologna, in 1838 in Ravenna and in 1840 in Pisa, and was a friend of Grove.[6] He charged two strips of platinum with hydrogen and oxygen, respectively, and was able to measure a voltage, when he immersed the strips in acidified water. Unlike Schoenbein and Grove, however, he did not give an explanation and, therefore, was not able to understand how the effect occurred.

Similar experiments had been carried out by Schönbein.[53] In a letter to the editors of the (British) Philosophical Magazine and Journal, he says that he had read a paper containing an account of the results of his research on the voltaic polarization of solid and fluid bodies before the scientific meetings that took place at Bale and Fribourg, some months previously. This letter carried the heading: "On the Voltaic Polarization of Certain Solid and Fluid Substances by Professor Schönbein". He said that it was to have been published soon in the Bibliotheek Universitat and in Poggendorf's Annalen. In the bulk of this letter, he described 14 experiments and at the end of it presented five conclusions. Schönbein finished his letter: "*we are entitled to assert that the current in question is caused by the combination of hydrogen with (the) oxygen (contained dissolved in water) and not by contact. I am, Gentlemen, yours, etc. Bale, Dec. 1838. C.F. Schönbein*". It was published January 1839.

This current Schönbein refers to and a potential in the carefully executed experiments was measured only with platinum as well as with hydrogen and an oxidant – oxygen or chlorine. The above sentence is the first written identification of the fuel cell effect, as Bossel called it.[20] Thus, Schönbein discovered the fuel cell effect. His picture is to be seen in Figure 13.

1.2.2 From Volta and Ritter to Schöenbein and Grove

A very important discovery, which at the same time was an invention, is Volta's pile. From this an apparatus, a device, piece of equipment or engine could be developed which could produce electricity with remarkable power. As we have described already, this 'engine' opened electrochemistry on the one hand and electrical power generation on the other. Volta was certainly a great scientist – and an ingeniously practical man – but he was not the theoretician to understand what was happening in his galvanic cell. This

Figure 13. Christian Friedrich Schoenbein. (Reproduced by permission of Edgar Fahs Smith Collection, University of Pennsylvania Library.)

was reserved for Ritter, who explained the voltaic series of metals, described the reactions of water electrolysis and discussed many other processes – also with his scientific friend Goethe.[56]

It would be interesting to describe and discuss the reflections of Schoenbein. He was the first to publish results of experiments about the 'fuel cell effect', a term which was introduced by Bossel. As one can see from his paper in January, 1839, he not only showed the voltage of the hydrogen/oxygen chain but also spoke of a current. He told this for the first time to an audience in lectures in France and Switzerland in late summer of 1838.[59, 61] Similar findings were published in an addendum of a letter to the editor by Grove 1 month later in February 1839.[62] Bossel analyzed the two papers and came to the conclusion that the content of the addendum to the letter was completely different from the main part of the letter and, consequently, the title of the letter consisted of two parts: "*On voltaic series*" and "*the combination of gases by platinum*".

The title in full reads: "*On Voltaic Series and the Combination of Gases by Platinum. By W.R. Grove, Esq. M.A. Swansea, Dec. 14, 1838, Gentlemen...*" This letter to the editor was published in the Philosophical Magazine in February, 1839. In the main part of the letter, Grove described battery combinations with copper and iron and copper with zinc, and he closed this part of the letter like closing a letter: "*I cannot refrain from expressing, with much diffidence, a hope that these experiments may be thought worthy of verification and extension by those older in practice, abler than myself. I remain, Gentlemen, yours, etc., W.R. Grove*". But this was not the end of the letter because he immediately continued: "*P.S. Jan. 1839. I should have persued these experiments further, and with other metals,*

158 *Part 4: Fuel cell principles, systems and applications*

but I was led aside by some experiments ...I will, however, mention one which goes a step further than any hitherto recorded; and affords, I think, an important illustration of the combination of gases by platinum". In this second half of the whole letter of three pages, Grove described his experiments with two platinum electrodes immersed half in acidified water and half in hydrogen and oxygen, respectively. He closed the letter with the sentence: *"I hope, by repeating this experiment in series, to effect decomposition of water by means of its composition."*

This is a similar statement as Schoenbein's at the end of his paper 1 month earlier. Therefore, it is not understandable that Euler asserts that Schoenbein did not recognize the fact that the electrochemical reaction of hydrogen with oxygen is the origin of the fuel cell effect. By the way, Schoenbein called it "*Sauer-Wasserstoff-Säule*" (oxygen/hydrogen pile) and Grove called it "*gaseous voltaic battery*", which Euler himself reports. It is probable that Euler put too much value on the fact that Schoenbein, being a thorough scientist, also discussed the contact theory, which would not need oxygen for the explanation. In the course of his experiments, Schoenbein was finally convinced of the importance of the electrolytical conduction between the electrodes. After all, he was also a scientific friend of Faraday, with whom he had an intense correspondence for many years if not decades. In 1834, Faraday introduced the terms: electrode, electrolyte, anode, anion, cathode and cation.[63–65]

Schoenbein attended the annual meeting of the Royal society in Birmingham in the summer of 1839, in order to present the invited lecture "Notices Of New Electro-Chemical Researches". There he received a grant of £40. There he met Grove, and they established a life long friendship. This is evident by reading the many letters that they exchanged. The complete correspondence between the two scientists has recently been published by Bossel,[20] the organizer of the European Fuel Cell Forum in Switzerland (two international meetings on topics in the field of fuel cells every year in Luzern, Switzerland). Bossel writes: *"For the first time, one can participate in the exchange between the two outstanding scientists of the past century (the nineteenth), who, together, should be credited with the creation of the fuel cell."*

In Figure 14, the first cell for Grove's gaseous chain is shown; it consists of a glass vessel with three necks; the outer two in the form of longer tubes; these contain a strip of platinum each. The cell is filled with dilute sulfuric acid completely. By carrying out electrolysis, the two tubes will be filled with hydrogen and oxygen, respectively. After interruption of the wire reaching the battery, a potential is measured by a galvanometer, which stems from the hydrogen/oxygen chain. The figure was not found in any of Grove's papers, but in a book by Peters from 1897.[66]

Figure 14. First cell for Grove's gaseous chain. (Reproduced by permission of VARTA, Aktiengesellschaft, Hanover.)

The main working field of Grove was the development of batteries (apart from his theoretical reflections, which will be discussed later). Therefore, he spent most of his laboratory time in the search for new systems for batteries. Thus, approximately at the same time as his first publication on the 'gaseous battery' experiment, he published his experiments about a zinc/nitric acid battery: "On A New Voltaic Combination".[67] A second publication the same year was titled: "On A Small Voltaic Battery Of Great Energy".[68] He used amalgamated zinc, a mixture of nitric acid with sulfuric acid and platinum electrodes. Two years later, R.W. Bunsen has substituted the platinum electrodes with carbon.[69]

In 1842 Grove published what may be called his first paper about a working fuel cell: "On a Gaseous Voltaic Battery".[70] Figure 15 is taken out from that publication.

As can be seen from the picture, he used four cells in series and carried into effect what he promised in his first publication: He carried out electrolysis of water with the current generated by the 'fuel cell battery'.

At this point, it may be permitted to make a remark regarding the volumes of hydrogen and oxygen in the tubes containing the electrodes of the fuel cells. Contrary to the opinion of several scientists, we think that they are correctly drawn and not the other way round, when taken into consideration that the tubes were completely filled with the gases. On operation of the fuel cell, the amount

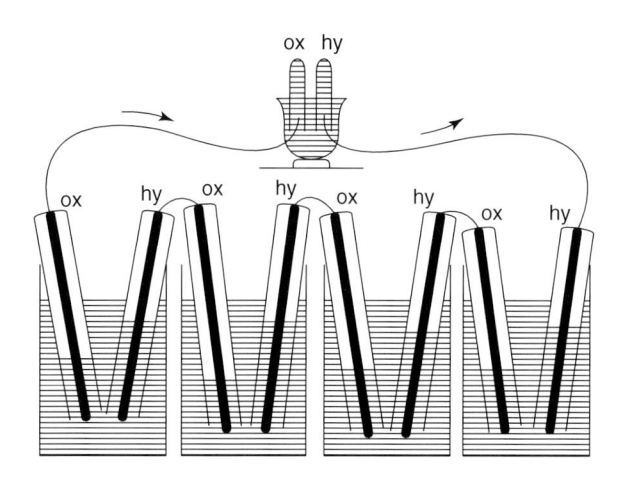

Figure 15. Grove's gaseous voltaic battery (1842). (Reproduced from Liebhafsky and Cairns (1968).[8])

of hydrogen consumed is twice the amount of oxygen. Consequently, the liquid level has to be twice as high in the hydrogen tube compared with the oxygen tube. The other way around would refer to the electrolysis reaction.

This publication already contains an important observation: a gaseous electrochemical reaction needs a triple phase border, see the third paragraph of that paper:

"As the chemical or catalytic action in the experiment detailed in that paper, could only be supposed to take place, with ordinary platina foil, at the line or water-mark, where the liquid, gas and platina met, the chief difficulty was anything like a notable surface of action".

Therefore, he enlarged the borderline by extending the surface of the platinum by applying electrolytically-formed platinum black. The highest number of cells he connected in series was 50; and with this battery he was able to feed an electric arc between two carbon electrodes.

Grove continued the experiments with his gaseous voltaic battery and, at the same time, improved the apparatus. The results were published in 1843[71] and 1845[72] and further. According to Bossel, a complete fuel cell electricity generator was described in the 1845 paper,[73] including the generation of hydrogen from zinc and sulfuric acid. Therefore, Bossel called this paper the start of fuel cell technology. In my view, in addition, another part of the technology is of great importance, namely the use of electrodes only partly immersed in the electrolyte, whereas the other part of the electrode protruded into the air, thus, forming the first air electrodes. This construction is to be seen from the figures in the paper under consideration (see Figure 16).

The 1845 paper was published in the "Philosophical Transactions", and it was 12 pages (DIN A4) long with one page of drawings. The experiments again were carried out very accurately, starting with number 31 and ending with 53, altogether 23 experiments – the last one is a description of the whole apparatus.

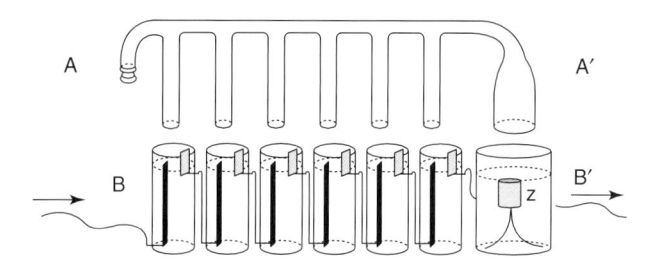

Figure 16. Grove's six-cell hydrogen/air gaseous voltaic battery (1845). (Reproduced from Liebhafsky and Cairns (1968).[8])

The paper also contains a list of substances which could be used as fuels and, likewise, another list of oxidants for the cathode. So, at this stage of his research, Grove already had treated all aspects of the fuel cell. Later on, he also tried carbon as a fuel.

As a result of all the experiments, Grove established a list of the electrochemical relations of substances, called 'voltaic excitants',[73] which are listed in Table 1.

In the last experiments (no. 53), Grove describes his Figures 4 and 5 (see Figure 16). *"I have never thought of the gas battery as a practical means of generating voltaic power, though in consequence of my earlier researches, which terminated in the nitric acid battery, having had this object in view... There is, however, a form of gas battery, which I may here describe, which, where continuous intensity or electromotive force is required... appears to me to offer some advantage over any form of battery hitherto constructed... We have now a gas battery, the terminal wires of which will give the usual voltaic effects, the atmospheric air supplying an inexhaustible source of oxygen, and the hydrogen being renewed as required by the liquid rising to touch the zinc; by supplying a fresh*

Table 1. Grove's list of the electrochemical relations of substances called 'voltaic excitants'.

Positive electrode	Negative electrode
Chlorine	Camphor
Bromine	Essential oils
Iodine	Olefiant gas
Peroxides	Ether
Oxygen	Alcohol
Deutoxide of nitrogen	Sulfur
Carbonic acid	Phosphorus
Nitrogen	Carbonic oxide
Metals which do not decompose water under ordinary circumstances	Hydrogen
	Metals which decompose water

piece of zinc when necessary, it thus becomes a self-charging battery, which will give a continuous current; no new plates are ever needed, the electrolyte is never saturated, and requires no renewal except the trifling loss from evaporation, which indeed is lessened, if the battery be in action, by the newly composed water...I have one of ten cells constructed, shown at Figure 5, which succeeds perfectly, giving sparks, decomposing water, etc., and is ever ready for use".

From this publication, it can be seen that Grove has already described a fuel cell system including the fuel supply for the generation of electricity. Sir William Robert Grove, Esq., Barrister-at-law, M.A., F.R.S., V.P.R.I (Vice President of the Royal Institution), born on 11th July, 1811, in Swansea, Glamorgan, Wales, and died 1st August, 1896, in London, UK, was a British physicist and lawyer, Lord justice of Britain's high court and Queen's Counsellor and, finally, Professor of Experimental Philosophy at the London Institution (see Figure 17).[74] Besides being an experimental physicist he also was a philosopher and theorist – for example, regarding the correlation and conservation of forces (earlier than Mayer and Helmholtz!) – and in addition to his invention of the two-fluid electrochemical cell (zinc/nitric acid battery), he achieved for the first time the thermal dissociation of atoms within a molecule. He brought steam into contact with a strongly heated platinum wire and showed that it is decomposed into hydrogen and oxygen in a reversible reaction. After retirement in 1887, he resumed his scientific studies.

To conclude the description of Grove's career, let us state that he is famous for the invention of the technical fuel cell and, therefore, ought to be found in the Encyclopaedia Britannica.[74] This is actually the case, and his results described above are mentioned in it, including the Grove cell, but this is the two-fluid zinc battery. There is nothing about the fuel cell, even not in the currently updated Internet version![75]

1.2.3 Researchers after Grove until Mond, Langer and Ostwald

Although Grove had already described a complete fuel cell system in 1845 using the oxygen from the air as the depolarizer (though expression was not yet known), Vergnes declared this system exactly as in his invention 15 years later and had it patented.[76] In 1849, Beetz experimented with various gases and found out that these obey a voltage row similar to that of the metals.[77] Furthermore, he investigated the influence of the diffusion of the dissolved gases on the polarization of the electrodes.[78]

Coal

The first scientist who thought about coal being used as a fuel in his electrochemical cell was Grove in 1854.[79] This meant it was no longer a gaseous cell. The first, who is known to have suggested a cell for the application of coal as an electrode, is Edmond Becquerel, son of Alexandre Cesar Becquerel and father of Henry Becquerel. He described his experiments in a few sentences in his book about electricity in 1855.[80] A glowing rod of charcoal was immersed in molten potassium nitrate in a platinum crucible. Thereby, the carbon rod constituted the negative electrode. Twenty-two years later, Pierre Jablochkoff[81] described a very similar experiment without mentioning Becquerel's research. Euler came to the conclusion that Jablochkoff had not known of Becquerel's short publication, being within a chapter of the book mentioned.[6] He also used iron as the material for the crucible. Euler also describes in his book the results of Adolphe Archereau in 1883 – contained in his British patent no. 1027[82]. This researcher used for the first time alkali hydroxides and alkali carbonates and melted them in a crucible that was lined with a sheet of copper, and he then realized that oxygen was needed for the positive electrode. This short paragraph was introduced here because the findings are of interest for the low temperature cell development as well.[4]

Figure 17. Sir William Robert Grove, Esq., Barrister-at-law, M.A., F.R.S., V.P.R.I, Lord justice and Queen's Counsellor, Professor of Philosophy and Physics. (Reproduced by permission of VARTA, Aktiengesellschaft, Hanover.)

Westphal and Scharf
Regarding his gaseous cell, Grove did not publish anything really new after his splendid paper of 1845. Also Schoenbein occupied himself with other topics, and Faraday had already finished working on voltaic cells. Thus, it took nearly 40 years before work on the gaseous Grove cell was seriously taken up again. Lord Raleigh made some experiments with platinum fabric instead of smooth or powdered platinum and he also achieved a current with coal gas, which was published in 1883.[83] Westphal, 1880, and Scharf, 1888, made inventions, which they only described in patents. Westphal was the first to recognize the possibility of achieving a high efficiency with such cells.[84] He wrote: *"The small percentage, which results by the conversion of the coal stored power into electricity by using an engine and a dynamo machine, points to a larger yield, if the two bridging machines could be omitted"*. This was written 14 years earlier than the famous statement of Ostwald!

Furthermore, Westphal invented the porous electrode and at the same time gave an explanation for its way of working. He wrote: *"You can press the gases, which are under pressure, under water through the diaphragms, which are covered outside by a conducting material, or through diaphragms, which consist completely out of porous conductive material, in such a way that they come out in the form of small bubbles while touching the liquid and the metal at the same time"*. Westphal was able to apply water gas, generator gas or coal gas. The next to take a big step forward was Paul Scharf. The patent he filed dealt with a multi-fuel cell stack for the first time.[85] The electrodes were flat plates in a round shape – as an example – and put together in an alternating way, between each cell a gas tight plate. This may also have been the first presentation of a stack according to the filter press principle. All hydrogen chambers and all oxygen chambers were connected and the electrolyte chambers between the electrodes were connected too, so that the electrolyte could be circulated by pumping. But the stack did not have a series connection of the cells, it had a parallel connection of the cells, which means that all negative electrodes and all positive electrodes were combined together, respectively.

Mond and Langer
Further improvements were undertaken by Ludwig Mond and Charles Langer after they had collected 15 papers about fuel cells published after the first Grove paper. In 1889, they discovered that they had to increase the surface of triple contact (three-phase zone: electrode, electrolyte and gas).[86] In addition to the application of porous electrodes therefore, they introduced a matrix that soaked the electrolyte, preventing the electrodes from flooding. Their cells operated on hydrogen and oxygen at 0.73 V and a current density of $6.5\,mA\,cm^{-2}$. According to Appleby, the phosphoric acid fuel cells (PAFCs) of today are in principal similar to the Mond and Langer cells, except in the use of modern materials. They also wanted to use cheaper hydrogen and used 'water gas' made from coal (the Mond gas process). But they soon gave up because they discovered poisoning of the platinum by carbon monoxide, still a problem today. However, Mond and Langer should be remembered because they introduced the name 'fuel cell'. Furthermore, they realized that at the voltage of 0.73 V, they achieved a useful effect of nearly 50% of the total energy contained in the hydrogen. Therefore, Appleby summarized that Mond and Langer's concept contained all the elements of the modern low temperature fuel cell (LTFC) for the first time, which could be taken as a reason for a celebration as being the 50th anniversary after Grove.

Ostwald
In 1894, a new impulse rose among researchers, who wanted to generate continuous, electrical energy from a 'battery', and which was given through a lecture by Wilhelm Ostwald (see Figure 18). Ostwald was the Professor for Physical Chemistry in Leipzig and was an expert in energy and was fond of promoting electrochemistry. He was born in Riga on 2nd September, 1853, and died in Großbothen near Leipzig on 14th April, 1932. He founded the "Zeitschrift für Elektrochemie" (Journal of Electrochemistry) and was a co-founder of the Bunsen-Gesellschaft für Physikalische Chemie.[87, 88] The paper mentioned was the manuscript of a speech before an audience of engineers, who he considered in general as fighters against poor efficiency.

Figure 18. Wilhelm Ostwald. (Reproduced by permission of VARTA, Aktiengesellschaft, Hanover.)

162 *Part 4: Fuel cell principles, systems and applications*

As part of this speech he stated: "*I don't know whether everybody everywhere imagines, what an incomplete thing...our essential source of energy is, which we use, I refer to the steam engine. We obtain no more than 10%...of the energy of the burning coal in form of mechanical work. We certainly know that heat cannot be converted into mechanical energy completely, but we can calculate the fraction...And also in the light of this fact we find that we use only one seventh of the energy that is convertible...The total huge temperature difference between the furnace room and the boiler is lost...But thermodynamic machines are not the only ones...The maximum of the energy is theoretically totally independent from the route on which the conversion is carried out. Could we, hence, convert the chemical energy of the fuel material into mechanical energy in such a way that heat is not involved, then we would not be bound to the inconvenient high temperatures and could gain the whole amount without putting up with that inconvenience. The way how to solve that biggest of all technical problems, the procurement of cheap energy, this way has to be found through electrochemistry. If we had a galvanic cell, which supplies immediate electrical energy from coal and the oxygen of the air, in an amount, which, to a certain extent, is in relation to the theoretical value, in that case we stand before a technical revolution, against that of the invention of the steam engine would vanish. Think about, when taking into account the uncomparable convenient and flexible distribution that is permitted by the electrical energy, the appearance of our industry sites would change. No smoke, no soot, no steam vessel, no steam engine, even no fire, because fire is needed only for a few processes, which cannot be carried out by means of electricity and which decrease from day to day*".

From then on new attempts were made to attain a fuel cell, but although the theoretical background was very important and totally correct, the hint to the 'coal fuel cell' was misleading the motivated community because the process engineering and reactivity of solid substances is too difficult and too slow. This, by the way, had already been shown by Borchers in 1894, who gave many reasons for solid coal being unsuitable as a fuel for fuel cells.[89] Thus, it took further a 40 years until gaseous fuels were preferred. Nevertheless, the way to get there was scientifically interesting.

Wilhelm Ostwald had a large house as his private home in Großbothen near Leipzig, Germany. Although his sphere of action was at the university in Leipzig, he built his home into a large laboratory and library and called it the house of energy. Keeping the name, it is, nowadays, a center of studies and conferences and, naturally, an exhibition of Ostwald's work in science and philosophy.

1.3 The first boom of fuel cell development

1.3.1 Jacques, Haber, Nernst, Baur, Gläßner, Taitelbaum and other pioneers

Jacques

The reader may wonder why we are discussing high temperature fuel cells (HTFCs) again, which we will briefly do in the following section. The reason is two-fold or even three-fold, if you like. The researcher wanted to convert coal directly into electricity. On the one hand, coal is an attractive fuel and on the other hand, there was this suggestion from an eminent authority (Ostwald), which was much too optimistic. So was W.W. Jacques, who intended to use huge fuel cells to power an ocean ship. That is why he published this proposal in Harper's Magazine (see Figure 19). However, it should be mentioned that the third reason is a quite remarkable one. For the first time, he built cells on a kilowatt scale. Jacques cells were large and had a high performance. His cells were 1.2 m high and 0.3 m

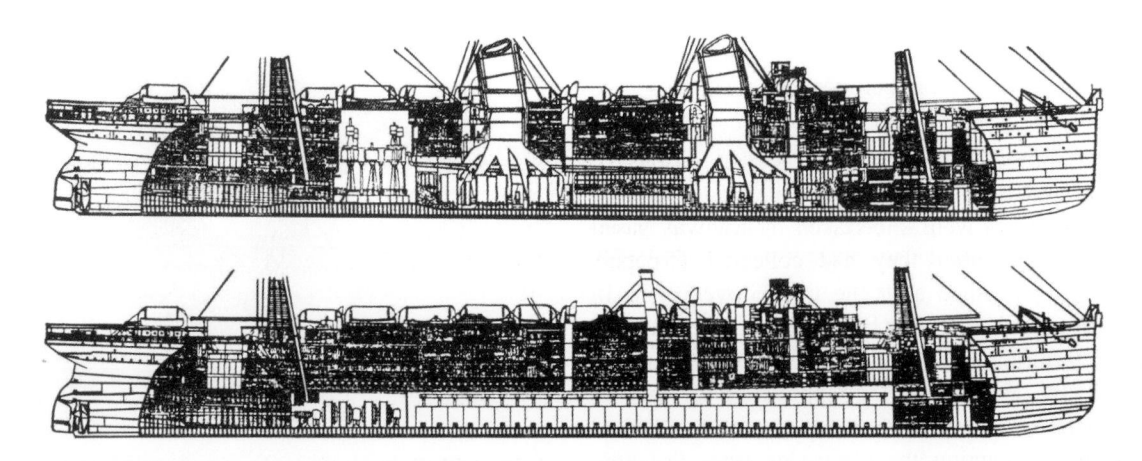

Figure 19. Jacques' vision of powering a ship with large fuel cells. (Reproduced by permission of VARTA, Aktiengesellschaft, Hanover.)

in diameter. They consisted of cylindrical iron pots which were filled with liquid molten potassium hydroxide and operated at 450 °C. The cell delivered a power of ca. 300 W; thus, the current density amounted to about 100 mA cm^{-2} at a voltage of 1.0 V. On operation, air was blown through the molten potassium hydroxide by the walls of the iron pot, which constituted the cathode. Jacques, by the way, was an electrician at the Bell Telephone Company in Boston, MA, USA; the article he wrote for Harper's Magazine was titled "Electricity Direct from Coal".[90] But the Jacques' cell with molten potassium hydroxide had to fail because of carbonate formation and corrosion problems. It was more or less a large uneconomical primary cell.[9]

VARTA

Detailed investigations were carried out by Liebenow and Strasser[91] in the laboratories of the Akkumulatorenfabrik AG in Hagen, Germany, in 1897, which had been founded by Adolph Müller in 1888. Later the company was renamed VARTA (the well-known battery company). Further investigations by Haber and Brunner[92, 93] in 1904 showed which reactions occurred and why the cell was not suitable.

Haber

Haber and his school have continued working with the hydrogen/oxygen-chain (they called it 'Knallgaskette'), but used electrolytes for higher temperatures, for instance molten glass, which is not our topic here (and which were not successful in the end). Others also tried gas cells at higher temperatures. Among those was Emil Baur, who devoted all his life to fuel cells (see next section).

Redox cells

Together with Gläßner,[94] Baur constructed the first redox fuel cell in 1903: Ce(IV) compounds in aqueous alkaline solution were reduced by sugar, which were then oxidized to CO_2. The reduced cerium compound was then electrochemically oxidized using oxygen from the air afterwards. Another one of Baur's group investigated the reduction of concentrated sulfuric acid with carbon or carbonaceous materials. This was Taitelbaum, who, in addition, made a thorough study of redox pairs, also in aqueous solutions.[95] On the other hand, he also tried molten potassium hydroxide at 370–390 °C in an iron crucible, which was passivated by molten Saltpeter. Using coke, Jungner had already tried reactions with concentrated sulfuric acid earlier than Taitelbaum.[96] Jungner was, by the way, the inventor of the nickel/cadmium accumulator in 1899.[39]

Nernst

Redox phenomena were also studied by Walter Nernst, who was a student of Ostwald. His Habil thesis dealt with the electrode potentials of ions, which were also important for the fuel cell.[97] Later he became professor at the university of Berlin in 1906. Nernst is famous for his equation for the equilibrium potential in ionic solution. Researchers at that time were caught in the idea of potentials and not in reaction velocity. The equilibrium thermodynamics diverted from the investigation and improvement of the kinetics of the reaction. That is why progress with fuel cell reactions was slow in the early years of the twentieth century.[98] Another student of Ostwald was Smale, who studied gas chains.[99, 100]

But going back to the research work of Nernst, he discovered the oxygen ion conductivity of certain oxides at higher temperatures (Nernst glower), which are the basis of the solid oxide fuel cell (SOFC) (see below).[101] Regarding his investigations into redox processes, he had a contract with the VARTA company in 1913 to assist them in building up fuel cell activity. He suggested using additives such as thallium, titanium or cerium, but his suggestions were not effective enough.[5] Among other things, he also suggested activating the oxygen electrode by adding some chlorine to the air. The resulting hydrogen chloride could be regenerated to chlorine by applying the Deacon process.[102] Nernst was born in Briesen, Westpreussen, Germany, on 25th June, 1864, and died on 18th November, 1941.

Alkaline electrolyte

The application of an aqueous alkaline electrolyte took place relatively late. In the first book about batteries by Hauck, published in 1883, alkaline batteries were not mentioned.[103] Whereas it went without saying for F. Peters, to speak about alkaline batteries in his book in the year 1897.[66] But still some more time had to pass until fuel cells with an aqueous alkaline electrolyte were described. Reid filed a patent in 1902[104] and Noel filed one in 1904.[105] They used hydrogen and air; Reid also vaporized liquid fuels in an oven.

1.3.2 Baur

Emil Baur took the suggestions of Ostwald seriously and was really excited for fuel cells. He pioneered in several areas of the fuel cell field, but has, nevertheless, worked in a number of other fields in physical chemistry. He was born in Ulm, Germany, on 4th August, 1873, and died in Zürich on 14th March, 1944. In 1907, Baur went as Professor to the University of Braunschweig and 1911 to the ETH Zürich. His life story can be found in the publication by his famous co-worker W.D. Treadwell,[106] whose father and brother were also chemists. Baur was active at the ETH Zürich until the summer of 1942. He occupied himself with nearly all types of fuel cells, especially HTFCs. Of these, just two types may be mentioned, one with a molten or pseudo solid electrolyte and one with really solid electrolyte. In the first field, his co-workers included H. Ehrenberg (Über neue Brennstoffketten[107]) and W.D. Treadwell (Ausführungsformen von Brennstoffketten bei hohen Temperaturen[108]). Together with H. Preis, he invented the first HTFC with a solid oxide electrolyte of the zirconium type. In the year 1937, they published a cell with which carbon could be electrochemically converted via an integrated gasification step inside the cell.[109, 110] By the way, 2 years earlier Schottky who was a pupil of Nernst, had theoretically investigated the ionic conductivity of solid electrolytes in 1935.[111] Only today has the development of integrated SOFC systems been taken up again.

Baur was also quite active in the low temperature field of fuel cells. A fundamental paper was the one about electrolytic dissociation.[112] His work about oxygen and air electrodes was important.[113, 114] Together with Tobler, he published a thorough investigation about the state-of-the-art in the whole field of fuel cells.[1] Their classified groups of cells and electrodes are shown below.

Anodes (fuel electrodes)
1. (a) the direct carbon cell with an aqueous electrolyte (seven different cells have been stated, starting with Davy[57] in 1802: $C/H_2O/HNO_3/C$);
1. (b) direct carbon cells with molten electrolytes (eight different cells);
2. (a) indirect carbon cells with carbon anodes (ca. 20 different cells with molten electrolytes);
2. (b) SO_2 cells (five different cells with concentrated sulfuric acid);
2. (c) oxidation-reduction cells (= redox systems) (nine different cells with potential transferring substances, containing a diaphragm);
3. (a) Fuel gas cells at ambient temperature (four different types of electrodes and seven different types of operation of the cells are described, starting with Grove[62]);
3. (b) the hydrogen electrode (16 different cells are collected, again starting with Grove);
3. (c) carbon monoxide electrodes (four different cells are described);
3. (d) other fuel gas electrodes at ambient temperature (two doubtful cells have been listed; and it was interesting that the UNION-ELEKTRZITÄTSGESELLSCHT BERLIN had suggested hemoglobin as an oxygen catalyst in 1902);
3. (e) gas electrodes at higher temperature (25 combinations have been proposed with lead/lead oxide anodes, copper/copper oxide anodes, iron/iron oxide anodes, oxyhydrogen chains and other gas chains at higher temperatures);
4. regenerative 'deposition cells' (eight types of metal/air cells have been proposed. The first was Nernst, who suggested zinc as a solution electrode;[115] even the sodium chain has been proposed);
5. fuel chains of special kind;
5. (a) gaseous electrolyte;
5. (b) organic electrolyte;
5. (c) chains, which deliver current by cooling.

Cathodes (air electrodes)
1. At ambient temperature;
1. (a) platinum air electrodes (four authors are cited);
1. (b) the practice of carbon air electrodes (the first carbon/air electrode was invented by Fery in 1917 and applied in a whole zinc/carbon cell[116, 117]);
1. (c) carbon cathodes (seven cells have been cited);
1. (d) theory of the carbon/air electrode (interesting discussion about the carbon electrode being an oxide electrode like the platinum oxygen electrode; one further statement is, for instance that good catalysts for the decomposition of hydrogen peroxide give good catalysts for the oxygen electrode/cathodic depolarization);
1. (e) air electrodes from non-platinum metals (silver, copper, lead dioxide);
1. (f) air electrodes with potential transfer (redox substance) (at first Davy was cited and then about ten cells are listed in the form of chains; furthermore, copper oxide cathodes were discussed);
2. cathodes of fuel cells at high-temperature (silver, platinum, iron);
3. attachment: chlorine as a depolarizer (five cell types have been briefly described, partially involving the Deacon process).

From this they concluded, surprisingly, that for the time being, the most prospective cell type is one with an aqueous alkaline electrolyte with a carbon/air electrode and a metal anode.

Consequently, Baur developed metal air cells using cadmium or iron electrodes, but he did not really recognize the advantages of a zinc/air cell, for example. He took a number of patents, especially together with Treadwell. Another interesting result of his work is the investigation of organic substances, e.g., amino acids, glycol and alanine.[118, 119] On the other hand, it is questionable whether Baur thought about the possibility of using them as fuel in a fuel cell.

Referring again to the eminent survey paper, it should be stated that Baur and Tobler discussed more than 200 literature citations – and 'discussed' means that they involved their own ideas, so that the interested researcher can profit a lot from reading the survey thoroughly even today. On the other hand, it has to be stated that many dispositions in the whole field of fuel cells had been already made before 1933, but there was a fundamental lack of knowledge, which was corrected after Baur and Tobler.

Baur and Tobler's air electrode for the low temperature application was described by Tobler.[120] It could be also used as hydrogen electrode and was in that respect the first hydrophobic hydrogen electrode. The cylindrical gas electrode is shown in Figure 23.

In Figure 20, a picture of Baur is shown. Five years before his death, Baur wrote a kind of resume of his scientific life:[121] *"Here, we have one of those...cases, where the theoretical knowledge has been hurried on in advance of the experimental one. We...also possess the mental connection. No wonder that the electrochemists were full of hope in 1894...How strange that nearly 50 years have passed since then without getting nearer to what we are aiming at. Should the electrochemists themselves have lost their belief? It almost seems to be so. Nevertheless, in the meantime work proceeded. – There are numerous possibilities...They all have to be tested thoroughly. It would be nicest to operate with aqueous electrolyte at ambient temperature or at about 200 to 300 °C with molten sodium hydroxide or pure sulfuric acid. In this temperature region, however, nothing...can be achieved with catalysts. Both, the oxygen as well as the fuel gas electrode suffer...from too low a reaction velocity.*

Figure 20. Emil Baur. (Reproduced by permission of VARTA, Aktiengesellschaft, Hanover.)

One has to recourse to high temperatures, which immediately implies the biggest technical difficulties. In that case, the electrolyte has to be a molten salt, for instance potassium or sodium carbonate...The molten salts had to be sucked into porous bodies out of magnesia.[122]*...One has to step back further and avoid all kinds of melts. Only cells with solid electrolyte are stable in the end for any length of time...The battery [fuel cell] has to possess the highest operational stability and reliability. Well, here technology should show, what it can perform. For if...only 50% of the heat of combustion would appear in the electric control room, in that case, a revolution of the energy economy of the world would occur".*

Today, more than 50 years have passed since Baur's vision, nearly 110 years since Ostwald's challenge and about 160 years since Grove and Schoenbein's discovery. How many more decades are still needed?

1.3.3 The first air electrodes

From Grove to Jungner

Already Grove had recognized that "*a notable surface of action*" must be available for the gases so that they can react with the water. Certainly, he did not have a clear idea about the kind of reaction that had to happen, but he was sure that the goal had to be to increase the size of the surface area. He tried to platinize the platinum electrodes. The result was that the electrolyte spread over the surface by capillary action, as he himself stated. Though this phenomenon, he wanted to expose an extended surface to the gaseous atmosphere. However, he succeeded only partially because the pores were full.

The concept of enlarging the reacting surface as a means of increasing performance was taken up by Westphal and later by Mond and Langer, as was mentioned above. But success was limited because they had not been able to hydrophobize the surface with the objective of providing a triple zone for gas and liquid on the electrode (catalyst) surface. The first, who invented the possibility of making the carbon surface hydrophobic by treatment with paraffin or similar substances was Jungner.[123, 124] He had found the first porous hydrophobic air electrode.

Schmid

The significance of the hydrophobicity seems not to have been clear at that time because the following researchers did not make any use of it. One was Alfred Schmid, who produced his thesis in Basel, Switzerland, with A.L. Bernoulli in 1922 (see Figure 21). The title read (translated) "Construction of a New Diffusion Hydrogen Electrode and Measurements with Hydrogen/Halogen Chains with Diffusion Electrodes". In 1923, he wrote the book[125] (translated)

Figure 21. Alfred Schmid. (Reproduced by permission of VARTA, Aktiengesellschaft, Hanover.)

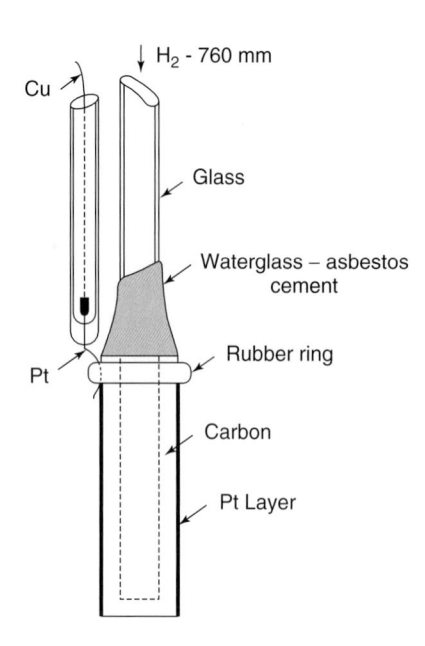

Figure 22. Schmid's hydrogen diffusion electrode. (Reproduced from Liebhafsky and Cairns (1968).[8])

"The Diffusion Gas Electrode". Although Schmid was of the opinion that his diffusion gas electrode was suitable for any kind of gas, he decided not to develop the oxygen electrode further because he thought that the hydrogen electrode did have the higher polarization. In addition, he was of the opinion that the prospect of the LTFC was much less than that for the high temperature cell. On the other hand, he saw an interesting application in the hydrogen/chlorine cell, which he transferred to practice by developing such a thing.[126] In this paper, he described a chlorine-detonating gas cell with electrodes out of porous carbon, which contained a hydrogen electrode with platinum and delivered a current density of $20\,mA\,cm^{-2}$. Nevertheless, Schmid made, by this hydrogen/chlorine cell, an important contribution to fuel cell development because he developed a gastight double layer hydrogen electrode, which was quite active. For the completion of the fuel cell, Schmid intended to apply the Deacon process for the regeneration of the chlorine corresponding to Nernst's proposal.

Schmid's electrode is shown in Figure 22. He used arc lamp coal, although this type of coal did not have the high electrical conductivity he was aiming for. The electrode consisted of a cylindrical porous graphite tube with a porous layer of electrodeposited platinum on its external surface. Hydrogen entered through a concentric cylindrical boring that ended a little bit above the bottom of the electrode. The pores in the platinum were intended to be smaller than those in the graphite and thus a dual layer, dual porosity electrode was formed. In this, the inner layer, the graphite tube had larger pores for the access of gas and collection of current, and the outer layer, made from platinum, had smaller pores so it acted as an electrocatalyst and as a gas/electrolyte seal. Schmid carefully examined the platinum layer microscopically and calculated an average pore density of $1050\,pores\,mm^{-2}$, and an average pore diameter of $16\,\mu$. The average thickness was about $0.2\,\mu$ and the total plated area about $10\,cm^2$, the amount of platinum was about $0.4\,mg\,cm^{-2}$. The coarse pores of the carbon tube were freed from the electrolyte by the gas pressure. A pore diameter of about $10\,\mu$ corresponds to a capillary pressure of $0.13\,bar$. Therefore, the pore diameter of the platinum layer must have been smaller than $16\,\mu$. The fine pores of the carbon may have been $<1\,\mu$ and completely filled with electrolyte. At any rate, Schmid had thus developed the first hydrophilic double-layer electrode. He used it as a hydrogen electrode in hydrochloric acid and achieved a current of $200\,mA$. As the active area depends on the depth of immersion, it can only be estimated. So the current density may have been as high as $50\,mA\,cm^{-2}$, a remarkable result at room temperature. The carbon/chlorine electrode showed only a negligible polarization.

Schmid did not spend his entire life at university. He was born on 29th January, 1899, in Mühlhausen, Alsace. After his study at the University of Basel he stayed there and became a professor of physical chemistry in 1928. But by 1932 he had already left the university and gone into industry, where he turned to other subjects. In 1965, he was in a very bad accident and died in 1968.

Nyberg, Heise and Schumacher

Nyberg also used the hydrophilic diffusion electrodes as oxygen electrodes. The first hydrophobic air electrode in practical use was developed by Heise and Schumacher in 1932.[127, 128] They aimed at a zinc amalgam battery with concentrated sodium hydroxide as the electrolyte. The cathode operated on air and consisted of a mixed carbon electrode. Non-graphitic carbon was baked with powdered charcoal to obtain a high porosity (50–65%). In addition, the electrode was impregnated with a solution of water-repelling substances, such as paraffin. The penetration of the electrolyte into an electrode treated in such a way could be kept below $2\,\text{mm A}^{-1}$. Taking into consideration the fact that this hydrophobic electrode was the first one at all for continuous use at room temperature, it showed a remarkable current density, namely $5\,\text{mA cm}^{-2}$.

Baur and Tobler

At the same time, namely in 1933, as when Baur and Tobler had carried out their literature study about the state-of-the-art, which was described above, Tobler was developing a voluminous carbon/oxygen electrode, which also could be used as a hydrogen electrode (see Figure 23). The electrode consists of a sack of cotton filled with granulated graphite, so that the grains are in intimate contact with one another and with a central current collector; the sack rests on a glass frit, which is connected to a glass tube. The figure shows two different sizes of electrodes, which are immersed in strong caustic solution at room temperature.

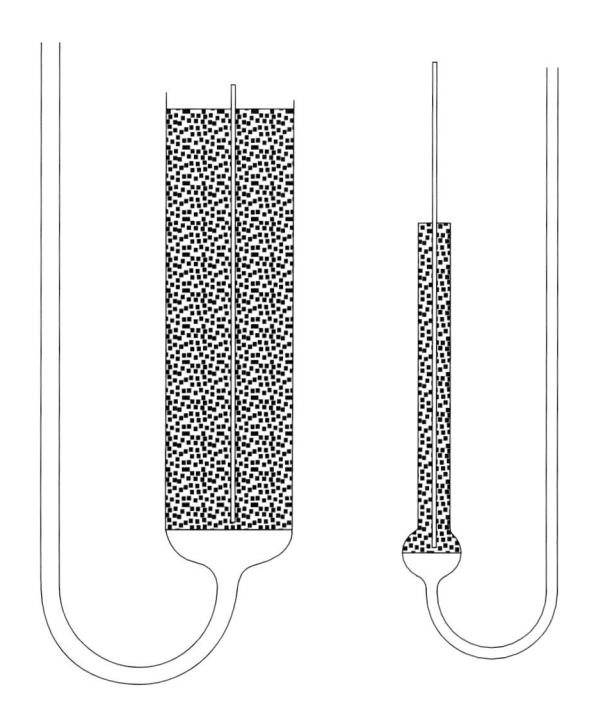

Figure 23. Tobler's carbon air electrode. (Reproduced from Liebhafsky and Cairns (1968).[8])

Tobler realized that the wetting of the graphite electrode by the electrolyte is an important factor. Therefore, he tried to achieve controlled wetting of the electrode by the electrolyte by proper impregnation of the graphite with paraffin. But the porous granules must not be completely repellent.

The advantages of these electrodes were that different electrocatalysts could be studied. Regarding the hydrogen electrode activity, platinum was found to be the most effective catalyst. The platinum was produced in place by the decomposition of chloroplatinic acid. The best additive for the cathode proved to be iron derived from blood or another source. The best performance achieved with hydrogen and air was a current density of $7\,\text{mA cm}^{-2}$, although it is difficult to relate the current to a suitable geometric area of the electrode because the electric field is horizontal.

Apart from the knowledge gained about the wetting characteristics and catalyzing, this development ran into a dead-end because the shape of the electrode could never be improved for the reduction of ohmic loss.

Berl

The next step in the investigation of oxygen electrodes – especially those made from carbon – was the work of Berl, which soon received the designation of being 'classic'. He found out that the reduction of oxygen persued the two-electron mechanism, i.e., the reaction product is hydrogen peroxide instead of water.[129] The cathodic reduction of oxygen stops at the peroxide stage, and water is only formed by the following reaction, namely by the catalytic decomposition of the peroxide (in alkaline solution of the peroxide anion):

$$O_2 + H_2O + 2e^- \longleftrightarrow HO_2^- + OH^-$$
$$HO_2^- \longleftrightarrow \tfrac{1}{2}O_2 + OH^-$$

The consequence for the activity of the electrode is that you have to look for a catalyst that has a good peroxide decomposition activity. This was found for instance in platinum, which is at the same time a good electrocatalyst.

Other carbon electrodes

On the other hand, there are some carbon examples which already have a quite high activity themselves, especially after certain treatments. Vielstich used such carbon electrodes for an alkaline methanol cell.[130]

Instead of platinum, also non-platinum catalysts were used, e.g., transition metal oxides, which are also quite active in both respects. Such catalysts were discovered first by Kordesch and Marko, who prepared their electrodes by impregnating carbon with a solution of nitrates and subsequent heating.[131] Most active was the cobalt spinel $CoAl_2O_4$. The nitrates were decomposed by heating and an aggressive gas, NO_2, was formed, which attacked the surface of the carbon and oxidized it and, in this way, increased it.

Metallic electrodes
The cathode of the first alkaline cell, which was developed by Reid,[104] consisted of sheets of steel, which contained many holes, where the oxygen bubbled through. The next metal electrode seems to have been developed by Niederreither, who at the same time applied pressure to the whole cell.[132] He too used sheet metal, which was perforated or pierced, so that the oxygen could get through. The oxygen electrode consisted of nickel or nickelated iron with a cover made from platinum or carbon, which served as a catalyst.

After Niederreither, Bacon was the first to use metal electrodes and higher pressure and both at the same time. The electrodes were made from nickel gauze, which was, in the case of the oxygen electrode, heated in air in order to produce an oxidized surface. The performance of the first cell, which was finished in 1939, was already quite good. It was operated with 27% potassium hydroxide solution at 13 mA cm^{-2} and 0.89 V at a pressure of about 500 p.s.i. and a temperature of 100 °C.[133]

1.3.4 The 1930s and 1940s

Niederreither
One possibility to increase the concentration of the reacting gases at the electrode surfaces is to raise the pressure because by this, solubility of the gases in the liquid will be increased. The first to recognize this was Niederreither,[132] who filed a patent claiming that the feature increased the pressure up to 600 atm. His electrodes consisted of iron or nickel, which were catalyzed and operated in potassium or sodium hydroxide solution. The hydrogen electrodes were covered with platinum or palladium and the oxygen electrodes with carbon or platinum. Although the pressure improved the performance of the cells remarkably, Niederreither at the same time pointed out that it may become dangerous if the pressure is raised too high, as the solubility of the gases in the electrolyte could increase so much that ignitable gas mixtures could result by diffusion.

Carbon electrodes catalyzed with platinum could also be used at both sides of the cell. Using these electrodes (diameter = 3 cm), he built a cell and operated it at a temperature of up to 200 °C and a pressure of up to 200 atm. Temperature and pressure had to correspond with each other in such a way that the vapor pressure is always smaller than the gas pressure. He developed a pressure regulating system for a pressure container, which contained two groups of fuel cells, which were electrically connected in series. He also used this pressure regulating system for a pressure electrolyzer for the production of hydrogen and oxygen.[134] This pressure regulating system included another pressurizing system with nitrogen, which was to be used if parts of the total equipment started to leak. Such a system, by the way, was also introduced into the Apollo fuel cell system.

Figure 24. Johann Niederreither. (Reproduced by permission of VARTA, Aktiengesellschaft, Hanover.)

Niederreither (see Figure 24), who was born on 25th October, 1899, extended his ideas to an energy storage system by operating of an oxygen/hydrogen gas battery (a fuel cell) together with an electrolyzer.[135, 136] This had already happened in 1932 and has since then had been proposed by others several times. He was even of the opinion that one and the same cell could be used alternatively for water decomposition and power production. At any rate, the system could also be used for the distribution of energy by hydrogen pipelines.

Bacon
The most successful fuel cell pioneer has been Francis Thomas Bacon (see Figure 25). He was born on 21st December, 1904, went to Eton College and Cambridge University, where he received a BA degree in 1925. After that he became an employee at the company C.A. Parsons & Co and stayed there until 1940. In the year 1932, he happened to read a German publication about an energy

Figure 25. Francis Thomas Bacon. (Reproduced by permission of VARTA, Aktiengesellschaft, Hanover.)

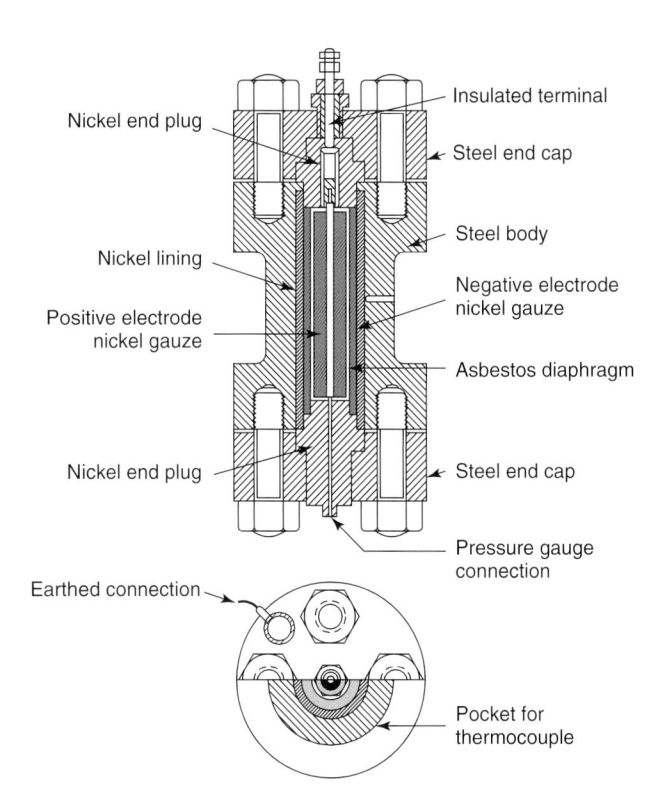

Figure 26. First high pressure cell, a reversible cell, designed by Bacon in 1938. (Reproduced by permission of VARTA, Aktiengesellschaft, Hanover.)

storage system that consisted of a pressure electrolyzer and a hydrogen engine.[6] Incidentally, at the same time Niederreither was filing his patent about a storage system, which was composed of a pressure electrolyzer together with a fuel cell instead of an engine. As a consequence, Bacon started his reflections about such a system. In 1937, he even proposed an apparatus which could be used alternatively as an electrolyzer and a fuel cell. But the board of the company declined, and nevertheless after 1 year Bacon, had a cylindrical high pressure cell produced in the workshop (see Figure 26).[133] This cell was a kind of accumulator – it could withstand a pressure of up to 210 bar and a temperature of up to 200 °C. First he operated the cell as electrolyzer, until a pressure of about 160 bar was reached – at a temperature of only 100 °C. After that the cell was discharged with a current density of 13 mA cm^{-2} until the final discharge voltage was reached, which was fixed by him at quite a high level, namely 0.89 V. During discharge, naturally, the pressure decreased to about 75 bar. Thus, Bacon's first fuel cell, a single cell, which was at the same time a reversible cell, already had a remarkable performance in 1938. The electrodes were made from nickel gauze and activated by alternately oxidizing and reducing in air and hydrogen, respectively. A tube of asbestos served as a separator – it was soaked with 27% potassium hydroxide solution. As the data for the dimensions of the cell have not been published, Euler has estimated them have had a length of about 300 mm, a diameter of about 150 mm, a weight of about 35 kg and the geometric area of the electrodes to be about 150 cm^2, so a current of about 2 A and a power of about 2 W was produced.

In spite of this outstanding result, the Parsons company was still not interested in the development of fuel cells, so Bacon left the company and went on request of Dr. Charles H. Merz, one of the two owners of an engineering office, to King's College of the University of London, where he continued his work under the guidance of Prof. A.J. Allmand, who was very much interested in fuel cells and held a lecture about this topic. Bacon constructed and built an apparatus with two cells, one fuel cell and one electrolyzer, both operating under pressure. The cells were connected with four tubes, so that two independent electrolyte circulations were possible. The electrodes consisted of six layers of nickel gauze. Under the operation of the electrolyzer, hydrogen and oxygen gases dissolve in the electrolyte, anolyte and catholyte, respectively, which are led separately to the fuel cell. By the end of 1941, Bacon had to interrupt these experiments and stayed as Temporary Experimental Officer in a test station of the British Admirality during the war. In 1946, on behalf of the semi-governmental Electrical Research Association, he went to the University of Cambridge and was taken up in the Department of Colloid

Science under Sir Eric Rideal. As Rideal had already had a publication about fuel cells,[137] he promoted Bacon's work, who at first repeated the double cell arrangement. The optimal conditions were found to be a pressure of 75 bar at a temperature of 240 °C, which resulted in a performance of the fuel cell of 81 mA cm^{-2} at a voltage of 0.65 V. The same year, Bacon received a Master of Arts from Cambridge University, and in 1947, he became Member of the Institute of Mechanical Engineers.

Then in 1949, the work was continued in the Department of Metallurgy under Prof. G. Wesley Austin. Here, Bacon now turned to the fuel cell alone and developed a new electrode, namely a sintered electrode, made out of nickel powder from the Mond Nickel Co. He invented the double-layer sintered electrode (see Figure 27), and the patent was owned by the British Electrical & Allied Industries Research Association.[138] In the main claim, one can read: *"that the pore size of a layer on the gas side is greater than that on the electrolyte side, so that on the layer side, the pores are sufficiently constricted for the difference in gas pressure across the electrode to force the electrolyte out of the pores on the gas side, but to prevent the gas bubbling through into the electrolyte"*. A current density of 1076 mA cm^{-2} at a voltage of 0.6 V was achieved with this type of cell – the highest value ever and not considered possible at the time.

After that he transferred to the Department of Chemical Engineering at Cambridge University, which was managed by Prof. T.R.C. Fox. Here, a coverage for the cathode was developed because it had to be protected against corrosion. This was achieved by a layer of nickel oxide, doped with lithium, which had, contrary to the pure nickel oxide, a good electrical conductivity. Bacon filed a patent for this electrode, which also belonged to the British Electrical & Allied Industries Research Association.[139] The new electrodes were a 1/16th of inch thick and had a diameter of 5 in. The battery consisted of six cells and contained concentrated potassium hydroxide solution as the electrolyte, with the concentration ranging between 37 and 50%. Normal operating conditions were: temperature = 200 °C and pressure = 400 p.s.i., which resulted in a power of 150 W in continuous operation, corresponding to a power density of 0.2 W cm^{-2}. A single cell of this type is shown in Figure 28. The battery was shown in London in 1954 and Bacon published his first report about his work.[140]

By 1955 work at the University of Cambridge had finished, but in 1956 it had already been resumed at the company Marshall of Cambridge. In this year, Bacon gave his second paper before the World Power Conference,[141] and he started the construction of a 40 cell battery with 10 inch electrodes. In 1957, the National Research Development Corporation (NRDC) of The UK took over the financing. So Bacon finished the first electrode with a 25 cm diameter (see Figure 29), and put this first large cell into operation in 1958 for several years. With these electrodes, he built a 40 cell battery of the filter press type, which could deliver 6 kW at a current of 250 A corresponding to a current density of 700 mA cm^{-2} at 200 °C and a pressure of 38 bar (see Figure 30). The electrolyte was 37% potassium hydroxide, the cell voltage 0.6 V and the open circuit voltage 1.15 V. The fuel cell stack was used for a forklift truck and for welding, and a demonstration was given in August, 1959.

With this large stack, it was no longer possible to remove the reaction water by thermosyphonic action in the circulating hydrogen steam mixture. Therefore, a small hydrogen blower was installed (below the stack in Figure 30). Moreover, a pressure difference of 0.15 bar between the electrolyte and the gas chambers had to be maintained.[142]

Bacon delivered a speech at the symposium of the Gas and Fuel Chemistry Division of the American Chemical Society meeting in Atlantic City in September, 1959. At this meeting he explained: *"It has always been hoped that some specialized application will arise – one for which a fuel cell is particularly suited. In this connection, the possible use of fuel cells in satellites and space vehicles is of great interest. Then, when further experience has been obtained, it should be possible to enter the commercial field in competition with storage batteries. It would seem that fuel cells of this type are most suitable for traction purposes, both road and rail. The combination of a battery and direct current series-wound motor provides an ideal propulsion unit for many types of vehicle, the limiting factor thus far being the weight of the battery"*.[133]

The American company Pratt & Whitney Aircraft applied for licenses at the end of 1959 and obtained them in 1961, when the work had been finished at Marshall of Cambridge.

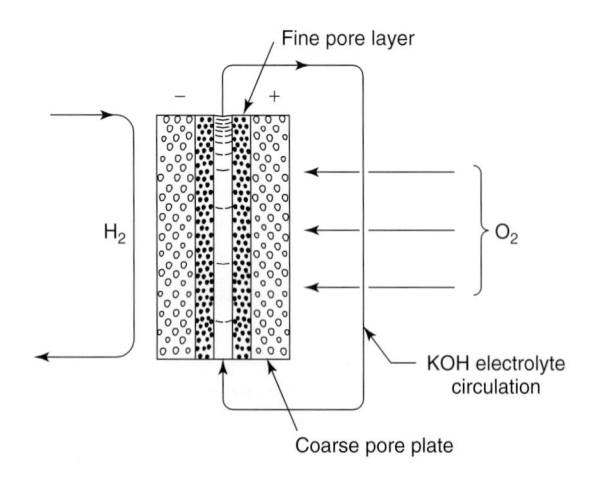

Figure 27. The first double-layer double porosity electrode. (Reproduced from Adams *et al.* (1963).[142])

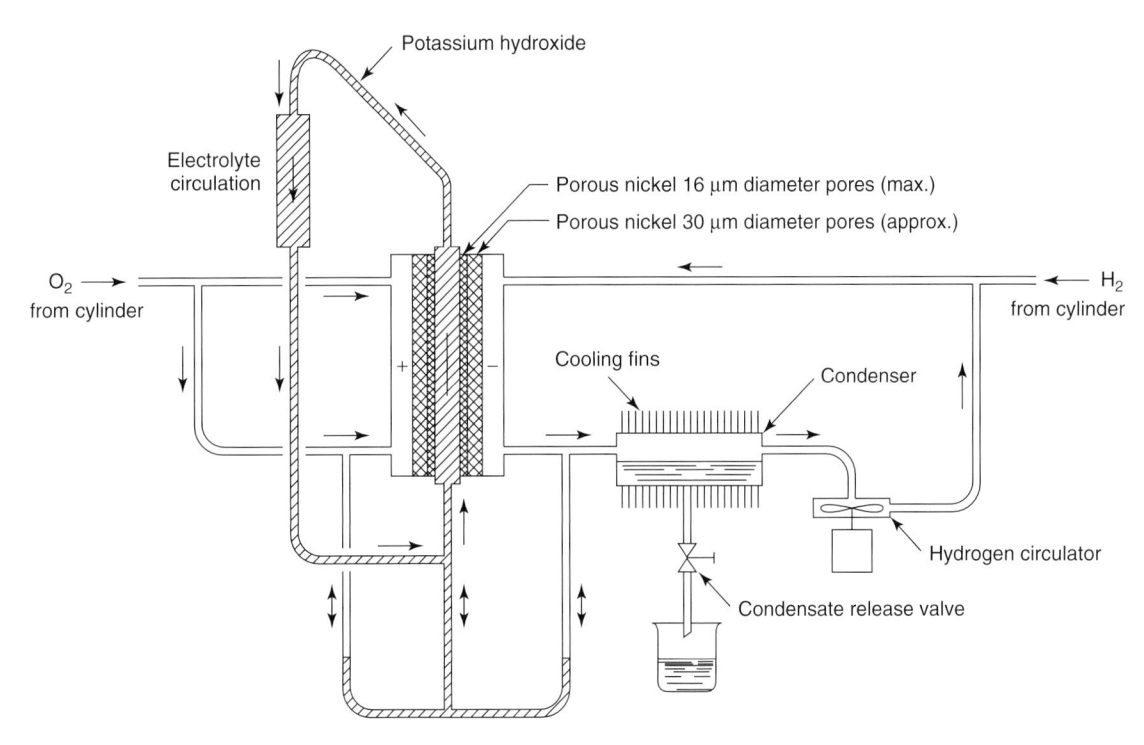

Figure 28. A single high pressure fuel cell with double layer electrodes and pressure balancing and hydrogen circulation systems. (Reproduced from Vielstich (1970)[185] by permission of Wiley-Interscience.)

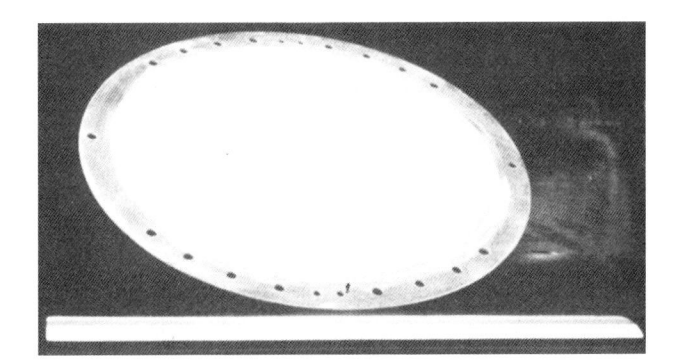

Figure 29. Bacon's 25 cm diameter double-layer electrode. (Reproduced from Bacon (1960).[133])

Pratt & Whitney had received an order from the National Aeronautics and Space Administration (NASA) to develop fuel cells for the Apollo mission. In order to reduce the weight of the stack, the laboratory of Pratt & Whitney increased the concentration of the potassium hydroxide solution up to 75%, which is still liquid at a temperature of 200–230 °C and the partial pressure of water over this solution is only 0.5 bar (the total pressure was adjusted to 4 bar), which made it possible to reduce the weight remarkably. The first flight using the fuel cell was in 1966. In 1962, a new company was founded in Basingstoke, Hampshire, UK, with the name Energy Conversion, where Bacon acted as a consultant. The objective was to use fuel cells in marine technology but finally, in 1970 work there was brought to

Figure 30. A 40 cell stack mounted on a trolley. (Reproduced from Bacon (1960).[133])

an end. A kind of continuation of the work took place at the Atomic Energy Research Establishment in Harwell, UK. From then on the high pressure route was left.

Bacon received several decorations for his work. In 1965, the Royal Society presented him with the S.G. Brown Medal; in 1967, he was awarded the Order of the British Empire (Civil Division) and in 1969, he was given the British Silver Medal of the Royal Aeronautical Society. Bacon died in 1992.

Davtyan

Oganes K. Davtyan began his research in the early 1940s in the G. Krzhizhanovsky Institute for Power Research (Energeticheskij Institut) of the USSR Academy of Sciences in Moscow. He investigated different aspects of the fuel cell problem, and in 1947, published his Ph.D. thesis in the form of a monograph "The Problem of Direct Conversion of Chemical Energy of Fuels into Electrical Energy" (Publishing House of the USSR Academy of Sciences, Moscow, Leningrad, 144 pages). This monograph is without a doubt the most comprehensive treatment of this subject published before the 'great boom' in fuel cell development in the 1960s. It includes a detailed analysis of the literature on this subject up to 1937, as well as the results of the author's investigations in the fields of LTFCs and HTFCs.

Studying performances of room temperature hydrogen/oxygen fuel cells with aqueous solutions, Davtyan for the first time emphasized the fact that these performances depended mainly on the catalytic activity of the materials used to prepare the hydrogen and oxygen electrodes. He compared the catalytic activity of different electrodes for hydrogen ionization, for hydrogen evolution, as well as that for catalysts used for the hydrogenation of organic compounds. He noted a distinct parallel between these activities for different reactions. The same parallel was found for the catalytic activity of electrodes for oxygen ionization and oxygen evolution.

He investigated the activity of different materials for hydrogen and oxygen electrodes in alkaline solutions. The best results for the hydrogen electrode were obtained with a silver deposit on activated carbon (about 10% silver), which was prepared by soaking the carbon samples with a solution of silver salts and the subsequent reduction of these salts by hydrogen gas at 200 °C. For oxygen electrodes, the highest activity was found for electrodes with nickel deposits (about 10%) on activated carbon prepared by a similar procedure. The activity of this electrode was even higher than that for electrodes containing platinum black. Using such hydrogen and oxygen electrodes, he assembled hydrogen/air cells with a 35% potassium hydroxide solution as the electrolyte. The open circuit voltage of these cells was 1.11 V, and the voltages at different current densities were 0.8 V (at 0.6 A dm^{-2}), 0.74 V (at 1 A dm^{-2}) and 0.5 V (at 2 A dm^{-2}). After improving the process of electrode preparation, he achieved current densities up to 3.5 A dm^{-2} at voltages of 0.75–0.8 V.

At the same time, he noted the main drawback of room temperature fuel cells with alkaline electrolytes, viz. the necessity to use very clean hydrogen and oxygen gases. Therefore, he considered the impossibility of using hydrogen prepared by the conversion of hydrocarbons or other organic fuels and also the impossibility of using air for the cathode without thorough (and expensive) purification from traces of carbon dioxide in order to prevent carbonization of the electrolyte solution. Due to this, he expressed his opinion that in the future alkaline hydrogen fuel cells (in combination with water electrolyzers and huge gas holders) should be applied to large-scale energy storage installations for the purpose of load-leveling in energy systems, including systems utilizing wind energy – a topic that was largely discussed 20–30 years after the publication of Davtyan's book.

The major part of Davtyan's investigations were devoted to the problem of using carbon monoxide in HTFCs with solid electrolytes. Noting that at temperatures higher than 500–800 °C, the main problem is not that of the catalytic activity of the electrodes (which at these temperatures is always high enough). He focussed mainly on investigating different types of solid electrolytes that could be used in fuel cells working at high temperatures. He investigated the conductivity and the mechanical properties of a great number of solid electrolytes based mainly on the use of monacite sand – a fossil found in the Ural region in Russia and containing 3–4% ThO_2 and some small amounts of oxides of metals of the cerium group. Other ingredients used in different ratios were sodium carbonate, sodium silicate, calcium oxide, tungsten trioxide, quartz and clay. The best results were obtained with an electrolyte of composition: 43% $Na_2Ca_2CO_3$ + 27% heat-treated monacite + 20% WO_3 + 10% Na_2SiO_4. The electrolyte was prepared by mixing and heating (melting) the ingredients at 1200 °C up to cessation of bubble evolution. Then the highly viscous melt was poured into a preheated receptacle, pressed and slowly cooled. The specific resistance of the electrolyte at 600° was 20 Ω (Ω cm) and after a subsequent heat treatment was 6 Ω. At the temperature of 600°, the electrolyte in appearance was quite solid (although it is possible that some liquid components remained in the framework of the ceramic matrix). With this electrolyte, carbon monoxide/air fuel cells were assembled. The negative electrode for carbon monoxide oxidation contained a mixture of 60% iron oxide, 20% powdered iron and 20% clay. This blend was heat treated in an air atmosphere for 2–3 h at 800–900°. The positive air electrode was prepared in a similar way from a blend of 60% Fe_3O_4, 20% Fe_2O_3 and 20% clay. Prior to the assembly of the cell, the surfaces of the electrolyte and of both electrodes were polished in order to ensure a good contact between them. The performances of these cells were investigated in the temperature range between 500 and 900°. At a current density of 2 A dm^{-2}, the voltage of the cells was 0.6 V at 500° and 0.8 V at 700°. The latter figure corresponds to an overall efficiency of energy conversion of 58%. The thickness of the cells was 20 mm (3 mm for each electrode, 4 mm for the electrolyte and 5 mm for each gas chamber). The calculated energy density for a 100 V battery of such high temperature cells at a working temperature of 700° and current density of 3 A dm^{-2} is 5 kW m^{-3}. The author noted that in some instances a cracking of the solid electrolyte occurred, especially at the very beginning of their discharge or after the discharge, during the cooling down period. This cracking was the result of sharp temperature changes. In the absence of such changes, the cell functioned for up to 20 h without any signs of electrolyte cracking.

In conclusion, Davtyan noted that despite some successes achieved in the field of fuel cells, a large amount of problems remained, which required further and extensive investigations. He expressed his opinion that the problem of fuel cells is one of the most important topics that needed to be solved in the immediate future. It can be seen from this brief review that in 1947, O.K. Davtyan anticipated many of the problems with which investigators were confronted 2–3 decades later.

In the 1960s, O.K. Davtyan was professor of physical chemistry at the State University in Odessa (now in the Ukraine). There he organized a little group (about eight persons) and continued some investigations in the field of fuel cells. His subsequent fate is not known. Apparently, he returned to his home country of Armenia. He died in about 1975.

1.3.5 The early 1950s

Justi and Winsel

Eduard Wilhelm Leonhard Justi was born in Hong Kong, China, on 30th May, 1904. He studied physics, chemistry, geology and mathematics at the universities of Kiel, Marburg and Berlin, where he received his Ph.D. in 1929 and became professor in 1942. Since 1938, he has been Regierungsrat at the Physikalisch-Technische Reichsanstalt. In 1944, he became professor at the University of Posen and in 1946 at the Technical University of Braunschweig, where he founded the Institut of Technical Physics, which he managed until his retirement. His picture can be found in the internet (see Figure 31). By the way, he was proud of the forename Eduard, because EDU is the abbreviation of Energie Direkt Umwandlung, which was a favorite expression of his.

Justi had started already his work on fuel cells in Berlin and published his first experimental results in 1948 in his book about the conductivity of solids.[143] He worked on various subjects in the field of energy[144] but worked continuously on fuel cells throughout his life until he died in Braunschweig, Germany on 16th December, 1986. His first fuel cell was similar to the one of Baur,[121] namely a

Figure 31. Professor Eduard Wilhelm Leonhard Justi. (Image supplied to G. Sandstede by Professor Justi.)

HTFC. It consisted of an iron wire as the anode, magnetite as the cathode and sodium carbonate as the electrolyte contained in a cup of steel. The second report on fuel cells by Justi was again published in a book.[145] At first, he continued experiments with the high temperature cells and filed patents, together with his co-workers K. Bischoff, H. Spengler and H.H. Keese. Then he started his work with low temperature cells with liquid electrolytes, e.g., using 6 N potassium hydroxide. Together with H.J. Thuy and A. Winsel, he developed a new type of carbon diffusion electrode and by thermal treatment the pore structure improved.[146]

Parallel to the carbon electrodes, completely new electrodes out of nickel had been developed and a first patent about these double skeleton catalyst electrodes had been filed by Justi, Scheibe and Winsel.[147] Double skeleton catalyst electrodes in German is Doppel Skelett Katalysator Electrode and is, therefore, abbreviated to DSK electrode. With this type of electrode they had found the opportunity to avoid using platinum as a catalyst because they used Raney nickel. Raney had developed these catalysts for the hydrogenation of organic compounds in 1925, and Justi and Winsel introduced them into electrochemistry as electrocatalysts. In order to produce the catalyst, nickel powder is mixed together with aluminium powder in a stoichiometry of 1 : 2 or more. After heating in argon, a reaction starts and gives rise to melting. After cooling down, the regulus has to be crushed and ground. This powder is the pre-catalyst and has to be mixed with carbonyl nickel powder and a pore-forming powder, which can be dissolved, then pressed and heated for sintering. After this has been carried out, the aluminium can be dissolved in potassium hydroxide solution. The electrode must not be dried afterwards because the Raney nickel is self-ignitable. In order to get the right double structure, different layers have to be formed and put together before the sintering.[148]

The Raney method has also been employed for the cathode. First a Raney/silver alloy was produced, which was sintered together with nickel powder. The main problem which had to be solved was the fact that the Raney/silver alloy was not brittle enough. Various procedures were tried by Justi's co-worker, K.H. Friese, and finally it was found that the melting of the alloy under a cover of calcium chloride increased the brittleness sufficiently. Also the addition of other metals, e.g., manganese, to the mixture, which was melted into the Raney/silver alloy, was successful.[149]

The performance of the DSK electrodes can be seen in Figure 32 (the data for the whole cell are included).[150]

Winsel

Justi's second book on fuel cells was published in 1962.[151] The co-author was A. Winsel, with whom Justi had the closest co-operation throughout his life. At first, Winsel stayed at the University of Braunschweig and later became the head of the Research Centre of VARTA in Kelkheim near Frankfurt am Main, where he continued the development of the DSK system – double skeleton Raney catalyst electrodes.[152, 153]

August Winsel was born on 23rd August, 1928. He studied physics at the Technical University of Braunschweig and received his Ph.D. there in 1957. Winsel developed complete fuel cell systems of the AFC type and had used them for the propulsion of a forklift by 1967. Furthermore, he invented the Eloflux system: A slow circulation of an electrolyte through the cells, which are arranged according to the filter press type, preventing concentration gradients within the cells. As well as fuel cell arrangements, Winsel and his co-workers developed components of fuel cell systems, e.g., alkaline direct methanol fuel cell (DMFC), water electrolysis, electrodialysis for the removal of carbon dioxide and water, inert gas control, electrolyte reconcentrators, carbon dioxide washer, ammonia splitter and air electrodes for chlorine electrolysis.[152–158] After retirement from the VARTA company Prof. Winsel continued his professorship at the University of Kassel full-time.

Early researchers from Justi's group were Grüneberg and Spengler, who built a dissolved fuel cell similar to that of Vielstich using ethylene glycol dissolved in potassium hydroxide solution[159] (see Figure 33).

Kordesch

Karl Kordesch was born 1922 in Vienna, Austria. He studied Chemistry and Physics at the University of Vienna

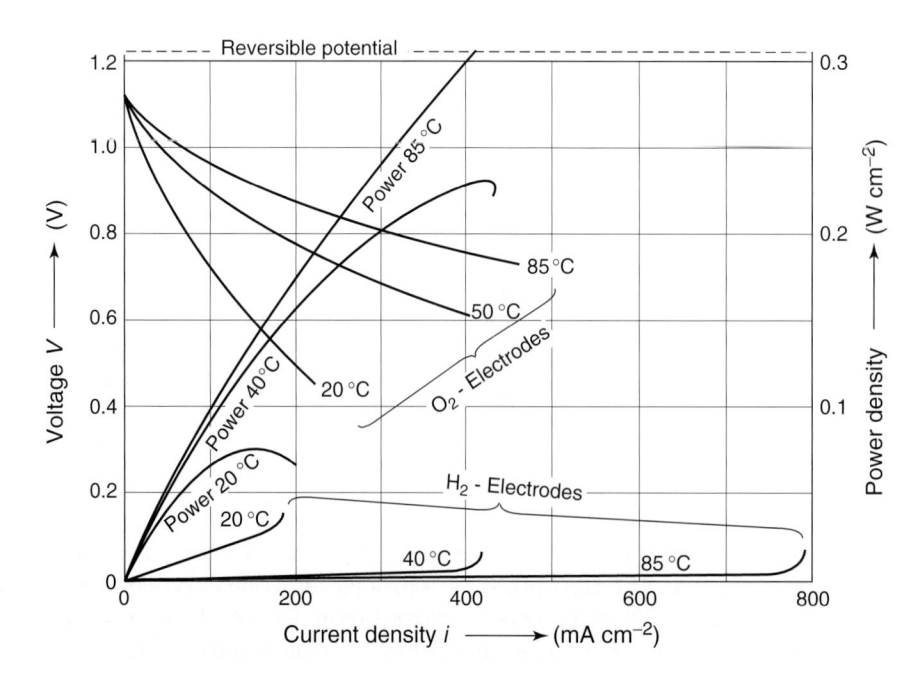

Figure 32. Performance of a hydrogen/oxygen cell of a double skeleton of Raney catalyst electrodes. (Reproduced by permission of VARTA, Aktiengesellschaft, Hanover.)

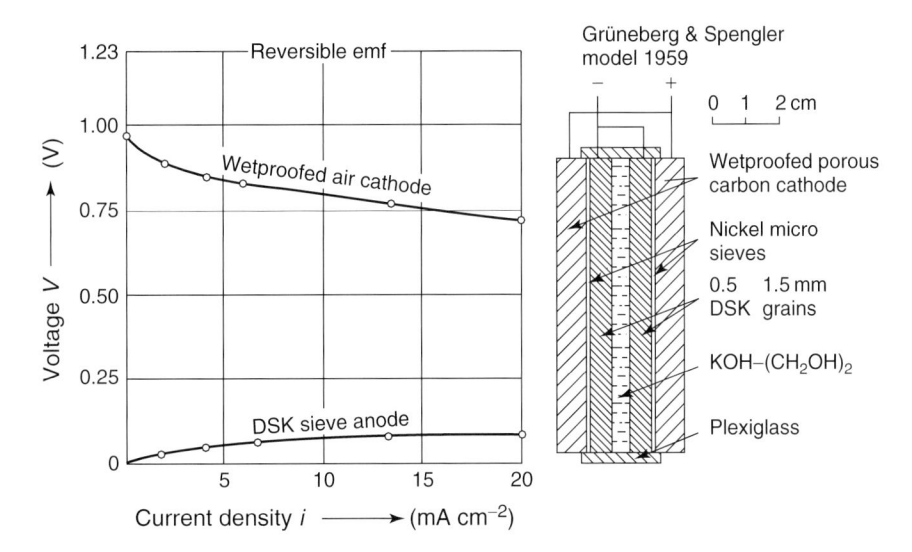

Figure 33. Alkaline glycol/air cell for room temperature with diagram of its performance. (Reproduced by permission of VARTA, Aktiengesellschaft, Hanover.)

and graduated in 1948 with a PhD. As a university assistant at the Physical Chemical Institute in Vienna he worked together with various companies, developing a zinc/air battery and was granted several patents. In the early 1950s, Kordesch worked on alkaline fuel cells (AFCs) with hydrophobic electrodes of the carbon type.[160] He invented the spinell catalyst for the oxygen electrode.[131, 161] In 1953, he was invited to continue his work at the US Signal Corps as a scientific staff member and moved with his family to the USA. There, he worked on several electrochemical systems and electronic circuits, resulting in more patents.

In 1955, Dr. Kordesch joined Union Carbide Corp. to work at the Research Laboratory in Parma, Ohio, USA. As the battery group leader, he continued the development of different batteries. The alkaline manganese dioxide/zinc primary cell was one of his major projects. Union Carbide at that time was the manufacturer of Eveready Batteries in the USA. The new alkaline system rapidly replaced the zinc/carbon (LeClanche) manganese dioxide battery between 1960 and 1970 in all Western countries. Dr. Kordesch obtained many of the basic patents on the compositions and design matters. The alkaline manganese primary battery is now the leading consumer battery and is known as the Eveready product under the name of 'Energizer', but is also produced worldwide by Duracell, Rayovac, Varta, Sony and many other companies.

During his time at Union Carbide, Dr. Kordesch, as contract manager contributed to the development of hydrogen fuel cells for the US space program, the US Navy and General Motors.[162] In the early 1970s, he changed an Austin A40 into a fuel cell city car for his personal use and operated it for several years on public roads with an Ohio license number (see Figure 34). Again, several patents and scientific publications were the result.

In recognition of his work, the US Electrochemical Society awarded Dr. Kordesch the Gold Medal for Outstanding Achievements (Vittoria de Nora). Several other awards of technical societies followed in Europe also. In 1977, Dr. Kordesch was called by the president of Austria to fill the chair of a full professor at the Technical University of Graz, serving as a director of the Institute for Inorganic Chemistry. He continued his work on alkaline batteries and fuel cells at the institute in Austria. Many publications and two more books were the result.[163] Besides his fuel cell work, the opportunity of recharging the alkaline manganese dioxide/zinc battery (which was already tried at Union Carbide) looked promising. After a few changes (patents) and with the help of Austrian government and some industrial grants, rechargeability was brought to a high level. However, it was not possible to commercialize the rechargeable alkaline battery in Europe, partly due to the resistance of the established battery industry, which favored the single use-primary battery, which had by that time become of universal use in the electronic field. Therefore, in 1986, together with Wayne D. Hartford, he founded a company called Battery Technologies (BTI) in Canada. A pilot plant production was established and negotiations about licensing this product were initiated. Rayovac, in Madison, WI, USA, became the first BTI licensee and started to manufacture all four cylindrical types of the rechargeable alkaline manganese dioxide/zinc (RAM™) Battery under the trade name 'Renewal'. In the meantime, over 50% of the small consumer Ni/Cd batteries in the USA had been replaced by

Figure 34. Kordesch's electro-car, an altered Austin 40 with a 6 kW hydrogen/air fuel cell system: (A) electrolyte removal, (B) air tubes, (C) exhaust, (D) hydrogen tubes and water disposal. On top is carbon dioxide absorption tank. (Reproduced by permission from Kordesch, "Brennstoffbatteries," Springer-Verlag, New York (1984).[349])

the new system, which found a great consumer acceptance, partly due to the popular advertising of Rayovac.

Dr. Kordesch is still heading a research group in Austria. While he retired in 1992, as professor emeritus he is entitled to guide students during their doctorate studies and can use all the facilities of the Institute at the Technical University in Graz and the Technical University in Vienna for life.

Vielstich

In the 1950s, Wolf Vielstich, working at Ruhrchemie/Oberhausen together with Justi, Winsel and Grüneberg from Braunschweig University, carried out investigations on the fundamentals of fuel cells and applied questions.

One step forward in obtaining one of the first views on electrochemical systems was the application of cyclic voltammetry to solid metal electrodes, using the first fast response potentiostat.[164] Therefore, he showed the first CVs for small organic molecules[165, 166] as well as for the oxidation of hydrazine[164] at platinum electrodes. Of special interest was the observation that the partial coverage of lead on platinum increased the rate of HCOOH oxidation by two orders of magnitude. This first experiment showing a upd effect is given in Figure 35.[167]

Small organic molecules such as methanol or formic acid perform at platinum much better in alkaline than in acidic media. Therefore, in the 1960s, fuel cell applications using alkaline electrolytes were tested. Because of carbonate formation, the fuel/electrolyte mixtures had to be renewed periodically. Porous nickel electrodes with and without

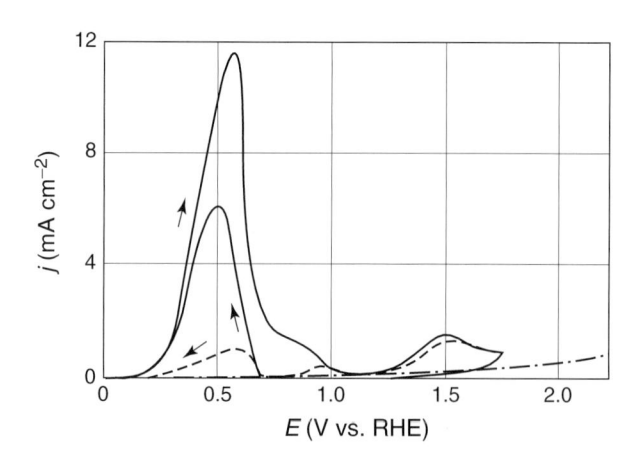

Figure 35. Cyclic voltammograms for formic acid oxidation at smooth, polycrystalline platinum, without addition of lead ions (dotted lines) and with lead ions in the electrolyte (full lines); 1 M HCOOH/1 M H_2SO_4, $5 \times 10^-$ M Pb^{2+}, 50 mV s^{-1}. (Reproduced from Schwarzer and Vielstich (1969).[167])

Pt or Pd/Pt[168] had been used as anodes and hydrophobic carbon electrodes (also with silver additions) as air cathodes.[169] Glycol and methanol have the special feature of high Ah capacities per unit volume and per unit weight. A total of 1000 Ah was available from 1 l of 6.2 M methanol solution. With glycol as the fuel, the terminal voltage of 0.8 V at 2 mA cm^{-2} and 20 °C decreased over 3000 h of discharge to 0.6 V.

With methanol as well as with methanol/formate mixtures, the energy supply to signal stations like buoys has been probed.[170] With formate as the fuel and it having only two electrons per molecule, gives it the advantage of being less volatile than methanol and so can be used at temperatures up to 100 °C. In addition, formate oxidation performs well at lower temperatures as well. Such a 20 W (40 W) methanol/formate air cell (see Figure 36) served a high altitude relay station in the upper Valais, Switzerland, for 6 months during the winter of 1965/1966.[170] Only one charge of fuel/electrolyte was used.

Problems with refilling the cells finally stopped the development of alkaline cells using liquid carbon containing compounds as fuel. A few years ago, in about 1986, a commission of the European Community decided to start a common project on methanol cells using acid electrolytes, concentrating on the severe electrocatalytic problems ahead. The first experiments of Petrii[171] with a platinum/ruthenium catalyst came into the focus of the new approach. The laboratory cells were first like the original alkaline versions. The development of high power hydrogen/oxygen fuel cells with proton conducting polymer membranes as electrolyte at the same period of time, finally led to today's DMFC as a PEMFC (see **Direct methanol fuel cells (DMFC)**, Volume 1).

Already in the early 1960s, Vielstich suggested the use of fuel cell type electrodes for other energy devices as well as for industrial electrolysis. With the nickel DSK (double skeleton) electrodes of the alkaline hydrogen/oxygen cell of Justi and Winsel came the idea of using the high hydrogen content (about one hydrogen atom per nickel atom) as an anode in a rechargeable battery.[172] For Raney nickel, there is a hydrogen capacity of 450 Ah kg^{-1}. About 100–200 Ah kg^{-1} can be reversibly taken up and released. Figure 37 shows the charge and discharge curves of a nickel hydride/nickel oxide accumulator as well as the respective curves of the positive and negative plate. The average charging voltage is

Figure 36. Left: Television station over Visp (Switzerland). Energy supply by a 20 W (40 W) methanol/formate air fuel cell from Brown, Boveri a. Cie, Baden, Switzerland. Right: Battery of 16 individual fuel cells of box form suitable for combination to form large units. Twenty of batteries (320 cells) were used at ca. 3000 m altitude for about 7000 h. (Reproduced from Vielstich (1970)[170] by permission of Wiley-Interscience.)

applications. For special applications in marine environments, however, an aluminum/air cell has been suggested as an isolated power source.[174]

In a number of electrolytic processes, the terminal voltage can be reduced if the usual hydrogen evolution electrode is replaced by an oxygen dissolution electrode. The replacement is equivalent to the connection of a hydrogen/oxygen cell in series with the electrolysis cell. The lesser voltage must be paid for by the loss of hydrogen as a product. Figure 38 shows as an example the combination of a chlorine anode with an oxygen cathode for the chlorine/alkali electrolysis via the amalgam process.[175] Some years ago. Hoechst was developing industrial size cells according to this idea, now Bayer is following this interesting new procedure.[176]

The energy-saving concept can be applied to diaphragm cells, and in addition to HCl electrolysis also. The saving in cell voltage is given by the difference in the working potential of hydrogen evolution and oxygen reduction, i.e., it results in about 1 V. A pilot plant with a capacity of 10 000 t year^{-1} has been built by Bayer, at Leverkusen. The unit, comprised of 72 cells, could be operated up to 5 kA m^{-2}, with a normal load of 4 kA m^{-2}. The energy consumption was 180 kWh tCl$_2$ at nearly 100% A h^{-1} efficiency, the chlorine quality being 99.9%. The first data is shown in Figure 39.[176]

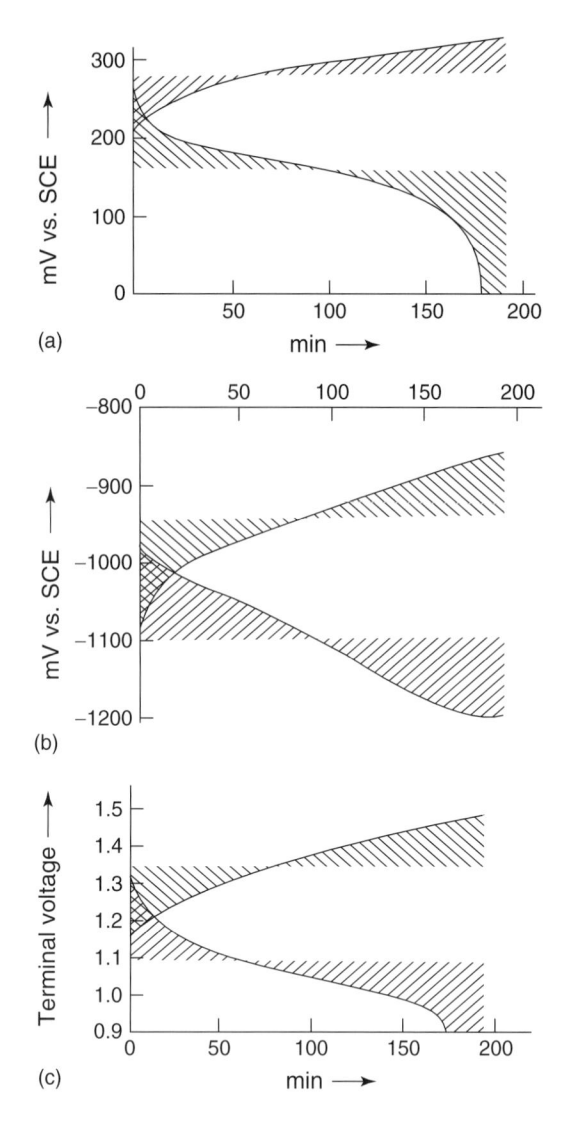

Figure 37. Charge and discharge of a nickel oxide battery electrode. (a) A Raney nickel skeleton electrode, (b) an alkaline accumulator consisting of both these electrodes, (c) electrode area = 12 cm^2, 6 M potassium hydroxide, 10 mA cm^{-2}. (Reproduced from Vielstich (1970)[172] by permission of Wiley-Interscience.)

1.3–1.4 V and the discharge voltage is near 1.1 V. The problem of this first example of a NiMeH battery is the cycle life. The nickel skeleton is not stable at the charge and discharge of hydrogen. The system had to wait for the introduction of stable nickel alloys such as LaNi$_5$.

The development of oxygen electrodes for high current densities is of interest also for metal/oxygen primary cells, such as zinc/air or zinc/oxygen. Due to its high Ah capacity per weight of almost 3000 Ah kg^{-1}, aluminum also is an interesting anode material. High current densities can be obtained in an alkaline electrolyte,[173] but problems of corrosion and gel type reaction products hinder industrial

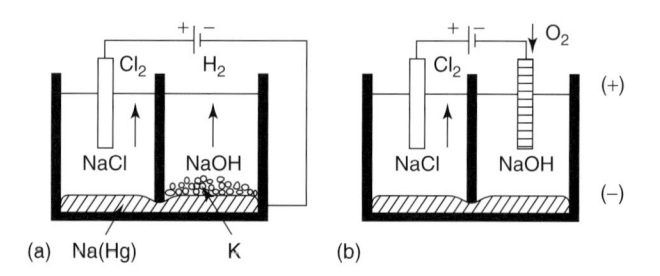

Figure 38. Chlorine/alkali electrolysis by the amalgam process. (a) Catalytic decomposition of the amalgam in the denuding cell and (b) amalgam/oxygen cell for purposes of reducing the cell voltage to obtain the Cl$_2$ product. (Reproduced from Vielstich (1970)[185] by permission of Wiley-Interscience.)

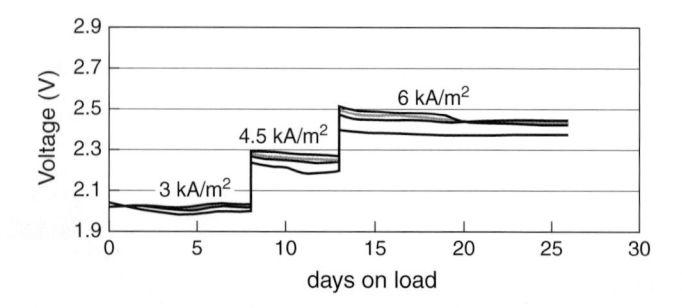

Figure 39. NaCl electrolysis with oxygen dissolution at the cathode. (Reproduced from Gestermann (2002).[176])

1.4 The final start of fuel cell development in the 1950s and 1960s

1.4.1 The first groups in the UK

As described in Section 2.2, the fuel cell is an English invention but about 40 years after Grove, C. Westphal a German, took up the research. The next step was carried out again by Englishmen. In 1889, Mond and Langer introduced a matrix for the uptake of the electrolyte. According to Appleby, this was the earliest start leading to the PAFC of today.

Later fuel cell research in the UK was done by two research groups. Firstly, Bacon's work at the University of Cambridge between 1937 and 1940 has to be mentioned. In 1940 and 1941, he continued his work after a transfer to King's College in London but he had to stop because *"it soon became apparent that a practical conclusion could not be achieved under war-time conditions"*.[177] During those years, he had been able to experiment with hydrogen/oxygen cells, first with an acid and then with an alkaline electrolyte and replacing the platinum electrodes by nickel ones. By the end of 1941, he had succeeded in connecting two cells together. In his summary about Bacon's work, which we described in Section 3.4, Schaeffer[11] remarks: *"After the war, in 1946, Bacon was able to interest the Electrical Research Association, at that time a co-operative industrial research organization financed by the British government as well as industry, in becoming a sponsor of his work. He resumed his fuel cell work at Cambridge University. After redesigning the cell (gaseous hydrogen and oxygen were used this time, instead of these gases being dissolved in the electrolyte of the fuel cell), improving the components and acquiring more money and manpower (in the early 1950s the team consisted of seven people), a six-cell 150 W fuel cell could be demonstrated at an exhibition in London in 1954. Industrial firms, however, showed no interest, but Bacon later admits that the industrialists had a point. In those days there was no idea what the use these devices could be: 'The best I could say at that time was that you should electrolyze water, transfer the hydrogen and oxygen to pressure vessels on a vehicle, and generate electrical energy in a stack of cells on board the vehicle for the purpose of propulsion. I shall never forget the shout of uproarious laughter with which this was greeted! I could of course see this point of view"*.[178]

Work at Cambridge University was discontinued in 1955, but Bacon resumed his work in 1956 at Marshall, a company which repaired and modified aircrafts, under sponsorship of the newly formed NRDC. The fuel cell staff at Marshall grew to 14. It was decided to build a 40-cell fuel cell system, producing 5 kW, which goal was successful and in fact surpassed in August, 1959.

Schaeffer reported that another fuel cell research activity in the UK started in 1953 at the Sondes Place Research Institute under the leadership of H.H. Chambers. The object of this work was to develop fuel cells to operate on conventional fuels, such as town gas (synthetic gas from coal), natural gas and vaporized liquid hydrocarbons. Sponsorship could be obtained by the Ministry of Power and the Central Electricity Authority (later the Central Electricity Generating Board). This work started with an analysis of the Davtyan cell. The claims of Davtyan could not be sustained, but during the experiments insights were produced into what was seen by the researchers as the basic problems that had to be resolved before a satisfactory design of a fuel cell could be made. They developed an electrolyte of molten carbonates and proposed and tested two designs at the cell level. Results of these experiments were reported at the first fuel cell conference in 1959.[179]

As Schaeffer continued, the industry remained skeptical towards fuel cells, also after the demonstration of Bacon's fuel cell system in 1959. In 196, Bacon's employer, Marshall, decided to discontinue its support. Most of Bacon's team members went to US companies. Despite Marshall's decision, NRDC remained interested in fuel cells, and in early 1962, it succeeded in enlisting the support of three industrial firms (Guest, Keen and Nettlefold, British Petroleum and British Ropes) with whom a new company, Energy Conversion Limited (ECL) was formed. This company also incorporated the work of Chambers and Tantram in HTFCs. During the 1960s, ECL continued its research and development activities on fuel cells but did not find any commercial application. In 1971, it was disbanded, the patents of the fuel cell technology went to the Atomic Research Center at Harwell, UK. A new company, Fuel Cells Limited, was set up, but this last effort ended in 1973.[178]

Another important UK player in the fuel cell community was Shell, which focused on low temperature, low pressure fuel cells with alkaline electrolytes, using hydrogen and hydrazine as fuels[180] and (sulfuric) acid electrolyte fuel cells (ACFCs) that used methanol as the fuel. Also the combination of a fuel reforming unit in which methanol and steam were converted into hydrogen and carbon dioxide, and a 5 kW hydrogen/air alkaline electrolyte fuel cell was developed, under contract with the Ministry of Aviation.[181]

A third company in the UK that was involved in fuel cell research was the battery company Chloride. Its R&D included studies of the basic principles of electrodes of various kinds, adsorption and catalysis, and the building of practical cells that consumed hydrogen, hydrazine, methanol and hydrocarbon fuels.[180] Most of the work was carried out on dissolved fuel cell systems at low-temperature and atmospheric pressure.[8] A light tow truck equipped with a 3.5 kW fuel cell was used at the Chloride laboratories for several years.[182]

1.4.2 The first groups in the USA

Perhaps the first significant fuel cell effort in the USA was that of W.W. Jacques, who developed a rather large fuel cell that used coal as the fuel and an electrolyte of molten potassium hydroxide. His work was reported in Harper's Magazine in 1896.[90] There is little else reported of fuel cell activity in the US until efforts began in the late 1950s at a number of laboratories, including Consolidation Coal, National Carbon, General Electric (GE), Leesona, Allis-Chalmers and Aerojet. In 1959, Allis-Chalmers demonstrated a 20 kW fuel cell-powered tractor, causing much excitement in the field.

Those laboratories reporting work at the Power Sources Conference (Red Bank, NJ) in 1957[183] and in the 1961 at the American Chemical Society (ACS) Fuel Cell Symposium[184] included GE Research Laboratory, United Aircraft, Allis-Chalmers, Lockheed Aircraft, the Institute of Gas Technology, Leesona Moos Laboratories, California Research Corp., National Carbon Corp. and Ionics. By 1963, several other groups were reporting results, including TRW, Battelle, Columbus, Texas Instruments, the US Army laboratory at Ft. Monmouth (New Jersey, USA), American Cyanamid, Westinghouse and Esso Research and Engineering. A photograph of some of the investigators of this period is shown in Figure 40. Many other companies and laboratories were engaged in fuel cell work, but did not report their results. It was estimated that over 50 US companies were involved in fuel cell R&D.

A variety of fuel cell types were being pursued by the groups above. The GE group developed the proton exchange membrane fuel cell (PEMFC) invented by W.T. Grubb of the GE Research Laboratory, initially using sulfonated polystyrene membranes and later Nafion membranes. These cells used platinum black electrodes bonded to the membranes. Cells were stacked to form 1 kW units which were the first fuel cells in space, as the electrical power source for the Gemini spacecraft, in 1965.

Molten carbonate fuel cells (MCFCs) were being developed in the 1960s by Consolidation Coal, the Institute of Gas Technology, Texas Instruments and a few others. The zirconia electrolyte fuel cell was under development by Westinghouse and GE.

Alkaline electrolyte fuel cells were being pursued at Pratt & Whitney, National Carbon, Leesona and Allis-Chalmers. The Pratt & Whitney AFC was developed based on the Bacon fuel cell but was engineered for space application and was used in the Apollo space program. A derivative of this system is now used in the space shuttle.

Work on the direct oxidation of organic fuels in fuel cells was very active in the mid-1960s, with several labs reporting results on a wide variety of fuels, including alcohols, unsaturated hydrocarbons and saturated hydrocarbons containing up to 20 carbon atoms per molecule. These included the California Research Corp, GE, Esso Research and Engineering and American Cyanamid. It was established that a variety of organic fuels could be oxidized in acidic electrolytes to carbon dioxide and water but the performance was insufficient for practical applications and the precious metal electrocatalyst requirements were too high.

This period of enthusiastic investigation in the 1960s was followed by a decrease of financial support for fuel cell R&D in the US. The 1970s and much of the 1980s was a relatively quieter period for fuel cell development in the US.

1.4.3 The first groups in Russia (and other countries of the former USSR)

One of the first Russian publications in the field of fuel cells was in 1941. It was a paper in the Journal of the USSR Academy of Sciences, "Science and Life" (No. 6, pp. 22). Pavel M. Spiridonov, an engineer who was working in the laboratory of Prof. A.N. Frumkin at the L. Karpov Institute of Physical Chemistry in Moscow, reported about investigations on hydrogen/oxygen fuel cells. He noted the possibility of using nickel instead of platinum as a material for the hydrogen electrode in alkaline solutions. On a perforated steel plate, a carbon/nickel mixture was pasted. For the oxyen electrode, activated carbon was pasted on the steel plate. At room temperature, the voltage of such a cell was 0.75 V at a current density 30 mA cm^{-2}. A battery

Figure 40. Photo of several fuel cell researchers active in the 1960s: left to right, W.T. Grubb, E.J. Cairns, G.H.J. Broers, B.S. Baker, I. Trachtenberg and L.W. Niedrach. (Image supplied by E. J. Cairns.)

of six such cells produced a current of 1.5 A at 4.5 V (an efficiency about 60%).

At about the same time, Davtyan began his studies on low temperature and HTFCs and had in 1947 published the first monograph on fuel cells. (See Section 3.4.). In the L. Karpov Institute, Dr. R.Ch. Burstein continued investigations of different types of carbon air electrodes for iron/air primary cells for many years.

Up to about 1958, there were no other Russian publications directly connected with fuel cells, but in many places investigations on different fundamental aspects had been performed which later were connected to the problem of fuel cells, e.g., investigation of the kinetics of the electrochemical reactions of oxygen reduction and hydrogen oxidation (Moscow State University, 1948–1953).

After the publication in 1959 of the paper by F.T. Bacon about the successful testing of his 500 W hydrogen/oxygen battery, the amount of R&D work in Russia (as well as in many other countries) sharply increased the 'fuel cell boom'. By decision of the USSR Academy of Sciences in 1960, a scientific council on fuel cells was formed. Prof Alexander N. Frumkin, then director of the Academy's Institute of Electrochemistry in Moscow, was appointed to be the head of this council and remained so until his death in 1976. The main task of this council was to co-ordinate and to promote all investigations in this field carried out in the Institutions of the Academy of Sciences in Universities, as well as in industrial centers.

A significant amount of fundamental and applied research in the field of LTFCs was performed at the Institute of Electrochemistry of the Academy of Sciences in Moscow. Dr. R.Ch. Burstein (who some years earlier joined the staff of this institute) was appointed head of the fuel cell laboratory. In her laboratory, processes at Raney nickel hydrogen electrodes (Dr. A.G. Pshenichnikov) and at oxygen electrodes with silver catalysts (Dr. M.R. Tarasevich) were investigated, and in about 1961, the first prototype of a 100 W alkaline hydrogen/oxygen fuel cell with such electrodes was assembled and tested over several months. Great attention was devoted to investigations of the behavior of porous gas electrodes. A group of physicists headed by Dr. Y.A. Chizmadzhev developed the first comprehensive theory of the kinetics of the processes in porous gas electrodes, and in 1971 published the first monograph on this subject. In Dr. V.S. Bagotsky's laboratory, a group headed by Dr. Y.B. Vassiliev began extensive studies of the mechanism of electrochemical oxidation of methanol and of other organic substances in order to assess the possibility of developing methanol/oxygen fuel cells.

The center of research of HTFCs was in the Institute of high temperature electrochemistry of the Ural branch of the Academy of Sciences in Sverdlovsk (now Yekaterinburg) (director Prof. S.V. Karpachov). In Dr. S.F. Palguyev's group, different types of solid electrolytes were developed, while in Dr. G.K. Stepanov's group, processes in molten carbonate electrolytes were investigated. In both groups, prototypes of HTFCs with a power up to 100 W were assembled and investigated.

Among these universities a group at the Moscow Power Institute (now Power University) headed by Dr. N.V. Korovin must be noted. In this group, in the early 1960s, the process of hydrazine oxidation was investigated and prototypes of hydrazine/oxygen fuel cells were developed.

At the University of Chemical Technology in Dnepropetrovsk (Ukraine), Dr. O.S. Ksenzhek's group performed numerous investigations in the field of macrokinetics in porous electrodes. Theoretical calculations in this field were also performed in Dr. J.G. Gurevich's group at the Institute for Heat and Mass Exchange of the Belorus Academy of Sciences in Minsk.

In the early 1960s, investigations on fuel cells were under way in different industrial institutes. In the Moscow Institute for Power Sources (director Dr. N.S. Lidorenko), different prototypes of hydrogen/oxygen fuel cells with alkaline electrolytes and with ion exchange membranes (IEMs) were designed and tested. In the Institute for Carbonaceous Products near the Moscow Region (director Dr. A.S. Fialkov), hydrogen/oxygen fuel cells with thin electrodes based on carbonized and graphitized cloths were developed.

The most sophisticated version of an industrial 1–2 kW fuel cell battery was developed by Dr. Y.L. Golin at the Ural Electrochemical Combined Plants near Sverdlovsk (Ekaterinburg) – an institution of the Nuclear Power Ministry. This battery was intended for use in the Russian version of a space shuttle, Buran, and passed successfully all the preliminary test procedures but was never practically used as the Buran program was discontinued.

1.4.4 The first groups in Germany

1880–1950

Although this discussion should start with the middle of the 20th century, it helps to understand the historical perspective better if the beginning of fuel cell development in Germany is briefly considered. About 40 years after Grove's invention in London the first important progress was done in Germany (leaving aside the investigations of the German Beetz and the Frenchmen Edmond Becquerel and Pierre Jablochkoff and others.) In 1880, C. Westphal recognized the possibility of achieving a high efficiency with gaseous voltaic cells,[84] 14 years earlier than Ostwald. Furthermore, Westphal invented the porous electrode. Eight years later, the next person who to make a big step forward was Paul Scharf, who filed a patent that dealt with a multi-fuel cell stack.[85] After him, the next to have an important influence on the development of fuel cells was Ostwald. He was a professor of physical chemistry in Leipzig, Germany, and was a great expert in energy (see Section 2.3). At the turn of the century, Nernst and Haber, well-known physico–chemists, dedicated considerable work to various types of fuel cells. Baur occupied himself with HTFCs and LTFCs. As he changed from the University of Braunschweig to the ETH Zürich, he may be considered Swiss. Apart from Nernst, Schottky and Wagner dealt with the structure of ionically conductive oxides, which led to the development of the SOFC.

The first one to propose the low temperature gaseous fuel (hydrogen) fuel cell with a liquid (alkaline) electrolyte was E. Justi in Braunschweig (Bacon was earlier but he himself said that his cell was not a real low temperature cell). Together with Winsel, he also dealt with dissolved fuel, methanol and ethylene glycol. At about the same time, Vielstich occupied himself with the dissolved fuel cell. And also the same time, the Austrian Kordesch started his work with gaseous and liquid fuels, preferring cells with carbon electrodes.

Technical University of Braunschweig
At the end of the 1940s, Justi started with HTFCs as had Davtyan and Baur earlier. With his group at the technical university in Braunschweig, Justi introduced the Raney method into the electrocatalyst preparation in the early 1950s. He specifically developed electrodes with Raney nickel and Raney silver[151] and adopted the double layer electrode structure with fine pores towards the electrolyte, so that with a little overpressure the gas could not flow through the electrode and the electrolyte could not fill the coarse pores at the back. This layer contained the electrocatalyst also, the fine pores of which were partly filled. Therefore, the whole electrode was a DSK electrode, with the electrodes being free of platinum (see Figure 41). The fine pore layer of carbonyl nickel, the coarse pore layer of carbonyl nickel powder and grains from the Raney alloy before activation can be seen. The right-hand side of the picture shows the Raney grains after activation, as indicated by the cracks in the grains.

University of Bonn
Vielstich became a professor in Bonn. He and his group studied mainly liquid reactants.[185] A principal cell of that time is shown in Figure 42. Later work dealt with electrochemical mass spectrometry in order to identify the intermediates and the reaction products of organic fuels.

Ruhrchemie
This company carried out fuel cell research quite early, see for instance the work of Spengler and Grüneberg[159] and Figure 33.

VARTA, Kelkheim
This company belongs to the great battery companies of the world and also started very early with fuel cell R&D. They took over the principle of Justi and Winsel, and Prof. Winsel became the head of VARTA's R&D center near Frankfurt

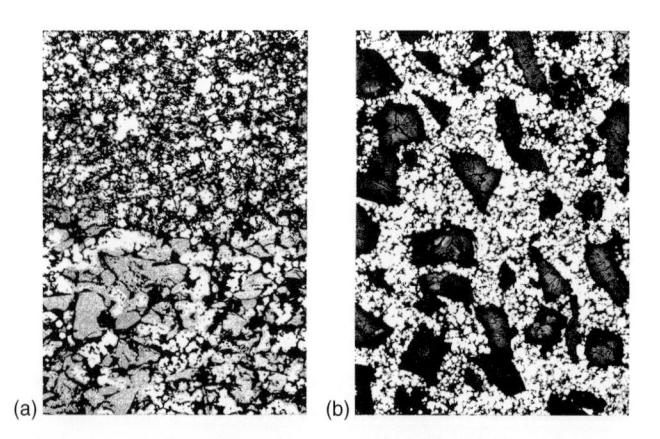

Figure 41. (a) Ground section of a DSK electrode according to Justi and Winsel, magnified to 500×. (b) Raney catalyst layer after leaching of the aluminium. (Reproduced by permission of VARTA, Aktiengesellschaft, Hanover.)

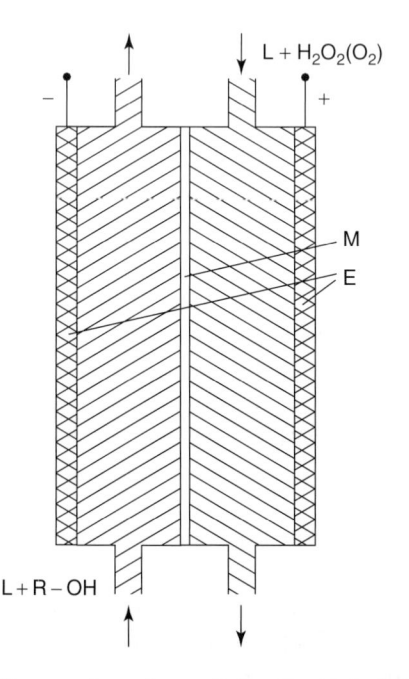

Figure 42. Construction of a cell for liquid fuel (e.g., alcohol R-OH) and dissolved oxidant H_2O_2 (M = membrane, L = electrolyte). (Reproduced from Vielstich (1970)[185] by permission of Wiley-Interscience.)

am Main after a while. By the mid-1960s, AFC batteries had been built. Figure 43 shows a forklift produced in 1966 with a 1.5 kW hydrogen/oxygen fuel cell, which had a peak performance of 5 kW.

There was a large R&D staff at VARTA, who built the fuel cell systems according to a modular design. In Figure 44 the basic module completely poured in plastic can be seen. It consists of ten cells for hydrogen/oxygen and delivers a power of 100 W at 60 °C, the weight is 3 kg and the measurements are $75 \times 140 \times 207$ mm.[186] They also developed hydrazine and methanol batteries (see for instance Figure 57). About 10 years ago, VARTA stopped everything in the field of fuel cells.

Battelle Institut in Frankfurt am Main
In the late 1950s, Sandstede formed two groups in the electrochemistry section of the physics department, one to deal with LTFCs and one for HTFCs. Within 3 years, a big program could be carried out, after the first planar SOFC was invented, which was able not only to run on hydrogen and air but also on hydrocarbons with the help of an integrated reformer. The results were presented at the 1962 meeting of the International Society of Electrochemistry (ISE, and at that time CITCE) in Rome.[8, 187] Apart from H. Binder and A. Köhling, A. Isenberg and later W. Baukal were the main co-workers on this project.

In the low temperature group, at first Raney platinum metal catalysts were developed with these results being published from 1961 onwards, for example being presented at the meetings of the Electrochemical Society (ECS), ACS, ISE (CITCE) and others. The following independent findings may be mentioned:

- alkaline methanol/air cells;[188-190]
- acid methanol/air cells;[191-193]
- a platinum/ruthenium electrocatalyst for methanol and carbon monoxide;[192]
- platinum covered with sulfur and selenium as an electrocatalyst for formic acid and carbon monoxide (see Figures 45 and 46);[194-198]
- hydrocarbon/air cells;[199-202]
- hydrophobic hydrogen/air cells with a Raney nickel/palladium catalyst and with a silver (from silver carbonate) catalyst;[188-190, 203]
- tungsten carbide (WC) anode catalyst;[204-206]
- metal chelates, especially cobalt porphyrine, as an oxygen catalyst;[207-211]
- thiospinells as oxygen catalysts.[212, 213]

The fuel cell work was brought to an end about 15 years ago.

Siemens
In the early 1960s, Siemens also started in the field of fuel cells. They also worked in cooperation with VARTA using Raney nickel and Raney silver as catalysts. Looking for a simpler design, von Sturm and colleagues arrived at an electrode structure, which they called a 'supported electrode'. The principle can be seen in Figure 47. Although the electrode consists of many layers, they are combined in

Figure 43. Forklift powered by a 1.5 kW hydrogen/oxygen VARTA fuel cell. (Reproduced by permission of VARTA, Aktiengesellschaft, Hanover.)

Figure 44. VARTA fuel cell module of 100 W. (Reproduced by permission of VARTA, Aktiengesellschaft, Hanover.)

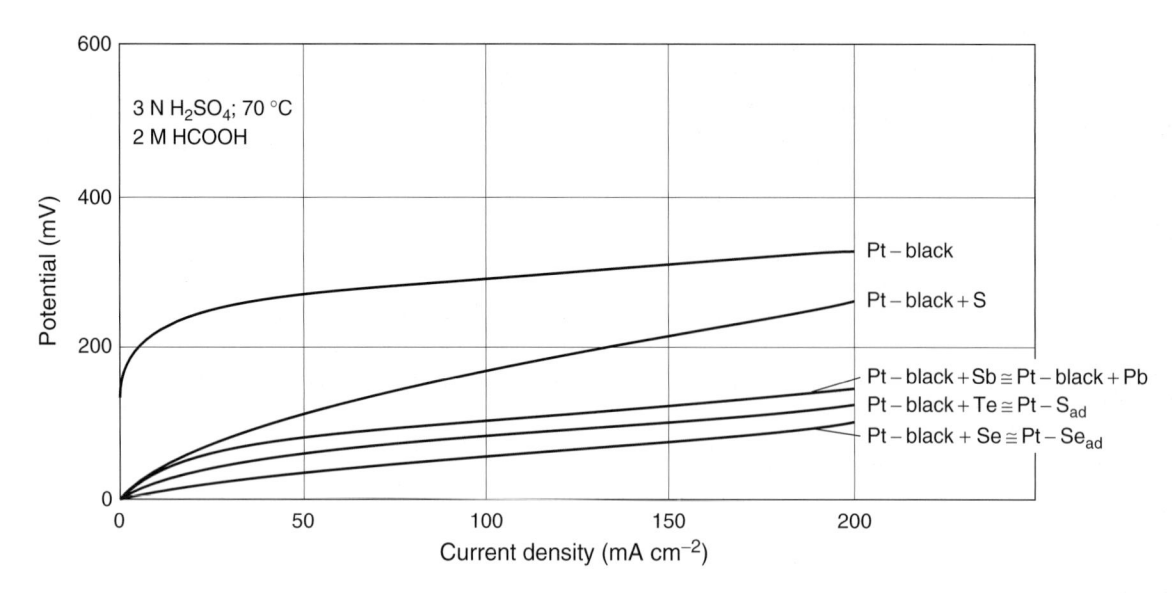

Figure 45. Stationary galvanostatic current/voltage curves for formic acid oxidation on platinum black and on platinum black mixed with various elements as well as on platinum with a 50% sulfur or selenium adsorbate. (Reproduced by permission of University of Washington Press.)

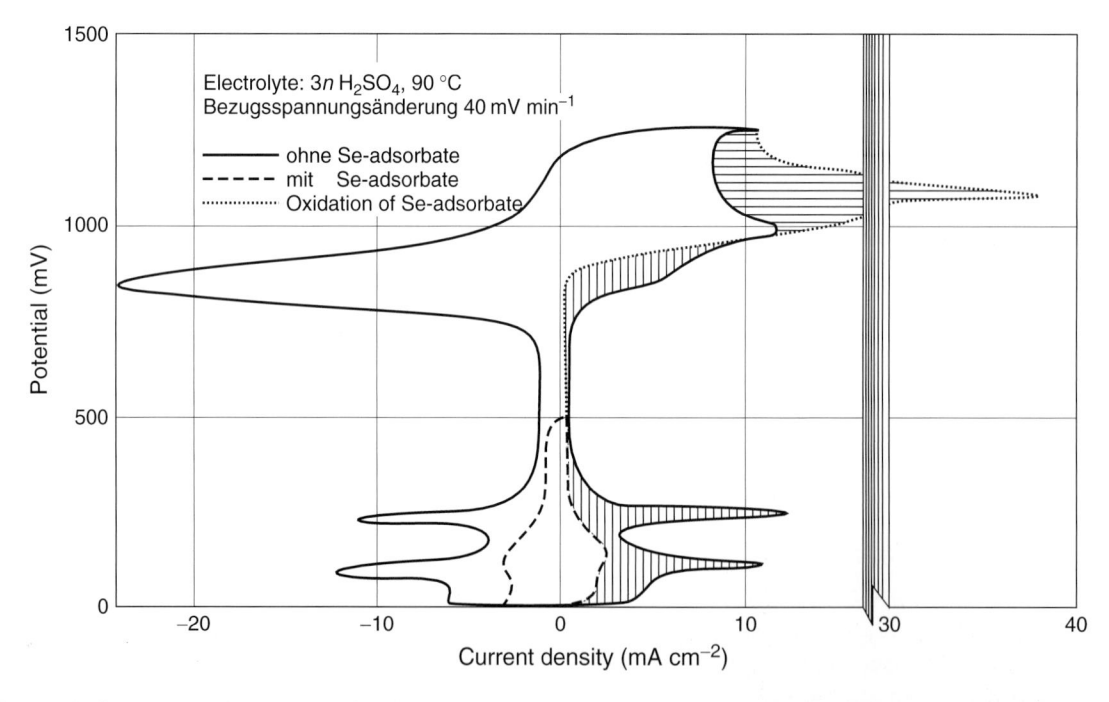

Figure 46. A periodic current/voltage curve of a platinum electrode without and with a selenium adsorbate and potentiodynamic curve of the anodic oxidation of the selenium adsorbate. (Reproduced by permission of University of Washington Press.)

a simple way. On the outside of the cell is a bipolar plate, while the inside is covered by a nickel screen, on which a layer of coarse grains of catalyst (Raney nickel) is cast. This layer is covered by a diaphragm (in the beginning asbestos, which later became, for example, polypropylene). Then the channel in the middle is formed by a waved thin metal sheet. The electrode on the other side is built in the same way, using Raney silver and the nickel screen to connect to the bipolar plate of the next cell.[214, 215]

In 1965, Siemens equipped a boat with a 500 W fuel cell according to their approach (see Figure 48).[186] Later they built 6 kW units (see Figure 49) for the propulsion of a submarine.[216] Later still, Siemens changed to the proton exchange fuel cell (PEFC), which they licensed in 1984

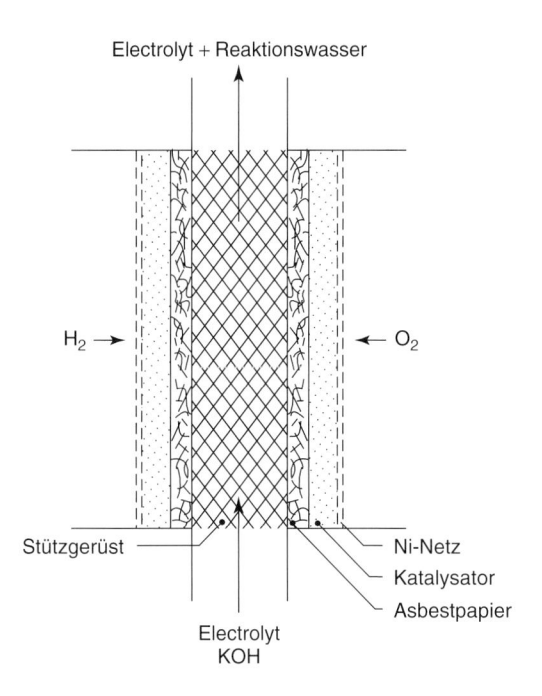

Figure 47. Siemens' supported electrode. (Reproduced by permission of VARTA, Aktiengesellschaft, Hanover.)

Figure 48. Boat with a 500 W Siemens fuel cell (1965). (Reproduced by permission of VARTA, Aktiengesellschaft, Hanover.)

from GE and from then on used for submarines of the German Navy, quite a number of which had already been built.

Von Sturm and Richter were also active in research on inorganic catalysts but up to this time were not satisfied with the results. Later, younger electrochemical managers Mund and Waidhas were occupied with the intensive developmental work for the acidic methanol cell, the biggest problem of which was namely the cross-over of the methanol to the cathode, thus decreasing its potential had not yet been solved. Nevertheless, $50\,mW\,cm^{-2}$ had been achieved so far.[217, 218]

Siemens had also dealt with the SOFC and developed an interesting planar system, which will not be described

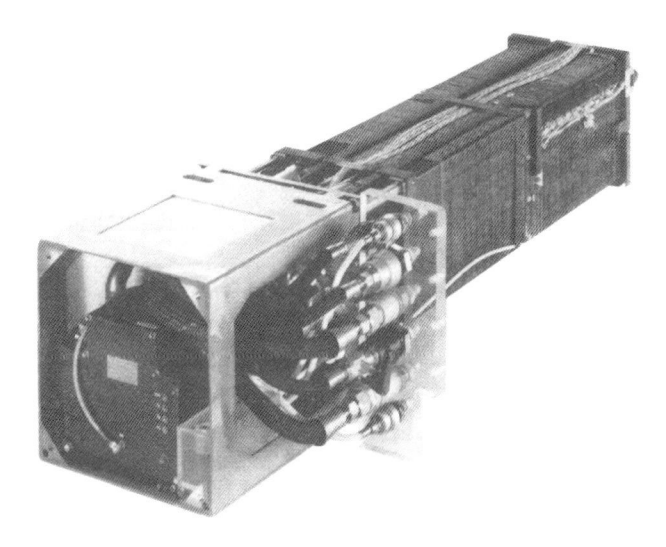

Figure 49. Siemens 6 kW AFC unit. (Reproduced by permission of Hüthig GmbH & Co. KG.)

in this chapter. On the other hand, it has to be stated that they stopped work for the time being because after buying Westinghouse Energy Operations, they adopted its tubular system.

BBC/ABB

The Research Center in Heidelberg occupied itself with SOFC development for many years. The main researcher was F.J. Rohr and the main manager was Groß. They started in 1963 with a tubular concept and stopped about 10 years ago with a planar concept.[219, 220]

University of Greifswald

At the same time, about 1960, as Sandstede in West Germany carried out his developmental work with a first SOFC, in East Germany Möbius began to investigate zirconium oxide, carry out oxygen potential measurements and, finally, build cells. Only many years later was it that they got to know of each other. Möbius wrote a historical survey in Ref. [19].

Technical University of Dresden

Only Klaus Wiesener's fields of activity shall be discussed. He also investigated metal chelates, especially porphyrines, for cathodic oxygen reduction. By applying a certain temperature treatment, he could achieve an important activation.[221, 222]

Robert Bosch, Stuttgart

As will be mentioned in Section 5.5, the Bosch research center was interested in methanol cells in particular H. Jahnke developed and investigated phthalocyanines as oxygen electrode catalysts.[223]

AEG–TELEFUNKEN

The research center in Frankfurt am Main was quite active in catalyst research and was also successful. Böhm and Pohl were a few months earlier than the Battelle team with the discovery of WC as an anode catalyst, who discovered it independently.[224, 225] With this catalyst they developed a hydrogen/air cell of the PAFC type.

DECHEMA

The Deutsche Gesellschaft für Chemische Technik und Biotechnologie (Society for Chemical and Biotechnology, DECHEMA), is comprised of the chemical industry and chemical science and engineering and has established many research committees, one of which is called 'Electrochemical Processes'. Until 1973, the chairman of this committee was H. Fischer (from Karlsruhe) and afterwards for 25 years G. Sandstede (from Frankfurt am Main). In addition, DECHEMA is the organizer of the ACHEMA, the largest chemical and chemical engineering exhibition of the world takes place every 3 years. Until 1995, the managing director of DECHEMA had been D. Behrens and from then on G. Kreysa was in charge.

In 1974 and 1975, the DECHEMA committee Electrochemical Processes carried out a study on electrochemistry under the supervision of D. Behrens and G. Sandstede (see Ref. [226]), including a chapter called "Electrochemical Energy Conversion and Storage (batteries and fuel cells)" by F. Barz, H. Binder, H. Böhm, W. Fischer, G. Sandstede, F. von Sturm, W. Vielstich and G. Walter.[227] In 1985, the DECHEMA committee Electrochemical Processes carried out a study about hydrogen technology, again under the supervision of D. Behrens and G. Sandstede (see Ref. [228]), including a chapter about fuel cells by H. Böhm, J.A. A. Ketelaar, K. Kordesch, G. Richter, G. Sandstede, W. Vielstich and G. Walter.[229] In 1986, the DECHEMA held a meeting (called the Tutzing Symposium) in Tutzing about the hydrogen economy and the conference proceedings were edited by G. Collin and G. Sandstede,[230] and included a paper by W. Vielstich.[231]

Since the early 1960s, member's lectures, who came from all over Germany, at the yearly sessions of the committee for Electrochemical Processes partly consisted of topics relevant to fuel cells. The leaders of the Electrochemical Group at the DECHEMA Institute were – in succession – E. Heitz, G. Kreysa and K. Jüttner, who have all been involved in fuel cell research.

In the 1980s and 1990s, further symposia took place in Tutzing at the annual meetings of the DECHEMA and at the ACHEMA on hydrogen and fuel cells or on topics that had something to do with fuel cells.

It may be pointed out that DECHEMA started the study of a 100 MW pilot project for hydrogen production in Canada and its transfer to Europe in 1986.

LBST

The Ludwig Bölkow Systemtechnik (LBST) was founded by the pioneer of aviation and – among other things – hydrogen, Ludwig Bölkow, who has just celebrated his 90th birthday. This company has been active in the application of hydrogen and solar energy for more than 20 years and has dealt with fuel cells to a greater extent since the late 1980s. R. Wurster, for instance, first studied water electrolysis[232] in 1984 and later also covered fuel cells in the form of comprehensive studies, which can be found on the Internet, either via LBST or DWV information (Deutscher Wasserstoffverband) or HYWEB. He also is the project leader of the Euro/Quebec Hydro/Hydrogen Project (see Section 4.7.2).

Technical University of Darmstadt

In the period when fuel cell support decreased, i.e., in the mid-1970s and early 1980s, Prof. Hartmut Wendt organized a hydrogen program for the European Union (EU). He himself participated by carrying out fundamental research on water electrolysis.[233, 234] Later he wrote a letter of reminders and warnings concerning the lagging behind of Germany in the whole field of fuel cells. Consequently, the government resumed its support. He himself, consequently, changed from research into electrolysis to that in fuel cells and was successful in electrolyte research and electrocatalysis.[234, 235] In addition, H. Wendt also founded the fuel cell committee of the Verein Deutscher Ingenieure.

HOECHST AG, BASF AG

At HOECHST, research was carried out in replacing the hydrogen electrode of the chlorine/alkali electrolysis by an oxygen-using air electrode. This silver containing hydrophobic electrode was developed with VARTA and was called a Silflon electrode. In the course of this research, a so-called falling film electrode was developed. Then, at the end of the 1980s, a complete alkaline falling film fuel cell was developed. This cell, which used hydrogen and oxygen, was very active and stable and achieved a performance of several amps per cm^2. However, HOECHST changed to the PEFC.

Being a large catalyst producer the BASF started the development of small reformers for methanol.

Mannesmann, Messer-Griesheim, Linde, LURGI, Südchemie

These are companies which have been active in many gaseous chemical processes, including the erection of plants for hydrogen or methanol production or for hydrogen storage. All of them have also dealt with fuel cells, more or less, as have other similar companies in the world.

DLR; FhG, FZJ, FZK, ISET, ZSW
The Deutsches Zentrum für Luft- und Raumfahrt (German Center for Aviation and Space Exploration, DLR) has had its fuel cell activities in the Institute for Thermodynamics in Stuttgart since the early 1980s. At first it was just alkaline systems, then SOFCs and also PEFCs. The first researchers were Fischer, Nitsch, Schnurnberger and Henne.

The Fraunhofer Gesellschaft has had fuel cell activities in the Institute for Solar Energy Systems (ISE) in Freiburg for decades. The main researchers were K. Ledjeff-Hey (he died on 1st June, 1999) and Angelika Heinzel, who is now professor in Duisburg. They dealt with the MCFCs, PEFCs and whole hydrogen energy systems. The work is continuing.

The Forschungszentrum Jülich (research center) (FZJ) formerly (KFA) established a new institute for energy process technology in the 1980s, the director of which was first von Sturm, after him Stimming and then Stolten. They have been dealing mainly with SOFCs and PEFCs.

The Institute for Solar Energy Technology in Kassel has worked (O. Führer) with A. Winsel since the 1980s.

The Zentrum für Sonnenenergie und Wasserstoffforschung (Center for Solar Energy and Hydrogen Research) in Stuttgart und Ulm has been active in fuel cell and battery research since the 1980s. Garche the director and Jörissen have been involved in technology development for many years.

Dornier, MTU, Daimler-Benz and the automobile industry
The aviation company Dornier, now belonging to Daimler-Chrysler, started with high temperature electrolysis (Dönitz, Erdle) and then switched to SOFCs. The turbine company now also belongs to Daimler-Chrysler, and it has been very intensely active in the MCFC development. After having been active in hydrogen research for decades Daimler-Benz jumped into fuel cells in the early 1990s. Everybody knows the activities in the automobile industry: Opel/GM, MAN, VW, Ford, Neoplan.

The gas industry, electric utilities and house installation industry
Some have already a long history, as many companies have been more or less intensely involved in fuel cells research and demonstration.

Teaching industry
Small companies which started one or two decades ago with developing and selling tiny fuel cells and the like to schools, colleges, universities and similar institutions should not be forgotten. They are very important. It goes without saying that this also holds for other countries.

1.4.5 The first groups in Japan

The first fuel cell in Japan in 1935
Apart from USA and outside Europe, by far the greatest fuel cell developers are Japanese institutions with respect to fundamental as well as applied research. The first overview on the fuel cell situation in Japan was given by T. Takahashi from the Department of Applied Chemistry, Faculty of Engineering, University of Nagoya,[236] which was repeated in Ref. [237] The first publication in Japan was about a single research project dealing with a HTFC.[238]

Until 1960, no further publications appeared. In this year, fuel cell R&D started, which was motivated without a doubt, by activities in Europe and the USA. Whereas in the US work on fuel cells was aimed at the development of a power system for the application in space, such as the Gemini, Apollo and Lunar excursions, investigators in Japan studied and developed fuel cells with the object of producing them on a commercial scale. But there was no definite focus for concentrating the energies of fuel cell research for about the first ten years.

Various types of fuel cells have been studied independently at a number of universities and companies. The results have been reported at the symposium on batteries, which have been held annually under the auspices of the Chemical Society of Japan and the Electrochemical Society of Japan. The sixth symposium on batteries was held in October, 1965, at Nagoya University and 27 papers on fuel cells were read.

After the first phase of fuel cell studies, which ended in the last couple of years of the 1960s, there was a slowdown in activities in the fuel cell area. Then, in 1973, the oil crisis effected a completely new situation in the energy field and new motivation for fuel cell research resulted (see Section 4.5.3).

The first phase of fuel cell research in Japan
Fundamental studies on low temperature hydrogen/oxygen fuel cells with aqueous electrolytes started in 1960 and were carried out by a large number of institutions. At Nagoya University, the oxygen electrode was investigated theoretically. At Kyoto University, Tokyo Institute of Technology, the Government Industrial Research Institute of Osaka (GIRIO), the Kobe Denki Company and the Japanese Storage Battery Company, fundamental research on carbon electrodes were carried out with respect to wet-proofings and catalysts. Sintered porous nickel electrodes were investigated as cathodes and anodes at Nagoya

University, Tokyo College of Science and the Mitsubishi Electric Corp. Furthermore, the Tokyo Shibaura Electric Company developed a new type of hydrogen electrode based on palladium.

A practical low temperature hydrogen/oxygen fuel cell battery with an alkaline electrolyte was developed by the Yuasa Battery Company, the Matsushita Electric Industrial Company and the Japanese Storage Battery Company. In 1963, the Yuasa Battery Company displayed a forklift powered with a 1 kW fuel cell battery at the sixth International Trade Fair held in Tokyo.

IEM fuel cells were studied at Osaka University and the Sanyo Electric Company, and a multi-cell type battery with an IEM equilibrated with 1 N NaOH was built in 1964. Like other institutions in the world, hydrazine fuel cells were investigated by the Kobe Denki Company and Kyoto University.

Extensive research dealt with the methanol fuel cell. The Tokyo Institute of Technology developed a methanol electrode out of copper and nickel, which was heated so that methanol was decomposed to produce hydrogen. Investigators from the Kobe Denki Company, together with Kyoto University, studied methanol electrodes and oxygen electrodes and found methanol electrodes made from porous carbon activated by a platinum catalyst give the best results. Anodic oxidation of methanol in alkaline solution was investigated at the GIRIO and the Tokyo Shibaura Electric Company. By application of the potential-sweep technique, it was observed that formaldehyde was the most reactive, formate the least and methanol in between. At Osaka University, a new methanol electrode was developed.

Also practical methanol cells were developed. The Japanese Storage Battery Company constructed a 24 W (24 V, 1 A) methanol/air fuel cell battery using a sintered nickel plate activated by platinum as a methanol electrode. And, the Matsushita Electric Industrial Company succeeded in operating a methanol/air fuel cell battery as a buoy light source in the mid-1960s.

Furthermore, sodium amalgam fuel cells were studied basically at Kyoto University, and a 15 V, 150 A sodium amalgam fuel cell battery using 0.4% Na/Hg as the fuel was constructed by the Japanese Storage Battery Company. The GIRIO investigated redox type fuel cells. Fundamental studies of the hydrogen/chlorine fuel cell were carried out at Kyoto University. Biochemical reaction cells were basically studied at Tokyo Institute of Technology. Also at this institute humic acid was extracted from coal and used as fuel for the purpose of developing a coal fuel cell capable of operating at ambient temperature.

HTFCs with alkali carbonate electrolytes were studied at the Nagoya University, Osaka University and the Mitsubishi Electric Corporation, whereas those with solid electrolytes were dealt with by researchers at Tohoku University and Nagoya University.

Also, consumable electrode fuel cells were studied. The Japanese Storage Battery Company constructed a 168 W (24 V, 7 A) zinc/air fuel cell battery, applying porous carbon as an air electrode.

All these above mentioned studies and developments were carried out within between 1960 and 1965. The literature has been cited in both, the "Sixth and Seventh Status Report on Fuel Cells", by the US Army Mobility Equipment Research and Development Center, Fort Belvoir, Virginia, USA.[237, 239]

The phase of keeping interest alive in fuel cell technology in the early 1970s
Around 1970, research activities slowed down. Although some interesting scientific results and technical information had been produced, the developmental stage had not progressed enough for the market needs for the use of fuel cells to be foreseen. However, the situation changed during the so-called oil crisis. Oil prices rose drastically and the worldwide balance of supply and demand was changed. Therefore, the Japanese Government as well as industry decided to develop new types of energy and the Agency of Industrial Science and Technology (AIST) initiated the 'sunshine project', which was followed by the 'moonlight project'.

Besides several types of fuel cells, high temperature gas turbines, advanced batteries, stirling engines and high-efficiency heat pumps were proposed for development with the aim of practical application. The moonlight program started in 1981 with the main focus of technological development for the PAFCs. But also research and development for the MCFCs was conducted, and, in addition, SOFCs and AFCs were also pursued to some extent.

The Moonlight Program
Already by 1973, the Tokyo Gas Company and the Osaka Gas Company had dealt with the PAFC and participated in the TARGET project, which was launched by the American Gas Association, by demonstrating field tests of 12 kW units and later also 40 kW units. The objective of the developmental work within the moonlight program was to build and test two 1 MW PAFC power plants, one operating under pressure and the other one at atmospheric pressure. Four companies started in 1981 and carried out the work, Fuji Electric, Mitsubishi Electric, Hitachi and Toshiba and finished in 1987. Testing followed in 1988 and 1989.

Other companies, for instance Sanyo and Mitsubishi Heavy Industries, developed PAFCs without the support of the Government. Sanyo made cross-licensing agreements with ERC in the USA with respect to their air-cooled PAFCs. Besides the government, support came also from the electric and gas utilities, which partly had their own fuel cell programs. The Tokyo Electric Power Company (TEPCO) ordered a 4.8 MW plant from the United Technology Corporation (UTC) in the US. In 1986, Toshiba joined UTC and formed with them a new company: International Fuel Cells (IFC), which developed the 200 kW units under the name PC-25. Thus, Toshiba did the engineering of all the PAFCs of the Japanese sites. The Japanese oil industry started its own fuel cell research at its Petroleum Energy Center and, in addition, a number of Japanese firms, like Tokai Carbon and Kobe Steel or Showa Shell, Kobe Steel and Babcock-Hitachi, as far as fuel processing equipment was concerned conducted research on components of the fuel cell or systems.

Besides the main focus of PAFC development, the R&D for the MCFC as well as research on the SOFC and some work for the AFC belonged to the Moonlight Program as well.

Largest fuel cell power plant of the world in 1991
The development and engineering of a 11 MW PAFC power plant was carried out by IFC in cooperation with Toshiba in the years 1985–1987. Imagine, the cell stacks are of a size of 670 kW and there is only one reformer for the 11 MW plant. TEPCO ordered the plant from Toshiba in 1988. It was assembled and installed on the premise of the Goi Thermal Power Station along the Tokyo Bay and achieved its full operation in April, 1991. The co-generation plant demonstrated a gross a.c. power efficiency of 43.6% at the rated power of 11 MW.

R&D for MCFCs
The number of companies dealing with the MCFC R&D was also considerable: Fuji Electric, Hitachi, Ishikawajima-Harima Heavy Industries (IHI), Mitsubishi Electric and Toshiba – all within the moonlight program. In addition, the Central Research Institute for the Electric Power Industry (CRIEPI) created a central testing location if the stacks of the industrial companies and the GIRIO (GIRIO) carried out basic research on materials for MCFC components.

The first aim of the program, which was to have 10 kW stacks in operation by 1987, was more or less achieved. Then the Government selected Hitachi and IHI to continue with the development and to construct a 1 MW demonstration plant by 1995 that would have an efficiency of 45%.

Research on SOFCs
Compared to the massive activities by the Japanese industrial companies and government to develop PAFCs and MCFCs, SOFC efforts were relatively small. Only two research institutes, GIRIO and the Electrotechnical Laboratory of AIST, were funded by the government to carry out some basic research, which also was part of the moonlight program.

On the other hand, the two largest gas utilities, Tokyo Gas and Osaka Gas were very interested in SOFC technology, which they showed by buying each a 3 kW SOFC system from Westinghouse in the US in 1986.

The moonlight program and follow-up
Under the frame of the moonlight program, Fuji Electric constructed a 7.5 kW alkaline hydrogen/oxygen fuel cell in 1985. Then the support was stopped and Fuji stopped its AFC activities, which they had been performing for about 20 years, in order to concentrate on the other fuel cell types.

The R&D programs for the PAFCs and MCFCs and also others was managed, technically and financially, by the New Energy Development Organization (NEDO) under the supervision of AIST, as mentioned at the beginning. Literature describing the many projects can be found in Refs. [11, 163, 240–243].

At the end of the 1980s, the development of components, stacks, reformers, inverters and also system technology was continued under a number of national and private projects. Then, in 1991 and 1992, the Ministry of International Trade and Industry (MITI) initiated two projects aimed at providing early market entry for PAFCs. MITI subsidized further projects about new types of PAFCs and MCFCs through NEDO.

1.4.6 The first groups in other European countries

Austria
Around 1950, Kordesch was working on fuel cells at the University of Vienna. He prepared hydrophobic carbon electrodes with catalysts for oxygen reduction. These were cobalt/aluminium spinells. Furthermore, the carbon tubes or plates were steam-activated to enlarge the active surface. The carbon/hydrogen electrodes contained small amounts of platinum catalysts and could also oxidize alcohols and aldehydes at ambient temperatures. Kordesch says that these cells were delivering relatively high current densities with hydrogen at ambient temperatures, confirming Baur's predictions made in 1933.[244]

Kordesch had transferred the technology to Union Carbide by 1951, and he himself went there in 1955 – work in Austria being taken up by Christoph Fabjan at the time. Later, in 1977, Kordesch came back and formed a new group on fuel cells in Graz. Breiter also returned to Vienna.

Belgium
This country saw the demonstration of one of the first fuel cells at the 1958 Brussels World Fair where the US National Carbon Company had laboratory models on

display there. But the Belgians waited for about 15 years to really begin their fuel cell work, when that of the others went down. Several companies had formed a consortium (Association Belge pour l'Étude et le Développement des Piles à Combustible (ABEPAC)) that involved in fuel cell research and development.

After several years of work, H. Van Den Broeck started his company called ELENCO in January, 1976. They concentrated on alkaline hydrogen/air AFCs and developed a stack technology suitable for mass production.

Right from the beginning, they concentrated on traction applications, and around 1980 carried out a study for a fuel cell bus.[245] One result of the study was the following: *"The overall cost per km of a hybrid fuel cell bus is 10–20% higher than that of a diesel bus but it is lower than that of battery bus and somewhat lower than that of a trolley bus, if normal bus frequencies of 3–4 buses per hour are considered"*.

Bulgaria
E. Budevski is professor at the Central Laboratory of Electrochemical Power Sources of the Bulgarian Academy of Sciences in Sofia. This institute is well known for his fundamental and applied work in electrochemical energy conversion, including battery development. Budevski's overview of the activities in research and the application of fuel cells in Bulgaria may be summarized as follows:[246] *"Structure and electrocatalysis of the oxygen electrode has been investigated. Also, metal chelates as oxygen electrode catalysts have been developed including a heat treatment. Complete practical cells have been developed, for instance zinc/air batteries and also aluminium/air batteries"*.

Denmark
The Energy Research Laboratory of the Odense University in Odense has dealt with solid electrolytes for fuel cells and other electrochemical devices. The electrolyte is called a solid-state protonic conductor (SSPC). In 1985 J. Jensen gave an overview on the work at the workshop in Ravello, Italy, organized by UNESCO together with the Commission of the European Communities.[247] They investigated various materials for the electrolyte, such as hydrated acids, amphoteric oxide hydrates, clays like montmorillonite and ion exchanged ceramics like β-alumina. All these proton conductors show interesting results, which have to be improved.

A well-known chemical company, which has been active in the production of hydrogen for fuel cells on-site or on-board for a long time, is Haldor Topsoe in Lyngby.

France
The first person who was known to have suggested a kind of fuel cell, a continuous voltaic cell for the application of coal as an electrode, was Edmond Becquerel, son of Alexandre Cesar Becquerel and father of Henry Becquerel. As has been mentioned already, he described his experiments in a few sentences in his book about electricity in 1855[80] (see Section 2.3.1). It goes without saying that this could not have had any success. Maybe this was the reason that researchers in France did not try to deal with fuel cells again for about a century.

The first genuine attention was paid to the subject by the Comité d'Action Scientific de la Defense National (CASDN). They called for a meeting of l'Institut Francais du Pétrole (IFP), Gaz de France and Electricité de France (EdF) and others. At this meeting and other occasions, the concern was put forward that France would lag behind with respect to fuel cell development. Therefore, a concerted action was demanded, which is in general in the responsibility of an independent organizational body, the Délégation Générale à la Recherche Scientifique et Technique (DGRST). At the beginning of 1960, the Comité de Conversion d'Energie was set up and allocated a great part of its funds to fuel cells. In 1965, this committee was replaced by a new committee, namely the Comité d'Electrotechnique Nouvelle, belonging to the DGRST as well. From 1970 onwards, funding slowly decreased and ceased totally around 1975. The situation in France between 1959 and 1981 is described in detail by G.J. Schaeffer.[11] As a consequence of the oil crisis in 1973, EdF decided to change predominantly to nuclear power.

Therefore, in France also, 1959 was an important year for the development of fuel cells as in other countries. The Laboratoire d'Electrolyse of the Centre National de la Recherche Scientifique (CNRS) dealt with electrode structure and electrocatalysis for hydrogen and hydrocarbon electrooxidation. IFP carried out extensive studies on the electrocatalysis of methanol and hydrocarbons. Furthermore, they were going to construct a 25 kW hydrogen/oxygen fuel cell in co-operation with the Regie Renault Companies. Other institutions, which carried out fuel cell research, were Compagnie de Telegraphie Sans Fil, the Societe les Piles Wonder, Gaz de France, Societe des Accumulateurs Fixes et de Traction and the aforementioned EdF and GdE. L'office National Industriel de l'Azote occupied themselves with LTFCs as well HTFCs. In addition, the company Alsthom was very active in the field. Towards the end of the 1960s, further companies took up work in this area too. Le Carbone Lorraine, Ugine Carbone and Compagnie Generale d'Electricité (CGE) may be mentioned.

Detailed reports about the institutions and their programs as well as their results at that time can be found in the "Fifth Status Report on Fuel Cells 1965", the "Sixth Status Report on Fuel Cells 1967" and the "Seventh Status Report on Fuel Cells 1972", issued by the US Army Electronics Command,

Fort Monmouth NJ, USA.[239, 248, 249] In 1965, the IFP issued a book "Les Piles à Combustible", which was written by 42 authors from the IFP and other institutions.[250] In the 1970s, the fuel cell scene in France slowly but surely decreased. From 1975 onwards, only IFP and Alsthom remained involved in fuel cell research. In 1974, a contract research organization on electrochemistry, SORAPEC, was established, which took up some fuel cell work. IFP stopped its activities on fuel cells in 1981. That was the date when the French fuel cell story practically ended. From 1981 to 1994, there was only some fundamental electrochemical work at universities and at the Commission à la l'Energie Atomique. In the early 1990s, the later version of the Energy commission of DGRST, the Agence Francaise pour la Maitrise de l'Energie, pointed out that France was getting behind. Consequently, the companies and other institutions resumed their work on fuel cells and, in addition, the three car manufacturers got involved in European fuel cell programs from 1994 onwards.

Italy

Although the birth of the electrochemical cell or even of electrochemistry or at least of electrochemical energy generation had taken place in Pavia and Como in Italy (Alessandro Volta is well known), there is little to say about further contributions to the history of fuel cells. Only in the 1970s was there some development here, when work on fuel cells started at CISE in Milano. The funding came from the Italian Electricity Board and later also from AEM, ENEA, CNR and industrial partners. A brief description of these Italian activities can be found in Ascoli and Redaelli's paper.[251] Since 1983, research studies on various aspects of advanced fuel cells have been conducted. Among other subjects, they dealt with MCFC electrode materials and SOFC electrolyte materials. Furthermore, PAFC systems have been studied.

In addition, it should be mentioned that the fundamental research of S. Trasatti and his group at the University of Milano is well known.[252]

Sweden

There are two main institutions in Sweden that have carried out fuel cell R&D: the Department of Chemical Technology of the Royal Institute of Technology (KTH) and the University of Lund. The groups at KTH and Lund work in close contact. The emphasis at Lund is on basic phenomena, whereas the KTH group is more applications-oriented. The group in Lund has been involved in research on macrocycle electrocatalysts for oxygen reduction since the early 1980s. In addition, they have been concerned with perovskites for the same purpose. The group at KTH has also studied the perovskites and pyrolyzed macrocycles. Activities in Sweden have been summarized by O. Lindstrom.[253] He stated, for instance, Sweden has been an alkaline corner of the fuel cell world since 1960.

In the middle of the 1960s, ASEA produced a 200 kW prototype fuel cell system for a submarine. It was to run on ammonia (certainly together with a splitter). However, after an accident during the acceptance test of the 200 kW installation in 1968, work at ASEA was discontinued.

Sweden is a country very much based on an electrical energy economy. The per capita consumption of electrical energy is twice as high in Sweden as in the USA. As the Swedish parliament has decided to phase out nuclear power by the year 2010, fuel cell power plants could be an alternative. That is the reason for further AFC development. Other projects at that time (1985) were the direct methanol/air fuel cell and the second generation iron/air battery, according to Lindstrom.

The Netherlands

In this section, we would like to refer at the beginning to the author of the comprehensive study on the history and future of fuel cells, G.J. Schaeffer[347] and cite a few statements: *"In the Netherlands, fuel cell research started in 1949, when a Dutch undergraduate student at the technical university of Delft, W.C. Mulder, wrote his M.Sc. thesis that consisted mainly of a review of the state of the art of fuel cells. He concluded with a negative conclusion on the technical and economic possibilities of fuel cells. Only the most recent cells (those with solid electrolytes) seemed to show any promise. Mulder's supervisor, Prof. Hoogland, was less pessimistic and according to Broers (1969) it was on his instigation that in September 1950 a PhD. research project was started under the auspices and sponsorship of the Central Technical Institute of the National Institute on Applied Scientific Research (TNO). The PhD. project was performed by G.H.J. Broers at the Laboratory for Electrochemistry of the University of Amsterdam under supervision of Professor J.A.A. Ketelaar. In the first three years of this research Broers examined the electrolyte used by Davtyan and generated the insight that first of all the electrolyte was not a solid but a melt at the working temperature of $700\,°C$; and, second of all that it did not remain invariant but changed over time (as did the electrodes)...On the basis of this insight, Broers developed his own fuel cell. In his first cells, he used a molten carbonate electrolyte that was supported by a sintered magnesia diaphragm as the electrolyte. Between 1953 and 1958, he was able to gradually improve the cell...As materials for the electrodes, Broers ended up with a silver air electrode and a nickel fuel electrode. Broers concluded this period of research by his dissertation in 1958 with the title 'High-Temperature Galvanic Fuel*

Cells'... *Broers and Ketelaar's work in Amsterdam would become very influential in later years*". TNO stopped the support in 1969, as they were of the opinion that it would take at least 10 years to find cheaper materials in order to arrive at an economical fuel cell.

There was a second group active in the field of MCFCs, namely the Technical University of Eindhoven. They developed a simple structure with little sealings. International Nickel Ltd's 16/7 Cr/Fe/Ni alloy 600 proved to be an excellent material for the electrolyte container of the cell. In the mid-1970s, another group of the Eindhoven University dealt with fundamental studies on the reduction of oxygen on metallo complexes, such as phthalocyanines and porphyrines.

Apart from the activities of TNO, DSM had started work on hydrogen fuel cells with alkaline electrolytes, which was continued after 1970.

Furthermore, Shell Research was active with catalyst research for methanol fuel cells.

In 1983, the Ministry of Economic Affairs set up a commission for electricity production policy and strategy, which came to the conclusion that fuel cells for co-generation are the first to become economical. The detailed conclusions are summarized by the chairman of the commission, E. Barendrecht, in Ref. [254].

Other European countries
Unfortunately, for the sake of time, the other European fuel cell activities could not be searched thoroughly enough to present accurate results.

1.4.7 The first groups in other countries of the world

Brazil
In 1985, there were four institutes active in research of fuel cells in Brazil. This was reported by R.H. Topke of the Energy Department of the Financiadora de Estudos e Projetos (Federal Financing Agency) and E.R.Gonzalez from the Instituto de Fisica e Chimica de Sao Carlos (IFQSC).[255] The other three institutes were the Department of Chemistry of the Universidade Federal do Ceará, a group of the Universidade Federal do Rio de Janeiro and the Instituto de Pesquisas Technológicas. Furthermore, there were four other financing agencies, the Conselho National de Desenvolvimento Cientifico e Technológico (Brazilian National Research Council), the Cordencao de Aperfeicoamento de Pessoal de Ensino Superior (Coordination of Post-Graduation), Secretaria de Tecnologia Industrial (Secretary of Industrial Technology) and Fundo de Incentive à Pesquisa Técnico-Cientifica/Banco do Brazil (Banco do Brazil).

In 1980, the first Brazilian Meeting on Batteries and Fuel Cells was held at the University of Ceará. The work relating to fuel cells, including PAFCs, in the Department of Chemistry comprised of the study of components and materials as well as the evaluation of methanol and ethanol as fues. The team was composed of six researchers at that time.

The group of the Universidade do Rio de Janeiro, which comprised of four researchers at that time, was interested in AFCs starting in 1975, and a prototype of 150 W with 670 cm^2 electrodes was put into operation in 1983. The activities of a group of about seven researchers of the Instituto de Pesquisas Technológicas has concentrated on catalyst research since 1979. They developed copper and zinc alloys for the production of hydrogen from methanol until 1984. The IFQSC electrochemistry group, which consisted of eight researchers, started with the development of a 1 kW PAFC in 1982.

The production of ethanol for automotive use amounted to 8 billion l in 1983 and more than a million cars were running on pure ethanol. Fuel cells could be used for co-generation and could provide the heat for the production process of ethanol.

With an area of about 8.5 km^2 and with 130 million inhabitants, the development of appropriate R&D technique were still a great challenge for Brazil in the 1980s. Nevertheless, the effort put into the development of fuel cells shows the environmental efforts in Brazil.

Canada
Nearly everybody in the renewable energy field is familiar with a name from Canada: Ballard Power. They were the first to develop high-power membrane fuel cells. But there is another name in Canada that is at least known in the field of hydrogen energy – the Hydrogen Industry Council – which is located in Montreal. This is a group of representatives of from industry that have to do with energy in general and hydrogen energy in particular. They have been active for nearly 30 years to bring together researchers and application-minded people to promote the production and use of hydrogen, also for instance in the fuel cell. In 1986, the Hydrogen Industry Council held a meeting called "The Hydrogen Link Conference", at which representatives of various countries talked about hydrogen, because Canada has huge resources of water and nuclear energies (and also coal and oil sands). Therefore, electrical energy, especially hydroenergy, is cheap in Canada, which could be used to produce hydrogen by water electrolysis. The conference proceedings can be found in Ref. [256] and the article by Sandstede about the situation in Germany in Ref. [257].

DECHEMA together with Dr. Gretz of the EU and R. Wurster of LBST and Sandstede and others developed a project called the "One Hundred Megawatts of Hydrogen Produced in Canada and Transferred to Europe", which was supported by a great number of companies and after about 10 years was postponed[258, 259] (it could be re-activated soon). A spin-off of this project is the Island project – "Hydrogen Energy from Water and Geothermal Power". These projects are certainly contributions to the overall subject "Where does the hydrogen for fuel cells come from?"

Meanwhile, the Ballard Power Company has developed into the largest fuel cell manufacturer in the world with Daimler-Chrysler and Ford as associates and other companies as partners. We should bear in mind that Ballard was the first to think of the terrestrial applications of the PEFC and has moved this forward in his company Ballard Power, which further developed the PEFC under contract from the Canadian Department of National Defense in the 1980s, beginning in 1983.[260] They succeeded in improving the membrane and electrode assemblies (MEA)s of the PEFC, achieving several amps per cm^2.

China
Regarding China, we also refer to the workshop "Fuel Cells, Trends in Research and Applications" presented in Ravello, Italy, in 1985. Chuan-Sin Cha from the Department of Chemistry of the Wuhan University gave a report at that workshop about fuel cell activities in China.[261] He described two types of fuel cell systems, which they seem to have started in the 1970s.

The first system was designed for use in manned spacecraft and was called the "4001 Fuel Cell System". It runs on hydrogen and oxygen, which is supplied in liquid form. The normal output is 300–500 W and 700 W at the maximum. Cha reported that the fuel cells are of the asbestos membrane type. The cells have plastic-bonded carbon electrodes with a noble metal electrocatalyst and an alkaline electrolyte. The whole system weighs about 50 kg, not including the weight of the liquid hydrogen and oxygen containers. He concluded that there was no information as to when such a system will actually be utilized for flight into outer space.

The second system Cha described is a 200 W indirect ammonia/air fuel cell system, which had been developed in the laboratory of the University of Wuhan. This system was intended to be used as a power station for terrestrial applications in remote areas. More attention was, therefore, being paid to minimizing fabrication costs and to improve the reliability of the system during prolonged and even unattended operation. A preliminary report was published in 1979 and a more detailed paper was presented at the Fifth World Hydrogen Energy Conference held in Beijing, China, in 1985. Ammonia cracking is the most favored method for hydrogen generation in the case of small fuel cell systems. It may be cracked at about 700 °C to give a gas mixture containing 75% hydrogen and 25% nitrogen with no constituents harmful to the fuel cell catalyst and electrolyte. The vapor pressure of liquefied ammonia is about 10 bar at ambient temperature. Theoretically, 13% of the energy of the hydrogen generated is necessary for sustaining the cracking. Cha reported about ageing problems but is confident to have solved them.

Furthermore, he reported about metal/air cells. The silver catalyst in the zinc/air cells, which are on the market, will soon be replaced by heat-treated cobalt tetra phenyl porphyrine (CoTPP) supported on carbon, which is cheaper and more reliable. Zinc/air batteries of 1000 Ah with high performance plastic-bonded fuel cell type air electrodes are widely employed as power sources for isolated signal lamps. Aluminum/air and iron/air batteries were developed in several Chinese institutes.

India
A report on the situation of fuel cells in India was given at the same workshop where the reports on Brazil and China and other countries had been presented. J.D. Pandya from the Tata Energy Research Institute in New Delhi first gave an overview about the general energy situation in 1985, and afterwards discussed the status of fuel cell research.[262] India is a rural country with over 80% of its population living in sparsely inhabited villages. The total number of villages amounts to 585 000, of which are 297 000 electrified. Another figure of interest is the number of wells. The total number is 16 million, of which are about 5 million are energized, whereas the target to energize amounts to 11 million by the year 2000. This shows the potential for fuel cells.

In 1985 there were not many groups in India engaged in fuel cell research, although the first work was initiated in 1974 by the Central Fuel Research Institute in Dhanbad. Mukherjee and co-workers developed catalyst impregnated carbon electrodes for low temperature hydrogen/oxygen fuel cells. Rao and co-workers of the Indian Institute of Technology in Kharagpur studied the electrochemical oxidation of formamide in an acidic medium. Shukla and co-workers from the Indian Institute of Science in Bangalore were developing low-cost charcoal-based air electrodes for direct/indirect ammonia/air fuel cells. Venkatesan and co-workers at the Central Electrochemical Research Institute in Karaikudi planned to build a 1 kW low temperature alkaline hydrogen/oxygen fuel cell and fabricated single cells having electrodes of the three layer DSK type with areas of 15 and 30 cm^2. They also were interested in high temperature molten carbonate and solid electrolyte cells.

Govil and co-workers from the Tata Institute of Fundamental Research in Bombay dealt with bio-fuel cells. They have designed coenzyme-immobilized anodes and have succeeded in chemically linking flavine adenine dinucleotide (FAD) via a carbon bridge to PTFE-bonded carbon black and have subsequently immobilized glucose oxidase on the FAD-modified electrodes.

Banerjee at the University of Delhi initiated a research program to investigate the suitability of various low temperature hydrogen/oxygen fuel cells for a low-cost power supply. Vasudevan and co-workers at the Indian Institute of Technology in New Delhi plan to synthesize a variety of electrocatalysts such as iron, cobalt and nickel phthalocyanines and study their activity. Pandya, Varshney and Copalan of the Tata Energy Research Institute in New Delhi plan to investigate electrode kinetics of fuel cell reactions.

The first ten years of fuel cell research created a lot of motivation, so that this historic beginning forms a sound basis for many applications to come, taking into consideration that India is especially suited for decentralized energy generation.

Other countries of the world
From Argentina to Zealand, there are many other countries active or at least interested in fuel cells. Because of the limited time budget we have to stop rather arbitrarily or because historical information is rather scarce.

1.5 History of R&D for various types of LTFCs

1.5.1 AFC

Fuel cells with alkaline electrolytes are attractive because of the fact that the alkaline electrolyte is much less corrosive than a strong acidic electrolyte. This allows much greater latitude in the selection of electrocatalysts and materials of construction. In addition, the conductivity of the alkaline electrolyte is nearly as high as that of a strong acid electrolyte. A disadvantage of the alkaline electrolyte is its lack of invariance when used with organic fuels or with air (containing 350 ppm of carbon dioxide).

Francis T. Bacon, a British engineer, appreciated the advantages of the alkaline electrolyte and worked for over 25 years developing a high power density AFC operating on hydrogen and oxygen at elevated temperatures ($\sim 200\,^\circ$C) and pressures (~ 28 atm).[142] The electrolyte was typically 45% potassium hydroxide and the main material of construction was nickel. Bacon used dual-layer porous nickel electrodes for hydrostatic stabilization of the gas/electrolyte interface within the porous electrodes. The oxygen electrode had a coating of lithiated nickel oxide on its surface, which had electronic conductivity and acted as the electrocatalyst. A notable feature of this electrode is that it was able to achieve the reversible potential for oxygen reduction. The hydrogen electrocatalyst was nickel.

Bacon first developed single cells, then scaled them up to 12.5 cm in diameter and built small cell stacks. Finally, he scaled the cells up to 25 cm in diameter and built a 40 cell stack, which was demonstrated publicly in 1959.[142] Cell performance data from 1960 are shown in Figure 50. A photograph of the 40 cell system, which was operated at 7–10 kW, is shown in Figure 51.[142]

Bacon's technology was very impressive and represented a large advance in fuel cell technology and engineering. Other companies worked on the further developments of Bacon's system. These included Leesona Moos and Pratt & Whitney (now United Technologies). Pratt & Whitney adapted Bacon's system for operation in space and supplied the fuel cell systems for the Apollo spacecraft, as shown in Figure 52.[263] This represented the second fuel cell for use in space (after the GE PEMFC system used in the Gemini spacecraft).

In the 1950s, Eduard Justi in Germany was developing a somewhat different AFC based on the use of Raney nickel as the electrocatalyst for the hydrogen electrode and Raney silver for the oxygen electrode.[149, 151] Operating at 85 °C, these cells could deliver about 300 mW cm^{-2} of power. This system was further developed by the Brown, Boverie and Cie company in Germany. Systems of hundreds of watts were demonstrated.

In the USA, Kordesch and co-workers at the Union Carbide Corporation developed porous carbon electrodes for alkaline electrolyte fuel cells.[162, 264] Initially, the electrodes were porous carbon tubes with electrocatalysts deposited in the pores. Subsequently, flat plate electrodes (~ 6 mm thick) were used and finally, thin, porous nickel supported electrodes were developed.[264] The electrocatalyst layer was porous carbon with electrocatalysts added. For the anode, platinum was used and for the cathode, a variety of electrocatalysts was evaluated, including perovskites, silver, platinum and others. As might be expected, the life limitation was the accumulation of solid potassium carbonate in the pores of the air electrode. The lifetime was extended greatly by scrubbing the carbon dioxide from the air before it was admitted to the cathode.

The Union Carbide cells were scaled up to about 1000 cm^2, and modules of cells were used in various demonstrations, including a fuel cell powered van built by General Motors. Union Carbide also built a battery fuel cell hybrid automobile that was operated by Kordesch.[162, 264]

An interesting early demonstration of a vehicle powered by an AFC was that of Allis-Chalmers. This system used

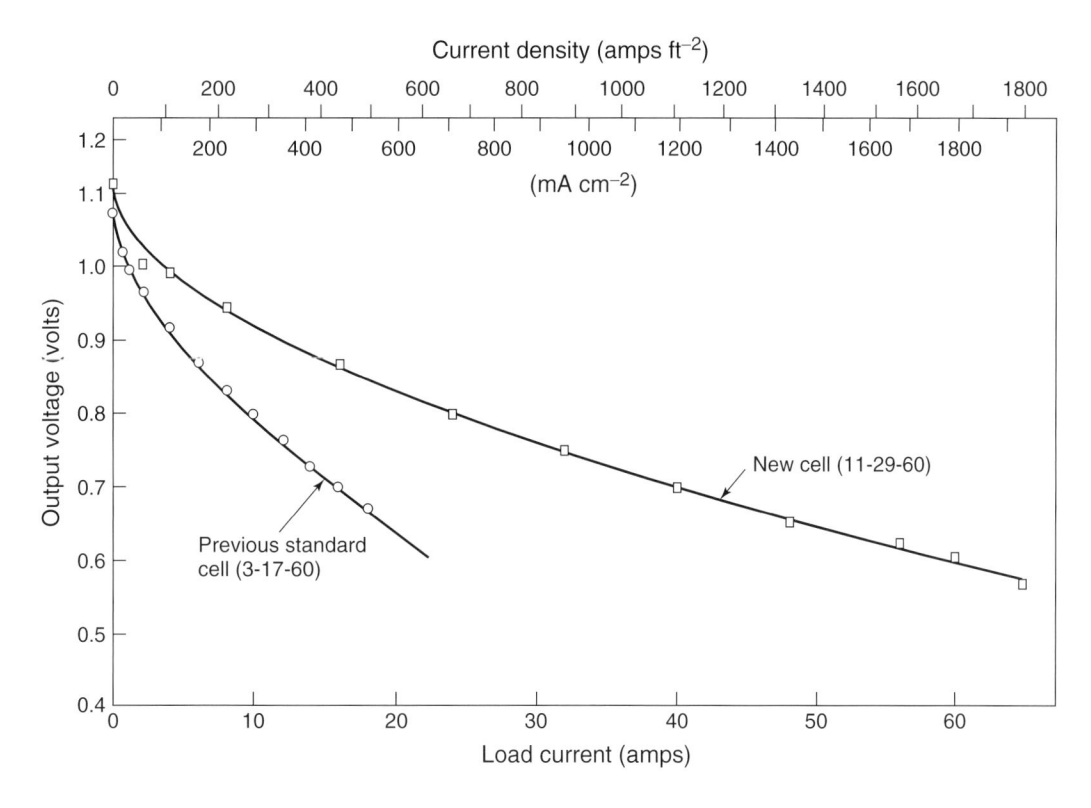

Figure 50. Bacon cell performance. (Reproduced from Reference [142].)

Figure 51. Photo of Bacon 40-cell stack. (Image supplied to E. J. Cairns by F. Bacon.)

an alkaline electrolyte immobilized in a porous matrix. A 20 horsepower system operating on hydrogen was installed in a tractor and demonstrations of the tractor plowing up a Milwaukee golf course were given in the fall of 1959.

The most enduring application of the AFC is its use in space. It has served as the on-board power source for the Apollo spacecraft and the space shuttle. It is still the system of choice for the space shuttle program.

1.5.2 ACFC

The abbreviation 'ACFC' is a new one although the acid fuel cell (acid electrolyte fuel cell), i.e., the fuel cell with an acid electrolyte was the first one that had been invented. It operated on hydrogen and oxygen. In 1842, W.R. Grove published his second article about fuel cells, which he called "Gaseous Voltaic Batteries" and showed a picture of this battery for the first time, a four-cell battery, which was of carrying out electrical work, namely electrolyzing water (see Figure 15). The cell contained by means of sulfuric acid acidified water. The abbreviation "ACFC" is the beginning of a row:

- ACFC = acid fuel cell;
- PAFC = phosphoric acid fuel cell;
- PEFC = proton exchange fuel cell or polymer electrolyte fuel cell;
- BEFC = bromide electrolyte fuel cell;
- FEFC = fluoride electrolyte fuel cell (hydrofluoride acid fuel cell);
- CEFC = hydrochloride acid fuel cell.

Figure 52. Photo of Apollo fuel cell. (Image supplied to E. J. Cairns by C. C. Morrill.)

Therefore, keeping the systematics AFC should be changed to ALFC (alkaline electrolyte fuel cell).

Although Grove had already recognized that the reaction needed *a "notable surface of action"*, it took nearly 40 years to translate this idea into practice. In 1880 Westphal invented the porous electrode and explained it as an increase of the three-phase contact zone for the reaction of the gases. In his patent,[84] he wrote: *"The gases, which are under pressure, can be pressed through diaphragms, which are covered on the outside with conducting material or which consist wholly of conducting porous material, so that they, being in the form of small bubbles, are in contact with the liquid and the metal at the same time"*. He also talks of porous tubes and of using water gas, generator gas or coal gas, whereby he recognized the advantage of the acid electrolyte. Another inventor developed a multi-cell battery with porous electrodes, which was built in 1888 according to the filter press principle.[85] Scharf's other feature had been the introduction of a porous non-conducting disk for the soaking of the electrolyte.[6] The cells were electrically connected in parallel. For using an acid electrolyte, Scharf can also be designated the forerunner of the membrane cell.

Two other inventors, Mond and Langer, have already been acknowledged for having built the first practical cells with porous electrodes.[9] They published their results in a scientific journal, opposite Westphal and Scharf, in 1889. Nevertheless, Mond and Langer developed practical cells also using an acid electrolyte soaked in a non-conducting matrix between the electrodes and operated them with carbon monoxide as well.[86]

In about 1910, Walther Nernst developed a hydrochloric acid cell for hydrogen and chlorine, which could be regenerated according to the Deacon process.[6] Alfred Schmid also developed a hydrochloric acid cell.[125] Elton Cairns has thoroughly investigated cells with hydrofluoric acid and even used hydrocarbons as a fuel.[265]

Sulfuric acid is said to be not very suitable as an electrolyte for fuel cells at least at higher temperatures because it will be reduced giving sulfurous acid or even sulfur or hydrogen sulfide, which can poison the platinum catalyst.[266] As Gerd Sandstede discovered that this need not be taken so seriously, at least if Raney catalysts of noble metals are used.[267] Thus, it is possible to anodically oxidize completely saturated hydrocarbons and a hydrocarbon fuel cell was developed and demonstrated at the ACHEMA exhibition in Frankfurt am Main in 1967.

Acid fuel cells with non-platinum metal catalysts were also developed. Tungsten carbide WC is, for instance, an electrocatalyst that merits much more attention. Böhm and Pohl used this catalyst in phosphoric acid,[224] while Sandstede and co-workers used it with sulfuric acid. A cell with a WC electrode, coupled to a reformer for methanol was demonstrated at the ACHEMA exhibition in Frankfurt am Main in 1970.[205]

The attractive features of trifluoromethane sulfonic acid (TFMSA) were investigated by Appleby and Baker.[268] This acid is characterized as a super acid: it has a high ionic conductivity, good thermal stability, higher solubility of oxygen than in non-fluorinated acids and a low degree of adsorption of the anion on the electrode surface. To overcome the problems encountered with TFMSA, instead of fluorinated sulfonic acids fluorinated carboxylic, phosphoric and antimonic acids were investigated.[269]

The ACFC branches, so to speak, into various families of fuel cells:

- a matrix for the electrolyte at higher temperatures: PAFC;
- a proton exchange membrane as an electrolyte: PEFC;
- a soluble fuel (methanol): DMFC;
- a proton conducting inorganic membrane at higher temperatures;
- a hydrogen carbonate electrolyte (see Section 5.8).

All these types of fuel cells have the advantage that they do not need to deal with carbon dioxide, be it as a natural component in the air, in the reformate of the organic fuel materials or resulting in the electrochemical conversion of organic fuel substances, in any case, carbon dioxide will be rejected.

1.5.3 PAFC

Although the carbon dioxide problem can also be solved in alkaline systems, it is easier to start with a system that does not have such a problem. Therefore, after the first successful applications of the ALFC, simultaneously the development of the PAFC was started in the 1960s. In order to achieve a high enough conductivity, the temperature had to be raised up to $200\,°C$. In principle, this may be a disadvantage but it is, however, an advantage because the water can be easily rejected and the heat can be easily extracted, so that the PAFC is an ideal co-generation system. Furthermore, the higher temperature means that less carbon monoxide is adsorbed on the surface of the platinum electrocatalyst. On the other hand, it is not so easy to stabilize the electrolyte at a higher temperature. The stabilization could be achieved by applying a matrix material. The best one proved to be silicon carbide, which is kind of a hydrophilic reservoir for the electrolyte.

In the late 1960s, development expenditure increased because industry created the TARGET program (a team to advance research for gas energy transformation). The program was initiated by a group of gas and electric utilities, the principal objective of which was the development of small combined heat and electricity power plants for residential use. Altogether 65 experimental units of $12.5\,kW$ natural gas-fuelled, PAFC power plants were manufactured in the laboratories of the Pratt & Whitney Aircraft Company. They were operated by utilities across the United States, Canada and Japan between 1971 and 1973. This was the first large scale attempt of a practical terrestrial application of the fuel cell.

Historical reviews of PAFC developments can be found in a number of publications.[7, 243, 266, 270] Following TARGET, several development programs were sponsored by the ERDA (which is now called Department of Energy, DOE), the Electric Power Research Institute and the Gas Research Institute (GRI) in the USA. At first, the construction and operation of 48 units of $40\,kW$ PAFC cogeneration plants was supported by the GRI and DOE, starting from 1976. The contractor was the UTC and later IFC, belonging to UTC and Toshiba. This plant was called Fuel Cell Generator-1 (FCG-1) and the project joined nine electric utilities. After the successful operation of these units, a 1 MW demonstration power plant was built in the UTC Power Systems Division facility in South Windsor, Connecticut, in 1977. Also this rather large unit went successfully through the testing procedures and, consequently, a scale-up was made to the $4.5\,MW$ size for a demonstration plant at a Consolidated Edison site at East 15th Street and the Franklin D. Roosevelt Drive in Manhattan. The original plan was that the plant would start running in 1978 but installation and testing were slowed down by licensing regulations and fire department uncertainties concerning untested aspects of the new technology. The partially unforeseen pressure testing led to damages and, therefore repeatedly delayed the start of the operation and when finally everything was completed after 5 years, the fuel cell stacks had reached their lifetime end – they had dried out.

The second $4.5\,MW$ demonstrator was ordered from UTC by the TEPCO in 1980. It was located at the Goi Power Plant on Tokyo bay and since the cell stacks had more electrolyte inventory than the New York stacks, the shelf-life problems encountered in New York did not occur. This plant operated from 1983 to 1985 successfully, achieving an overall degree of efficiency of about 37%. Based on operation and construction experience, IFC and Toshiba jointly developed an improved, semi-conventional 11 MW PAFC plant. It began operation in April 1991 and achieved an electrical efficiency (HHV) of 41% on an average.

It is interesting to persue the specific amount of platinum the electrodes needed. In the mid-1960s, ca. 10 mg cm^{-2} was necessary and the electrodes were made from PTFE-bonded Pt-black supported by a tantalum mesh screen. The electrolyte was supported by glass fiber paper. In the mid-1970s, a carbon support for the catalyst was introduced and the amount of platinum went down to 0.25 mg cm^{-2} for the anode and 0.50 mg cm^{-2} for the cathode. As a catalyst carrier, Vulcan XC-72, a conductive oil furnace black, product of the Cabot Corp., the surface area of which amounted to 220 m^2 g^{-1}, was employed. The support for the electrodes was changed to carbon paper and for the electrolyte to PTFE-bonded SiC. Since then only minor improvements have been carried out, for instance the amount of platinum for the anode was reduced further down to 0.1 mg cm^{-2}.

In the 1990s, UTC/IFC concentrated on manufacturing 200 kW units and sold about 200 of them. Meanwhile other companies also invested in the development of PAFCs in the USA and in Japan. After reaching their first goal, namely building of a demonstration unit, most of them discontinued further work, partly because UTC had already covered the market and secondly, the market was not at all as large as originally assumed, probably resulting from a very slow cost and prize reduction. Furthermore, other types of fuel cells were ripening and it was too difficult to recognize which system would serve best the requirements. The competitive systems include the MCFCs, SOFCs, PEFCs and ALFCs.

1.5.4 PEFC

When Willard T. Grubb invented the PEMFC in 1957,[271–273] he called it a ion exchange membrane (IEM).[274] Later the GE changed the name into solid polymer electrolyte (SPE), which was their trademark because the striking characteristic of the membrane was that it was a gas-tight polymer with ionic conductivity. A polymeric support as a matrix for the electrolyte had already been used. The abbreviation used was SPFC or SPEFC. It was only later that the electrolyte property came to the forefront. As an intermediate step, the designation read PMFC (polymeric membrane fuel cell), and finally we arrived at PEFC (polymer electrolyte or polymer electrolyte membrane (PEM) or proton exchange membrane, which could also be abbreviated to PEMFC – this has the advantage of sounding fluent when spoken). Other expressions are the PEM cell or membrane cell.

The invention was also investigated by colleagues at GE,[275] (see Figure 53) and immediately after the invention became known, others followed.[276, 277] GE wanted to use this type of fuel cell for application to the Gemini space flight, the first manned space project. The main problem in the beginning was the low current density because of the poor contact of the electrodes to the membrane; therefore, strong acid was used to provide a better contact between the adjacent membrane and the surface of the catalyst. But there was a great disadvantage in that the reaction water was not completely neutral. Consequently, methods were sought that created an intimate contact between the catalyst and the membrane surface. Those methods were the in situ precipitation of the catalyst onto the membrane.

The first IEM was not stable enough against higher temperatures. The material used in the Gemini program was polystyrene sulfonate. The membrane showed a hot spot effect from drying, leading to irregular conductivity over the surface. The wet spots got hotter and hotter till they

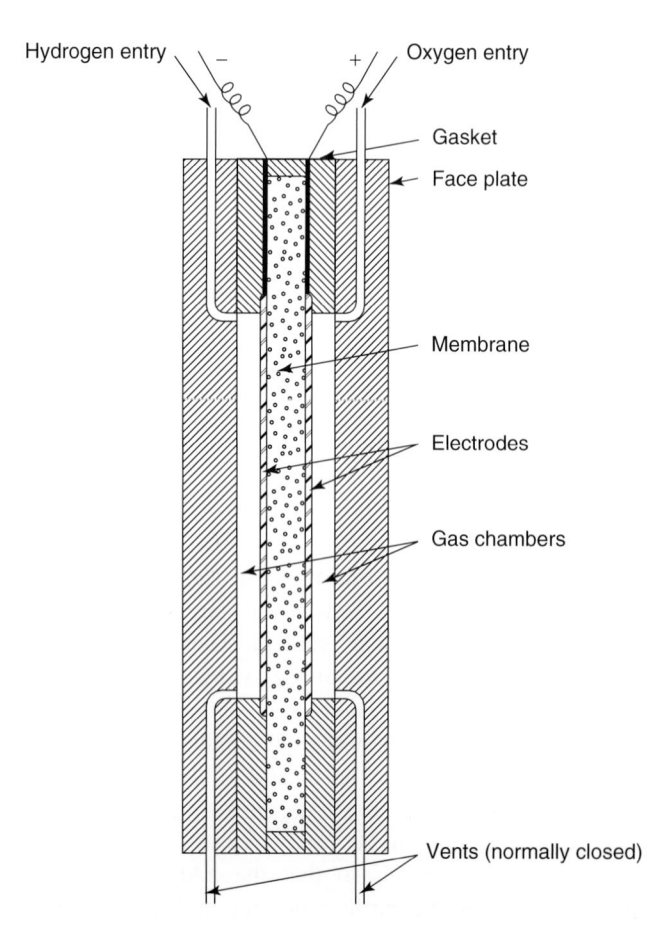

Figure 53. Basic cell structure of the PEFC. (Reproduced from Vielstich (1970)[185] by permission of Wiley-Interscience.)

finally melted and, thus, destroyed the cell. Later DuPont invented Nafion – a perfluorinated product: sulfonated polytetrafluorethylene – which was used in the chlorine/alkali electrolysis for producing chlorine and alkali hydroxide. This material has been used since the early 1970s.[278–281] It has very attractive properties: a thermal stability at temperatures higher than 100 °C and also a strong chemical stability without losing the pure ionic conductivity. Consequently, it had one disadvantage of having a high cost. This holds still today with only a few groups seeming to have achieved producing competitive membranes.

GE built a 1 kW power plant for each of the seven space flights in the early 1970s. The second GE PEFC unit was a 350 W module, which powered the Biosatellite spacecraft in 1969. For this cell, a Nafion membrane manufactured by DuPont was used. The power density of the cell was about 90 W ft^{-2}, whereas it was for the Gemini cells it was about 50 W ft^{-2}.

Another development achieved is that present PEFCs do not use any electrolyte other than the hydrated membrane itself. In the basic cell, the proton-conducting membrane is sandwiched between two platinum-impregnated porous electrodes, whereas the backs of the electrodes are made hydrophobic by coating with a PTFE suspension. By this wet-proof coating, a path for gas diffusion to the catalyst layer is provided. This unit is called a MEA (see Section 6.3).

On the basis of the same principle, GE also developed a SPE electrolysis cell using the same Nafion membranes.[282] We can call it PEEC (EC = electrolysis cell) or SPEC. Combining both cells, we get the PEM electrolysis fuel cell system, which in orbit, facing the sun, uses the current from the solar cells to generate hydrogen and oxygen and during shady periods provides electric energy from the fuel and oxidant for the station. Meanwhile, other companies have also developed PEECs for terrestrial applications.

In the mid-1980s, GE SPE technology (fuel cells and electrolyzers) was transferred to UTC–Hamilton Standard, and a license for the basic know-how for PEFCs was granted by GE to Siemens AG in Germany during 1983–1984, which was to have been used for as a power source for an air-independent submarine.

Terrestrial applications of the PEFC has been forwarded by the Canadian company Ballard, which further developed the PEFC under contract from the Canadian Department of National Defense in the 1980s, beginning in 1983.[260] They succeeded in improving the MEAs by soaking with dissolved PTFE, achieving several amps per cm^2. Ballard also tried other membranes with the Dow Chemical membrane, MK IV, and they improved the stability up to 120 °C and the pressure up to 700 kPa. In 1993, Daimler-Benz came and took the stacks for driving a car and showed the world that the exhaust was only pure drinking water (attention and warning because of the missing minerals!).

In the 1980s, other companies in the USA began developing PEFCs as well, including Ergenics Power Systems, the A.F. Sammer Corp., Engelhard and Treadwell. Then LANL did basic studies and fundamental research, becoming the greatest promoter of the PEFC over the years. In the late 1980s, organizations such as IFC (a joint venture between UTC and Toshiba), General Motors (Delco Remy, GM Research), Giner Inc. and Texas A&M in the USA and also DeNora in Italy and Japanese companies started development.

In 1985, Ballard achieved a further significant development for the total system of the fuel cell with reformer. They included a selective oxidation process for the removal of carbon monoxide from the reformate, thus making the use of methanol or hydrocarbons as a fuel possible. Furthermore, alloy catalysts were developed, which can cope with about 100 ppm of carbon monoxide in the fuel gas.

1.5.5 DMFC

Nearly all designations for fuel cells are based on the type of the electrolyte – there are only very few exceptions and the first one is the DMFC. But again one researcher is to be admired for being the first – William Grove. In 1845, he had already screened a long list of organic substances with respect to their activity as fuels, which he called 'voltaic excitants' including alcohol.[73] The next researcher was Taitelbaum, who converted the chemical energy of liquid fuels (petroleum, stearic acid and starch)[95] into electricity. But the possibility of the direct electrochemical oxidation of methanol was at first investigated by E. Müller.[283] On the other hand, the real beginners started in the 1950s and early 1960s: Kordesch and Marko,[284] Bloch and colleagues,[285] Justi, Winsel, Grüneberg and group,[151] Vielstich and co-workers,[130, 286] Sandstede and team,[191, 200] Schlatter,[287] Foust and Sweeney[288] and Williams and co-workers[289].

Because methanol is an attractive fuel, several research groups tried to develop a fuel cell for the direct conversion of this liquid fuel, as can be seen from the above. But there is one obstacle for a high power density like hydrogen and that is the complicated reaction mechanism, which results in a steep polarization of the methanol anode. This is demonstrated in Figure 58, shown at the Detroit meeting of the ECS 1961.[191] Therefore, intensive research followed in order to explain the reaction mechanism. An early result was found by Breiter[290] and by Bagotsky.[291] It can be summarized by:

$$CH_3OH + 3Pt \longrightarrow Pt_3COH + 3H^+ + 3e^-$$

$$3Pt + 3H_2O \longrightarrow 3Pt - OH + 3H^+ + 3e^-$$

$$Pt_3COH + 3Pt - OH \longrightarrow CO_2 + 2H_2O + 3Pt$$

Further investigations concerning the mechanism of the anodic oxidation of methanol, form aldehyde and formic acid were carried out by Binder, Frumkin, Iwasita, Mund, Parsons, Sandstede, Stimming, Vielstich, Waidhas, Winsel and others. In the early 1960s, empirical research arrived at alloy electrocatalysts, for instance platinum/ruthenium was found by Heath,[292] Frumkin[293] and Petrii, Sandstede, Vielstich, and Williams. Figure 54 shows an example of the dependence of the activity on the composition of the electrocatalyst.[192, 267]

Figure 55 shows the influence of the alloying of platinum with ruthenium on a methanol electrode in a hydrogen carbonate solution[267]. The activity is nearly as high in this equilibrium electrolyte as in an acid electrolyte. For further work with buffer electrolytes see Section 5.8.[294]

A big problem is the fact that methanol is transferred to the cathode by diffusion and additively while the cell is operating. This is called methanol cross-over. This is not only the case in a liquid electrolyte but also in a membrane electrolyte because the proton is not only hydrated with water molecules but also solvated with methanol molecules. Thus, the higher the current, the more methanol is transferred to the cathode. This problem has not yet been completely solved. Nevertheless, a number of companies and other research groups have developed methanol fuel cell modules. By in 1965, Shell researchers in the UK had developed a 40-cell 300 W module[289] with an acid electrolyte for methanol and air (see Figure 56).

As a further example, Siemens' 100 W demonstration unit may be mentioned.[295, 296] In addition, the work of IFC should be pointed out,[297] as they investigated the methanol cross-over. At any rate, they achieved a current density of 270 mA cm^{-2} at a voltage of 0.5 V at 100 °C with a platinum/ruthenium catalyst for the anode.[163]

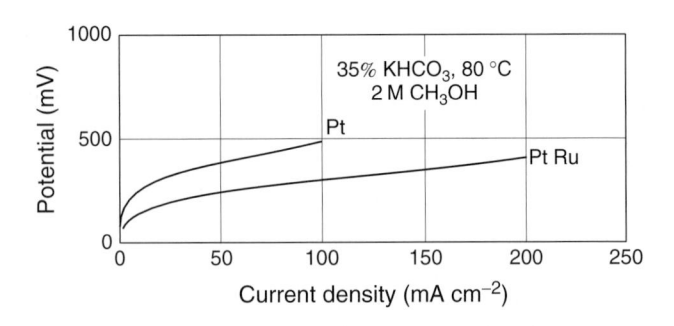

Figure 55. Potential/current density plots showing the increase in electrocatalytical activity of the methanol electrode with Raney platinum in an equilibrium electrolyte potassium hydrogen carbonate solution. (Reproduced from Binder (1965).[192])

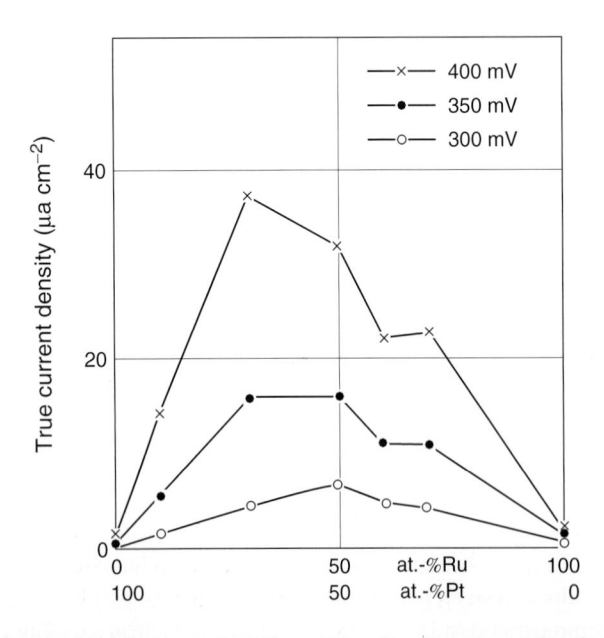

Figure 54. Electrocatalytical activity for the electrochemical methanol conversion in terms of true current density as a function of the composition of the platinum/ruthenium catalyst. (Reproduced by permission of University of Washington Press.)

Figure 56. A 40-cell 300 W direct methanol/air module with an acid electrolyte (Shell Research Ltd., 1964). (Reproduced from Vielstich (1970).[356])

Besides ruthenium, other alloying components have also been found that enhance the activity of platinum. These are Sn, Ti, Re and Mo, and have been discussed by McNicol.[298] Also Mo added in various ways gives an interesting activity enhancement.[198, 299]

Alkaline methanol fuel cells were also developed. There was one company that was occupied very early and quite intensively with AFC systems: Allis-Chalmers in the US who, besides the hydrogen tractor in 1959, developed the first methanol module in 1962.[151, 163] At the same time, Justi and Winsel and the VARTA company also developed methanol cells (see Figure 57).[186] Vielstich and the BBC Co. have been mentioned already.[185]

Two other groups in Germany may be mentioned: the Battelle (Frankfurt) team (Sandstede and co-workers) and Robert Bosch GmbH (H. Jahnke, W. Ilge and co-workers). Figure 58 shows the voltage/current density plots of a methanol/air cell with Raney Pd/Ag/PE anodes and Ag (from Ag_2CO_3)/Ni/PTFE cathodes of Battelle (Frankfurt).[188]

Similar electrodes were incorporated in the complete system of the Robert Bosch GmbH, which can be seen from Figure 59. It is a methanol/oxygen fuel cell system with alkaline electrolyte and has a performance of 100 W.[185]

The electrolyte of the alkaline methanol cells has to be regenerated because the carbon dioxide formed during the operation will be dissolved in the electrolyte converting it to carbonate. This can be done by dialysis processes, as was shown by Winsel.[155] However, a practical apparatus has so far not yet been manufactured.

Figure 57. Methanol/oxygen fuel cell system with alkaline electrolyte delivering 140 W at 60 °C of VARTA AG (1967). (Reproduced by permission of VARTA, Aktiengesellschaft, Hanover.)

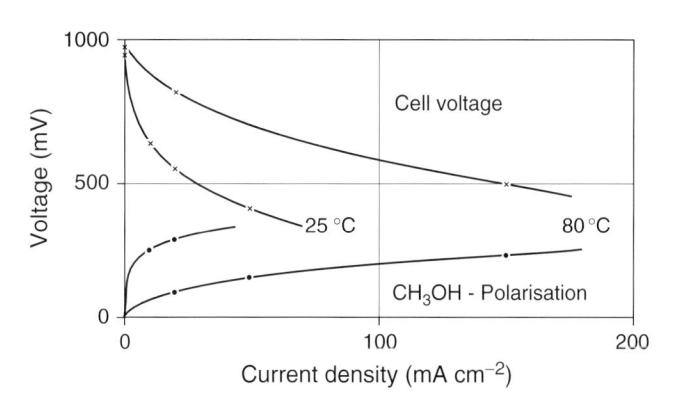

Figure 58. Performance of a methanol/air cell with Raney Pd/Ag/PE anodes and Ag (from Ag_2CO_3)/Ni/PTFE cathodes from Battelle (Frankfurt, 1961). (Reproduced from Vielstich (1970)[185] by permission of Wiley-Interscience.)

Figure 59. Methanol/oxygen fuel cell system with an alkaline electrolyte delivering 100 W at 65 °C (Bosch, Germany, 1963). (Reproduced from Vielstich (1970)[357] by permission of Wiley-Interscience.)

1.5.6 Other dissolved fuels for dissolved fuel fuel cells

Organic substances
Of the numerous number of organic substances other than methanol or hydrocarbons, a restricted number can be selected: ethanol, propanol, ethylene gycol, glycerol, higher and cyclic alcohols, formic acid and formate.

For special applications we refer to other chapters of this book. The historical viewpoints can be found in the books of Vielstich[185] as well as those of Justi and Winsel,[151] and concerning formic acid we refer to Sandstede's book.[267]

Inorganic substances
The most comfortable fuel is hydrazine. Therefore, many groups have developed cells, modules and complete systems, e.g., for traction. A good catalyst for the electrode is nickel with a small amount of palladium. Unfortunately, hydrazine turned out to be possibly a dangerous substance, being toxic. That is why Allis-Chalmers, Alstom, ESSO, EXXON, Monsanto, Shell and VARTA stopped working with it. But it must be said that a lot of knowledge has been accumulated which can be used in similar fields. Kordesch developed a hydrazine/air motorbike (see Figure 60).

It is powered with 800 W hydrazine/air fuel cell and 16 Ah Ni/Cd battery with 2 kW peak – the top speed is 35 miles h^{-1} and it has a range of 120 miles gal^{-1} of hydrazine (64%). The bike started in Cleveland and ran from there in 1968 on for several years through the town and country. Now it can be operated on the Graz University campus in Austria. Incidentally, it has not yet been definitely decided whether hydrazine is toxic itself or whether there is a side product dissolved in it that is responsible for its toxicity. Therefore, times may change.

Alkali metal borohydrides are strong reducing agents, and are quite stable in alkaline solutions in the absence of metal catalysts. Vigorous hydrogen evolution occurs in acid solution if metals are present. Potassium borohydride is a good fuel for portable fuel cells. More information can be found elsewhere.[185]

1.5.7 Other gaseous fuels for gaseous fuel cells

General remarks
Although hydrogen has been the exclusive fuel for fuel cells, there has always been the desire and intent to use more conventional, inexpensive fuels that are more easily stored. Electrochemical kinetics, however, has not made this an easy task. Early research has shown that a wide variety of fuels, including hydrocarbons and even solid carbon can be electrochemically oxidized to carbon dioxide and water. The rates are too low to be commercially viable but the hope remains that more active electrocatalysts will be found.

Ammonia
R.A. Wynveen[300] was one of the early investigators of direct oxidation of ammonia in a fuel cell. It is necessary to employ a strongly alkaline electrolyte for this fuel as it will react with an acidic electrolyte and is quite

Figure 60. Professor Karl Kordesch's motorbike powered with a 800 W hydrazine/air fuel cell and a 16 Ah Ni/Cd battery with a 2 kW peak. (Reproduced from Vielstich (1970)[359] by permission of Wiley-Interscience.)

soluble in dilute aqueous electrolytes. Wynveen used a strong potassium hydroxide electrolyte and porous carbon electrodes catalyzed with platinum black. It was found that the only product detected in the anode exit was nitrogen and the amount was 95–98.5% of that expected for oxidation to nitrogen and water. The performance was low: 10–25 mA cm^{-2} at 0.3 V for operation at 30–80 °C The open circuit voltage was 0.45–0.55 V.

Noting that the overpotentials for ammonia oxidation were high at low current densities, Eisenberg increased the operating temperature to as much as 300 °C, using a molten hydroxide electrolyte.[301] He was able to obtain open circuit voltages of about 1.0 V and current densities near 30 mA cm^{-2} at 0.5 V, using porous nickel electrodes.

Significant performance improvements for the direct ammonia fuel cell were made at the GE Company.[302–304] Cairns and co-workers reported current densities up to 600 mA cm^{-2} at 0.5 V and 140 °C for a platinum loading of 51 mg cm^{-2} and a slightly lower performance at 120 °C and at a Pt loading of 1 mg cm^{-2}. McKee *et al.* investigated the use of Ir and Pt/Ir alloys for direct ammonia oxidation. Ir and Pt/50%Ir reduced the anode overvoltage by 0.1 V at operating current densities of 100 mA cm^{-2}. Figure 61 shows these results.[304]

In addition to the direct use of ammonia in a fuel cell, McKee *et al.* reported on the catalytic decomposition of ammonia at temperatures in the range of 400 °C using supported Ir catalyst. The elevated temperature could be easily sustained by bleeding up to 15% oxygen into the ammonia stream. Thus, there is a design choice to be made: direct oxidation or catalytic decomposition of the ammonia.

It appears that ammonia can be an acceptable fuel for some applications. An attractive feature is the ability to store the ammonia as a liquid and feed it as a gas. An alkaline electrolyte is necessary and carbon dioxide scrubbing of the air will be required.

Hydrocarbons
The first use of hydrocarbons in fuel cells was that reported by W.R. Grove in 1874.[305] He successfully oxidized ethylene on a platinum electrode in dilute sulfuric acid at room temperature. Because the performance was much lower than that of hydrogen, the use of hydrocarbon fuels was not pursued further. Essentially no other work with hydrocarbons in fuel cells was reported until 1959, when Young and Rozelle[306] reported some electrochemical activity with ethylene and acetylene on platinum (supported on porous carbon) in 40% potassium carbonate at room temperature. This report and that of Allis-Chalmers on the electrochemical oxidation of propane in a 20 kW fuel cell system spurred great activity into the pursuit of the direct hydrocarbon fuel cell. (In the meantime, the Allis-Chalmers claim was retracted.)

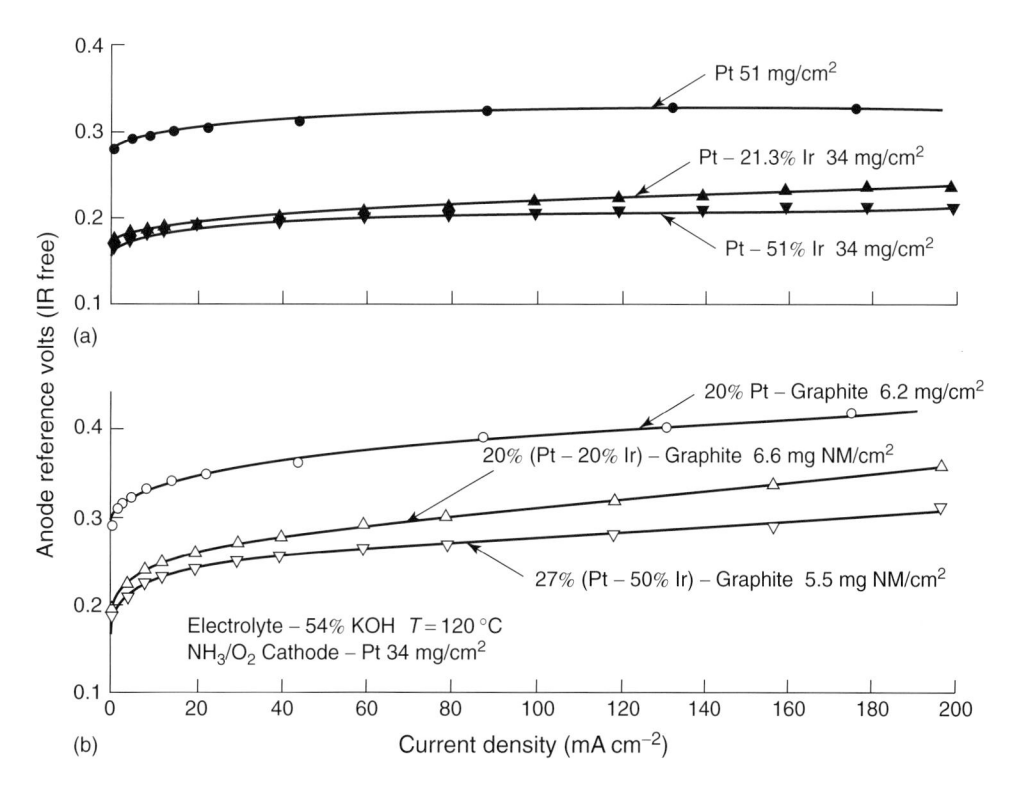

Figure 61. Ammonia anode performance. (Reproduced from McKee *et al.* (1968).[304])

In 1961, researchers at the California Research Corp. presented results for the anodic oxidation of ethylene and propane to carbon dioxide on platinum in 5 N H_2SO_4 at 25 °C.[307] Later in 1961, Young and Rozelle reported the performance curves for C_2 to C_4 hydrocarbons on carbon-supported Pt in 6 M H_2SO_4 at 90 °C. They also found that these hydrocarbons yielded about 50% of the desired carbon dioxide.[308] Schlatter[287] presented performance results for many organic fuels in both acidic and alkaline electrolytes using Pt/C electrodes and temperatures below 100 °C. The current densities achieved were less than 10 mA cm^{-2}, values too low for practical applications.

In 1962, investigators at the GE Research Laboratory reported the oxidation of various hydrocarbons using Pt black electrodes and a H_2SO_4 electrolyte at 65 °C[309] or an ion-exchange membrane electrolyte up to 85 °C.[310] At this point, the number of laboratories investigating and developing direct hydrocarbon fuel cells expanded very rapidly. Many major petroleum companies moved quickly into this field. There were 13 papers on hydrocarbon fuel cells presented at just one ACS symposium in 1965.[311]

Ongoing work explored the use of various electrolytes with the highest performance being achieved with an acidic fluoride electrolyte of CsF + HF.[265] The performance of propane on Pt in CsF + HF at 150 °C is shown in Figure 62. Other work was performed with a variety of fuels, including those that are liquid at operating temperature. Okrent and Heath of Esso Research[312] presented results for a five-cell stack operating in liquid decane at 150 °C with a 14.7 M H_3PO_4 electrolyte and 50 mg cm^{-2} Pt black electrodes with an active area of about 10 × 10 cm. The performance is shown in Figure 63.

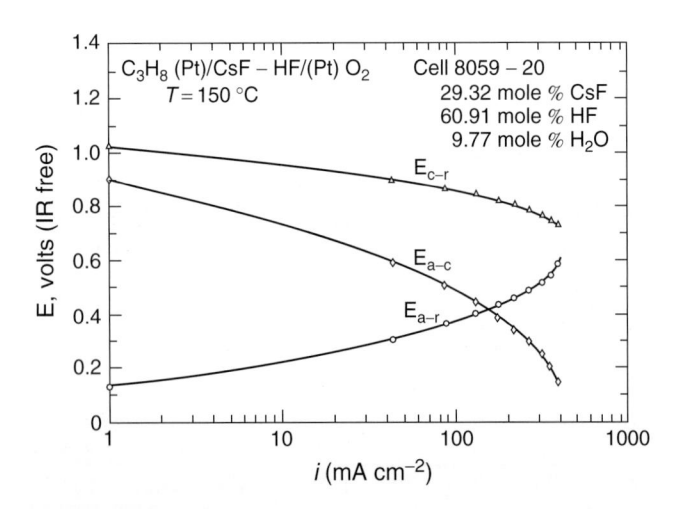

Figure 62. Performance of propane on Pt in a CsF/HF electrolyte. (Reproduced from Cairns (1965)[265] by permission of Academic Press.)

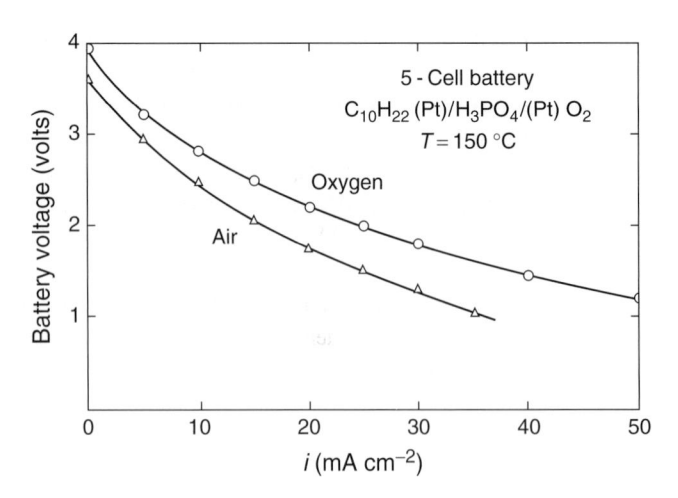

Figure 63. Performance of a decane fuel cell stack. (Reproduced from Okrent and Heath (1969).[312])

As work proceeded with direct hydrocarbon oxidation, it became clear that during operation the anode electrocatalyst accumulated a carbonaceous residue that hindered operation and caused cycling of the potential of the anode to high values necessary for the oxidative removal of the residue. Thus, it was not possible to achieve steady operation with sulfuric or phosphoric acid electrolytes. In contrast, Cairns found that HF-based electrolytes did not exhibit this behavior, presumably due to the lack of anion adsorption in the fluoride electrolytes.[265]

A more complete history and detailed analysis of hydrocarbon fuel cell R&D can be found in Refs. [8, 313]. An interesting result was obtained by the German Battelle team using Raney platinum, which is shown in Figure 64.[200, 201] This electrocatalyst has the same activity for ethane and propane but also methane gives a relative high current

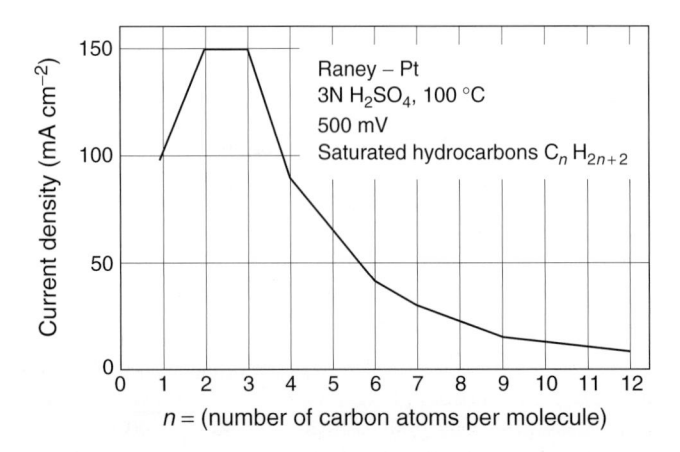

Figure 64. Rate of electrooxidation of saturated hydrocarbons at Raney platinum as a function of the number of carbon atoms. (Reproduced by permission of University of Washington Press.)

density. The long chain saturated hydrocarbons can also be anodically oxidized completely to carbon dioxide.

As a result of the high level of research activity during the 1960s, it was shown that very high electrocatalyst loadings and elevated temperatures were necessary to achieve current densities above about $100\,\text{mA}\,\text{cm}^{-2}$ at useful potentials. This performance was too low for commercial exploitation and interest in direct hydrocarbon fuel cells waned. Later, interest developed in the use of reformers with hydrocarbon fuels to prepare hydrogen fuel.

1.5.8 Other electrolytes for LTFCs

Another electrolyte system that was investigated in the early 1960s in connection with cells that operate on air (containing carbon monoxide$_2$) or organic fuels is that of an aqueous solution of alkali carbonates and bicarbonates. These electrolytes must be carefully chosen, since some of the carbonates and bicarbonates have relatively low solubilities, which can cause the plugging of the pores of the electrodes as water is lost or the relative amounts of carbonate and bicarbonate change in response to the operating conditions of the cell. The salts with the highest solubilities are those of caesium and rubidium. The advantages of these electrolytes include:

- low corrosion rates compared to strong acids and bases;
- invariance at steady state (rejection of carbon dioxide);
- ability to support the complete oxidation of a variety of organic fuels.

Disadvantages are:

- lower conductivity than strong acids or bases;
- more difficult water management under some operating conditions (temperatures approaching 200 °C);
- lower reaction rates than strong acid electrolytes.

Cairns and co-workers demonstrated that the performance, invariance and endurance of fuel cells using carbonate electrolytes with a selection of organic fuels, including ethylene, ethane and methanol.[294, 314]

1.6 Special historical considerations

1.6.1 The role of space

Fuel cells are particularly attractive for use in remote locations where the energy per unit mass of a system is important. Batteries could provide up to about $100\,\text{Wh}\,\text{kg}^{-1}$, not counting the support system. Fuel cells plus the fuel and oxidant supplies could provide several times as much energy per unit mass, making them an important power supply for space applications.

The first real application of fuel cells was as an onboard electrical power supply for the Gemini space capsule in 1964.[315, 316] This system was based on the PEM fuel cell developed by the GE Company. It used a polystyrene sulfonic acid membrane and platinum black electrocatalyst. Each cell was about $20 \times 20\,\text{cm}$ and operated at $30\,\text{mW}\,\text{cm}^{-2}$. The cells were assembled into stacks of 32 cells connected in series. Six of these stacks were assembled into two cylindrical housings (three stacks each) and provided a total power of up to 2 kW. Figure 65 shows a cutaway view of one of the cannisters. The hydrogen and oxygen were stored cryogenically as liquids in spherical containers separate from the fuel cell stacks. This system proved to be very rugged and reliable but did not offer a very high specific power ($2000\,\text{W}\,66\,\text{kg}^{-1} = 30\,\text{W}\,\text{kg}^{-1}$).

An interesting feature of this system was the water management scheme for zero gravity operation. Flat wicks were used to soak up and remove the liquid water from the cells. The wicked water was transported by capillary action to a porous ceramic that was well wetted by the water. A pressure differential across the porous ceramic forced the liquid water through the pores into a storage container. This product water was used as drinking water by the astronauts.

The second fuel cell system to be used in the space program was the alkaline system based on the Bacon cell, as developed by the United Aircraft Corp.[317] Three modules of 31 fuel cells each were used for this system. The stacks operated at 3.5 atm. and 200 °C. The three stacks, each packaged separately, as shown in Figure 66 provided a total power of 2.3 kW and weighed 100 kg.

The AFC of United Technologies has been improved significantly since the 1960s and is currently used in the space shuttle. It now uses a gold-based electrocatalyst in the oxygen electrode. This is a highly reliable system and can operate for thousands of hours with little performance change.

It is to be expected that fuel cells will continue to play a key role in space exploration because of their unique combination of characteristics, including the use of lightweight reactants at high efficiency and the provision of a low thermal management load and a very high specific energy. The fact that low cost is not a requirement has been important in the early use of fuel cells in space.

1.6.2 The role of military requirements

Soon after the announcements of Bacon and Allis-Chalmers in 1959, there was great interest in fuel cells as power sources for military use. Both the US Army and the US Navy[279] initiated programs for the development of

Figure 65. Cutaway view of one cannister of the Gemini fuel cell system. (Reproduced by permission of Schenectady Museum.)

Figure 66. Photograph of the Apollo fuel cell power plant together with the 1 kW fuel cell power plant of the Lunar Excursion Model of 1964. (Reproduced from Reference [181].)

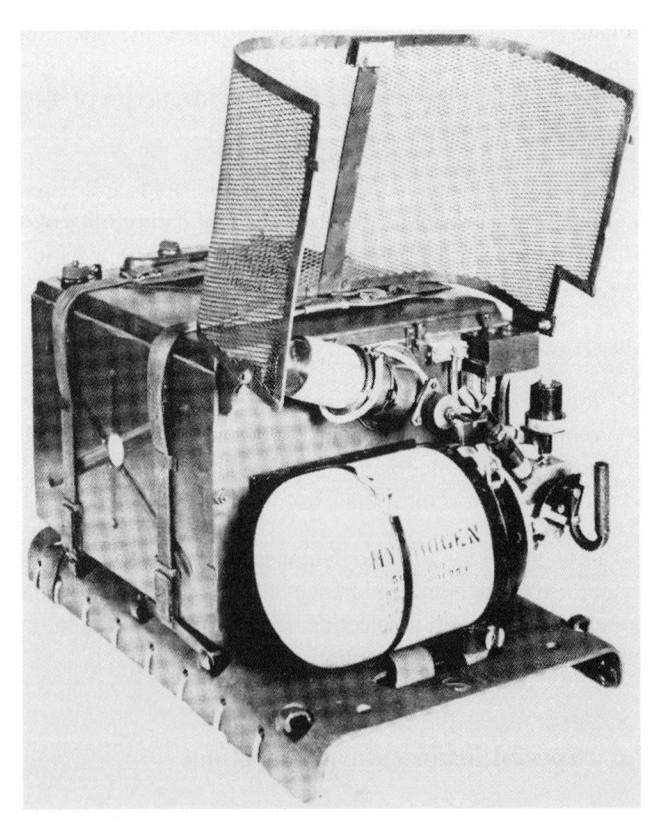

Figure 67. Photo of the GE 200 W backpack. (Reproduced from Oster and Chapman (1963).[318])

fuel cell systems for portable applications. Oster and Chapman[318] presented a detailed description of a 200 W portable hydrogen/air system developed for the Bureau of Ships and the Army Signal Corps. This unit could be operated on compressed hydrogen or on hydrogen from a hydride. A photo of this unit is shown in Figure 67.[318]

The US Army expanded its program to include a range of sizes of fuel cells from 1 to 15 kW, as described by Huff and Orth.[319] The attractive features of fuel cells for military use included silence, low-thermal signature and high-fuel efficiency. There was a strong interest in the use of logistic fuels, such as JP-4, so several programs on the direct oxidation of hydrocarbons were sponsored for several years. In addition, several units that included reformers were demonstrated.[319]

It was the interest of the US Army that provided the main driving force for the study of direct hydrocarbon oxidation in the US. Okrent and Heath reported the development of a liquid-fed decane/air fuel cell system that produced 5 W of power,[312] and Luksha and Weissman[320] reported on the anodic oxidation of multicomponent liquid hydrocarbon fuels, simulating military logistic fuels, in phosphoric acid on platinum black electrodes. The fuels could be oxidized to carbon dioxide and water but the performance was too low for practical applications. In the late 1960s, the sponsorship of fuel cell R&D by the US military decreased significantly.

In more recent years, interest in fuel cells for military applications has increased again, due to the increase in demand for electrical power for the soldier. The modern soldier has more electronics used for communication, surveillance and weapons systems. Batteries cannot fulfill all of the requirements and fuel cells are being pursued again. In addition to power for the foot soldier, electric power for vehicle propulsion is of a higher priority now. There is still a need for the use of logistic fuels but some compromise may be possible.

In the 1960s, there was little commercial interest in fuel cells, so the government-sponsored programs were very important for the development of fuel cells. Today, there is a very large worldwide commercial interest in fuel cells for communications and transportation applications, as well as stationary power use. In comparison, the military market is very small and, thus, it has a smaller influence on the development of the field.

1.6.3 From IEMs and PTFE to MEAs

W.T. Grubb of the GE Research Laboratory was the first to recognize the utility of the IEM as an electrolyte in a fuel cell.[271, 272] The initial performance of such cells was low, due to limited contact between the membrane and the electrocatalyst. Improvements were made by Grubb and by Niedrach,[321] raising the performance by improving the electrode structure and by bonding the electrode and its current collector to the membrane. The membrane materials first used were commercial materials, such as Permion membranes, followed by sulfonated phenol/formaldehyde polymers, then polystyrene divinyl benzene sulfonic acid, which were later fluorinated.[322]

During this same time period, the small-scale cells (11.4 cm^2) were scaled up and the first of the modern-style MEAs were produced by Cairns et al.,[280, 323] in sizes up to 15 × 15 cm. The principle of current collection using bipolar plates as contacts to the electrocatalyst layer and the stacking of cells in series was demonstrated with these cells, as was heat removal using extended cooling fins. A stack of such cells without the fins is shown in Figure 68.

At this point, GE produced the fuel cell system for the Gemini spacecraft,[316] completing the evolution from invention to application in just about eight years. Some steps in this evolution are shown in Figure 69.

In 1966, GE began using duPont Nafion (a random co-polymer of tetrafluoro ethene and PSEPVE, a vinyl ether sulfonyl fluoride) as the membrane.[324–326] It has exhibited excellent mechanical properties and chemical stability and is the standard fuel cell membrane today. The structure of the MEA has undergone various changes over the years, including the use of carbon-supported electrocatalysts bonded to carbon cloth or paper, which is in turn bonded to the membrane, and the gradual change to thinner electrode structures, including those based on unsupported electrocatalysts, as introduced by Cairns and Douglas.[280, 323] Recent changes to the membranes include the use of reinforced membranes for greater strength at small thicknesses.

Figure 68. Cairns–Douglas IEM fuel cell stack. (Reproduced by permission of Schenectady Museum.)

Figure 69. Evolution of the IEM cell to the Gemini system. (Reproduced by permission of Schenectady Museum.)

1.6.4 Theory of porous electrodes

Since the solubility of hydrogen and oxygen in electrolyte solutions is low, in hydrogen/oxygen fuel cells porous electrodes are used that are partly filled with the solution and partly with the respective gas. These electrodes contact the electrolyte with their front face and the gas space with the other (rear) face (Figure 70).

As the part of the pore space that is filled with the solution reactants dissolved in the solution are supplied by diffusion to the inner reaction sites, an electric current (an ion flow) is carried from the reaction sites to the front face. The reacting gas is mainly supplied not by diffusion through the liquid but by flow in the gas channels. The electrochemical reaction occurs mainly beneath thin electrolyte films forming at the walls of the gas pores close to the places where the gas and liquid pores meet, viz. near the so-called three-phase boundaries. The reacting gas is moved by diffusion through these thin electrolyte films (Figure 71).

Porous electrodes of this type are called gas-diffusion electrodes. The first hydrogen diffusion electrodes were described in 1923 by E. Müller (Germany). Refined versions of porous gas-diffusion electrodes were proposed in 1959 by F.T. Bacon (UK) and in 1962 by E. Justi and A. Winsel (Germany).

Porous electrodes have large true surface areas, S_{true}, of the inner surface as compared to their external (geometric)

Figure 70. Schematic of a porous gas diffusion electrode in an electrochemical cell.

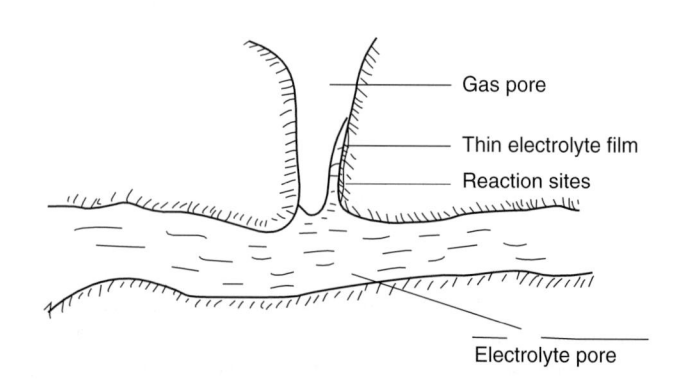

Figure 71. A thin electrolyte film in a gas pore near the three-phase boundary.

surface, S_{geom}. In the following, the specific surface area per unit of external area is denoted as $\gamma\ (= S_{true}/S_{geom})$ and that per unit of the electrode's volume as $s_V\ (= S_{true}/V)$. Under conditions of uniform work of the full internal surface area, the value of the net (overall) current density of the porous electrode, i_{por}, would be γ times larger than the current density, i_{sm}, of a smooth electrode of the same geometric size working under the same conditions. In practice this is not so and the value of i_{por} is much lower. The ratio

$$h = \frac{i_{por}}{\gamma i_{sm}} (h < 1) \quad (1)$$

is called the efficiency factor of the porous electrode.

In order to explain this discrepancy and to allow the calculation of values of the efficiency factor in the 1950s and 1960s, different theories of processes in porous electrodes were developed by O.S. Ksenzhek (Ukraine), Yu.A. Chizmadzhev (Russia) and J.S. Newman (USA).

Porous electrodes are systems with distributed parameters. Different points within the electrode are not equally accessible to the electrode reaction. Due to limited values of the effective diffusion coefficients, D_{eff}, of the reactants in the electrolyte present in the pores, concentration gradients are formed and due to limited values of the electrolyte's effective conductivity, σ_{eff}, ohmic potential gradients in the electrolyte are also established. For both these reasons the local current density is different at different depths, x, of the porous electrode. It is largest at the front surface of the electrode ($x = 0$) and falls with increasing depth within the electrode.

Mathematical calculations of the local current density distribution in the depth of the electrode when taking into account simultaneous diffusion processes, ohmic losses and electrode polarizations are rather difficult. In the particular case when only ohmic drops are present but concentration gradients are absent and when the electrode polarization ΔE is proportional to current density, the current density distribution, i_x, can be described by equation:

$$i_x = i° \cosh\left[\frac{(d-x)}{L_{ohm}}\right] \cosh\left(\frac{d}{L_{ohm}}\right) \quad (2)$$

where

$$L_{ohm} = \left[\left(\frac{RT}{nFi°}\right)\sigma_{eff}s_V\right]^{1/2} \quad (3)$$

has the dimension of length and is called the characteristic length of the ohmic process, d is the thickness of the electrode and $i°$ is the exchange current density of the reaction.

The net overall current density, i_{por}, can be computed by integrating the local volume current density at depth $x s_V i_x$ over the electrode thickness:

$$i_{por} = s_V \int_0^d i_x dx = i_0 L_{ohm} s_V \tanh\left(\frac{d}{L_{ohm}}\right) \quad (4)$$

Hence we find

$$h = \tanh\left(\frac{d}{L_{ohm}}\right) \quad (5)$$

In another particular case, when for fuel cell electrodes liquid reactants are used and when taking into account only concentration gradients but disregarding ohmic potential gradients an equation for the distribution of the process inside the electrode identical to equation (2) is valid but in which the rate of attenuation depends upon the characteristic length of the diffusion process

$$L_{diff} = \left(\frac{nFD_{eff}}{s_V k}\right)^{1/2} \quad (6)$$

where k is the rate constant of the reaction.

1.6.5 Historic development of zinc/air systems

The first developments of zinc/air batteries can be found more than 100 years ago. When it became known that a partial conversion of oxygen from the air takes place in zinc/manganese ore batteries in the Lechlanche cells, some people in the mid-1840s remembered the work about gaseous cells by Sir William Grove and they came to the conclusion that oxygen from the air must have been cathodically reduced.

The first patent of a current source in which oxygen from the air is converted originated from Maiche in 1878.[327] Zinc/oxygen batteries with ammonium chloride solution as the electrolyte were developed independently by W. Cohen[328] and T.H. Nash[329] around 1891. However, industrial production of these cells started by C. Fery[330, 331] only in 1915. By 1894, W. Walker and F.H. Wilkins[332] carried out experiments with zinc/oxygen (air) cells containing an alkaline electrolyte. The manufacturing of these was initiated by R. Oppenheim from 1929.[333, 334]

Decisive progress with respect to the performance of this type of battery was achieved when it was recognized that

- that active carbon would improve the cathode;
- that deposition of catalysts would accelerate the cathodic oxygen reduction; and
- later that the electrodes could be formed out of several layers.

Both world wars stimulated the development of the primary zinc/air battery because carbon as an electrode material was much more easily obtainable than manganese ore. As a result, the zinc/air battery used to be frequently called a makeshift of the Lechlanche battery, which was absolutely wrong.

Over recent decades, R&D has predominantly concentrated on the better performing alkaline systems. Since the begin of the 1960s, many attempts have been undertaken to develop rechargeable zinc/oxygen (air) batteries mainly for military applications and for electrotraction – not so much for consumer batteries, for which primary zinc/air batteries are in particular suitable.

Mainly two problems have to be overcome. One is that the stability of a bifunctional oxygen electrode is not sufficient. The second is that the zinc electrode suffers from a shape change resulting from repeated charging. Simple solutions for these problems are to use an additional electrode for oxygen evolution during charging and the regeneration of the zinc outside of the battery, which delivers the current.

Gulf General Atomic Company developed a 20 kW battery for electrotraction with a mechanical exchange of the zinc anodes.[335] Analogous developments at Leesona Moos[336] led to experimental batteries with an energy content of up to 1.2 kWh, which were predominantly earmarked for military applications. Investigations of batteries with mechanically exchangeable electrodes without electrolyte circulation were also carried out at General Motors. Such a battery with an energy content of 35 kWh was tested in a vehicle.[337] Later on, at the beginning of the 1970s, experiments followed at Sony in Japan,[338] at Battelle,[339] at Citroen[340] and at the CGE.[341] With these experiments attempts were made to oxidize the circulating zinc powder electrolyte suspension at an anode current collector to zinc oxide, which could be regenerated externally. Of this work the most advanced seemed to be that of CGE, where the a zinc dust/electrolyte suspension was pumped through tube-like collector nets. This idea of such a zinc powder/slurry battery was taken up by Foller[342] in the 1980s. Despite exceedingly positive statements of the developing groups, the commercialization of these batteries has not yet been tried.

Over recent years, progress has been achieved in the development of electrocatalysts and electrode support materials, so that activity and corrosion stability have improved. The results can be found in reviews that have been issued recently.[343–345]

REFERENCES

1. E. Baur and J. Tobler, *Z. Elektrochem., Angew. Physik. Chem.*, **39**, 169 (1933).
2. W. Ostwald, 'Elektrochemie, Geschichte Lehre', Verlag von Veit & Comp, Leipzig (1896).
3. W. Ostwald, 'Electrochemistry: History and Theory', Amerind for the Smithsonian Institution and the National Science Foundation, New Delhi (1980).
4. H. C. Howard, 'Direct Generation of Electricity from Coal and Gas (fuel cells)', in "Chemistry of Coal Utilization", H. H. Lowry (Ed), John Wiley & Sons, New York, Vol. 2, Chapter 35, pp. 1568–1585 (1945).
5. J. Euler, 'Aus der Frühzeit der Galvanischen Brennstoffelemente', VARTA AG, Frankfurt am Main (1966).
6. K.-J. Euler, 'Entwicklung der Elektrochemischen Brennstoffzellen', Verlag Karl Thiemig, München (1974).
7. J. A. A. Ketelaar, 'History', in "Fuel Cell Systems", L. J. M. J. Blomen and M. N. Mugerwa MN (Eds), Plenum Press, New York, pp. 19–35 (1993).
8. H. A. Liebhafsky and E. J. Cairns, 'Fuel Cells and Fuel Batteries', John Wiley & Sons, New York (1968).
9. A. J. Appleby, *J. Power Sources*, **29**, 3 (1990).
10. A. J. Appleby, *J. Power Sources*, **29**, 267 (1990).
11. G. J. Schaeffer, 'Fuel Cells for the Future', Twente (1998).
12. 'Energy, Electricity, Other Energy Topics', in "Encyclopædia Britannica" (1999).
13. Encyclopædia Britannica (2002). **http://www.britannica.com**
14. G. Sandstede, *Bunsen Magazin*, **2**, 66 (2000).
15. G. Sandstede, *Eur. Fuel Cell News*, **7**(4), 3 (2000).
16. O. K. Davtyan, 'The Problem of Direct Conversion of Chemical Energy of Fuels into Electrical Energy', Publishing House of the USSR Academy of Sciences, Moscow, Leningrad (1947).
17. W. F. Schnurnberger, 'From Vision to Technology: The History of Fuel Cells', Presented at the Annual Meeting of the International Society of Electrochemistry, Pavia, Italy, International Society of Electrochemistry (1999).
18. L. Schlapbach, 'History of Electrochemistry', Presented at the Electrochemistry Symposium, Villigen, Switzerland, Paul Scherrer Institut, G. G. Scherer (Ed) (2002).
19. H.-H. Möbius, *J. Solid State Electrochem*, **1**, 2 (1997).
20. U. Bossel, 'The Birth of the Fuel Cell', Presented at the European Fuel Cell Forum, Oberrohrdorf, Switzerland (2000).
21. 'Fuel Cells: Collecting History with the World Wide Web', Smithsonian institution (2001). **http://americanhistory.si.edu/csr/fuelcells/index.htm**
22. W. Jaenicke, '100 Jahre Bunsen-Gesellschaft 1894–1994', Steinkopff, Verlag, Darmstadt (1994).
23. M. V. Lomonossow, in 'Encyclopædia Britannica' (1999). ,
24. J. Teichmann, 'Energietechnik und Kulturgeschichte', München (2002).
25. J. R. Mayer, *Liebigs Ann. der Chemie*, **42**, 233 (1842).
26. W. R. Grove, 'The Correlation of Physical Forces', Longmans Green, London (1846).

27. W. R. Grove, 'Die Wechselwirkung der Physischen Kräfte', Verlag von Julius Springer, Berlin (1863).
28. W. R. Grove, 'Die Verwandschaft der Naturkräfte. Mit einem Anhange, enthaltend die Rede des Autors. Über den ununterbrochenen Zusammenhang in der Natur', Verlag von Friedrich Vieweg und Sohn, Braunschweig (1871).
29. 'History of energy' (2001). **http://www.swifty.com/apase/charlotte/histe.html**
30. 'California Energy Commission, Today in Energy History' (2002). **http://www.energy.ca.gov/m+pco/history.html#month**
31. 'History of Electricity' (2002). **http://www.codecheck.com/pp_elect.html**
32. 'Driven to drive, Automobile History' (2002). **http://www.trentyne.com/driventodrive/autohistory.htm**
33. 'Automotive History – A Chronological History', Antique Automobile Club of America (2002). **http://www.aaca.org/history**
34. 'The Origin of the Automobile' (2002). **http://www.geocities.com/MotorCity/Lot/3248/hist01.htm**
35. 'Zur Geschichte von Batterien' (1997). **http://www.elektroauto-tipp.de/battgesch.html**
36. 'Geschichte von Elektroautos' (1997). **http://www.elektroauto-tipp.de/eautog1.html**
37. G. Sandstede, 'Verbindung der Physikalisch Orientierten Naturwissenschaften mit der Gesellschaft durch den Physikalischen Verein', in "Der Weg der Wahrheit", P. Eisenhardt, F. Linhard and K. Petanides (Eds), Georg Olms Verlag, Hildesheim, Zürich, New York, pp. 261–274 (1999).
38. G. Sandstede, 'Beiträge des Physikalischen Vereins zur Entwicklung von Technik und Naturwissenschaft', in "Jahresbericht der Wissenschaftlichen Vorträge", Frankfurt am Main, Physikalischer Verein, Druckerei und Verlag Otto Lembeck, Vol. 169, pp. 141–166 (1993).
39. E. W. Jungner, 'Nickel/Cadmium Accumulator', Germany, DRP 110210 (1899).
40. K. H. P. Bienek, 'Die Siemensstadt – 200 Jahre elektrische Batterie' (2001). **www.kbtext.de/batt.htm**
41. F. Didik, 'History and Directory of Electric Cars from 1834–1987' (2001). **http://www.didik.com/ev_hist.htm**
42. 'Econogics, Some EV History, History of Electric Cars and other Vehicles' (2000). **http://www.econogics.com/ev/evhistry.htm**
43. P. A. Hughes, 'A History of Early Electric Cars' (1996). **http://www.geocities.com/Athens/Crete/6111/electcar.htm**
44. A. Mittasch, 'Döbereiner, Goethe und die Katalyse', Hippokrates-Verlag Marquardt & Cie, Stuttgart (1951).
45. M. J. Kirschner, 'Oxygen', in "Ullmann's Encyclopedia of Industrial Chemistry", VCH Verlag, Weinheim, Vol. A18 (1986).
46. P. Häussinger, R. Lohmüller and A. M. Watson, 'Hydrogen', in "Ullmann's Encyclopedia of Industrial Chemistry", VCH Verlag, Weinheim, Vol. A13 (1986).
47. G. Sandstede, *Technik Bayern/Hessen*, **6/99**, 19 (1999).
48. O. Krätz, 'Goethe und die Naturwissenschaften', Callwey, München (1998).
49. 'Balloon', in "Encyclopædia Britannica" (1999). **http://www.britannica.com/search?query=balloon&ct=&fuzzy=N**
50. J. Verne, '*The Mysterious Island*' (1874).
51. G. B. Kauffmann, *Platinum Metals Rev.*, **43**, 122 (1999).
52. P. Walden, *Z. f. Angewandte Chemie*, **43**, 792, 847 and 864 (1930).
53. C. F. Schoenbein, *Phil. Mag. (III)*, **14**, 43 (1839).
54. V. Engelhardt, '*Die Elektrolyse des Wassers*', Verlag von Wilhelm Knapp (1902).
55. H. Berg and K. Richter (Eds), "Entdeckungen zur Elektrochemie, Bioelektrochemie und Photochemie von Johann Wilhelm Ritter", Akad. Verlagsges, Geest u. Portig, Leipzig (1986).
56. L. Dunsch, 'Geschichte der Elektrochemie', VEB Deutscher Verlag für Grundstoffindustrie, Leipzig (1985).
57. H. Davy, *Nicholson's J. Nat. Phil.*, **144**, (1802).
58. N. Gautherot, *J. Physique*, **56**, 429 (1802).
59. C. F. Schoenbein, *Schweiz. Ges.*, 82 (1838).
60. C. Matteucci, *Compt. Rend. Acad. Sci., Paris*, **7**, 741 (1838).
61. C. F. Schoenbein, *Compt. Rend. Acad. Sci., Paris*, **7**, 741 (1838).
62. W. R. Grove, *Phil. Mag.*, **14**, 127 (1839).
63. M. Faraday, *Pogg. Ann.*, **32**, 301 (1834).
64. M. Faraday, *Pogg. Ann.*, **32**, 401 (1834).
65. M. Faraday, *Pogg. Ann.*, **32**, 481 (1834).
66. F. Peters, 'Die Primär und Secundär Elemente', A. Hartleben's Verlag, Wien, Pest, Leipzig (1897).
67. W. R. Grove, *Phil. Mag.*, **14**, 388 (1839).
68. W. R. Grove, *Phil. Mag.*, **15**, 287 (1839).
69. R. W. Bunsen, *Pogg. Ann.*, **55**, 265 (1842).
70. W. R. Grove, *Phil. Mag. (III)*, **21**, 417 (1842).
71. W. R. Grove, *Ann. Phys. Chem.*, **48**, 202 (1843).
72. W. R. Grove, *Phil. Mag. (III)*, **24**, 268 (1844).
73. W. R. Grove, *Phil. Transactions (I)*, 351 (1845).
74. 'Grove, Sir William Robert', in "Encyclopædia Britannica" Extended Series, London (1997).
75. 'Grove, Sir William Robert', in Encyclopædia Britannica (2002). **http://www.britannica.com**
76. M. Vergnes, US Patent, 28,317, May 15 (1860).
77. F. W. H. V. Beetz, *Pogg. Ann.*, **77**, 493 (1849).
78. F. W. H. V. Beetz, *Pogg. Ann.*, **90**, 42 (1853).
79. W. R. Grove, *Phil. Mag.*, **8**, 399 (1854).
80. E. Becquerel, 'Traité d'Electricité, Librairie de Firmin Didot Frères', A. C. Becquerel and E. Becquerel (Eds), Paris, (1855).
81. P. Jablochkoff, *Compt. Rend., Paris*, **85**, 1052 (1877).
82. A. Archereau, British Patent, 1,027 (1883).

83. Lord Rayleigh, *Proc. Cambridge Philos. Soc.*, **4**, 198 (1883).
84. C. Westphal, 'Apparat zur Erzeugung elektrischer Ströme', German Patent, DRP, 16.12.1880, ausgeg. 29.06.1883, DRP 22393 (1880).
85. P. Scharf, German Patent, DRP, 16.08.1888, DRP 48446 (1888).
86. L. Mond and C. Langer, *Proc. Roy Soc.*, **46**, 296 (1889).
87. W. Ostwald, *Z. f. Elektrotechnik Elektrochem.*, **1**, 81, 122 (1894).
88. W. Ostwald, *Z. Physik. Chem.*, **15**, 409 (1894).
89. W. Borchers, *Z. F. Elektrotechnik Elektrochem.*, **1**, 484 (1894).
90. W. W. Jacques, *Harper's Mag.*, **96**(559), 144 (1896); W. W. Jacques, *Harper's Mag.*, May (1897).
91. C. Liebenow and L. Strasser, *Z. Elektrochem.*, **3**, 353 (1897).
92. F. Haber and L. Brunner, *Z. Elektrochem.*, **10**, 697 (1904).
93. F. Haber and L. Brunner, *Z. Elektrochem.*, **12**, 78 (1906).
94. E. Baur and A. Gläßner, *Z. Elektrochem.*, **9**, 534 (1903).
95. I. Taitelbaum, *Z. Elektrochem., Angew. Physik. Chem.*, **16**, 286 (1910).
96. E. W. Jungner, *Electrochem. Industry*, **6**, 256 (1908).
97. W. Nernst, *Z. F. Physikalische Chem.*, **1**, 129 (1889).
98. J. O. M. Bockris and S. Srinivasan, 'Fuel Cells Their Electrochemistry', McGraw–Hill, New York (1969).
99. F. J. Smale, *Z. f. Physikalische Chem.*, **14**, 577 (1894).
100. F. J. Smale, *Z. f. Physikalische Chem.*, **16**, 562 (1895).
101. W. Nernst, *Z. f. Elektrochem.*, **4**, 41 (1899).
102. W. Nernst, 'Adding Chlorine to the Air Electrode', Germany, N 13403 and later DRP 259241 (1913).
103. W. P. Hauck, 'Die Galvanischen Batterien', A. Hartlebens Verlag, Wien, Pest, Leipzig (1883).
104. J. H. Reid, 'Verfahren zur Erzeugung von Elektrizität', US Patent, 736,016, May 17 (1902).
105. P. G. L. Noel, 'Generateur Electrique à Gaz', French Patent, 350,110, August 4 (1904).
106. W. D. Treadwell, *Helv. Chim. Acta*, **27**, 1302 (1944).
107. E. Baur and H. Ehrenberg, *Z. Elektrochem., Angew. Physik. Chem.*, **18**, 1001 (1912).
108. E. Baur, W. D. Treadwell and G. Trümpler, *Z. Elektrochem.*, **27**, 199 (1921).
109. E. Baur and H. Preis, *Z Elektrochem., Angew. Physik. Chem.*, **43**, 727 (1937).
110. E. Baur and H. Preis, *Z Elektrochem., Angew. Physik. Chem.*, **44**, 695 (1938).
111. W. Schottky, *Wiss. Veröff. Siemens-Werke*, **14**(2), 1 (1935).
112. E. Baur, *Z. Elektrochem.*, **12**, 725 (1906).
113. E. Baur, *Z. Elektrochem.*, **39**, 168 (1933).
114. E. Baur, *Z. Elektrochem.*, **40**, 249 (1934).
115. R. Lorenz, *Z. Elektrochem.*, **4**, 395 (1897).
116. C. Féry, *Elektrotechn*, **39**, 298 (1918).
117. C. Féry, 'Piles primaires accumulateurs', Bailliére et fils, Paris (1925).
118. E. Baur, *Z. Elektrochem.*, **41**, 794 (1936).
119. E. Baur, *Z. Elektrochem.*, **43**, 821 (1937).
120. J. Tobler, *Z. Elektrochem.*, **39**, 148 (1933).
121. E. Baur, *Bull. Schweiz ETV*, **30**(17), 478 (1939).
122. E. Baur and W. D. Treadwell, Swiss Patent, 78,591 (1917).
123. E. W. Jungner, 'Hydrophobieren von Potassium Hydroxidele mit Paraffin', Swedish Patent, June 17 (1919).
124. E. W. Jungner, 'Hydrophobieren von Potassium Hydroxidele mit Paraffin', German Patent, DRP 348, 393 (1920).
125. A. Schmid, 'Die Diffusionsgaselektrode', Ferdinand Enke, Stuttgart (1923).
126. A. Schmid, *Helv. Chim. Acta*, **7**, 370 (1924).
127. G. W. Heise and E. A. Schumacher, *Trans. Electrochem. Soc.*, **62**, 383 (1932).
128. G. W. Heise and E. A. Schumacher, *Trans. Electrochem. Soc.*, **92**, 173 (1947).
129. W. G. Berl, *Trans. Am. Electrochem Soc.*, **83**, 253 (1943).
130. W. Vielstich, 'Brennstoffelemente', Verlag Chemie, Weinheim (1965).
131. K. Kordesch and A. Marko, *Österr. Chemiker Z.*, **52**, 125 (1951).
132. J. Niederreither, 'Verfahren zur Erzeugung Elektrischer Energie durch Betreiben von Gasketten', Germany, DRP 648940, March 30 (1932).
133. F. T. Bacon, 'The High Pressure Hydrogen/Oxygen Fuel Cell', in "Fuel Cells, Symposium of the ACS", Atlantic City, Reinhold Publishing Corp., New York and Chapman & Hall, London, G. J. Young (Ed), Vol. I, pp. 51–77 (1960).
134. J. Niederreither, *Z. VDI 92*, 995 (1950).
135. J. Niederreither, 'Verfahren zum Speichern und Verteilen Elektrischer Energie', German Patent, DRP 648,941, March 24 (1932).
136. J. Niederreither, 'Method of Producing, Storing and Distributing Electrical Energy by Operation of Gas Batteries, Particularly Oxy-Hydrogen Gas Batteries and Electrolyzers', USA Patent, 2,070,612, March 18 (1933).
137. E. K. Rideal and U. R. Evans, *Trans. Faraday Soc.*, **17**, 466 (1921).
138. F. T. Bacon, 'Improvements Relating to Galvanic Cells and Batteries', British Patent, 667,298, June 8 (1949).
139. F. T. Bacon, 'Improvements Relating to Electric Batteries', British Patent, 725,661, Jan 16 (1953).
140. F. T. Bacon, *Beama J.*, **6**, 61 (1954).
141. F. T. Bacon and J. S. Forrest, 'British Work on Fuel Cells', Presented at the Fifth World Power Conference, Vienna (1956).
142. A. M. Adams, F. T. Bacon and R. G. H. Watson, 'The High-Pressure Hydrogen/oxygen Fuel Cell', in "Fuel Cells", W. Mitchell, Jr (Ed), Academic Press, New York, pp. 129–192 (1963).

143. E. W. Justi, 'Leitfähigkeit und Leitungsmechanismus Fester Stoffe', Verlag Vandenhoeck und Ruprecht, Göttingen, pp. 327–334 (1948).
144. E. W. Justi, 'Die Energie- und Rohstoffquellen der Zukunft', in "Wie Leben wir Morgen?" Alfred Kröner Verlag, Stuttgart (1957).
145. E. W. Justi, 'Stand und Aussichten der Reversiblen Erzeugung Elektrischer Energie', in "Brennstoffelementen", Akademie der Wissenschaften und Literatur, Mainz, pp. 9–24 (1956).
146. E. W. Justi, H. J. Thuy and A. W. Winsel, German Patent, DBP 957,491 (1957).
147. E. W. Justi, W. Scheibe and A. W. Winsel, 'Double Skeleton Catalyst Electrodes', German Patent, DBP 1,019,361, Oct. 10 (1954).
148. E. W. Justi, M. Pilkuhn, W. Scheibe and A. W. Winsel, 'Hochbelastbare Wasserstoff-Diffusions-Elektroden für Betrieb bei Umgebungstemperatur und Niederdruck', (High Drain Hydrogen-Diffusion Electrodes Operating at Ambient Temperature and Low Pressure), Verlag der Wissenschaften und der Literatur in Mainz bei Franz Steiner Verlag, Wiesbaden (1959).
149. H. M. Dittmann, E. W. Justi and A. W. Winsel, 'DSK Electrodes for the Cathodic Reduction of Oxygen', in "Fuel Cells", Presented at the Symposium of the ACS, Chicago, Reinhold Publishing Corp., New York and Chapman & Hall Ltd., London, Vol. II, pp. 133–142 (1963).
150. E. W. Justi and A. W. Winsel, J. Electrochem. Soc., **108**, 1073 (1961).
151. E. W. Justi and A. W. Winsel, 'Kalte Verbrennung', (Cold Combustion), Franz Steiner Verlag, Wiesbaden (1962).
152. A. W. Winsel and G. J. Richter, 'Alkaline Fuel Cells', in "Electrochemical Hydrogen Technologies", H. Wendt (Ed), Elsevier, Amsterdam (1990).
153. A. W. Winsel, 'Galvanische Elemente', in "Brennstoffzellen", Ullmann (1976).
154. K. Rühling, C. Fischer, O. Führer and A. W. Winsel, 'Elektrolyse- und Brennstoffzellen in Eloflux-Technik', DECHEMA Monographie (1991).
155. A. W. Winsel, 'Stoff- und Wärmeflüsse in Brennstoffzellen am Beispiel der Akalischen H_2/O_2 Brennstoffzelle', Presented at Ulms Electrochemical Talks, Ulm, Germany (1999).
156. A. W. Winsel, 'Eloflux-Zellen für die Wasserelektrolyse, für Brennstoffzellen und ihre Einsatzmöglichkeiten bei der Energiespeicherung', DECHEMA Monographie, Frankfurt am Main, **92**, pp. 21–43 (1982).
157. A. W. Winsel (Ed), Elektrochemie in Energie- und Umwelttechnik, Frankfurt am Main, DECHEMA Monographie, 124 (1991).
158. K. Rühling, 'Das Brennstoffzellen-System der VARTA Batterie AG und seine Weiterentwicklung', VARTA (1990).
159. H. Spengler and G. Grüneberg, 'Galvanische Brennstoffelemente bei Raumtemperatur', DECHEMA Monographie, Frankfurt am Main, **38**, pp. 579–599 (1960).
160. K. V. Kordesch, J. Electrochem. Soc., **125**, 77C (1978).
161. K. Kordesch and F. Martinola, Monatsh. Chem., **84**, 39 (1953).
162. K. V. Kordesch, 'Low Temperature Hydrogen/oxygen Fuel Cells', in "Fuel Cells", J. Mitchell Will (Ed), Academic Press, New York, London, pp. 329–370 (1963).
163. K. Kordesch and G. Simader, 'Fuel Cells and Their Applications', VCH, Weinheim (1996).
164. W. Vielstich, Z. Instrumentenk., **71**, 29 (1963).
165. W. Vielstich, Presented at the Electrochemical Soc. Indianapolis Meeting, Indianapolis, USA, J. Electrochem Soc., vol. 108, Ext. Abstract No. 113 (1961).
166. A. Kutschker and W. Vielstich, Electrochim. Acta, **8**, 985 (1963).
167. E. Schwarzer and W. Vielstich, 'Journées d'Etude des Piles a Combustible', Presented at the 3rd International Fuel Cell Symposium, Bruxelles, SERAI, Presses Academiques Europeennes, Revue Energie Primaire Vol. III, pp. 220–229 (1969).
168. H. Schmidt and W. Vielstich, Z. f. Analytische Chem., **224**, 84 (1967).
169. W. Vielstich, 'Fuel Cells', Wiley-Interscience, London, pp. 271–281 (1970).
170. W. Vielstich, 'Fuel Cells', Wiley-Interscience, London, pp. 459–468 (1970).
171. O. A. Petrii, I. Podlovchenko, A. N. Frumkin and H. Lal, J. Electroanal. Chem., **10**, 253 (1965).
172. W. Vielstich, 'Fuel Cells', Wiley-Interscience, London, pp. 172–178 (1970).
173. D. W. Gibbons, E. J. Rudd and D. Gregg, Proceedings of the 36th Power Sources Conference, IEEE, New York, pp. 132 (1994).
174. M. Ritschel and W. Vielstich, Electrochim. Acta, **24**, 885 (1979).
175. W. Vielstich, Chem. Ing. Techn., **34**, 346 (1962).
176. F. Gestermann, Ges. Deutscher Chemiker, Monography, 23 (2002).
177. F. T. Bacon, Electrochim. Acta, **14**, 569 (1969).
178. F. T. Bacon, J. Electrochem. Soc., 7C (1979).
179. G. J. Young (Ed), 'Fuel Cells', Reinhold Publishing Corp., New York and Chapman & Hall, London (1960).
180. M. Barak, Adv. Energy Conversion, **6**, 29 (1966).
181. K. R. Williams (Ed), 'An Introduction to Fuel Cells', Elsevier, Amsterdam (1966).
182. A. McDougall, 'Fuel Cells', The MacMillan Press, London (1976).
183. Presented at the 11th Annual Power Sources Conference, PSC Publication Committee, Red Bank (1957).
184. E. Gorin and G. J. Young (Eds), Presented at the Symposium on Recent Advances in Fuel Cells, American Chemical Society, Washington DC (1961).
185. W. Vielstich, 'Fuel Cells', Wiley-Interscience, London (1970).
186. H. H. v. Döhren and K. J. Euler, 'Brennstoffelemente', VDI Verlag, Düsseldorf (1971).

187. H. Binder, A. Köhling, H. Krupp, K. Richter and G. Sandstede, *Electrochim. Acta*, **8**, 781 (1963).
188. G. Sandstede, 'Galvanische Brennstoffzellen', DECHEMA Monographien, Frankfurt, **49**, pp. 303–342 (1964).
189. H. Binder, A. Köhling, W. H. Kuhn, W. Lindner and G. Sandstede, 'Hydrogen and Methanol Fuel Cells with Air Electrodes in Alkaline Electrolyte', in "From Electrocatalysis to Fuel Cells", University of Washington Press, Seattle, London, pp. 131–41 (1970).
190. H. Behret, H. Binder and G. Sandstede, 'Development of Fuel Cells – A Materials Problem', in "Materials Science in Energy Technology", G. G. Libowitz and M. S. Wittingham (Eds), Academic Press, New York, pp. 381–426 (1979).
191. H. Krupp, H. Rabenhorst, G. Sandstede, G. Walter and R. McJones, *J. Electrochem. Soc.*, **109**, 553 (1962).
192. H. Binder, A. Köhling and G. Sandstede, 'The Anodic Oxidation of Methanol on Raney-type Catalysts of Platinum Metals', in "Hydrocarbon Fuel Cell Technology", B. S. Baker (Ed), Academic Press, New York, pp. 91–102 (1965).
193. H. Binder, A. Köhling and G. Sandstede, 'Effect of Alloying Components on the Catalytic Activity of Platinum in the Case of Carbonaceous Fuels', in "From Electrocatalysis to Fuel Cells", G. Sandstede (Ed), University of Washington Press, Seattle, London, pp. 43–58 (1970).
194. H. Binder, A. Köhling and G. Sandstede, *Adv. Energy Conversion*, **7**, 471 (1967).
195. H. Binder, A. Köhling and G. Sandstede, *Adv. Energy Conversion*, **7**, 112 (1967).
196. H. Binder, A. Köhling and G. Sandstede, *Nature*, **214**, 268 (1967).
197. H. Binder, A. Köhling and G. Sandstede, *J. Electroanal. Chem. Interfacial Electrochem.*, **17**, 111 (1968).
198. H. Binder, A. Köhling and G. Sandstede, 'Platinum Catalysts Modified by Adsorption or Mixing with Inorganic Substances', in "From Electrocatalysis to Fuel Cell", G. Sandstede (Ed), University of Washington Press, Seattle, London, pp. 59–79 (1970).
199. H. Binder, A. Köhling and G. Sandstede, 'Anodic Oxidation of Methane, Ethane, Propane, Butane and Carbon Monoxide in Phosphoric Acid at 110 °C', Presented at the ECS Fall Meeting, Battery Division of the Electrochemical Society, Washington DC, in "Extended Abstracts", pp. 25–28 (1964).
200. H. Binder, A. Köhling, H. Krupp, K. Richter and G. Sandstede, *J. Electrochem. Soc.*, **112**, 355 (1965).
201. G. Sandstede, *Chem. Ing. Technik*, **27**, 632, 782 (1965).
202. H. Binder, A. Köhling and G. Sandstede, 'Anodische Oxydation von Kohlenwasserstoffen in Säuren', Presented at Journees Interantional d'Etude Piles a Combustible, Bruxelles, SERAI, Revue Energie Primaire I, pp. 74–81 (1965).
203. G. Sandstede, 'Elektrochemische Brennstoffzellen', in "Fortschritte der Chemischen Forschung", E. Heilbronner, U. Hofmann, K. Schäfer, G. Wittig and F. Boschke (Eds), Springer–Verlag, Berlin, **8**, pp. 171–221 (1967).
204. H. Binder, A. Köhling and G. Sandstede, 'On the Performance of WC Electrodes upon Oxidation of Hydrogen, Hydrazine and Formaldehyde', Presented at the American Chemical Society Meeting, American Chemical Society, New York, Preprints of Papers, **13**(3), 99 (1969).
205. K. Von Benda, H. Binder, A. Köhling and G. Sandstede, 'Electrochemical Behavior of Tungsten Carbide Electrodes', in: "From Electrocatalysis to Fuel Cells", G. Sandstede (Ed), University of Washington Press, Seattle, London, pp. 87–100 (1970).
206. H. Binder, A. Köhling, W. H. Kuhn, W. Lindner and G. Sandstede, *Energy Conversion*, **10**, 25 (1970).
207. H. Alt, H. Binder, A. Köhling and G. Sandstede, 'Organic Cathodes with Air Regeneration', in "From Electrocatalysis to Fuel Cells", G. Sandstede (Ed), University of Washington Press, Seattle, London, pp. 333–343 (1970).
208. H. Alt, H. Binder, A. Köhling and G. Sandstede, *J. Electroanal. Chem. Interfac. Electrochem.*, **31**, App. 19 (1971).
209. H. Alt, H. Binder and G. Sandstede, *J. Catalysis*, **28**, 8 (1973).
210. H. Behret, W. Clauberg and G. Sandstede, 'Berichte der Bunsengesellschaft' *f. Physikalische Chemie*, **83**, 139 (1979).
211. H. Behret, H. Binder, G. Sandstede and G. G. Scherer, *J. Electroanal. Chem.*, **117**, 29 (1981).
212. H. Behret, H. Binder and G. Sandstede, 'Inorganic and Organic Non-Noble Metal Containing Electrocatalysts for Fuel Cells', in "Electrocatalysis", M. W. Breiter (Ed), The Electrochemical Society, Princeton, NJ, pp. 319–338 (1974).
213. H. Behret, H. Binder and G. Sandstede, *Electrochim. Acta*, **20**, 111 (1975).
214. F. Von Sturm, H. Nischik and E. Weidlich, *Ingenteur Digest.*, **5**, 52 (1966).
215. F. Von Sturm, 'Elektrochemische Stromerzeugung', Verlag Chemie, Weinheim (1969).
216. K. Strasser, 'Die alkalische Siemens-Brennstoffzelle in Kompaktbauweise', in "Brennstoffzellen", VDI-Buch 996, pp. 25–46 (1990).
217. H. Grüne, K. Mund and M. Waidhas, 'Fuel Cells for Transportation', Workshop ELsS (1994).
218. M. Baldauf, W. Lager, W. Preidel and M. Waidhas, 'Direkt-Methanolbrennstoffzellen', in "Brennstoffzellen, Entwicklung, Technologie, Anwendung", K. Ledjeff-Hey, F. Mahlendorf and J. Roes (Eds), C.F. Müller Verlag, Heidelberg, pp. 77–100 (2001).
219. F. J. Rohr, 'Festelektrolyt für Brennstoffzellen', DE, 1967 DE-P 1 611, pp. 704 (1967).
220. F. J. Rohr, 'Hochtemperatur-Brennstoffzellen mit Zirkonoxid-Festelektrolyten', in "Elektrochemische Energietechnik – Entwicklungsstand und Aussichten", Bundesminister für Forschung und Technologie, Germany, Bonn, 264, pp. 179 (1981).
221. K. Wiesener, J. Garche and W. Schneider, 'Elektrochemische Stromquellen', Akademie Verlag, Berlin (1981).
222. K. Wiesener and D. Ohms, 'Electrocatalysis of the Cathodic Oxygen Reduction', in "Electrochemical Hydrogen Technologies", H. Wendt (Ed), Elsevier, Amsterdam, pp. 63–135 (1990).

223. H. Jahnke and M. Schönborn, 'Journees International d'Etude Piles a Combustible', Bruxelles, SERAI, Revue Energie Primaire III (1969).
224. H. Böhm and F. A. Pohl, *Wiss. Ber. AEG-Telefunken*, **41**, 46 (1968).
225. H. Böhm and F. Fleischmann, 'Phosphorsaure und Schwefelsaure Brennstoffzellen', in "Brennstoffzellen, Stand der Technik, Entwicklungslinien, Marktchancen", H. Wendt and V. Plzak (Eds), VDI-Verlag, Düsseldorf, pp. 82–98 (1990).
226. D. Behrens and G. Sandstede, 'Elektrochemische Prozesse, BMFT/DECHEMA', D. Behrens and G. Kreysa (Eds), Chemische Technik – Rohstoffe, Prozesse und Produkte Band 3, Bonn, Frankfurt am Main (1975).
227. F. Barz, H. Binder, H. Böhm, W. Fischer, G. Sandstede, F. Von Sturm, W. Vielstich and G. Walter, 'Elektrochemische Energieerzeugung und -speicherung', D. Behrens and G. Sandstede (Eds), Elektrochemische Prozesse, DECHEMA, Frankfurt am Main, pp. 287–327 (1975).
228. D. Behrens and G. Sandstede (Eds), 'Wasserstofftechnologie – Perspektiven für Forschung und Entwicklung', DECHEMA-Studie, Frankfurt am Main (1986).
229. H. Böhm, J. A. A. Ketelaar, K. Kordesch, G. Richter, G. Sandstede, W. Vielstich and G. Walter, 'Brennstoffzellen', in "Wasserstofftechnologie – Perspektiven für Forschung und Entwicklung", D. Behrens (Ed), DECHEMA-Studie, Frankfurt am Main, pp. 243–284 (1986).
230. G. Collin and G. Sandstede (Eds), 'Wasserstoffwirtschaft – Herausforderung für das Chemie-Ingenieurwesen', DECHEMA, Frankfurt am Main, Dechema-Monogr., 106 (1987).
231. W. Vielstich, 'Brennstoffzellen als Energiewandler in einer Wasserstoffwirtschaft', in "Wasserstoffwirtschaft – Herausforderungfür das Chemie-Ingenieurwesen", G. Collin and G. Sandstede (Eds), Tutzing, DECHEMA, VCH-Verlagsges., 106, pp. 299–314 (1986).
232. G. Sandstede and R. Wurster, 'Water Electrolysis and Solar Hydrogen Demonstration Projects', in "Modern Aspects of Electrochemistry", R. E. White, J. O. M. Bockris and B. E. Conway (Eds), Plenum, New York, **27**, pp. 411–515 (1995).
233. C.-J. Winter and J. Nitsch (Eds), 'Wasserstoff als Energieträger', Springer Verlag, Berlin (1986).
234. H. Wendt, 'Electrochemical Hydrogen Technologies', Elsevier, Amsterdam (1990).
235. H. Wendt and V. Plzak (Eds), 'Brennstoffzellen, Stand der Technik, Entwicklungslinien, Marktchancen', VDI-Verlag, Düsseldorf (1990).
236. T. Takahashi, *J. Electrochem. Soc., Jpn.*, **34**, 60 (1966).
237. T. Takahashi, 'Research and Development of Fuel Cells in Japan', in "Sixth Status Report on Fuel Cells" G. R. Frysinger (Ed), US Dept. of Commerce, Office of Tech. Serv., US Army Electronics Command, Fort Monmouth NJ, Report No. AD 665353, 159–170 (1967).
238. S. Tamaru and K. Ochiai, *J. Chem. Soc. Jpn.*, **56**, 92, 103 (1935).
239. J. R. Huff, 'Seventh Status Report on Fuel Cells', US Army Mobility Equipment Research and Development Center, Fort Belvoir, Virginia, 2039 (1972).
240. K. Kishida, 'Status and Interest of Japanese Industrial Development of Fuel Cells', in "Fuel Cells: Trends in Research and Application", A. J. Appleby (Ed), Hemisphere Publishing Corp., Washington, pp. 173–190 (1987).
241. H. Sato, 'Status of Fuel Cells R&D and Prospects for their Application in Japan', in "Fuel Cells: Trends in Research and Application", A. J. Appleby (Ed), Hemisphere Publishing Corp., Washington, pp. 233–245 (1987).
242. R. Anahara, 'PAFC Plants in Japan', in "Brennstoffzellen", K. Ledjeff (Ed), C.F. Müller Verlag, Heidelberg, pp. 45–62 (1995).
243. R. Anahara, 'Research, Development and Demonstration of Phosphoric Acid Fuel Cell Systems', in "Fuel Cell Systems", L. J. M. J. Blomen and M. N. Mugerwa (Eds), Plenum Press, New York, pp. 271–343 (1993).
244. K. V. Kordesch and J. C. Tambascodeoliveira, 'Fuel Cells', in "Ullmann's Encyclopedia of Industrial Chemistry", VCH Verlag, Weinheim, A12, pp. 55–83 (1989).
245. H. Vandenbroeck, 'Application of Fuel Cells to Large Vehicles', in "Fuel Cells: Trends in Research and Application", A. J. Appleby (Ed), Hemisphere Publishing Corp., Washington, pp. 281–287 (1987).
246. E. Budevsky, 'Status of Research and Application of Fuel Cells in Bulgaria', in "Fuel Cells: Trends in Research and Application", A. J. Appleby (Ed), Hemisphere Publishing Corp., Washington, pp. 69–90 (1987).
247. J. Jensen, 'Materials Research for Advanced Solid-State Fuel Cells at the Energy Research Laboratory, Denmark', in "Fuel Cells: Trends in Research and Application", A. J. Appleby (Ed), Hemisphere Publishing Corp., Washington, pp. 145–159 (1987).
248. S. J. Bartosh, J. M. Mcdonagh and W. G. Taschek, 'Fifth Status Report on Fuel Cells', US Dept. of Commerce, Office of Tech. Serv., US Army Electronics Command, Fort Monmouth, NJ, AD 620114 (1965).
249. G. R. Frysinger, 'Sixth Status Report on Fuel Cells', US Dept. of Commerce, Office of Tech. Serv., US Army Electronics Command, Fort Monmouth, NJ, AD 665353 (1967).
250. Institut-Francais-Du-Petrole (Ed), 'Les Piles á Combustible', Editions Technip, Paris (1965).
251. A. Ascoli and G. Redaelli, 'Long Term Testing of an Air-cooled 2.5 kW PAFC Stack in Italy', in "Fuel Cells: Trends in Research and Application", A. J. Appleby (Ed), Hemisphere Publishing Corp., Washington, pp. 51–58 (1987).
252. D. Ohms, V. Plzak, S. Trasatti, K. Wiesener and H. Wendt, 'Electrode Kinetics and Electrocatalysis of Hydrogen and Oxygen Reactions', in "Electrochemical Hydrogen Technologies", H. Wendt (Ed), Elsevier, Amsterdam, pp. 1–135 (1990).
253. O. Lindstrom, 'Status of Research and Applications of Fuel Cells in Sweden', in "Fuel Cells: Trends in Research and Application", A. J. Appleby (Ed), Hemisphere Publishing Corp., Washington, pp. 191–202 (1987).
254. E. Barendrecht, 'Status of Present and Intended Research on Fuel Cells in the Netherlands', in "Fuel Cells: Trends in Research and Application", A. J. Appleby (Ed), Hemisphere Publishing Corp., Washington, pp. 59–62 (1987).

255. R. H. Topke and E. R. Gonzalez, 'Status of Research and Applications of Fuel Cells in Brazil', in "Fuel Cells: Trends in Research and Application", A. J. Appleby (Ed), Hemisphere Publishing Corp., Washington, pp. 275–279 (1987).

256. The Hydrogen Link Conference March 1986, Hydrogen Industry Council, Montreal, **1** (1986).

257. G. Sandstede, 'Hydrogen in Industrial Chemical Processes in Germany', Presented at the Hydrogen Link Conference, Montreal, Hydrogen Industry Council, Montreal, Vol. I, pp. 138–202 (1986).

258. J. Gretz, J. P. Baselt, O. Ullmann and H. Wendt, *J. Hydrogen Energy*, **15**, 419 (1990).

259. J. Gretz, J. P. Baselt, O. Ullmann and H. Wendt, 'The 100-MW Euro/Quebec Hydro-Hydrogen Pilot Project', in "Wasserstoff-Energietechnik II", VDI-Verlag, Stuttgart, VDI-Berichte (1989).

260. D. S. Watkins, K. Dircks and D. Epp, 'Canadian Solid Polymer Fuel Cell Development', Presented at the 33d International Power Sources Symposium, Cherry Hill, NJ, The Electrochem. Soc., pp. 782–791 (1988).

261. C.-S. Cha, 'Research and Applications of Fuel Cells Techniques in the People's Republic of China', in "Fuel Cells: Trends in Research and Application", A. J. Appleby (Ed), Hemisphere Publishing Corp., Washington, pp. 95–98 (1987).

262. J. D. Pandya, 'Status of Research and Applications of Fuel Cells in India', in "Fuel Cells: Trends in Research and Application", A. J. Appleby (Ed), Hemisphere Publishing Corp., Washington, pp. 203–208 (1987).

263. C. C. Morrill, personal communication (1966).

264. G. E. Evans, 'Status of the Carbon Electrode Fuel Cell Battery', in "Fuel Cells – a CEP Technical Manual", American Institute of Chemical Engineers (AIChE), New York, pp. 6–10 (1963).

265. E. J. Cairns, 'Fluoride Electrolytes for Saturated Hydrocarbon Fuel Cells', in "Hydrocarbon Fuel Cell Technology", B. S. Baker (Ed), Atlantic City, New Jersey, Academic Press, New York, pp. 465–483 (1965).

266. S. Srinivasan, B. B. Davè, K. A. Murugesamoorthi, A. Parthasarathy and A. J. Appleby, 'Overview of Fuel Cell Technology', in "Electrochemistry of Fuel Cells", L. J. M. J. Blomen and M. N. Mugerwa (Eds), Plenum Press, New York, pp. 37–72 (1993).

267. G. Sandstede (Ed), 'From Electrocatalysis to Fuel Cells', University of Washington Press, Seattle, WA (1972).

268. A. J. Appleby and B. S. Baker, *J. Electrochem. Soc*, **125**, 404 (1978).

269. K. Kinoshita, F. R. McLarnon and E. J. Cairns, 'Fuel Cells: A Handbook', US Department of Energy, Morgantown, WV, DOE/METC88/6096 (1988).

270. A. J. Appleby and F. R. Foulkes, 'Fuel Cell Handbook', Van Nostrand Reinhold, New York (1989).

271. W. T. Grubb, in 'Proceedings of the 11th Annual Battery Research and Development Conference', Asbury Park, NJ, PSC Publications Committee, Red Bank, NJ, pp. 5 (1957).

272. W. T. Grubb, US Patent No. 2,913,511 (1959).

273. W. T. Grubb, *J. Phys. Chem.*, **63**, 55 (1959).

274. D. S. Watkins, 'Research, Development and Demonstration of Solid Polymer Fuel Cell Systems', in "Fuel Cell Systems", L. J. M. J. Blomen and M. N. Mugerwa (Eds), Plenum Press, New York, pp. 493–530 (1993).

275. W. T. Grubb and L. W. Niedrach, *J. Electrochem. Soc*, **107**, 131 (1960).

276. E. Joachim and W. Vielstich, *Electrochim. Acta*, **244**, 3 (1960).

277. J. Perry, Jr, Presented at the 14th Annual Power Sources Conference, Atlantic City, NJ, pp. 50 (1960).

278. 'Acid Electrolyte Fuel Cell Program, Final Report', General Electric, NASA, Contract NAS 9-12332 (1973).

279. E. A. Oster, Presented at the 14th Annual Power Sources Conference, Atlantic City, NJ, PSC Publications Comm., Red Bank, NJ, pp. 59 (1960).

280. E. J. Cairns, D. L. Douglas and L. W. Niedrach, *A.I.Ch.E.J.*, **7**, 551 (1961).

281. L. W. Niedrach and W. T. Grubb, 'Ion Exchange Membrane Fuel Cells', in "Fuel Cells", J. Mitchell Will (Ed), Academic Press, New York, London, pp. 253–298 (1963).

282. A. C. Erickson and L. J. Nuttal, 'Development Status of a Regenerative Fuel Cell System for Orbital Energy Storage', Presented at the 18th IECEC, pp. 1519–1524 (1983).

283. E. Müller, *Z. Elektrochem.*, **28**, 101 (1922).

284. K. Kordesch and A. Marko, *Oesterr. Chemiker Ztg.*, **52**, 125 (1961).

285. O. Bloch, M. Prigent and J. C. Balaceanu, Presented at the Indianapolis Meeting, Indianapolis, The Electrochemical Society (1961).

286. W. Vielstich, 'Alcohol-Air Fuel Cells – Development and Applications', in "Hydrocarbon Fuel Cell Technology", B. S. Baker (Ed), Atlantic City, NJ, Academic Press, pp. 79–90 (1965).

287. M. J. Schlatter, 'Fuel Cell Intermediates and Products', in "Fuel Cells", G. J. Young (Ed), Presented at the Symposium of the ACS, Chicago, Reinhold Publishing Corp., New York, Chapman & Hall Ltd., London, Vol. II, pp. 190–215 (1963).

288. C. W. Foust and W. J. Sweeney, 'Carbonaceous Fuels', in "Fuel Cells", J. Mitchell Will (Ed), Academic Press, New York, London, pp. 371–388 (1963).

289. K. R. Williams, M. R. Andrew and F. Jones, 'Some Aspects of the Design and Operation of Dissolved Methanol Fuel Cells', in "Hydrocarbon Fuel Cell Technology", B. S. Baker (Ed), Atlantic City, NJ, Academic Press, New York, pp. 143–149 (1965).

290. M. W. Breiter, 'Electrochemical Processes in Fuel Cells', Springer–Verlag, Berlin (1969).

291. V. S. Bagotsky, Y. B. Vasiliev and O. A. Khazova, *J. Electroanal. Chem.*, **81**, 229 (1977).

292. C. E. Heath, 'Journees Int. d'Etude Piles a Combustible', SERAI, Bruxelles, Revue Energie Primaire I, pp. 99 (1965).

293. A. N. Frumkin, Presented at the International Symposium on Batteries, Brighton, UK, pp. 4 (1964).

294. E. J. Cairns and D. C. Bartosik, *J. Electrochem. Soc.*, **111**, 1205 (1964).

295. M. Waidhas, 'Methanol-Brennstoffzellen', in "Brennstoffzellen", K. Ledjeff (Ed), C.F. Müller Verlag, Heidelberg, pp. 137–156 (1995).

296. K. Ledjeff-Hey, F. Mahlendorf and J. Roes (Eds), 'Brennstoffzellen, Entwicklung, Technologie, Anwendung', C.F. Müller Verlag, Heidelberg (2001).

297. D. L. Maricle, B. L. Murach and N. L. Vandine, 'Direct Methanol Fuel Cell Stack Development', Presented at the 36th Power Sources Conference, pp. 99 (1994).

298. B. D. McNicol, 'Direct Methanol/Air Systems', in "Power Sources for Electric Vehicles", B. D. McNicol and D. A. J. Rand (Eds), Elsevier, Amsterdam, pp. 807–838 (1984).

299. H. Binder, A. Köhling and G. Sandstede, *Energy Conversion*, **11**, 17 (1971).

300. R. A. Wynveen, 'Preliminary Appraisal of the Ammonia Fuel Cell System', in "Fuel Cells, Symposium of the ACS", G. J. Young (Ed), Reinhold Publishing Corp., New York, Chapman & Hall, London, pp. 153–167 (1963).

301. M. Eisenberg, Presented at the 18th Annual Power Sources Conference, PSC Publications Committee, Red Bank, NJ (1964).

302. E. J. Cairns, E. L. Simons and A. D. Tevebaugh, *Nature*, **217**, 780 (1968).

303. E. L. Simons, E. J. Cairns and D. J. Surd, *J. Electrochem. Soc.*, **116**, 556 (1968).

304. D. W. McKee, A. J. Scarpellino, Jr, I. F. Danzig and M. S. Pak, *J. Electrochem. Soc.*, **116**, 562 (1968).

305. W. R. Grove, 'The Correlation of Physical Forces', 6th Edition, Longmans Green, London (1874).

306. G. J. Young and R. B. Rozelle, 'Catalysis of Fuel Cell Electrode Reactions', in "Fuel Cells, Symposium of the ACS", G. J. Young (Ed), Reinhold Publishing Corp., Atlantic City, New York, Chapman & Hall, London, pp. 23–33 (1960).

307. R. P. Buck, L. R. Griffith, R. T. Macdonald and M. J. Schlatter, 'Proceedings of the 15th Annual Power Sources Conference', PSC Publications Committee, Red Bank, NJ, pp. 16–21 (1961).

308. G. J. Young and R. B. Rozelle, 'Low temperature Electrochemical Oxidation of Hydrocarbons', in "Fuel Cells, Symposium of the ACS", G. J. Young (Ed), Reinhold Publishing Corp., Chicago, New York, Chapman & Hall, London, pp. 216–224 (1963).

309. W. T. Grubb, 'Proceedings of the 16th Annual Power Sources Conference', PSC Publications Committee, Red Bank, NJ, pp. 31–34 (1962).

310. L. W. Niedrach, *J. Electrochem. Soc.*, **109**, 1092 (1962).

311. B. S. Baker (Ed), 'Hydrocarbon Fuel Cell Technology', Academic Press, New York (1965).

312. E. H. Okrent and C. E. Heath, 'A Liquid Hydrocarbon Fuel Cell Battery', in "Fuel Cell Systems II", B. S. Baker (Ed), Presented at the Symposium of the Division of Fuel Chemistry of the ACS, Chicago, Illinois, American Chemical Society, Washington DC (1969); *Adv. Chem.*, **90**, 328 (1967).

313. E. J. Cairns, 'Advances in Electrochemistry and Electrochemical Engineering', C. W. Tobias (Ed), Wiley-Interscience, New York, pp. 8 (1971).

314. E. J. Cairns and D. I. MacDonald, *Electrochem. Technol.*, **2**, 65 (1964).

315. T. K. Johnson, 'Proceedings of the 18th Annual Power Sources Conference', PSC Publications Committee, Red Bank, NJ, pp. 25–27 (1964).

316. J. H. Russell, 'Proceedings of the 19th Annual Power Sources Conference', PSC Publications Committee, Red Bank, NJ, pp. 33–38 (1965).

317. C. C. Morrill, 'Proceedings of the 19th Annual Power Sources Conference', PSC Publications Committee, Red Bank, NJ, pp. 38–41 (1965).

318. E. A. Oster and L. E. Chapman, 'Description of a Fuel Cell Power Pack Operating on Hydrogen and Ambient Air', in "Fuel Cells", G. J. Young (Ed), Presented at the Symposium of the ACS, Chicago, Reinhold, New York, vol. 2, pp. 168–181 (1963).

319. J. R. Huff and J. C. Orth, 'The USAMECOM-MERDC Fuel Cell Electric Power Generation Program', in "Fuel Cell Systems II", B. S. Baker (Ed), Presented at the Symposium of the Division of Fuel Chemistry of the ACS, Chicago, Illinois, American Chemical Society, Washington DC (1969); *Adv. Chem.*, **90**, 315 (1967).

320. E. Luksha and E. Y. Weissman, 'Electrochemical Oxidation of Multicomponent Hydrocarbon Fuels III. Relative Reactivities in the Electrochemical Oxidation of Hydrocarbon Fuel Components', in "Fuel Cell Systems II", B.S. Baker (Ed), Presented at the Symposium of the Division of Fuel Chemistry of the ACS, Chicago, IL, American Chemical Society, Washington, DC, pp. 200–222 (1969).

321. L. W. Niedrach, 'Proceedings of 13th Annual Power Sources Conference', PSC Publications Committee, Red Bank, NJ, pp. 120–121 (1959).

322. *Chem. Eng. News*, **43**, 65 (1965).

323. D. L. Douglas and E. J. Cairns, US Patent, 3,134,696 (1964).

324. Dupont, press release, Sept. 12 (1969).

325. *C&E News*, Aug. 27, pp. 15 (1973).

326. W. G. Grot, *Macromol. Symp.*, **82**, 161 (1994).

327. I. Maiche, French Patent, 127,069 (1878).

328. W. Cohen, British Patent, 15,407 (1891).

329. T. H. Nash, *Electrician*, **26**, 718 (1891).

330. C. Féry, *Electrician*, **80**, 320 (1917).

331. C. Féry, German Patent, DRP 440,806 (1924).

332. W. Walker and F. H. Wilkins, US Patent, 524,229 and 524,391 (1894).

333. K.-J. Euler, 'Batterien und Brennstoffellen', Springer-Verlag, Berlin, Heidelberg (1982).

334. B. Siller, 'Luftsauerstoffelemente', VDI Verlag, Düsseldorf, VARTA-Fachbuchreihe Bd., 5 (1968).

335. R. R. Shipps, 'Proceedings of the 20th Annual Power Sources Conference', Atlantic City, NJ, pp. 86–88 (1966).

336. S. M. Chodosh, B. Jagid and E. Katsulis, 'Power Sources 2', D. H. Collins (Ed), Pergamon Press, London, pp. 423 (1970).
337. R. R. Witherspoon, E. L. Zeitner and H. A. Schulte, 'Proc. 6th Intersoc. Energy Conv. Eng. Conf.', Boston, pp. 96 (1971).
338. H. Baba, US Patent, 3,555,032 and 3,560,262.
339. Battelle Memorial Institute, French Patent, 2,096,046.
340. D. Doniat, K. Beccu and A. Porta, German Patent, 2,125,576.
341. A. J. Appleby and M. Jacquier, *J. Power Sources*, **1**, 17 (1976).
342. P. C. Foller, *J. Appl. Electrochem.*, **16**, 529 (1986).
343. O. Haas, S. Müller and K. Wiesener, *Chem. Ing. Tech.*, **68**, 524 (1996).
344. F. R. McLarnon and E. J. Cairns, *J. Electrochem. Soc.*, **138**, 645 (1991).
345. D. Kuller and A. Belanger, 'Proc. 38th Ann. Power Sources Conf.', Terry Hill, NJ, pp. 282 (1998).
346. G. J. Schaeffer, Fuel Cells for the Future, Twente, p. 349 (1998).
347. G. J. Schaeffer, Fuel Cells for the Future, Twente, p. 353 (1998).
348. W. Vielstich, Fuel Cells, Wiley-Interscience, p. 231 (1970).
349. K. Kordesch, Brennstoffbatterien, Springer-Verlag, Wien, New York, p. 150 (1984).
350. H. H. von Döhren and K. J. Euler, Brennstoffelemente, 6. Auflage, VDI Verlag, Düsseldorf, p. 35 (1971).
351. O. Führer and A. Winsel, *Alkalische Brennstoffzellen in Europe*, in Brennstoffzellen, Entwicklung, Technologie, Anwendung, K. Ledjeff, HF-Hey, Mahlendorf, Roes (Eds), 2nd Edition, Hüthig GmbH & Co. KG, Heidelberg, pp. 43–60 (2001).
352. H. H. von Döhren and K. J. Euler, Brennstoffelemente, 6. Auflage, VDI Verlag, Düsseldorf, p. 52 (1971).
353. H. H. von Döhren and K. J. Euler, Brennstoffelemente, 6. Auflage, VDI Verlag, Düsseldorf, p. 132 (1971).
354. W. Vielstich, Fuel Cells, Wiley-Interscience, London, p. 193 (1970).
355. G. Sandstede (Ed), From Electrocatalysis to Fuel Cells, University of Washington Press, Seattle, WA, p. 51 (1972).
356. W. Vielstich, Fuel Cells, Wiley-Interscience, London, p. 270 (1970).
357. W. Vielstich, Fuel Cells, Wiley-Interscience, London, p. 458 (1970).
358. H. H. von Döhren and K. J. Euler, Brennstoffelemente, 6. Auflage, VDI Verlag, Düsseldorf, p. 86 (1971).
359. W. Vielstich, Fuel Cells, Wiley-Interscience, London, p. 475 (1970).
360. G. Sandstede (Ed), From Electrocatalysis to Fuel Cells, University of Washington Press, Seattle, WA, p. 19 (1972).
361. S. U. Falk and A. J. Salkind, *Alkaline Storage Batteries*, John Wiley & Sons, Inc, New York (1969).

Chapter 13
History of high temperature fuel cell development

H. Yokokawa and N. Sakai
National Institute of Advanced Industrial Science and Technology, Ibaraki, Japan

1 HISTORICAL SURVEY OF DEVELOPMENTS OF HIGH TEMPERATURE FUEL CELLS

1.1 Historical overview

Molten carbonate fuel cells (MCFCs) and solid oxide fuel cells (SOFCs) are often discussed within the same framework as high temperature fuel cells, and there are actually many common features against low temperature fuel cells such as phosphoric acid fuel cells (PAFCs) or polymer electrolyte fuel cells (PEFCs). Historically, development of high temperature fuel cell stacks started quite later when compared with the low temperature fuel cells.[1–4] This is because the high temperature fuel cells were not appropriate for space and/or military applications. For the space application, hydrogen was used as a convenient fuel, whereas the strongest interest in the high temperature fuel cells was on the direct use of carbon-containing fuels, such as coal related fuels. In this sense, the oil crisis in 1974 changed the situation of high temperature fuel cells. Natural gas became one of the major fuels to be used in the high temperature fuel cells. It was also strongly required to improve the conversion efficiency from fuel to electricity.

Although R&D of both MCFC and SOFC started from 1960s, there appeared to be some differences between MCFCs and SOFCs[2–4]. In the beginning, general comparisons between MCFC and SOFC lead to a thought that MCFCs were more favored than SOFCs; this was probably based on expected advantages of MCFCs in its lower operation temperature (650 vs. 1000 °C) and in the wider "available" choice of materials. At that time, there was a belief that the higher the temperature, the more difficult it would become to find appropriate materials for fuel cells. Only precious metals such as Pt were used for electrodes and interconnections in SOFCs in the beginning. Thus, MCFCs attracted much more interest, and effort for the development of MCFCs stacks was made by the Institute of Gas Technology (IGT) and the United Technology Corporation, and then later by the Energy Research Corporation (ERC) and the General Electric Corporation (GE).[5] On the other hand, only the Westinghouse Power Corporation (WHPC) paid strong and persistent interest to SOFCs.[6, 7]

In the early 1980s, however, the basic materials selection by WHPC had progressed successfully for SOFCs.[8] In addition, Dornier also reached a similar materials selection for water electrolyzers.[9] This success in materials selection gave a new competitive feature between MCFCs and SOFCs. During the 1980s and 1990s, intensive and extensive investigations were made on materials and stack developments on MCFCs and SOFCs.[10] During this period of time, the difference in strategy for developing stacks became clearer between MCFCs and SOFCs. Until the early 1980s, the cell performance of small MCFC stacks had been very much improved through efforts of many developers such as IGT, ERC, the International Fuel Cells Corporation (IFC) and GE. In addition, almost all material problems and electrolyte management issues associated with MCFC also appeared during this period.[5, 11–14] In SOFCs, a sealless tubular stack was eventually successfully developed in the early 1980s. This epoch-making achievement was due to breakthroughs that enabled the provision of simultaneous solutions for technological issues for design, fabrication and materials. Just after WHPC proposed the sealless tubular stacks, Argonne National Laboratory in the USA analyzed the efficiency of SOFCs compared with PAFC and MCFCs. The conclusion reached was that for installation without a bottoming cycle, the efficiency of a SOFC power plant was approximately the same as a PAFC power plant but about 15% less than a MCFC power plant.[15] This statement reflected some of the atmosphere at that time about high temperature fuel cells. This is based on the following merits and demerits that (1) the theoretical value for open-circuit potential OCV shows about 100 mV higher at 650 °C for the MCFC operation temperature than 1000 °C for SOFCs; (2) the concentration polarization loss is higher at 1000 °C. On the other hand, (3) the activation polarization loss can be expected to be low in SOFCs; (4) the merits of high temperature heats in anode-exit gas can be more advantageous in SOFCs; (5) gaseous circulation related with CO_2 is not energetically favored in MCFCs. Since the materials issue was not critical any more for SOFCs, the conversion efficiency and other operational characteristics became key features to be compared between MCFCs and SOFCs.

From the late 1980s, MCFCs started the development of large and tall stacks towards MW size as schematically shown in Figure 1. This was a big step to target the commercialization of the MCFC system, which was initiated in Japan and the USA. In a sense, this was a reasonable decision, because the superiority of MCFCs was still believed to be as described above. Furthermore, as shown in Figure 1, the scale-ups were made rapidly. Even so, it should be recognized that there remained some materials issues limiting life and efficiency. Needless to say, life and efficiency are the most important characteristics for fuel cell systems. For MCFCs, expected efficiency was high, but the efficiency actually obtained was low. To improve them, materials investigations were also made in parallel to stack development.

The development of the SOFC system with the tubular design has been accelerated from the late 1980s by WHPC with support from the US Department of Energy. Their strategic efforts were focused on the confirmation of the long-term stability and the high efficiency of the SOFC systems.[16, 17] The most important impact of their earlier achievement is the confirmation of the stability and compatibility of the materials. They showed that stability is not governed by temperature alone. Since SOFC has no liquids, the kinetic reaction rate is extremely small compared with MCFCs. This makes SOFC materials more stable than MCFC materials, despite its higher operation temperature. Another important feature of WHPC's SOFCs is confirmation of the high efficiency for natural gas-fueled SOFCs, which is sufficiently competitive with that for the analogous MCFCs. In addition to WHPC, several different strategies appeared in SOFCs in the late 1980s.[18] One is to adopt the wet-sintering process to reduce the fabrication costs, another being attempts of utilizing metal interconnects to reduce the material cost and improve the thermal cycle characteristics. There was also strong interest in planar stacks for their high power density. Although it took more time, planar stacks have been successfully constructed in several 10 kW size as shown in Figure 1.

One of the major objectives for developing high temperature fuel cells is to achieve the highest efficiency from carbon-containing fuels to electricity. This is the common target among natural gas fueled energy converters. Rapid advance in gas turbine technology in particular has changed the situation about the conversion efficiency; that is, the advanced combined cycles of gas and steam turbines in several hundred megawatt sizes have achieved 50% (higher heating value (HHV)) with a 1300 °C class gas turbine (GT) and 53% (HHV) for a 1500 °C class GT. In this sense, the development of the hybrid system of fuel cells and gas turbines has become the key technology for high temperature fuel cells. For such a combination, high pressure operation is needed for the fuel cell side. Although developing large stacks was early in the development of MCFCs, the first hybrid system was tested on a SOFC/GT system in 2000. This is due to the fact that a pressurized operation for MCFCs had not yet been well established because of the shorting caused by nickel dendrites originated from the NiO cathode dissolution, which is enhanced under high CO_2 partial pressure. As a result, the SOFC system was recognized for high efficiency as high temperature fuel cells earlier than MCFC systems.

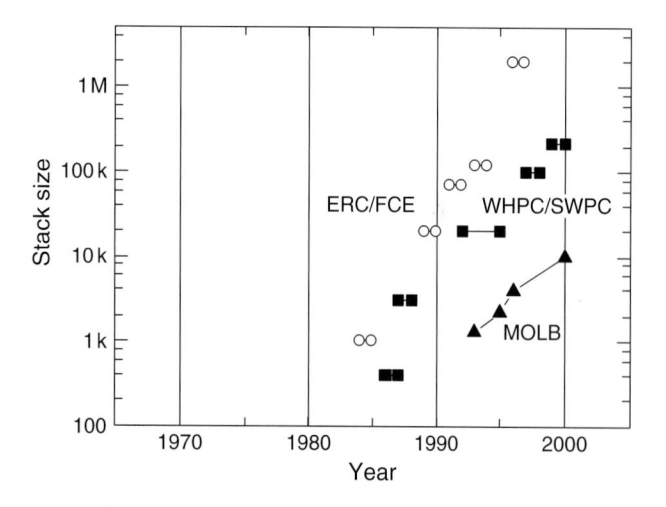

Figure 1. The chronological developments of MCFC and SOFC stacks. Typical sizes of MCFC stacks (Energy Research Corporation/Fuel Cell Energy) and SOFC stacks (Westinghouse Power Corporation/Siemens Westinghouse Power Corporation and MOLB-type by Mitsubishi Heavy Industry) are plotted.

From the mid-1990s, new situations appeared in energy conversion. That is, the environmental issue and the deregulation of power generation, distribution and retail. To reduce CO_2 emission, focus was placed mainly on natural gas (or biogas in the near future) rather than on coal related gas. Furthermore, the distributed power generation makes middle-size power generators more attractive; on-site generation makes it possible to provide electricity and heat simultaneously and to improve the total efficiency of utilizing fuels. All these factors combined lead to SOFCs being more favorable. For example, in the early 1990s, SOFCs were regarded as inappropriate for large power plants because a plausible SOFC system was considered as several tens of megawatts or less. In recent years, however, interest in SOFCs has been renewed because of their appropriate size for distributed generators. Even in MCFCs that should be operated as a base load with less frequent thermal cycles, there appeared a new strategy to establish the medium size of stacks and then to combine them into larger systems after achieving cost reduction in the medium size systems.

Environmental issues make it attractive and necessary to develop fuel cells applicable to automobiles. From the quick start-up and fast load-following characteristics, PEFCs have attracted much interest for transportation applications and, therefore, intensive efforts have been focused on PEFCs. Even so, in 1998, BMW, Renault and Delphi proposed the utilization of a SOFC as the auxiliary power unit (APU). Since the main driving force is supplied with combustion engines, SOFCs for APU uses the same fuels as the engines. This makes it possible to introduce fuel cells without changing the infrastructure of the fuels. The application of SOFCs as APUs is also based on the advantages of high temperature fuel cells against the low temperature fuel cells. The reforming process is simpler and the efficiency is higher. On the other hand, SOFCs have weak points as transportation applications; start-up is not rapid and the power density per volume or per weight is still low. Thus, these are new challenges for SOFCs to overcome such weak points.

Developments in the early stages have been well documented in many textbooks.[1–4, 7] From the 1980s, many arguments on stack development, materials and modeling of MCFCs and SOFCs have been made in several series of international symposia; MCFC technology,[19–23] SOFCs,[24–31] high temperature solid oxide electrolyte[32] and the European SOFC forum.[33–36]

In what follows, a brief historical development will be described for MCFCs and SOFCs. In Sections 2 and 3, the common features between MCFC and SOFC are described from the viewpoint of high temperature materials science and irreversible thermodynamics treating mass transfer under gradients of thermodynamic quantities developed inside stacks. In Section 4, stack development of MCFCs is described from the molten chemical point of view. In Section 5, stack development of SOFCs is described in relation to the technological means to overcome mechanical weakness. Finally, a comparison is made and future perspectives are given for high temperature fuel cells in Sections 6 and 7.

1.2 MCFCs

Initial attempts were made mainly in Europe and the USA to find electrolytes at high temperatures in a mixture of solid and liquid and then in molten salts. Alkali carbonates were soon selected as suitable electrolytes because other salts could not provide the steady state electrochemical reaction; when cations are used as the carrier, one of salt components is consumed or accumulated in electrode areas. When anions are used as the carriers, salts reacted with carbon dioxides that are one of the reaction products. Thus, it was natural to use molten carbonates. For molten carbonates, the cathode and the anode reactions can be written as follows:

$$0.5O_{2(g)} + CO_{2(g)} + 2e^- \longrightarrow CO_3^{2-} \qquad (1)$$

$$CO_3^{2-} + H_{2(g)} \longrightarrow CO_{2(g)} + H_2O_{(g)} + 2e^- \quad (2)$$

Note that other species than carbonate ions are not involved in the electrode reactions. A total reaction is given as follows:

$$H_{2(g)} + 0.5O_{2(g)} \longrightarrow H_2O_{(g)} \qquad (3)$$

In order to conserve CO_2, it is necessary to transfer the CO_2 evolved at the anode to the cathode, where CO_2 is required. This circulation of CO_2 was regarded as one of weak points of MCFCs; this makes the system complicated. In stacks, molten carbonates were used as a free, paste electrolyte or an electrolyte matrix. As materials for paste electrolyte or matrix, MgO, Al_2O_3 and alkali aluminates were tested. As cathode materials, Ag or CuO were used, and Ni, Pd and Ag were used initially. After pioneering work by Baur et al. in 1920 and Davytan in 1960, Broers and co-workers[37] investigated MCFCs in the 1960s and provided the basis of modern MCFC technology. Figure 2 shows their stack. It operated for more than 1 year.

From 1959, IGT in the USA started MCFC development to facilitate the use of natural gas. A 32 cell stack was constructed in 1964 using paste electrolyte and Ag and Ni electrodes. It followed the TARGET program organized by the American Gas Association, which was initiated in

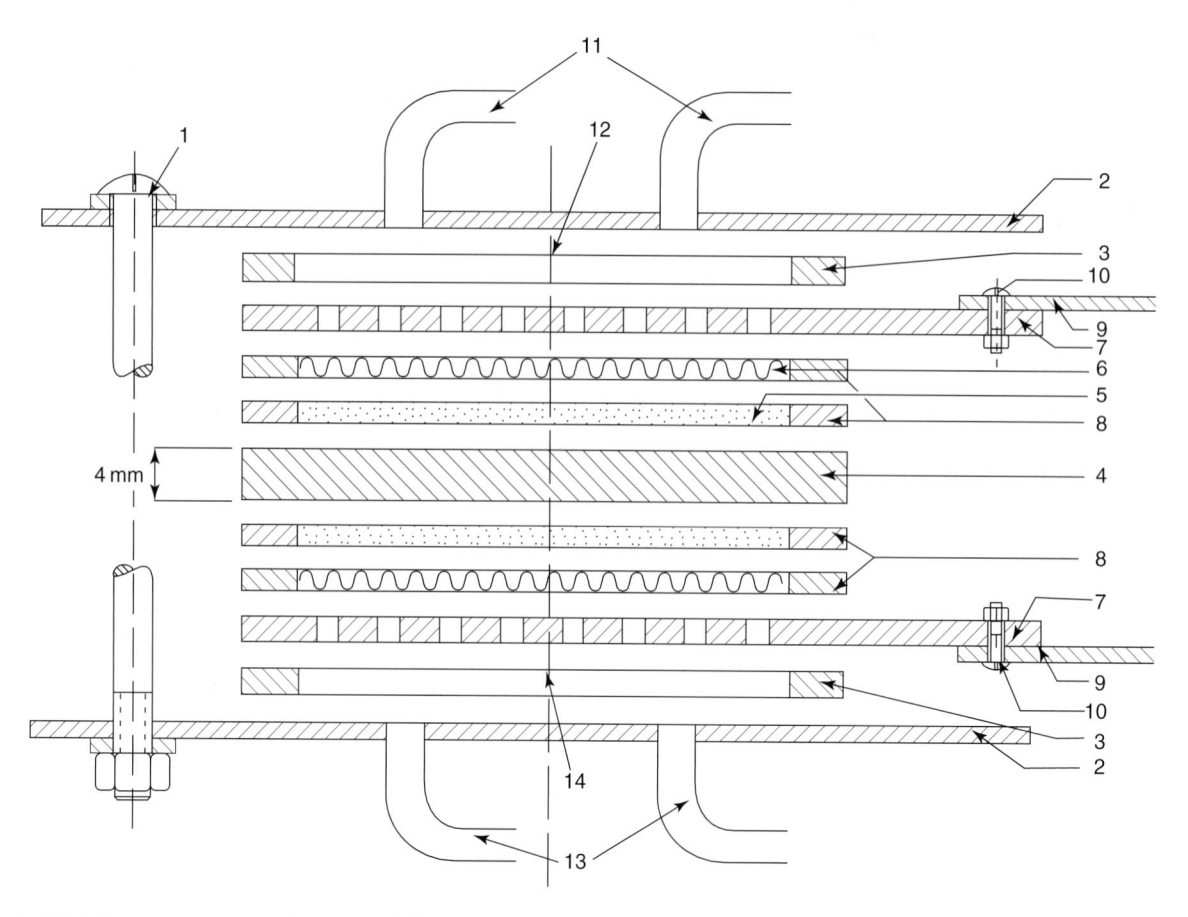

Figure 2. MCFC stacks in the early stages of development with an electrolyte matrix to hold the electrolyte. 1, screw with mica insulation rings; 2, steel covers; 3, asbestos; 4, impregnated disk; 5, thin layers of metal powders; 6, silver wire gage; 7, firm perforated stainless disk; 8, gaskets of mica; 9, terminal wire; 10, screw; 11, pipes for circulating gases; 12, central line; 13, pipes for circulating gases; 14, central line. (Reproduced from Broers and Ketelaar (1960),[37] copyright The American Chemical Society.)

1967 with the United Technology Corporation (later IFC with the Toshiba Corporation) as the main contractor. From 1975, the USA Department of Energy and Electric Power Research Institute (EPRI) started to support the MCFC program.

Problems appearing in this early stage are fundamental. For example, a severe problem associated with shorting occurred due to the Ag (or CuO) dissolution from the cathode. Dissolved Ag migrate to the anode side and was reduced by dissolved hydrogen; as a result, dendrites are formed and in turn grew towards the cathode side, leading to shorting. This was most severe when a free electrolyte was used without an electrolyte matrix because the convection promoted the dendrite formation. Whenever a paste electrolyte or an electrolyte matrix was used, this tendency weakened. Later, lithiated NiO was adopted as the cathode instead of Ag or Cu because this effect was small. As the electrolyte, the Li_2CO_3/K_2CO_3 system and the $Li_2CO_3/Na_2CO_3/K_2CO_3$ systems were widely used but many stack developments started with use of the Li_2CO_3/K_2CO_3 system.[5]

When a paste electrolyte or an electrolyte matrix was used, the molten carbonates are confined inside electrolyte region due to capillary forces. Even when an electrolyte matrix was used, the alkali carbonate escaped out of the matrix region due to the creepage of electrolyte. For small test cells, frequent additions of the electrolyte were needed to continue the cell operation. Electrode performance was not good in the early stage. This was due to flooding of the electrolyte caused by the large surface tension. To overcome this issue, control of pore size[38] in the electrolyte and electrodes was essential, as shown in Figure 3. To confine the electrolytes inside the matrix, the pore size in the electrolyte should be fine. Low performance was also as a result of sintering of the electrode. Sealing was another big problem because gas leakage took place from gaskets. Fine cracks in the matrix led to failure. In some cases, fire occurred when hydrogen penetrates through the cracks to the cathode.

Improvements in the problems appearing at this stage had been intensively made in 1980 by Energy Research

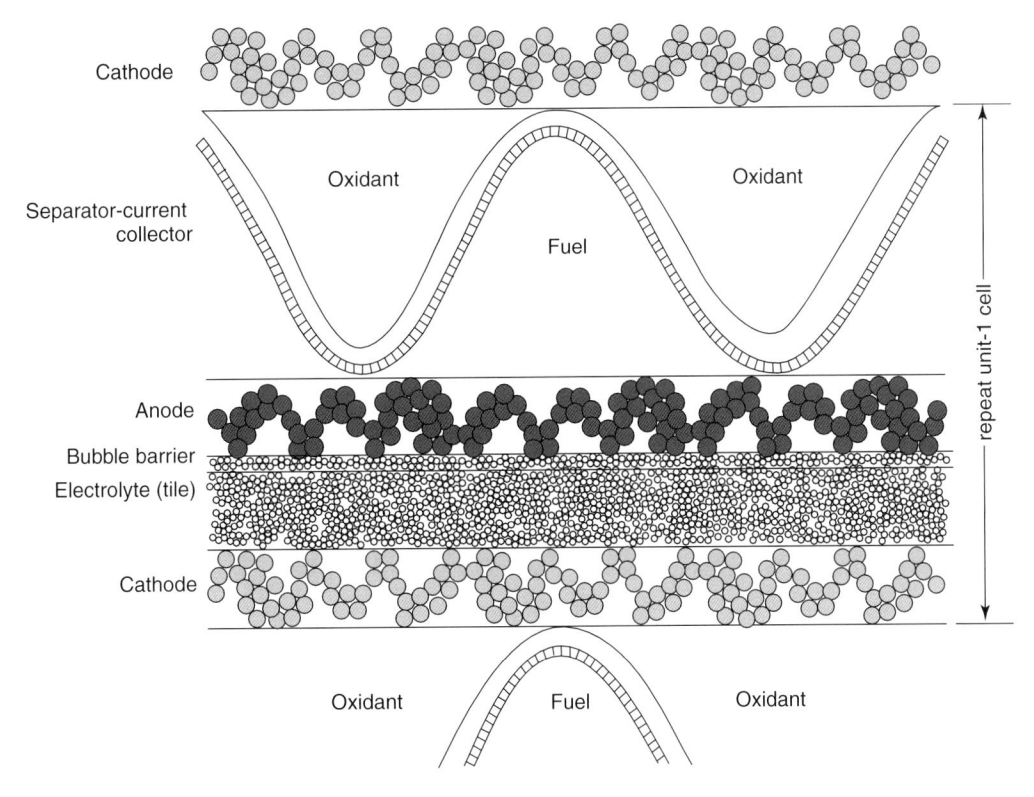

Figure 3. Schematic diagram illustrating the MCFC stacks consisting of electrolyte matrix with fine pores, porous cathode and anode and bipolar plates.[13] The electrolyte is held with the capillary forces in the matrix. (Reproduced by permission of the Risø National Laboratory.)

Corporation (ERC (later Fuel Cell Energy Corporation)), General Electric Corporation (GE), IGT, IFC and other stack developers as follows:

1. Thermal cycling characteristics: It was essential to strengthen the electrolyte matrix materials. Attempts to add spherical ceramic particles or fibers were successful in strengthening the matrix materials.[12] Improvements in the matrix materials made it possible to fabricate larger cells with larger active areas.
2. Gas crossover: Basically, gas crossover was prevented by controlling the pore size distribution in the matrix. In addition, a bubble barrier with fine pores was used as illustrated in Figure 3.[13]
3. Sinterability and creep of Ni anode: The anode exhibited deformation due to creep under compression.[14] To inhibit creep and sintering, Cr or Al was added to the Ni anode. Added metallic elements react with the Li_2CO_3 component to form $LiCrO_2$ and $LiAlO_2$. These oxides were stable in the molten carbonate and inhibited Ni sintering.
4. Corrosion: Stainless steels 316 (or 310) were used for current collector and metal interconnect. Corrosion occurred in both the cathode and anode compartments. Corrosion was more severe in the anode, and Ni clad was found to be effective.[13]
5. Wet-seal: Gas seal was the most important part in stacks. In many MCFCs, the wet-seal approach was adopted to obtain gas-tightness by sandwiching the electrolyte matrix with bipolar plates. Severe corrosion occurred due to the formation of local cells between the anode and air outside the stacks.[39, 40] Coating of aluminum on the wet-seal parts provided a nonconductive film made of $Al_2O_3/LiAlO_2$ to present the local corrosion.
6. Electrolyte redistribution:[41] It had been found that significant change in electrolyte composition occurred inside the stacks. Two related phenomena were involved. One was the migration beyond many cells in tall stacks, the other was the redistribution between the cathode and the anode sides inside one cell. This kind of migration can take place from various kinds of driving forces. It was, however, clarified that this occurred due to the difference in the cation diffusion in the molten carbonates.
7. NiO cathode dissolution: In 1981, it was found that even for the NiO cathode, shorting occurred due to the dissolution, migration and recrystallization in matrix materials.[42] Efforts to overcome this major issue

are still ongoing. The NiO solubility is about 20×10^{-6} mole fraction at $P(CO_2) = 1$ atm at 923 K.[43] Many attempts are still being made to overcome this issue. Attempts have been made to develop an alternative cathode having a lower solubility; this is not succeeded because stable compounds do not show a good performance. Other attempts have been made to lower the NiO solubility by changing the electrolyte composition.

8. Electrolyte loss causes severe degradation in performance because this seriously affects the electrode/electrolyte/gas interface. This is a result of the following phenomena: cathode lithiation (Ni is installed, oxidized and lithiated before operation); anode lithiation (Cr or Al reacts with the Li component); hardware corrosion ($LiFeO_2$, $LiCrO_2$, K_2CrO_4 are formed); vaporization (the K component reacts with water vapor to form $KOH_{(g)}$); electrolyte migration. Electrolyte loss still remains as one of the major issues to determine the stack life.

As given in Figure 1, the development of MCFCs started a new era from the late 1980s; that is, scale up of stacks and establishment of stack and system engineering. This direction was first adopted in Japan. Investigations in MCFCs in Japan started from the early 1980s on fundamental cell and stack technology and moved to test on 10 kW stacks. In the late 1980s, a strategic decision was made to start a scale-up program.

In the USA, stack and system engineering and scale up were also initiated in a similar manner. The major developers were the Energy Research Corporation and MC Power: The Energy Research Corporation adopted an external-manifolded and internal-reforming system. From the early stages, they proposed the internal reforming scheme to improve efficiency. Later, direct internal reforming (DIR) and also indirect internal reforming were tested. A 250 kW test facility was built in Danbury (Connecticut, USA). The first 250 kW test was made for 2500 h. They scaled up 10-ft^3 area stacks and tested 20 kW tall stacks. At Santa Clara, they constructed a 2 MW system consisting of 16 sets of 125 kW stacks and operated for 500 h. The efficiency of HHV 40% (LHV 44%) was obtained for 75% fuel utilization. MC-Power was formed in 1987 supported by IGT, IHI and Bechtecl to scale up and commercialize MCFC stacks based on the cell technology and the internally manifolded heat exchanger stack concept developed by IGT. In 1993, they constructed 20 kW stacks and 250 kW stacks at UNOCL in 1994–95. In 1997, a 250 kW system was tested at Miramar Naval Air Station and 1 MW was tested in 1998; this is an internally manifolded, externally reforming system. Although they built a plant having the ability to manufacture 4 MW per year, they stopped activities in 2000.

In Japan, the MCFC Research association was established in 1987 to develop 100 kW in the first half and 1000 kW at the end of the second half of a 10 year program. The following are MCFC stack developers involved in the Japanese MCFC project; Hitachi, Toshiba, Mitsubishi Electric Company, Ishikawajima–Harima Heavy Industries Co., Ltd. (IHI) and Sanyo. In 1988–93, two 100 kW class external reforming pressurized MCFC stacks were built. From 1993, design, manufacture, construction and installation of 1000 kW were started. IHI and Hitachi constructed 250 kW stacks with internal manifolds and an external reforming system as parts of 1 MW test stacks at Kawagoe (Mie, Japan). The Mitsubishi Electric Company constructed an advanced internal reforming type 200 kW stacks in Amagasaki (Osaka, Japan).

In Europe, investigations on MCFCs have been made by the Dutch group, MTU Friedrichshafen, and Ansoldo. The Dutch MCFC program was initiated in 1986. The Netherlands Energy Research foundation (ECN) started by acquiring the basic technology from IGT. They constructed a 10 kW stack in early 1990. From 1996, the Dutch Fuel Cell Corporation made a consortium consisting of British Gas, Gaz de France, ECN and Sydkraft. They aimed at develop "second generation" (advanced DIR-) MCFCs, which should be cost effective and have simple, improved lifetime and reliability. This was called the SMARTER system (concept). Another European consortium comprising of the German company, MTU Fiedrichshafen, Ruhrgas AG and RWE Energie AG as well as the Danish companies Elkraft A.m.b.H. and Haldor Topsøe A/S started the MCFC program from 1992 with a partner of ERC. From a system analysis, they proposed the Hot Module system combining all components of a MCFC system operating at a similar temperature and pressure into a common thermally insulated vessel. From 1997, they started its first full size test plant in Dorsten. MOLCARE is an R&D program on MCFCs led by ANSALDO Ricerche from 1992. A 100 kW proof-of-concept finished in the middle of 1999.

These activities have made big contributions to the commercialization of MCFC systems. Even so, there remain several issues to be surmounted. These are:

1. Although MCFCs have the advantage of the high OCV value, the current density is limited to low values. This increases the facility cost.
2. Heat management is important to realize the high efficiency. The internal reforming process and the total heat management of a whole system including gas circulation is extremely necessary.
3. High-pressure operation is also required to achieve high efficiency. For this purpose, the most important issue is

to reduce the NiO dissolution. Many efforts have been made to clarify the mechanism of shorting caused by the NiO dissolution. There should be still a need to adopt a systematic approach to optimizing the MCFC materials for high-pressure operation.

1.3 SOFCs

Zirconia based solid oxide conductors were found by Nernst in his investigation on "glower" in 1899. Practical attempts of SOFCs were made by Baur and Pries in 1937 after they discarded molten salts in their search for an appropriate electrolyte.[3] They investigated mainly on ZrO_2-based ceramics. By using a zirconia electrolyte, oxygen can be transported from one side to the other in the form of the oxide ion. The oxide ion can react with hydrogen to form water vapor.

$$0.5O_{2(g)} + 2e^- \longrightarrow O^{2-} \quad (4)$$

$$O^{2-} + H_{2(g)} \longrightarrow H_2O_{(g)} + 2e^- \quad (5)$$

A total reaction is given as follows:

$$H_{2(g)} + 0.5O_{2(g)} \longrightarrow H_2O_{(g)} \quad (6)$$

Although the total reaction is the same as that in a MCFC, the electrochemical reaction is simpler in a SOFC and there is no need to circulate carbon dioxide. Even so, solid oxide electrolytes investigated showed a relatively low ionic conductivity. This lead to the necessity of testing SOFCs around 1000 °C.

Construction of SOFC stacks was initiated with a planar design similarly to other types of fuel cells. Immediately after some attempts, efforts were focused on a tubular design. Initially, WHPC tested two segmented-cell-in-series types:[44–46] one is the "Bell and Spigot" type, the other being the "banned" type. The former "Bell and Spigot" type was also adopted by Dornier for a water electrolyzer.[9] The banned type was also adopted by the Electrotechnical Laboratory and later by Mitsubishi Heavy Industry Corporation (Nagasaki, Japan) (see Figure 4).

In the beginning, fabrication of thin electrolyte and low contact resistance between the electrolyte and electrodes was the most important concern. Even in the segment cell types, there remained the big technological issues about how to connect cell to cell without inducing thermal stress, chemical degradation, etc. This was closely related with design. Despite the difficulty in stack fabrication, system cost/benefit considerations made for power generation and industrial cogeneration showed promising results.

Although the zirconia-based electrolyte was used from the initial stage, there was a limited allowance for selecting electrode materials and interconnect materials. In 1960, Pt electrodes were used for the cathode and anode. Interconnect was also made of platinum. Later, the cathode changed to Pr-ZrO_2/SnO/In_2O_3, the anode being Ni/yttria stabilized zirconia (YSZ). As interconnects, Mn-doped cobalt chromite was tested. Even after such improvements in the electrode and interconnect, results in performance were not satisfactory. This was one of the reasons why SOFCs were regarded as being unattractive compared with MCFCs. The history of R&D on SOFCs in the early stage was well summarized by H.-H. Möbius.[81]

In early 1980, Isenberg[8] in WHPC made big breakthroughs to overcome the issues of thermal stress and contact resistance by adopting a sealless tubular design (Figure 5) to be fabricated with the electrochemical vapor deposition (EVD) technique (for details, see below and Figure 22). The cell components are listed in Table 1.[16] The fabrication procedure was started with Ca-stabilized zirconia as the supporting tube. A 500–1000 μm thick porous film of lanthanum strontium manganite was first coated on this supporting tube. Dense films of 50 μm thick YSZ and of 30 μm lanthanum chromium magnesium oxide as interconnect were deposited by the EVD technique. This technique provides a clever way of fabricating dense films onto porous electrolyte. Subsequently, Ni or Co was slurry coated on the tubes and finally fixed with YSZ by the EVD technique. Tests on cells and cell-bundle generator modules showed excellent results. The development of the stacks was supported by the USA's Department of Energy, the Gas Research Institute (GRI) and also by Japanese consortiums composing of electricity utilities, gas companies and the New Energy Development and Industrial Technology Organization (NEDO).

Since the WHPC tubular stacks have many advantages, a module test was successfully scaled-up in 1984 to a 400 W stack test (24 cells, 2000 h) by the Tennessee Valley Authority and 3 kW stack tests by Tokyo gas and Osaka gas in 1987 and GRI in 1988. In addition, the long-term test was made on tubes fabricated in the late 1980s. Results after 70 000 h showed an extremely low degradation as shown in Figure 6.[17] The demonstration of the tubular SOFC system was made by step-wisely increasing size as follows: tests on 20 kW were made in 1989–91 in Japan. A 100 kW stack was tested by the Dutch and Danish consortium EDB/ELSAM from January 1998 as an atmospheric SOFC combined heat and power (CHP) system in Westervoort (The Netherlands) for more than 14 000 h. The 53.6% direct current (d.c.) LHV efficiency or the 45.9% alternating current (a.c.) LHV efficiency was achieved and excellent load following capability was demonstrated. The

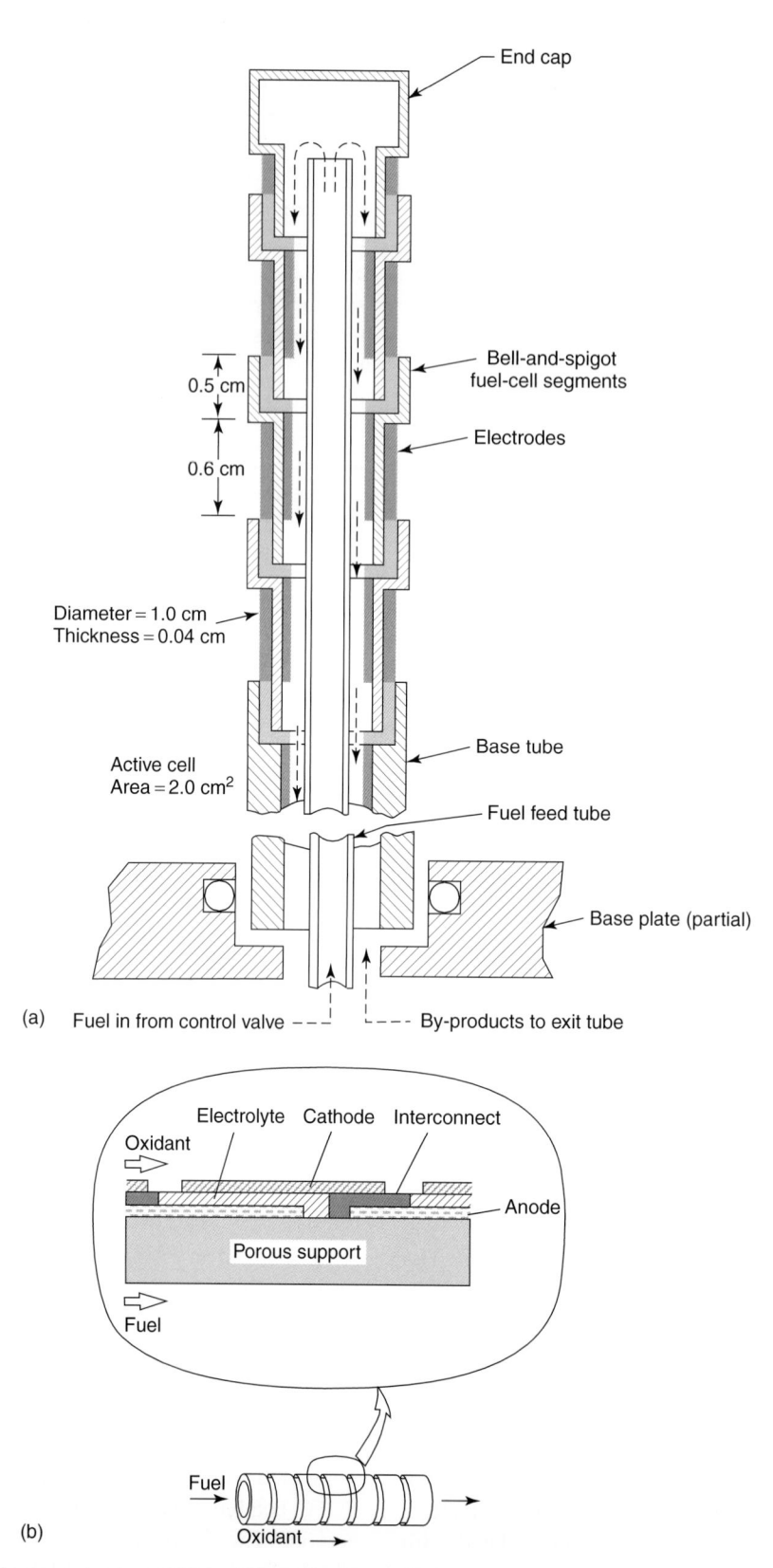

Figure 4. Segmented-cell-in-series type SOFC stacks in the early stage. (a) "Bell and Spigot" type. (b) "Banned" type. (Reproduced from Minh and Takahashi (1995)[7] by permission of Elsevier Science.)

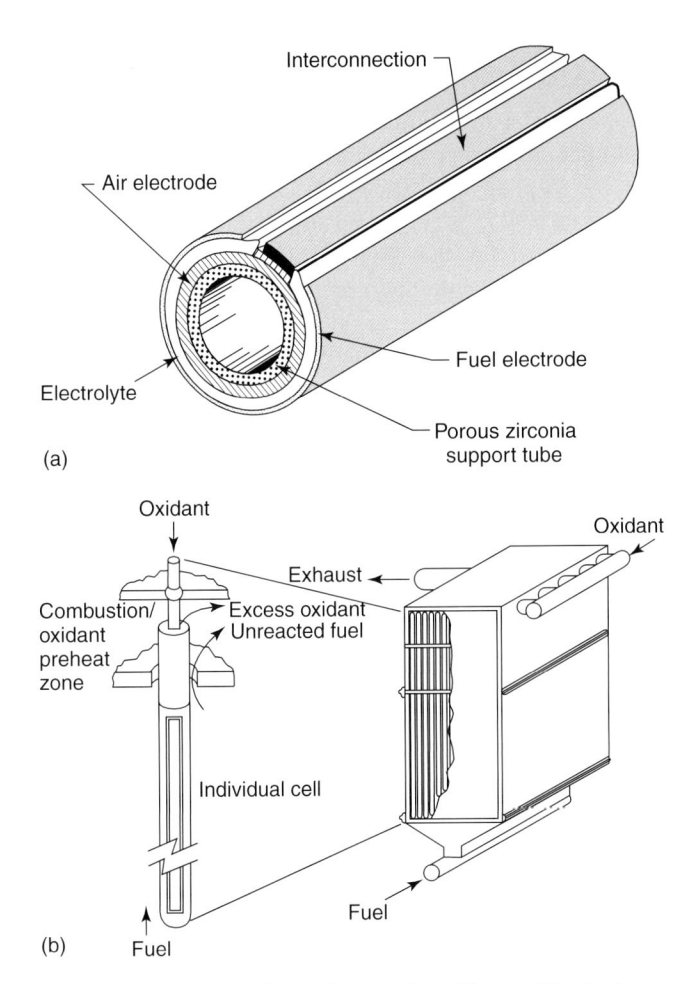

Figure 5. Sealless tubular design developed by the Westinghouse Power Corporation. (a) Tube with stripe interconnect and (b) tubular design with gas manifold. Details of cell components and fabrication processes are listed in Table 1. (Reproduced from Minh and Takahashi (1995)[7] by permission of Elsevier Science.)

Table 1. Cell components, materials and fabrication processes of the early sealless tubular SOFCs developed by WHPC in the 1980s.[a]

Component	Materials	Thickness	Fabrication process
Support tube	(Zr, Ca) O_2	1.2 mm	Extrusion-sintering
Air electrode	(La, Sr) MnO_3	1.4 mm	Slurry coat-sintering
Electrolyte	(Zr, Y) O_2	40 μm	EVD
Interconnection	La (Cr, Mg) O_3	40 μm	EVD
Fuel electrode	Ni/(Zr, Y) O_2	100 μm	Slurry coat/EVD

[a]The corresponding electrical conductivities and thermal expansions are shown in Figure 31.

Southern California Edison tested a 220 kW pressurized SOFC/micro-gas turbine generator (MTG) power system as the world's first hybrid cycle system in 2000. As of autumn 2001, the following plans have been announced for a 220 kW SOFC/CHP atmospheric pressure system in Canada, a 300 kW PSOFC/MTG system in Europe, two 1000 kW PSOFC/MTG systems with the Environmental Protection Agency's Environmental Science Center and with European commission.

In parallel to the development and demonstration of the SOFC system, improvements have been made in the design, fabrication and materials by WHPC. These are adoption of air electrode support tubes, modification of the cathode/electrolyte interface to improve low temperature performance, modification in geometrical design and

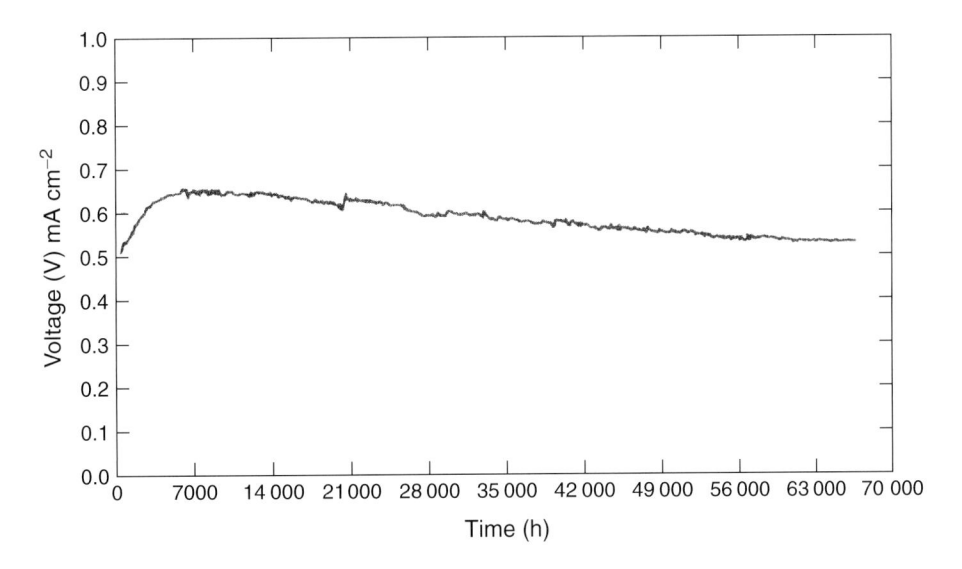

Figure 6. Long-term performance of the sealless tubular cells fabricated and tested by the Westinghouse Power Corporation. (Reproduced from Singhal (1997)[17] by permission of The Electrochemical Society Inc.)

fabrication process change from the EVD process to the plasma spray technique.[16, 17]

The success of the tubular SOFC by WHPC in the early 1980s formed a strong impact on other groups involved in high temperature electrochemical energy conversion. Despite the success by WHPC, there remained room for other groups to believe that competition can be made in developing other types of SOFCs. The first of such groups was the Argonne National Laboratory, who recognized well the weak points of the tubular cells; that is, high cost and low power density.[47] To overcome these disadvantages, they proposed the concept of monolithic cells in the mid-1980s.[48] Their stack design is shown in Figure 7. This contained the following new ideas in stack developments:

(i) The monolithic "honeycomb-type" structure is adopted as the stack design. This shortens the electrical path per unit cell.
(ii) To fabricate SOFCs, the ceramic technology that has been well developed in the electroceramic fields will be utilized. Significant cost reduction in fabrication can be expected against the EVD process.
(iii) To fabricate stacks in a smaller number of steps, the cell components are sintered simultaneously.
(iv) This stack concept makes it possible to apply SOFCs to various fields, such as residential use or transportation use.

The idea and technology of the monolithic cells were transferred to Allied Signals (later, Honeywell).

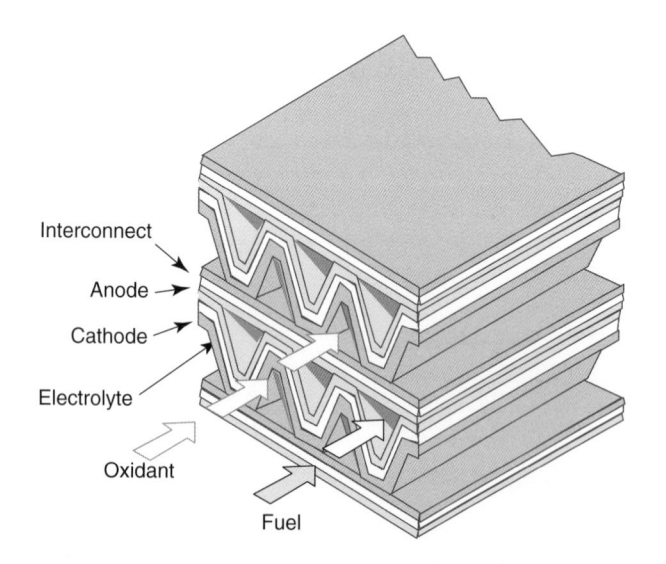

Figure 7. Monolithic cells based on a "honeycomb type" design proposed by the Argonne National Laboratory. "Monolithic" implies that the all cell components are united into one structure and that cells should be fabricated by one simultaneous sintering step. (Reproduced from Minh and Takahashi (1995)[7] by permission of Elsevier Science.)

The establishment of the sealless tubular design and the proposal of the monolithic cells stimulated activities in Europe and Japan. Dörnier already started the work on the high operating temperature electrolysis (HOT ELLY) process by adopting the segmented-cell-in-series type tubular cells from 1975.[9] They also used similar cell materials to those used for the WHPC's fuel cells, although the dopants of manganite and chromite are different between the two groups. That is, a Ni/YSZ cermet anode, doped lanthanum manganite cathode, yttria (>10%) doped zirconia electrolyte and doped lanthanum chromite ring. From long term operation, they had already found at that time that to have no degradation of YSZ, more than 10% doping is needed. For fuel cells, they adopted a planar design by using essentially the same materials for electrolyzers. Ceramatec (later SOFCo was formed in 1994) in the USA started the work on planer cells in 1986. After the fundamental investigation on materials, fabrication technology and design, a 1.4 kW stack was successfully operated with 85% fuel utilization.

In Japan, on the basis of the fundamental work carried out on the wet/sintering method in the national laboratories, the NEDO project started in 1989, the first target being a small 100-W size. The Murata manufacturer focused on the co-firing of cell components of planar cells. TOTO started to fabricate tubular cells by the wet/sintering process. Mitsubishi Heavy Industry worked on planar cells as well as tubular cells. In the early 1990s, progress had been made on materials chemistry associated with the adoption of the wet/sintering method; particularly, in the sintering[49] of a lanthanum chromite based interconnect, and the reactions[50] between the perovskite oxide cathode and YSZ were clarified physico–chemically. These provided the basis of optimizing the composition of the SOFC materials in fabrication of the respective tubular/planar cells. Actually, TOTO succeeded in fabricating sealless tubular cells by the wet process and obtaining a reasonably good performance.[51] Recently, Mitsubishi Heavy Industry (Kobe, Japan) succeeded in constructing the mono-block layer build (MOLB) type of SOFC stacks.[52] In one sense, this can be regarded as the monolithic cell fabricated by several sintering steps instead of one. These achievements in the development of planar design were largely due to the maturing of the oxide interconnect technology. In other words, without good interconnect technology, no planar stacks can be constructed. In this sense, the interconnect technology is a key in SOFCs technology.

Use of a metal interconnect is another challenge in fabrication of SOFC stacks. Those constructing stacks with a metal interconnect were Z-tek in the USA, Siemens (Erlangen, Germany) and the Julich Research Center in Germany, Sulzer Innotec in Switzerland and Sanyo and Fuji Electric in Japan. Siemens started from 1990 by developing, with Plansee, a Cr-base alloy, $Cr_5FeY_2O_3$, having the same thermal expansion coefficient as YSZ.[53] They tested a 1 kW stack in 1994 and a 10 kW stack in 1995. This was stopped shortly after Siemens bought WHPC to form Siemens Westinghouse Power Corporation (SWPC). Ceramic Fuel Cells Limited in Australia was formed in 1992 and carried out studies on planar stacks. Recently, they scaled up to 25 kW; natural gas (80–85 utilization) + air (50%) at 850 °C. After operating for 400 h, the stacks failed from excessive leaks due to various aspects of the stack design including manifold. Sulzer-Innotec (and Sulzer Hexis from 1996) in Switzerland developed planar stacks for commercial building and smaller industrial plants from 1988. Field tests were made in Switzerland, Germany, Japan and Spain in 1999–2001. Global Thermoelectric started planar stacks based on an anode-support cell with a metal interconnect in 1997. They provided one 1 kW stack to Delphi who started work on applications of SOFCs to the automobile field. They also constructed a 2 kW residential CHP system and obtained an electrical conversion efficiency of 36% d.c. and 32% a.c. LHV in 2001.

Use of metal interconnect gave rise to many arguments about the applicability of SOFCs as follows:

(i) Lowering operational temperature and its relation to conversion efficiency;
(ii) cost reduction in materials and fabrication;
(ii) durability against frequent thermal cycles;
(iii) applicability to automotive use.

Summarizing the progress of the SOFC stack development, the following can be pointed out:

1. The sealless tubular cells and the planar cells with oxide interconnects can be called the first generation SOFCs. Such stacks are composed of thermodynamically stable materials in the fuel cell environments and the other cell components, so therefore, essentially, there are no problems with materials. In other words, the long-term stability is well established. This achieved efficiency confirmed the earlier estimate. The benefit of low activation polarization loss and the high temperature heat availability compliment the lower theoretical efficiency and the larger concentration polarization loss. Further enhancement of efficiency can be expected by improving heat management associated with gas circulation and combination with other engines.
2. Stacks with metal interconnects can be categorized as the second generation SOFCs. Since thermodynamically nonequilibrium materials, namely metal interconnects, are used, technological issues becomes more complicated. For example, to improve the durability of alloys, a coating can be made; however, this in turn increases the fabrication cost and reduces the advantages of low material costs for metal interconnects. Optimization should be, therefore, be made as a whole among the materials selection, fabrication and stack design. In view of this, the apparent features of stack development of the second generation SOFCs are similar to those of MCFCs, although the corrosion rate is much more severe in MCFCs. Lowering the temperature from 1000 °C to 600–700 °C also makes the second generation SOFCs similar to MCFCs from the materials and the chemical point of view.

2 MCFCs AND SOFCs: THEIR SIMILARITY FROM A HIGH TEMPERATURE, MATERIALS SCIENCE VIEWPOINT

MCFCs and SOFCs have many common features as high temperature fuel cells when compared with the low temperature fuel cells, PAFC and PEFC. The main common features of the high temperature fuel cells are the following:

1. SOFCs and MCFCs both use electrolytes that transport the charged species associated with the oxidant, namely, the oxide ion (O^{2-}) or carbonate ion (CO_3^{2-}). This provides a wide fuel flexibility. For the low temperature fuel cells, hydrogen is the only fuel that can be transported through the electrolyte to the cathode side where the electrochemical oxidation takes place to produce water. When use is made of other fuels, such as natural gases, the reforming into CO and H_2 has to be done around 800 °C endothermically, and those heats required are supplied from combustion of some parts of fuels. This makes the fuel-to-electricity conversion efficiency low. On the other hand, natural gas, coal-related gases and other hydrocarbons have been tested from the start of investigations into high temperature fuel cells. In principle, if an appropriate anode would be available, any fuels can be used in high temperature fuel cells. Practically, however, the Ni anodes that are used for both MCFCs and SOFCs are not tolerable against carbon deposition in most cases; this makes it necessary to reform fuels with water or CO_2 into $H_2 + CO$.
2. Since high temperature electrochemical reactions are based on rather fast electrode reactions, there is no need to use precious metals for the electrodes. This makes high temperature fuel cells much more attractive from the materials-cost point of view.

3. The electrode overpotentials contribute to only a small part to the performance loss. For MCFCs and the intermediate temperature SOFCs, cathode overpotentials tend to become larger with decreasing temperature. Even so, the magnitude of such overpotentials is much smaller than that of PEFCs.
4. When the heat required for reforming can be provided from the heat generated in an exothermic electrochemical reaction inside the stacks, the total conversion efficiency becomes quite high. At this point, there are some differences between MCFCs and SOFCs because typical reforming is made between MCFC operation temperature (650 °C) and SOFC operation temperature (1000 °C).
5. Tolerance of electrodes against impurities is high for high temperature fuel cells. Particularly, CO can be used as a fuel, although CO is poisonous for platinum anodes in the low temperature fuel cells.

The following disadvantages originate from high temperature operation:

1. With increasing temperature, it becomes difficult to make the materials properties able to meet the requirements for high temperature operation because of the general trend of increasing reactivity among materials with increasing temperature. Precious metals such as platinum, gold or silver are often used because of their chemical inertness or chemical stability, not because of their catalytic activity. This is contradictory to the second advantage given above.
2. Working life is limited by the wearing out of materials. This is particularly true when metallic materials are used in corrosive environments because they are not thermodynamically stable in the fuel cell's environment.
3. Since high temperature fuel cells use ceramics in their stacks, the thermal stress originated from mismatch in thermal expansion should be avoided in fabrication and cell operation. This makes it necessary to have good performance against thermal cycles.

The technological issues are, therefore, how to overcome these expected disadvantages without losing the many excellent advantages.

1. To overcome the first disadvantage, it is highly necessary and recommended to properly use the equilibrium thermodynamics. Actually, advances in materials thermodynamics makes it possible to analyze the thermodynamic stability of cell components in the fuel cell environment and also the materials' compatibility. Particularly, the complicated chemical equilibrium calculations and the construction of various kinds of stability diagrams (or chemical potential diagrams) have provided useful information about the reactivity and stability of materials in MCFCs[54] and SOFCs.[49, 50, 55, 56] By using these facilities, it is not difficult to find out the proper materials that can be stable in fuel cell environments. It should be also noted that due to high reactivity at high temperatures it is quick to find out wrong materials and, therefore, makes stability issues clear. This leads to good interaction between theoretical predictions and experimental examinations.
2. To ensure the long life of the materials and stacks, it is necessary to make appropriate materials selection to avoid reactions or retard reaction rates. One important technology is therefore the coating of well-designed materials on alloys. Another possible way of controlling reactions is management of mass transfer. Heterogeneous reactions do not proceed unless reactants and/or products are transferred to/from reaction sites. In view of this, the flows of gases or liquids, if any, should be carefully designed. Diffusion in substances is also important. This is driven by concentration gradients, chemical potential gradient and gradients of other thermodynamic variables. Such phenomena can be described within the framework of the irreversible thermodynamics with the aid of associated data, such as solubility, diffusion coefficients and so on.
3. To control the thermal expansion or the volume change as a function of the oxygen potential, it is necessary to dope other compounds in a substantial amount. Usually, volume-related properties do not sensitively depend on an impurity-level amount. This suggests the necessity of controlling the volume by adopting systematic approaches based on the key physico–chemical understanding of the materials.

As a whole, the materials' thermodynamics play the crucial roles in stack development.

3 EQUILIBRIA, STABILITY AND MASS TRANSFER ASSOCIATED WITH HIGH TEMPERATURE FUEL CELL MATERIALS, MATERIALS PROCESSING AND FUEL PROCESSING

To understand the historical aspects of materials development for high temperature fuel cells, it is essential to know the thermodynamic features behind the respective attempted efforts. Since no description on material thermodynamics is given in this volume of the "Fuel Cell Handbook", the stability of fuel cell materials and related fundamental properties are summarized here by adopting MCFC and SOFC materials as examples.

3.1 Thermodynamic features for redox and acid/base relations

The electrochemical materials are optimized in their electrical and catalytic properties. Those properties are related to the particular valence state of the component ions or with defect properties regulated by valence control. Thus, the valence stability of metal oxides is the key to achieving high ionic or electronic conductivities together with chemical stability in the fuel cell environments. Figure 8 shows the stable oxygen potential region for metal and metal oxides at 1273 K. In the anode atmosphere of SOFCs, Fe, Co, Ni and Cu are present as the metal. At 923 K of the MCFC operation temperature, about the same stability is realized.

The stability of double oxides or more complex oxides can be characterized in terms of the stabilization energy defined as follows;

$$AO_{m/2} + BO_{n/2} \longrightarrow ABO_{(m+n)/2} \quad (7)$$

$$\delta(ABO_{(m+n)/2}) = \Delta G_f^\circ(ABO_{(m+n)/2}) - \{\Delta G_f^\circ[AO_{m/2} + \Delta G_f^\circ(BO_{n/2})]\} \quad (8)$$

When double oxides with transition metal oxides are concerned, the large ions having higher valence numbers tend to have stronger stabilization energies. For example, perovskite oxides, AMO_3, have the largest stabilization when the ideal ionic configuration for the perovskite-type lattice is achieved (namely, the tolerance factor, t, derived from the ionic radii becomes unity as shown in Figure 9). This makes it possible to stabilize the unusual valence state (for example, Ni^{3+}, Co^{3+}, Fe^{4+}) in the perovskite lattice. In other words, the redox of transition metal ions is determined from the valence stability as well as the stabilization energy of double oxides. For the alkali metal transition metal double oxides, the stabilization energy exhibits significant dependence on alkali metal ion size. The lithium double oxides have small stabilization, whereas the potassium analogues have a larger stabilization (see Figure 10).

For solid solutions or mixed molten carbonates, the stability can be expressed in terms of the interaction parameters that are used to represent the nonideal part of the Gibbs energy:

$$G^{total} = G^{ideal} + G^{excess} \quad (9)$$

$$G^{ideal} = \sum_i x(M_i)\{G^\circ(M_i) + RT \ln x(M_i)\} \quad (10)$$

$$G^{excess} = \sum_i \sum_j x(M_i)x(M_j) \times \{A_{ij}^0 + A_{ij}^1(x(M_i) - x(M_j))\} \quad (11)$$

where $x(M_i)$ is the composition of the constituent, M_i, and A_{ij}^0, A_{ij}^1 are the first and second interaction parameters in the sub regular solution. Optimization of interaction

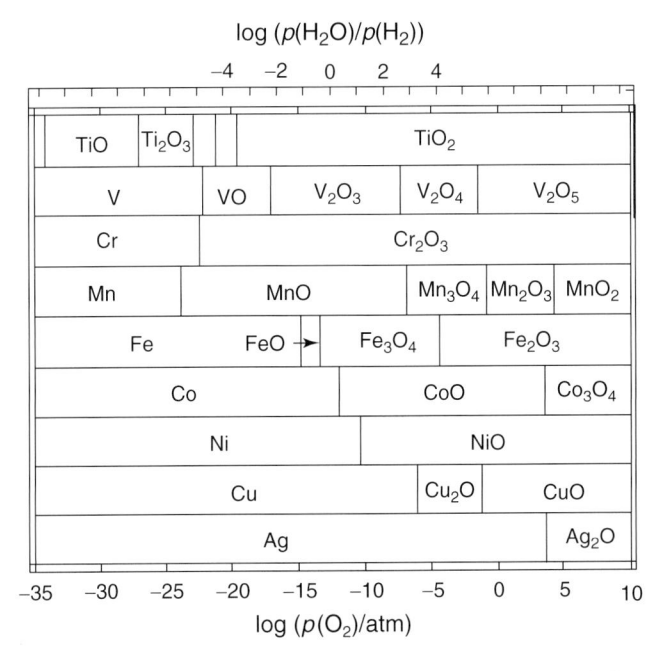

Figure 8. Valence stability of metal/metal oxides at 1273 K. Anodes are selected from those metallic elements in an anode atmosphere; Co, Ni, Cu, Ag are important. Interconnects should be one valence state over the oxygen potential range for fuel cells; Cr_2O_3 (TiO_2) is critical.

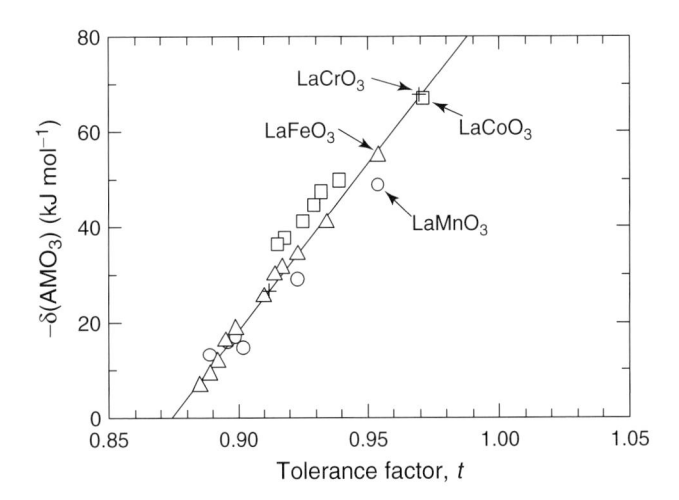

Figure 9. Stabilization energy of lanthanum transition metal perovskite oxides from the constituent oxides (La_2O_3 and transition metal oxides). The crystallographic matching is best when the tolerance factor (t) is unity.

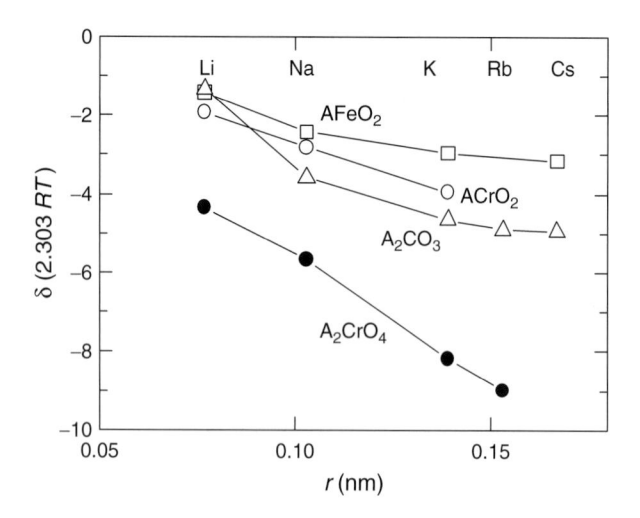

Figure 10. Stabilization energy of alkali metal compounds as a function of the ionic size of alkali ions. The alkali metal ions with the largest ionic size have the largest stabilization.[58]

parameters was made for molten carbonates[57] and for zirconia-based ceramics.[55, 56] (See Figure 11).

The stability of materials in various environments is determined from the relative magnitude of the thermodynamic properties. For example, the reaction of metal oxides with carbon dioxide to form metal carbonates can be discussed in terms of the differences in the enthalpy between oxides and carbonates:

$$M_2O_{(s)} + CO_{2(g)} \longrightarrow M_2CO_{3(s)} \qquad (12)$$

$$\Delta_r G°[\text{eq. (12)}] = \Delta_f G°(M_2CO_3) - [\Delta_f G°(M_2O)$$
$$+ \Delta_f G°(CO_2)] = \Delta_f H°(M_2CO_3)$$
$$- [\Delta_f H°(M_2O) + \Delta_f H°(CO_2)] - T\{\Delta_f S°(M_2CO_3)$$
$$- [\Delta_f S°(M_2O) + \Delta_f S°(CO_2)]\} \qquad (13)$$

In Figure 12, the enthalpy difference for the respective elements is plotted against the enthalpy change for oxide formation. Elements in the upper region tend to exist as the carbonate, the lower region being as the oxide. In the left-lower corner, elements will exist as metals. This diagram provides the fundamental nature of the elements in fuel cell environments. In MCFCs, the molten electrolyte consists of Li_2CO_3, Na_2CO_3 and K_2CO_3; these elements are located in the upper part. Those elements used in MCFC electrodes are found in the lower part. Similar diagrams for oxides vs. chlorides or hydroxides vs. carbonates can be set up to see the appropriate nature of elements in selecting or changing the component in particular materials. The relative stability between the hydroxide and carbonate is also important because MCFCs are operated in the presence of water vapor in both the anode and cathode. In SOFCs, the relative stability between the chloride and oxides becomes crucial in fabrication using the vapor deposition technique. Metal chloride vapors are used as starting materials and they are deposited as a result of oxidation with oxygen or water vapors. There are thermodynamic limits on the applicability of this technique to oxide film preparation. The same considerations can be made on the behavior of oxides against HCl impurities.

The reactivity among oxides can be well interpreted in terms of the acid/base concept. On the other hand, the reactivity can be also discussed in terms of the valence stability of metal oxides and the stabilization of complex oxides. For example, the interaction of NiO with alkali carbonate can be written as follows;

$$NiO_{(s)} \longrightarrow Ni^{2+}_{(in\ carbonate)} + O^{2-}_{(in\ carbonate)} \qquad (14)$$

$$NiO_{(s)} + CO_{2(s)} \longrightarrow Ni^{2+}_{(in\ carbonate)} + CO_3^{2-}\,_{(in\ carbonate)}$$
$$\longrightarrow NiCO_{3(in\ carbonate)} \qquad (15)$$

where the NiO dissolution is regarded as acidic because the oxide ion is formed on dissolution. In other words, the NiO dissolution can be regarded as a reaction with CO_2 to form dissolved nickel carbonate. This can explain the temperature dependence and the CO_2 partial pressure dependence of the solubility as shown in Figure 13.[43] This NiO dissolution can be also discussed from the electrolyte side. An alkali carbonate, A_2CO_3, decomposes into the alkali oxide and carbon dioxide.

$$A_2CO_3 \longrightarrow A_2O + CO_{2(g)} \qquad (16)$$

Furthermore, the alkali oxide can dissociate to form oxide ions.

$$A_2O \longrightarrow 2A^+ + O^{2-} \qquad (17)$$

Thus, the NiO solubility is related with the alkali carbonate through the CO_2 equilibrium partial pressure or the concentration of the oxide ions.

$$NiO + A_2CO_3 \longrightarrow NiCO_3 + Li_2O \qquad (18)$$

The NiO solubility is governed by the concentration of oxide ions, $[O^{2-}]$. Among A_2CO_3, the concentration of $[O^{2-}]$ is highest in Li_2CO_3 because the energy difference between the carbonate and oxide is the smallest as shown in Figure 12. The basicity of Li_2CO_3 is, thus, the highest, and as a result the NiO solubility is smallest in Li_2CO_3. Therefore, the NiO solubility decreases with increasing concentration of Li_2CO_3 as shown in Figure 14.[43] The acid/base considerations can be easily related with the

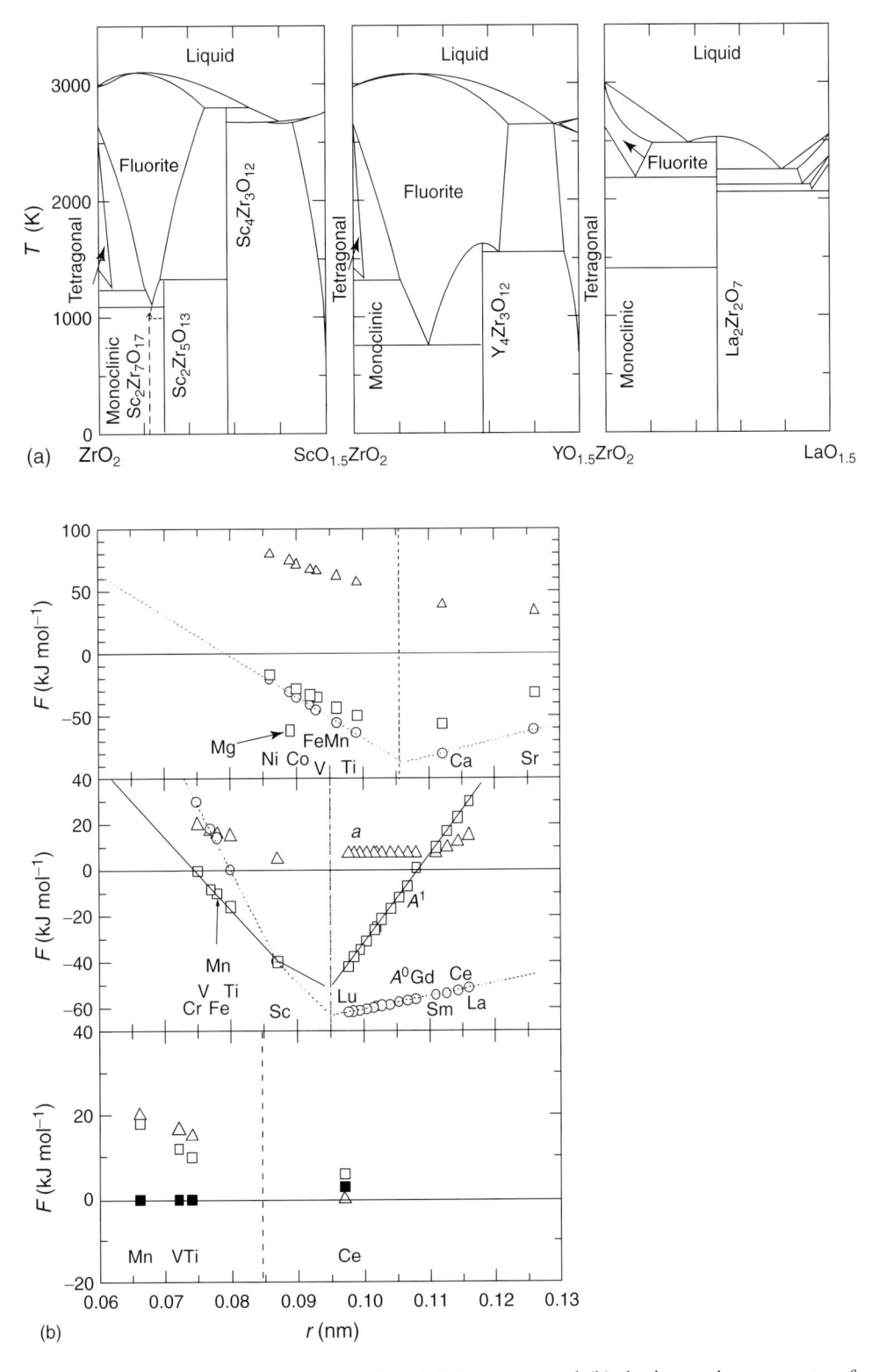

Figure 11. (a) Phase diagrams for the $ZrO_2/MO_{1.5}$ (M = Sc, Y, La) systems and (b) the interaction parameters for the cubic phase in the $ZrO_2/MO_{n/2}$ (n = 2, 3, 4) systems. Good regularity is obtained between the interaction parameters and the ionic size. By using those interaction parameters, the solubility of transition metal oxides in YSZ were evaluated (see Figure 18).[55, 56]

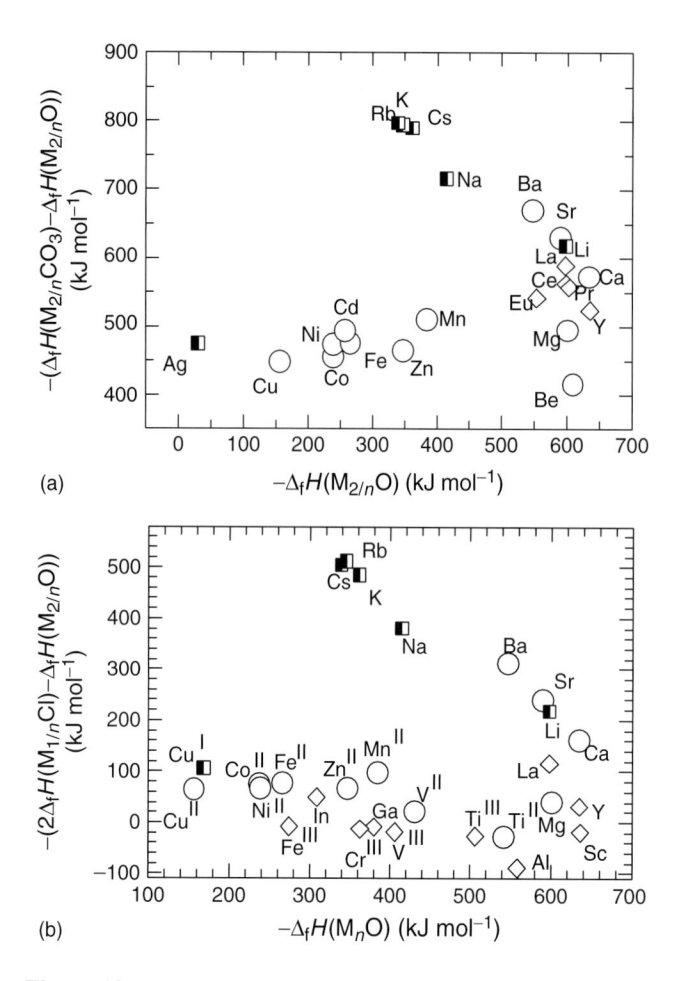

Figure 12. Relative stability among the various chemical forms for respective elements. Elements are shown with the enthalpy difference as a function of the enthalpy of the particular chemical form. (a) $[\Delta_f H(A_{2/n}CO_3) - \Delta_f H(A_{2/n}O)]$ vs. $\Delta_f H(A_{2/n}O)$ and (b) $2[\Delta_f H(A_{1/n}Cl) - \Delta_f H(A_{2/n}O)]$ vs. $\Delta_f H(A_{2/n}O)$. In this plot, the upper parts are the stable regions for metal carbonates (a) or chlorides (b), the lower parts being for the oxides. Data are taken from the thermodynamic database MALT2.[59]

considerations in terms of the stabilization energy. This provides the quantitative basis for the acid/base concept.

3.2 Thermodynamic nature of electrolytes

3.2.1 Molten carbonates

The major components of molten carbonates are Li_2CO_3, Na_2CO_3 and K_2CO_3 as shown in Figure 15. The thermodynamic stability of carbonates can be roughly evaluated by the stability of the respective constituent carbonates. Figure 12 shows that among Li_2CO_3, Na_2CO_3 and K_2CO_3, the Li_2CO_3 component has the strongest tendency to decompose into Li_2O and $CO_{2(g)}$. This is consistent with

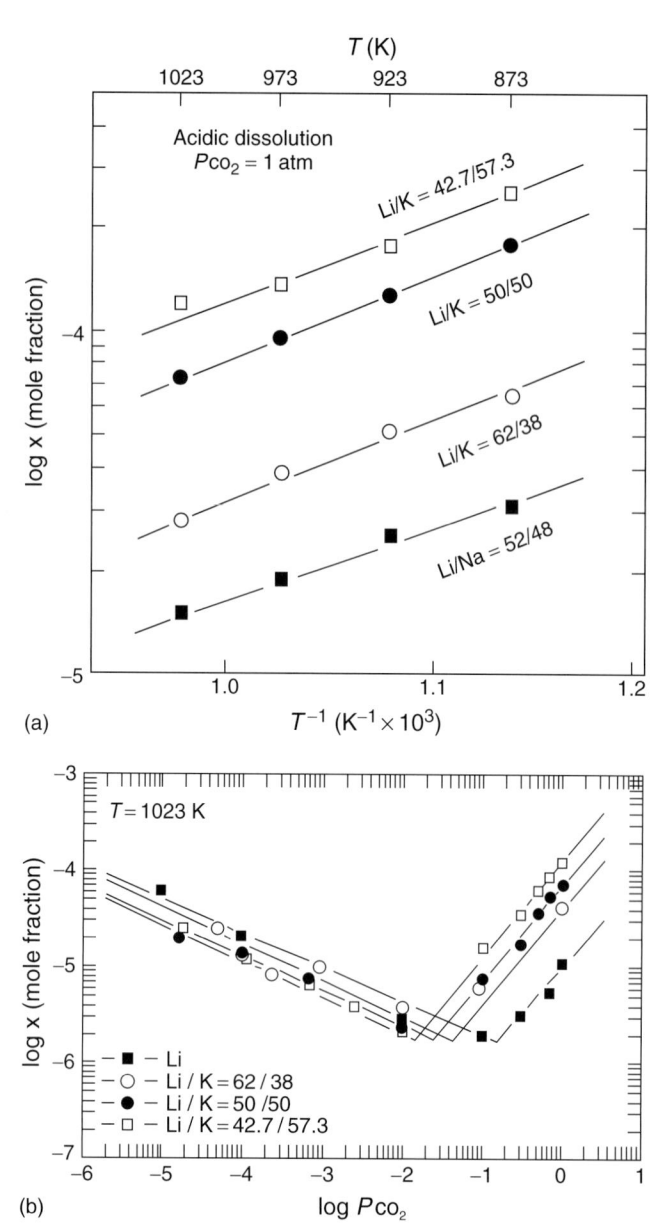

Figure 13. NiO solubility in the Li_2CO_3/K_2CO_3 mixture. (a) temperature dependence and (b) CO_2 partial pressure dependence. (Reproduced from Ota et al. (1995)[43] by permission of The Electrochemical Society Inc.)

the sequence of the basicity of the alkali carbonates determined from the observed solubility by Ota[43] as follows:

$$Li_2CO_3 > Na_2CO_3 > K_2CO_3 \qquad (19)$$

The reactivity of mixed carbonates with other oxides is related to these properties of the components. The Li_2CO_3 component dominates reactions with many compounds. For example, Al_2O_3, which was used as an electrolyte matrix material in the early stages, reacts with the Li_2CO_3

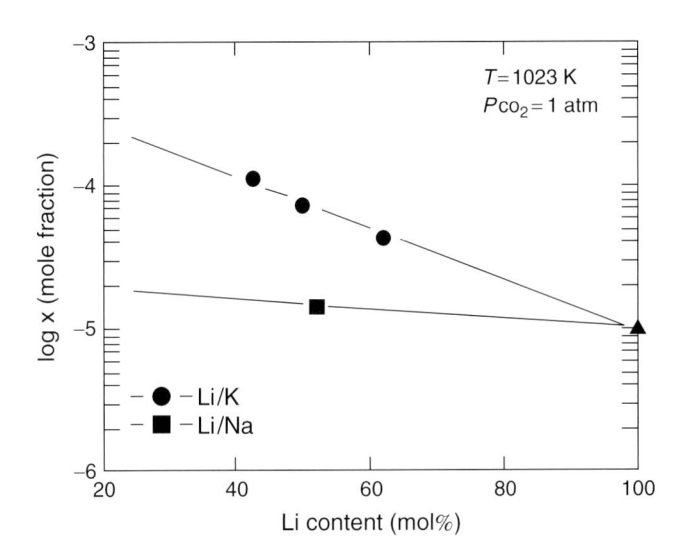

Figure 14. NiO solubility in molten carbonates as a function of the concentration of Li_2CO_3. (Reproduced from Ota et al. (1995)[43] by permission of The Electrochemical Society Inc.)

component to form $LiAlO_2$ to evolve carbon dioxide.

$$Li_2CO_3 + Al_2O_3 \longrightarrow 2LiAlO_2 + CO_{2(g)} \qquad (20)$$

Similarly, Fe_2O_3 and Cr_2O_3 react with the Li_2CO_3 component to form $LiFeO_2$ and $LiCrO_2$. For the case of Cr_2O_3, reactions with the K_2CO_3 component can take place dependant on the oxidative atmospheres.[58]

The stabilization energy of K_2CO_3 is much larger than that of Li_2CO_3. This implies that K_2O has the stronger reactivity with acidic substances. Thus, the reactivity of K_2O will be important for reactions with other acidic compounds; for example, the K_2CO_3 component reacts with Cr_2O_3 to form a chromate, K_2CrO_4.[58]

$$K_2CO_3 + 0.5Cr_2O_3 + 0.75O_{2(g)} \longrightarrow K_2CrO_{4(l)} + CO_{2(g)} \qquad (21)$$

In the acid/base theory, this can be regarded as basic dissolution of Cr_2O_3. Similarly, the K_2CO_3 reacts with H_2O to form the hydroxide gaseous species.

$$K_2CO_3 + H_2O_{(g)} \longrightarrow 2KOH_{(g)} + CO_{2(g)} \qquad (22)$$

Although the liquid KOH formation reaction is not favored compared with the liquid LiOH formation reaction, the KOH gas formation is favored. As a result, volatilization occurs mainly in the form of $KOH_{(g)}$. This causes the decrease in the electrolyte amount and also a change in the electrolyte composition. The volatilization decreases with increasing pressure. In a cathode atmosphere, there exists peroxide, O^- and O_2^-. These species have been discussed in the cathode reaction mechanism. These properties should be given from the thermodynamic data for A_2O_2 or AO_2.

The solubility of the transition metal oxides is determined by two factors: the carbonate formation energy vs. the formation energy of the double oxide with Li_2O or K_2O. For example, Ag, CuO and NiO do not react with Li_2O to form

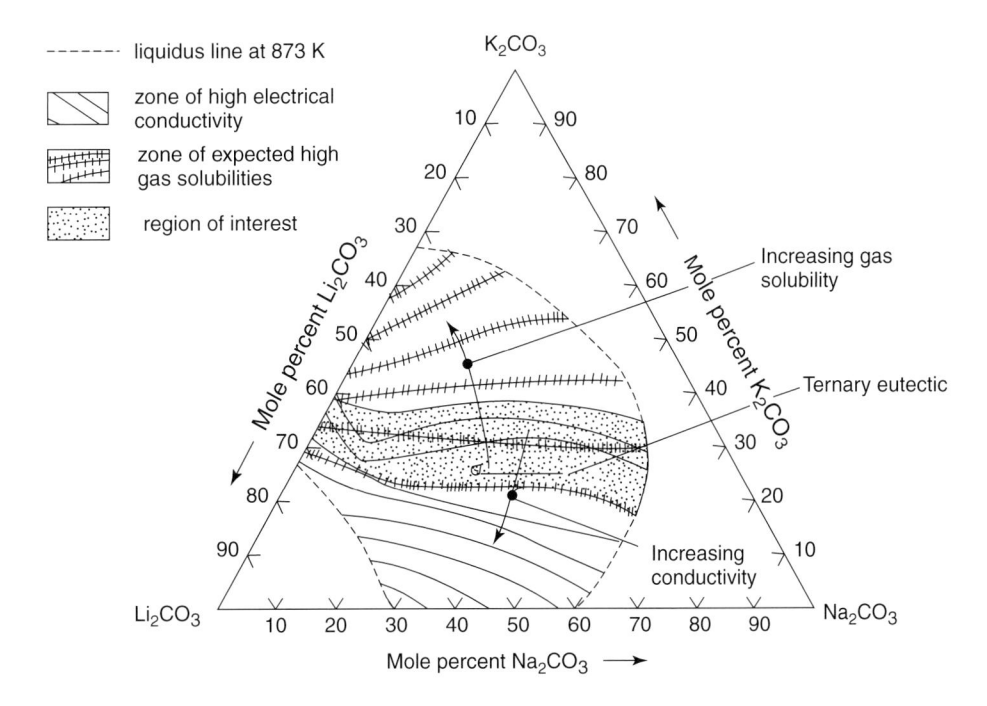

Figure 15. Triangle diagram for the $Li_2CO_3/Na_2CO_3/K_2CO_3$ system at 873 K. Gas solubility and electrical conductivity are shown in the diagram. (Reproduced from Selman (1986)[5] by permission of Elsevier Science.)

the double oxide in the fuel cell environment, whereas Fe, Co and Mn form the $LiMO_2$ phase or similar compounds as a result of reaction with the Li_2CO_3 component. In such a case, the solubility is determined by the dissolution behavior of $LiMO_2$. The dissociation reactions of $LiMO_2$ can be written as

$$LiMO_2 + O^{2-} \longrightarrow Li^+ + MO_3^{3-} \quad (23)$$

$$LiMO_2 + CO_3^{2-} \longleftarrow Li^+ + MO_3^{3-} + CO_{2(g)} \quad (24)$$

The temperature dependence and the CO_2 partial pressure dependence of the solubility are controlled in terms of the acid/base properties. For the acidic dissolution, the solubility increases with increasing CO_2 partial pressure or with decreasing temperature.

The solubility dependence on the oxygen potential is determined by the difference in valence between the solid state and the dissolved state. For Ag,

$$2Ag + O_{2(g)} + CO_{2(g)} \longrightarrow 2Ag^+ + CO_3^{2-}$$

$$2Ag^+ + H_{2(\text{in electrolyte})} + 2CO_3^{2-} \longrightarrow 2Ag$$

$$+ 2H^+ + 2CO_{2(g)} + 2OH^- \quad (25)$$

The solubility decreases significantly with decreasing oxygen potential because Ag exists as Ag^+ in the molten state. For NiO, it does not change within the NiO stable oxygen potential region, whereas it changes for the Ni stable region in a similar manner to that of Ag. These behaviors are shown schematically in Figure 16.

There are only a few compounds that are stable (or have little soluble) in the MCFC environment, and these are MgO, $SrTiO_3$ and $LiAlO_2$. Attempts have been made to use these compounds as electrolyte matrix materials.

3.2.2 Solid electrolyte

The fluorite phase of pure ZrO_2 is stable only above 2662 K; the tetragonal phase is stable between 1445 and 2662 K, the monoclinic phase being below 1445 K.[55, 56] When yttria, Y_2O_3, is doped to more than 10%, the stability region of the fluorite phase becomes down to 1273 K, as shown in Figure 17. Another important feature of the large stabilization is that the chemical potential of the yttria component is maintained at extremely low values.[59, 60] This leads to the lower reactivity of the yttria component. Y_2O_3 is a rather basic oxide so that its reactivity with carbon dioxide and water vapor in the anode or with acidic oxides, such as Al_2O_3 or SiO_2, should be carefully examined. So far, it has been confirmed that the yttria

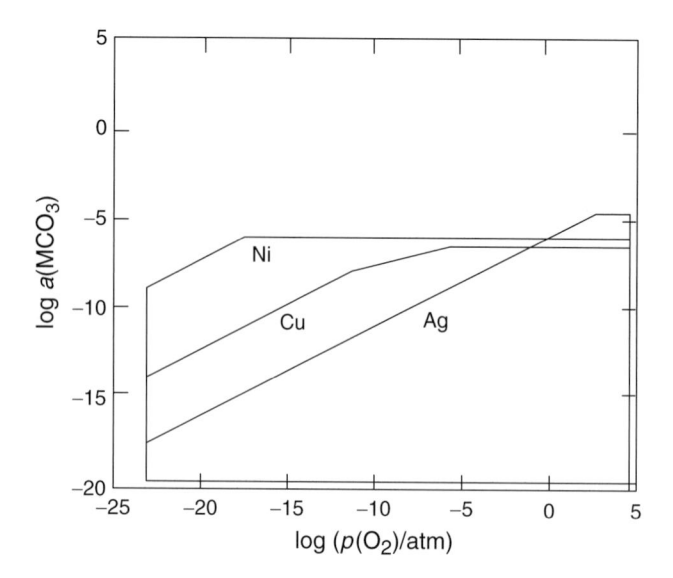

Figure 16. Schematic behavior of the solubility of dissolved MCO_3 (M = Ni, Cu, Ag_2), which is represented tentatively by the oxygen potential dependence of the chemical potential of MCO_3 in equilibrium with the respective stable metal/metal oxides in molten carbonates at 923 K and $p(CO_2) = 0.1$ atm.

component in YSZ is not reactive against such substances. V_2O_5 is, however, harmful because of its reactivity with the yttria component.

$$0.075\ V_2O_5 + (Zr_{0.85}Y_{0.15})O_{1.925} \longrightarrow 0.15\ YVO_4 + 0.85\ ZrO_{2(\text{monoclinic})} \quad (27)$$

As a result of extraction of the yttria component, zirconia is destabilized to the monoclinic phase, causing a destructive reaction.

The solubility of the transition metal oxide in the fluorite structure has been measured for many systems. At a glance, it is hard to see the regularity among the experimentally determined solubility. Even so, some regularities have been recognized from the thermodynamic considerations[55, 56] as shown in Figure 18. The solubility is determined by two factors: the stability of the transition metal ions in the fluorite phase and the stabilization energy of the transition metal oxide in double oxides with the components of YSZ. The former is well interpreted in terms of the valence of the transition metal ions and their relative size to the critical dopant size determined for the respective valences given in Figure 11.

The oxygen potential dependence, which was thermodynamically evaluated from the obtained interaction parameters, was found to be consistent with the accumulated experimental data on solubility. In the SOFC operation region, the interesting phenomena appear as follows: the solubility changes drastically when the valence changes.

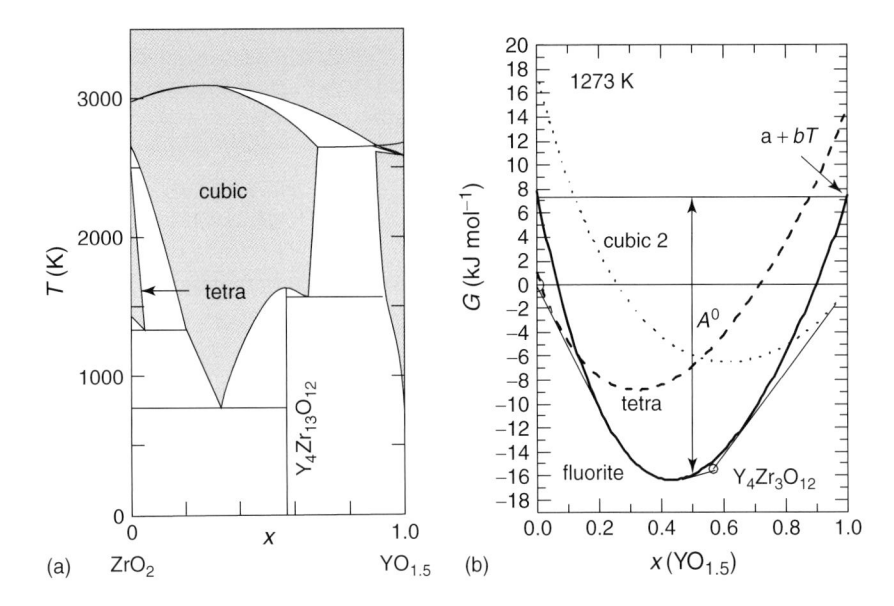

Figure 17. (a) Phase diagram of the $ZrO_2/YO_{1.5}$ system[55] and (b) related stabilization of cubic phase given in terms of the total Gibbs energy and the chemical potential of $YO_{1.5}$.[60] The significant lowering of the chemical potential of $YO_{1.5}$ corresponds to the chemical stability of YSZ against various atmosphere and other cell components.

For example, the solubility of Ti increases with decreasing oxygen potential because the trivalent or divalent ions show more stability in the fluorite structure due to their large ionic sizes. The behavior of Mn and Fe also can be understood in terms of their valence stability and ionic size. For Ni and Co, the solubility is constant within the MO (M = Ni, Co) stable region, but decreases rapidly in the metal regions. This is the same behavior as in molten carbonates as illustrated in Figure 16.

The temperature dependence of the solubility in YSZ is usually simple; that is, it increases with increasing temperature.

3.3 Thermodynamic stability of cathode and anode materials

3.3.1 Thermodynamic stability of electrode materials in the mixed alkali carbonate melt

The solubility of Ag, Cu and Ni in the anode atmosphere is extremely small, leading to a good chemical stability as anode major components. Ni is often doped with Al or Cr to improve the morphological stability and creep resistivity. Those elements, Al or Cr, react with Li_2CO_3 to form $LiMO_2$. For the cathode, lithiated NiO is used in the major MCFC stacks. In the early stages, Ag or Cu were used but soon were abandoned because of shorting caused by dissolution. Figure 16 gives an explanation on why Ag or Cu cathodes causes shorting due to the dissolution/dendrite formation mechanism. Since the oxygen potential dependence of solubility is steep, this provides a large driving force for precipitation as a metallic dendrite immediately after dissolution. Transition metal oxides with a higher valence react with the electrolyte to form stable compounds.

In Figure 19, the stability of some transition metal oxides is shown for the A/M/C/O (A = alkali, M = transition metal) systems at 923 K. The lithium compounds, such as $LiMO_2$ (M = Fe, Co) and Li_2MnO_4, are stable. The SrO-based perovskite oxides, $SrMO_3$ (M = Fe, Co, Ni), are generally unstable against reactions with the Li_2CO_3 component to form $LiMO_2$ and $SrCO_3$. This is understandable in view of the magnitude of the stabilization energies shown in Figures 10 and 12. The La_2O_3-based perovskites, $LaNiO_3$, show some stability, whereas $(La,Sr)MO_3$ (M = Ni, Co) is unstable because of the highly reactive nature of the SrO component and the unusual valence in Ni and Co in perovskites. Cathode materials should be also stable against reactions with matrix material, for example, $LaNiO_3$ was found to react with $LiAlO_2$ to form $LaAlO_3$ and lithiated NiO.

3.3.2 Thermodynamic stability of electrodes in contact with YSZ

The interactions of the electrode materials with YSZ can be characterized in terms of interdiffusion across the interface and the formation of the zirconia-based complex oxides or the yttria-based complex oxides. Perovskite oxides have attracted much interest as cathodes because of their mixed conductive feature and their ability to stabilize the unusual valence state. The reactions of the zirconia component in

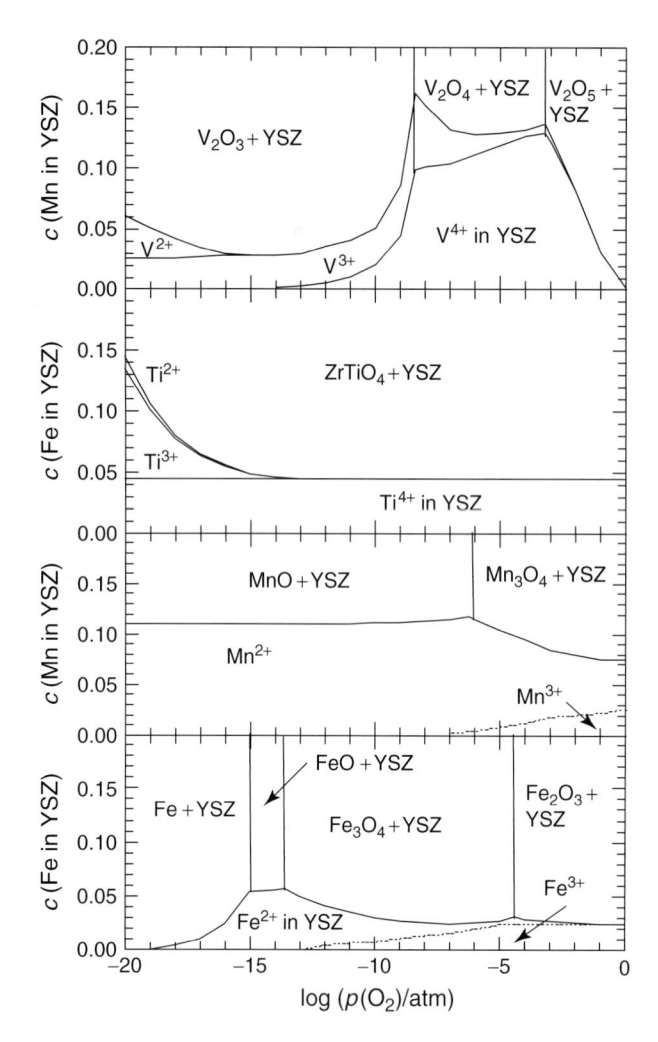

Figure 18. Oxygen potential dependence of solubility of transition metal oxides in YSZ at 1273 K. The solubility is determined by the valence stability of transition metal ions and their stabilization in the cubic fluorite structure. When the difference in valence number appears between the fluorite phase and the equilibrated phase, the solubility exhibits the large oxygen potential dependence.[56]

YSZ with the La_2O_3 component in perovskites, however, were found to be critical. For example, $LaCoO_3$, which was first investigated among the perovskites oxides, was found to react with the zirconia component to decompose into lanthanum zirconates and cobalt oxide as follows:

$$LaCoO_3 + ZrO_2 \longrightarrow 0.5\ La_2Zr_2O_7 + CoO + 0.25O_{2(g)} \quad (28)$$

This can be characterized as the destructive reduction reaction. On the other hand, $LaMnO_3$-based perovskites also exhibit some reactivity with YSZ, although this is not destructive but an only small part of the perovskites reacts to change the cation nonstoichiometry.

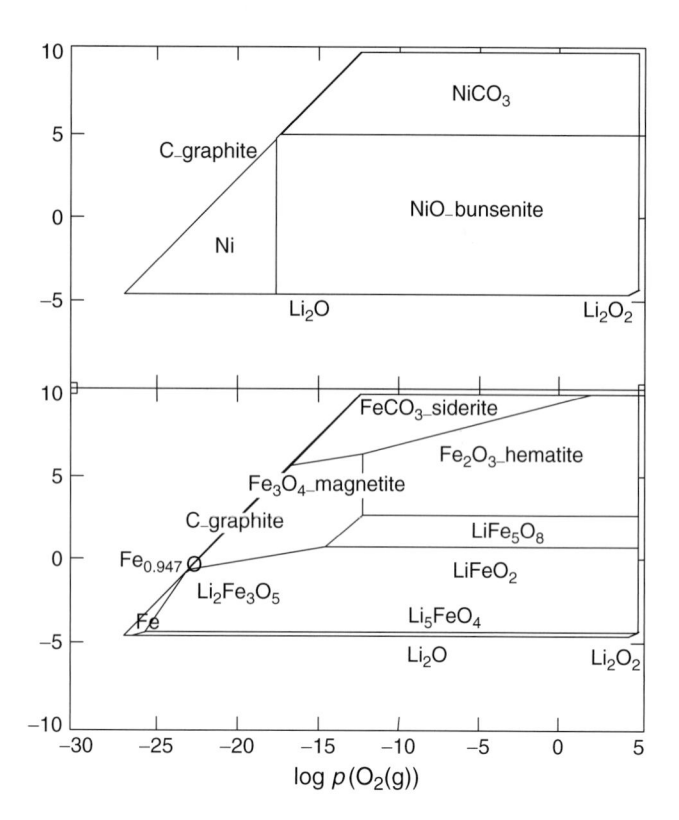

Figure 19. Chemical potential diagram for the Li/M/C/O (M = Ni, Fe) system under the condition of $\alpha(L_2CO_3) = 1$ at 923 K. This shows the stable chemical form of Ni or Fe in the carbonates in the $\log p(CO_2)$ vs $\log p(O_2)$ plot. The lower and the left hand sides are a thermodynamically inaccessible region because of the decomposition of metal carbonates. (Reproduced from Yokokawa et al. (1993)[58] by permission of The Electrochemical Society Inc.)

$$LaMnO_3 + y(ZrO_2 + 0.75O_{2(g)}) \longrightarrow$$
$$yLa_2Zr_2O_7 + La_{1-y}MnO_3 \quad (29)$$

These features have been well understood from thermodynamic considerations as shown in Figure 20.[50]

For anode materials, reactions with YSZ are not severe during operation. When fabrication is carried out in air, the NiO dissolution occurs slightly into YSZ. After being in contact with fuels, NiO particles as well as dissolved NiO will be reduced to form Ni. It can be expected that the dissolved NiO can be effective in fixing Ni particles on the YSZ surface and will give some durability against coagulation and other degradation.

3.4 Thermodynamic stability of interconnect materials

Similarly to the electrolyte materials, the interconnect materials have to meet the chemical stability requirements in previously shown in Figure 8, so that use of the Cr element is necessary for both oxide and metal interconnects. On the other hand, the Cr hexavalent state is also stable in gases, liquids and solids. This makes the interconnect chemistry complicated.

As oxide interconnects for SOFCs, $CoCr_2O_4$ and $NiCr_2O_4$[8] were tested in the early stage of development without success. This suggests that the stability of the complex oxides is determined by the weakest oxide against attacks; in the above case, CoO and NiO in the double oxides are reduced in the reducing atmosphere, although the Cr_2O_3 component is stable against the reduction. As the oxide interconnect, only a few oxides can be candidates; for example, $LaCrO_3$-based oxide and AMO_3 (A = Ca, Sr; M = Ti, Zr)-based oxides can be selected from the stability point of view. Among them, $LaCrO_3$ has been selected from their relatively high electrical conductivity of doped systems. Many attempts have been made to find an appropriate kind and amount of dopants to optimize the characteristic features on electrical conductivity, thermal expansion, chemical volume expansion and sinterability. Good results were obtained only for thermodynamically stable dopants.

For MCFCs, only metal interconnects have been attempted, probably because electron conductive perovskites are unstable in molten carbonates. For SOFCs, attempts have been also made to use metal interconnects by lowering operating temperature. Since metal interconnects are oxidized in both the cathode and anode environments, the technological issue is about the growth rate of oxide scales and their electrical resistivity. When the electrical current passes through across the oxide scales, the chemical and morphological stability is in many cases a trade off to higher electrical conductivity. To have a stable oxide scale, the dense Cr_2O_3 layer without defects will be the best. However, the electrical conductivity of such a scale is expected to be very small. For MCFCs, oxide scales are formed as a result of the oxidative reaction with the alkali carbonates. Iron tends to react with Li_2CO_3 to form $LiFeO_2$ and Cr tends to react with K_2CO_3 to form K_2CrO_4 in the cathode environment and to react with L_2CO_3 to form $LiCrO_2$ in the anode environment as shown in Figure 21. In an anode atmosphere, corrosion is generally more severe than in a cathode atmosphere, suggesting significant effects from the water vapor.

In an anode atmosphere, Ni and Co are thermodynamically stable so that these materials can be used effectively in SOFCs and MCFCs. In MCFCs, corrosion in the anode compartment is so severe that Ni is usually clad on alloys. In tubular SOFC stacks by SWPC, Ni is used as a material for bundling cells. For this purpose, cobalt alloys can be also used despite its higher price.

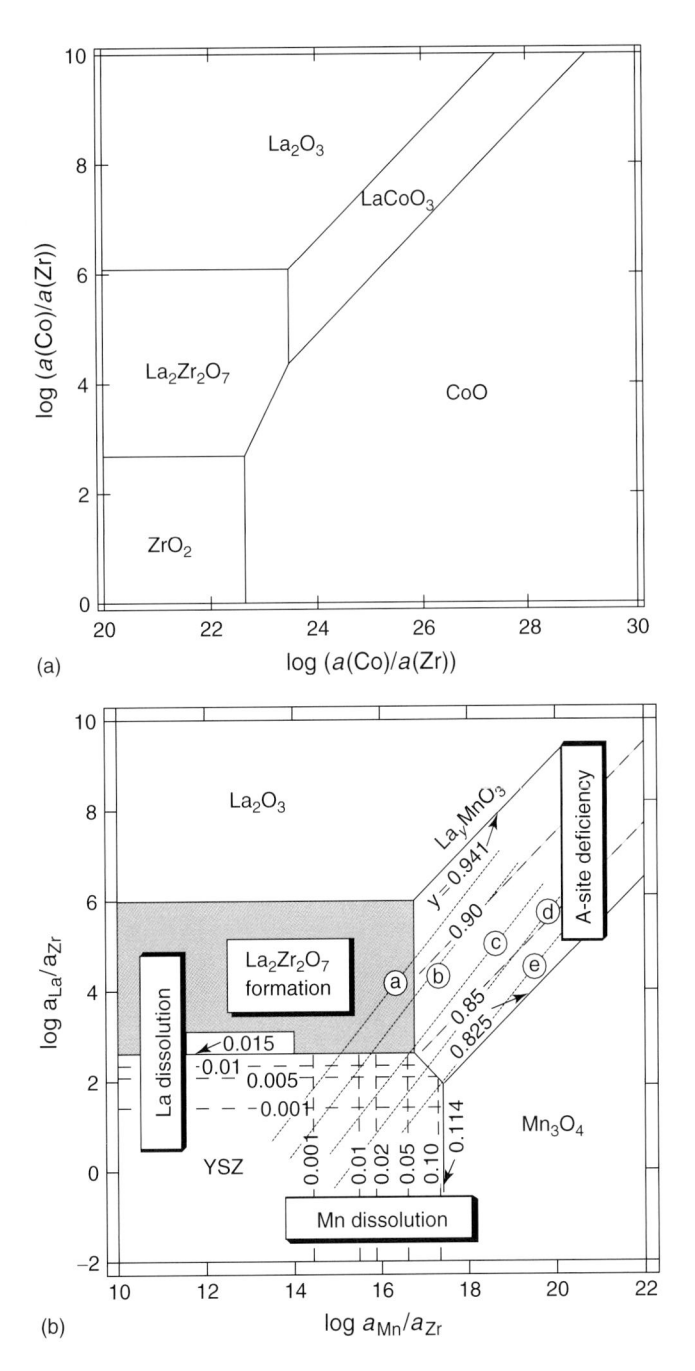

Figure 20. Chemical potential diagram for the La/Zr/M/O (M = Co, Mn) system under the condition of $p(O_2) = 1$ atm at 1273 K. The thermodynamic stability is not realized for the $LaCoO_3/ZrO_2$ interface (a) but for the $LaMnO_3/ZrO_2$ interface (b), except the nearly stoichiometric region for the $La_2Zr_2O_7$ formation. (Reproduced from Yokokawa et al. (1991)[49] by permission of The Electrochemical Society Inc.)

both MCFCs and SOFCs. Attempts have been made to use both oxides and metals. From the materials thermodynamic point of view, the chemical behavior of Cr is crucial. Chromium trivalent ions have a strong chemical stability, as

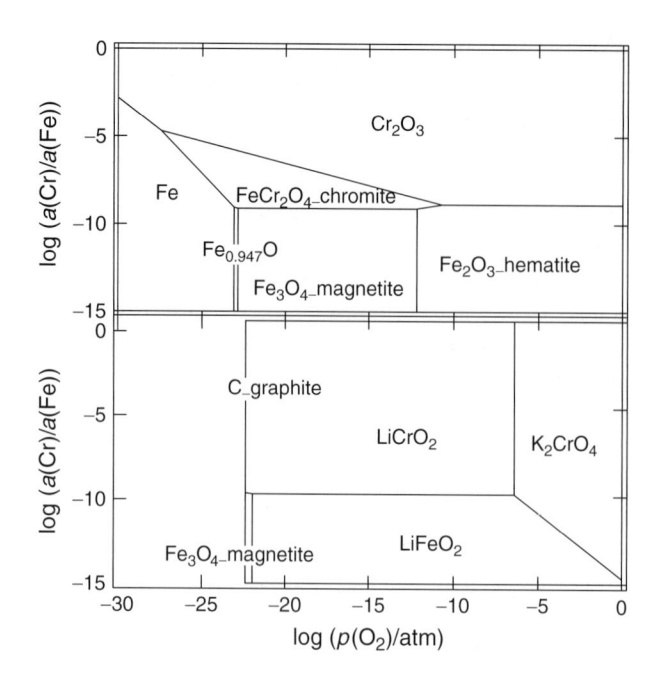

Figure 21. Chemical potential diagram for the Li/K/Fe/Cr/C/O system at 923 K under the conditions of $\alpha(Li_2CO_3) = 0.34$ and $\alpha(K_2CO_3) = 0.14$. This is compared with the corresponding diagram for the Fe/Cr/O system at 923 K. (Reproduced from Yokokawa et al. (1993)[58] by permission of The Electrochemical Society Inc.)

3.5 Mass transfer under the chemical potential gradients

There can be many different kinds of driving forces for mass transfer in fuel cells. These are:

1. Chemical diffusion due to concentration or chemical potential gradients is important at any interface with other components.
2. Electrical potential gradient: the charged particles have the additional driving force from the electrical field. Cations tend to move to cathode area. Particularly, a large gradient appears inside the electrolyte. Furthermore, edge parts of cells, where current density exhibits a steep change, have an additionally large gradient.
3. Oxygen potential gradient: between the anode and the cathode atmospheres, there appears a large oxygen potential gradient in electrolyte, interconnects and associated sealing parts, if any. This provides the driving forces for oxygen permeation through electrolyte or interconnect. In SOFCs, oxygen permeates as a form of oxide ions and electrons (or holes). In MCFCs, dissolved hydrogen permeates through electrolyte.
4. Within the local equilibrium approximation, the chemical potentials of metallic elements in oxides tend to have the inverse gradient to that of the oxygen potential. This is the driving force for cations to migrate to the cathode side and for anions to migrate to the anode side. When there are many cations, the redistribution of cations occurs as a result of the difference in diffusion coefficients. This is called kinetic decomposition or kinetic demixing. Such demixing can appear whenever the thermodynamic variables have a steep gradient and the diffusion coefficient is not negligible. In SOFCs, demixing seldom occurs due to the extremely low cation diffusivity. An exceptional case is the Ca diffusion through grain boundaries in air sintered (La,Ca)CrO$_3$.[61] In MCFCs, this phenomenon is popular because the cation diffusion coefficients are rather high because of liquidity. As a result, the K$_2$CO$_3$ component tends to be concentrated in the cathode side.[62] The effects of additives, such as alkaline earth carbonates, have been tested in order to improve this.
5. In some reaction mechanisms, overpotentials correspond to a change in the oxygen potential. This suggests that the distribution of the oxygen potential depends on the overpotential or the current density.
6. Temperature gradient: temperature distribution is mainly determined from the gas flow scheme (co flow, cross flow and counter flow), and flow rate in addition to the endothermic reforming process.
7. Difference in particle size or surface curvature: difference in surface tension gives rise to the driving force for mass transfer from small particles to large particles
8. For MCFC, convection, if any, is also one of the big driving force for mass transfer in stacks.

To know the oxygen potential distribution developed in fuel cell materials is the key when the mass transfer inside the fuel cell materials is examined. In the SOFC electrolytes, the magnitude of the electron and hole conductivities as a function of oxygen partial pressure makes it possible to know the distribution of the electrical potential, oxygen potential and other (electro-)chemical potentials of other species. This also makes it possible to evaluate the loss originated from the Joule effect and also from oxygen permeation via bipolar diffusion of oxide ions together with electrons/holes.[63] The latter can be regarded as shorting due to nonzero electron currents.

In MCFCs, the situation is much more complicated. The important information to be known is the solubility and mobility of hydrogen, oxygen and CO$_2$ gases and also those properties for other ionic species such as O^{2-}, OH$^-$, O$_2^-$, etc. Since the solubility of hydrogen is larger than other gases, it is recognized that the dissolved hydrogen comes close to the cathode side at the beginning of the operation, providing the oxygen potential profile has a steep change near to the cathode area.[64, 65]

3.6 Key mass transfer mechanisms in fabrication and cell operations–dissolution/deposition mechanism–evaporation/re-condensation mechanism

On solidification from liquids or gases, crystal growth sites are determined from the surface tension. When particles are separated, bigger particles grow faster. When particles are in contact with each other, growth occurs at the concave parts at the interparticles. This phenomenon is common among the solid/gas interface and the solid/liquid interface. In SOFCs, the most successful case is the EVD technique to fabricate a dense YSZ layer on a porous cathode adopted by SWPC as illustrated in Figure 22.[8] Deposition occurs selectively at the concave parts to make the surface film flat automatically. In the $LaCrO_3$-based interconnect, this phenomenon causes a serious problem of poor sinterability; the chromium vapor, $CrO_{3(g)}$, tends to deposit at the interparticle parts as Cr_2O_3 and prevents further densification.[49] In MCFCs, this phenomenon occurs in many cases. The most severe case is the deposition of nickel ions as nickel metals on the electrolyte matrix. Deposition occurs preferably at the bigger particles or at the interparticles/concave parts to assist the formation of the electrical path. This phenomenon takes place on the electrolyte matrix and electrodes,[38] leading to changes in pore size and pore size distribution as shown in Figure 23.

3.7 Thermodynamic properties related to the reforming process

One of the advantages of the high temperature fuel cells is the utilization of carbon containing fuels without

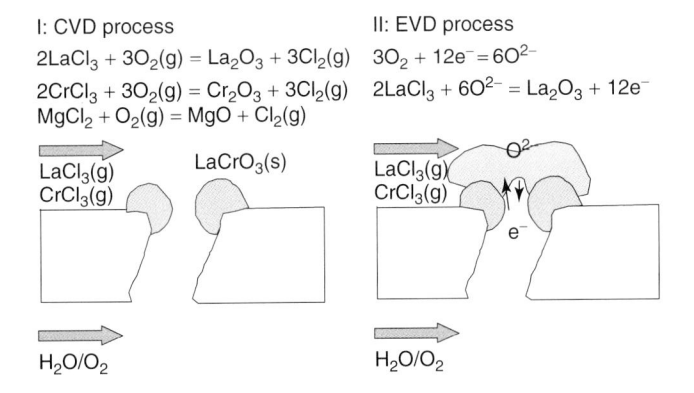

I: CVD process
$2LaCl_3 + 3O_2(g) = La_2O_3 + 3Cl_2(g)$
$2CrCl_3 + 3O_2(g) = Cr_2O_3 + 3Cl_2(g)$
$MgCl_2 + O_2(g) = MgO + Cl_2(g)$

II: EVD process
$3O_2 + 12e^- = 6O^{2-}$
$2LaCl_3 + 6O^{2-} = La_2O_3 + 12e^-$

Figure 22. A schematic illustration for the EVD process.[65] (a) In the normal chemical vapor deposition process, reactions occur at the places where the gaseous reactants encounter. (b) In the subsequent EVD process, the reaction proceeds with a diffusion of oxide ions and electrons in the deposited substance. This diffusion process becomes the rate-determining step to ensure that the film becomes homogeneous in thickness.

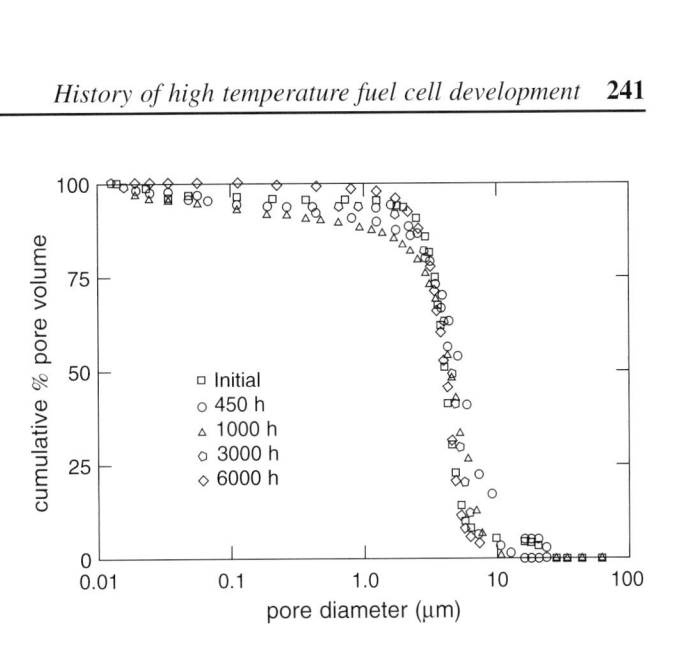

Figure 23. Size distribution change in MCFC materials. Ni/Cr plack at 923 K in a standard fuel environment; 34% H_2, 24% CO_2, 42% N_2. (Reproduced from Iacovangelo (1986)[38] by permission of The Electrochemical Society Inc.)

complicated fuel processes. For a large power generation system, the maximization of efficiency is the most important requirement. One of the important factors in determining the total efficiency is heat management of the reforming process. In SOFCs, it is not difficult to use heat evolved inside the stacks for the reforming process. In MCFCs, the external reforming and the internal reforming give rise to an appreciable change in efficiency; the external reforming occurs at 800–850 °C, and, therefore, heat should be supplied from the combustion of the remaining fuels. This is completely against the advantages of the high temperature fuel cells. In this sense, the external reforming MCFC has no greater advantages over the low temperature fuel cells. The usefulness of internal reforming has been well recognized from the early stages of the MCFC development particularly by ERC.

In the internal reforming of MCFCs, the reforming process is made at the same temperature as the fuel cells and, therefore, heat can be supplied from the heat generated from the electrochemical reactions. At lower temperatures, the methane conversion rate is limited to a small extent compared with 800–850 °C. However, by increasing the oxidation rate of fuels, the remaining methane is reformed further due to a shift in the equilibrium.

Generally speaking, carbon deposition during the reforming process and the fuel cell operation should be avoided. When steam is used, a steam to carbon ratio is selected of around three for MCFCs and two or less for SOFCs. This difference comes from the differences in the total electrochemical reactions and the operating temperature. In MCFCs,[57] the composition of the anode compartments changes as a result of oxidation by CO_3^{2-}.

$$CH_4 + nH_2O + 2(O_2 + 2CO_2) \longrightarrow 5CO_2 + (n+2)H_2O \quad (30)$$

On the other hand, in SOFCs, fuels are always oxidized by oxygen (oxide ions).

$$CH_4 + nH_2O + 2O_2 \longrightarrow CO_2 + (n+2)H_2O \quad (31)$$

These are shown in Figure 24 as different composition lines in the C/H/O triangle diagram, where the composition region for carbon deposition is given with a parameter of temperature. Despite the higher steam to carbon (S/C) ratio in MCFCs, the carbon activity in fuels can be higher because the composition becomes close to the carbon deposition region. This suggests that carbonization of alloys is more severe in MCFCs.

For SOFCs, some challenges have been made to introduce methane directly into the fuel cells. One is to use the fuel cell for the partial oxidation of methane at temperatures of 800–1000 °C.

$$CH_4 + 0.5O_2 \longrightarrow CO + H_2 \quad (32)$$

Although carbon deposition occurs at the anodes, these can be also electrochemically oxidized. The second approach is the direct oxidation at lower temperatures around 600 °C.

$$CH_4 + 2O_2 \longrightarrow CO_2 + 2H_2O \quad (33)$$

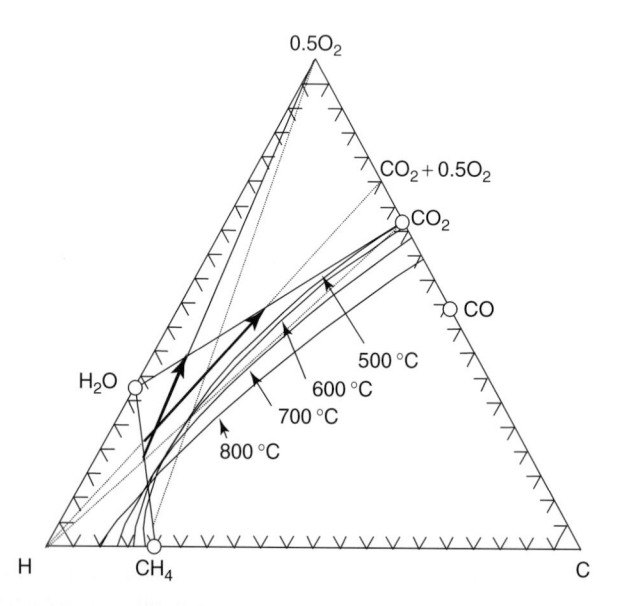

Figure 24. Triangle diagram for the C/O/H system for comparison of the reforming process between MCFCs and SOFCs. The curved lines for 500–800 °C are carbon precipitation regions at a total pressure of 10 atm. The composition of the anode atmosphere is plotted for S/C = 3 for MCFCs and S/C = 2 for SOFCs.

Although the direct electrochemical reaction is not plausible, the complicated electrochemical and thermal reactions can proceed to form CO_2 and water vapor. These new approaches are of interest from the viewpoint of heat management. When the reforming process is made for hydrocarbon fuels, more heat will be emitted in the electrochemical reaction, although some heat is utilized in the reforming process. For this purpose, heat has to be transferred from the cells to the reforming sites. When the oxidation is introduced as in equations (32) and (33), the heat to be transferred will be decreased. This may improve the situation of heat management.

4 DEVELOPMENT OF MCFCs

The molten carbonate electrolytes have a great advantage over solid electrolytes as the liquid materials have generally a higher ionic conductivity and a lower electron conductivity. In addition, several advantages can be derived also from the fluidity of electrolytes:

(i) Low contact resistance can be easily obtained due to the wetting of electrolyte.
(ii) Construction of cells with a large active area can be made by adopting electrolyte matrix materials (electrolyte immobilizing solid phase, tiles).
(iii) It would be easy to achieve a gas seal by adopting a sandwich structure of the electrolyte matrix and bipolar plate.

On the other hand, disadvantages also come from the liquidity of electrolyte:

(i) Large volume changes take place on melting or freezing of the electrolyte. This will damage the electrolyte matrix materials.
(ii) Even a small amount of dissolved materials can cause serious troubles because of the rather high diffusion coefficients of the cations as well as the carbonate ions.
(iii) A high corrosion rate and relatively high vapor pressure due to $KOH_{(g)}$.
(iv) The need for electrolyte management: electrolyte loss due to creepage/migration, corrosion and vaporization.

The main technological issues in MCFC developments can be summarized, therefore, as the establishment of an appropriate electrolyte management system to overcome above disadvantages without losing the merits of a liquid electrolyte.

4.1 Selection of electrolyte composition

In the early stages, the Li_2CO_3/K_2CO_3 mixture and the $Li_2CO_3/Na_2CO_3/K_2CO_3$ mixture were used as the electrolytes. In Figure 15, the gas solubility and the electrical conductivity are shown for the $Li_2CO_3/Na_2CO_3/K_2CO_3$ ternary carbonate system.[5] Since the cell performance was higher in the Li_2CO_3/K_2CO_3 mixture, the main effort in the construction of the MCFC stacks were shifted to the Li_2CO_3/K_2CO_3 mixture. As will be described below, the composition of the molten carbonate electrolyte affects, through many physicochemical properties, the cell performance and the long-term performance decay. In this sense, the composition should be determined from optimization by analyzing these possible effects in a systematic manner. The ionic conductivity is high for systems containing Li_2CO_3 or Na_2CO_3. The gas solubility, which is important for electrode reactions, is high in the Li_2CO_3/K_2CO_3 mixture. The basicity is higher in the Li_2CO_3/Na_2CO_3 system.[43] This system has attracted much interest due to the necessity of avoiding shorting due to the NiO dissolution and resulting dendrite formation of metallic Ni. One factor that makes the situation complicated is the effects of the presence of minor species in electrolyte such as OH^-, O_2^{-}, O^-.[5]

4.2 Key issues in materials selection in MCFCs: solubility of oxides in molten salts

The most important criterion for selecting the MCFC materials is whether or not the materials are stable in the selected molten electrolyte. Chemical reactions with the electrolytes and solubility into the electrolytes are well interpreted in terms of acid/base relations and the related thermodynamic properties.[54, 57, 58]

4.2.1 Matrix materials

Matrix materials confine the electrolyte in its pores. Therefore, this material should meet the following requirements:[5, 12]

1. Appropriate interstice size distribution. This should be stable during the operation, and also the resulting capillary forces should be sufficient to retain the electrolyte even in the presence of the pressure differentials between the anode and cathode gases.
2. The mechanical stability. The above electrolyte retention property is maintained by the capillary forces in the fine pores. When a crack is formed in such tiles, gas crossover through the cracks can take place, resulting in a fire of the fuels without generating electricity.
3. Chemical compatibility with electrolyte. For example, Al_2O_3 was abandoned because of the reaction with the Li_2CO_3 to form $LiAlO_2$.
4. Dimensional change should be small.
5. The volume fraction should be at a minimum to allow the maximum volume fraction of the electrolyte to be achieved and to provide the highest ionic conductivity.

The important properties of matrix materials are associated with the dissolution/reprecipitation process. This causes the Ostwald ripening of particles or a similar phenomena because solubility also depends on surface tension. When there are differences in the convexity of the solid surface, dissolution occurs from the more convex part to the less convex or concave parts. For example, MgO was used as an electrolyte material in the early stages. However, crystal growth due to this mechanism was found to be fast and uncontrollable. Instead of MgO, $LiAlO_2$ has often been used as an electrolyte matrix. The crystal growth rate depends on the solubility of $LiAlO_2$. Since the $LiAlO_2$ dissolution is known as basic reaction,

$$LiAlO_{2(s)} + O^{2-}_{(in\ electrolyte)} \longrightarrow Li^+_{(in\ electrolyte)} + AlO_3^{2-}{}_{(in\ electrolyte)} \quad (34)$$

The solubility and resulting crystal growth increases with increasing temperature, with decreasing CO_2 pressure and with increasing Li_2CO_3 content (activity), as shown in Figure 25, which compares the solubility of Ni[43] and $LiAlO_2$[66] as functions of the Li_2CO_3 concentration. In addition, $LiAlO_2$ shows polymorphism: α-, β- and γ-phases have quite different densities of 3401, 2610 and 2615 kg m^{-3}, respectively. During the cell operation, α-$LiAlO_2$ undergoes the transformation to γ-$LiAlO_2$ through the dissolution/reprecipitation mechanism. Since the density of the α-phase is larger than the γ-phase, this transformation gives rise to the enlargement of the pore volume in the matrix materials, leading to a decrease in the fills of the electrolyte. It is also found that a large partial pressure of H_2O would increase the solubility of $LiAlO_2$ and, therefore, increase crystal growth and the rate of transformation to γ-$LiAlO_2$.

The use of spherical ceramic particles or fibers was tested to strengthen the electrolyte composite with success. However, an electrolyte thickness of more than 1 mm is required for this purpose. This requirement makes it difficult to reduce the Joule loss. A highly ionic, conductive electrolyte is preferred.

As alternative matrix materials, GE tested the usefulness of $SrTiO_3$.[67] An improvement against $LiAlO_2$ can be expected in crystal structure (no transformation in $SrTiO_3$), water insensitivity and chemical stability, although $BaTiO_3$

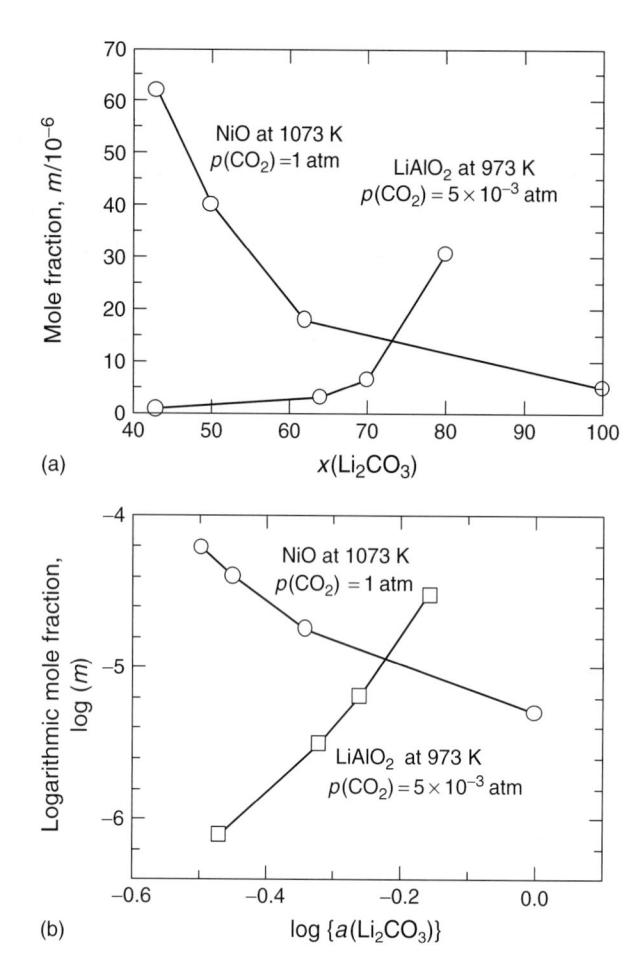

Figure 25. Comparison of the solubility of NiO[43] and LiAlO$_2$[66] in (Li$_{1-x}$K$_x$)$_2$CO$_3$ mixed carbonates plotted against the concentration (a) or plotted in log vs. log plot (b).

and MgAl$_2$O$_4$ reacted significantly, SrTiO$_3$ is stable. Since the electronic conductivity is negligibly small, this is appropriate for matrix materials. Unfortunately, this material is unstable against crystal growth. Crystal size doubled after a period of 250–420 h.

4.2.2 Cathode

Dissolution of cathode materials could lead to the most serious problems in the cell performance as follows:[5]

(i) Change in porosity and pore size distribution. The electrode/electrolyte/gas interface is affected so that the cathode performance is degraded.
(ii) Shorting occurs as described above.

Shorting was found first for Ag and Cu cathodes. The lithiated NiO cathode showed a much higher stability than Ag and Cu, so then it became the major cathode in 1970s. However, in 1981, shorting was observed even for the NiO cathode. Since then, many investigations have been made on the solubility of NiO and related properties in various kinds of electrolytes, in addition searching for alternative cathode materials. As alternative cathodes, the oxides of Mn, Fe, Co and Ni have been focused on because of their possible high electrical conductivity. The observed stable double oxides are Li$_2$MnO$_3$, LiFeO$_2$, ZnO, LiTiO$_3$, Li$_3$TaO$_4$, LiCrO$_2$, MgO, K$_2$WO$_4$, Li$_3$VO$_4$ and Li$_2$SnO$_3$. On the other hand, the following compounds have stability problems: the perovskite oxides, (La, A)MO$_3$ (A = alkali earth), are almost unstable in molten carbonates; SrFeO$_3$ is also unstable. However, the lithium transition metal oxides are stable so many efforts have been made with such compounds. So far, their performance is not as satisfactory as compared to the lithiated NiO.

4.2.3 Anodes

The solubility of metals in electrolytes is extremely small so that there are few materials problems. The major problems are sintering and creep of Ni. To improve these characteristics, Cr or Al is added to the Ni. These metals are oxidized and react with the Li component in the electrolyte to form Cr$_2$O$_3$/LiCrO$_2$ and Al$_2$O$_3$/LiAlO$_2$. Those dispersed oxides prevent the nickel sintering.[14]

As alternative anodes, oxides anodes such as doped MnO, CeO$_2$ and LiFeO$_2$, were tested but showed chemical instability. Among metal alternative anodes, Cu/Al has not got good creep strength. Anode materials should have a good compatibility with cladding materials on current collectors. When a copper-based anode is used with a Ni clad, significant diffusion between Cu and Ni is observed. Cu clad should be adopted when an anode consists mainly of Cu.

4.2.4 Bipolar plates/current collectors

Corrosions in molten electrolytes and the electrical conductivity across the oxide scale are crucial characteristics. When molten salts are involved, high temperature corrosions become severe. In this sense, corrosions of alloys with molten carbonates are the most severe material problem. It is well known that corrosions are more severe for those alloys in contact with both molten carbonates and oxidative atmosphere than those alloys immersed in the molten carbonates. Usually, corrosion products consist of many compounds and form a morphologically complicated arrangement. A decrease in the electrical conductivity across the oxide scale leads to significant degradation of cell performance. In addition, corrosions of alloys affect cell performance in many ways. For example, as a result of corrosion, the lithium component is lost from the molten carbonates. This contributes to the electrolyte loss and results in a lowering of the electrode performance. Furthermore, the ionic conductivity decreases with decreasing levels of the lithium component.[13]

The corrosion of the alloys is a combined effect of the oxidation of the alloy elements and the reaction with the molten carbonates. The most important feature of the corrosion in molten carbonates is that Cr_2O_3 is not protective in the cathode because Cr_2O_3 is oxidized by peroxide ions to form chromate ions.

Although many alloys have been tested for corrosion behavior, the 316L and 310S alloys are widely used as current collectors and bipolar plates. For the 316L alloy, the corrosion products in the cathodes are $LiFeO_2$ and $LiCrO_2$ or a K_2CrO_4-rich chromate melt. Corrosion becomes less severe when the 310S alloy with a higher Cr content is used. Although corrosion products are not only Cr_2O_3 but also other complicated oxides, the basic features of oxide scales are about the same; that is, when the stability is better, the electrical conductivity is worse.

Corrosion in anode environments is much more severe. Figure 26 shows the sequence of oxide formation and the thickening behavior.[14] In an anode environment, chromate is not produced, but Cr_2O_3 and $LiCrO_2$ are formed. $LiFeO_2$ tends to be formed as an outer scale, whereas FeO and Fe/Cr spinel form inner scales. Carburization takes place also deep inside the alloys.

Corrosion of the 316L alloys in anodes depends on fuel utilization; that is, the corrosion is enhanced with increasing fuel utilization. It can be ascribed to increasing CO_2 partial pressure because corrosion depends on CO_2 pressure. One possible explanation can be given in terms of acidity. At higher CO_2 pressures, the acidity of the electrolyte increases and leads to the increase of solubility of the corrosion products. It is also found that at high CO_2 pressures, the inner $LiFeO_2$ and the outer $LiCrO_2$ are not dense but porous, so are less protective. On the other hand, 310S alloys with high Cr and high Ni contents do not show any dependence on fuel utilization. However, electrical conductivity is not good. Attempts to use Fe/Cr/M (M = Co, Mn, Ni) were made to provide improvements in corrosion resistivity, loss of electrolyte and electrical conductivity.

Corrosions in an anode atmosphere are so severe that there has been no appropriate method other than Ni clad. Ni clad can also partly prevent carburization of the alloys.

4.3 Development of MCFC stacks

When a large and tall stack consisting of many cells with a large effective electrode area (1000–10 000 cm^2) started to be constructed, there arose several technological problems to be solved. The most important point is the dimensional stability. For this purpose, the electrolyte matrix should be strong against the thermal stress and volume change on melting/freezing. In addition, anode materials should

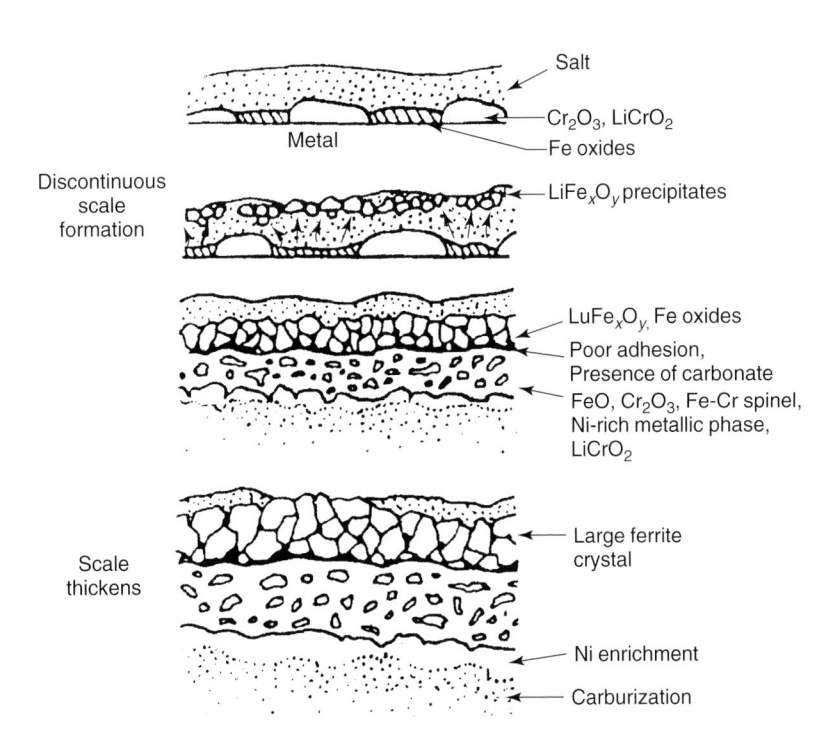

Figure 26. Oxide scale on alloys in anode environments in MCFC stacks. (Reproduced from Yuh et al. (1995)[14] by permission of Elsevier Science.)

be strong against creep either. The nondoped Ni anodes deformed significantly under a compressive load of stacks in such a way that they deformed by 50% within several days. Since anodes play an important role in holding extra electrolyte to be utilized after the electrolyte loss, the creep of anodes directly leads to a lowering of the cell performance. For long term use, the change in pore distribution or sintering is an important phenomena for the dimensional stability.

Sealing is the essential part in the construction the stacks. This gives rise to two major problems, electrolyte migration[62] and corrosion in the wet seal part.[39]

In a tall stack consisting of more than 100 cells built in the 1980s, a significant migration of electrolyte was observed. This is due to the creepage of electrolyte out of the effective cell parts to the side seal parts. When electrolytes in the respective cells are confined and separated from each other, not a large voltage is applied to the electrolyte. However, electrolytes came to the side seal parts and this electrically combined all the electrolytes. As a result, a large voltage is applied to the electrolyte and subsequently a large driving force for migration across many cells appears.

In addition, corrosion at the wet seal parts became a problem.[39] Seals of MCFC stacks were made by direct contact of the anode and the cathode plates on matrix materials containing the electrolyte. Although this wet seal is good for preventing gas leaks, this provides the environment for electrochemical corrosion. This is illustrated in Figure 27.[40] In the vicinity of the wet seals, there are three different gases: in the cathode atmosphere, the anode atmosphere and the air outside the seals. Inside the cells, CO_2 exists in both the cathode and anode, but does not in the air outside the cells. Furthermore, the distribution of water vapor is also different in these atmospheres. This is the driving force for corrosion in the wet seals. Since the electrolyte and metals are in direct contact in this area, there are several possibilities of local electrochemical cells. In addition, depending on the current density in the cells, the gaseous composition of the anode atmosphere and the distribution profile of the electrical potential and the chemical potentials of oxygen, carbonate ion and hydroxide ions change in the wet seal part. This leads to complicated features of corrosion due to local cells. The best way of avoiding this corrosion is to cut the electrical path in the local electrochemical paths. For this purpose, an aluminizing alloy in the wet seal parts provides a good coating of Al_2O_3.

The characteristic features to be improved in MCFC stack development can be summarized as cell performance and stack life:

1. Cell performance: Against the expectation in the early stages, high cell performance of MCFCs has not been obtained. The main factors for the decreasing performance originate from the large Joule loss in the electrolyte and from the large cathode over potential. The former is related to the rather large thickness of the electrolyte. The latter is related to the fact that CO_2 is involved in the cathode reaction as given in equation (1). A recent investigation clarified that the dissolution of CO_2 gas into molten carbonates is a rate-determining step.[68]

2. Life is limited mainly by the voltage decay in the operating stacks. This is determined by the following factors:
   (i) electrolyte loss (creepage and vaporization loss);
   (ii) NiO dissolution and resulting shorting;
   (iii) sintering and creep of porous electrodes;
   (iv) corrosion;
   (v) degradation of $LiAlO_2$;
   (vi) thermal cycling: volume change in melting and freezing;
   (vii) electrolyte redistribution.

Since some of these factors are described already above, electrolyte loss and shorting will be described below.

4.3.1 Electrolyte loss

Electrolyte loss occurs as a result of reactions with metallic elements (Cr or Al) during fabrication and of creepage and vaporization during operation. In addition to degradation of cell performance, electrolyte loss caused some problems in stack operation. Vaporized species were deposited in low temperature regions and caused blockage of the gaseous flow. In reformers, catalysts were damaged by the transported electrolytes. This vaporization/recondensation process continues via mass transfer. This, therefore, depends on many factors, such as temperature distribution, gaseous flow or other operational conditions. The vaporization occurs mainly in the form of KOH gas as written in equation (20). Thus, electrolyte composition changes due to migration may affect the vaporization rate.

4.3.2 Shorting

Shorting for the NiO cathode was found 1981. Before that time, the lithiated NiO was recognized as a much better cathode than Ag or CuO. Figure 16 implies that the solubility of NiO is not so small compared with those for Ag or CuO but the oxygen potential dependence is different. For Ag and CuO, solubility depends strongly on the oxygen potential. In the vicinity of the cathode, some oxygen potential gradient is developed. This suggests that

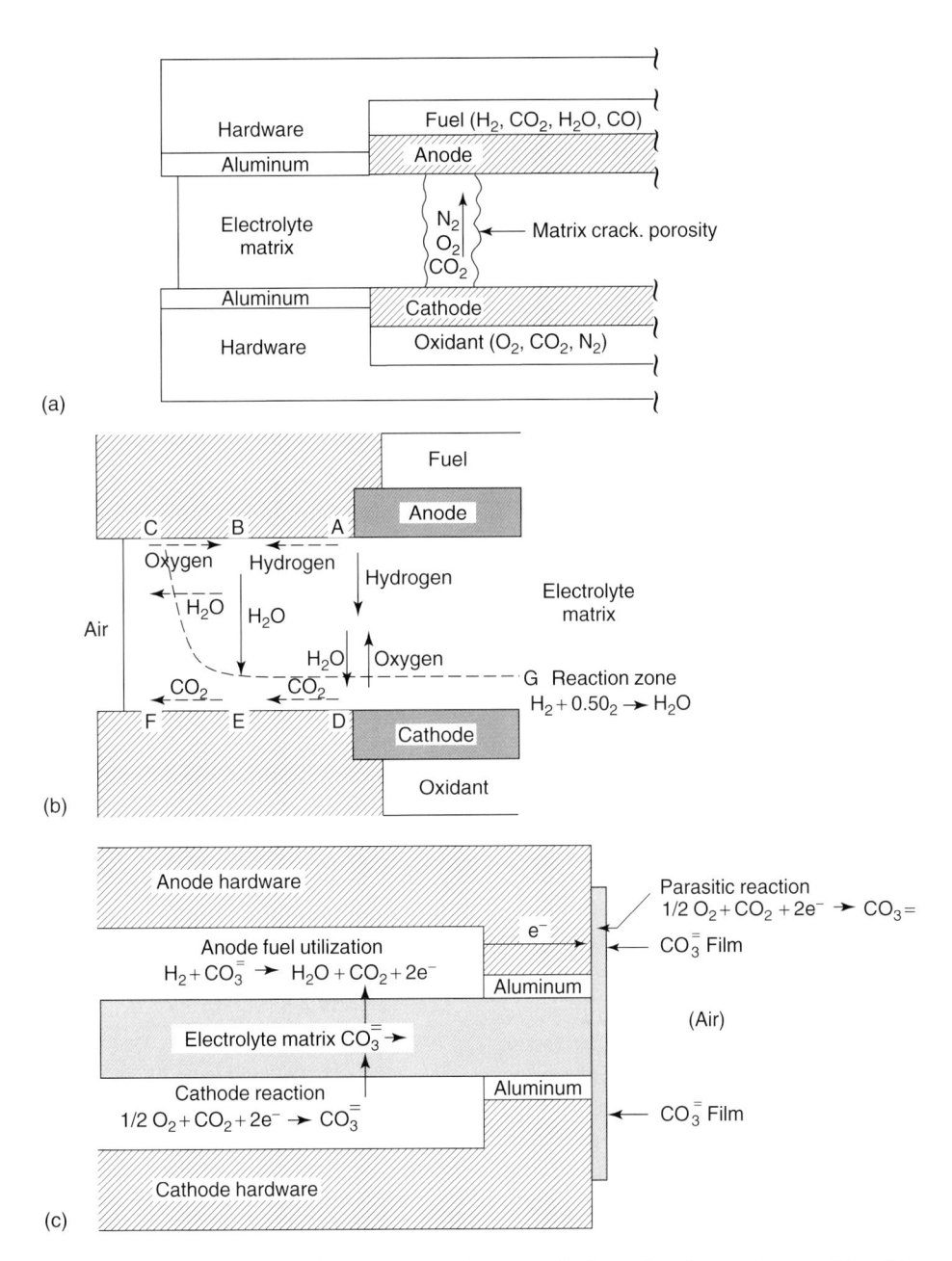

Figure 27. Parasitic Reactions in MCFCs. (a) Gas crossover due to crack formation in matrix materials, (b) gas flux in wet seal corrosions and (c) O_2 reduction on anode cell hardware in the presence of electrolyte Ni creep. (Reproduced from Iacovangelo and Jerabek (1986)[40] by permission of The Electrochemical Society Inc.)

the dissolved cathode can be easily precipitated shortly after some diffusion. This facilitates further dissolution. As a result, the shorting can easily take place. On the other hand, the solubility of NiO changes only in the reducing atmosphere. In the vicinity of the cathode, therefore, an immediate driving force for precipitation is not developed. This can explain the difference between NiO and Ag.

For the shorting due to the NiO cathode, detailed analysis has been made by experiments and simulation.[42] Figure 28 shows the experimental results on the distribution of deposited NiO particles in the electrolyte matrix at the operation time of 115, 334, 1124 and 3427 h. In the early stages of the cell operation, Ni was deposited at a position close to the cathode/matrix interface. This is explained by the difference in the solubility of hydrogen and oxygen.[64] Since the solubility of hydrogen is much higher, the reducing area expands closely to the cathode. The peak position in the distribution moved to the central part. This suggests that the oxygen potential distribution changes due to the changes in concentration of minor species, such

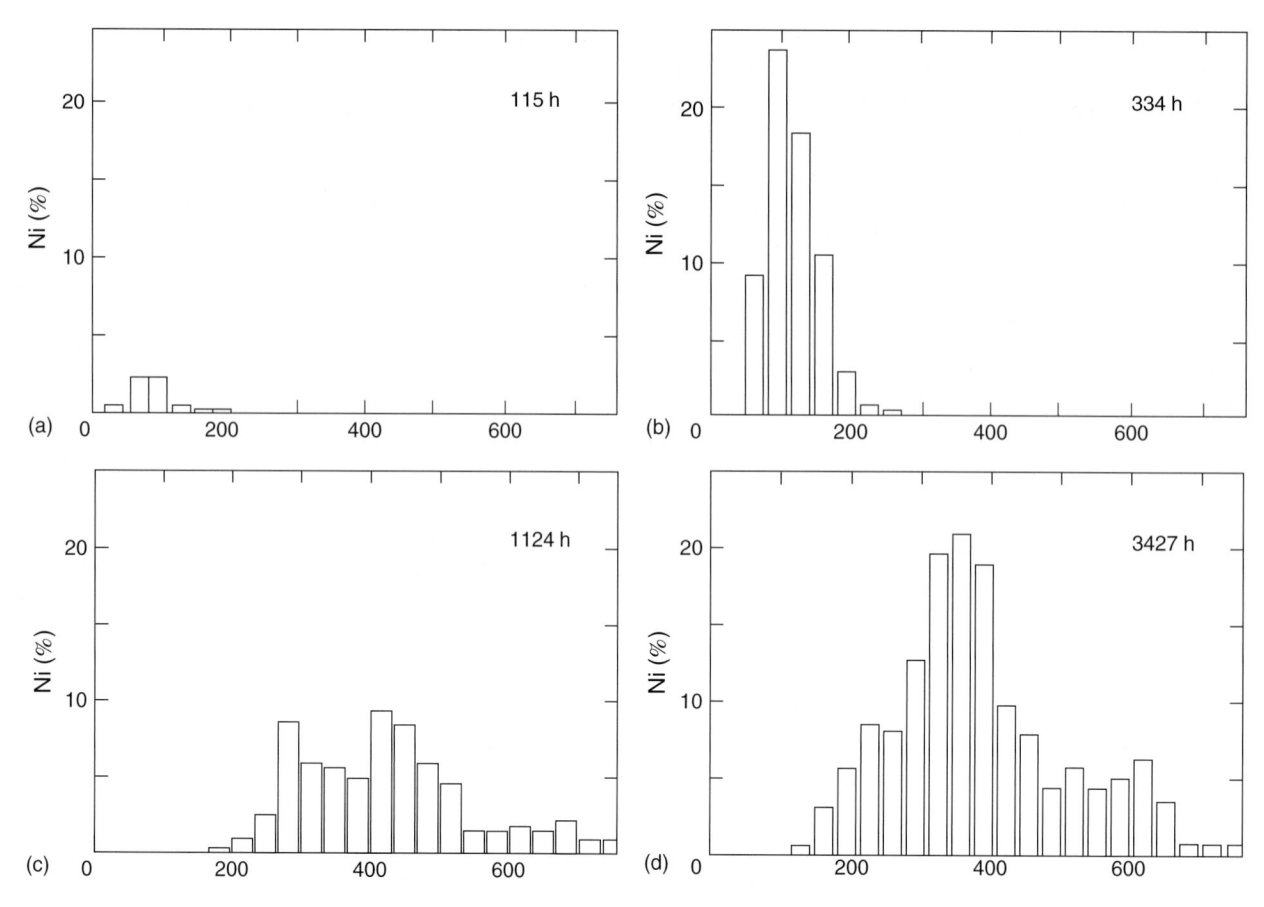

Figure 28. Deposited Ni particle distribution in the electrolyte matrix from the cathode to the anode interfaces with a parameter of operation time. (Reproduced from Mugikura *et al.* (1995)[42] by permission of The Electrochemical Society Inc.)

as O_2^-, O_2^{2-}, etc. Deposited Ni particles can dissolve again when the oxygen potential increases. Another factor for the moving of the peak position is the electrolyte redistribution. The NiO solubility depends on the Li_2CO_3 concentration. An increase in the K_2CO_3 component at the cathode side leads to higher NiO solubility. This may cause the change in distribution. A short circuit path can be setup by connecting deposited Ni particles. Shorting time depends on matrix thickness, carbon dioxide pressure and the NiO solubility. The matrix thickness should be determined from considering the cell performance and mechanical strength. Therefore, it is not practical to avoid shorting by thickening the electrolyte matrix. For cell performance and conversion efficiency, it is essential to operate MCFC at high pressures. The main efforts, therefore, have been focused on the search for other electrolyte compositions and the development of alternative cathodes.

4.3.3 Change in electrolyte composition

NiO solubility depends on the basicity of the electrolyte.[66] With increasing Li_2CO_3 content (activity), NiO solubility decreases. When the Li_2CO_3/Na_2CO_3 electrolyte system instead of the Li_2CO_3/K_2CO_3 system is used, NiO solubility decreases as expected from the basicity discussed by Ota *et al.*[66] Thus, many efforts have been made on the Li_2CO_3/Na_2CO_3 system. NiO solubility was found to be small and, therefore, shorting time is confirmed to be longer and is possible to operate at high pressures. Corrosion of 316L in the Li/Na system was severe at low temperatures but this can be improved by oxidizing with H_2O before exposure to the cathode atmosphere. Addition of other electrolyte components has been investigated.

4.3.4 Efficiency

Conversion efficiency is the most important factor for fuel cells. As described above, MCFC has some problems in cell performance; that is, the current density cannot be increased. In other words, at the low current densities, the voltage of MCFC stacks is high so that a high energy conversion efficiency can be expected for the low current density operation. However, the actually obtained efficiency is low so far. This is apparently due to the nonelectrochemical parts. These factors are:

(i) high fuel cell utilization operation cannot be made without energy loss;
(ii) heat management of gaseous flow is not appropriate;
(iii) for the reforming process, heats or fuels are utilized in an inappropriate manner.

From the early stages of development, ERC showed a strong interest in the direct fuel cell, that is, internal reforming. This is the proper selection in order to achieve high efficiency. External reforming loses fuels by just burning to provide the heat required for the reforming process. In internal reforming, however, heat can be supplied from the inside stacks. Furthermore, ERC proposed the gaseous flow from the anode to cathode to provide the CO_2. This gives better heat management and results in a better thermal efficiency. This, however, leads to the introduction of a large amount of water vapor in the cathodes. A similar idea was extended to develop the "hot module", which collects all facilities necessary to be operated at high temperatures into one hot module. This is an improvement to heat management outside the stacks.

5 DEVELOPMENT OF SOFCs

5.1 Key issues in the materials selection in SOFCs: chemical reactivity and mismatching in volume expansion

The advantages of solid electrolytes are the following:

(i) long stability of materials;
(ii) low reactivity and slow diffusivity.

On the other hand, disadvantages originate from the "solid" nature of the electrolyte:

(i) technological difficulty for obtaining good contact amongst materials;
(ii) ceramic materials are fragile so that it is difficult to handle large cells particularly in a planar design.

Since SOFCs consist of all solid components, the interface contact resistance is one of the most important technological issues. To achieve a good contact between dissimilar solid materials, it is essential to bond two materials in a chemically strong manner by means of high temperature heat treatment or by physical activation in dry processes. During such a fabrication process, however, chemical reactions or inter-diffusion may occur across the interface. Therefore, the chemical stability, the morphological stability and the compositional stability should be carefully examined when materials and associated fabrication techniques will be selected. The valence number or the valence stability is the convenient idea in such materials selection.

When the materials are selected and then fabricated into stacks, matching volume change becomes important to reduce the thermal and mechanical stresses. Volume and its expansion usually do not sensitively depend on minor components. Thus, doping or mixing in an appreciable amount becomes important. Volume change on transformation of the crystal structure and its dopant dependence should be known. In a similar manner, the volumes of materials can change on shift of the redox equilibrium with changing oxygen potential. For example, a large volume change is observed when rare earth doped ceria, $(Ce, Ln)O_2$, is reduced to have a large number of Ce^{3+} ions. This is usually called chemical volume expansion.

Material cost is also a big issue. Since the material cost is high for lanthanum or yttrium, the amounts used should be carefully examined. There are two approaches to reduce the cost. The first one is to replace them with cheaper materials without losing performance. A typical case is to use lanthanum concentrates instead of pure lanthanum in preparing perovskite materials, AMO_3 (A = La, Ln; M = Mn, Cr). The second approach is to increase the cell performance by using more expensive materials that can increase the performance significantly. The use of Sc-stabilized zirconia instead of YSZ is an example of this.

5.2 SOFC materials: requirements and selection

Although development for SOFC materials did not progress in the 1960s and 1970s, the basic selection of the electrolyte and electrodes was made in the beginning of the 1980s. In the 1980s and 1990s, refinement of the electrode materials and the development and testing of interconnects were performed with great success.

5.2.1 Electrolyte

The ionic conductivity of the solid electrolyte is smaller than that of the liquid electrolyte (PAFC, MCFC). Thus, higher conductivity as well as stability is the main criteria for selecting solid electrolytes. From the beginning, YSZ has been utilized because of its high ionic conductivity and stability.[55, 56] Although rare earth doped ceria[69] exhibits a higher conductivity than YSZ, this electrolyte has not been widely used because of its high contribution of electrons in the reducing atmosphere. At lower temperatures, such as 600 °C, the effect due to the electron conductivity becomes weakened compared with at 1000 °C.

The oxygen permeation through the electrolyte is important in maintaining high efficiency. The major factor in the determination of oxygen permeation is the electron or hole conductivity.

5.2.2 Cathode

In the early stages, other cathodes than lanthanum manganite-based perovskite oxides were tested. However, almost all stack developers now use this cathode. The main reasons are:[50, 56]

1. As oxide cathodes, the mixed (ionic and electronic) conductive nature is preferable. This suggests that perovskite oxides based on transition metal oxides are good candidates. Among perovskite oxides, lanthanum cobaltite has an excellent conductivity and catalytic activity. However, the thermal expansion coefficient is much higher that of YSZ and reactions with YSZ are severe, leading to degradation at 1000 °C. To decrease the thermal expansion coefficient, Sr-doping is effective. However, the reactivity of the Sr component with YSZ is much higher so that $SrZrO_3$ formation will give rise to degradation. To use $LaCoO_3$, therefore, it is necessary to use nondoped $LaCoO_3$ below 900 °C with an appropriate method to moderate the mismatch in thermal expansion.
2. $LaNiO_3$ shows a similar trend to that of $LaCoO_3$. The reactivity with YSZ is more significant.
3. $LaFeO_3$ does not react with YSZ but its performance is not good enough. Sr-doping enhanced the reactions with YSZ.

Although $LaMnO_3$ is rather stable, there remained some issues about the reactivity with YSZ. As practically usable cathodes, the SrO-doped and the CaO-doped lanthanum manganites have been developed by WHPC and Dörnier, respectively. The thermal expansion coefficient can be controlled by the amount of dopants.

5.2.3 Anode

Initially, Ni and Co were tested as anodes. Co showed a stronger tolerance against sulfur. However, Ni became common because of its price and good activity. Since Ni exhibits a slightly higher thermal expansion coefficient (12×10^{-6}), the compatibility with YSZ becomes the key issue in constructing stacks. One common way to moderate this mismatch is to use a cermet anode consisting of ceramic YSZ and metallic nickel. Another issue is sintering of the Ni particles, leading to degradation.

5.2.4 4 Interconnect

Compared with the electrolyte and electrodes, the selection of interconnects remains undetermined; however, there have been many ideas proposed about the development of interconnect. Interconnect chemistry and its related technology is one of the most important technological issues in SOFCs. This is not only the matter of material development but is also deeply concerned with the strategic issues in constructing stacks.

From the chemical point of view, lanthanum chromite based perovskite oxides are the most promising candidates despite their poor air sinterability. The first technological issue associated with the $LaCrO_3$ interconnect was on the fabrication process for making a dense film of this material. Although the oxide interconnect was initially fabricated by the EVD process, the air sinterability is essential for the establishment of SOFC stacks. The second issue was related to the thermal expansion coefficient. Since $LaCrO_3$ has a lower thermal expansion coefficient (9.4×10^{-6}) than that of YSZ (10.6×10^{-6}), the alkaline earth oxide is doped to increase the thermal expansion coefficient (see Figure 29).[70] Alkaline earth oxide doping is also preferable for increasing hole conductivity. However, the chemical volume expansion due to the reduction of Cr^{4+} also increases with increasing alkaline earth oxide content (see Figure 30).[71] There are several ways to overcome this mismatch in thermal and chemical volume expansion, and the respective methods adopted in different strategies will be described.

The use of a metal interconnect is another big issue concerning interconnect materials, stack design and heat management.[53] Since the high temperature corrosion in the air and anode atmosphere is severe at high temperatures, such as 1273 K, adopting a metal interconnect is inevitably accompanied with lowering the operation temperature (1073 K is one typical temperature).

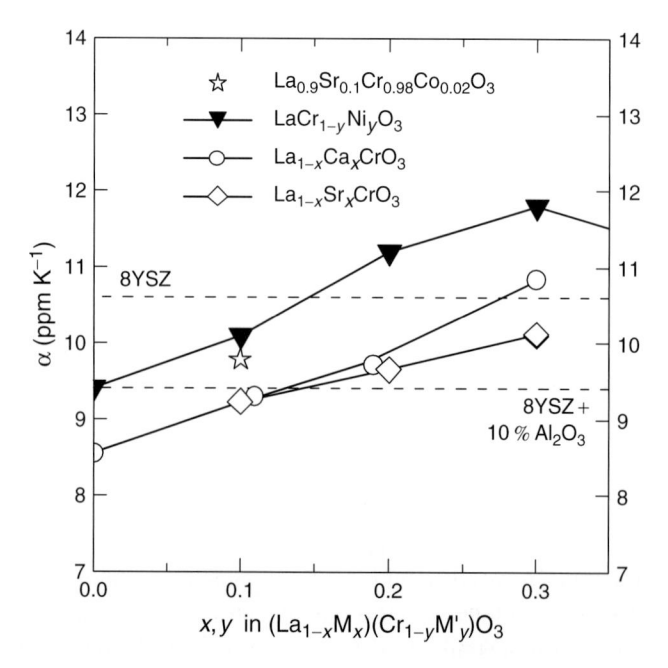

Figure 29. Thermal expansion coefficient of doped $LaCrO_3$ and comparison with YSZ and Al_2O_3 with added YSZ.[70]

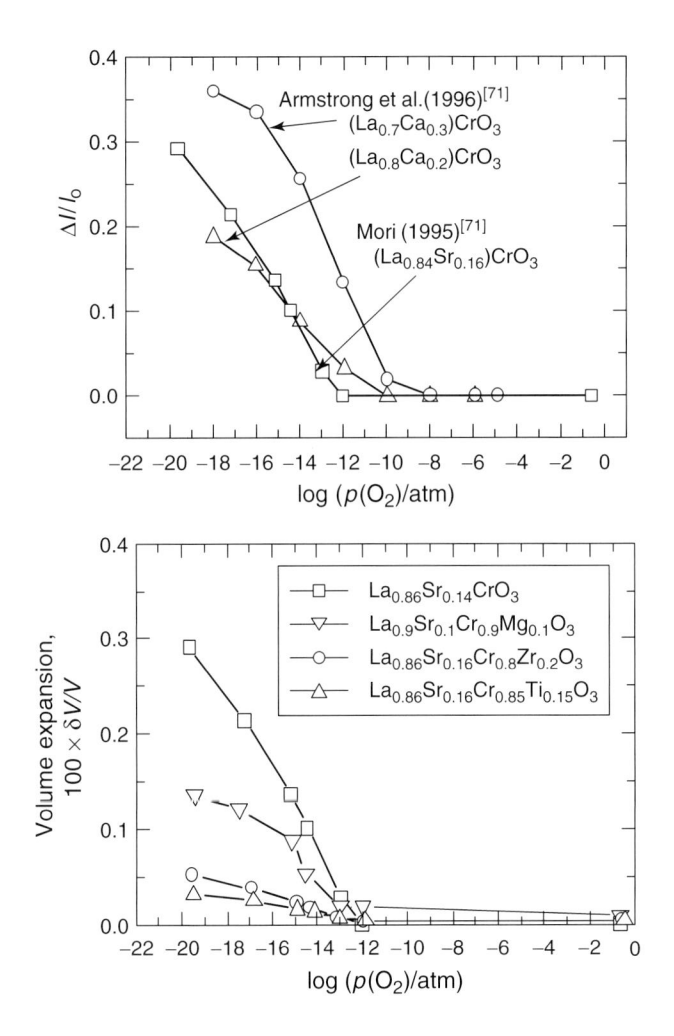

Figure 30. Chemical volume expansion of doped $LaCrO_3$.[71]

From the stack configurational point of view, there is essentially no difference between the oxide and the metal interconnects. However, the choice of the oxide or the metal interconnects affect many features in stack development. For example, metals cannot be fired in air with other oxide components so this affects the chronological sequence of stack fabrication. Although metal interconnects have a lower chemical stability, they exhibit excellent electrical and thermal conductivities and also have a good thermal shock resistance.

5.3 Characteristic features of materials, stacks developments in SOFCs

The developments of the materials and fabrication techniques cannot be separated from those of stack design. In this sense, the biggest technological issue is about the fundamental design, that is, tubular vs. planar. As will be described below, a variety of cell/stack designs have been proposed. These are closely related with the fabrication techniques adopted for selected materials in a selected design.

One issue is the chronological sequence of fabricating stacks. There are several ways:

(i) starting with the cathode: this is adopted in the WHPC tubular cell;
(ii) co-firing of cathode/electrolyte/anode;
(iii) starting with anode: anode-supported thin electrolyte.

Such a sequence becomes important when a particular interface exhibits extremely temperature-sensitive characteristics, the YSZ/lanthanum manganite is one of such interfaces. There is a general trend that when an electrode is fabricated together with the densification of electrolyte, the electrode activity gives a higher performance.

5.4 Seal-less tubular stacks by WHPC and its relation to stack development issues

The most important development in SOFC materials, fabrication technique and design was made by workers in WHPC around the early 1980s.[8] They invented the seal-less tubular stacks shown in Figure 5. The cell components and this fabrication technique in the earliest stage are given in Table 1. The key ideas about the stacks are:

1. a long stripe shape is adopted for the interconnect/electrolyte connection to minimize the interface length (see Figure 5(a));
2. by using a one-ended tube together with an alumina inner tube (see Figure 5(b)), it is possible to eliminate the parts to be sealed;
3. as cathode materials, the Sr-doped $LaMnO_3$, $(La_{0.84}Sr_{0.16})MnO_3$, is coated on calcium stabilized zirconia.
4. to fabricate a dense film of YSZ (10% Y_2O_3 stabilized zirconia) and 5–10% MgO-doped $LaCrO_3$, $La(Cr_{0.95}Mg_{0.05})O_3$, the EVD)technique was adopted (see Figure 22);
5. Ni is used as an anode. Ni (or Co) is also used as a thermodynamically stable cell-to-cell contact (see Figure 8). As a result, fuels flow outside the cells.

The disadvantages of this tubular cell were recognized as follows:

1. Although the EVD process is excellent in fabricating a good quality of cells, the fabrication cost is quite high.
2. Researchers in the Argonne National Laboratory[47] made analysis on this design and found that the major contribution to energy efficiency loss comes from the in-plane resistance of the cathode. According to this

criticism, WHPC made the modification of adopting the air electrode support tubular cells to reduce the in-plane resistance in the cathode. Still, this problem remains for tubular cells.

3. Since the EVD process was adopted, only MgO doped $LaCrO_3$ can be fabricated because of the thermodynamic limit as will described later. The thermal expansion coefficient of $La(Cr,Mg)O_3$ is lower than that of YSZ as shown in Figure 31.[17] From the thermal stress point of view, this mismatch between the interconnect and the electrolyte is one weak point.

Despite the disadvantages given above, the sealless tubular cells made epoc-making breakthroughs. Thus, it will be useful to discuss the importance of the sealless tubular stacks in relation to the technological issues.

Although the planar design is common in many fuel cells, initial attempts on planar SOFCs encountered severe problems associated with crack formation and sealing. In the planar design, there are wide areas to be sealed. In addition, mismatch in the thermal expansion coefficient gives rise to bending or crack formation. Even when crack formation is overcome, bending will also lead to a severe situation for side seals and electrical contact between two plates.

In view of this difficulty in fabricating planar stacks, it was not surprising that the main efforts were focused on tubular designs from the beginning. An initial stack design was the "Bell and Spigot" type or "banned" type as shown in Figure 4. Both can be categorized as a segmented-cell-in-series type. Generally speaking, the tubular is much stronger than the planar type ceramics. Even so, there were still severe problems of gaseous leakage and interface contacts among materials. The following can be pointed out as improvements made by WHPC:

1. The electrolyte film is fabricated by the EVD technique (see Figure 22). This makes it possible to fabricate a thin and dense film on supporting tubes. Since any pores in the electrolytes and interconnects can be closed during the EVD process, almost perfect gas tightness can be achieved. This process also makes it possible to prepare electrodes with an excellent performance.

2. The electrolyte to interconnect configuration. One of the most interesting points in the tubular design is that the electrolyte and interconnect materials are not fully overlapped but form a connected series. In other words, the cathode and anode are connected to the interconnect in the smallest area. This connection is not favored for the electrical path resistance because the in-plane resistance increases.[47] However, it is favored for the mechanical stability. For example, in the early WHPC tubular cell, the EVD-YSZ film is connected to the EVD-$La(Cr_{1-x}Mg_x)O_3$ film. Although matching in the thermal expansion is not completely enough, as shown in Figure 31, this tubular cell is rather stable against thermal cycles. This can be ascribed to the mechanical allowance of the electrolyte/interconnect dense tubes.

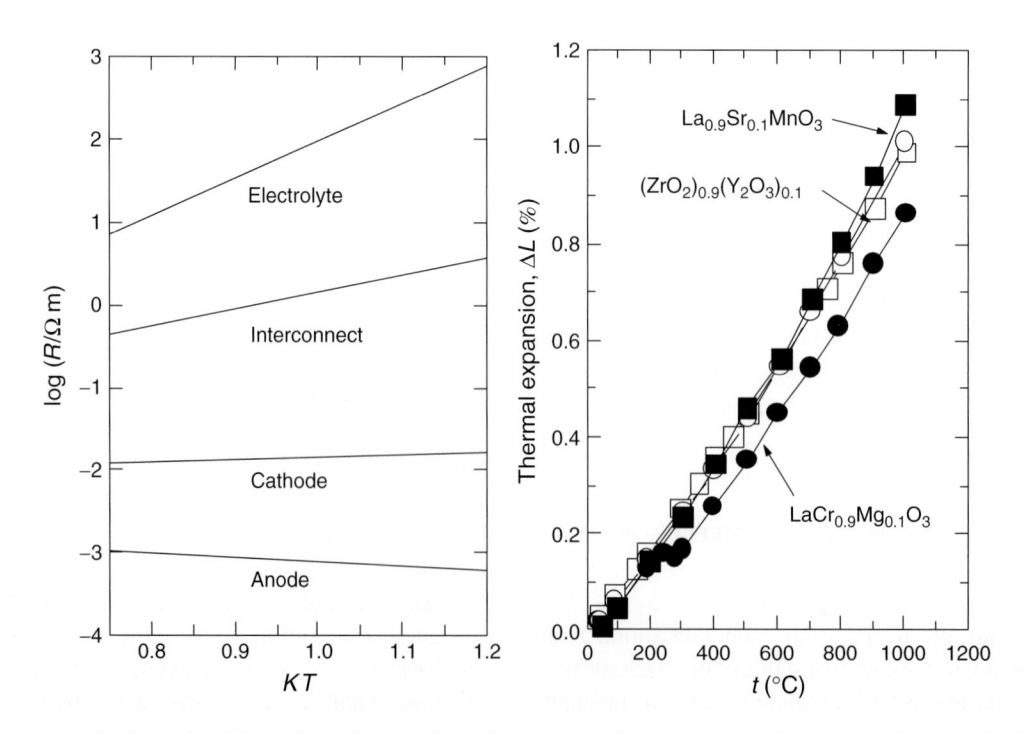

Figure 31. Fundamental characteristics of materials used in the early sealless tubular cells developed by WHPC. (a) electrical conductivity and (b) thermal expansion coefficient.

3. Cell-to-cell connection is made outside the tubes. This makes it possible to connect in the fuel environment using metals. Bundles consisting of several tubes are connected with each other by Ni felt. This can act as a buffer for absorbing any slight deformation of the tubular cells. This structure makes the stacks strong against the large temperature distribution along the tubes. On the other hand, the segmented-cell-in-series type tubular stacks have no such an allowance because the metallic materials area are used in a confined area. In view of this, this makes it difficult to make modules out of segmented-cell-type cells.

5.5 Various attempts after WHPC's achievement

In the mid-1980s, the Argonne National Laboratory proposed a monolithic concept[48] (Figure 7) for fabricating planar cells. This was challenging in the following ways:

1. They proposed a way of overcoming some disadvantages of the sealless tubular stacks. The power density of the tubular stacks is not high, whereas the honeycomb-like monolithic design makes it possible to increase the power density. This actually makes it not strange to discuss the possibility of SOFCs to be applied to automobiles and other transportation applications.
2. To eliminate thermal stress, they emphasized the matching thermal expansion coefficients.
3. They also proposed to adopt a co-firing technique to make stacks into a monolithic structure using only one sintering step. This needs to match the sintering behavior as well as the thermal expansion of all the cell components to be co-fired. Adoption of the co-firing process makes it possible to reduce the fabrication costs drastically. Note that the EVD process is unrealistically expensive as a fabrication process in a commercialization stage.

The monolithic cells were investigated in the late 1980s in the Argonne National Laboratory and were followed by Allied Signal. These activities in the USA in the early 1980s stimulated the interest of many groups in Japan and Europe. From the end of the 1980s, there appeared four new technological strategies:

1. Planar cells were constructed by developing the oxide interconnect. This can be categorized as similar attempts to the monolithic cells. The main efforts have been made in the development of the wet/sintering process. The major developers are Ceramatic (later SOFCOs) in the USA, Dörnier in Germany, Mitsubishi Heavy Industry Kobe (MOLB), Murata manufacturer, Osaka Gas, Fuji, Mitsui Shipenary in Japan and Risø National laboratory in Denmark.
2. Tubular cells were constructed with an oxide interconnect by the wet process. This was challenged by ceramic companies in Japan like TOTO. Since the design modification is not needed, efforts have been focused on the development of the fabrication technique.
3. Construction of segmented-cell-in-series type stacks with some modifications: despite some difficulties, the segmented-cell-type cells have some advantages. Mitsubishi Heavy Industry (Nagasaki) adopted banned type cells and continued the investigation in this type. Asea Brown Bovari A.G. (ABB) proposed a new design similar to the segmented-cell-in-series type. The cells are fabricated on Ca-stabilized zirconia support plate, which makes an air inside/fuel outside system.[72, 73] Rolls Royce also proposed a similar stack but based on an air outside/fuel inside system.
4. Construction of cells with metal interconnects: Attempts to use metal interconnects have been made by many groups with different ideas. One group are focused on small systems based on a round type, they are Z-tek, Fuji and Sulzer. The other groups are based on square types, they are Siemens (Erlangen), Sanyo, Ceramic Fuel Cells Limited and Global thermoelectric.

In addition to these attempts, WHPC has made also their own efforts to reduce material and fabrication costs and to modify designs to improve efficiency.

5.6 Wet/sintering process for stacks with oxide interconnect

WHPC's achievement in developing SOFC stacks stimulated some groups to apply the wet/sintering process for the fabrication of tubular cells. There were three materials issues associated with the wet/sintering process: (1) the chemical reaction between lanthanum manganite and YSZ, (2) the preparation of cermet anodes by the slurry coat technique and (3) the air sinterability of the $LaCrO_3$-based interconnect.

Figure 32 shows the composition range of $LaMnO_3$ that can be separated into two regions: the nearly stoichiometric region exhibits the reaction with YSZ to form $La_2Zr_2O_7$, whereas the MnO-rich (La-deficient) region can be in equilibrium with ZrO_2.[50] $La_2Zr_2O_7$ formation can be avoided when the La-deficient $LaMnO_3$ is used as the cathode. The interaction between the La-deficient lanthanum manganite and YSZ at temperatures higher than 1473 K lead to the degradation of cathode characteristics

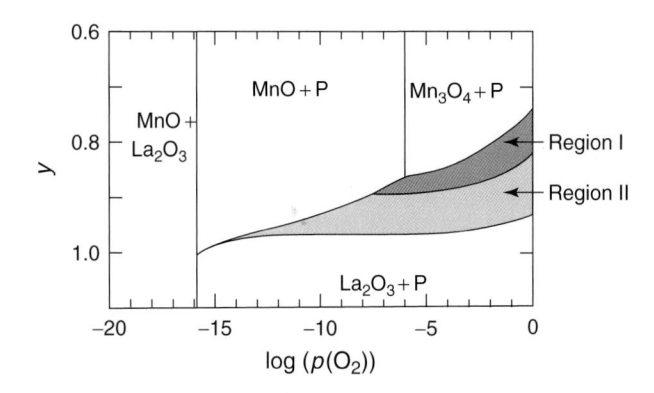

Figure 32. The reactivity of LaMnO$_3$ with YSZ. In the nearly stoichiometric composition region (Region II), a small amount of the La$_2$O$_3$ component in LaMnO$_3$ reacts with ZrO$_2$ to form La$_2$Zr$_2$O$_7$, whereas the La-deficient La$_{1-y}$MnO$_3$ (Region I) does not react with ZrO$_2$ but Mn dissolution to ZrO$_2$ occurs.[50]

due to the drastic change in the microstructure as shown in Figure 33. To overcome this problem, there are two ways. One is to keep the higher heat treatment temperature to ensure the long-term stability without losing high performance. The second way is to complete the sintering process of a YSZ film at temperatures lower than 1473 K with highly active fine particles of YSZ.

The second issue is the performance of an anode prepared by sintering a mixture of NiO and YSZ and subsequently being reduced to a Ni/YSZ cermet in situ inside the cells. It was found necessary to carry out a preheat treatment of mixtures and a subsequent firing with the electrolyte at high temperatures of around 1400–1500 °C, respectively.

It appeared that to achieve a good performance by cells prepared by the wet/sintering process, it is better to first fabricate the anodes at high temperatures and then the cathodes at lower temperatures. This is a typical sequence.

The third important issue was the air sinterability of the oxide interconnects. The most interesting achievement was that for Ca-doped LaCrO$_3$, good sintering was achieved by adopting a Ca-excess (Cr-deficient) composition (see Figure 34).[49] The Ca excess has a two-fold effect: One is a decrease in the Cr vapor pressure to prevent the vaporization/recondensation mechanism, which creates Cr$_2$O$_3$ as a result of the decomposition of LaCrO$_3$ (the presence of Cr$_2$O$_3$ prohibits further densification). The second effect is to form liquids (calcium chromates) as a sintering aid. These only exist in the intermediate stage of sintering and will disappear into grains after the densification process. The chemical volume expansion of Ca-excess LaCrO$_3$ (Figure 30) is, however, so large that it was clarified that this material cannot be applied to planar cells due to crack formation or bending. When the Sr-doped LaCrO$_3$ was adopted, the Sr-excess composition could not be used for air sinterability because La$_2$O$_3$ is formed as a second phase at the grain boundaries, resulting in disintegration due to the large volume change due to hydroxide formation with water. Good sintering of the Sr-doped LaCrO$_3$ is achieved by additional doping, such as a small amount of V$_2$O$_5$ or CoO in the Cr site.

The thermal expansion of LaCrO$_3$ is smaller than that of YSZ, and about 30% doping of Ca or Sr-doping can improve the mismatch as shown in Figure 29. Such a heavily alkaline earth doped LaCrO$_3$ exhibits a large chemical volume expansion as shown in Figure 30. This dilemma can be overcome by the following two ways.

One way is to adopt the low doping level (10%) of alkaline earth oxide to reduce the chemical volume expansion. To match the volume expansion coefficient, alumina is added to YSZ. This approach was adopted by Osaka Gas/Murata manufacturers and seemingly by Dörnier. The former group constructs the connection-type planar cells, which consist of unit cells made with one cathode/electrolyte/anode tri-layered plate combined with an oxide interconnect plate having grooves that are contact with the cathode in the electrolyte layer side as shown in Figure 35.[73, 74] The air channels are formed inside the cells. Fuel flows outside so that Ni felt can be used as a cell-to-cell connection in a similar way to the sealless tubular stacks.

The second way is to adopt the high concentration (30%) of alkali earth oxide for matching thermal expansion. To avoid a large chemical volume expansion, additional doping of Ti or Zr was attempted into the Cr-sites of (La,Sr)CrO$_3$. This was adopted by MHI Kobe for MOLB-type stacks.[52, 71] Such doping lowers the concentration of Cr^{4+} in the oxidative atmosphere and, therefore, the

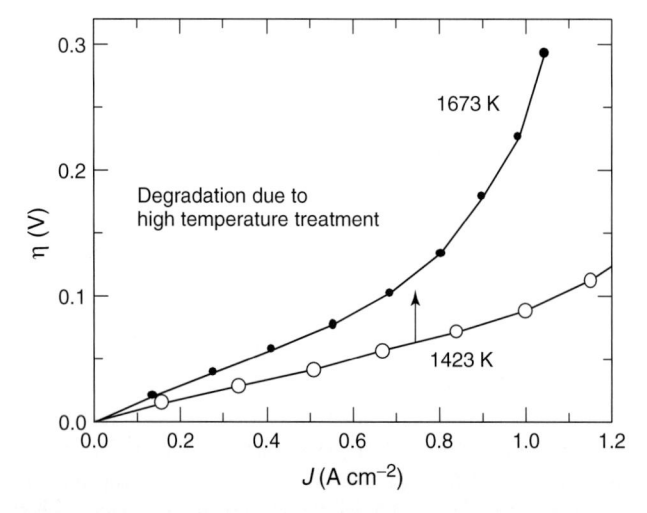

Figure 33. The difference in the La-deficient lanthanum manganite cathode performance between 1673 and 1423 K for the applied temperature on YSZ. (Reproduced from Mori et al. (1990)[73] by permission of The Electrochemical Society of Japan.)

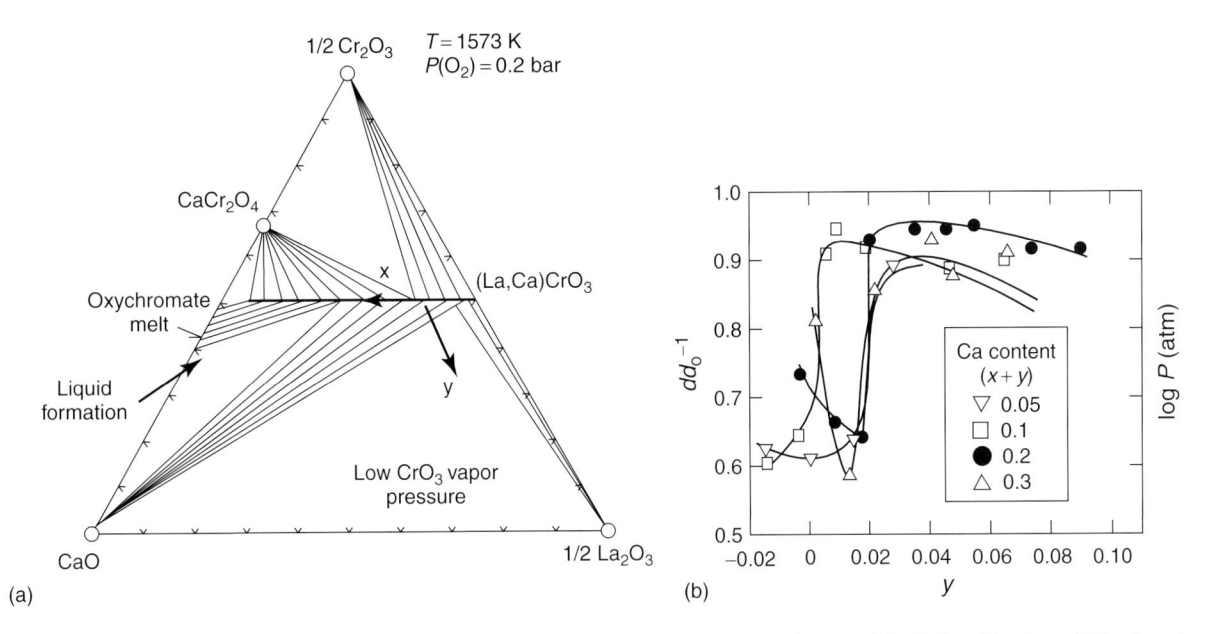

Figure 34. Phase relations in the La/Ca/Cr/O system in air and the air sinterability of Ca-doped $LaCrO_3$. Air sinterability is enhanced by the decrease in chromium vapor pressure and by the formation of the liquid phase during sintering for the Ca-excess (Cr-deficient) $(La,Ca)CrO_3$ perovskites. (Reproduced from Sakai *et al.*, *Solid State Ionics*, **40/41**, 394 (1990) by permission of Elsevier Science.)

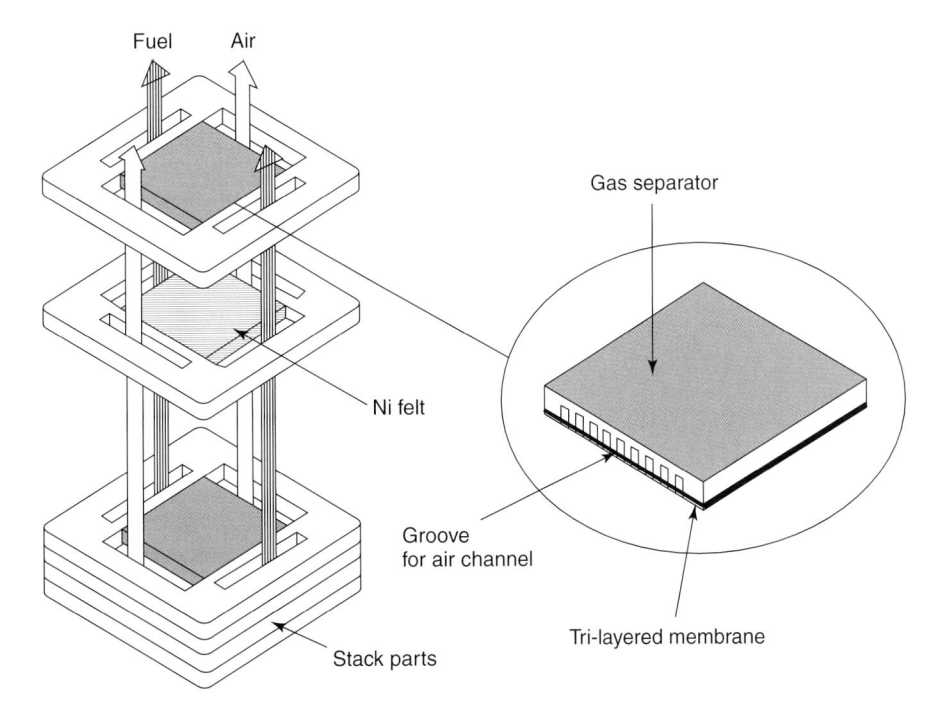

Figure 35. OG type planar cells. Electrolyte plates and interconnect plates are firmly connected. (Reproduced from Chikagawa *et al.* (1994)[74] with permission by The Solid Oxide Fuel Cell Society of Japan.)

chemical volume change due to reduction of the Cr tetravalent ions. Such doping, however, lowers the hole conductivity, which is govern by the Cr^{4+} ions.

For both cases, it is essential to have tight bonding between the electrolyte and interconnect. Thus, the geometrical configuration of these two components is important for the magnitude of mismatch allowed in volume expansion.

Note that the development of monolithic cells was stopped due to the difficulty of simultaneous sintering of the cell components. For the densification of the $LaCrO_3$-based interconnect, liquid substances should be formed during sintering process.[61] However, such liquids needed in sintering will escape through the interconnect layer to the electrode layers due to capillary forces. The electrodes should be porous but will be densified in the presence of the liquids. Unless the $LaCrO_3$ interconnect is used, simultaneous sintering is impossible. Instead of one step sintering, MHI Kobe fabricated the stacks into the monolithic structure by several steps. These respective planar stacks have similar features to those of the WHPC's sealless tubular cells in the following ways:

1. The Osaka gas type planar stacks[74] consist of combined unit cells that have air inside. This allows the use of nickel felt to connect the cells in fuel environments and gives some flexibility in stacking cells even for planar stacks. On the other hand, the interconnect and the electrolyte are tightly connected at wide areas so that a small mismatch in volume expansion gives rise to the stress concentration in contrast to some allowance given in the sealless tubular. For this purpose, the volume change on transformation of $LaCrO_3$ at around 500 K should be avoided by appropriate doping of $LaCrO_3$.
2. In MOLB-type stacks, on the other hand, the honeycomb-like structure was adopted. A dimple-type of waved electrolyte plates make it possible to manually handle such electrolytes. The interconnect plate is flat so as not to cause concentration of stress. The electrolyte and interconnect are connected only at the top and the bottom of the dimples. This can absorb any slight deformation in electrolyte. The interconnect plate is flat so that chemical volume expansion can be avoided carefully. It is interesting to note that despite the planar design, the monolithic cell and the MOLB-type SOFCs do not have a complete connection between the paralleled electrolyte and interconnect. This must provide greater strength against thermal cycles (see Figure 36).

5.7 New challenges in tubular designs

Improvements by SWPC have been made on the fabrication technique, materials selection and also stack design.[17]

1. The adoption of the air electrode supported cell will now be discussed. This is a natural extension, since the largest contribution to the Joule loss came from the in-plane resistance of cathode (65%).[47] One way to overcome this issue was to adopt a thick cathode substance instead of a thin cathode substance on a supported tube.

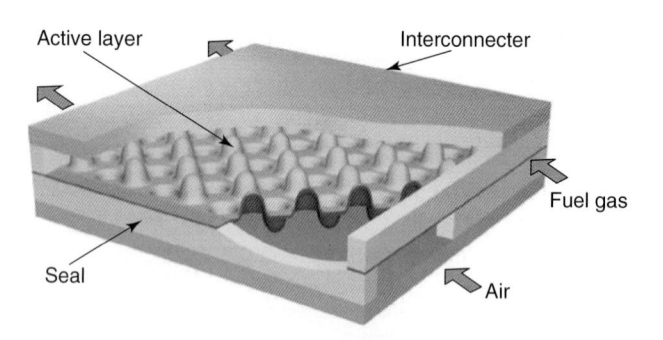

Figure 36. MOLB type planar cells. This can be regarded as one advanced type of the monolithic cells. The parts where the electrolyte plates and the interconnect plates are connected becomes narrower compared with the monolithic cells. This leads to greater strength against thermal cycles. (Reproduced by permission of Mitsubishi Heavy Industry.)

2. Adoption of the spray technique to coat the Ni particles in the electrolytes.
3. Adoption of the plasma spray technique for the deposition of the $LaCrO_3$-based interconnect and electrolyte.
4. Enlargement of the tubular cells. Since these tubular cells are cathode-supported, enlargement of the cells can be made by increasing the size and thickness of the cathode by keeping the thickness of the electrolyte.

These attempts are primarily to reduce the fabrication and materials costs.

To reduce the fabrication costs, it is essential to apply the wet/sintering process. This has been mainly made for planar stacks as described above. However, the same technique can be applied for fabricating the tubular cell. Particularly, TOTO[51] adopted the Ca-excess $(La,Ca)CrO_3$ and the A-site deficient lanthanum manganite $(La,Sr)_{1-y}MnO_{3-d}$ for the sealless tubular cells. They succeeded in firing the A-site deficient $LaMnO_3$ with YSZ at high temperatures (about 1673 K) without degradation. To apply the wet/sintering process to the sealless tubular cells, it should start with the cathode supporting tubes to be fired with the electrolyte at relatively high temperatures (1673 K). This gives rise to significant improvements for tubular cells:

(i) fabrication cost will be significantly reduced,
(ii) the electrical conductivity of the A-site deficient lanthanum manganite is higher so that the Joule loss contribution should be reduced,
(iii) the mismatch in the thermal expansion coefficient between YSZ and the interconnect $(La_{0.8}Ca_{0.21})CrO_3$ was much improved so that the thermal cycling characteristics became much better, as given in Figure 37.[51]

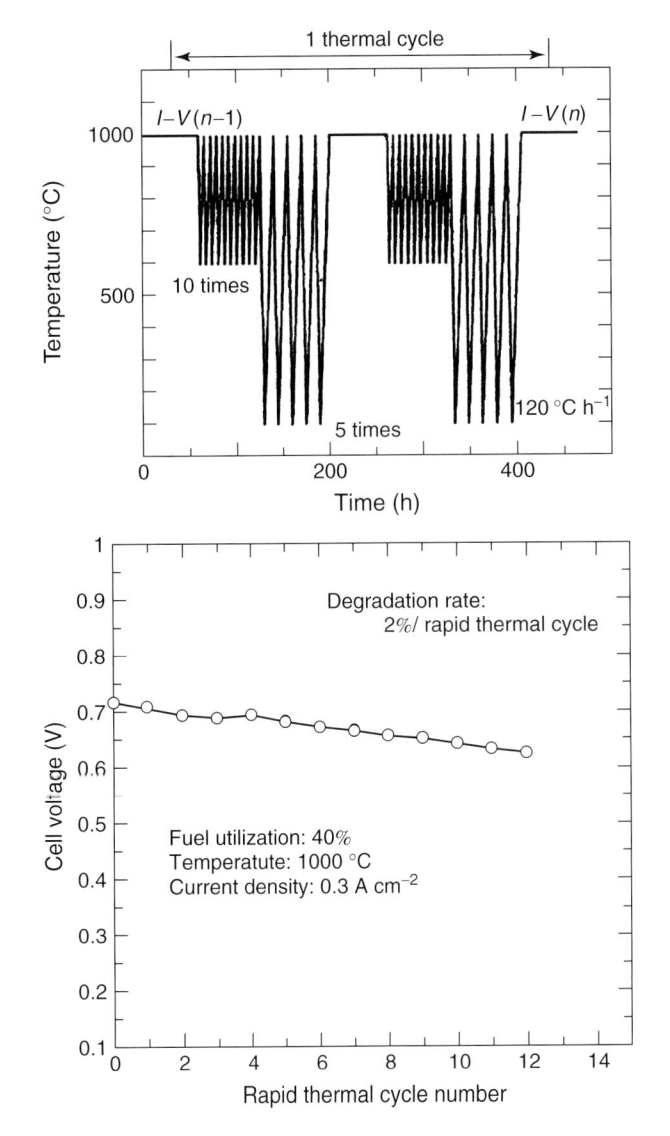

Figure 37. Thermal cycle test for TOTO cells. In TOTO cells, the thermal expansion coefficient mismatch has been improved very much and leads to low degradation during accelerated thermal cycle tests. (Reproduced from Takeuchi et al. (1999)[51] by permission of The Electrochemical Society Inc.)

5.8 Development in segmented-cell-in-series type tubular and related stacks

MHI Nagasaki fabricated the segmented-cell-type cells by adopting the plasma spray technique with following materials: $LaCoO_3$ for the cathode, YSZ for the electrolyte, Ni/YSZ for the anode and Ni/Al alloys for the interconnect. They constructed 10 kW high pressure stacks and confirmed that these cells are stable in the frequent thermal cycle test and in a long durability test. These are surprising results in the sense that these materials have large mismatches in the thermal expansion coefficients – that of $LaCoO_3$ is 20×10^{-6}. They also used metal interconnects without severe problems. Recently, they adopted the wet-sintering method to fabricate the segmented-cell-type cell. They adopted a lanthanum manganite cathode and a non-$LaCrO_3$ but titanate-based oxide interconnect.[75] This development of the interconnect is a new challenge because it has been long believed that there is no good non-$LaCrO_3$ oxide interconnect because the sinterability and the electrical properties are a trade off. In other words, titanate-based perovskites are stable and well sinterable but not good electronic conductors. They enhanced the electrical properties by the doping technique. Even though the conductivity on the cathode side is poor, this is similar situation to the Ni/Al alloy interconnect.

There have been made several attempts in modifying the segmented-cell-type tubular cells. ABB and Rolls Royce proposed a cell-in-series on flat tubes, as shown in Figure 38. Rolls Royce made some improvements to make a thin but well-conductive cathode by adopting a compositionally gradient structure between cobaltite and manganite.

5.9 Planar stacks with metal interconnects

Many technological issues arose about the utilization of metal interconnects, some of them remain still unresolved. These are:

(i) thermal expansion mismatch with YSZ;
(ii) stability vs. conductivity of the oxide scale;
(iii) Cr poisoning of the cathode;

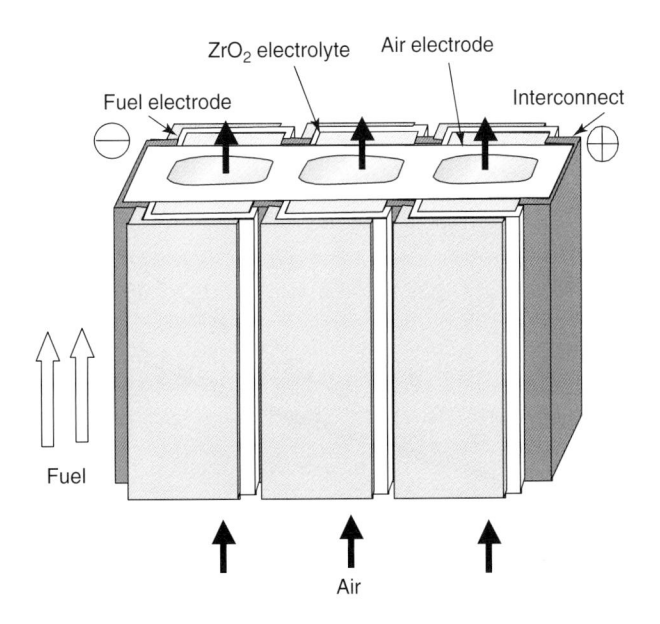

Figure 38. ABB stack design categorized as the modified segmented-cell-in-series type design.[72]

(iv) interface resistivity;
(v) sealing.

In the beginning, several attempts were made to allow the mismatch in the thermal expansion coefficients between YSZ and the metal interconnect by using soft seal materials containing silica and other glass forming oxides. Reactions of such sealing glasses and metal interconnects were severe. Migration of the sealing glass occurred, leading to performance degradation. The thermal cycling characteristics were not good enough on the sealing parts. This approach was not successful.

The main efforts were focused on developing new alloys having the same thermal expansion coefficient as YSZ. This was made by Plansee and Siemens. The resulting alloy was Cr5Fe1Y$_2$O$_3$.[53] More recently, Hitachi metals developed a Fe-based alloy, whereas Sumitomo Special Materials developed Fe/Cr/W alloys, which have similar thermal expansion coefficients to YSZ (Figure 39).

The oxidation of alloys cannot be avoided. To develop alloys suitable for SOFC applications, materials design of the oxide film on alloys is important. One way is for the alloy design to form an appropriate oxide scale by optimizing the alloy composition. Coating of well-characterized oxides is another way.

1. Cr-rich alloys provide a continuous oxide scale made of Cr$_2$O$_3$. This is in many cases dense and protective against further oxidation. On the other hand, such a stable and protective Cr$_2$O$_3$ layer has a small number of defects so that the electrical conductivity is low. When Cr content is small, oxide scales consist of the inner Cr$_2$O$_3$-rich layer and the outer Fe-rich layer. This Cr$_2$O$_3$ is usually porous and allows further diffusion of the Fe component through the oxide scale. This is not protective and corrodes faster.
2. When Cr$_2$O$_3$ covers the surface of interconnects, the Cr vapor starts to evolve from the surface as a result of reaction of Cr$_2$O$_3$ with oxygen or water vapor to form CrO$_{3(g)}$ or CrO$_2$(OH)$_{2(g)}$. This chromium vapor attacks the cathodes and leads to severe degradation.[76]

To avoid chromium poisoning of the cathode performance, several attempts have been made:

1. The use of a chromium getter: Sanyo used La$_2$O$_3$ as a Cr getter. La$_2$O$_3$ reacts with chromium vapor to form LaCrO$_3$. However, La$_2$O$_3$ is hygroscopic and easily reacts with water vapor to form La(OH)$_3$. This is not appropriate for SOFCs. Other developers used electron conductive perovskites, such as LaCoO$_3$.
2. Coating the electron-conductive oxides can be a much better solution to this problem. LaMnO$_3$-based or LaCrO$_3$-based perovskite oxides were attempted. The chemical properties of the interface between the perovskite oxides and alloys have been examined to see a change in electrical conductivity. In many cases, for the use of (La, Sr)MO$_3$, electrically conductive strontium chromates are formed at the interfaces so that the interface resistivity does not increase in the initial period of testing time. However, after several thousands of hours, the resistivity starts to increase due to the formation of a scale consisting of Cr^{3+} containing oxides. As fabrication technique, the vacuum plasma spray, has been tested with good result on adhesion.[77] However, a more cost-effective fabrication process is needed to coat the electronic conductive oxides onto the alloys.
3. A recent investigation by Tokyo gas[78] has revealed that Cr poisoning depends strongly on the electrolyte/electrode combinations. Sanyo already revealed that the degradation due to Cr poisoning strongly depends on the performance of lanthanum manganite cathodes on YSZ; that is, with increasing overpotential, degradation increases, leading to catastrophic damages. When the electrode and electrolyte change to ceria and lanthanum cobaltite such Cr poisoning disappeared. This suggests one possible way of avoiding Cr poisoning.

The most interesting and important finding by Sanyo is that the Cr poisoning does not decrease with lowering temperature. This is due to the fact that as the temperature decreases, the overpotential of the cathode increases and the driving force for Cr poisoning will increase. On the other hand, with decreasing temperature, the vapor pressure of the chromium oxide vapor will decrease. In the presence of

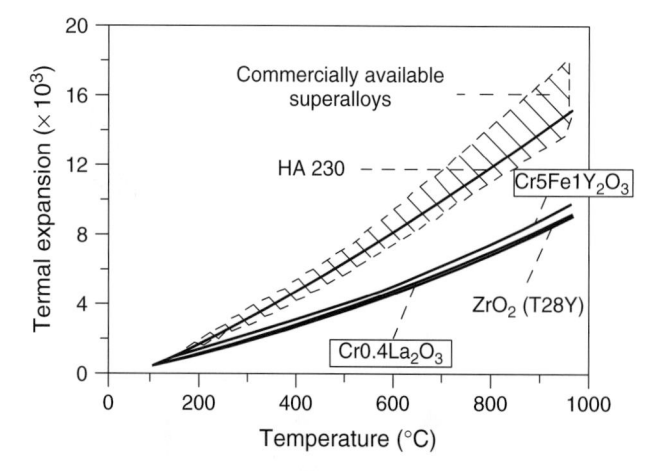

Figure 39. Thermal expansion coefficients of alloys tested for SOFC interconnects. (Reproduced from Quadakkers *et al.* (1994)[53] by permission of European Fuel Cell Forum.)

water vapor, however, the total vapor pressure increases and the temperature dependence is moderate. In other words, the effect of water becomes significant with lowering temperature. A similar effect of water on cathode reaction mechanisms has also been pointed out recently.

In the case of the metal interconnect, it has become popular to use a conducting layer between the metal interconnects and electrodes, mainly because the high temperature heat treatment for those interfaces, including metals, should be made in an inert atmosphere and, therefore, is not appropriate from a cost performance viewpoint.

From the viewpoint of the fabrication costs, it is highly necessary to use commercial alloys without coating. However, in the cathodes, Cr poisoning and the high resistivity of the stable oxide scale require a coating of perovskite oxide. In the anodes, the enhanced oxidation due to water vapor in the fuel environments after high utilization has been well recognized.[79] Furthermore, in the fuel environments even at low fuel utilization, carburization may have become a problem when methane or other hydrocarbons are present without being fully reformed. This suggests it might be necessary to use clad or coating on the anode side as well. In such a case, diffusion between the clad materials and alloys becomes important.

Many efforts have been made to lower the operation temperature of the SOFCs. From the Joule loss constraint, there can be only few ways in selecting electrolytes:

1. Anode support zirconia-based electrolyte: Since zirconia-based electrolytes have a rather high activation energy for the temperature dependence of the ionic conductivity, it is necessary to fabricate a thinner film of electrolyte. One way is to use anode-supported cells.
2. Ceria-based electrolytes:[69] When electrolyte supported cells are fabricated for intermediate temperature SOFCs, alternative electrolytes with a higher ionic conductivity should be developed. Most common electrolytes are rare earth-doped ceria. The materials problems of ceria-based electrolytes have been recognized as follows: (i) high electron conductivity due to the reduction of Ce^{4+} to Ce^{3+}; (ii) chemical volume expansion also due to the reduction; (iii) segregation of dopant along the grain boundaries; and (iv) mechanical strength is not enough,
3. Lanthanum gallate-based electrolyte: Ishihara[80] found that the $(La, Sr)(Ga, Mg)O_3$ perovskite showed a higher conductivity than that of YSZ. Since the magnitude of the electron and hole conductivities is low enough, this can be used as a SOFC electrolyte. The issues to be improved are (i) high vapor pressure in the reducing atmosphere; (ii) low mechanical strength; and (iii) fast interdiffusion with other components.

The stack development for intermediate temperature SOFCs has been focused on the anode-supported cell. The technological merits and demerits can be summarized as follows:

1. One merit is that cathode can be applied after the fabrication of the anode/electrolyte. This makes it possible to adopt a low temperature heat treatment for the cathode.
2. One demerit is due to the fact that the anode consists of two materials (NiO and YSZ) having different thermal expansion coefficients. Furthermore, on reduction of NiO to Ni, the volume changes drastically. When compared with the cathode support, the fabrication of anode support has many more difficulties.

For intermediate temperature SOFCs, improvement in the electrode activities is also a big issue. Particularly, the cathode shows a rapid decrease in activity with decreasing temperature. Although $LaMnO_3$ is recognized as a nonmixed conductive material and is not appropriate as a cathode at lower temperatures, recent attempts have revealed that a composite cathode consisting of $LaMnO_3$ + YSZ shows a good activity. As mixed conductors, $(La, Sr)(Co, Fe)O_3$ cathodes have been developed by Imperial college. Although the electrode activity is good, this material reacts with YSZ. For the prevention of such reactions, a thin layer of ceria has been successfully inserted between the YSZ electrolyte and LSCF cathode.

Construction of stacks with metal interconnects have been made by many groups. Siemens (Erlangen) and Ceramic Fuel Cells Corp. built several 10 kW modules. Sulzer Hexis has focused on a small system.

Anode supported cells have been constructed by Allied Signal, Julich Research Center, ECN and Global Thermoelectric. Particularly, Global Thermoelectric has constructed the planar cells consisting of a $LaMnO_3$/YSZ composite cathode, anode substrate, YSZ thin electrolyte and metal interconnect. They also have adopted a compression seal to build several kilowatt planar modules.

5.10 Other important topics in SOFC materials, processing, and stack developments

5.10.1 Electrolyte

Some attempts have been made to use partial stabilized zirconia (PSZ, 3% Y_2O_3 doped zirconia) instead of fully stabilized zirconia. This makes it easy to handle in electrolyte plate in fabrication, although the ionic conductivity is lower than YSZ. In many cases, the cathode performance is systematically lower when PSZ is used with the same cathode. This is due to the change in the crystal structure on manganese dissolution into PSZ.

5.10.2 Interconnect

The EVD process could not be applied to fabricate (La, Sr)CrO$_3$. This was attempted by WHPC with failure due to the thermodynamic limitation as given in Figure 12(b). This feature suggests that such materials as (La, Sr)CrO$_3$ will be weak against attack from impurities containing chlorine. Many attempts have been made to improve the air sinterability. One is co-firing of LaMnO$_3$/LaCrO$_3$. In some composition ranges, co-sintering can be achieved. However, the extraordinarily fast diffusion remained in the grain boundaries of perovskites, and as a result of diffusion under an oxygen potential gradient, the materials became completely porous.

5.10.3 Anode

To obtain the mechanical stability against the thermal and atmospheric cycles, Dörnier added Ceria (5 vol.%) to Ni/YSZ cermets. Their anode survived even after exposure to air and subsequent reduction of Ni/NiO. They explained that a small amount of ceria helps to bridge nickel particles. Murata added ZiSiO$_4$ to cermet to adjust the thermal expansion mismatch with YSZ. CRIEPI added coarse YSZ particles to a normal mixture of fine YSZ and NiO. This gave rise to an improvement in volume change behavior on reduction.

5.10.4 Other materials problems associated with mass transfer under cell operation

When internal reforming is taken place in cells, the anode temperature is lowered due to the endothermic nature of the reforming reaction. As a result, the poisoning effect of silica appears. Silica is reduced to form SiO$_{(g)}$ in the anode atmosphere. When the anode is cooled, the attack by SiO$_{(g)}$ becomes significant. This is analogous phenomena to Cr poisoning in the cathode. Similarly, use of silicone rubber should be avoided because silicone rubber emits silicon-containing species under a continuous flow.

6 COMPARISON BETWEEN MCFCs AND SOFCs IN STACK DEVELOPMENT

One of the advantages of MCFCs is use of a liquid electrolyte having such a high ionic conductivity. On the other hand, the thickness of the electrolyte layer is in the order of 1–2 mm in the state-of-art stacks. This should be compared with 50 μm of YSZ electrolyte thickness in the SOFC stacks. The width of the MCFC electrolyte is determined mainly by the following reasons:

(i) NiO dissolution and the resulting shorting depend on the width of the electrolyte matrix. To make the stack life reasonably longer, it is essential to have the thick layer of electrolyte.
(ii) The large cells with a large active area need to have a strengthened electrolyte matrix.

When compared with SOFCs, the large stack size of MCFCs is one of their advantages. Since the electrolyte melting/freezing process is a high barrier for MCFC to be stable against frequent thermal cycles, this leads to an idea that MCFCs should be operated as the base load without frequent thermal cycles. Enlargement of the cell size is therefore consistent with this idea. This means that to establish the advantage of making large area cells and to surmount the disadvantage of NiO dissolution, the electrolyte matrix is widened.

The lower operation temperature (650 °C) of MCFCs, derived from the rather high ionic conductivity, is also one of advantages against SOFCs. This makes it possible to use the metal interconnect and, therefore, to reduce material costs. On the other hand, the corrosive nature of the molten carbonates makes it necessary to use coating or cladding in many places. These are needed to fabricate a complicated structure of materials and, therefore, increase the fabrication costs. Furthermore, the cathode performance is not so good compared with the expectation that the contact between electrolyte, electrode and gas is easily established and the reaction rate can be expected to be fast. Recently, this was confirmed to be due to the slow dissolution of CO$_2$ gas into the molten carbonates.

The main technological issues associated with MCFC stacks have been resolved physicochemically, and some appropriate methods have been adopted for surmounting the corrosion of the metal interconnect, dissolution of the cathode and the resulting shorting and electrolyte loss. Even so, these still remain the major factors dominating the life and cost of MCFC stacks.

For MCFCs, it can be said that the advantages so gained are offset by a whole series of disadvantages. Efforts to overcome the disadvantages will be still needed, including development of materials and fabrication techniques.

The historical sequence of SOFC developments is somewhat different from that of MCFCs. For SOFCs, the ionic conductivity of YSZ at 1273 K is one order of magnitude lower than that of the molten carbonate at 923 K. This makes it necessary to operate SOFCs at higher temperatures. In addition, it is technologically required to examine the feasibility whether or not the thin YSZ electrolyte can be fabricated into stacks. This technological challenge has been successfully overcome. For example, 50 μm thick YSZ can be successfully fabricated by the EVD process as well as by the inexpensive wet/sintering method. This clearly shows that the disadvantage of the low ionic conductivity of a solid electrolyte can be surmounted by fabrication techniques. On the other hand, the main difficulty of fabricating SOFC stacks questions how to keep good electrical contacts among cell components in a gas-tight structure. WHPC adopted the physical fabrication process to overcome this difficulty to show the technological feasibility of SOFC stacks. Actually, their stack, after being fabricated successfully with the electrical contacts in a sealless structure, exhibits excellent long-term stability. When change is made in a fabrication technique, many technological issues have to be re-examined. For example, change to the wet/sintering process required confirmation concerning whether or not the electrode activity does not degrade after high temperature heat treatment. Furthermore, optimization has to be made for sintering behavior, thermal expansion and chemical expansion in fuel cell environments.

In view of these features, the most important stage of the SOFC stack development appears even in the smallest stack size. For the tubular cells, several kilowatts to several tens of kilowatts in size is a critical size, whereas for the planar, the several kilowatt size is very important and needs to have a longer period of time. During this period of time, materials, materials processing and cell design have to be investigated from the same strategy. Among the many technological key issues, the selection of the interconnect materials is the most important. This is because the interconnect should be dense and has to be connected firmly with the YSZ electrolyte. In this sense, the presently well-established SOFC stacks, which can be defined as the first generation SOFC stacks, can be categorized in terms of the oxide interconnect used and the fabrication method, as describe above.

SOFCs with metal interconnects can be categorized as the second generation. The developing pattern of the second generation of SOFCs is not similar to the first generation SOFC but similar to that of MCFCs in a sense that the SOFC stacks can be fabricated even thought almost all materials problems have not been fully resolved. In another sense, the long stability issue remains unresolved. In addition, similarity can be also found in the physicochemical properties of alloys in the cathode and anode atmospheres.

7 FUTURE TRENDS IN MATERIALS AND SYSTEM DEVELOPMENTS

Advantages of the high temperature fuel cells against the low temperature fuel cells should be further developed.

In such future developments, the following features in the electrochemical conversion methods should be considered together with other characteristics:

1. The 100% fuel utilization cannot be achieved. Except for pure hydrogen fuels for PEFC or PAFC, fuels cannot be consumed completely in fuel cells to avoid the rapid voltage loss due to the Nernst term. This is one weak point of fuel cell systems.
2. Fuel cells can be characterized as one of membrane reactors. This implies that scale-merit is small. The facility and materials costs tend to be high compared with thermal engines, which have large scale-merits. On the other hand, the high efficiency can be expected even for small systems. This in turns implies the importance of heat management in a small system, which significantly affects the efficiency. Furthermore, mass transfer associated with the reactions in the fuel cells is dominated by only one chemical species so that it makes it feasible to treat reaction products from the membrane reactor point of view. For example, reaction products exited from the anode compartment do not contain nitrogen so it makes it easy to separate CO_2 from the reaction products. Condensation of water vapor in the reaction gas leads to concentration of CO_2.
3. Electrochemical cells as key technology for the conversion among electricity, chemical energy and heat. Fuel cells provide electricity and heat from fuels. Heat is usually a secondary product. However, there are a few fuel cells that can provide electricity and chemicals from fuels and heat by adopting a specific reaction such as the partial oxidation of methane (equation (32)). This shows some possibility that the electrochemical energy conversion can provide better utilization of energy by means of the mutual conversion among electricity, chemicals and heat. Fuel cell is free from Carnet's constraints about the limit of the conversion from heat to work. In this sense, there can be many ways to utilize heat and chemical energy by adopting various electrochemical reactions.

On the basis of the above characteristics of the high temperature fuel cells, the following should be the important points in the future development of fuel cell systems.

7.1 Higher conversion efficiency

To maximize the conversion efficiency, the combination of fuel cells and gas turbines is essential. Generally speaking, gas turbines and other thermal engines have a weak point in that energy is lost during the combustion process, whereas fuel cells can generate electricity in the oxidation process but have some difficulty in utilizing 100% of fuels. In view of this, the two systems are complementary to each other in maximizing the conversion efficiency. Here, however, the two systems have different features on scalability. For gas turbines, the highest efficiency can only be achieved in the largest size at a high operation temperature, whereas the large scale-merit cannot be expected for fuel cells as described above. There are two ways: One is based mainly on the large size gas turbines combined with SOFC as a topper, which has attracted interest in the Japanese electricity utilities. The other is based on an expectation that the conversion efficiency of fuel cells alone is higher than that of large gas turbines, leading to a system consisting mainly of fuel cells together with small gas turbines. Conceptual work on "ultra fuel cell" by the USA Department of Energy is based on this idea. To maximize the efficiency, in "ultra fuel cells" it is assumed that gas exhausted from fuel cells should be small in amount and high in temperature. For this purpose, inside fuel cells, temperature distribution should be large to ensure heat exchange to cool the systems with a small amount of gases. This implies that the requirements for future fuel cells should be (i) to exhibit a high performance over a wide temperature range and (ii) to be mechanically strong against the temperature distribution. At the current stage of the technology, these requirements are severe for MCFCs and for planar SOFCs with oxide interconnects. For tubular SOFCs, the stability against the temperature distribution has been improved, and as a result the flow ratio of air to fuel becomes smaller. Even so, it will be hard to cover a wide range of operating temperatures, particularly down to 600 °C.

Even for the case where high temperature fuel cells are used as stand-alone, heat management is important to obtain high efficiency. Particularly, MCFCs have to have a good heat management system for reforming and gaseous circulation. Heat management for small SOFCs is also important. In a large SOFC system in a size of several hundreds of kilowatts or larger, it is not so difficult to install the pre-reforming unit with a gas circulation system to obtain the effective utilization of high temperature heat and reaction products, water and CO_2. In a small SOFC system, however, it becomes harder to maintain the good heat management associated with the reforming process. Gas circulation makes the small SOFC system complicated. In such a small system, another method of achieving the high efficiency should be examined. One possible way is direct the electrochemical oxidation of methane and other hydrocarbons.

7.2 Fuel flexibility

For the high temperature fuel cell systems, the most important advantage against the low temperature fuel cells is that a wide range of fuels, particularly carbon containing fuels, can be utilized without difficulties. In the near future, utilization of natural gas will be primarily important for stationary applications. In addition, gasoline, diesel and other liquid hydrocarbon fuels will be important for transportation applications. As a long-term target, coal related gases, methane to be produced from biogases or anaerobic digester gases will become crucial.

The high temperature fuel cell systems consist of fuel processing, fuel cells and the remaining fuel optimization process. When the fuel is changed, all the features of the fuel cell system have to be examined. Particularly, impurities and carbon deposition are examined. For MCFCs, the impurity effect in coal gases has been systematically examined. It becomes important to manage the complicated fuel processing and the remaining fuel optimization process.

The key points in fuel flexibility should be therefore the following:

(i) The ease fuel processing in context of the energy conversion?
(ii) The kinds of materials problems that occur when a simpler fuel processing system is adopted?

7.3 Separation of CO_2

One merit of the electrochemical cells is the possible removal of CO_2 out of the reaction products. For SOFCs, CO_2 can be separated after burning has been carried out with oxygen rather than air. For this purpose, the membrane reactor can be utilized. Even for exhausted gas from fuel cells, the remaining fuels make the oxygen potential low enough to provide a driving force for oxygen to be transported from the air into the remaining fuel area. Out of the resulting mixture of CO_2 and water vapor, CO_2 can be removed effectively. This has been and will be conducted by Praxiair/SWPC and Shell hydrogen/SWPC.

7.4 Wider applications

The first-generation tubular SOFC is suitable to use for stationary power generation as a base load supplier. One promising way of the SOFC future is to establish power generators that produce electricity with the highest conversion efficiency in combination with gas turbines.

Since the SOFC efficiency does not depend on stack size, the determining factor is how to manage heat flow and chemical flows in the combined cycles. There are great possibilities for even small SOFC systems of several megawatts to achieve the high efficiency. For industrial applications, a large size facility is required. In this sense, from the beginning of their development, MCFCs have been recognized as appropriate for industrial cogeneration purposes. In addition, there is a great possibility to use MCFCs in the co-production of chemicals in the chemical industry.

For the utilization of SOFCs in other applications, such residential use or automobile applications, further developments will be needed in materials, processing and design. For such an application, the system size will not be big. Instead, rapid start-up characteristics and fast load following behavior are required. For these new applications, new materials are highly hoped for the electrolytes, electrodes and interconnects. Particularly, in many cases, innovative utilization of metals is highly required in many respects. For example, the thermal shock resistivity of ceramic components can be improved by adopting materials with high thermal conductivities and low thermal expansion coefficients, or by adopting a fine structure with short relaxation times for thermally equilibration. When YSZ was selected as an electrolyte, improvement could not be made in the thermal properties of YSZ. An improvement should be achieved (1) by adopting an appropriate design or (2) by using YSZ with other materials appropriate for this purpose. From the former point of view, microtubes or honeycomb type structures are interesting. From the latter point of view, the use of metals is crucial because metals are good thermal conductors.

7.5 Further developments in materials technology

To further develop high temperature fuel cell systems, materials innovation is crucial. To develop one material, it is also essential to confirm the material's compatibility with other materials. Furthermore, equilibrium properties and kinetic properties are both critical for the future development of high temperature fuel cells. Recent development of high temperature fuel cells is based on time consuming investigations on phase relations, solubility determinations, examinations of chemical stability, etc. In future investigations, therefore, it is highly necessary to establish sophisticated, systematic ways of utilizing the accumulated physicochemical properties of materials and performance data in fuel cell environments. Concerning the equilibrium properties, progress has been achieved for the fuel cell materials. It has become not difficult to predict the stability of materials in fuel cell environments and materials interfaces. The kinetic properties are, however, in a stage of accumulating the properties for different materials. It is highly hoped to establish a more elegant way to predict the performance of fuel cell materials during operation. Advances in such a field will help to widen the applications of the high temperature fuel cells.

REFERENCES

1. US Department of Energy, 'Fuel Cell Handbook,' 5th edition (2000).
2. B. S. Baker (Ed), 'Hydrocarbon Fuel Cell Technology,' Academic Press, New York (1965).
3. F. Jones, 'Fuel Cells with Molten Carbonate Electrolyte', in "An Introduction to Fuel Cells", K. R. Williams (Ed), Chapter 7, Elsevier (1966); J. G. Smith, 'Solid Oxide Electrolytes', in "An Introduction to Fuel Cells", K. R. Williams (Ed), Chapter 8, Elsevier (1966).
4. W. Vielstich, 'Fuel Cells Modern Processes for the Electrochemical Production of Energy', Verlag–Chemie (1965) English version, translated by D. J. G. Ives, Wiley Interscience, London (1970).
5. J. R. Selman, *Int. J. Energy*, **153** (1986).
6. J. T. Brown, *Int. J. Energy*, **209** (1986).
7. N. Q. Minh, *J. Am. Ceram. Soc.*, **76**, 563 (1993); N. Q. Minh and T. Takahashi, 'Science and Technology of Ceramic Fuel Cells', Elsevier, Amsterdam (1995).
8. A. O. Isenberg, *Solid State Ionics*, **3/4**, 431 (1981); A. O. Isenberg, 'Cell Performance and Life Characteristics of Solid Oxide Electrolyte Fuel Cells', in "Proceedings of the Conference on High Temperature Solid Oxide Electrolytes", pp. 5–15, August 16–17 (1983); 'Anion Conductors', F. J. Salzano (Ed), Brookheaven National Laboratory, Vol. 1, BNL51728 (1983).
9. W. Dönitz, E. Erdle, W. Schäfer, R. Schamm and R. Späh, 'Status of SOFC Development at Dornier,' in "Proceedings of the 2nd International Symposium on Solid Oxide Fuel Cells", F. Grosz, P. Zegers, S. C. Singhal and O. Yamamoto (Eds), Commission of the European Communities, Brussels and Luxembourg, pp. 75–84 (1991).
10. A. J. Appleby, 'Fuel Cell Technology: Status and Future Prospects,' *Energy*, **21**, 521 (1996).
11. A. J. Appleby, 'Molten Carbonate Fuel Cell Technology – Status and Future Prospects,' in "Proceedings of the International Fuel Cell Conference", NEDO, pp. 141–144 (1992).
12. L. M. Paetsch, J. D. Doyon and M. Farooque, 'Review of Carbonate Fuel Cell Matrix and Electrolyte Developments,' in "Proceedings of the 3rd International Symposium on Carbonate Fuel Cell Technology", D. Shores, H. Maru, I. Uchida and J. R. Selman (Eds), The Electrochemical Society, PV93-3, pp. 89–105 (1993).
13. D. A. Shores, 'Materials Problems and Solutions in the Molten Carbonate Fuel Cell', in "High Temperature

Electrochemical Behaviour of Fast Ion and Mixed Conductors", F. W. Poulsen, J. J. Bentzen, T. Jacobsen, E. Skou and M. J. L. Østergård (Eds), Risø National Laboratory, pp. 137–150 (1993).

14. C. Yuh, R. Johnsen, M. Farooque and H. Maru, *J. Power Sources*, **56**, 1 (1995).

15. D. T. Hooie and E. H. Camara, 'Onsite Industrial Applications for Natural Gas-Fueled Fuel Cells', Fuel Cell Seminar, pp. 182–185 (1985).

16. S. C. Singhal, 'Solid Oxide Fuel Cell Development At Westinghouse', in "Proceedings of the 2nd International Symposium on Solid Oxide Fuel Cells", F. Grosz, P. Zegers, S. C. Singhal and O. Yamamoto (Eds), Office for Official Publications of the European Communities (1991).

17. S. C. Singhal, 'Recent Progress in Tubular Solid Oxide Fuel Cell Technology', in "Solid Oxide Fuel Cells", U. Stimming, S. C. Singhal, H. Tagawa and W. Lehnert (Eds), The Electrochemical Society, Pennington, NJ, PV97-40 (1997).

18. H. Yokokawa, N. Sakai, T. Horita and K. Yamaji, *'Fuel Cells – From Fundamentals to Applications'*, **1**(2), 1 (2001).

19. J. R. Selman and T. D. Claar (Eds), 'Proceedings of the Symposium on Molten Carbonate Fuel Cell Technology', The Electrochemical Society, Pennington, NJ, PV 84-13 (1984).

20. J. R. Selman, H. C. Maru, D. A. Shores and I. Uchida (Eds), 'Proceedings of the 2nd International Symposium Molten Carbonate Fuel Cell Technology', The Electrochemical Society, Pennington, NJ, PV 90–16 (1990).

21. D. Shores, H. Maru, I. Uchida and J. R. Selman, (Eds), 'Proceedings of the 3rd International Symposium on Carbonate Fuel Cell Technology', The Electrochemical Society, Pennington, NJ, PV 93-3 (1993).

22. J. R. Selman, I. Uchida, H. Wendt, D. A. Shores and T. F. Fuller (Eds), 'Proceeding of the 4th International Symposium on Carbonate Fuel Cell Technology', The Electrochemical Society, Pennington, NJ, PV 97-4 (1997).

23. I. Uchida, K. Hemmes, G. Lindbergh, D. A. Shores and J. R. Selman (Eds), 'Proceedings of the 5th International Symposium on Carbonate Fuel Cell Technology', The Electrochemical Society, Pennington, NJ, PV 99-20 (1999).

24. O. Yamamoto, M. Dokiya and H. Tagawa (Eds), 'Proceedings of the International Symposium on Solid Oxide Fuel Cells', Nagoya, Japan, Nov.13–14, 1989, Science House, Tokyo, Japan (1989).

25. S. C. Singhal (Ed), 'Proceedings of the 1st International Symposium on Solid Oxide Fuel Cells', The Electrochemical Society, Pennington, NJ (1989).

26. F. Grosz, P. Zegers, S. C. Singhal and O. Yamamoto (Eds), 'Proceedings of the 2nd International Symposium on Solid Oxide Fuel Cells', Commission of the European Communities, Brussels and Luxembourg (1991).

27. S. C. Singhal and H. Iwahara (Eds), 'Proceedings of the 3rd International Symposium on Solid Oxide Fuel Cells', The Electrochemical Society, Pennington, NJ, PV 93-4 (1993).

28. M. Dokiya, O. Yamamoto, H. Tagawa and S. C. Singhal (Eds), 'Proceedings of the 4th International Symposium on Solid Oxide Fuel Cells', The Electrochemical Society, Pennington, NJ, PV95-1 (1995).

29. U. Stimming, S. C. Singhal, H. Tagawa and W. Lehnert (Eds), 'Proceedings of the 5th International Symposium on Solid Oxide Fuel Cells', The Electrochemical Society, Pennington, NJ, PV 97-40 (1997).

30. S. C. Singhal and M. Dokiya (Eds), 'Proceedings of the 6th International Symposium on Solid Oxide Fuel Cells', The Electrochemical Society, Pennington, NJ, PV 99-19 (1999).

31. H. Yokokawa and S. C. Singhal (Eds), 'Proceedings of the 7th International Symposium on Solid Oxide Fuel Cells', The Electrochemical Society, Pennington, NJ, PV2001-16 (2001).

32. F. J. Salzano (Ed), 'Proceedings of the Conference on High Temperature Solid Oxide Electrolytes, Aug. 16–17, 1983', Brookhaven National Laboratory Associated Universities, Vol. 1 (1983).

33. U. Bossel (Ed), 'Proceedings of the of First European Solid Oxide Fuel Cell Forum, Lucerne, Switzerland', European SOFC Forum, Switzerland, Oct. 3–7 (1994).

34. B. Thorstensen (Ed), 'Proceedings of the 2nd European Solid Oxide Fuel Cell Forum', Oslo, Norway, European SOFC Forum, May 6–10 (1996).

35. P. Stevens (Ed), 'Proceedings of the Third European Solid Oxide Fuel Cell Forum, Nantes, France', European Fuel Cell Forum, June 2–5 (1998).

36. A. J. McEvoy (Ed), 'Proceedings of the Fourth European Solid Oxide Fuel Cell Forum, Lucerne, Switzerland', European Fuel Cell Forum, July 10–14 (2000).

37. G. H. J. Broers and J. A. A. Ketelaar, *Ind. Eng. Chem.*, **52**(4), 303 (1960).

38. C. D. Iacovangelo, *J. Electrochem. Soc.*, **133**(11), 2410 (1986).

39. R. B. Swaroop, J. W. Sim and K. Kinoshita, *J. Electrochem. Soc.*, **125**, 1799 (1978); R. A. Donado, L. G. Marianowski, H. C. Maru and J. R. Selman, *J. Electrochem. Soc.*, **131**, 2535, 2541 (1984).

40. C. D. Iacovangelo and E. C. Jerabek, *J. Electrochem. Soc.*, **133**(2), 280 (1986).

41. H. R. Kunz, *J. Electrochem. Soc.*, **134**, 105 (1987).

42. Y. Mugikura, T. Abe, S. Yoshioka and H. Urushibata, *J. Electrochem. Soc.*, **142**(9), 2971 (1995).

43. K. Ota, S. Mitsushima, S. Kato, S. Asano, H. Yoshitake and N. Kamiya, *J. Electrochem. Soc.*, **139**(3), 667 (1992).

44. E. F. Sverdrup, C. J. Warde and A. D. Glasser, in 'From Electrocatalysis to Fuel Cells', G. Sandstede (Ed), University of Washington Press, Seattle, WA, p. 255 (1972).

45. D. H. Archer, L. Elikan and R. L. Zahradnik, 'The Performance of Solid-Electrolyte Cells and Batteries on CO/H_2 Mixture; A 100-Watt Solid-Electrolyte Power Supply', in "Hydrocarbon Fuel Cell Technology", B. S. Baker (Ed), Academic Press, New York, pp. 51–75 (1965).

46. A. O. Isenberg, in 'Proceedings of the Symposium on Electrode Materials and Processes for Energy Conversion and Storage', J. D. E. McIntyre, S. Srinivasan and F. G. Will (Eds), The Electrochemical Society, Pennington, NJ, p. 682 (1977).

47. D. C. Fee, S. A. Zwick and J. P. Ackerman, 'Solid Oxide Fuel Cell Performance', in "Proceedings of the Conference on

High Temperature Solid Oxide Electrolytes", "Anion Conductors", F. J. Salzano (Ed), Brookhaven National Laboratory, Vol. 1, pp. 29–38 (1983).

48. D. C. Fee, P. E. Blackburn, D. E. Busch, T. D. Claar, D. W. Dees, J. Dusek, T. E. Easler, W. A. Ellingson, B. K. Flandermeyer, R. J. Fousek, F. C. Mrazek, J. J. Picciolo, R. B. Poeppel and S. A. Zwick, 'Monolithic Fuel Cell Development', "Presented at the Fuel Cell Seminar, Tuscon, Arizona", pp. 40–43, Oct. 26–29 (1986).

49. H. Yokokawa, N. Sakai, T. Kawada and M. Dokiya, *J. Electrochem. Soc.*, **138**, 1018 (1991); N. Sakai, T. Kawada, H. Yokokawa, M. Dokiya and T. Iwata, *J. Mater. Sci.*, **25**, 4531 (1990).

50. H. Yokokawa, N. Sakai, T. Kawada and M. Dokiya, *J. Electrochem. Soc.*, **138**(9), 2719 (1991); H. Yokokawa, N. Sakai, T. Kawada and M. Dokiya, *Denki Kagaku*, **57**, 829 (1989).

51. H. Takeuchi, H. Nishiyama, A. Ueno, S. Aikawa, M. Aizawa, H. Tajiri, T. Nakayama, S. Suehiro and K. Shukuri, 'Current Status of SOFC Development by Wet Process', SOFC VI, S. C. Singhal and M. Dokiya (Ed), The Electrochemical Society, PV 99-19, pp. 879–884 (1999).

52. Y. Sakaki, A. Nakanishi, M. Hottori, H. Miyamoto, H. Aiki and K. Kakenobu, 'Development of MOLB Type SOFC', SOFC VII, H. Yokokawa and S. C. Singhal (Ed), The Electrochemical Society, PV2001-16, pp. 72–77 (2001).

53. W. J. Quadakkers, H. Greiner and W. Köck, 'Metals and Alloys for High temperature SOFC Application', in "Proceedings of the First European Solid Oxide Fuel Cell Forum, Luzerne, Switzerland", U. Bossel (Ed), Oct. 3–7, pp. 525–541 (1994).

54. H. S. Hsu and J. H. Devan, *J. Electrochem. Soc.*, **133**, 2077 (1987).

55. H. Yokokawa, N. Sakai, T. Kawada and M. Dokiya, 'Phase Diagram Calculations For ZrO₂ Based Ceramics: Thermodynamic Regularities in Zirconate Formation and Solubilities of Transition Metal Oxides', in "Science and Technology of Zirconia V", S. P. S. Badwal, M. J. Bannister and R. H. J. Hannink (Eds), The Australian Ceramic Society, The Technomic Publication Co., pp. 59–68 (1993).

56. H. Yokokawa, 'Phase Diagrams and Thermodynamic Properties of Zirconia Based Ceramics', in "Zirconia Engineering Ceramics: Old Challenges – New Ideas", E. Kisi (Ed), Trans Tech Publications, pp. 37–74 (1998).

57. J. M. Sangster and A. D. Pelton, 'Critical coupled Evaluation of Phase diagrams and Thermodynamic Properties of Binary and Ternary Alkali Salt Systems', in "Phase Diagrams for Ceramists", The American Ceramic Society, Westerville, OH, Vol. VII (1989); see also: C. W. Bale, A. D. Pelton and J. Melançon, 'Calculation of Thermodynamic Equilibria in the Carbonate Fuel Cell' Centre du Department Technologique, Ecole Polytechnique de Montreal, Vol. II (1981).

58. H. Yokokawa, N. Sakai, T. Kawada, M. Dokiya and K. Ota, *J. Electrochem. Soc.*, **140**(12), 3565 (1993).

59. The Thermodynamic Database Task Group, 'The Thermodynamic Database MALT2', Kagaku-gijutsu sha, Tokyo, Japan (1991).

60. H. Yokokawa, N. Sakai, T. Horita, K. Yamaji, Y.p. Xiong, T. Otake, H. Yugami, T. Kawada and J. Mizusaki, *J. Phase Equilibria*, **22**(3), 331 (2001).

61. N. Sakai, T. Tsunoda, I. Kojima, K. Yamaji, T. Horita, H. Yokokawa, T. Kawada and M. Dokiya, 'Material Transport via the Grain Boundary of Lanthanum Chromites', in "Ceramic Interfaces", H.-I. Yoo and S.-J. K. Kang (Eds), IOM Communications, Vol. 2, pp. 135–156 (2001).

62. R. H. Arendt, *J. Electrochem. Soc.*, **129**, 942 (1982).

63. T. Kawada and H. Yokokawa, 'Materials and Characterization of Solid Oxide Fuel Cells', in "Electrical Properties of Ionic Solids", J. Nowotny and C. C. Sorrell (Eds), Trans Tech Publications, pp. 187–248 (1997).

64. C. E. Baumgartner, *J. Am. Ceram. Soc.*, **69**(2), 162 (1986).

65. C. E. Vallet and J. Braunstein, *J. Electrochem. Soc.*, **126**, 527 (1979).

66. H. Sotouchi, Y. Watanabe, T. Kobayashi and M. Murai. *J. Electrochem. Soc.*, **139**(4), 1127 (1992).

67. R. H. Arendt, *J. Electrochem. Soc.*, **129**, 979 (1982).

68. W. H. A. Peelen, K. Hemmes and G. Lindbergh, *J. Electrochem. Soc.*, **147**, 2122 (2000).

69. B. C. H. Steele, *Solid State Ionics*, **134**, 3 (2000).

70. N. Sakai and I. Yasuda, *Kikan Kagaku Sousetsu*, **49**, 210 (2001).

71. K. Mori, H. Miyamoto, K. Takenobu and T. Matsudaira, in 'Proceedings of the 3rd European Solid Oxide Fuel Cell Forum', P. Stevens (Ed), European SOFC Forum, Switzerland, pp. 179–183 (1998); T. R. Armstrong, J. W. Stevenson, L. R. Pederson and P. E. Raney, *J. Electrochem. Soc.*, **143**, 2919 (1996).

72. G. R. Heath and R. F. Singer, 'Development of An Advanced 1-kW Solid Oxide Fuel Cell Prototype – Status Report of the European Joint Program', in "Proceeding of the Second International Symposium on Solid Oxide Fuel Cells, Athens, Greece", Commission of European Communities, July 2–5, pp. 55–66, (1991).

73. M. Mori, N. Sakai, T. Kawada, H. Yokokawa and M. Dokiya, *Denki Kagaku*, **58**, 528 (1990).

74. O. Chikagawa, A. Shiratori, M. Iha, O. Yokokura, H. Takagi, Y. Sakabe and K. Akagi, 'Solid Oxide Fuel Cells', U. Stimming, S. C. Singhal, H. Tagawa and W. Lehnert (Eds), The Electrochemical Society, PV 97-40, pp. 204–211 (1997); K. Akagi, A. Siratori, M. Iha and O. Chikagawa, 'Development of Planar SOFC by Easy-to-fabricate Stack Structure', in "Proceedings of the 3rd Symposium on Solid Oxide Fuel Cells, Japan", The Solid Oxide Fuel Cell Society of Japan, pp. 21–24 (1994).

75. H. Hisatome, S. Takatsuki, K. Tomida, T. Nishi, H. Tukuda, A. Yamashita, T. Hashimoto and K. Kosaka, 'Development of Tubular Type SOFC by Sintering Method', in "Proceedings of the 6th Symposium on Solid Oxide Fuel Cells, Japan", The Solid Oxide Fuel Cell Society of Japan, pp. 29–32 (1997).

76. S. Taniguchi, M. Kadowaki, H. Kawamura, T. Yasuo, Y. Akiyama, Y. Miyake and T. Saitoh, *J. Power Source*, **55**, 73 (1995); K. Hilpert, D. Das, M. Miller, D. H. Peck and R. Weiß, *J. Electrochem. Soc.*, **143**, 3642.

77. C. Gindorf, K. Hilpert, H. Nabielek, L. Singheiser, R. Ruckdäschel and G. Schiller, 'Chromium Release from Metallic Interconnects with and without Coatings', in "Proceedings of the 4th European Solid Oxide Fuel Cell Forum", A. J. McEvoy (Ed), European Fuel Cell Forum, Switzerland, pp. 845–854 (2000).

78. Y. Matsuzaki and I. Yasuda, *J. Electrochem. Soc.*, **148**, A126 (2001).

79. K. Honegger, A. Plas, R. Diethelm and W. Glatz, 'Solid Oxide Fuel Cells VII', H. Yokokawa and S. C. Singhal (Eds), The Electrochemical Society, PV 2001-16, pp. 803–810 (2001).

80. T. Ishihara, H. Matsuda and Y. Takita, *J. Am. Chem. Soc.*, **116**, 3801 (1994).

81. H.-H. Möbius, *J. Solid State Electrochem.*, **1**, 2 (1997).

Chapter 14
Hydrogen/oxygen (Air) fuel cells with alkaline electrolytes

M. Cifrain and K. Kordesch
Graz University of Technology, Graz, Austria

1 GENERAL PRINCIPLES

Alkaline fuel cells (AFCs) normally use a mobilized or immobilized aqueous solution (30–45 wt%) of potassium hydroxide (KOH) as an electrolyte. Also, sodium hydroxide (NaOH) is possible but has some disadvantages, in particular the much lower solubility of sodium carbonate compared to potassium carbonate (see below).

Briefly, the electrochemical reactions inside an AFC are

$$2H_2 + 4OH^- \longrightarrow 4H_2O + 4e^- \quad \text{anode reaction}$$
$$O_2 + 4e^- + 2H_2O \longrightarrow 4OH^- \quad \text{cathode reaction}$$
$$2H_2 + O_2 \longrightarrow 2H_2O \quad \text{overall reaction}$$

Hydroxyl anions diffuse continuously from the cathode to the anode and water diffuses the opposite way. Most of the reaction water leaves on the anode side, but a small amount of reaction water is removed via the electrolyte loop, if a circulating system is considered.

The oxygen reduction in alkaline environments is more favourable than in acid environments, i.e., the voltage drop is lower.[1] It yields the highest voltage at comparable current densities, leading to a higher efficiency of the system. Consequently, the use of smaller amounts of noble metal electrocatalysts and the use of non-noble metal catalysts is more favourable in an AFC than in any other system. AFCs also operate well at room temperature and have a good cold start capability.

However, a major disadvantage is the CO_2 sensitivity of the alkaline electrolytes as carbonates are formed. In strong alkaline environments, the solubility of carbonates is rather poor. This leads to the formation of carbonate crystals, capable of blocking electrolyte pathways and electrolyte pores:[2]

$$CO_2 + 2OH^- \longrightarrow CO_3^{2-} + H_2O$$

As a consequence, the use of AFCs has been restricted to special applications, where no CO_2 is present, for example in space vehicles, where pure hydrogen and pure oxygen are used. The formation of carbonates is one of the causes of the poor long-term stability of AFCs.

It is obvious that this problem is much more critical in systems with an immobilized electrolyte than in systems with circulating electrolytes, as the saturation point of the carbonate is reached much faster and a regeneration of the electrode pores (like a washing procedure with acetic acid) is impossible. In the case of a circulating electrolyte, small amounts of CO_2 can be tolerated, as a circulated liquid electrolyte has a small capacity to absorb CO_2. When using air as oxidant, which contains 0.03% CO_2, the removal of a higher amount of CO_2 is easily possible using a simple absorbing tower (with soda lime or amines). Keeping this fact in mind, many earthbound applications are conceivable, such as vehicular applications (e.g., buses, cars, etc.) or in stationary applications (e.g., power units for houses or farms).

The removal of the 0.03% carbon dioxide from the air can be accomplished by chemical absorption in a tower filled with soda lime, for example. An industrial temperature swing absorption/desorption process using amines is very efficient, but currently not available for small units. On the fuel side, alkaline systems use high purity hydrogen from electrolysis, reformers or ammonia crackers.

The carbon dioxide scare is partly the reason why phosphoric acid fuel cells have made such an inroad into fuel cell technology since the 1970s and is also the reason for the switch to the proton exchange membrane fuel cell (PEMFC) systems in recent years.

Alkaline systems operate well at room temperature, yield the highest voltage (at comparable current densities) of all the fuel cell systems and their cell and electrodes can be built from low-cost carbon and plastics. Because of a good compatibility with many construction materials, AFCs can achieve a long operating life (15 000 h have been demonstrated). The choice of catalysts available for alkaline cells is greater than for acidic cells, which are currently limited to platinum group metals and tungsten carbides.

Contrary to alkaline matrix cells (Figure 1), AFCs with a circulating electrolyte (Figure 2) establish a reliable barrier against reactant gas leakage from the electrodes and can use the electrolyte as a cleaning medium. Accumulated impurities and carbonates can easily be removed. The main advantage of the circulating electrolyte is its use as a cooling liquid and as a water removal vehicle.

Some systems use the circulating electrolyte to carry the reaction water to cooling subsystems and separate water vaporizing units (regenerators) outside the stacks. In such systems, the reactants – hydrogen and oxygen – can be supplied dead-ended. Care must only be taken that inerts building up with time are removed by a continuous or periodic purging of the gases.

Another way of water removal is by evaporation through the electrodes into excess reactant gas streams. Vaporization cooling is an additional bonus. In AFCs (contrary to acidic cells), the reaction water is produced at the hydrogen

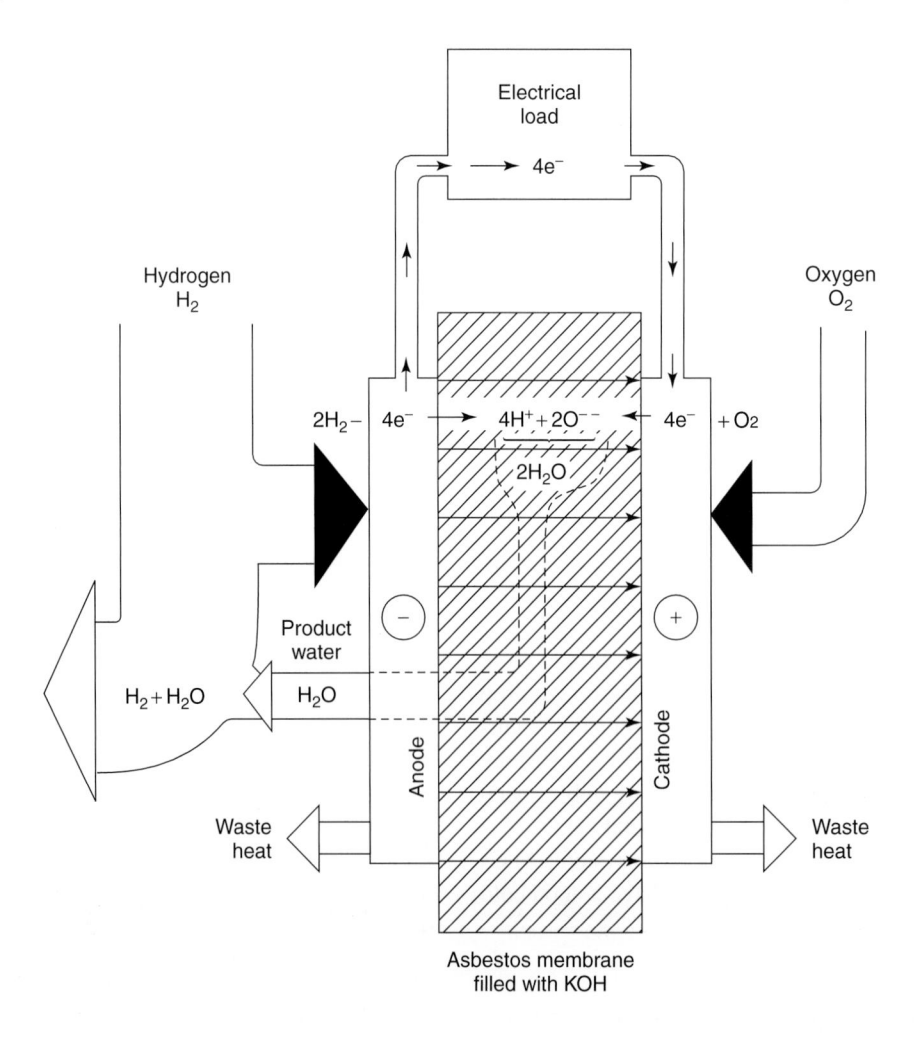

Figure 1. Principle of the AFC with a static electrolyte.

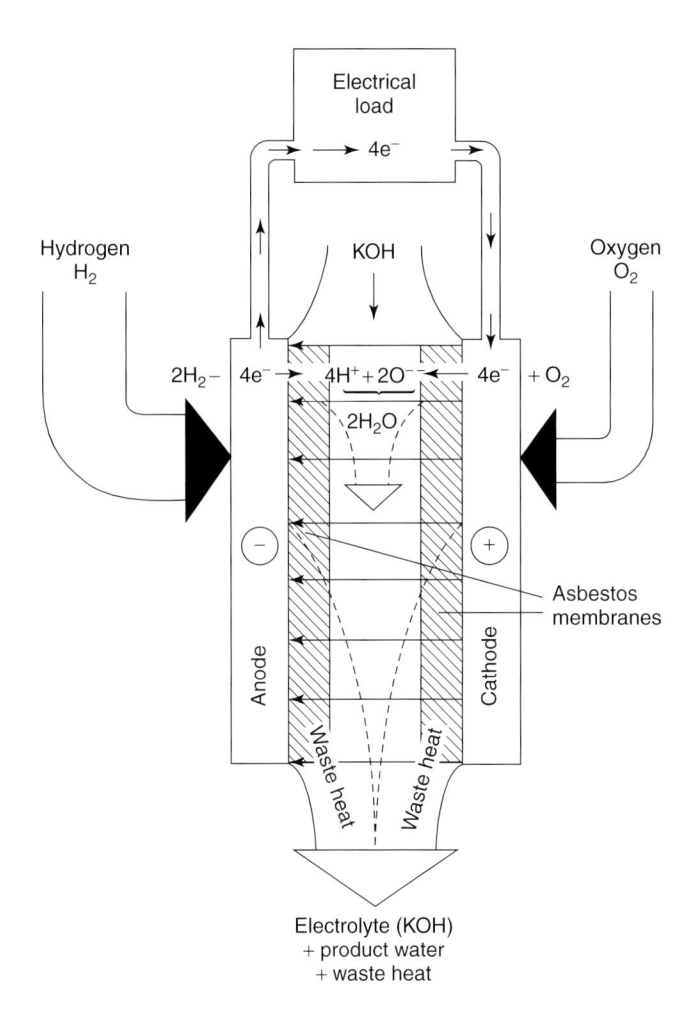

Figure 2. Principle of the AFC with a circulating electrolyte.

electrode (the anode). Therefore, the electrolyte concentration inside the anode structure is lower than in the bulk electrolyte. This increases the water vapour pressure and, thus, its evaporation rate. The excess hydrogen stream is obtained by circulating and cooling the hydrogen by means of a gas pump or by mounting a gas jet into the gas loop between the condenser and the gas port leading into the stack. In the latter case, the gas circulation rate becomes automatically proportional to the gas consumption. Oxygen circulation is not as effective as hydrogen circulation because the electrolyte concentrates in the cathode and its water vapour pressure is lower than that of the bulk electrolyte. A schematic of a dual/loop water removal system with circulating jets is demonstrated in Figure 2.

In air-operated fuel cells, excess air must be blown continuously through the cathodes to remove nitrogen from the air manifolds that would otherwise block them. The increasing necessity for carbon dioxide removal is the price that must be paid for this simple way of water and nitrogen removal without any condensers.

The disadvantage of circulating electrolyte systems, in which the electrolyte is fed through a common manifold into a multi-cell stack, is the occurrence of parasitic currents, through the ionically conductive liquid electrolyte junctions of the series-connected cells and parallel electrolyte flow: these currents must be minimized.

Regarding the R&D activities in the field of AFCs, the international scientific community has mainly changed the areas of investigation. The main interest now lies in the fields of PEM, molten carbonate and solid oxide fuel cells. Over the last few years, research into AFCs has mainly been carried by some European research teams and some in USA/Canada. The next section gives an overview of AFC systems already developed.

The AFCs developed for remote applications (space, undersea, military) are not strongly constrained by cost. On the other hand, the consumer and industrial markets require the development of low-cost components to successfully compete with alternative technologies. Much of the recent interest in AFCs for mobile and stationary terrestrial applications has addressed the development of low-cost cell components (carbon-based porous electrodes).[3, 4]

It is the KOH electrolyte, reaction kinetics and CO_2 sensitivity that are the only real common factors among different AFCs. Pressure, temperature and electrode structure vary greatly between designs and will, therefore, be discussed in specific sections below.

2 DEVELOPED AFC SYSTEMS

2.1 The Bacon fuel cell

Francis Thomas Bacon (1904–1992) at Cambridge (UK) began experimenting with alkali electrolytes in the late 1930s, settling on potassium hydroxide instead of using the acid electrolytes. KOH performed as well as the acid electrolytes and was not as corrosive to the electrodes he used. In fact, he demonstrated the first viable power unit. This monopolar 5 kW fuel cell was operated with pure hydrogen and oxygen and was equipped with porous nickel anodes and lithiated porous nickel oxide cathodes. The electrolyte was circulating 30 wt% aqueous potassium hydroxide. Strictly speaking, this system was a medium-temperature fuel cell system, as it was operated at 200 °C, at a pressure of 5 MPa, to prevent boiling of the electrolyte.[5, 6]

Bacon avoided expensive noble metal catalysts and decided to take nickel electrodes, as nickel also has a high catalytic activity for hydrogen oxidation. Therefore, nickel powder was sintered to obtain a rigid structure. To improve the establishment of the three-phase zone, two different sizes of nickel powers were used. The electrolyte wetted the fine pores easily because of strong capillary forces and the wider pores stayed free of the electrolyte. No stable wet-proofing agent such as polytetrafluoroethylene (PTFE) was available at that time, the three-phase reaction zone inside the electrodes was maintained by carefully regulated, differential gas pressure. The lithium salts on the surface of the cathode reduced the oxidation of the nickel.

This design was improved to a bipolar system (hydroxcell) by the Patterson-Moos Research Division of Leesona Corp.

2.2 The Allis/Chalmers manufacturing fuel cell system

On the basis of the Bacon fuel cell system, Allis/Chalmers built the first large vehicle equipped with a fuel cell. It was a farm tractor powered by a 15 kW stack consisting of over 1000 cells. With this 15 kW of power, the tractor generated enough power to pull a weight of about 1.5 tons.[7]

Together with the US Air Force, Allis/Chalmers have maintained a research program for some years, building a fuel cell powered golf cart, submersible and forklift. They concentrated on catalyst-coated porous sintered nickel plaque electrodes mounted in a bipolar assembly. The anodes and cathodes contained platinum palladium catalysts, and the KOH electrolyte was immobilized, absorbed in a sheet of microporous asbestos. Nickel-plated magnesium plates were used as bipolar plates. At the beginning, the system was equipped with a water condenser in the circulated hydrogen stream. Improvements over the years 1962–1967 led to a static water vapour control method for removing the reaction water, which followed load changes more quickly, as the matrix had a slowing down effect on the water equilibrium. This method included an additional moisture removal membrane on the anode side. The new concept (see Figure 3) became a model for many asbestos-matrix fuel cell constructions, e.g., in cells for many National Aeronautics and Space Administration (NASA) contracts.[8]

Fuel cells of 1.5 kW were built and tested. The operational temperature was maintained between 50 and 65 °C by conduction along the bipolar assembly with forced air cooling of the protruding cell edges.

2.3 The UCC fuel cell systems[9, 10]

Union Carbide Corporation (UCC) produced porous baked carbon plate electrodes and reported excellent power densities in the early 1960s. The electrodes were made by mixing

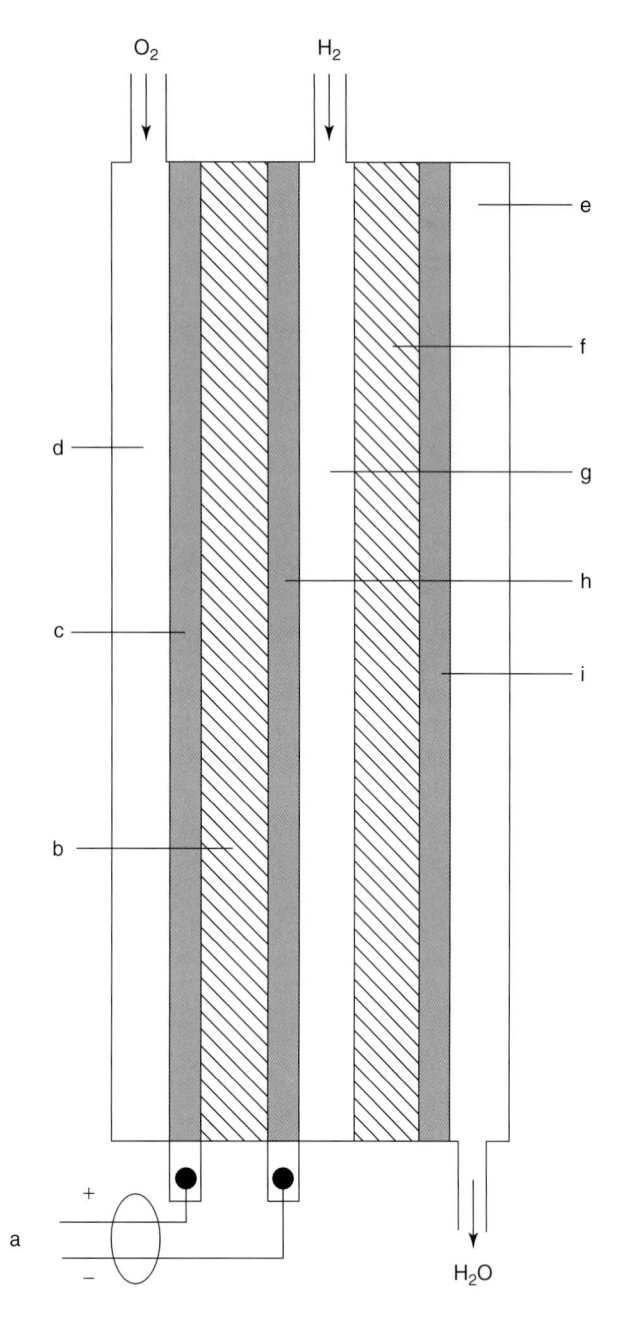

Figure 3. Allis/Chalmers static water vapour control. (a) Electrical power, (b) electrolyte, (c) porous oxygen electrode, (d) oxygen cavity, (e) moisture removal cavity, (f) moisture removal membrane, (g) hydrogen cavity, (h) porous hydrogen electrode, (i) porous support plaque.

filler (petroleum coke flour) with binder (coal-tar pitch), baking to carbonize the pitch and firing at high temperatures of 1000–1700 °C. The pore structure was determined by the size of the filler particles and the holes produced by gas escaping during carbonization. The holes were of very irregular shape. The conductivity depended on physical structure and on the degree of graphitization produced by prolonged firing at high temperatures. The electrodes were usually activated by burning out with air, steam or CO_2, which produced a sharp rise in the internal area of over a few percent of burnoff. Although these electrodes had some catalytic activity as oxygen cathodes, they were usually catalyzed by plating or impregnating with metal or metal oxide catalysts. For hydrogen oxidation, the activity without catalysts was always too low. UCC electrodes had given current densities over $1000\,mA\,cm^{-2}$ for short-term operations – a prototype of the bipolar construction was used by the US navy.

Anyway, these electrodes had two major disadvantages. Firstly, they flooded slowly and irreversibly. An increase of pressure, like in the sintered metal electrodes, did not restore their performance. It was assumed that a flooding of micropores was the reason, therefore, the electrodes were often wet-proofed with hydrocarbons to retard flooding. Secondly, the electrodes were brittle and had to be relatively thick (3–4 mm) to be sufficiently strong and to give a long life as slow flooding occurred (concentration and ohmic overvoltage). Another minor disadvantage was the slow oxidation of the carbon, but at temperatures of 80 °C this carbon corrosion still allowed such electrodes to be used for many years.

It was obviously desirable to combine the catalytic properties of activated carbon with the advantages of Teflon bonding to obtain active, thin, wet-proofed electrodes. In 1961, the first electrodes were constructed with a layer of active carbon powder on a layer of inactive carbon powder with highly hydrophobic plastic bonding on a layer of porous nickel containing an embedded nickel screen. The electrodes were flexible, lightweight and had a total thickness of less than 1 mm. The carbon powders used were pitch-bonded or sugar-bonded carbon flour ground to less than 20 μm in diameter. They became known as "fixed zone" electrodes.

At 70 °C, the initial cell performance on H_2 and O_2 with 12 M KOH was 0.8 V at $400\,mA\,cm^{-2}$ (IR free), with most of the overvoltage occurring at the cathode. However, the anode deteriorated within 200 h due to mechanical instabilities. The nickel backing had no effect on performance, providing it was strong, thin and permeable to gas. The sugar-bonded flour gave electrodes more resistance to failure. The performance of a cell improved over the first 2 weeks of use. Changing from 70 °C to room temperature greatly decreased the performance of the cathode and also decreased the anode performance at higher current densities. Both air and hydrogen electrodes showed an increased rate of overvoltage at current densities between 100 and $200\,mA\,cm^{-2}$, indicating concentration polarization effects (electrolyte not circulated).

All UCC systems used a circulating potassium hydroxide electrolyte for concentration control and heat exchange. The reaction water was removed by transferring the water vapour through the porous electrodes (anodes and cathodes) into the two excess gas streams circulated by jet pumps. The electrolyte temperature of 65–75 °C assured a self-regulation of the operating conditions as a function of the load profile up to current densities of $100\,mA\,cm^{-2}$. The materials used in 1970 did not allow operating temperatures of 80–85 °C, at which the hydrogen loop alone would have sufficed.

UCC demonstrated a fuel cell-powered mobile radar set for the US army and a fuel cell-powered motorbike (alkaline hydrazine/air fuel cell) and drew up plans for an undersea base that would run on fuel cells.

The largest fuel cell built by UCC was a 150 kW unit for the "Electrovan" of General Motors (GM) in 1967 (Figure 4). This system consisted of 32 modules with a top output of 5 kW each. The six-passenger van was designed for usage of H_2 and O_2, both in liquid form. Hence, this car was only used on company property. Anyway, although it had a driving range of 200 km and a top speed of $105\,km\,h^{-1}$, the overall system was much too heavy (3400 kg), the lifetime was poor (1000 h after first start-up) probably due to cell reversals in this high voltage system (400 V). Also, the liquid gases were objectionable. GM did not continue working on fuel cell technology.

Finally, in 1970, K.V. Kordesch built a 6 kW H_2/air fuel cell–lead acid battery hybrid passenger vehicle (an Austin A-40, see Figure 5) and drove it for 3 years on public roads. The fuel cell system was mounted in the rear compartment together with all of the necessary accessories for automatic start-up and shut-down, including nitrogen purging.[11–13] The lead battery is situated in the front next to the motor. For simplicity, the original four-gear shift transmission with clutch was retained (it proved very beneficial for up-hill driving). "Gas pedal" type speed control is achieved by field-winding switching. The early history of electric vehicle batteries and hybrids is well documented.[14] Fuel cell types and performances are described in several books.[15, 16]

The use of a fuel cell for vehicle propulsion poses problems regarding safe handling, short start-up time, avoidance of overloading and cell reversal, automatic controls in case of operating failures (electrolyte loss, overheating, etc.) and shut-down procedures. Several of these problems disappear if a fuel cell battery is operated in parallel with a secondary battery. The start-up time can be stretched to several minutes while the equipment (or vehicle) is operated during that time from the secondary battery. Overloading is effectively prevented because the secondary battery takes over at the minimum voltage determined by the fuel cell characteristics. Figure 6 shows the test polarization curves.[17] Seven 12 V lead/acid batteries are connected in series (42 cells)

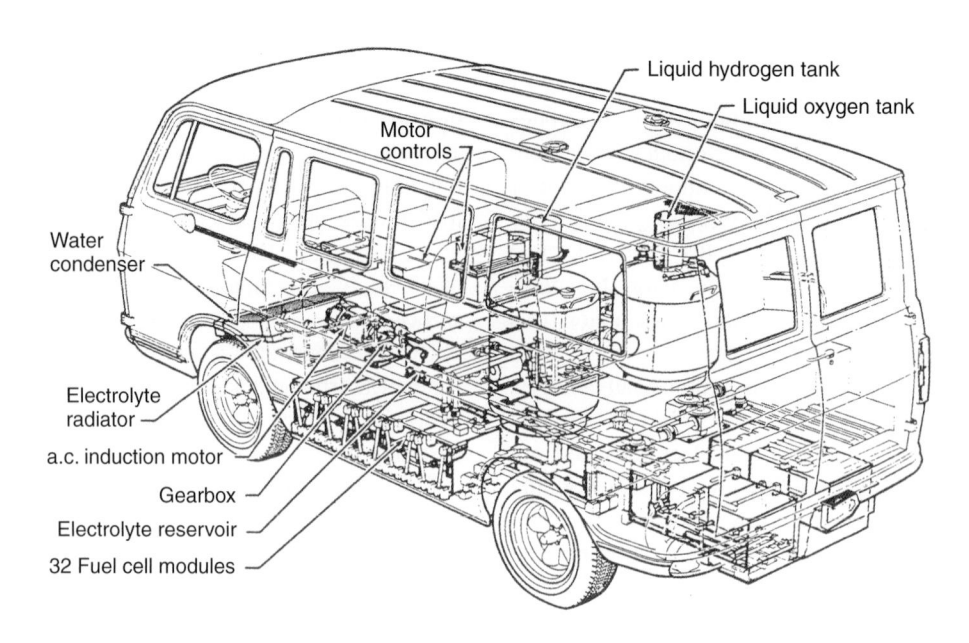

Figure 4. A phantom view of the General Motors Co. 'Electrovan'.

Figure 5. The converted Austin A-40 of K.V. Kordesch. On the roof the six hydrogen bottles were mounted, the fuel cell was in the rear and the lead/acid batteries were in the front of the car.

and this bank is connected in parallel with a hydrogen/air fuel cell battery consisting of 15×8 cell modules in series (120 cells). At low load levels, the fuel cell battery reaches a higher voltage than the lead/acid battery and is, therefore, able to charge the secondary battery. However, when the fuel cell battery is not in operation, it contains no hydrogen and has a far lower voltage than the lead/acid battery. The latter would now electrolyze water at the fuel cell electrodes and thereby cause wasteful and harmful operation. To prevent this, a diode is inserted into the connecting leads to permit current to flow in one direction only. This is one of the innovations permitting shut-down of the fuel cell battery and emptying of gases (and electrolyte) during non-operating periods. This feature prolongs the lifetime of the fuel cells.

During start-up, however, hydrogen reaches the cells at different times and causes the activated cells to reverse the still inactive cells. This results in permanent damage to the "driven" or reversed cells. The reversed cells can be "righted" only by individually applying a pulse current (in the charging direction), a very time-consuming procedure. The difficulty can be avoided simply by applying the lead/acid battery voltage to the fuel cell battery (while the anode compartments are filled with nitrogen) by means of a resistor to bridge the diode during that period causing a small charge current to flow, which "polarizes" the cells in the "right" direction. The nitrogen gas supply added to the fuel cell system to blanket the anodes when required can also be used for safety protection in case of an emergency.

Figure 7 shows the block diagram of the hydrogen/air fuel cell power plant as used in the automobile (the nitrogen supply is not shown). The hydrogen/air battery contains $120\times$ single H_2/air cells of the "duplex" or "bicell" construction. Eight such cells are arranged into a module. Hydrogen is stored as compressed gas in six cylinders on top of the car. The cylinders are made by the Pressed Steel Tank Company (Model No. 8RC 1500T). These lightweight tanks are made for aircraft use, scuba diving and similar purposes.

The gas regulator and the manifold parts are standard and are fitted with the required heat and overpressure safety plugs on the cylinder valves. At the main regulator (outside the car), the pressure is reduced to 30 or 50 p.s.i.g. and the gas is piped through the main solenoid valve to the

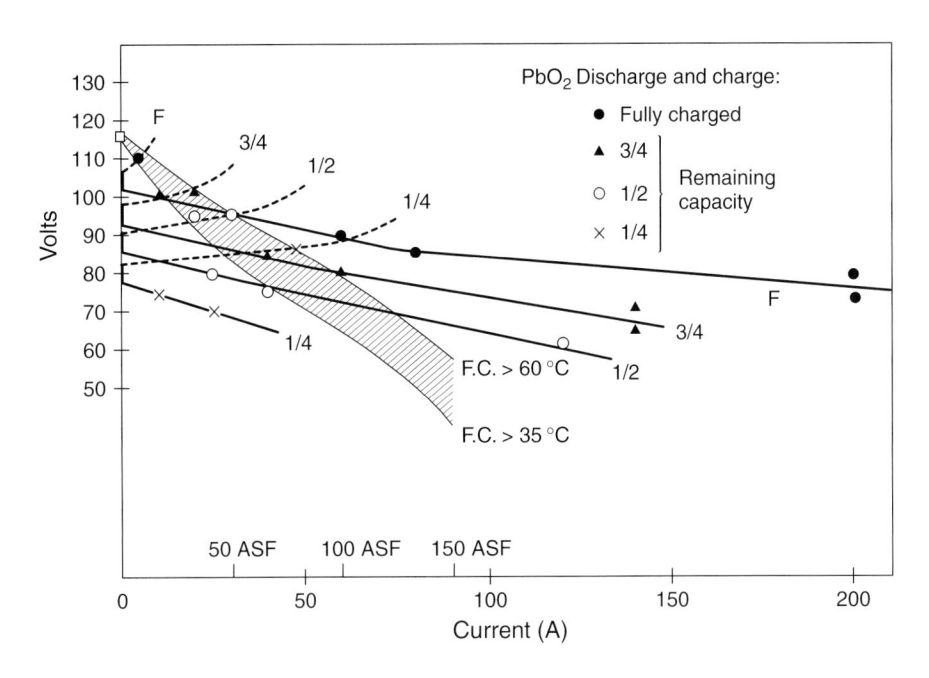

Figure 6. Test polarization curves of a hybrid automotive power plant. (Reproduced from Kordesch et al. (2000).[17])

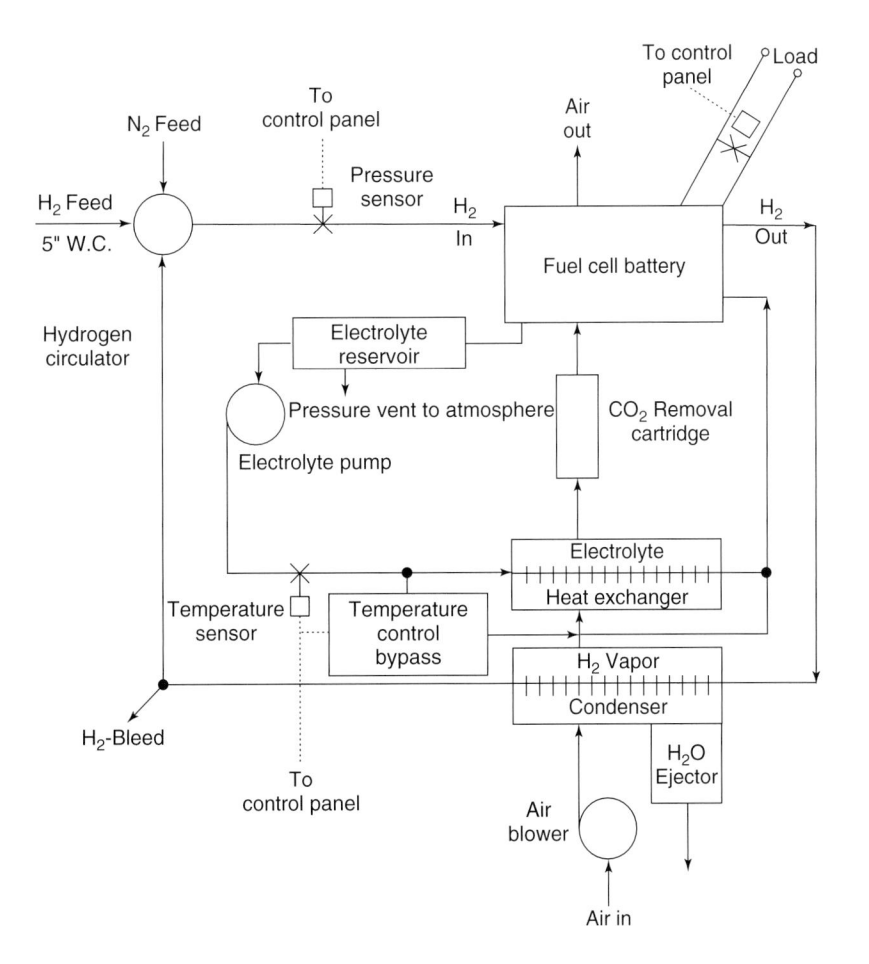

Figure 7. Block diagram of hydrogen/air fuel supply of the Austin A-40 (see Figure 5). See text for a detailed explanation.

low-pressure regulator that delivers hydrogen at a pressure range of 2–12 in. water column (w.c.) at a high flow rate. Stainless steel traps, placed after the battery, collect any water before the hydrogen returns to the circulator pump. This pump is able to produce a 1 in. w.c. differential pressure across the stacks at 10–20 times the theoretical gas circulation rate at the 3 kW level (60 ft^3 h^{-1}). This amount of circulation is sufficient to remove all the water produced at 50 A ft^{-2}.

A hydrogen-bleed valve (adjustable between 0 and 5 ft^3 h^{-1}) serves as a constant inert removal device. It is usually set at 1 to 1% ft^3 h^{-1} hydrogen. A rapid bleed valve is activated during the start-up and shut-down sequences to exchange the battery atmosphere rapidly (N_2 for air, H_2 for N_2 and N_2 for H_2). The maximum gas flow-through rate is ~15 ft^3 min^{-1} at a 10 in. w.c., which allows the exchange atmosphere to sweep the manifolds in a very short time should an emergency arise.

The hydrogen passes through a heat exchanger (water cooled) before it returns to the circulator pump (see Figure 7). There, water vapour condenses out and passes through an automatic drain valve into the water collector tank (10 l capacity) located under the car.

Nitrogen is stored in a cylinder mounted under the battery. The tank is of the same type as the hydrogen cylinders, but smaller (40 ft^3), and can be filled from the right rear side of the car. Nitrogen flow control is achieved by means of a solenoid valve that is energized from the dashboard sequence switch.

Air is moved through the battery by a vacuum cleaner-type blower. The air intake is in the car but the blower is underneath (outside) the car to reduce noise. The pressure side of the blower is connected to the CO_2 scrubber (mounted on top of the battery) with a 3 inch flexible hose. From the CO_2 scrubber, the air is directed into three air-intake manifolds, one for each battery block. The moisture-containing air is exhausted to the outside. The blower speed is changed by depressing the footswitch in accordance with needed power requirements. With a 1 in. w.c. backpressure across the stacks, the blower can deliver 30–50 ft^3 min^{-1} of air (maximum on full voltage). This is ten times the stoichiometric requirement on high level load (60 A). However, to conserve the CO_2-scrubber capacity, it is normally run from lower voltage taps (24 and 32 V) while still satisfying the average load of 30 A and at the same time removing water vapour to such an extent that the H_2/water removal circuit has only to carry 50% of its designed capacity. This versatility assures that a satisfactory KOH normality will be maintained under all conditions (high outside temperature and high humidity are the most unfavourable conditions).

The feature of taking the air from the inside of the car and exhausting it to the outside of the car is an important safety feature because it reduces the danger of small hydrogen leaks.

The CO_2 scrubber is filled with 10 kg of soda lime. According to laboratory tests it should last at least 500 miles of average driving. Additions of Baralyme and layers of indicating Lithadsorb in the compartments close to the battery end serve as a booster and warning indicator. Intermittent driving modes seem to permit far better bed utilization than earlier continuous tests have indicated. It is interesting to note that fuel cells pick up CO_2 from the air with "inefficiency" after about 80% of the CO_2 has been removed. This becomes even more pronounced when the electrolyte has a concentration under 9 N.

The air scrubber is mounted horizontally and five vertical buffer plates prevent bypassing of unscrubbed air when the soda lime has settled. The best particle size seems to be the 3–5 mesh range to combine low back pressure with satisfactory scrubbing.

It is not possible to operate all circuits from the 12 V accessory battery because the drain would be too high. The best solution for maintaining a separate ground is to charge the accessory battery from the fuel cell system by means of a direct current (d.c.)/d.c. inverter with an 84–90 V input and a 12–15 V output.

The seven 12 V batteries are all located under the hood, two in front of the motor and five above it. The batteries were commercially available batteries manufactured by the Globe Union Corporation.

The purpose of this project was to demonstrate (a) that an electric automobile could be powered by a fuel cell/lead/acid battery hybrid system and retain the advantages of both battery types, and (b) that such a system can be easily started and shut down and is no more complicated than any other assembly of batteries with such common accessories as pumps, motors and simple electrical circuits.

These facts were demonstrated, and the electric automobile showed in driving tests that it behaved essentially like the gasoline engine powered automobile, except that its top speed was somewhat limited.

2.4 The NASA Apollo fuel cell system

In the first manned space flight program of the NASA, the Mercury Program, the capsule was battery powered. From the Gemini Program onwards, fuel cells were used because the flights became longer, but Gemini used an acidic proton exchange membrane (PEM) system. Due to its higher efficiency as well as problems with the membranes (holes), NASA decided to use an alkaline system for their moon flight program, Apollo. So, in the early 1960s, Pratt & Whitney Aircraft licensed the Bacon patents and won the NASA contract to power the Apollo spacecraft with alkaline cells (see Figure 8).

Therefore, they activated the sintered nickel electrodes with platinum metal catalysts at high loading (up to 40 mg cm^{-2}) to boost performance in spite of lowering the operating pressure to 0.3 MPa. At this pressure, a much more concentrated electrolyte had to be used to prevent boiling (85% KOH), hence a high temperature (over 100 °C) was necessary to keep it liquid. The nominal operating temperature was between 200 and 230 °C, which was detrimental to the life of the electrolyte circulation pumps.

The electrodes were 2.5 mm thick and circular with a diameter of 200 mm. Each single cell was individually packed between nickel sheets, whereby 31 cells were stacked together and connected electrically in series, then three stacks were connected in parallel. The nominal rating of one stack was 1.5 kW (overload = 2.3 kW) and its weight was 109 kg. The water was separated in the hydrogen recirculation loop in a water condenser and a gas liquid separator. Waste heat was removed using circulating nitrogen in the stack containment.

Of the 92 fuel cells that had been delivered by Pratt & Whitney, 54 had been used (including nine flights to the moon, the three Skylab and the Apollo-Soyuz missions).

2.5 The Varta Eloflux system

The main difference of this system, described 1965 by Wendtland and Winsel, compared to the other systems is that the electrolyte is pressed perpendicular to the electrodes through the whole stack and to the gas flows.[18, 19] Therefore, the electrodes and the diaphragms contained a interconnecting system of narrow pores, in which the electrolyte was dispersed. The bigger pores were filled with pressurized hydrogen or oxygen, which flew independently of the electrolyte. Due to the different pore sizes, the gases could not enter the narrow pores. Anyway, the diaphragm perfection is critical as any gas cross leakage would cause a malfunction of the cell. The water was taken out by the circulated electrolyte using an external water vaporizer (or dialytic reconcentrator) as the system could either be operated as a fuel cell or an electrolysis cell.

2.6 The NASA space shuttle orbiter fuel cell system

NASA selected AFCs for their space shuttle fleet, as well as for the Apollo program, mainly because of their power generating efficiencies that approached 70%. AFCs also provide drinking water for the astronauts. The shuttle systems were based on the Apollo system technology and consisted of 32 cells with 465 cm^2 of active area

Figure 8. A PC3A-2 power plant for the NASA Apollo program.

each. Each shuttle was equipped with three 12 kW stacks (maximum power rating = 436 A at 27.5 V; emergency overload capability = 16 kW). Hence, compared to the old Apollo systems, the initial shuttle systems supplied eight times the power by weighing 18 kg less. They operated at 92 °C and 0.40–0.44 MPa, hence they were low-temperature systems.

The stacks had a bipolar configuration with lightweight, silver-plated magnesium foils as the bipolar plates also aiding the heat transfer. The electrolyte was 35–45 wt% immobilized, aqueous KOH in an asbestos separator. The anode was PTFE-bonded carbon loaded with a 10 mg cm^{-2} Pt/Pd (ratio 4 : 1) loading, pressed on a silver-plated nickel screen. The cathode consisted of a gold-plated nickel screen with 10 wt% Pt (related to 90% Au). Water was removed via the anode gas in a condenser and a centrifugal separating device. Cooling chambers, filled with a circulated heat-exchanging liquid, provided a constant temperature. In addition, each cell had an electrolyte reservoir plate made of porous sintered nickel with pierced holes to compensate for electrolyte changes during different load profile operations. Unfortunately, these plates represented 50% of the total stack weight.

Further improvements in cell design led to thinner carbon anodes, a butyl-bonded potassium titanate matrix, lightweight graphitic and metallized plastic electrolyte reservoir plates, gold-plated perforated nickel foils as electrode substrates, a polyphenylene sulphide edge-framing material for the cells and electroformed nickel foil replacing the gold-plated magnesium plates. Hence, the cell thickness decreased from 4.4 to 2.4 mm.

The United Technologies Corporation (UTC, in contract with International Fuel Cells, IFC) developed a system of which the specific power and energy density superseded that of any other galvanic system, while lifetimes of smaller units were demonstrated up to 15 000 h. The new fuel cells for the shuttles, however, will be PEMFCs.

2.7 The Elenco fuel cell system

This monopolar system, developed in the 1970s, was operated with a circulating 6.6 N KOH electrolyte, which was also used as coolant as well as a heating medium.[20] The nominal temperature was 65–70 °C. Because of the relatively low electrolyte concentration it was possible to remove the reaction water on the cathode side of the cells, as long as air was used. Otherwise, or at higher current densities, both gas loops could be used for water removal.

The anodes and cathodes were rolled into multilayer carbon gas diffusion electrodes with PTFE as the binding material. The layers were pressed into a supporting nickel mesh and the gas side of the electrodes were covered by a porous hydrophobic PTFE foil. The electrodes (size = 17 × 17 cm^2, thickness = 0.4 mm) were mounted on to injection moulded frames and 24 cells were stacked using a vibration welding method. Due to the low temperature and the very small amount of noble metal catalyst that was used (0.15–0.3 mg cm^{-2}), the current densities were low (0.7 V at 100 mA cm^{-2} and 65 °C, H$_2$/air).

Elenco investigated systems in the 1.5 kW, 15 kW and 50 kW range for different applications, e.g., small power generators, a VW van, etc. The largest installation (52 kW) was a mobile power supply for a trailer for the Belgian Geological Service.

Besides efforts to obtain a contract for the European space shuttle HERMES (see below), Elenco participated in the EUREKA Bus Project, supplying the fuel cell system. According to the design parameters of the 80 passenger demonstration bus, the power was delivered by a hybrid, whereby the AFC had a power output of 80 kW with a parallel Ni/Cd buffer battery which worked at 775 V and 80 Ah. The nominal power of the hybrid system was estimated to be 180 kW (800 V).

In 1994, the EUREKA Bus Project was stopped, and at end of April 1995, Elenco announced that they we discontinuing their fuel cell business.

2.8 The Siemens fuel cell system[21]

In the 1970s, Siemens in Erlangen (Germany) developed a AFC system for hydrogen and oxygen and also the manufacturing technology. The 7 kW fuel cell (49 V, 143 A) was fitted with 70 cells with nickel electrodes, 340 cm^2 in size, operating at 420 mA cm^{-2} (nominal) in 6 N KOH at 80 °C. The gas pressures were 0.2 MPa. The system was bipolar with a circulating electrolyte, but was fitted with asbestos diaphragms on every electrode to prevent gas leakage on the electrolyte side.

The anode consisted of catalyzed Raney nickel. To produce this type of electrode, aluminium and nickel were mixed in a way that no true alloy is formed, but as a microcrystalline structure, where both metals exist side by side. Then, the aluminium was dissolved in a strong alkali, which leaves behind a porous nickel material with a very high surface area. The pore size and pore distribution depended on the ratio of Ni/Al in the mixture (normally 1 : 1) and, of course, on the particle size. Due to the high hydrogen/oxidation catalytic effect of the dispersed nickel, no additional catalyst was needed for the anodes, but they were sometimes doped with Ti, which had a stabilizing effect (reducing material loss during operation).

The cathodes consisted of an active carbon matrix with Teflon and asbestos with silver as the catalyst, which was doped with Hg, Bi, Ni and Ti for stabilizing reasons. When the system was operated using air as the oxidant, approx. 60 mg Ag cm^{-2} was used.

The electrolyte was used for stack cooling and for water removal also. Both took place in one separate unit, the electrolyte regenerator. It simply looked like a fuel cell. Water evaporated through an asbestos membrane, thereby cooling the electrolyte, and condensed on parallel cooling plates.

In the early 1990s, Siemens built AFCs fitted with Raney nickel electrodes for submarines.

2.9 The Russian Photon fuel cell system

Not very much is known about Russian fuel cell systems. During the investigation of different foreign systems for the manned space ship HERMES from European Space Agency (ESA),[22] a flight model of the Russian fuel cell power plant "Photon" was tested in 1993 at the European Space Research and Technology (ESTEC) fuel cell test facility. Photon was built for the Russian space plane "Buran", but this project has already been stopped.

The Photon fuel cell system, manufactured in 1991, consisted of eight 32-cell stacks, all connected in parallel. The active electrode area was 176 m^2. The 40 wt% KOH electrolyte was immobilized in an asbestos matrix and the reaction water was removed in a hydrogen loop. The stack was operated at 100 °C at gas pressures of 1.2 MPa.

The tests, investigated at ESTEC, showed the excellent behaviour of the system concerning stable temperature control, water removal and electrochemical behaviour even at high loads. However, no lifetime data is available and the high noble metal loading of 40 mg cm^{-2} Pt restricted this system to military and space applications.

2.10 The ESA/Dornier fuel cell system for HERMES

The prime contractor for the development of the fuel cell power plant for the manned space shuttle HERMES (HFCP, sometimes FCPP) of the ESA was Dornier Systems. In the beginning (1984), there were four subcontractors involved in electrode and stack research: Siemens, Varta, Elenco and Sorapec.

Siemens (Germany) concentrated on a system with a mobile electrolyte and matrix system. The electrodes were nickel (anode) and silver-based (cathode). The mobile system again had an electrolyte reconcentrator unit and asbestos membranes as described above. Due to severe gas bubble problems (removal of bubbles is difficult in zero gravity, μg) Siemens removed one asbestos membrane and used the second one as electrolyte compartment (immobilisation). This greatly increased the performance (voltage gain = 40 mV at 200 mA cm^{-2}), mainly because of a decrease in ohmic loss over the smaller distances between the electrodes. Other advantages were an increased system simplicity and increased safety (no electrolyte circuit). Water removal was now on the anode side but caused tremendous problems when changing currents by electrode flooding or dry-out. These being problems that are well known in today's PEMFCs.

Varta (Germany) based their research on their Eloflux system (see above) and double-layer skeleton electrodes (DSKs, similar to Bacon's electrodes, see above). Several companies and universities were involved, e.g. Hoechst AG, the University of Kassel and the Fraunhofer Institute, Freiburg (all Germany). The anodes were of the Raney nickel type doted with various metals like platinum, palladium, gold, silver and mercury, although the usage of noble metals is not necessary due to the high activity of Raney nickel. The oxygen electrodes were silver-based DSK electrodes. Varta also introduced a diffusion gap evaporator, which consisted of porous hydrophilic membranes at the electrolyte and the product water sides with a hydrogen-filled diffusion gap in between.

Elenco (Belgium) built their thin carbon-based and PTFE-layered electrodes for a system version with a mobile electrolyte. Much work was expended in decreasing the KOH leaking, especially through the cathodes ("weeping"), like using a thicker PTFE layer (290 μm) or using ammonia bicarbonate as a pore former. For the prepared electrodes, an excellent electrochemical behaviour was reported: 887 mV at a loading of 200 mA cm^{-2}, a pressure of 4 bar and a temperature of 80 °C for lifetimes over 2000 h.

Sorapec developed thin electrodes for matrix systems but did not play an important role in the overall HFCP project.

It has to be noted that the technologies described above were state of the art in December 1988, when Dornier made a controversial decision. Based on their own set of rules[23] for the system selection, they decided to continue with the Siemens immobile system despite the numerous advantages and the progress of the Elenco system at that time. However, Elenco became a subcontractor of Siemens.

Until 1993, when the HERMES program was also stopped for various reasons, Siemens worked on the stack itself, building five-celled stacks for testing purposes. The final stack as specified by Dornier should have consisted of 104 cells, supplied in parallel by the reactants oxygen and hydrogen.[24, 25] The main challenge concerns the fine control of the content of water removed according to the stack working conditions, i.e., avoiding dry-out of the stack matrix on the one side and the risk of flooding of the stack on the other side. Three tasks were given to Elenco, namely the gas purge subassembly, the cooling subassembly and the development of a "pseudo stack back-up technology". Thereby, they concentrated on further decreasing the cathode-weeping and on stack construction (mainly material science). In any case, the electrode performance which was reported by Siemens, could not be obtained by Elenco.

2.11 The Hoechst falling film fuel cell system

Hoechst developed a Chlor-alkali electrolysis cell also capable of working as a fuel cell with a very low energy demand. These membrane (Siflon®) cells had an active area of $1-3 \, m^2$ and contained silver as the catalyst and PTFE (Hostaflon®). At 2.6 MPa reactant gas pressure, 80 °C, $300 \, mA \, cm^{-2}$ and with NaOH as the electrolyte, only 20-mV degradation was measured after 3 years of operation.[26]

2.12 The Swedish small scale stack technology

A secret project of the Swedish Navy to develop an AFC system at the University of Stockholm (Sweden) ended with an accident because of burning of the oxygen electrodes. To continue, Lindström tried in 1993 to use this technology for small bio-fuel cells in India, units with a 500 kW output for a village power supply were planned.[27–30] Obviously this project was discontinued after Lindström died.

However, Lindström published a very critical assessment of fuel cell technology, which indicates that AFC technology has been under-evaluated over recent years. He indicates in his works that most of the problems have returned but now occurring in the PEM systems in contrast to the AFC systems,[31, 32] and this is especially true in the case of the so-called CO_2 syndrome. In contrast to the fuel cell community, Lindström was convinced that the carbon dioxide can be removed cost-effectively from the hydrogen feed (which is practised in every ammonia plant).

2.13 The Austrian fuel cell system

In 1985, a program based on the Union Carbide electrode technology was started at the University of Technology in Graz (Austria) to develop a low-cost alkaline bipolar fuel cell system whose components should be mass

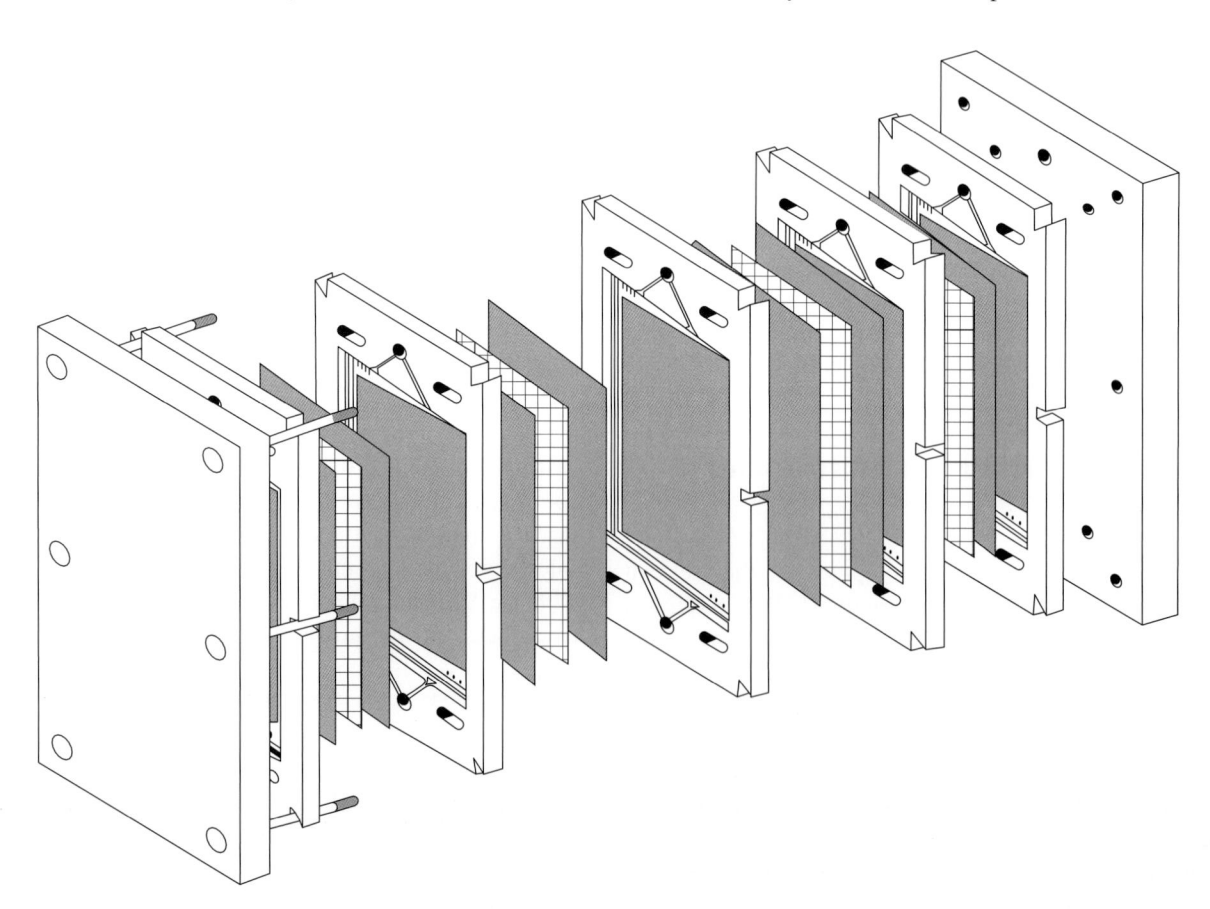

Figure 9. Bipolar AFC battery concept developed at the Technical University Graz.

producible (see Figure 9). The all-carbon electrodes were PTFE-bonded carbon black rolled on a carbon foil. They were operated in air and hydrogen and only used a minimum amount of noble metal catalyst. Extruded carbon/polypropylene plates were used as bipolar plates and manifolds.[33]

2.14 The Apollo Energy Systems Inc. fuel cell

At the end of 1997, Apollo Energy Systems Inc. (AESI, former Electric Auto Corporation, EAC) started a R&D program at the University of Technology in Graz (Austria) to develop mass producible low-cost AFC electrodes based on the thin PTFE-bonded carbon electrodes of the Union Carbide Corporation and several R&D projects in Graz with a circulating potassium hydroxide electrolyte, but using today's raw materials.[34–38] Within 4 years, the current density was doubled compared to UCC technology (maximum = 400 mA cm^{-2} using air and hydrogen, both at ambient pressure, 75 °C and 0.5 mg Pt cm^{-2}). In addition, two functional modules were built, fitted with bipolar 21-cell stacks (first generation, 10 A × 15 V = 150 W nominal) and monopolar 21-cell stacks (second generation, 15 A × 15 V = 225 W nominal), as it turned out that choosing the bipolar system with all its constructional problems is useless under nominal stack currents of 60 A. The size and the weight of the monopolar stack was approximately half of that of the bipolar stack.

REFERENCES

1. K. Kinoshita, 'Electrochemical Oxygen Technology', John Wiley & Sons, New York, Chapter 2, pp. 19–112 (1992).
2. K. Kordesch, J. Gsellmann and B. Kraetschner, *Power Sources*, **9**, 379 (1983).
3. K. Kordesch, 'Gibt es Moeglichkeiten zur Herstellung und Verwendung von 'Low Cost–Low Tech' Zellen?' in "Brennstoffzellen: Stand der Technik Enwicklungslinien Markchancen", H. Wendt and V. Plzak (Eds), VDI–Verlag, Duesseldorf, pp. 57–63 (1990).
4. K. Kordesch, 'The Choice of Low-Temperature Hydrogen Fuel Cells: Acidic or Alkaline?', Presented at Hydrogen Energy Progress–IV, in "Proceedings of the 4th World Hydrogen Energy Conference", California, USA, T. N. Veziroglu, W. D. V. Vorst and J. H. Kelley (Eds), Vol. 3, pp. 1139–1148, 13–17 June (1982).
5. F. T. Bacon, British Patent 667,298 (1950); DAS 1,025,025 (1954); British Patent 725,661 (1955).
6. H. v. Döhren and J. Euler, 'Der heutige Stand der Brennstoff elemente', Akadem. Verlagsges., Frankfurt (1965).
7. E. Justi and A. Winsel, 'Kalte Verbrennung–Fuell Cells', Steiner Verlag, Wiesbaden (1961).
8. J. L. Platner, D. Ghere and P. Heiss, Presented at the US Army 19th Annual Power Sources Conference (Symposium), pp. 32–35 (1965).
9. L. G. Austin, 'Fuel Cells', Scientific and Technical Information Division, NASA, US Government Printing Office, Washington DC (1967).
10. K. V. Kordesch and G. Simader, 'Fuel Cells and Their Applications', Wiley, Weinheim, New York, Tokyo (1996).
11. K. Kordesch, J. Gsellmann, M. Cifrain, S. Voss, V. Hacker, C. Fabjan, T. Hejze and J. Daniel-Ivad, *J. Power Sources*, **80**, 190 (1999).
12. K. Kordesch, Union Carbide Corp., Parma Res. Lab., CRM 244, Aug. 11 (1970).
13. K. Kordesch, 'City Car with H_2/Air Fuel Cell and Lead Battery', Presented at the 6th IECEC Congress, SAE-Paper No. 719015 (1971).
14. K. Kordesch, 'Lead Acid Batteries and Electric Vehicles', Marcel Dekker (1977).
15. K. Kordesch, 'Brennstoffbatterien', Springer–Verlag (1984) (in German).
16. K. Kordesch and J. Oliveira, 'Fuel Cells', in "Ullmanns Encyclopedia", VCH, Weinheim, Vol. A-12, pp. 55–83 (1989).
17. K. Kordesch, V. Hacker, J. Gsellmann, P. Enzinger, M. Cifrain, R. Aronson, G. Faleschini, M. Muhr and K. Friedrich, *J. Power Sources*, **86**(1–2), 163 (2000).
18. A. Winsel, 'The Eloflux Fuel Cell System', DECHEMA-Monogr. 92, pp. 1885–1913 (1983).
19. A. Winsel and R. Wendtland, US 3,597,275 (1971).
20. H. van den Broeck, 'Commercial Development of Alkaline Fuel Cells', Presented at the CEC Italian Fuel Cell Workshop, Taormina, Italy, pp. 73–77 (1987); The Commission of the European Communities, Brussels, Belgium (1988).
21. Siemens AG, Final Report '20 kW Brennstoffzellenanlage in Kompaktbauweise', parts 1-7, T 80-055 to T 80-059, Erlangen (1980).
22. M. Schautz *et al.*, 'Results of Tests on the Photon Fuel Cell Generator', in "Proc. Europ. Space Pow. Conf.", Graz, Austria, Vol. 2, pp. 653–658, Aug. 23–27 (1993).
23. Dornier System GmbH, 'WP1800 System Selection', Hermes Fuel Cell Power Plant Phase C1, Technical Note, December (1988).
24. F. Baron, 'Fuel Cell for European Space Vehicle. Status of Activities', in "Proc. Europ. Space Pow. Conf.", Graz, Austria, Vol. 2, pp. 635–646, Aug. 23–27 (1993).
25. L. Blum, U. Gebhardt, J. Gilles and B. Stellwag, 'Development of an AFC for Space Use', in "Proc. Europ. Space Pow. Conf.", Graz, Austria, Vol. 2, pp. 647–652, Aug. 23–27 (1993).
26. K. H. Tetzlaff, R. Walz and C. A. Gossen, *J. Power Sources*, **50**, 311 (1994).
27. C. Myrén and S. T. Naumann, 'Fuel Processor Systems for Small Scale AFC Power Plants', Presented at the Fuel Cell Seminar, San Diego, CA, pp. 192–195 (1994).

28. O. Lindström, S. Schwartz, Y. Kiros, M. Ramanathan and A. Sampathrajan, 'Small Scale AFC Stack Technology', Presented at the Fuel Cell Seminar, San Diego, CA, pp. 323–324 (1994).
29. Y. Kiros, 'Tolerance of Gases by the Anode for the Alkaline Fuel Cell', Presented at the Fuel Cell Seminar, San Diego, CA, pp. 364–367 (1994).
30. J. Kivisaari, 'Influence of Impurities in the Anode and Cathode Gases on Fuel Cell Performance and Life – A Literature Overview', Presented at the Fuel Cell Seminar, San Diego, CA, pp. 436–437 (1994).
31. O. Lindström, 'A critical Assessment of Fuel Cell Technology', Presented at the Fuel Cell Seminar, San Diego, CA, pp. 297–298 (1994).
32. O. Lindström, 'A Critical Assessment of Fuel Cell Technology', Department of Chemical Engineering and Technology, Royal Institute of Technology Stockholm 1993, ISRN KTH/KT/FR, SE, ISSN 1101-9271 (1993/1994).
33. K. Kordesch *et al.*, 'Fuel Cell Research and Development Projects in Austria', in "Proceedings of the 7th World Hydrogen Energy Conference", Moscow (1988); T. N. Vézirglu (Ed.), Presented at the Hydrogen Energy Progress – VII, Pergamon Press, Oxford (1988).
34. K. Kordesch, J. Gsellmann and M. Cifrain *et al.*, 'Fuel Cells with Circulating Electrolytes and their Advantages for AFCs and DMFCs. Part 1: Alkaline Fuel Cells', Presented at the 39th Power Sources Conference, Cherry Hill, USA, pp. 108, June 12–15 (2000).
35. K. Kordesch, V. Hacker, J. Gsellmann and M. Cifrain *et al.*, *J. Power Sources*, **86**, 162 (2000).
36. K. Kordesch, M. Cifrain, T. Hejze, V. Hacker and U. Bachhiesl, 'Fuel Cells with Circulating Electrolytes', Presented at the 2000 Fuel Cell Seminar, Portland, Oregon, USA, pp. 432–435, Oct. 30–Nov. 2 (2000).
37. K. Kordesch, J. Gsellmann, M. Cifrain and V. Hacker *et al.*, 'Advantages of Alkaline Fuel Cell Systems for Mobile Applications', Presented at the 2000 Fuel Cell Seminar, Portland, Oregon, USA, pp. 655–658, Oct. 30–Nov. 2 (2000).
38. M. Cifrain, 'Thin Carbon-based Electrodes for Alkaline Fuel Cells with Liquid Electrolytes', PhD Thesis, Graz University of Technology, Austria (2001).

Chapter 15
Hydrazine fuel cells

H. Kohnke
i.H. Gaskatel, Kassel, Germany

1 INTRODUCTION

Hydrazine may be regarded as an ideal fuel. It is easily stored and transported as an aqueous solution. It has a high boiling point (115 °C) and is electrochemically oxidized at potentials close to the hydrogen potential. However, it is difficult to visualize hydrazine as a commercially attractive fuel, as it is toxic and quite expensive.

There are two ways to use the chemical energy of hydrazine in fuel cells. Either the fluid hydrazine is given into the electrolyte of the fuel cell where it is converted at special hydrazine electrodes or the hydrazine is converted in a external decomposer into a hydrogen/nitrogen rich gas.

2 HISTORY

In 1962, the Chloride Electric Storage Company described a hydrazine/oxygen fuel cell. At 20 °C, the current density was about 50 mA cm^{-2} at a cell voltage of about 0.6 V.[1] In 1963, the Allis-Chalmers Company developed a hydrazine fuel cell with palladium-catalyzed nickel as the hydrazine electrode and silver-catalyzed oxygen electrodes.[2] Kordesch developed hydrogen-driven vehicles (see Figure 1) in the 1960s. He used alkaline fuel cells and a lead acid accumulator as a back-up system. The problem with the hydrogen storage was first overcome by using hydrazine in his motorcycle.[3]

3 PROPERTIES OF HYDRAZINE

Pure hydrazine that is free from water is dangerous. It reacts with air and also with some metals and metal oxides. It is less dangerous to use mixtures of hydrazine and water. Commercially available are 15, 51 and 64% hydrazine/hydrate mixtures.

Furthermore, hydrazine is known for causing cancer and poisoning the liver and the erythrocytes. The maximum concentration allowed is about 0.1 ppm.[4]

4 HYDRAZINE AS A FUEL FOR FUEL CELLS

In fuel cell systems that must operate for long periods without any servicing, the advantage of liquid reactants over gaseous ones is that they can be stored in simple light weight containers. In addition, their energy density is much higher. Thus, the theoretical energy density for liquid hydrazine hydrate is 3560 Wh l^{-1} compared with 590 Wh l^{-1} for pressurized hydrogen at 200 bar.

As already mentioned hydrazine can be converted to hydrogen and nitrogen in a reactor and the gas mixture fed to the fuel cell, or the hydrazine is added to the electrolyte of the fuel cell and converted at special hydrazine electrodes.

4.1 The decomposer

4.1.1 Fundamentals

There are two reaction paths for hydrazine, if converted in an external reactor:

$$N_2H_4 \longrightarrow N_2 + 2H_2 + 50.66 \text{ kJ} \qquad (1)$$

the hydrazine flow vertically through the plates, which avoids mass transport resistance. The hydrogen generated inside the plates is led away by a cross-flow process in the plane of the plates. These biporous structures are known in other plate-shaped reactors, as described elsewhere.[5] Apart from the most efficient use of the catalyst, another advantage in comparing a tube reactor with a catalyst bed is that the pump which pumps the hydrazine to the reactor can be operated at lower pressures (10–14 N cm^{-2}). However, the hydrogen leaving the reactor is still at high pressure.

In demonstration units, four of these biporous catalyst plates with dimensions of 116×82 mm^2 were mounted into the decomposer. At a hydrazine pressure of about 1.2–1.5 bar, the pressure of the exhaust hydrogen gas was about 2 bar. As mentioned in Section 4.1.1, the rate of gas production is increased by increasing temperature, but the content of ammonia in the gas rises too. The following dependencies have been observed:

$$\text{gas evolving rate} \propto e^{\frac{-7500}{T}} \quad (3)$$

$$\text{amount of ammonia} \propto e^{\frac{-5000}{T}} \quad (4)$$

T is for temperature (K).

The total weight of the decomposer is about 0.75 kg and the overall dimensions are 0.47 l. At an inlet pressure of 1.2 bar and a temperature of about 60 °C, 905 l h^{-1} of gas are produced, suitable for a 1 kW fuel cell. The amount of ammonia in the gas is 8%, and 1% of the hydrazine leaves the reactor. If the decomposer is operated at the smallest mass flux and at only 30 °C, the content of ammonia and the remaining hydrazine decreases to less than 0.1% (see Figures 2 and 3).

Figure 1. Kordesch on the hydrazine fuel cell motorcycle. (Reproduced with permission from Karl Kordesch.)

The second path is:

$$3 N_2H_4 \longrightarrow 4 NH_3 + N_2 + 355 \text{ kJ} \quad (2)$$

and leads to a production of ammonia and should, therefore, be avoided.

These demands are met by Raney nickel catalysts with the lowest possible iron content because of the preferred NH_3 development at iron. This side reaction decreases at lower temperatures. At lower temperatures, the reaction rate is reduced. A way out is to choose optimized dimensions of the reactor. The developed gas mixture is then directly fed into a hydrogen/oxygen fuel cell.

4.1.2 Construction of the decomposer

The reactor for decomposition of the hydrazine can be, as many other reactors, a tube with a catalyst bed or like a lamellar reactor with catalytically active plates. Lamellar reactors are in principle of smaller size and the thermal conditions can be better controlled. A particularly good use of the reactor volume is achieved if the plates are made of porous materials. In this case, it is possible to pump

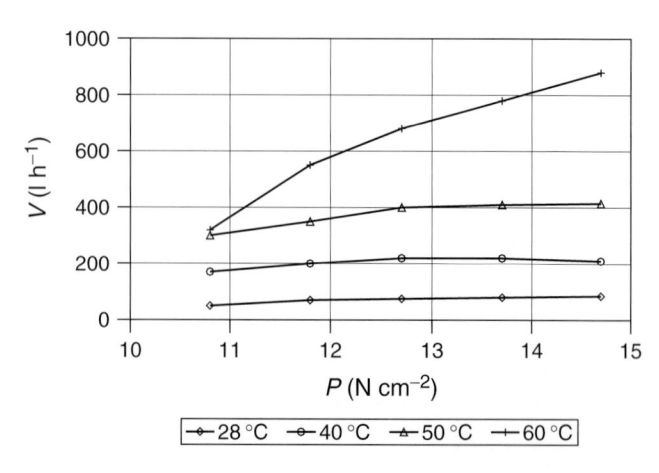

Figure 2. Hydrogen flow independent of the temperature and hydrazine pressure.

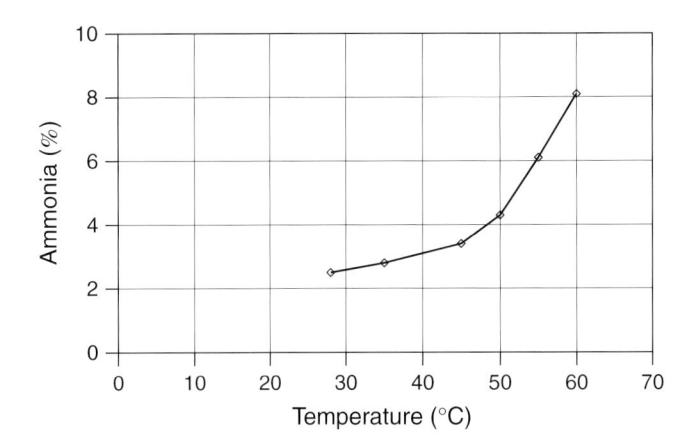

Figure 3. Ammonia content in the hydrogen flow (temperature-dependant).

4.2 The hydrazine fuel cell

4.2.1 Fundamentals

Hydrazine occupies a special place amongst other liquid fuels. It is the only liquid fuel of which reaction products do not react with alkaline electrolytes and, therefore, do not consume it. During the direct hydrazine conversion, the fuel is dissolved in the alkaline electrolyte and specific catalytically active electrodes are used in the reaction:

$$N_2H_4 + 4OH^- \longrightarrow N_2 + 4H_2O + 4e^-$$
$$\varphi' = -30\,\text{mV RHE} \tag{5}$$

At the cathode the following reaction takes place:

$$O_2 + 2H_2O + 4e^- \longrightarrow 4OH^- \quad \varphi = 1229\,\text{mV RHE} \tag{6}$$

Thus, the open circuit potential of a hydrazine/oxygen cell should be 1.56 V. But it is well known that the oxygen potential is reduced in real systems to about 1.1 V, and the rest potential of a hydrazine electrode is near the hydrogen potential (for details see Ref. [7]). Therefore, the open circuit voltage of a hydrazine/oxygen cell should be near 1.1 V.

4.2.2 Anode catalysts

A suitable catalyst for hydrazine should not support chemical dehydrogenation or the formation of ammonia. The metals and alloys with three and more unfilled positions in the d bands are well-known as dehydrogenation catalysts. Because of this, all metals with holes in the 3d band would be excluded, just as alloy catalysts that are composed of metals with three or more d band holes. Furthermore, the catalyst has to have a high hydrogen overpotential. Otherwise the electrode will produce hydrogen. This is the reason that Raney nickel is only of restricted use.

It has been shown that catalytic electrodes consisting of amalgamated nickel do not catalyze the self-decomposition of hydrazine to a significant extend – no ammonia is generated – but nevertheless the electrode can be loaded with high current densities. The best performance of the electrodes is observed if they contain 4 mg of mercury per cm^2.[6]

Most of the previously described hydrazine/oxygen cells with a lower voltage than 1.56 V could, therefore, be described as hydrogen/oxygen fuel cells because the anode catalyst reacts directly with the hydrazine as described in Section 4.1.1.

The activities of different catalysts have been demonstrated by Vielstich in half cell tests[7] (see Figure 4).

In general, one can assume that the energy efficiency decreases with increasing activity of the catalyst because of the side reaction referred to in equation (2).

4.2.3 Hydrazine fuel cells

In addition to the UCC hydrazine fuel cells mentioned above, Allis-Chalmers and the Chloride Electric Storage Company, there are several other known cells, for which the technical data are outlined in the following.

Alsthom
Alsthom developed a hydrazine/hydrogen peroxide fuel cell for submarine applications. The anode is composed of thin, non-porous metal foils covered with a cobalt catalyst. The total weight for the fuel cell is about 2 kg kW^{-1}.

Grüneberg et al.
In most cases the fuel cell's performance is limited by the cathode if air is the oxidant. But even with pure oxygen, the polarization at the cathode is more important than the polarization at the anode. It is well known that hydrogen peroxide as a liquid oxidant leads to very high current densities at the cathode.

Therefore, Dünger, Grüneberg and Wedding have developed a hydrazine/hydrogen peroxide fuel cell. The anode is flame-sprayed with Raney nickel, and the cathode is formed out of electrodeposited silver. The current/voltage performance is shown in Figure 5.

Liquids are pumped into the anolyte and catholyte circuit. The fuel cell module, composed of 12 cells in series, provides approximately a 1 kW peak load. The specific weight is 10 kg kW^{-1} and 6 l kW^{-1}. The hydrazine efficiency is 70%.[8]

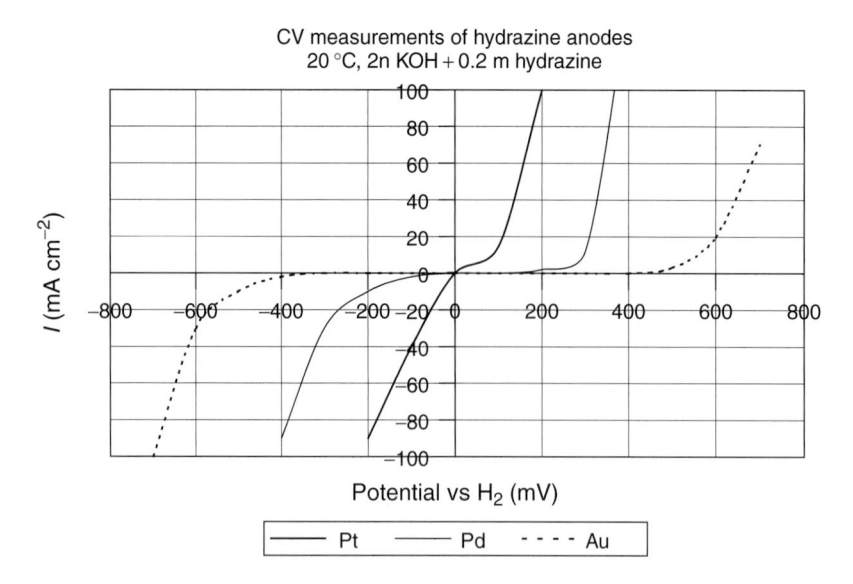

Figure 4. Potential of various hydrazine electrodes in half cell tests. (Reproduced from Vielstich (1965)[7] with permission of Verlag Chemie.)

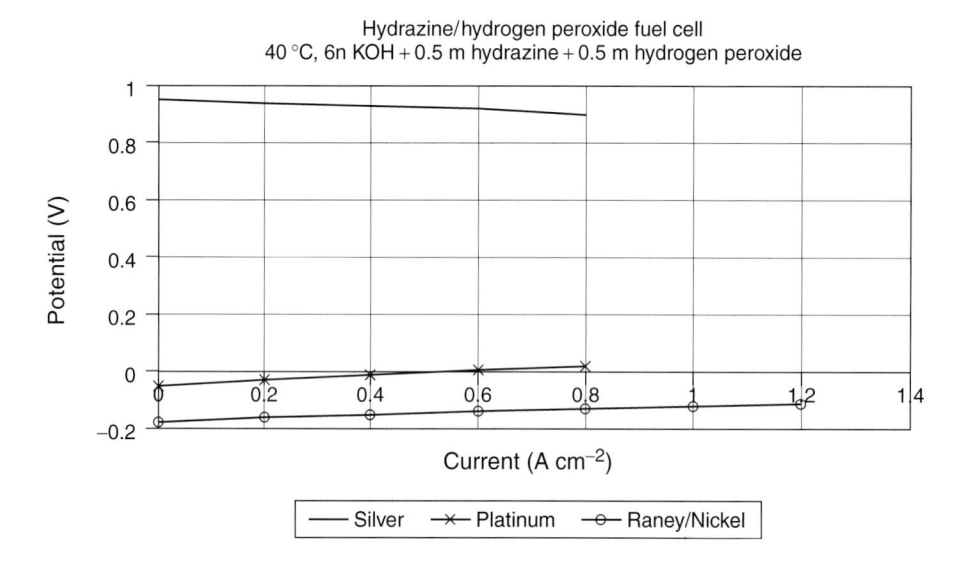

Figure 5. Half-cell data of the Grüneberg hydrazine/hydrogen peroxide fuel cell. (Reproduced from Grüneberg and Wedding (1967).[8])

Hitachi

Hitachi showed a 3 kW hydrazine/air fuel cell in 1978 for military applications. The cathode was a polytetrafluoroethylene-bonded carbon electrode loaded with several milligrams of palladium and platinum. The anode was a porous sintered nickel electrode with several milligrams of palladium catalyst. The single voltage was about 0.71 V, the current density was 80 mA cm^{-2}. The specific volume was 54 W l^{-1} and the specific weight was 63 W kg^{-1}.[9]

Matsushita electric industrial company

Hydrazine electrodes formed out of sintered nickel and palladium was added by electroless plating. The fuel cell stack, comprising of 18 unit cells in piled form and in which an aqueous solution of caustic potash has a specific gravity of 1.3, is used as an electrolyte and the hydrazine hydrate is used as a fuel. When continuous discharge is effected at 60 mA cm^{-2} with a fuel concentration of 30%, the reduction in voltage after a lapse of 8000 h is 6–8% of the initial voltage. The utilization efficiency of the fuel is about 60–70%. This cell has an excellent lifetime.[10]

Siemens

Instead of using noble catalysts like platinum and palladium, several other cheaper catalysts such as Ni$_2$B, Raney nickel and cobalt are also suitable. Because of the extremely negative potential of the anode, it can be assumed that the nickel catalyst does not only produce hydrogen out of the hydrazine but also converts the hydrazine directly (see Figure 6).

Siemens used an asbestos paper serving several functions in the fuel cell. It serves as an electrode spacer, as well as a diffusion barrier for the hydrazine. The diffusion of the fuel to the oxygen electrode permits a chemical short circuit resulting in the oxidation of the hydrazine to nitrogen, ammonia and water, thus, increasing the consumption of oxygen. Polarization at the cathode also increases due to the formation of mixed potentials.

The Faradaic efficiency of the hydrazine in a 10 cell unit is over 90% at a temperature of 40 °C and a hydrazine concentration of about 0.1 M. The cell also has an excellent lifetime, as the voltage drop in the first 2000 h is less than 90 mV.[11]

Yuasa

The anode of the Yuasa hydrazine fuel cell is a sintered body out of nickel and Ni_2B. The voltage of a single cell at a current density of $50 mA cm^{-2}$ in air is about 0.7 V. The power density, $45 kg kW^{-1}$, is, in comparison to the Alsthom fuel cell, quite low.

5 NEW APPLICATIONS

The advantage of fuel cells is that the electrochemical converter and the storage device are separated. This means that in long-term operations fuel cells could be a better proposition than batteries if the container for the fuel is not too heavy or too big. The fuels suitable for such long-term applications are methanol or hydrazine.

As mentioned above, the hydrazine fuel cell was of interest for underwater applications like submarines. Today other underwater applications include sensors for drift and oceanography. One new topic is the detection of methane at a depth of 2500–3000 m. Because the electric demand of the sensor is about 10 W and the operating time would be more than 1 year, fuel cells, especially the hydrazine fuel cell, have to be taken into account as one of the most favorable technologies for this purpose.

However, the toxicity of hydrazine must be considered in all applications.

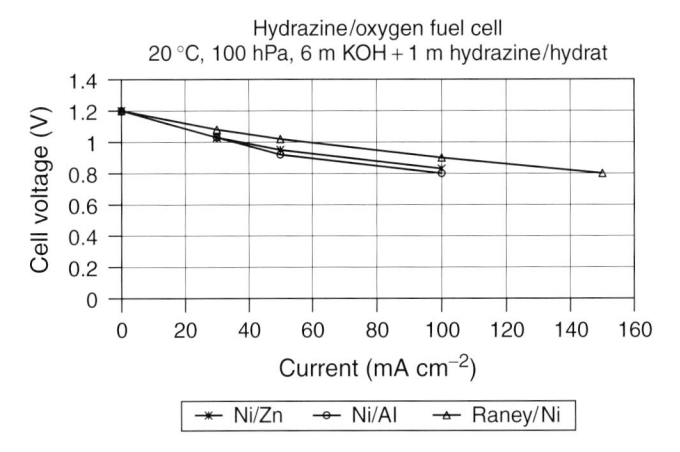

Figure 6. Cell voltage of the Siemens hydrazine fuel cell. (Reproduced from Collins.[11])

REFERENCES

1. G. R. Lomax and M. J. Gillibrand, in "Proc. 3rd Int. Symp. Batteries", Pergamon, pp. 221 (1962).
2. S. S. Tomter and A. P. Antony, 'Fuel cells', in "Chem. Eng. Prog. Tech. Manual", Am. Inst. Chem. Eng., pp. 22 (1963).
3. K. Kordesch and G. Simader, 'Fuel Cells and their Application', Wiley–VCH, Weinheim (1996).
4. D. Henscher, 'Gesundheitsschädliche Arbeitsstoffe – Toxikologisch – Medizinische Begründung der MAK Werte', Verlag Chemie, Weinheim (1972).
5. BMFT Report, EDU 202/68, Varta Battery AG (1973).
6. USP 3811949, Jung (1973).
7. W. Vielstich, 'Fuel Cells', John Wiley & Sons (1970).
8. G. Grüneberg and F. Wedding, Presented at the AGARD Meeting, Liège, Belgium (1967).
9. K. Tamura, *New Mater. New Processes*, **2**, 317 (1983).
10. US Patent 4,001,040, Jan. (1977).
11. D. H. Collins, 'Power Sources 3', Oriel Press, pp. 373ff, Sept. (1970).

Chapter 16
Phosphoric acid electrolyte fuel cells

J. M. King[1] and H. R. Kunz[2]

[1] Manchester, CT, USA
[2] Department of Chemical Engineering, University of Connecticut, Storrs, CT, USA

1 FUEL CELL APPLICATIONS

Phosphoric acid fuel cells have been applied to stationary power, tactical military power and motive power applications.

1.1 Stationary power

Stationary power applications have ranged from homes to commercial buildings. Types of installations included: combined heat and power; on-line back-up power; continuous, no-break power; operation as distributed generators in the electric utility network; and operation on the digester gas produced in wastewater treatment plants.[1]

For phosphoric acid fuel cells, home application has been limited to experimental evaluation in a field test conducted in the early 1970s as part of the Team to Advance Research in Gas Energy Transformation (TARGET) program, which was sponsored by a group of gas utilities and United Aircraft (the predecessor to United Technologies Corporation). Homes, commercial and industrial buildings from Calgary, Alberta in Canada to New England, the Midwest and Southwest of the United States, and in Japan installed the 12.5 kW unit. In this field test of the on-site fuel cell concept, the power plants operated independently of the utility network following the electrical load of the home automatically. These tests were limited to a few thousand hours of operation, but exposed the power plants to load profiles of different building types, altitude up to 1600 m (1 mile), weather extremes (Southern California to Chicago and New England) and to a range of natural gas compositions. Power plants were applied singly, and in groups of 2, 3 and 6 units. Total operating experience with 65 power plants exceeded 200 000 h.

Initial testing of combined heat and power application of fuel cells was conducted under sponsorship of the US Department of Energy and the Gas Research Institute in an activity during the mid-1980s. This program tested a number of 40 kW fuel cell power plants in commercial, government, health care and industrial applications.[2] Two of the power plants used in these tests are shown in Figure 1. The installation is at a health club in California; heat is recovered for use in the domestic hot water system of the building. Figure 2 shows the electrical and overall (electrical plus thermal) efficiency for one of the power plants in this installation over roughly a 1 year period of testing. The 40 kW field-testing included 53 power plants and over 350 000 h of testing. A few power plants operated in excess of 12 000 h.

The combined heat and power application constituted a majority of the early installations of the first commercial fuel cell power plant, the 200 kW PC25™ A, which was installed for the first time in April 1992. This power plant was manufactured by ONSI Corporation, a subsidiary of International Fuel Cells (in November 2000, ONSI was absorbed into IFC and production of the PC25 continues at UTC Fuel Cells). A photograph of the PC25 A is provided in Figure 3, which shows an installation at a hospital in New York where the heat is used for domestic hot water.

Handbook of Fuel Cells – Fundamentals, Technology and Applications, Edited by Wolf Vielstich, Hubert A. Gasteiger, Arnold Lamm.
Volume 1: *Fundamentals and Survey of Systems*. © 2003 John Wiley & Sons, Ltd. ISBN: 0-471-49926-9.

Figure 1. Two 40 kW power plants installed at a health club.

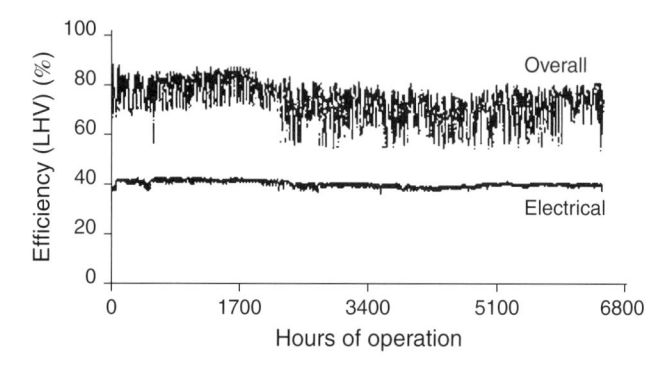

Figure 2. Electrical and overall efficiency of 40 kW power plants in a health club application (Southern California Gas Power Plant serial no. 8224).

Figure 3. Installation of PC25A power plant at a hospital.

Hotels, nursing homes, district heating systems in Europe and food processing installations are other examples where this power plant has been used in the combined heat and power application. A large activity by the US Department of Defense (DOD), managed by the US Army Construction Engineering Research Laboratory, has focused application in combined heat and power using improved models of this 200 kW unit, the PC25 B and PC25 C. In addition to the heat applications noted above, these installations include application to boiler feed water heating for central heating systems for military bases. Performance data on these DOD sites is available at http://www.dodfuelcell.com. Heat recovery experience with these units ranges from near zero heat recovery to recovery of heat energy at a rate over 750 000 BTU/h or over 220 kW thermal; many of the DOD sites for which heat recovery data are available show recovery of heat equal to 60–100 kW thermal.

As experience with the PC25 fleet has increased confidence in the power plant reliability, the power plant has been utilized for applications where power reliability is an important life safety or business objective. These applications are referred to as "premium power" or "assured power" installations. Several different system configurations have been utilized. The simplest configuration is operation of the power plant connected to the utility network with switching of the power plant to grid-independent power supply to a critical load when outages of the electric network occur. Hospitals are a good example of this type of application. The electrical configuration is similar to that for a standby engine generator set and a period of 5–10 s is required to switch the load to the grid-independent fuel cell mode. The advantage of the fuel cell in terms of reliability is that it is always on-line and its condition is therefore always known so that any problems with the back-up power source are identified and corrected immediately. Figure 4 is an example of an application that satisfies the needs of computers for "no break" power, defined as permitting an interruption of no more than 4 ms. The installation is at a regional mail-sorting center in Anchorage, Alaska. Frequent power interruptions of the electric network at this location disrupted the mail sorting equipment and required a work shift to recover. The fuel cells were installed in a manner that has eliminated the interruptions. At this location, the normal operation of the five fuel cells rated at 200 kW each is connected to the electric network. If there is an interruption of the electric network, a static switch disconnects the fuel cell installation from the grid and the five fuel cells operate in a grid-independent mode to provide critical power to the building. The switching action takes place in less than 4 ms so the computer operation is not affected. Since the critical load is 750 kW, there is one redundant fuel cell available and an outage would only occur with the simultaneous failure of the electric network and two of the five fuel cells. Other premium power applications include

Figure 4. 5 Power plant central power installation at a mail sorting center in Anchorage, Alaska.

use of the fuel cell as primary power with the electric network providing back up through a static transfer switch and use of redundant fuel cells and standby engine generators in parallel with the electric supply network providing power to an uninterruptible power system for extremely high power reliability.

Critical requirements for application of fuel cells in assured power applications include high electrical efficiency over a broad load range and rapid response to electrical load change. Figure 5 shows the electrical efficiency of the 200 kW power plant is nearly constant from 50% to 100% of rated power. Figure 6 shows output voltage dips only slightly for a few electrical cycles when power is increased instantaneously by 80 kW or 40% of rated load.

The efforts to develop fuel cells for on-site power described above were conducted by United Technologies Corporation and its partners and customers. Other efforts in this area were conducted by Fuji Electric Corporation and Mitsubishi Electric Corporation.

Interest in the use of fuel cells as distributed generators in the electric supply network began in the early 1970s. In this application, the fuel cell would avoid investment in transmission lines to serve a rapidly growing load and would provide reactive power to compensate for inductive loads on the system. The first experiments with operation in this manner occurred during the TARGET program in 1971 and 1972. Three of the 12.5 kW power plants were installed in parallel with the utility grid at a substation in Newark, New Jersey and two were installed in parallel with the utility grid in Enfield, Connecticut.

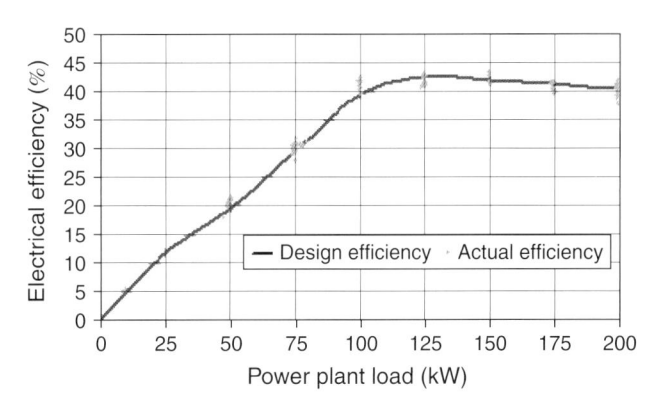

Figure 5. Efficiency versus electrical output for 200 kW power plant.

This experience with the 12.5 kW power plant stimulated investment by a group of utilities and United Aircraft to develop a megawatt scale power plant for installation at distribution substation locations. A result of this activity was fabrication and test of a 1 MW power plant in a breadboard configuration that operated in the experimental fuel cell facility in South Windsor, Connecticut from December 1976 through June 1977. This power plant, shown in Figure 7, was a pressurized unit capable of operation on either natural gas or naphtha fuel. The power plant had

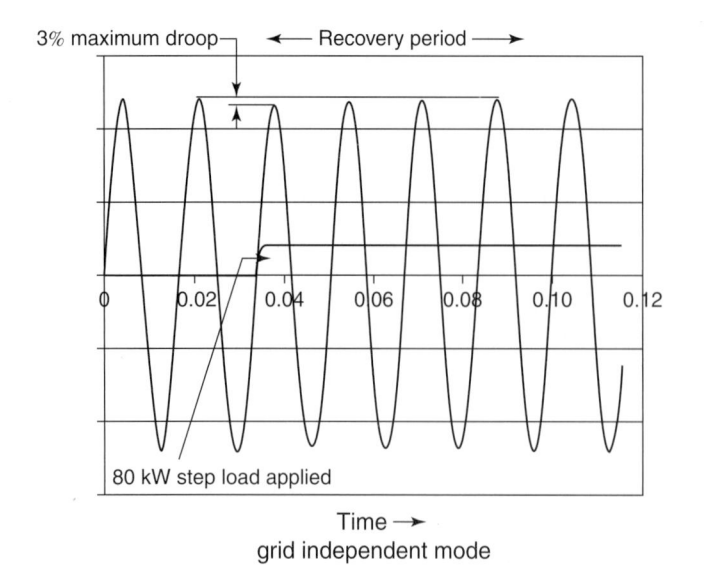

Figure 6. Nearly constant AC voltage output during a 40% step load increase.

Figure 7. 1 MW pilot fuel cell power plant.

20 cell stacks arranged electrically in two parallel strings of 11 stacks each. The 11 000 kW power plant had 18 cell stacks arranged in three parallel strings of six cell stacks each. The cell stacks in the 11 000 kW power plant are the largest PAFC stacks with each cell having an area of $0.93\,m^2$ ($10\,feet^2$) and a total of 492 cells per stack. One of these stacks was operated at an overload condition of 1.05 MW.

Additional efforts to demonstrate pressurized, multi-megawatt distributed generators were conducted by Japanese manufacturers with the support of the New Energy Development Organization (NEDO) during the 1990s. The effort to commercialize pressurized phosphoric acid fuel cell power plants at the megawatt scale was ultimately unsuccessful. Contributing technical factors include control of corrosion within pressurized cell stacks, control of temperature in pressurized reformers, system complexity associated with the turbochargers and the large amount of

Figure 8. 4.8 MW fuel cell power plant installation at Goi, Japan power plant.

six cell stacks shown in the foreground and a fuel processing system shown in the background. It was connected to a 13.8 kV distribution feeder through a power conditioner located in an adjacent room and demonstrated the operational capability and good environmental characteristics of fuel cells. This unit also stimulated the start of efforts to design fuel cell control systems to account for the control and protection concerns of electric utilities regarding distributed fuel cell generators. The operation of this one megawatt pilot plant led to development and test of a 4500 kW power plant and later an 11 MW power plant that are shown in Figures 8 and 9. These units were also pressurized with the cell stacks operating at 3 and 8 atmospheres, respectively. The 4500 kW power plant included

Figure 9. 11 MW power plant tested at Goi, Japan.

field erection required for megawatt scale power plants. A contributing business factor is that this application can not benefit from use of fuel cell heat and does not provide any premium power value to end use customers. Consequently, while the experimental results did show the way to overcoming the technical problems, the business case was not sufficient to continue the costly development activity.

Smaller distributed generators can provide additional value and one application of the 200 kW power plant at a police station in Central Park illustrates these values. While this location in the middle of New York City would seem to be urban and well served by the utility distribution network, this site is well within the park and was served by a very old underground electric line and a natural gas line. When modernization of the police station equipment required an increase in the electric service rating, it was found that the cost and environmental disturbance associated with providing the increased power by installing a new line were prohibitive. At the same time, the natural gas line was sufficient to permit operation of a 200 kW fuel cell so the fuel cell was installed and now provides all the power to the station operating independently from the utility grid. An engine generator set provides backup to minimize disruptions associated with fuel cell outages for infrequent planned or unplanned maintenance actions.

The final stationary application of phosphoric acid fuel cells has been operation on anaerobic digestion gas from waste-water treatment facilities or landfills. Experience with landfill gas is not encouraging for a number of reasons:

1. Generally, there is no use for power at the site so it must be sold into the network in competition with very low cost power from central stations at the several hundred megawatt level;
2. There are no customers nearby who could use the available heat;
3. The gas heating value per cubic foot is only 55% of that for natural gas;
4. Impurities in the gas require extensive clean up before introduction to the fuel cell power plant;
5. The gas supply is unreliable because weather and truck traffic interfere with the flow of gas to the power plant.

Experience with digester gas from wastewater treatment plants is much better because:

1. Both the electricity and heat can be utilized in the treatment plant;
2. The gas heating value is 65% of that for natural gas;
3. While impurities are greater than those in natural gas, they are more amenable to removal than those in landfills;
4. The gas supply is more reliable.

There are municipal wastewater applications in Europe, Japan and the United States; there are also installations for industrial wastewater plants at two breweries in Japan.

1.2 Tactical military applications

Since the early 1960s, the US Army has sponsored development of fuel cell power plants for use in tactical military applications. The reasons for this interest are several-fold:

1. Quiet operation combined with low temperature exhaust and low temperature components reduces the chance of enemy detection;
2. Engine generators are quite unreliable resulting in significant logistic support requirements;
3. High efficiency minimizes the weight of fuel required to complete a mission.

The effort to develop a tactical fuel cell began with evaluation of experimental fuel cell power plants based on alkaline fuel cell technology because this was the only technology sufficiently advanced for experimental evaluation in power plants. Figure 10 shows a 500 W power plant that operated on a sulfur-free, light petroleum product. A 3 kW version of the power plant in Figure 10 was used in the first experiment of a home powered by a fuel cell in 1968. These power plants made an important contribution to the development of fuel processing systems and overall power plant control. Palladium–silver separators were incorporated to provide pure hydrogen fuel to the cell stacks; however, after a number of efforts to clean ambient air sufficiently to permit extended operation failed, military interest moved to phosphoric acid fuel cells in the 1970s.

A 1.5 kW phosphoric fuel cell developed for the US Army used a thermal cracker to process logistic fuels such

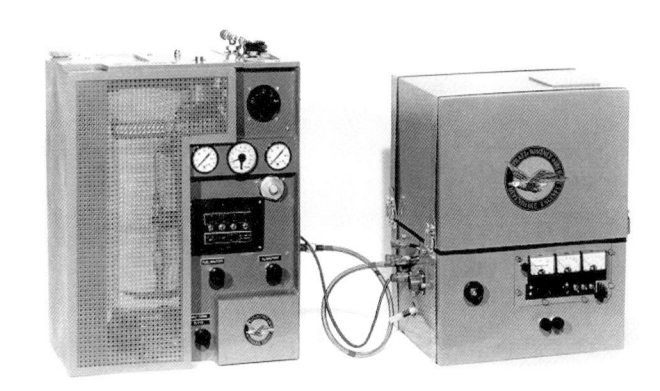

Figure 10. 500 W power plant for army experiments.

as JP-4 to a hydrogen-rich gas for consumption in the phosphoric acid fuel cell stack. Operation on commonly available logistic fuels has always been a key issue with tactical military power plants and the thermal cracker was an attempt to meet this requirement. In the thermal cracker, the hydrocarbon is thermally dissociated into hydrogen and carbon. The carbon deposits and is subsequently burned to provide the heat for thermal dissociation. The thermal cracker employed in this power plant had many problems with variations in gas purity, sealing and durability and this approach to the military requirement was unsuccessful. Another attempt at military fuel cell power plants involved use of methanol fuel. A 1.5 kW power plant using methanol fuel was fabricated and tested during the early 1980s. This development effort was marginally successful, but the fact that methanol is not readily available in the battlefield meant that final development of military fuel cells had to be based on use of petroleum fuels such as JP-4 or diesel.

1.3 Transportation applications

Phosphoric acid fuel cells have been studied for application to transportation for some time. In the 1990s, the Department of Transportation sponsored efforts managed by Georgetown University to develop phosphoric acid power plants for application to transit buses. Methanol was the fuel selected for this program. The first activity in this endeavor involved a Fuji cell stack integrated into a total power plant for a 9.14 m (30 ft) transit bus by H Power. The second activity involved development of a methanol-fueled power plant by International Fuel Cells, which was integrated into a 12.19 m (40 ft) transit bus.[3] The power plant installation in the rear of the bus is shown in Figure 11. The 100 kW direct current (DC) power plant occupies a volume of only 175 cubic feet or 1.75 cubic feet per kW. This compares with the DC portion of the 200 kW PC25 power plant that occupies a volume of 7.5 cubic feet per kW. The size reduction is a result of some changes in requirements, advances in technology and an optimization for the bus that put a premium on volume reduction. This power plant is rated at 100 kW and operates in a hybrid configuration with batteries and a motor drive to provide the steady state and overload power required by the bus. The transient response of the power plant to changes in load associated with an urban driving cycle is more than adequate as shown in Figure 12. Because polymer electrolyte membrane (PEM) fuel cells offer additional size reductions and operating flexibility in the transportation application and give promise of cost reductions associated with automobile production rates, they are now the preferred approach for all transportation applications.

Figure 11. 100 kW PAFC power plant in rear of 12.19 m (40 ft) transit bus.

2 COMPONENT EVOLUTION

2.1 Overall system

While the cell stack is at the heart of a fuel cell power plant, many other components and subsystems are required to provide a complete system that consumes fuel and generates useful alternating current power. These subsystems include the conversion processes for fuel and electricity. Hydrocarbon fuel is converted to a hydrogen-rich stream for use in the fuel cell by a fuel processor and the direct current electricity produced by the fuel cell is converted to alternating current at the voltage and frequency required by a power conditioner. These primary conversion systems are supported by ancillary components and subsystems to:

1. supply air for the fuel cell and fuel processor;
2. manage heat produced by the fuel cell stack and other systems;
3. manage water;
4. provide a weather-proof enclosure;
5. control the system and provide a man–machine interface.

Figures 13 and 14 show all these elements for the 200 kW PC25™ fuel cell power plant. Figure 13 shows the three conversion subsystems from left to right: fuel processing, cell stack, and power conditioner. The condenser is

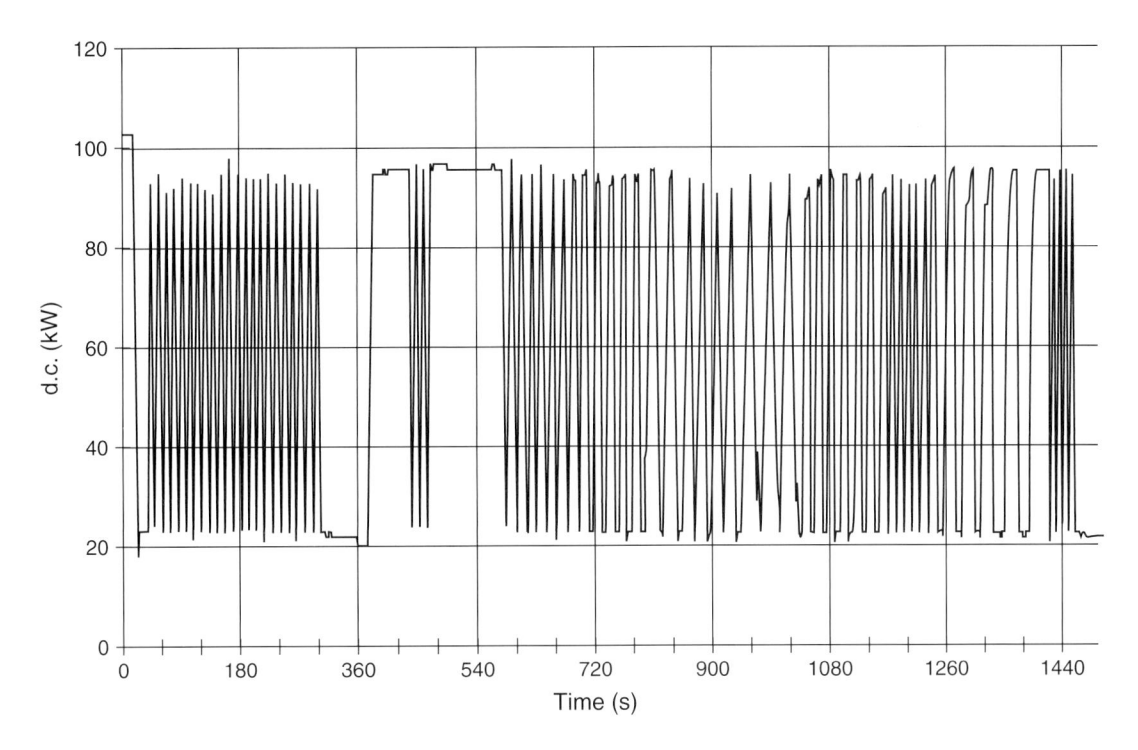

Figure 12. Transient response of power plant to changes in load associated with urban driving.

Figure 13. Fuel processor, cell stack, inverter and condenser of 200 kW power plant.

at the top in this photograph. Figure 14 shows the ancillary components and subsystems associated with air supply, thermal management, water management and control. The power plant is enclosed in a weather-proof cabinet and the cabinet has large doors permitting ready access for maintenance. The cell stack itself comprises less than 20% of the weight and volume of the overall power plant.

Individual subsystems of hydrocarbon fuel cell power plants and their evolution are discussed below.

2.2 Cell stack

The design of phosphoric acid fuel cell stacks must maximize the electrical power output for a given reactant

Figure 14. Ancillary components and subsystems of 200 kW power plant.

flow rate. In order to do this, the design must satisfy the requirements for uniform distribution of fuel and oxidant gases, removal of product water and removal of reaction heat as well as the physical phenomena identified in Section 4, particularly Sections 4.2.2 and 4.2.4.

Uniform distribution of reactants has been satisfied by an integrated consideration of reactant manifold design and manufacturing tolerances for the gas passages in the cell stack. These considerations have resulted in external manifolds for all practical phosphoric acid cell stacks that have been demonstrated in applications from 40 to over 600 kW per stack. Internal manifolding of a size sufficient to provide good distribution to each cell would be prohibitively expensive. External manifolding comes with the requirement for a good seal between the manifold and the cell stack and corrosion resistance for the manifold itself. Both these issues have involved considerable development effort and the success of these efforts is evident in the routine operation of 200 kW cell stacks for 30 000–40 000 h and more.

Product water removal is accomplished by evaporation into the exhausted cathode reactant stream. At the 200 °C operating temperature of current phosphoric acid fuel cells, this presents no problem. Since recovery of product water is a requirement for practical power plant operation, high cell stack oxygen utilization is an important factor in the power plant design.

Removal of cell stack waste heat through the reactant gases was attempted early in the evolution of phosphoric acid cell stacks, but this proved to be impractical because it resulted in excessive loss of electrolyte by evaporation. As an alternative, some developers used separate gas passages within the stack to remove heat, but this proved impractical because of parasite power and cost considerations. Single-phase silicone oils were also evaluated; stability problems and the necessity of an additional heat exchanger to generate steam required by the fuel processor resulted in abandonment of this approach. Since the mid-1980s, two phase cooling with water has been employed. This approach circulates pressurized water through the cell stack where low quality steam is generated. The water vapor is separated in an accumulator external to the cell stack and used for the fuel processor make-up steam. This has proven to have advantages with regard to system simplicity and transient response although it has presented challenges with regard to achieving uniform coolant distribution and in maintaining water quality at a level, which avoids cooling system deposits within the cell stack. Lack of proper design with respect to either of these effects can result in cell stack overheating and damage. While early cell stacks using this cooling technique required cooling system maintenance after only a few months of operation, current designs operate for more than 5 years with no maintenance or cell stack problems.

Phosphoric acid loss to evaporation has been minimized through careful design of the cooling system. Migration of the electrolyte from cell to cell has required aggressive steps to mitigate this migration in the cell stack design. Current designs can operate with the original electrolyte inventory for 40 000 h or more before a lack of acid is evident in any cell within the stack. Replenishment of the acid has been achieved successfully so extension of cell stack operation beyond 40 000 h may be possible with further development of replenishment techniques.

Material corrosion was a major factor in early phosphoric acid cell stacks, even with operation at temperatures as low as 135 °C. Improvements that involved higher graphitization temperatures to improve corrosion resistance as well as design changes to avoid imposing corrosive conditions have resulted in robust cell stack components for operation at 200 °C. Examination of cell stack components after 40 000 h operation in the ambient pressure conditions of the 200 kW PC25™ fuel cell power plant shows the structural integrity is satisfactory for many more hours of operation, which is consistent with results of experimental evaluation at the materials laboratory level. As is well known, operation of phosphoric acid cell stacks at pressurized operating conditions has been less successful. Corrosion issues at pressure have proven to be much more difficult. While technical solutions to these issues have been demonstrated, implementation has not been attempted because previously discussed issues identified for pressurized power plant operation make this power plant approach unattractive.

Current state of the art for phosphoric acid fuel cell stacks is embodied in the cell stack for the 200 kW PC25 fuel cell power plant. This stack consists of more than 250 cells, each with an area greater than 0.5 m². More than 200 of these stacks have been produced with total fleet operating time exceeding 5 million operating hours and individual cell stacks having been operated for more than 40 000 h.

2.3 Fuel processor

The fuel processor converts hydrocarbon fuels to a hydrogen-rich stream for use in the fuel cell stack. Hydrogen is also a major constituent of the gas produced by anaerobic digestion of wastewater; consequently, wastewater gas can also be used in fuel processors designed for natural gas. Phosphoric acid cells operating at the 200 °C level tolerate carbon monoxide concentrations of 1% or less and this establishes one requirement for the fuel processor. Another requirement is that the fuel processor be able to utilize widely available fuels. Natural gas is available in much of the developed world at competitive prices and delivery by pipeline to the end user is consistent with the use of fuel cells in on-site, distributed generator applications. Consequently, natural gas was selected as the fuel for initial application of stationary fuel cells. The primary constituent of natural gas is methane and, as delivered, impurity levels are quite low with the only impurity requiring removal usually being the sulfur compounds added for odorization. These odorants are either mercaptans or thiophanes and they can be removed in a low pressure hydrodesulfurization process, which converts the sulfur compounds to hydrogen sulfide which is then reacted with zinc oxide to form zinc sulfide. This makes the fuel suitable for reaction with steam over nickel based catalysts like those used in catalytic steam reforming in the chemical industry to provide a source of hydrogen. The steam used can be generated using the waste heat of the phosphoric acid cell stack so the fuel processor-cell stack combination can provide high efficiency levels. Consequently, the primary fuel processing approach used with phosphoric acid fuel cells has been catalytic steam reforming. After sulfur removal in the hydrodesulfurizer, the natural gas fuel is mixed with steam and heated to over 700 °C to form carbon monoxide and hydrogen. The gas is then cooled to approximately 250 °C and introduced to a shift converter catalyst bed where steam and carbon monoxide react to form carbon dioxide and additional hydrogen. There is a small amount (less than 1%) of carbon monoxide at the shift converter exit, but with current phosphoric acid cell temperatures at 200 °C this does not present a problem to the anode performance.

This process is used extensively in the chemical industry and is the primary method used to produce industrial hydrogen. As practised in the chemical industry, the process operates at constant output, efficiency is not important and size is of little consequence. Development of this process for stationary fuel cell power plants required an efficient process and integration with the fuel cell to utilize the steam produced by cooling the stack and the tail gas from the anode to provide the thermal energy needed by the reform reaction. Another change to improve efficiency involves regeneration of the exit gas from the reformer to provide a portion of the endothermic requirements for the inlet gas to the reformer and the initial stages of the reform reaction. Convective heat transfer was substituted for the radiant furnaces used in the chemical industry to reduce size and improve the rate of response to load changes. A number of heat exchange processes, the hydrodesulfurizer, and the shift converter were integrated to improve cost and efficiency, to reduce size, and provide inherent reduction of temperature. Control approaches that provide improved response to load change have been developed. The final result is a robust fuel processor that meets the requirements of stationary applications where natural gas is available and where the power plant operates continuously in response to ranging loads.

Alternative fuel processing approaches such as autothermal reforming or catalytic partial oxidation offer improvements in start time and energy and fuel flexibility which are needed to extend the applicability of fuel cell power plants in transportation, tactical military and stationary applications for intermittent use or where natural gas is not available. These alternative fuel processing approaches are being developed primarily for use with proton exchange membrane fuel cell power plants.

2.4 Power conditioning

Stationary fuel cell power plants must provide the alternating current power used throughout the world; it is impractical to consider direct current power since utilization devices have been designed since early in the 20th century to use alternating current (a.c.) power. This a.c. power is at higher voltage levels (120–480 V) than the direct current (d.c.) voltage levels produced by typical fuel cell stacks (50–200 V). Consequently, to have a useful device, the fuel cell power must be converted to a.c. and the voltage must be transformed to standard levels. The output power frequency and voltage regulation must match those that are common in the electric supply system and, for grid-connected power plants, the control and protection system must be consistent with proper operation of the electric utility network.

At the beginning of the efforts to develop stationary fuel cell power plants, the common approach to convert d.c. to a.c. power was through the use of motor-generator sets. This approach is large, expensive and inefficient. Solid state approaches using power semiconductor switches were in the infancy of development. It was possible to generate single or three phase square waves in a combination of bridges and the added output of the bridges could be filtered to produce a.c. current which closely approaches the sine wave from traditional equipment. It was clear that the solid state approaches would benefit from advances in semiconductor technology and soon surpass the characteristics of power conditioning based on motor generator sets and the development proceeded on this basis.

From the mid-1960s to the middle of the 1980s, power conditioner development for fuel cells over 5 kW concentrated on the use of silicon controlled rectifiers (SCRs), which are also known as thyristors. An SCR can be switched on through a low power gating signal, but turning it off requires removal of the load current and imposition of a reverse voltage. The reverse voltage is applied from electrical energy stored in an L-C circuit controlled by a separate SCR that is referred to as a commutation circuit for self-commutated power conditioners (inverters). In grid-connected situations, the reverse voltage could be provided by the electric network.

The self-commutated inverter versus line-commutated inverter comparison was very active in the 1970s. Line commutated inverters were less complex and required simpler SCRs, but they were not suitable for grid-independent applications, could not provide independent control of real and reactive power, and imposed additional reactive power supply requirements on the utility grid. Further, self-commutated inverters would benefit from advances in power semiconductors. Accordingly, by the mid-1970s fuel cell power conditioner development was firmly focused on self-commutated inverters.

SCRs did benefit from semiconductor development and their characteristics in terms related to cost and efficiency improved markedly. The ultimate development of this technology was realized in gate turnoff devices, which were employed in the 11 000 kW power plant. However, when higher power transistors and finally insulated gate bipolar transistors (IGBTs) became available in the mid-1980s, they afforded many advantages and the use of SCRs was discontinued. These advantages include:

1. They require much less parasitic power to perform the switching function;
2. Commutation circuits are not required thereby reducing complexity;
3. Fast switching is possible so that pulse width modulation can be used to generate a very close approximation to the sine wave.

Fast switching speed is also advantageous in controlling the inverter to prevent damage from overvoltage on the electric supply network (caused by lightning or switching actions). All current fuel cell power conditioners use IGBT switches in pulse-width modulated systems. Current focus is on reduction of losses and cost and on developing standards and designs to those standards that facilitate installation and connection to the electric supply network. That activity is currently centered in development of interconnection standards such as that underway by the Institute of Electric and Electronic Engineers in IEEE 1547.

3 SYSTEMS EVOLUTION

Ancillary systems of fuel cell power plants constitute about a third of the power plant volume and are responsible for most of the forced outages of the power plant. Consequently, a successful power plant design requires attention to these systems.

Thermal management and water recovery are relatively straightforward. The cell stack waste heat is provided, as noted above, in the form of low quality steam. Early power plants simply exhausted this steam and took in make-up water for the stack cooling in an open cycle system. This proved to be impractical because of the water consumption, variations in water quality that caused plugging of the cooling system and the inability to recover heat. Consequently, efforts after the early 1970s used closed cycle cooling systems. In these systems, the thermal management system separates the gaseous and liquid components of the low quality steam produced in the cell stack in a steam separator or accumulator, delivers the required amount of steam to the fuel processor and cools the water stream before reintroduction to the cell stack to cool the stack. Make-up water to accommodate delivery of steam for the reformer is recovered from the power plant exhaust in a condenser. This water is treated to remove carbon dioxide and other contaminants, as boiler feed water quality is necessary to avoid depositing material and blocking the cell stack cooling passages. In the 40 kW experiments in the mid-1980s, chemical cleaning to remove these deposits was required every 2000–3000 operating hours, which is not acceptable. This resulted in many changes in the cell stack cooling and water treatment system to eliminate these problems, and the cell stacks in the 200 kW power plant have clean cooling passages after 40 000 h of operation with no stack maintenance required.

The water and thermal management loops in a fuel cell power plant are subject to environmental extremes. The 40 kW power plant testing in the mid-1980s showed that systems which permit direct contact with ambient air are vulnerable to freezing even with imposition of control systems designed to avoid it. Consequently, a secondary loop containing antifreeze was imposed between the water loop and the power plant radiator to avoid these problems.

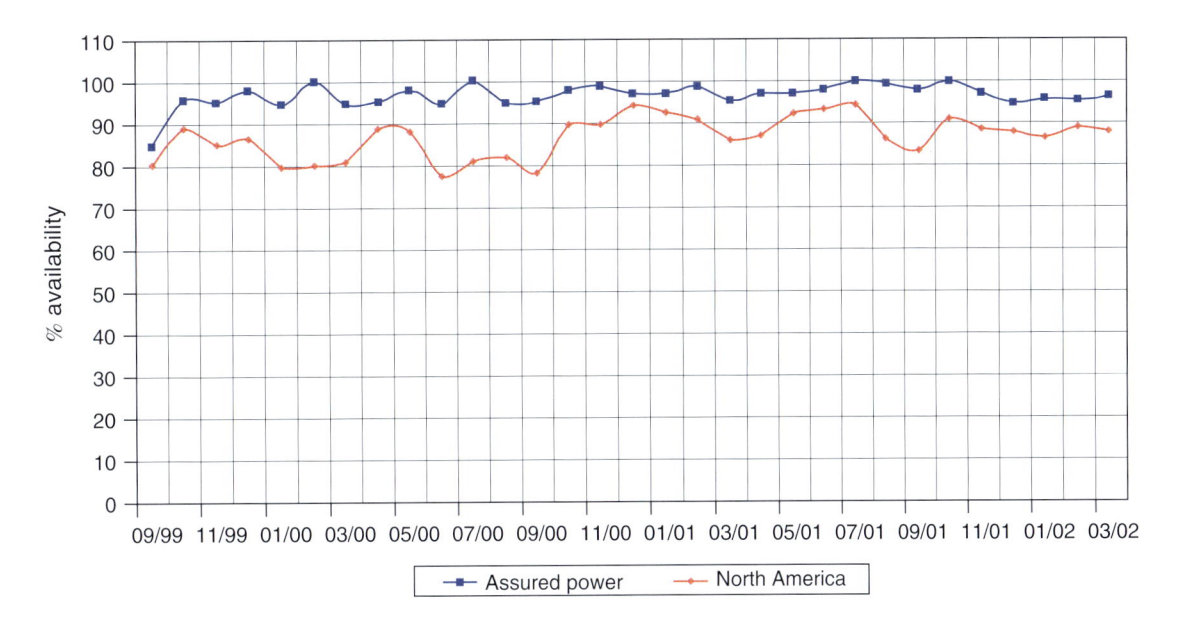

Figure 15. Recent availability experience with 200 kW fuel cell power plant.

The final ancillary system of interest is the controller. This system has evolved from a combination of thermo-mechanical and analog electrical control using pneumatically or hydraulically actuated valves to a system using digital control and electrically actuated valves. As computation power and communication technology has evolved, diagnostics and remote control and diagnostic capability have been added to the control system. This facilitates low cost maintenance for an unattended power plant that is essential to practical application of stationary fuel cell power plants.

The improvements in ancillary system components and rapid outage response and diagnostics associated with remote annunciation of failure and remote diagnostics provide the opportunity for the excellent power plant availability required in distributed power applications. Figure 15 shows that assured power installations with local stocking of spare parts and rapid service response achieve availability consistently in excess of 95% and as high as 99%. A less vigorous maintenance approach associated with a demonstration philosophy results in availability around 90%.

4 TECHNOLOGY EVOLUTION

4.1 Cell performance

Phosphoric acid fuel cells were selected for commercialization over alkaline electrolyte fuel cells because of the ability of phosphoric acid to reject carbon dioxide, and were selected over Nafion® because of phosphoric acid's low cost and ease of water management. At the time, prospects for performance and low membrane cost for PEM were not evident. This situation changed markedly during the 1990s. In the early period of investigation of fuel cells for ground based applications, the sulfuric acid cell was also considered, but issues of stability of the electrolyte caused this technology to be abandoned. Materials problems were much less severe than with molten carbonate or solid oxide fuel cells so they were considered later generation devices.

In conjunction with the development of phosphoric acid fuel cells, considerable research was performed to improve performance, reduce cost and increase endurance. Much of this technology provides foundations for subsequent proton exchange membrane fuel cell development. The earliest phosphoric acid cells operated at about 80 °C, whereas the current cells operate at about 200 °C. This increase in temperature was accomplished by the improvement in the materials of construction used in the fuel cell stack. The matrix material used to hold the electrolyte between the electrodes evolved from fiberglass to silicon carbide. The cathode platinum loading was initially at 20 mg cm^{-2} of platinum black but is currently 0.75 mg cm^{-2} of platinum alloy supported on a graphitized carbon black. The anode catalyst evolved from about 20 mg cm^{-2} of platinum black to various unsupported platinum-containing alloys and mixtures to provide carbon monoxide tolerance. Presently the anode uses 0.5 mg cm^{-2} of platinum supported on carbon. The sophistication of the fabrication of Teflon™-bonded electrodes was improved through the understanding of the colloid chemistry of their fabrication[4, 5] and electrode mathematical modeling.[6, 7] Carbon paper substrates and carbon separator plates evolved to provide suitable electrolyte management. The state-of-the-art of phosphoric acid fuel cells was published in 1977.[8]

4.1.1 Phosphoric acid

Many of the properties of phosphoric acid needed to develop a fuel cell power plant were not known when the first unit was built. Research was performed to determine many of these required properties.[9] Phosphoric acid was found to be quite stable from an electrochemical standpoint over the complete potential range of fuel cell operation. The loss of the acid was found to be very small due to evaporation. Reactant gas diffusion coefficients and solubilities were determined. Ionic conductivity and electrolyte concentration polarization were measured. The rate of the oxygen reduction reaction was found to be slower than with sulfuric acid because of phosphate anion adsorption and a lower acidity. The electrolyte was found to be suitable at 200 °C where the equilibrium concentration of the electrolyte exceeded 100 wt% based on molecular H_3PO_4.

4.1.2 Cathode catalyst

Platinum has long been recognized as a catalyst for the reduction of oxygen in acidic electrolytes. However, when the development of phosphoric acid fuel cells was begun, the activity had not been accurately measured. Experimental data were available for oxygen reduction on flat sheets of platinum that led to the speculation that excessive platinum quantities would be needed for significant fuel cell commercialization. Teflon®-bonded, platinum-black cathodes of about 20 mg cm^{-2} loading supported on screen were evaluated. The surface area of this catalyst was about 25 m^2 g Pt. A reduction in the quantity of platinum required supporting the platinum. The material to use as the support presented a problem because of the requirements that must be met by such a material. It had to be high in surface area, thermally stable, electrochemically stable, and an excellent electronic conductor. Conventional supports used in catalysis such as alumina could not be used because of poor electronic conductivity. Types of carbon black met the conductivity requirement but carbon has an oxidation potential about one volt below the cathode operating potential and would therefore be expected to corrode. To investigate the use of carbon as the support, Teflon®-bonded electrodes were fabricated and experiments were performed to determine the activity of platinum in a highly dispersed form. This study showed that the catalytic activity of supported platinum in high area form of about 100 m^2 g was significantly different from that of both flat sheet platinum and platinum black.

The polarization curve for oxygen reduction on flat-sheet platinum shows the existence of two distinct Tafel slopes.[10] At ambient temperature, a shallow one of about 60 mV decade^{-1} occurs at higher potentials above about 0.9 V relative to a hydrogen electrode in the same electrolyte (reference hydrogen electrode (RHE)) and a slope of twice this value at lower potentials. These two Tafel slopes have been explained due to a change in the adsorption characteristics of oxygen with electrode potential.[11] For high surface area platinum catalyst supported on carbon, only one Tafel slope is present with a value of 60 mV decade^{-1} at ambient temperature and 90 mV decade^{-1} at higher temperatures as 160 °C.[12] This single Tafel slope has been rationalized by a change in mechanism for the high surface area catalyst.[13] Experimental performance data for oxygen reduction on supported platinum from Ref. [8] are shown in Figure 16.

A change in catalytic activity has also been observed with a change in the catalyst surface area when the catalyst is supported on carbon.[14] As the catalyst surface area increases, the catalytic activity per unit surface area decreases. Here the activity is expressed as the current for oxygen reduction at 0.9 V vs. RHE. An increase in Tafel slope with the reduction in surface area is also evident. The catalytic activity was found to vary between the crystal faces of the platinum crystallite.[15] These observations sparked a great interest in the understanding of the oxygen reduction reactions on platinum. Alloys of platinum with non-noble metals were discovered that enhance the catalyst activity.[16] Work continues in this area using more advanced analytical techniques that help clarify the oxygen reduction mechanism. Summaries of oxygen reduction technology have been published.[17, 18]

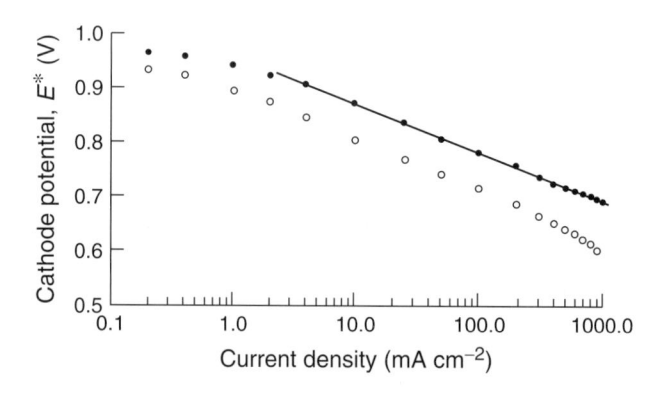

Figure 16. Cathode performance for 0.25 mg cm^{-2} Pt supported on Vulcan XC-72 at 160 °C in 96% H_3PO_4. Closed symbols, oxygen; open symbols, air. A straight line with a slope of 90 mV decade^{-1} has been drawn through the oxygen data. Data are for ambient pressure. (Reproduced from Kunz and Gruver (1975)[12] with permission from The Electrochemical Society, Inc.)

The activity of platinum was found to be reduced by an increase in the concentration of the phosphoric acid.[19] This effect is thought to be caused by the loss of water from the electrolyte lowering the activity of the proton that is involved in the oxygen reduction process.

Considerable effort has been expended to replace platinum using non-precious metal catalysts such as macrocycles.[20] However, not only must those materials be lower in cost than platinum, but they also must perform nearly as well as platinum and be stable. If their performance is less, other material costs in the fuel cell stack increase because of the lower current density of operation. No replacement for platinum has been found.

4.1.3 Anode catalyst

Fundamental studies have been performed to determine the mechanism of hydrogen oxidation on platinum and the effect of carbon monoxide in poisoning this reaction.[21] The alloying of platinum has long been known to enhance this catalyst for the oxidation of hydrogen containing carbon monoxide.[22] Platinum/ruthenium alloy supported on carbon was evaluated for use in phosphoric acid fuel cells and found to result in enhanced tolerance to CO.[23] Numerous other catalysts were evaluated and identified to result in enhanced tolerance. However, as the phosphoric acid fuel cell temperature was increased because of the availability of more stable materials, pure platinum was found to result in the best performance and was selected for use.

4.2 Cell decay

Several mechanisms of performance decay are present in the stacks of phosphoric acid electrolyte fuel cells. The most important ones are cathode catalyst degradation, electrolyte migration, cell contamination, and material corrosion.

4.2.1 Cathode catalyst degradation

The platinum cathode catalyst loses activity by both dissolution into the electrolyte and a loss of surface area per unit mass. Platinum has a solubility that increases with potential with the dissolved specie being the Pt^{2+} ion.[24] This ion migrates toward the anode in the cell and is plated as Pt metal by reduction with the hydrogen.

Platinum catalyst loses surface area per unit mass by three processes: sintering, dissolution–precipitation, and crystallite migration. Platinum black and platinum supported on carbon both lose specific area by the dissolution–precipitation process (Ostwald ripening) at high potential. Platinum black also loses area by sintering at lower potentials where the platinum solubility is negligible. This process occurs due to surface atom migration[25] and also results in a loss in catalyst porosity.[26] The cathode performance then decays due to both a reduced catalytic activity due to the loss of area and an increase in diffusional losses due to the loss in porosity. Platinum supported on carbon loses surface area by crystallite migration followed by crystallite aggregation and sintering at low potentials where the platinum solubility is low.[27, 28]

4.2.2 Electrolyte migration

Electrolyte migration can result in decay due to three mechanisms. First, the electrolyte can evaporate from the cell so the cell becomes deficient in electrolyte. The evaporating species are phosphorous oxides. Electrolyte reservoirs are frequently used in fuel cells that have liquid electrolyte to accommodate electrolyte volume changes with time and operating conditions. Secondly, electrolyte can migrate from the cathode side to the anode side of the cell due to the potential profile across an operating cell.[29] This migration can result in a flooding of the anode at high current density. This decay mode can be alleviated through the control of the pore size distribution of the anode and cathode components. The third electrolyte migration that can occur is through the bipolar separator plates. If these plates have a small porosity that fills with electrolyte with time, electrolyte can migrate through the plates from the negative end to the positive end of the stack due to the potential profile along a stack. The control of electrolyte migration has been very important in achieving the present low level of performance decay in phosphoric acid fuel cells.

4.2.3 Cell contamination

Harmful contaminants can enter the cell in the reactant gas streams. These contaminants can poison catalysts or react with the electrolyte to change its properties. For example, sulfur compounds present in fuel gas or the air can adsorb on the surface of the anode catalyst and harm its catalytic activity and hence the cell performance. Ammonia formed in a fuel processor can react with the electrolyte and result in an electrolyte concentration polarization that reduces the cell performance.[30]

4.2.4 Material corrosion

Carbon used in the stack is the primary component, other than the cathode platinum, that corrodes. The carbon support material used on the cathode side is electrochemically oxidized. This oxidation rate depends strongly on the potential of the electrode, the form of the carbon,[31] and the water concentration.[32] For Vulcan XC-72 that is commonly used, the interior of the prime particles is corroded because the corrosion occurs on the edges of basal planes and defects.[33] For this reason, this carbon has been heat treated at 2700 °C to increase the degree of graphitization.[34]

The carbon used in the separator plates and electrode substrates (backing or diffusion layers) has also indicated corrosion,[35] but the problem has been virtually eliminated by modified fabrication techniques.

REFERENCES

1. J. M. King and M. J. O'Day, *J. Power Sources*, **86**, 16 (2000).
2. '40 kW On-Site Fuel Cell Field Test Program', in "Final Technical Report for Gas Research Institute", Contract no. 5084-244-1077, (July 1986).
3. A. P. Meyer, J. M. King and D. Kelly, 'Progress in Development and Application of Fuel Cell Power Plants for Automobiles and Buses', SAE Paper 1999-01-0533.
4. S. Kratohvil and E. Matijevic, *J. Colloid Interface Sci.*, **57**, 104 (1976).
5. S. Kratohvil and E. Matijevic, *Colloids Interfaces*, **5**, 179 (1982).
6. R. Brown and L. A. Horve, *Electrochem. Soc. Proc.*, **67-1**, Abstract No. 203 (1967).
7. J. Giner and C. Hunter, *J. Electrochem. Soc.*, **116**, 1124 (1969).
8. H. R. Kunz, in 'Proceedings of the Symposium on Electrode Materials and Processes for Energy Conversion and Storage', The Electrochemical Society, Pennington, NJ, PV77-6 (1977).
9. J. F. Zemaitis, Jr, D. M. Clark, M. Rafal and N. C. Scrivner, 'Handbook of Aqueous Electrolyte Thermodynamics', Design Institute for Physical Property Data, pp. 409–414, pp. 675–682 (1986).
10. A. Damjanovic and V. Brusic, *Electrochim. Acta*, **12**, 615 (1967).
11. A. J. Appleby, *J. Electrochem. Soc.*, **117**, 328 (1970).
12. H. R. Kunz and G. A. Gruver, *J. Electrochem. Soc.*, **122**, 1279 (1975).
13. W. M. Vogel and J. M. Baris, *Electrochim. Acta*, **22**, 1259 (1977).
14. L. J. Bregoli, *Electrochim. Acta*, **23**, 489 (1978).
15. H. R. Kunz, *Proc. Electrocatal. Fuel Cell React.*, **79**, 14 (1978).
16. V. Jalan and E. J. Taylor, in 'Proceedings of the Symposium on Chemistry and Physics of Electrocatalysis', Electrochemical Society Proceedings, Pennington, NJ, PV84-12, pp. 546–553 (1984).
17. E. Yeager, *J. Mol. Catal.*, **38**, 5 (1986).
18. K. Kinoshita, 'Electrochemical Oxygen Technology', John Wiley & Sons, Chichester (1992).
19. H. R. Kunz and G. A. Gruver, *Electrochim. Acta*, **23**, 219 (1978).
20. K. A. Radyushkina and M. R. Tarasevich, *Sov. Electrochem.*, **22**, 1087 (1986).
21. W. Vogel, J. Lundquist, P. Ross and P. Stonehart, *Electrochim. Acta*, **20**, 79 (1975).
22. L. W. Niedrach, D. W. McKee, J. Paynter and I. F. Danzig, *J. Electrochem. Tech.*, **5**, 318 (1967).
23. P. N. Ross, K. Kinoshita, A. J. Scarpellino and P. Stonehart, *J. Electroanal. Chem.*, **63**, 97 (1975).
24. P. Bindra, S. J. Clouser and E. Yeager, *J. Electrochem. Soc.*, 1631 (1979).
25. P. A. Stonehart and P. A. Zucks, *Electrochim. Acta*, **17**, 2333 (1972).
26. K. Kinoshita, K. Routsis, J. A. S. Bett and C. S. Brooks, *Electrochim. Acta*, **18**, 953 (1973).
27. J. A. Bett, K. Kinoshita and P. Stonehart, *J. Catal.*, **35**, 307 (1974).
28. G. A. Gruver, R. F. Pascoe and H. R. Kunz, *J. Electrochem. Soc.*, **127**, 1219 (1980).
29. H. R. Kunz, *Electrochem. Soc. Proc.*, **99-14**, 191 (1999).
30. S. T. Szymanski, G. A. Gruver, M. Katz and H. R. Kunz, *J. Electrochem. Soc.*, **127**, 1440 (1980).
31. K. Kinoshita, 'Carbon, Electrochemical and Physicochemical Properties', Wiley-Interscience, New York (1988).
32. P. Stonehart and J. P. MacDonald, in 'Proceedings of the Workshop on the Electrochemistry of Carbon', Electrochemical Society, Pennington, NJ, PV84-5 (1984).
33. G. A. Gruver, *J. Electrochem. Soc.*, **125**, 1719 (1978).
34. K. Kinoshita, in 'Proceedings of the Workshop on the Electrochemistry of Carbon', Electrochemical Society, Pennington, NJ, PV84-5, pp. 273–290 (1984).
35. L. G. Christner, H. P. Dhar, M. Farooque and A. K. Kush, *Corrosion*, **43**, 571 (1987).

Chapter 17
Aqueous carbonate electrolyte fuel cells

E. J. Cairns

Lawrence Berkeley National Laboratory, and University of California, Berkeley, CA, USA

1 INTRODUCTION

Fuel cells with aqueous electrolytes commonly employ either strongly acidic (e.g., H_3PO_4 or H_2SO_4) or strongly alkaline (KOH or NaOH) electrolytes. These offer relatively high electrolytic conductivity and support high electrochemical reaction rates for hydrogen and oxygen. The acid electrolytes may also be used with organic fuels, as they reject CO_2, but they are also very corrosive, severely limiting the choice of electrocatalysts to noble metals and their alloys. Sodium or potassium hydroxide electrolytes cannot be used directly with ambient air or with organic fuels, since they react with CO_2 to yield carbonate, eventually converting the hydroxide electrolyte to a carbonate electrolyte. At the concentrations normally used, this results in the precipitation of sodium or potassium carbonate and/or bicarbonate, damaging the electrodes and rendering the cell useless. If this problem can be overcome, then aqueous carbonate electrolytes would be a possibility for applications in which the reduced corrosivity and moderate pH of aqueous carbonates would be important.

2 THE ISSUE OF INVARIANCE

A fundamental requirement of fuel cell electrolytes is invariance. They must maintain a constant composition during fuel cell operation, and must not enter into the net electrochemical reaction of the fuel cell. An invariant electrolyte should have the following features:

1. The fuel cell performance should not decrease over extended operating periods due to any changes in the electrolyte;
2. No spontaneous chemical reactions should occur between the electrolyte and the fuel, oxidant, or products;
3. The electrolyte should support complete electrochemical oxidation of the fuel (including carbonaceous fuels);
4. The solubility of the fuel and the oxidant in the electrolyte should be limited (to avoid chemical reaction at the opposite electrode);
5. The electrolyte must possess sufficient electrolytic conductivity;
6. The electrolyte should not react with any of the cell components.

As will become clear below, properly formulated aqueous carbonates can fulfill the invariance requirements, and can be used with a variety of fuels, including organic fuels.

3 ELECTROLYTE SYSTEMS EXAMINED FOR USE IN FUEL CELLS

The use of invariant aqueous carbonate electrolytes with organic fuels was reported by Gruenberg et al.,[1, 2] who made use of potassium carbonate/bicarbonate electrolytes, and by Cairns and Macdonald,[3] who made use of the highly soluble bicarbonates and carbonates of cesium and rubidium to avoid the precipitation problems indicated above. The solubilities of the cesium and rubidium salts are compared to those of the other alkali carbonates and bicarbonates in Figure 1. It is clear from

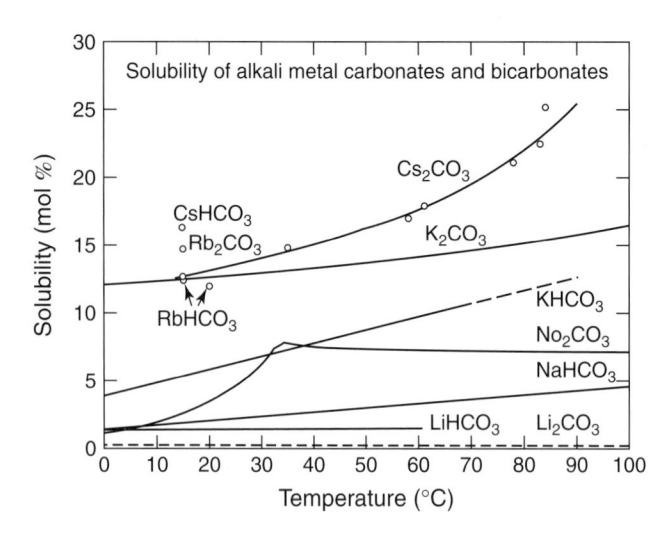

Figure 1. Solubilities of some carbonates and bicarbonates in water.

this figure that the high solubilities of the Cs and Rb salts can provide a highly concentrated electrolyte which allows operation at temperatures well above 100 °C at atmospheric pressure. This feature is important in connection with the use of the less-reactive organic fuels. The conductivities of these electrolytes are in the range $0.05-0.3\,\Omega^{-1}\,cm^{-1}$ in the composition and temperature range of interest.[4] Other relevant properties of these electrolytes are reported in Ref. [3]. The expectation was that these electrolytes would reach a steady-state composition that depends upon the partial pressures of water and carbon dioxide, resulting in an invariant electrolyte for a given operating condition.

4 PERFORMANCE ON CARBON-CONTAINING FUELS

Stable operation on several organic fuels, including ethylene, ethane, methanol, and carbon monoxide was reported for periods up to 200 h.[3] Detailed investigations of the operation on methanol were carried out, establishing stable performance and complete oxidation of methanol for a period of 500 h at 130 °C and atmospheric pressure. Figure 2 shows the performance of a $CH_3OH(Pt)/Cs_2CO_3/O_2(Pt)$ fuel cell at various temperatures.[5] The electrodes were platinum black-bonded with polytetrafluoroethylene (PTFE), and contained a platinum mesh current collector. The methanol was fed as a gas. The results of the 570 h endurance test are shown in Figure 3, which indicates that the performance was very stable, and improved slightly with time. The stoichiometric rate of carbon dioxide production was found at steady state.

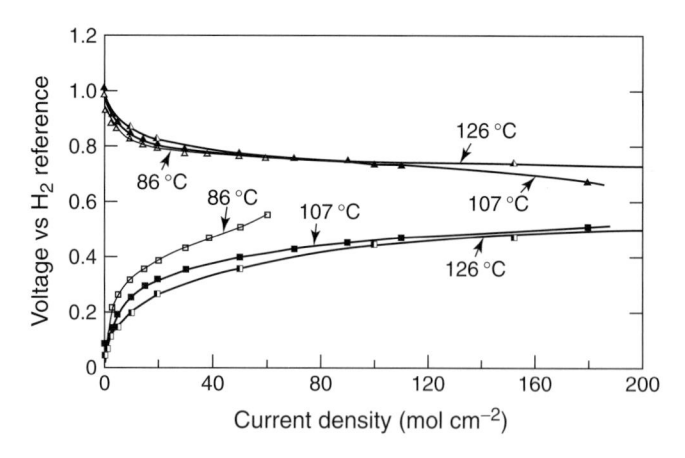

Figure 2. Methanol performance in the $CH_3OH_{(g)}(Pt)/Cs_2CO_3/O_2(Pt)$ fuel cell at various temperatures.

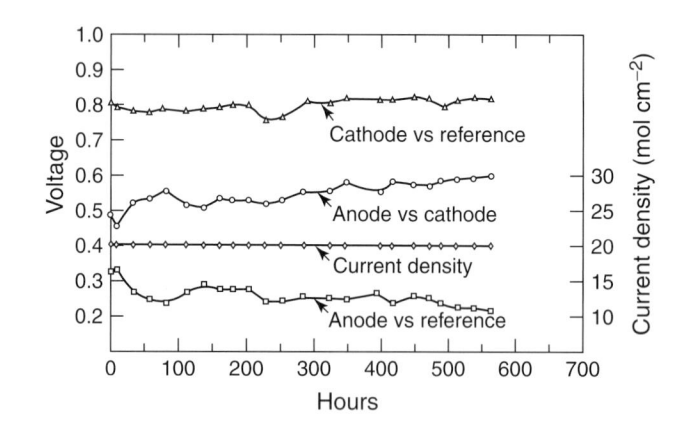

Figure 3. Endurance test of a $CH_3OH_{(g)}(Pt)/Cs_2CO_3/O_2(Pt)$ fuel cell operating at 130 °C.[4]

Based on the results of the electrochemical experiments and the material balances with chromatographic analysis of products, the following reaction scheme was proposed.

Anode reaction

$$CH_3OH + 3CO_3^= \longrightarrow 4CO_2 + 2H_2O + 6e^- \quad (1)$$

This reaction consumes carbonate ion, and must be compensated in some manner if the electrolyte is to be invariant. It was postulated that some of the product CO_2 is reabsorbed by the electrolyte, maintaining a balance between bicarbonate and carbonate in the electrolyte

$$3CO_2 + 3H_2O + 3CO_3^= \longrightarrow 6HCO_3^- \quad (2)$$

The bicarbonate ions can diffuse to the cathode where they can participate in a reaction that restores the carbonate inventory in the electrolyte.

Cathode reaction

There were two possibilities proposed for the cathode reaction, with equivalent overall results

$$\frac{3}{2}O_2 + 3H_2O + 6e^- \longrightarrow 6OH^- \quad (3)$$

followed by

$$6OH^- + 6HCO_3^- \longrightarrow 6CO_3^= + 6H_2O \quad (4)$$

Or the reaction could be simply the sum of the two reactions above

$$\frac{3}{2}O_2 + 6HCO_3^- + 6e^- \longrightarrow 6CO_3^= + 3H_2O \quad (5)$$

These six carbonate ions then can move to the anode where three of them react according to equation (1) and three according to equation (2). This balances the whole process, giving the following overall cell reaction

$$CH_3OH + \frac{3}{2}O_2 \longrightarrow CO_2 + 2H_2O \quad (6)$$

For reactions involving other organic fuels, corresponding reactions can be written, preserving invariance, and yielding only carbon dioxide and water as products. Experimentally it has been shown that no detectable products other than carbon dioxide and water were formed for several organic fuels.[3, 5]

More recently, the kinetics of oxygen reduction on platinum in aqueous carbonate electrolytes was investigated using rotating ring-disc methods, revealing that the reaction rate constant is quite high, even higher than for a hydroxide electrolyte.[6] The high rate constant is partially compensated by lower oxygen solubility in the carbonate electrolyte. This research was extended to the situation of porous Pt/C fuel cell electrodes (2.1 mg(Pt) cm^{-2}), wherein the results showed clearly the accumulation of hydroxide ions in the pores of the electrocatalyst layer (reaction (3)).[7] A more complete mathematical description of the operation of the oxygen reduction reaction in a fuel cell electrode with an aqueous carbonate electrolyte was reported by Striebel et al.[8] This model takes into account reaction (4), the kinetics, and mass transport of all of the species involved in the reaction. With no adjustable parameters, excellent agreement with the experimental data was shown, as can be seen from Figure 4.

This investigation showed that for oxygen reduction on typical carbon-supported Pt fuel cell electrodes in carbonate electrolytes, the important transport limitation is not due to oxygen, but hydroxyl ion. Figure 5 shows the calculated reaction rate distribution for 4 M K_2CO_3 electrolyte. In this situation, it would be beneficial to reduce the thickness of the electrocatalyst layer that is flooded with liquid electrolyte to allow a shorter diffusion path of hydroxide to the bulk electrolyte. At relatively high current densities, the effective reaction order of oxygen is one half.

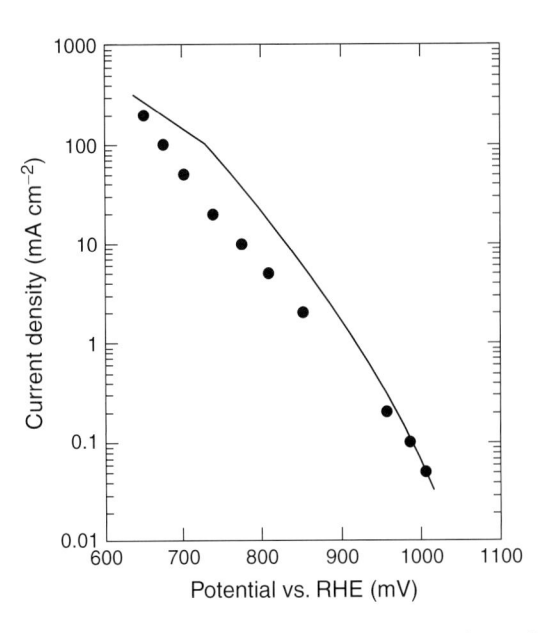

Figure 4. Comparison between calculated and experimental performance for oxygen reduction on a 0.5 mg Pt cm^{-2} electrode in 4 M K_2CO_3 at 25 °C.[7] Data points, experiment; curve, model results.

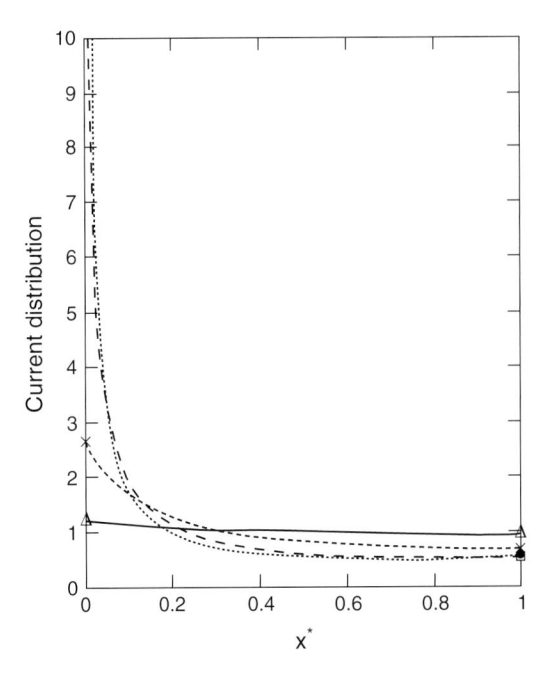

Figure 5. Calculated current distribution for oxygen reduction on a 0.5 mg Pt cm^{-2} electrode in 4 M K_2CO_3 at 25 °C.[7] △, i = 0.1 mA cm^2; X, i = 1.0 mA cm^2; □, i = 10 mA cm^2; ●, i = 100 mA cm^2.

5 ADVANTAGES AND LIMITATIONS OF FUEL CELLS WITH AQUEOUS CARBONATE ELECTROLYTES

Aqueous carbonate electrolytes offer advantages of lower corrosivity, allowing a wider range of electrocatalyst materials to be considered. Thus far, there has been little work on such electrocatalysts, although there is the possibility of perovskites or some metal oxides such as $La_{0.6}Ca_{0.4}CoO_3$, which are active for oxygen reduction in hydroxide electrolytes. The use of cesium or rubidium carbonate electrolytes offers the opportunity to operate above 100 °C at atmospheric pressure, gaining the opportunity for increased reaction rates, and rapid product water removal.

The advantages above are partially balanced by the need for careful water management, to avoid precipitation of solid phases. In addition, the conductivity of the carbonate electrolytes is lower than that of hydroxide or acidic electrolytes of the same concentration.

The progress made in the development of polymer electrolyte membrane (PEM) fuel cells has diminished the interest in aqueous electrolytes for use in fuel cells. This is true because of the relative simplicity of the PEM fuel cell, and the opportunity to have significantly thinner cells, leading to higher power per unit volume and unit weight.

6 CONCLUDING REMARKS

Aqueous carbonate fuel cells have shown some promise for the anodic oxidation of a variety of organic fuels to carbon dioxide and water. They can be operated at atmospheric pressure and temperatures up to nearly 200 °C, offering the advantages of improved kinetics and rapid water removal. In principle, any fuel or oxidant stream containing carbon dioxide can be accommodated, in contrast to the lack of such tolerance by hydroxide electrolytes. Because off their reduced corrosivity, a greater range of electrocatalysts can be considered for use in the carbonate electrolytes, as compared to acidic or hydroxide electrolytes.

REFERENCES

1. G. Gruenberg, M. Jung and H. Spengler, German Patent 1,146,562 (1958).
2. W. Vielstich, 'Fuel Cells', John Wiley & Sons-Interscience, p. 241 (1970).
3. E. J. Cairns and D. I. Macdonald, *Electrochem. Tech.*, **2**, 65 (1964).
4. E. J. Cairns, *Electrochem. Tech.*, **5**, 8 (1967).
5. E. J. Cairns and D. C. Bartosik, *J. Electrochem. Soc.*, **111**, 1205 (1964).
6. K. A. Striebel, F. R. McLarnon and E. J. Cairns, *J. Electrochem. Soc.*, **137**, 3351 (1990).
7. K. A. Striebel, F. R. McLarnon and E. J. Cairns, *J. Electrochem. Soc.*, **137**, 3360 (1990).
8. K. A. Striebel, F. R. McLarnon and E. J. Cairns, *Ind. Eng. Chem. Res.*, **34**(10), 3632 (1995).

Chapter 18

Direct methanol fuel cells (DMFC)

A. Hamnett

University of Strathclyde, Glasgow, UK

1 BASIC PRINCIPLES OF THE DMFC

Whilst the power efficiencies of common H_2/O_2 fuel cells continue to show steady improvements, there remain problems with the use of hydrogen as the active fuel: either this hydrogen must be obtained by in situ reformation of solid or liquid C–H fuels, such as petroleum, coal or methanol, or it must be pre-purified and stored as the gas under pressure, or in the form of an admixture with a metal alloy, carbon or some other absorbent. In all of these cases, either there is a considerable weight penalty or there is increased engineering complexity, adding to costs. It remains highly desirable to design a fuel cell that would directly oxidize a liquid fuel at the anode, but retain the high power/weight ratio of the solid-polymer electrolyte fuel cells described elsewhere. One such fuel cell, now under active development in Europe and the USA, is the direct methanol fuel cell, and the principles are shown in Figure 1. The anode reaction is

$$CH_3OH + H_2O = CO_2 + 6H^+ + 6e^- \quad (1)$$

and the cathode reaction is

$$\tfrac{3}{2}O_2 + 6H^+ + 6e^- = 3H_2O \quad (2)$$

The thermodynamic cell voltage associated with these two couples, assuming 1 M methanol and liquid water is 1.20 V.[1] Methanol possesses a number of advantages as a fuel: it is a liquid, and therefore easily transported and stored and dispensed within the current fuel network; it is cheap and plentiful, in principle renewable from wood alcohol, and the only products of combustion are CO_2 and H_2O. The advantages of a direct methanol fuel cell are: changes in power demand can be accommodated simply by alteration in supply of the methanol feed; the fuel cell operates at temperatures below ~150 °C so there is no production of NO_x, methanol is stable in contact with mineral acids or acidic membranes, and it is easy to manufacture; above all, the use of methanol directly as an electrochemically active fuel hugely simplifies the engineering problems at the front end of the cell, driving down complexity and cost.

2 SUGGESTED CONFIGURATIONS AND CELL DESIGNS

Figure 1 represents the most basic design, developed initially by Shell, and also by Exxon and Hitachi.[2] In all these cases, 1–2 M sulfuric acid was used as the electrolyte and unsupported platinum black was used as the electrocatalyst; the most significant development was a 50 W direct methanol fuel cell (DMFC) stack at Hitachi.

The end of the 1980s saw the introduction of solid-polymer electrolyte membranes for the first time into DMFCs, but this brought some substantial problems with the fabrication of electrodes that both optimized exposed catalyst area and allowed ionic and electronic conductivity to be maintained. It was realized that optimization of the complete electrode membrane structure was needed, rather than attempts to optimize each part separately, and different designs of the membrane–electrode assembly (MEA) have been reported by different groups.

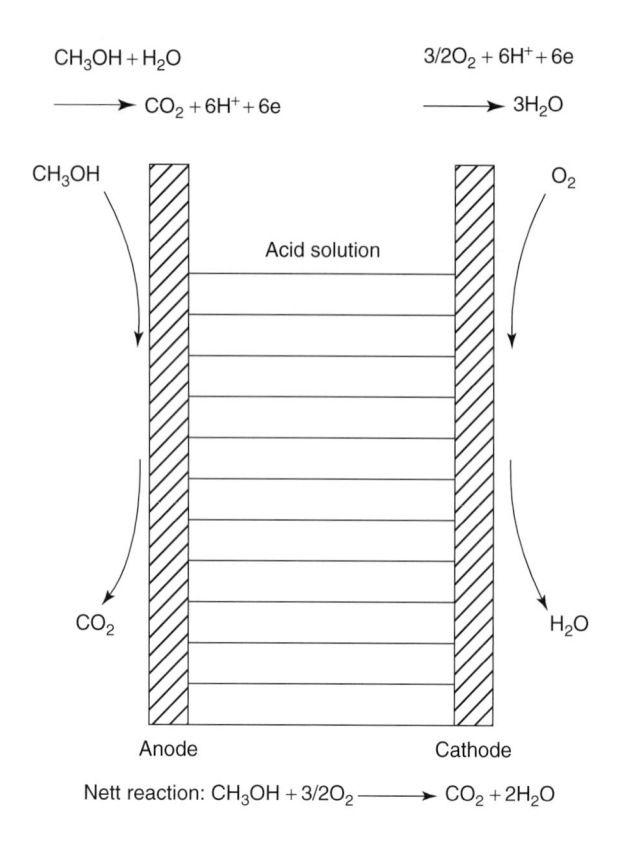

Figure 1. Schematic diagram of a simple direct methanol fuel cell with dilute aqueous acid as the electrolyte.

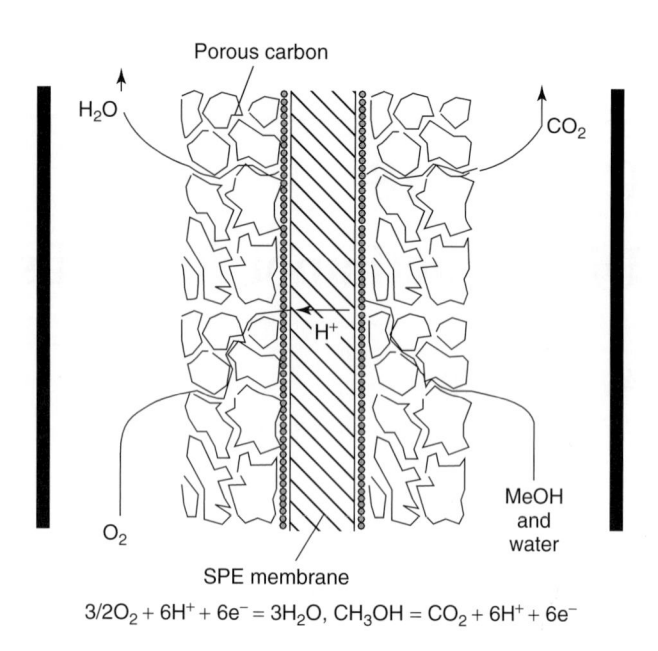

Figure 2. Realization of a direct methanol fuel cell with a solid-polymer electrolyte membrane as the electrolyte and impregnated porous carbon electrodes as anode and cathode.

The early MEAs resembled that of Figure 2, with catalyst impregnated carbon-polytetrafluoroethylene (PTFE) composites being used to form gas-diffusion electrodes, and Nafion ionomer sol being spread onto the electrode surface followed by hot-pressing.[3] Clearly the problem here is that much of the catalyzed carbon is, effectively, inaccessible electrochemically owing to lack of penetration of the Nafion. Furthermore, the PTFE, needed for structural support of the porous layer, tends itself to block access. This can be ameliorated by direct mixing of the electrocatalyst and the ionomer sol with the PTFE sol, or, as later discovered, even without PTFE.[4, 5] The main difficulty with this approach is that the ionomer sol does not penetrate into the smaller pores of the catalyst, leading to less than optimal use of the catalyst, and improvements have centered both on increasing the pore diameters and decreasing the size of the ionomer colloidal particles.

A second set of problems is associated with the fact that both air at the cathode and methanol at the anode produce exhaust gases. This is quite different from the normal H_2/O_2 fuel cell, where the only product is water. When air is fed to the cathode side, the oxygen reacts to form water but the nitrogen remains trapped in the pores of the electrode; the trapped nitrogen is a diffusion barrier for the incoming oxygen, and this results in mass transport losses. In addition, similar problems are encountered at the anode side as the methanol is converted to gaseous CO_2, which is, again, trapped in the pores. If very thin electrodes could be used, this would not be too much of a problem, but the higher catalyst loadings used in DMFCs imply the use of thicker electrode layers, with the need to control gas ingress and egress more carefully.

One approach is to retain the PTFE, mixing the electrocatalyst–ionomer mixture ultrasonically with the PTFE-carbon composite before spreading onto a carbon cloth or carbon paper current collector. This provides a series of pores to allow the nitrogen or CO_2 to escape[4] without effecting the utilization of the electrocatalyst. A second approach is try to modify the structure of the Nafion-bonded electrode by curing at temperatures above $\sim 150\,^\circ$C; this leads to degradation of the Nafion, increasing the hydrophobicity of the electrode layer.[6] A third approach is to use pore formers, such as $(NH_4)_2CO_3$; decomposition of this material gives wholly gaseous products, and it can be used to generate novel micro-structures.[7]

Once the MEA is fabricated, it can be incorporated into a complete cell, as shown schematically in Figure 3. The structure is derived from the standard plate-and-frame design, and is capable of operation at temperatures well in excess of $100\,^\circ$C and at pressures of up to 5 bar. Most modern DMFC designs are of a similar ilk.[8]

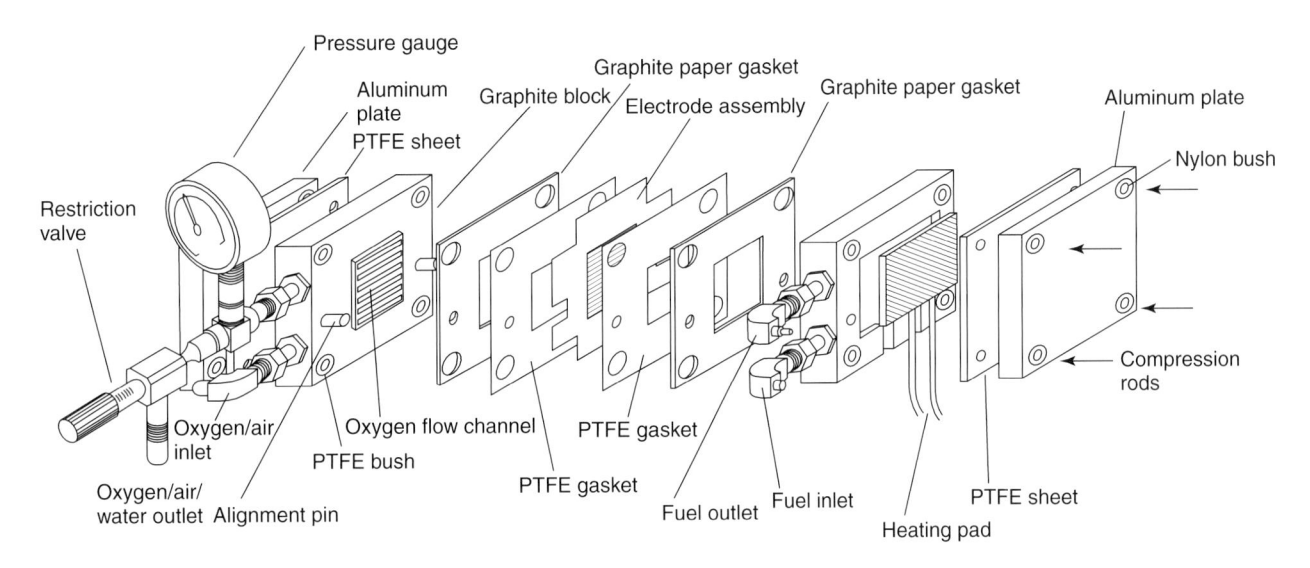

Figure 3. Construction of a practical methanol vapor-feed fuel cell, showing the membrane-electrolyte assembly and the modular components of anode and cathode.

3 PROBLEMS WITH THE DMFC

As indicated above, methanol can be used both directly and indirectly, the latter involving reformation of the methanol to hydrogen. Comparison of the two approaches is bedeviled by costing difficulties, but whilst the design and performance of reformers have shown real progress recently, it is important to recognize that there also remain problems in this area, even beyond the obvious cost implications. Reformation of methanol can be effected using steam:

$$CH_3OH + H_2O \longrightarrow 3H_2 + CO_2 \quad \Delta H = +131 \text{ kJ mol}^{-1} \quad (3)$$

a reaction that can be seen to be clearly endothermic, requiring heat input either from the waste heat of the fuel cell, which is not practicable if the fuel cell operating temperature is below $\sim 250\,°C$, or from combustion of the fuel in a separate burner. An alternative reformation procedure uses oxygen as partial combustant:

$$CH_3OH + \tfrac{1}{2}O_2 \longrightarrow 2H_2 + CO_2$$
$$\Delta H = -154.9 \text{ kJ mol}^{-1} \quad (4)$$

While this is now exothermic, the efficiency of fuel utilization is evidently reduced as only two molecules of hydrogen are produced for every molecule of methanol. Evidently, a judicious mixture of steam and oxygen would allow the reformer to operate at optimal fuel efficiency, but even if this could be achieved, modern reformers are still unable to produce hydrogen of sufficient purity for anode operation at lower fuel cell temperatures (below $\sim 200\,°C$) without either modification of the Pt anode, further secondary reformer stages designed to remove all but the last traces (<10 ppm) of CO from the fuel input, or further oxidation of the CO with an oxygen bleed using the Pt anode itself as a heterogeneous catalyst. The reason for this is that at lower temperatures, CO adsorbs strongly on Pt, poisoning the catalyst and preventing facile oxidation of hydrogen. As we shall see in a later section, adsorbed CO, CO_{ads}, an important intermediate in the electrochemical oxidation of methanol, is the main reason why the anode kinetics for direct electrochemical oxidation of methanol are so problematic.

The basic problems currently faced by the direct methanol fuel cell are: (a) the anode reaction has poor electrode kinetics, particularly at lower temperatures, making it highly desirable to identify improved catalysts and to work at as high a temperature as possible; (b) the cathode reaction, the reduction of oxygen, is also slow: the problems are particularly serious with aqueous mineral acids, but perhaps not so serious with acidic polymer membranes. Nevertheless, the overall power density of the direct methanol fuel cells is much lower than the $600+ \text{ mW cm}^{-2}$ envisaged for the hydrogen fuelled solid polymer electrolyte fuel cell (SPEFC); (c) perhaps of greatest concern at the moment is the stability and permeability of the current perfluorosulfonic acid membranes to methanol, allowing considerable fuel crossover, and, at the higher temperatures needed to overcome limitations in the anode kinetics, degradation of the membrane both thermally and through attack by methanol itself; (d) the fact that methanol can permeate to the cathode leads to poor fuel utilization, the appearance of methanol in the anode and cathode exhausts, and to the development of a mixed potential at the cathode, since conventional cathode catalysts are based on platinum, which is

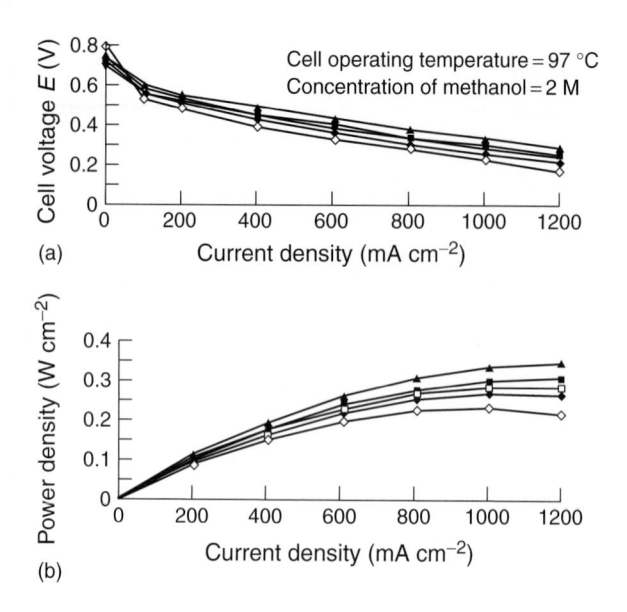

Figure 4. Performance of the cell of Figure 3 under varying oxygen-feed pressures: (◊) 1 bar, up to (▲) 5 bar.

highly active for methanol oxidation at the higher potentials encountered at the cathode. Experimentally, this problem has been tackled both by seeking alternative oxygen reduction catalysts and by increasing the Pt loading substantially; the latter clearly increases costs significantly.

In spite of these difficulties, the DMFC solid polymer electrolyte (SPE) does have the capability of being very cheap and potentially very competitive with the internal combustion engine, particularly in niche city driving applications, where the low pollution and relatively high efficiency at low load are attractive features. Performances from modern single cells are highly encouraging: an example from the Newcastle Fuel Cell Group is shown in Figure 4, and it can be seen that in oxygen, power densities of up to $0.35\,\mathrm{W\,cm^{-2}}$ are possible; in air a power density of $0.2\,\mathrm{W\,cm^{-2}}$ has been attained with a pressure of 5 bar.[8]

4 APPROACHES TO IMPROVED PERFORMANCE AT THE ANODE

4.1 Fundamental studies at platinum

The electrooxidation of methanol is not only an important reaction in its own right, since it is one of the primary reactions in modern fuel cell technology, but it has important generic characteristics, representing, as it does, a whole class of electrocatalytic processes. Although the literature was reviewed extensively by Parsons and Van der Noot[9] in 1988, and more recently by the author[10, 11] progress remains exceedingly rapid, largely driven by modern in situ spectroscopic and spectrometric techniques.

The chemisorption process of methanol on clean platinum surfaces has been found to be far faster than the steady-state oxidation rate for methanol, with the result that the initial electrooxidation current is found to decay rapidly with time. The origin of this decrease is now known to be the accumulation of adsorbed intermediates on the surface whose further oxidation to CO_2 is slow, and the rate differences allow us to separate the process of chemisorbate formation from that of the subsequent oxidation of the chemisorbed intermediates to CO_2. The nature of the intermediates can, in principle, be studied purely electrochemically by measuring separately the charge, Q_{ads}, required to chemisorb methanol, the charge Q_H required to cover the remaining the surface with adsorbed hydrogen (which can also be measured for the intermediate-free surface, as Q_H^o, and used to estimate the total surface area assuming a charge of 0.21 mC per true cm^2) and the charge Q_{ox} required completely to oxidize the adsorbed fragments, from which may be deduced n_{eps}, the number of electrons per site needed for the oxidation. However, these purely electrochemical experiments, whilst suggestive, have not proved definitive. The basic scheme for oxidation of methanol is shown below in the form of a scheme of squares:[12]

$$\begin{array}{ccccccc}
CH_3OH & \to & CH_2OH & \to & CHOH & \to & C-OH \\
 & & x & & xx & & xxx \\
 & & \downarrow & & \downarrow & & \downarrow \\
 & & CH_2O & \to & CHO & \to & CO \\
 & & & & x & & x \\
 & & & & \downarrow & & \downarrow \\
 & & & & HCOOH & \to & COOH \\
 & & & & & & x \\
 & & & & & & \downarrow \\
 & & & & & & CO_2
\end{array}$$

where the x symbolizes a Pt–C bond, and which summarizes the following reactions:

$$CH_3OH + Pt(s) \longrightarrow Pt-CH_2-OH + H^+ + e^- \quad (5)$$

$$Pt-CH_2-OH + Pt(s) \longrightarrow Pt_2CHOH + H^+ + e^- \quad (6)$$

$$Pt_2CHOH + Pt(s) \longrightarrow Pt_3COH + H^+ + e^- \quad (7)$$

$$Pt_3COH \longrightarrow Pt-CO + 2Pt(s) + H^+ + e^- \quad (8)$$

$$Pt(s) + H_2O \longrightarrow Pt-OH + H^+ + e^- \quad (9)$$

$$Pt-OH + Pt-CO \longrightarrow Pt-COOH \quad (9a)$$

or

$$Pt-CO + H_2O \longrightarrow Pt-COOH + H^+ + e^- \quad (9b)$$

$$Pt-COOH \longrightarrow Pt(s) + CO_2 + H^+ + e^- \quad (10)$$

Additional reactions described in the scheme above include

$$Pt-CH_2OH \longrightarrow Pt(s) + HCHO + H^+ + e^- \quad (11)$$

$$Pt_2CHOH \longrightarrow Pt(s) + Pt-CHO + H^+ + e^- \quad (12)$$

$$Pt-CHO + Pt-OH \longrightarrow 2Pt(s) + HCOOH \quad (12a)$$

or

$$Pt-CHO + H_2O \longrightarrow Pt(s) + HCOOH + H^+ + e^- \quad (12b)$$

$$Pt_3C-OH + Pt-OH \longrightarrow 3Pt(s) + Pt-COOH + H^+ + e^- \quad (13)$$

or

$$Pt_3C-OH + H_2O \longrightarrow 2Pt(s) + Pt-COOH + 2H^+ + 2e^- \quad (14)$$

The difficulties faced in analyzing a scheme of this complexity by purely electrochemical means are clearly formidable, and all the underlying assumptions made by earlier workers have now been challenged: (a) that the potential protocol adopted to clean the surface prior to each measurement does not alter the morphology of the surface; (b) that the surface, once cleaned, will remain clean indefinitely, save for such processes as we wish to study; (c) the only contribution to the anodic charge observed on stepping the potential from the anodic limit usually chosen for cleaning the surface ($\sim$1.2–1.6 V) is the chemisorption of methanol to form adsorbed fragments.

The initial measurements on polycrystalline Pt[13] suggested a three-electron intermediate, such as $Pt_3\equiv C-OH$, was the dominant adsorbed species, but later work suggested a more complex time dependence, with long times and high concentrations of methanol favoring Pt–CO.[14–16] The complex picture emerging from these studies on polycrystalline platinum impelled several groups to re-examine the problem using single-crystal platinum surfaces, with the aim of working with well-defined surfaces, which would be highly sensitive to any perturbation. This work established not only that the different low index surfaces of Pt had different activities towards methanol oxidation,[17, 18] but that methanol (at 0.1 M concentration) and CO both showed similar behavior towards oxidation on the different Pt surfaces, adding plausibility to the assignment of CO_{ads} as an important intermediate (since CO_{ads} is known to form when Pt is exposed to CO dissolved in the electrolyte). In addition, it was established that there was a considerable difference in activities for the Pt(111) surface in perchloric and sulfuric acids, with the latter showing strong anion adsorption, particularly on Pt(111),[19] that severely inhibited methanol oxidation. Finally, these studies established that for methanol oxidation, the behavior of a poly-faceted Pt-bead electrode can, to a first approximation, be described as the weighted sum of the activities of the three low index platinum surfaces.

These investigations also established that whilst values of n_{eps} close to 2 are found for all (100) derived surfaces in perchloric acid, suggesting that adsorbed linearly bonded CO dominates on these surfaces in the absence of any anion specific adsorption, adsorption of bisulphate ions reduces n_{eps} below 2, which can be ascribed either to simultaneous adsorption of both Pt_3C-OH and Pt–CO or, as the authors point out, to a mixture of linearly and *bridge* bonded CO, the latter having the structure Pt_2CO or even Pt_3CO for CO bridging two Pt atoms on the surface or adsorbed in a three-fold site. Further, for Pt(100), n_{eps} appears to rise from a value of about 0.9 at short adsorption times to a value close to 2 at higher times,[16] the very low value of n_{eps} at short times was again ascribed to the formation of bridge-bonded CO.

Intrinsic to much of this earlier work was the third assumption above, that the stable adsorbate, once formed, is also the dominant intermediate en route to the formation of the product CO_2. This is by no means obvious: the actual intermediate could be a species such as $Pt_3\equiv C-OH$ or some form of activated CO_{ads}, or indeed a third, unsuspected species, such as HCHO or HCOOH in the near electrode region. There is certainly good evidence that CO_{ads} is not a kinetically limiting intermediate at potentials above $\sim$0.6 V,[20] but the balance of evidence seems to be that at lower potentials, CO_{ads} is an increasingly important intermediate species, becoming dominant below $\sim$0.45 V.

The electrochemical studies summarized above point clearly to CO_{ads} as the dominant adsorbed species at long times and high methanol concentrations, and this conclusion has been confirmed by innumerable spectroscopic studies in the last two decades. The results of these studies can be summarized as:[11]

i. Infrared (IR) data usually show at least two peaks in the CO stretching region, assigned to linear and multiply bonded CO. Typical results are shown in Figure 5.[21]

ii. There are slight differences between the CO_{ads} derived from adsorption of CO and from methanol oxidation, possibly reflecting the presence of small amounts of such adsorbates as $Pt_3\equiv C-OH$. These differences are particularly marked in single-crystal studies on low index Pt faces, where the coverages of CO_{ads} derived

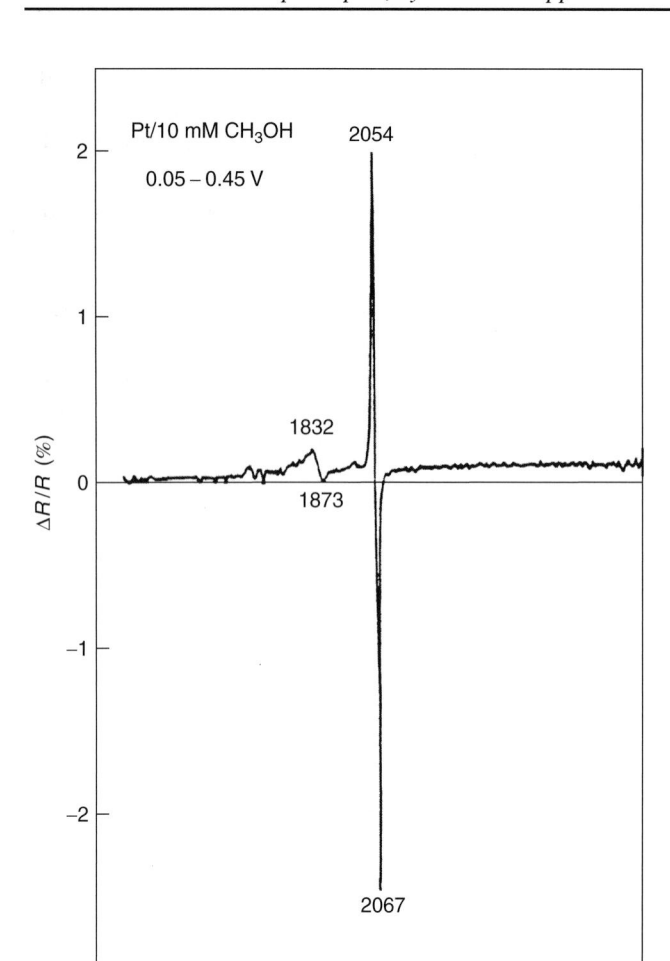

Figure 5. In situ FTIR differential reflectance spectrum of methanol oxidation on platinum showing surface-bonded singly and multiply bonded CO; the electrode was held sequentially at 0.5 and 0.45 V vs. NHE in a sulfuric acid/methanol electrolyte, and the resultant reflectance spectra at the two potentials subtracted from one another.

from methanol oxidation are significantly smaller than those derived from CO adsorption, and the ratio of bridging or multiply bonded species to the linear adsorbed form is also different, being generally smaller for methanol oxidation.

iii. There is a substantial shift in peak position as the potential and coverage are both increased, with $(\partial v_{c\equiv o}/\partial E)_\theta \sim 100\,\text{cm}^{-1}\,\text{V}^{-1}$.

iv. The increase in intensity of the CO_{ads} band with time for an electrode maintained at 0.4 V in 0.01 M methanol/0.5 M H_2SO_4 was found accurately to obey the expression $\theta_{ads} \sim \text{const.} + (1/\alpha f)\ln t$, consistent with Temkin activated adsorption.

v. Bridged CO_{ads} species appear more prevalent at lower potentials and coverages. For example on Pt(111) in 0.1 M $HClO_4$/CO, the results at a constant coverage of 0.65 show that below ~ 0.4 V, there is a very low frequency vibration ($\sim 1770-1790\,\text{cm}^{-1}$) which is assigned to triply bonded CO, and above 0.4 V, there is a jump to higher frequencies ($1840-1850\,\text{cm}^{-1}$) assigned to doubly bonded CO (however, see below), and in general for this surface, the proportion of terminal to bridged CO increases monotonically with θ.

vi. Evidence from a variety of spectroscopic and structural studies strongly suggests that at lower coverages, the CO does not adsorb randomly on the Pt(111) surface, but forms close-packed islands, a process that is particularly evident if a monolayer of CO_{ads} is formed and then partially oxidized. Under these circumstances, island formation is very marked. There is evidence that CO_{ads} island formation is less marked on Pt(100), where the islands appear kinetically more easy to dissociate. On Pt(110), island formation is complicated by the lifting of the normal reconstruction on CO adsorption. For methanol oxidation, the evidence suggests that CO_{ads} islands also form, but that they are smaller in extent.

vii. Support for this suggestion of island formation on single-crystal surfaces comes from isotope measurements: if a ^{13}CO monolayer is formed and then partially oxidized, and then ^{12}CO adsorbed, the ^{12}CO is preferentially oxidized to $^{12}CO_2$, suggesting that the CO islands, once formed, are not involved in subsequent nucleation and growth. In addition CO_{ads} sub-monolayers oxidize at lower potentials confirming the view that co-adsorbed water facilitates the oxidation of CO. It would appear that on the highly ordered Pt(111) surface, migration of CO_{ads} from the islands is relatively slow. There is evidence that relaxation of the islands is faster on Pt(100), and much faster on polycrystalline Pt.

viii. scanning tunneling microscopy (STM)/infrared reflection absorption spectroscopy (IRRAS) studies[22] of CO adsorption on Pt(111) from $HClO_4$ solution have shown that for saturation coverage, and for $E < 0.25$ V, a (2×2)-3CO adlayer forms possessing both terminal and triply bonded CO (in the ratio 1:2), and with an IRRAS spectrum showing peaks at 2066 and $1850\,\text{cm}^{-1}$ (with intensity 2:1). However, above 0.25 V, a surface phase transition takes place to yield a $(\sqrt{19} \times \sqrt{19})$-$R23.4°$ structure with 13 CO molecules per unit cell.

ix. On the Pt(533) surface, which can be thought of as [4(111) × (100)] in character, the step edge sites are preferentially occupied by adsorbed CO at low coverage, with terrace bonding only taking place at higher coverages. The step-edge CO$_{ads}$ shows relatively little shift in frequency with coverage, but the bridge-bonded CO$_{ads}$ on edge sites does show a more marked shift. As coverage increases, the terrace sites become populated, with the appearance of new IR bands.

x. Spectroscopic studies also confirm the stronger interaction of HSO$_4^-$ with the Pt surface, as well as suggesting that ordered HSO$_4^-$ overlayers have structures dependent on CO$_{ads}$ and other species adsorbed on the surface, the ratio $\theta_{HSO_4^-}/\theta_{CO}$ is *always* higher for methanol oxidation than CO oxidation.

xi. Other nonelectrochemical measurements have suggested, particularly at lower concentrations of methanol, that significant amounts of other adsorbed species are present during oxidation of methanol: differential electrochemical mass spectrometry (DEMS) data[23] strongly support the existence of Pt$_3$≡C–OH as an important adsorbate, as does electrochemical thermal desorption mass spectroscopy (ECTDMS).[24] Clearly, if≡C–OH is the intermediate formed at low coverage, it must also be very stable on the surface, since it differs little in energy from CO, a fact which would tend to rule it out as the active intermediate in any low potential parallel pathway.

xii. IR characterization of≡C–OH has proved difficult: the most convincing work[25] identified two small peaks at 1200 cm^{-1} and 1273 cm^{-1}, in addition to rather larger signals in the linear and bridged carbonyl region, the first being assigned to ν(C–O) of≡C–OH and the peak at 1200 cm^{-1} tentatively to δ(C–OH) of adsorbed =CHOH.

xiii. Study of Pt high index surfaces during methanol oxidation[26] has gained increasing importance as these surfaces are thought to be good models for both polycrystalline and particulate Pt. Edge-bonded and terrace-bonded linear CO$_{ads}$ is seen at much lower potentials on Pt(533) than linear CO$_{ads}$ on Pt(111), with presumably the terrace-bonded CO$_{ads}$ arising through migration.

xiv. Studies of CO$_{ads}$ on particulate Pt dispersed on carbon are very difficult, and bedeviled by technical and interpretational problems. There is evidence from earlier studies that CO$_{ads}$ is migrating from terrace to edge sites as the latter are preferentially oxidized to a mixture of HCOOH and CO$_2$,[27–29] with the former reacting further with methanol in solution to form methyl formate, a species identified both by IR and DEMS.

The electrochemical and spectroscopic studies on platinum have established that a complex partially ordered surface adlayer exists on single-crystal surfaces, and have shown that CO$_{ads}$ is accompanied both by adsorbed bridging CO species, by species such as≡C–OH, and by more or less ordered hydrogen-bonded anion adlayers. This complex surface cannot easily be studied by conventional electrochemical kinetics: Tafel slopes are found to depend critically on the state of the surface, on the concentration of methanol, and on the coverage of the surface by intermediates. Early studies[30] did, however, establish that methanol oxidation was inhibited by formation of an oxide layer on the Pt, and the kinetic law was consistent with oxidation of the methanol chemisorbate being rate-limiting, with the concentration of Pt–OH sites also being limiting at lower potentials: this implies that the rate-limiting step is:

$$\text{Pt–CO}_{ads} + \text{OH}_{ads} \longrightarrow \text{CO}_2 + \text{H}^+ + e^- \quad (15)$$

with the alternative

$$\text{Pt–CO}_{ads} + \text{H}_2\text{O} \longrightarrow \text{CO}_2 + 2\text{H}^+ + 2e^- \quad (16)$$

being suggested at higher potentials.

Combining electrochemical and in situ IR studies, rather similar data were reported for smooth Pt by Christensen and co-workers,[31] but for rough Pt the current becomes almost potential independent between 0.4 and 0.55 V, suggesting that roughening the Pt creates active Pt–OH$_{ads}$ sites and that migration of CO$_{ads}$ to these sites is rate-limiting. Similar conclusions as to the importance of CO$_{ads}$ diffusion to sites of OH$_{ads}$ adsorption were also reached by Gasteiger and co-workers[32] in their consideration of the catalytic activity of adsorbed Ru on Pt. The diffusion of adsorbed CO on both Pt and Ru in vacuo is actually quite facile: diffusion coefficients of order 10^{-9} cm^2 s^{-1} have been reported for both metals,[33] and whilst these will certainly be rather smaller in electrolyte solution, there is a high probability of the surface mobility playing a substantial role in methanol oxidation. This conclusion is also strongly supported by recent ^{13}C nuclear magnetic resonance (NMR) studies by Day and co-workers[34] who established that the spin-spin relaxation time, T_2, for ^{13}C could be modeled with a simple diffusion model, with an activation energy of 33 ± 8 kJ mol^{-1}, a result very similar to earlier data reported for methanol oxidation on both Pt and Pt/Ru.

IR studies have also shown that on potentiodynamic sweeps, the coverage of CO$_{ads}$ remains roughly constant on polycrystalline Pt electrodes pre-cycled in methanol up to a potential of 0.5 V, above which it falls rapidly, reaching zero above 0.7 V. For single-crystal Pt, the results are less clear cut: CO$_{ads}$ derived from methanol chemisorption on Pt(111) at 0.05 V[35] behaved in a manner similar to polycrystalline Pt, but CO$_{ads}$ derived from methanol chemisorption at 0.6 V was only removed from the surface at potentials in excess of 0.85 V, suggesting a much more stable species, which further studies suggested formed highly kinetically stable islands. The CO_{ads} formed at lower potentials appears to consist of a variety of islands of very different sizes and stabilities.

The implication of this is that the stability of CO_{ads} and the effective diffusion coefficient of CO on Pt will both depend very heavily on the experimental conditions employed. Even on polycrystalline Pt, it may be possible to generate very stable monolayers of CO_{ads} whose oxidation by potential stepping from lower potentials can only be effected by a nucleation mechanism in which an initial hole is created in the layer, further oxidation taking place by a growth mechanism similar to that encountered in film growth on electrodes. This model can be fitted quantitatively[36] with a mixed nucleation-diffusion equation. Note that if the size of the potential steps is extended into the oxide-formation region, nucleation of a Pt-O surface within the adsorbate layer, with chemical reaction taking place at the OH_{ads}/CO_{ads} boundary, may become rate-determining.[37]

The kinetic model emerging from the previous few paragraphs suggests that the details of the rate-limiting oxidation of CO_{ads} depend crucially, at lower potentials, on the extent to which the adlayer manages to order into more or less stable islands. That these islands are important comes not only from the structural data reviewed above, but also from isotope measurements. Studies of methanol oxidation below 0.5 V show that neither ^{13}CO nor $\equiv^{13}C-OH$ formed on the surface from electrosorption of $^{13}CH_3OH$ exchange on immersion in $^{12}CH_3OH$, even during oxidation of the $^{12}CH_3OH$ to $^{12}CO_2$.[38, 39] Not only does this strongly support the suggestion above that methanol adsorption leads to islands of adsorbed CO,[40] but it also suggests, in agreement with Christensen et al.,[31] that oxidation to CO_2 takes place only at the *edges* of these islands. This may well be the resolution of the enigma of the parallel reaction, and supports a very early suggestion[41] that CO_{ads} exists in two forms, one much less stable and more active than the other. Oxidation of methanol probably involves rapid chemisorption at higher potentials on the partially free surface of Pt to give CO_{ads} that migrates to the edges of islands to maintain their extension, but which is subsequently oxidized to form CO_2. The CO_{ads} at the centre of the islands thus remains unaffected, at least over short periods, by the oxidation of methanol, but the so-called active intermediate, is a CO_{ads} molecule at the edge of the island. This argument would not be significantly different even if the chemisorbed intermediate did not become completely dehydrogenated, provided it was sufficiently mobile to migrate to the edges of adsorbed CO islands rapidly compared to further oxidation to form CO_2.

There are, therefore, two kinetic models: the first is the oxidation of CO_{ads} at the edges of islands and the second is the migration of CO_{ads} to active sites on the Pt surface, at which oxidation can occur. Which of these dominates will depend primarily on the morphology of the surface: roughened surfaces are likely to inhibit stable island formation and to increase the number of active oxidation sites; it is likely that oxidation at these sites is through adsorbed Pt-OH, which, at low potentials, is otherwise present only at minute coverage. Better defined surfaces will favor island formation, and attack on the CO_{ads} at the edges of these islands is more likely to be by H_2O than OH_{ads}. If attack is by the latter, then the question arises as to the mobility of OH_{ads} on the surface; little seems to be known about this, but OH_{ads} spillover and migration from ad-atoms on the surface has been suggested on a number of occasions as providing a suitable kinetic model.[42] Whatever species is surface-mobile, the ideas above, whilst remaining tentative at this stage, do allow us to consider, within a coherent framework, the mechanism of promotion of platinum by other species as discussed below.

4.2 Platinum promotion

Platinum itself is not sufficiently active to be useful in commercial DMFCs, and there has been an intensive search for more active materials. This search has been illuminated by the mechanistic analysis given above; in particular, the realization that the catalyst must be both capable of chemisorbing methanol and oxidizing the resultant chemisorbed fragments (the concept of bi-functionality) has led to the search for materials that might be able to combine with platinum, with the latter acting to chemisorb methanol and the promoter then providing oxygen in some active form to facilitate oxidation of the chemisorbed CO.

The simplest method for enhancing oxidative activity is to generate more reticulated and active Pt surfaces on which Pt-O species will form at lower potentials. The drawback with this method of promotion is the poor long-term stability of the particles. A second approach has been the use of surface ad-atoms, deposited by under potential deposition (UPD) on the platinum surface. These have been studied extensively with most recent attention being on Sn[43] and Bi.[44] Several modes of operation have been suggested for these ad-atoms, including the blocking of hydrogen adsorption, the ability to modify the electronic properties of the platinum, the ability to act as redox centers, the blocking of poison adsorption sites or the ability of the ad-atom to induce Pt-O formation on neighboring sites, creating additional active sites for oxidation of CO_{ads}. Again there are issues of long-term stability, however.

A third type of promotion is the use of alloys of platinum with different metals, where the second metal forms a surface oxide in the potential range for methanol oxidation. This is the basis for studies of Pt/Sn, Pt/Ru, Pt/Os, Pt/Ir etc. Studies of Pt alloyed with the other noble metals has shown that Ru has by far the largest effect, giving a substantial catalytic advantage.[45, 46] The mechanism(s) whereby Ru promotes Pt have been the subject of intense study: the original suggestion by Watanabe and Motoo[43] that the mechanism is bifunctional, that methanol chemisorbs on Pt to form ≡C–OH or CO_{ads} and that these intermediates are oxidized by Ru–O(H) moieties, still remains the most plausible. Experimentally, almost any method of mixing Pt and Ru will give enhanced activity, provided Ru is present at the surface. Studies of Ru promotion on single-crystal Pt(111) suggest that the Ru segregates to form mono-atomic islands of roughly 1 nm diameter,[47] and it would seem likely that CO_{ads} diffusion to these islands would be rate-limiting. It should be noted that alloying Pt with Ru would be expected also to have an electronic effect, and IR data confirm this[48, 49] in these experiments, the IR stretching frequency increases on Pt/Ru as compared to Pt, which has been ascribed to reduced CO binding to the Pt as a result of Ru alloying. Confirmation of the importance of the presence of Ru on the surface has also come from ex situ observations;[50] whilst the presence of Ru in sub-surface layers has a substantial effect on CO_{ads} binding, no enhancement of electroactivity to methanol oxidation over Pt itself is observed unless Ru is present on the surface. In addition to Ru, Ir and Os have also been found to act as promoters, though less effectively than Ru,[45] with Au and Pd inhibiting the reaction. Amongst the nonnoble metals, electrodeposited Pt/Re alloys were reported to show substantial enhancement by Cathro,[51] a result that seems not to have been followed up. Given the effectiveness of binary alloy formation, particularly with Ru/Pt, attention has turned recently to ternary alloys, and Pt/Ru/Os[52] has shown unequivocal advantages.

The fourth type of promoter to be described in the literature is a combination of Pt with a base-metal oxide. Initially, these were reported by Hamnett et al.,[53] with Nb, Zr and Ta being found most active for Pt promotion, particularly at higher currents. However, although the effect on Pt itself is clear, attempts to use ZrO_2 to promote Pt/Ru were unsuccessful,[54] suggesting that the original mechanism for promotion by these oxides is similar to that for Ru itself. More recently, Shen and Tseung[55] have identified *hydrous* WO_{3-x} as a substantial promoter for activity when Pt is electrodeposited on its surface, and this was confirmed separately by Shukla et al. who reductively co-deposited Pt and WO_{3-x} onto a porous carbon substrate.[56] This promoter appears to operate by providing a surface redox site, cycling between W(IV) and W(VI). The development of hydrous W and Mo oxides as catalysts apparently operating somewhat differently from Ru has suggested that ternary catalysts of the form Pt/Ru/W or Pt/Ru/Mo might be developed,[57, 58] and these have proved moderately more active than Pt/Ru but not substantially so.

In addition to the electrodeposition or reductive deposition of Pt onto an oxide surface, it is possible to incorporate the Pt directly into many oxide structures, and these species have been examined recently. Perovskite-based oxides, in particular, $SrRu_{0.5}Pt_{0.5}O_3$, were found to give good current densities when used as electrodes in Nafion-based cells,[59] and similar results were reported for $Dy_xPt_3O_4$.[60] It is unclear how stable these oxides are in the longer term in acidic environments, and reductive decomposition through methanol adsorption to give finely divided platinum particles at the surface which act as the main catalytic sites, remains a plausible mechanism.

4.3 Preparation methods

Provided that both Ru and Pt are present at the surface, it does not seem to matter precisely how the catalyst is formulated, at least some enhancement of activity will be seen: methods used include bulk alloys, reductive co-deposition or electrodeposition of both metals, deposition of Ru on platinum particles, thermal decomposition of chlorides and chloro-acids, co-impregnation of metal salts on carbon paper, implantation of Ru into Pt by ion bombardment, formation of Raney alloys of Pt/Ru, or spray deposition from organic solvents. Deposition of the Pt/Ru catalyst by whatever means is usually followed by an activation step, and these steps are crucial in ensuring that both platinum and ruthenium are indeed present at the surface: as shown by McNicol and Short, for example,[61] activating the Pt/Ru in hydrogen at 300 °C gave a poorly active catalyst, since this treatment favors Pt on the surface, whereas activation in oxygen gives an active catalyst since this treatment favors segregation of some of the Ru to the surface to form oxides.

From the practical perspective, the primary purpose is to apply the Pt/Ru catalyst to high surface area carbon as the preliminary to manufacture of the MEA as described above. There are two basic methods: one utilizes solutions of precursors such as chlorides, carbonyls etc., that are impregnated onto carbon and then reduced using either a liquid-phase reductant such as N_2H_4 or $NaBH_4$, or a gas-phase reductant such as H_2. This method has the advantage that it can be used for multi-element catalysts, but it does require a high surface area carbon, and attempts to generate high metal loadings are often frustrated by aggregation of the catalyst particles.

A second route is the use of colloidal dispersions[62] based initially on sulfito chemistry; the basic chemistry involves formation of sulfito complexes in solution of the general form $[H_xPt(SO_3)_4]^{-(6-x)}$ and analogously for Ru. These complexes are then oxidized, the sulfito-ligands converted to sulfate, which do not coordinate under the conditions used; this leaves unstable $[Pt(OH)_n]^{-(n-2)}$ species that aggregate to form small (2 nm) colloidal particles. These are adsorbed onto carbon and reduced. Very careful pH control is essential if uniform 2 nm particles are to be obtained, but the advantage of this technique is that even at high loadings, the particles show little tendency to aggregate.

Once the catalyst has been prepared, fabrication of the MEA follows the routes outlined above. Each catalyst formulation will have its own hydrophobicity, and it is essential that the preparation techniques for each MEA are optimized separately: this is not a case of one size fits all. It is for this reason that comparison of different catalyst formulations is so difficult in the literature: varying the catalyst composition may appear to lead to improvements that are actually associated with failure to optimize the initial catalyst/electrode morphology and hydrophobicity, the additives giving a better catalyst by virtue, for example, of improved wetting. Even experienced chemists can be badly misled in this field.

5 OXYGEN REDUCTION

5.1 Oxygen reduction electrocatalysts

The reduction of oxygen is a key electrochemical reaction, and improved understanding of its mechanism and the development of more efficient electrocatalysts underlie much current fuel cell research.[63, 64] There is no doubt that the prospects for low temperature fuel cell commercialization would be transformed by the discovery and development of improved cathode catalysts in both conventional and DMFCs. In the case of DMFCs, of particular importance would be the development of so-called methanol tolerant cathodes: as indicated in section three above, methanol cross-over from anode to cathode is a major problem, and the electrooxidation of fuel at the cathode not only leads to poor overall fuel utilization in the cell, but reduces the operating voltage through generation of a mixed potential at the cathode. If electro-catalysts could be developed that were inactive to methanol oxidation but comparable to platinum in activity to oxygen reduction, this would represent a major step forward. Such catalysts are starting to appear, and are discussed in more detail below.

There are some four classes of oxygen electrocatalyst now known: the most familiar are the noble and coinage metals, particularly platinum and gold, which have been extensively investigated both as pure metals, nanoparticles[65] and alloys,[66–68] and as polycrystalline and single-crystal surfaces.[69–71] They show a wide variety of behavior, reflecting both the surface structure and the presence or absence of oxide films, but platinum, in particulate form dispersed on carbon, still shows the highest overall activity of any oxygen electrocatalyst in acid solutions.

A second class of electrocatalyst contains the macrocyclic derivatives of a wide range of transition-metal compounds.[72] The most well investigated are Co and Fe and amongst the ligands studied, porphyrins, phthalocyanines, tetra-azaannulenes and dimethylglyoxime derivatives are well established. The involvement of the metal centre in the electrochemistry is well established for this class; less clear is the involvement of ligand orbitals, particularly in view of the near degeneracy in energy between the HOMO on the ligand and the d-orbitals on the metal.[73, 74] Such complexes have even been found to be active following pyrolysis on carbon substrates, and mimetic derivatives, made by thermolysis of metal salts in a polypyrrole matrix, have also been found to be active.[75]

A third class of catalyst is derived from metallic oxides. Many oxides, particularly of the second and third row transition elements, show metallic conductivity,[76] usually derived from M–O–M bonding rather than direct M–M overlap, and can therefore be fabricated into electrodes without addition of a conducting matrix. Particularly in alkaline solution, a number of such oxides, including spinel, perovskite and pyrochlore structures[77] show remarkable activity for oxygen reduction, but in acid solutions, the activity declines substantially, and the stability of the oxide phase is also far less. There have been reports of oxide catalysis in acid solution, though it has been suggested that in a number of such cases, homogeneous rather than heterogeneous mechanisms may play a significant role, particular when the oxide is acting as a promoter.[78, 79] Metal oxides can, in principle, be stabilized to acid by immobilization in conducting polymers, and it has been shown that $Pb_2Ru_2O_{6.5}$ can be stabilized in Nafion[80] as can the spinel oxides in polypyrrole.[81]

Related to the oxides are members of the fourth class of electrocatalyst, based on transition-metal compounds with other nonmetallic counter-ions derived from the chalcogenides[82] and indeed from other nonmetal species. The chalcogenides are frequently highly stable, especially in combination with later transition-metals, and can be immersed in aqueous acid and held at high positive potentials without appreciable degradation. Studies of compounds from this latter class has only taken place within the last 10 years or so, but it is already clear that their activity towards oxygen reduction is remarkable.[83–89]

5.2 The mechanism of oxygen reduction

Our understanding of the mechanism of oxygen reduction remains incomplete, even on well-established catalysts such as the noble metals. Part of the reason for this lies in the fact that almost all our knowledge is derived from classical kinetic studies of reaction-rate dependence on concentration of O_2 and pH and on electrode potential. Modern spectroscopic measurements have proved much less rewarding: whilst some useful Fourier Transform infrared (FTIR) work has been carried out,[89] the fact that IR is blind to O_2 itself, and that species such as $O^-_{2\text{ads}}$ appear to have a rather low absorption cross-section and/or surface coverage, has meant that very few surface studies have been reported. Similar difficulties have been encountered with other techniques, and mechanistic details remain very obscure in many cases. What is clear is that there are certain critical steps in oxygen reduction, and that branching from these steps can determine whether reduction occurs to the two-electron products H_2O_2 or HO_2^-, or reduction takes place down to the normally desired four-electron product H_2O. The steps are:[90–92]

1. The physical access and possibly physisorption of dioxygen onto the catalyst;
2. Possible initial electron transfer from the interior of the electrode to a surface-localized redox center;[93]
3. Electron transfer to O_2, usually with immediately following[90, 94] or simultaneous[95–100] chemisorption of the super-oxide one-electron reduction product onto the catalyst in either the end-on or bridged configuration, perhaps with simultaneous proton tunneling in acid solution to form $HO_{2(\text{ads})}$;
4. As an alternative, the disproportionation of O_2^- or $HO_2^·$ in the outer Helmholtz layer to form HO_2^- or H_2O_2 may take place without chemisorption. The O_2 formed is then recycled;[99, 101]
5. Subsequent reduction of chemisorbed O_2^- to HO_2^- or H_2O may take place; the former may further catalytically disproportionate to form O_2 and H_2O, with recycling of the O_2.[102]

It will be apparent that the chemisorption step is the key process, and the evidence is now quite strong that this normally follows the first electron-transfer step or is concerted with it. Only in the most unusual cases, such as oxygen reduction on Au(100), has it been suggested that dissociative chemisorption of oxygen takes place at room temperature before net electron transfer, and even in this case the mechanism is controversial.[103, 104] This has very important consequences, not the least of which is that the first electron-transfer process is normally rate-limiting at all but the lowest overpotentials, and unless the O_2^- can be stabilized strongly by concerted chemisorption, reduction of O_2 will be severely kinetically impeded.[105, 106]

5.3 Improved oxygen electrocatalysts based on Pt for the DMFC

Pt remains the electrocatalyst of first resort for low-temperature fuel cell cathodes, and most investigations have either been carried out to improve the dispersion characteristics, as discussed above for MEA fabrication, or to alloy Pt with other metals. In this regard, high specific activities of Pt/Cr and Pt/Cr/Co alloy electrocatalysts for oxygen reduction as compared with that on platinum have been reported in H_2/air polymer electrolyte membrane fuel cells (PEMFCs).[107, 108] The formation of a tetragonal ordered structure upon thermal treatment, in the case of Pt/Cr/Co[109–111] leads to a more active electrocatalyst than that with a Pt face centered cubic (FCC) structure. The enhanced activity has been ascribed to a variety of causes, including interatomic spacing, preferred orientation, and electronic interactions, and considerable recent effort has been expended on fabricating Pt alloy particles of uniform small size using co-precipitation, impregnation and colloidal preparation procedures as described above.[112]

Clearly the main drawback of Platinum is its activity to methanol oxidation. In principle, since the methanol oxidation and oxygen reduction reactions make different demands in terms of surface structure, it ought to be possible to optimize the size of the Pt particles for the latter, and minimize their activity to the former reaction. In practice, however, this has proved exceptionally difficult; even the use of alloying elements, such as Co and Cr, has not proved very successful, partly as both these elements can be leached from Pt at high potentials to form fractal surfaces that show enhanced activity for methanol oxidation as well. Preparatively, dispersed mixtures of Pt and Cr and/or Co are impregnated onto carbon as above, but reduction and annealing requires higher temperatures, usually 700–900 °C rather than the 100–400 °C used for DMFC anodes. This alloying temperature allows the formation of true face-centered tetragonal alloys to take place, as well as increasing the mean particle size, the latter unfortunately decreasing the mass activity of the catalyst.

5.4 New methanol-tolerant oxygen reduction electrocatalysts

The search for oxygen electrocatalysts has, then, been the informed search for electrode materials that are both stable at high potentials in acid or alkaline solution (an exceptionally demanding requirement in its own right), and which can coordinate oxygen in its partially reduced form. The ternary metallic oxides (particularly those based on Ru and Ir), have proved particularly successful in this regard, especially in alkaline solution. In both cases, electron transfer leads either to direct M–O–O bonding or to the formation of O_2^- that can displace water or OH from a metal coordination site at the surface of the oxide. The strong coordination of the oxygen in this latter case to the Ru[90, 113] or other transition metal[114] leads almost entirely to 4–electron reduction. The role of the counter cation in ternary oxide catalysts has been the subject of considerable, if inconclusive, investigation. Control of the oxidation state of the main catalytic site, as in the perovskite $Pr_{1-x}Sr_xMnO_3$[115] and spinels $Cr_{1-x}Cu_xMn_2O_4$[116] can be important, and Egdell et al.[90] postulate an active site associated with the Pb/O/Ru centre in $Pb_2Ru_2O_{6.5}$. Goodenough et al.[117] also identify such A/O/Ru sites as significant; but more extreme views have been put forward, notably by Shukla et al.,[118] who suggest that in the pyrochlore ruthenates the active site might actually be the A cation, and by Widelov et al.[119] who find that activation of (apparently stoichiometric) $Bi_2Ru_2O_7$ requires the leaching of surface Bi^{3+} ions to create a high surface area Ru-rich interphase. However, the general consensus is that critical to the success of these catalysts is the fact that there is both a high density of states at the Fermi level, and that the surface sites are connected electronically to the bulk, but can sustain quasi-localized electron transfer. This can only be accomplished if the bandwidth is sufficiently narrow for Mott-Anderson localization to be possible in the surface but not in the bulk, and it is no accident that successful metallic oxide catalysts have been derived from species such as the manganese perovskite or Ru-based pyrochlores where the bandwidth is high enough to allow itinerant electron conduction in the bulk but where electrons at the surface can be conceived as localized. This configuration allows the active site to behave as an essentially multi-electron reductant, and this approach has also been the key to ensuring that macrocyclic complexes can facilitate the 4–electron reduction of O_2 by incorporating the active sites as dimeric species, in which the dioxygen molecule is accommodated as a bridging species, allowing 4–electron reduction to take place.[120, 121]

The necessity for the active site to be able to accommodate multi-electron reduction is well recognized in biology, and led Alonso-Vante and Tributsch[83] to propose that the Chevrel phases, based on the $Mo_6X_8^{n-}$ unit might be active in oxygen reduction as well as being stable under the hostile electrochemical conditions required for a successful cathode in fuel cells. Subsequent exploration by the Berlin Group showed that this conjecture was correct, and a wide range of Chevrel phases were synthesized and explored electrochemically;[122, 123] however, although this new class of material showed moderate oxygen reduction activity, the intrinsic reduction rates are significantly lower than Pt metal.

It has been recognized that to increase the rate of reduction of oxygen by these novel Chevrel phases, some means would have to be found of dispersing them onto a high-surface-area substrate such as carbon. It was discovered in Berlin[124] that heating a mixture of the relevant metal carbonyls and elemental sulfur or selenium in xylene and adding carbon to adsorb the products did give rise to an encouraging increase in activity. The Newcastle Group[86] were able to establish several key intermediates in this reaction, involving the formation both of xylenyl metal carbonyls and cluster carbonyls, and were also able to establish for the first time that the materials formed were not Chevrel phases, and did not appear to be related to such phases, having stoichiometries that were highly metal-rich, but were still active for oxygen reduction.

It has become clear that, whilst the highest activity was initially found for materials containing both Mo and Ru, the catalyst of nominal composition $Mo_2Ru_2S_5$ was unstable to the loss of Mo, and activation led to actual compositions typically of the type $Mo_2Ru_{11.4}S_9$ on untreated XC–72 and $Mo_2Ru_8S_{3.8}$ on sulfur-treated XC–72. Even though the Mo content was reduced, however, it still showed promotion effects, particularly at higher current densities; furthermore, whilst the Ru appeared to remain coordinated to S or Se, the Mo seemed to be coordinated to O, and it has been suggested that the Mo might be playing a role in either improving the wettability of the catalyst or possibly in coordinating oxygen during the early stages of electro-reduction as indicated above. Nonetheless, whilst Mo evidently does act as a promoter, the main catalysis does appear to be mediated by Ru/S or Ru/Se phases.[125]

There is now clear evidence that these mixed-metal chalcogenides are not only highly active to oxygen reduction but also inactive to methanol oxidation, and the Newcastle Group has incorporated catalysts of the general form $A_xB_yS_zO_?$, where the oxygen content is unknown, A is a metal from the earlier part of the transition group, such as Mo, W, and Re, and B is a metal from group VIII, such as Ru, Rh or Os, into quasi-optimized fuel cells. The results, shown in Figure 6[88] illustrate that activities not far from Pt can be attained in DMFCs under the right conditions.

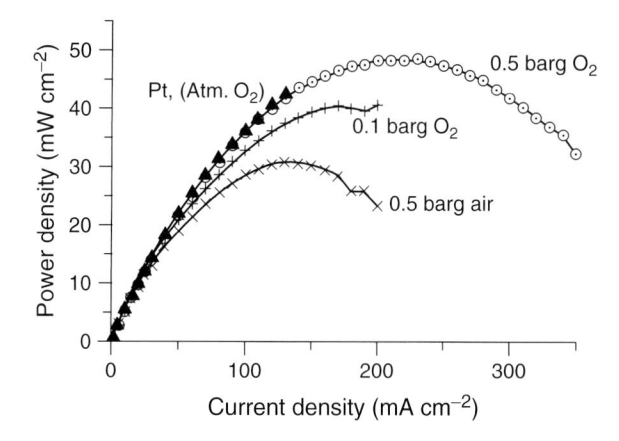

Figure 6. Performance of a liquid-feed DMFC at 90 °C, with 2 M methanol feed and ambient pressure O_2 using an E-TEK Pt:Ru (1:1) anode and a $Rh_2Ru_{11.8}S_{9.4}$ (1.5 mg cm^{-2}) cathode on Ketjen 600 carbon at various oxygen partial pressures. The performance on dispersed Pt at ambient oxygen pressure is shown as (▲).

Whilst these materials are extremely interesting, their structure and morphology remains obscure. Studies[88, 125–127] suggest two possible basic structural features in these catalysts: either that they are composed of extremely small nanoparticles of *metal* which are stabilized against dissolution and aggregation by surface adsorbed sulfur, carbonyl or other species, or that the particles are forms of metal-rich RuS clusters which do not form ordered arrays, and which are admixed with deposits of (possibly hydrous) RuO_x. The latter must be screened in some way, or at the least is only present in rather small quantities, since at the temperatures of operation of the fuel cells in Figure 6, ruthenium is moderately active for methanol oxidation, and very little such oxidation was observed with such catalysts.

The nature of the active site in these materials is even less well understood. Tributsch has suggested[128] that the active site is, in fact, a Ru *carbonyl* cluster on the surface of a Ru nanoparticle stabilized by S or Se. In support of this suggestion, active electrocatalysts have recently been reported in which there is no chalcogenide: simply by heating $W(CO)_6$ in xylene or $Os_3(CO)_{12}$ or $Ru_3(CO)_{12}$ in 1,2 dichlorobenzene[129] and admixing the product with graphite and paraffin oil has been found to generate catalysts of moderate activity. However, this has been strongly challenged: extended x-ray absorption fine structure (EXAFS) data show that a near O-free precursor, of approximate stoichiometry Ru_4Se_2, is profoundly modified in solution, with the clear formation of Ru–O species,[125] and diffuse reflectance infrared Fourier Transform (DRIFT) spectra of the catalysts show no signs of carbonyl moieties. The alternative suggestion of the Mo–O–Ru site as the active centre has received support from several areas, and clearly this is an important aspect of further investigation into these fascinating systems.

6 IMPROVEMENTS IN MEMBRANE PERFORMANCE

As indicated above, one of the most difficult aspects of the DMFC is the transport of methanol from the anode to the cathode compartments. The seriousness of this problem can be gleaned from Table 1 below, showing some recent work from the Newcastle Group. In this study, three cathodes were used, Pt, a Mo/Ru sulfide catalyst and a Re/Ru sulfide catalyst, the latter two being essentially inactive for methanol oxidation. Table 1 shows the measured concentration of methanol in the cathode exhaust in a liquid-feed

Table 1. Measured Methanol concentrations in the cathode exhaust from MEA-Pt, MEA-MoV and MEA-ReV under various operating conditions.

Anode feed concentration (mol dm^{-3})	Current density (mA cm^{-2})	Cell T (°C)	Methanol concentration in cathode exhaust (mol dm^{-3})		
			MEA-Pt	MEA-MoV	MEA-ReV
2	50	60	–	–	0.7
2	50	70	0.10	0.59	1.57
2	50	80	0.07	1.20	1.3
2	50	90	0.09	0.97	1.25
2	100	90	0.07	0.85	0.86
2	10	90	0.10	1.13	–
2	0	90	0.10	1.18	1.33
4	10	90	0.10	–	–
4	50	90	0.07	1.98	–
4	100	90	0.08	–	–
4	130	90	–	1.50	–

cell as a function of methanol concentration in the anode feed and other parameters. It is evident that very substantial permeation of the methanol is taking place, and that much of this is lost at Pt through parasitic oxidation.

The Nafion membrane used for these studies is typical of the perfluoro-sulfonic acid membranes developed by both DuPont and Dow, and typical structures are shown in Figure 7. Fuel efficiency losses can amount to as much as 20% under operating conditions,[129] and improved membranes are of considerable importance in this area. Reversion to a liquid-based electrolyte system is not attractive: solid-state proton-conducting membranes limit corrosion, reject CO_2 and have high conductivities, all of which are extremely valuable. The requirements otherwise are chemical and electrochemical stability and low methanol permeability ($<10^{-6}$ mol cm^{-2} min^{-1}), and several types of experimental membrane have been reported recently that start to meet these criteria: polyvinylidene fluoride,[130] styrene grafted and sulfonated membranes,[131] sulfonated poly (ether ether ketone) and poly (ether sulfone),[132] or zeolites gel films (tin mordenite)[133] and/or membranes doped with heteropolyanions.[134]

Whilst sulfonated poly (ether ether ketone) and poly(ether sulfone) electrolytes show good mechanical strength, even as 10–50 μm films, acceptable conductivities and reduced permeability to methanol, there is evidence that they are not stable to methanol, and that this instability increases with sulphonation. One approach is to cross-link the polymer chains, which should both increase stability and decrease methanol permeation, but this process also decreases the conductivity and careful compromise is necessary. An alternative approach is to use phosphoric-acid-doped or sulfuric-acid-treated polyacrylamid and polybenzoimidazole[129, 135, 136] membranes, though the stability of the acid to leaching in the anode feed is open to serious question.

Further developments include composite membranes, such as Nafion-silica or Nafion-zirconium phosphate composites with or without added heteropolyacid, that retain the desirable qualities of Nafion itself whilst decreasing methanol permeation rates.[129, 137–139] These membranes allow the operation of DMFCs at up to 150 °C, since the water retention properties of the Nafion are also increased, and similar properties are also shown by Nafion impregnated with silica/heteropolyacid or zirconium phosphate.

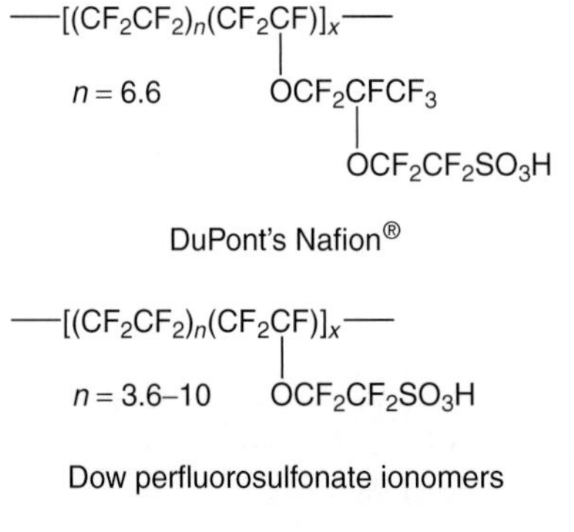

6.1 Performance characteristics of modern DMFCs

Experimental single cell DMFCs, in comparison to H_2/air PEMFCs, show the effects of both poor electrode kinetics at the anode and methanol cross-over. Open circuit potentials are usually ~150–200 mV lower, in the range 0.7–0.9 V, and on polarization, there is an immediate further potential loss of 0.15–0.2 V, following which an ohmic plateau is encountered. The performance is usually quoted at a standard cell voltage, often 0.5 V, which strongly depends on such operating conditions as working temperature, mass-transport conditions, use of air or oxygen at the cathode, pressure and fabrication conditions and catalysts used in the MEA.[129]

The best single cell (5–50 cm^2 active area) DMFC performances, achieved by various groups in the last five years, under high pressures, are 300–450 mW cm^{-2} and 200–300 mW cm^{-2} as maximum power density, in presence of oxygen and air feed at the cathode, respectively, as illustrated in Figure 4.[140–144] These performances have been obtained at temperatures close to or above 100 °C, with overall Pt loadings of 2–5 mg cm^{-2} and, in most cases, under pressurized conditions. At a cell voltage of 0.5 V, the best performances in presence of air as oxidant are 150–200 mW cm^{-2}. If the DMFC is operated at room temperature and under air breathing conditions, the performances become less than one-tenth of those achieved under the above conditions.

Although the initial driving force for the development of DMFCs was traction, there are substantial numbers of applications, particularly for portable equipment, that could also be addressed by the DMFC, as indicated in Table 2.[129]

Owing to the current high catalyst loadings for DMFCs, and attendant costs, attention has increasingly shifted to the application of DMFCs as attractive power sources for cellular phones, laptop computers, palm-held devices, etc., with power levels running up to a few kilowatts. This is the current limit of development of DMFC stacks, with Ballard power Systems in Canada reporting a 3 kW system utilized in an experimental one-car system by DaimlerChrysler.

Within Europe, a consortium led by Siemens AG, IRF A/S in Denmark and Johnson-Matthey in the UK is constructing a 25 W 16-cell stack, and has already built a

$$—[(CF_2CF_2)_n(CF_2CF)]_x—$$
$$n = 6.6 \quad\quad OCF_2CFCF_3$$
$$|$$
$$OCF_2CF_2SO_3H$$

DuPont's Nafion®

$$—[(CF_2CF_2)_n(CF_2CF)]_x—$$
$$n = 3.6–10 \quad\quad OCF_2CF_2SO_3H$$

Dow perfluorosulfonate ionomers

Figure 7. Chemical composition of perfluorosulfonic acid polymers used a proton conductors in solid-polymer membranes.

Table 2. Potential applications of DMFCs.[129]

Potential application	Field	Rated power (kW)	Overall efficiency requirements (%)	Specific power (W kg^{-1})
Transportation	Electromotive	20–50	35–45	350–500
Portable	Laptop microcomputers	0.05–0.1	20	50
	Cellular phones	0.001–0.003	20	30
Stationary	Residential	5–10	35–45	200

3 cell air-fed stack operating at 1.5 bar, but optimal performance was only obtained at very high air-flow rates. A 150 W stainless steel-based air-fed DMFC was also built in Europe under the support of the Joule Programme under the framework of the NEMECEL project.[129, 145] A rather similar stage has been reached by Los Alamos National Laboratory,[141] which recently reported a 50 W stack giving an operational efficiency of 37%.

Jet Propulsion Laboratory (JPL) has been actively engaged in the development of miniature DMFCs for cellular phone applications over the last two years.[129, 146] The power requirement of cellular phones during the standby mode is about 100–150 mW, whereas under operating conditions it can be in the 800–1800 mW range. In order that the proposed DMFC power source is compatible with the cellular phone, as for state-of-the-art lithium ion batteries, it is projected that the weight and volume be about 50 g and 50 ml respectively. Pure methanol will be fed from a cartridge and will be appropriately diluted prior to being fed to the anode. The products, water and carbon dioxide, will be contained within the package. Air will be delivered to the cathode by natural convection. Based on the results of the current technology, the JPL researchers predict that a 1 W DMFC power source, with the desired specifications for weight and volume and having an efficiency of 20% for fuel consumption, can be developed for a 10 h operating time, in between replacement of methanol cartridges.

Motorola[129, 147] has recently demonstrated a prototype of a miniature DMFC using a multi-layer ceramic technology for processing and delivering methanol and air to the fuel cell. In the current design, the MEA is mounted between two porous ceramic plates. The lower ceramic plate acts as the methanol feed, while the upper one provides for passive air delivery. In this prototype, four cells, connected in series in a planar configuration (5×5 cm^2 in area and 1 cm thick) exhibited average power densities between 15 and 22 mW cm^{-2}. The reasons for using four cells in series was that a potential of greater than 1 V for the DMFC is necessary to be used with a direct current (DC) to DC converter to step up the potential to about 4–5 V, which is required for charging the lithium ion battery.

7 THE FUTURE

The construction of stacks in the last few years has signaled the start of a more mature phase in the development of the DMFC. For long time a laboratory curiosity, excoriated by funding agencies as being wholly impractical, it is now a serious contender, particularly for the smaller scale applications discussed above. These applications include both civilian and military: anywhere in which the intrinsic weight advantages of the DMFC can be exploited will, in the next few years, be targeted. In the longer term, the undecided question is whether the DMFC will ever be a serious contender for motor vehicle traction. This remains quite uncertain: the high catalyst loadings and the continued reliance on expensive membranes with substantial cross-over problems and consequent exhaust clean-up difficulties are major hurdles. Some progress continues to be made in both areas: it is incremental, and it remains likely that breakthroughs, if they come, will probably arise in unexpected ways in unrelated areas. Nonetheless, the pieces of the jigsaw are mostly in our grasp; if this very promising technology is to develop, however, we will have to see substantial investment in the years to come.

ACKNOWLEDGEMENTS

I would like to acknowledge helpful conversations with Dr. Paul Christensen and Dr. Ashok Shukla, as well as discussions earlier with Dr. Wolf Vielstich and Dr. Theresa Iwasita. I would also like to acknowledge the many recent literature reviews on methanol electro-oxidation and DMFC fabrication; of particular value was the recent article by Professor Arico and co-workers[129] on new commercialisation possibilities for DMFCs.

REFERENCES

1. A. J. Bard, R. Parsons and J. Jordan, 'Standard Potentials in Aqueous Solution', Marcel Dekker, New York (1985).

2. B. D. McNicol, D. A. J. Rand and K. R. Williams, *J. Power Sources*, **83**, 15 (1999).
3. J. B. Goodenough, A. Hamnett, B. J. Kennedy, R. Manoharan and S. A. Weeks, *Electrochim. Acta*, **35**, 199 (1990).
4. A. K. Shukla, P. Stevens, A. Hamnett and J. B. Goodenough, *J. Appl. Electrochem*, **19**, 383 (1989).
5. M. S. Wilson and S. Gottesfeld, *J. Appl. Electrochem.*, **22**, 1 (1992).
6. A. S. Aricò, P. Creti, N. Giordano, V. Antonucci, P. L. Antonucci and A. Chuvilin, *J. Appl. Electrochem.*, **26**, 959 (1996).
7. A. Fisher, J. Jindra and H. Wendt, *J. Appl. Electrochem.*, **28**, 277 (1998).
8. M. P. Hogarth, P. A. Christensen, A. Hamnett and A. K. Shukla, *J. Power Sources*, **69**, 113, 125 (1997).
9. R. Parsons and T. Van der Noot, *J. Electroanal. Chem.*, **257**, 1 (1988).
10. A. Hamnett, *Catal. Today*, **38**, 445 (1997).
11. A. Hamnett, in 'Interfacial Electrochemistry', A. Wiechowski (Ed), Marcel Dekker, New York (1999).
12. V. S. Bagotzky, Yu. B. Vassilyev and O. A. Khazova, *J. Electroanal. Chem.*, **81**, 229 (1977).
13. V. S. Bagotzky and Yu. B. Vassilyev, *Electrochim. Acta*, **12**, 1323 (1967).
14. T. Biegler, *J. Phys. Chem.*, **72**, 1571 (1968).
15. S. Wilhelm, T. Iwasita and W. Vielstich, *J. Electroanal. Chem.*, **238**, 383 (1987).
16. A. Papoutsis, J.-M. Leger and C. Lamy, *J. Electroanal. Chem.*, **234**, 315 (1987).
17. C. Lamy, J. M. Leger, J. Clavilier and R. Parsons, *J. Electroanal. Chem.*, **150**, 71 (1983).
18. S. G. Sun and J. Clavilier, *J. Electroanal. Chem.*, **236**, 95 (1987).
19. H. Kita, Y. Gao, T. Nakato and H. Hattori, *J. Electroanal. Chem.*, **373**, 177 (1994).
20. H. Matsui and T. Hisano, *Bull. Chem. Soc. Jpn.*, **60**, 863 (1987).
21. K. Kunimatsu, *Ber. Bunsen-ges. Phys. Chem.*, **94**, 1025 (1990).
22. J. Villegas and M. J. Weaver, *J. Chem. Phys.*, **101**, 1648 (1994).
23. T. Iwasita and W. Vielstich, *J. Electroanal. Chem.*, **201**, 403 (1986).
24. J. Willsau and J. Heitbaum, *Electrochim. Acta*, **31**, 943 (1986).
25. S. Wilhelm, W. Vielstich, H. W. Buschmann and T. Iwasita, *J. Electroanal. Chem.*, **229**, 377 (1987).
26. T. Iwasita and F. Nart, *J. Electroanal. Chem.*, **317**, 291 (1991).
27. J. Shin and C. Korzeniewski, *J. Phys. Chem.*, **99**, 3419 (1995).
28. P. A. Christensen, A. Hamnett, J. Munk and G. L. Troughton, *J. Electroanal. Chem.*, **370**, 251 (1994).
29. J. Munk, P. A. Christensen, A. Hamnett and E. Skou, *J. Electroanal. Chem.*, **401**, 215 (1996).
30. V. S. Bagotzky and Yu. B. Vassilyev, *Electrochim. Acta*, **12**, 1323 (1967).
31. P. A. Christensen, A. Hamnett and G. L. Troughton, *J. Electroanal. Chem.*, **362**, 207 (1993).
32. H. A. Gasteiger, N. Markovic, P. N. Ross and E. J. Cairns, *J. Phys. Chem.*, **98**, 617 (1994).
33. V. J. Kwasniewski and L. D. Schmidt, *Surf. Sci.*, **274**, 329 (1992).
34. J. B. Day, P.-A. Vuissoz, E. Oldfield, A. Wieckowski and J.-P. Ansermet, *J. Am. Chem. Soc.*, **118**, 13 046 (1996).
35. N. Furuya, S. Motoo and K. Kunimatsu, *J. Electroanal. Chem.*, **239**, 347 (1988).
36. N. A. Hampson, M. J. Willars and B. D. McNicol, *Faraday Trans. Chem. Soc.*, **46**, 2535 (1979).
37. C. McCallum and D. Pletcher, *J. Electroanal. Chem.*, **70**, 277 (1976).
38. L.-W. H. Leung and M. J. Weaver, *Langmuir*, **6**, 223 (1990).
39. J. D. Roth and M. W. Weaver, *J. Electroanal. Chem.*, **307**, 119 (1991).
40. J. Lu and A. Bewick, *J. Electroanal. Chem.*, **270**, 225 (1989).
41. E. P. M. Leiva and M. C. Giordano, *J. Electroanal. Chem.*, **158**, 115 (1983).
42. B. E. Hayden, *Catal. Today*, **38**, 473 (1997).
43. M. Watanabe, Y. Furuuchi and S. Motoo, *J. Electroanal. Chem.*, **191**, 367 (1985).
44. S. A. Campbell and R. Parsons, *J. Chem. Soc. Faraday Trans.*, **88**, 833 (1992).
45. M. M. P. Janssen and J. Moolhuysen, *J. Catal.*, **46**, 289 (1977).
46. A. Hamnett and B. J. Kennedy, *Electrochim. Acta*, **33**, 1613 (1988).
47. E. Herrero, J. M. Feliu and A. Wieckowski, *Langmuir*, **15**, 4944 (1999).
48. T. Iwasita, F. C. Nart and W. Vielstich, *Ber. Bunsen-ges Phys. Chem.*, **94**, 1030 (1990).
49. T. Frelink, W. Visscher and J. A. R. van Veen, *Langmuir*, **12**, 3702 (1996).
50. J. C. Davies, B. E. Hayden and D. J. Pegg, *Surf. Sci.*, **467**, 118 (2000).
51. K. J. Cathro, *Electrochem. Technol.*, **5**, 441 (1967).
52. K. L. Ley, R. Liu, C. Pu, Q. Fan, N. Leyarovska, C. Segre and E. S. Smotkin, *J. Electrochem. Soc.*, **144**, 1543 (1997).
53. A. Hamnett, B. J. Kennedy and S. A. Weeks, *J. Electroanal. Chem.*, **240**, 355 (1988).
54. A. Hamnett, P. Stevens and G. L. Troughton, *Catal. Today*, **7**, 219 (1990).
55. P. Shen, K. Chen and A. C. C. Tseung, *J. Chem. Soc. Faraday Trans.*, **90**, 3089 (1994).
56. A. K. Shukla, M. K. Ravikumar, A. S. Aricó, G. Candiano, V. Antonucci, N. Giordano and A. Hamnett, *J. Appl. Electrochem.*, **25**, 528 (1995).

57. M. Gotz and H. Wendt, *Electrochim. Acta*, **43**, 3637 (1998).
58. K. Y. Chen, Z. Sun and A. C. C. Tseung, *Electrochem. Solid State Lett.*, **3**, 10 (2000).
59. J. H. White and A. F. Sammells, *J. Electrochem. Soc.*, **140**, 2167 (1993).
60. K.-I. Machida, M. Enyo, G.-Y. Adachi and J. Shiokawa, *Bull. Chem. Soc. Jpn.*, **60**, 411 (1987).
61. B. D. McNicol and R. T. Short, *J. Electroanal. Chem.*, **81**, 249 (1977).
62. H. G. Petrow and R. J. Allen, US Patent, 3,992,331 (1976).
63. A. Wiechowski, 'Interfacial Electrochemistry', Marcel Dekker, New York (1999).
64. S. Trassatti and M. M. Jaksić, *Electrochim. Acta*, **45**(25–26), (2000).
65. L. Geines, R. Faure and R. Durand, *Electrochim. Acta*, **44**, 1317 (1998).
66. T. Toda, H. Igarashi, H. Uchida and M. Watanabe, *J. Electrochem. Soc.*, **146**, 3750 (1999); **145**, 4185 (1998).
67. S. Mukerjee, S. Srinavasan, M. P. Soriaga and J. McBreen, *J. Electrochem. Soc.*, **142**, 1409 (1995).
68. S. Mukerjee, S. Srinavasan, M. P. Soriaga and J. McBreen, *J. Phys. Chem.*, **99**, 4577 (1995).
69. S. Strbac and R. R. Adzic, *J. Electroanal. Chem.*, **403**, 169 (1996).
70. S. Strbac and R. R. Adzic, *Electrochim. Acta*, **39**, 983 (1994).
71. G. Briscard, N. Bertrand, P. N. Ross and N. Markovic, *J. Electroanal. Chem.*, **480**, 219 (2000).
72. R. Jasinski, *J. Electrochem. Soc.*, **112**, 526 (1965).
73. S. Y. Ha, J. Park, T. Ohta, G. Kwag and S. Kim, *Electrochem. Solid State Lett.*, **2**, 461 (1999).
74. K. J. Moore, J. T. Fletcher and M. J. Therien, *J. Am. Chem. Soc.*, **121**, 5196 (1999).
75. W. Seeliger and A. Hamnett, *Electrochim. Acta*, **37**, 763 (1992).
76. J. B. Goodenough, *Prog. Solid State Chem.*, **5**, 39 (1971).
77. S. Trassatti, 'Electrodes of Conducting Metal Oxides', Elsevier, Amsterdam, Vols. A, B (1980).
78. H. C. Chiu and A. C. C. Tseung, *Electrochem. Solid State Lett.*, **2**, 379 (1999).
79. L. Gurban, A. Teze and G. Hervé, *C. R. Acad. Sci. II C*, **1**, 397 (1998).
80. J. M. Zen, R. Manoharan and J. B. Goodenough, *J. Appl. Electrochem.*, **22**, 140 (1992).
81. H. N. Cong, K. Abbassi and P. Chartier, *Electrochem. Solid State Lett.*, **3**, 192 (2000).
82. T. Biegler, D. A. J. Rand and R. Woods, *J. Electroanal. Chem.*, **60**, 151 (1975).
83. N. Alonso-Vante and H. Tributsch, *Nature*, **323**, 431 (1986).
84. N. Alonso-Vante, M. Giersig and H. Tributsch, *J. Electrochem. Soc.*, **138** 639 (1991).
85. O. Solorza-Feria, K. Ellmer, M. Giersig and N. Alonso-Vante, *Electrochim. Acta*, **40**, 567 (1995).
86. V. Trapp, P. A. Christensen and A. Hamnett, *J. Chem. Soc. Faraday Trans.*, **92**, 4311 (1996).
87. R. W. Reeve, P. A. Christensen, A. Hamnett, S. A. Haydock and S. C. Roy, *J. Electrochem. Soc.*, **145**, 3463 (1998).
88. R. W. Reeve, P. A. Christensen, A. J. Dickinson, A. Hamnett and K. Scott, *Electrochim. Acta*, **45**, 4237 (2000).
89. J. Brooker, P. A. Christensen, A. Hamnett, R. He and C. A. Paliteiro, *Faraday Disc.*, **94**, 339 (1992).
90. R. G. Egdell, J. B. Goodenough, A. Hamnett and C. C. Naish, *J. Chem. Soc. Faraday Trans.*, **79**, 893 (1983).
91. A. Damjanovic, *J. Electroanal. Chem.*, **355**, 57 (1993).
92. A. J. Appleby *J. Electroanal. Chem.*, **357**, 117 (1993).
93. D. W. Shoesmith, J. S. Betteridge and W. H. Hocking, *J. Electroanal. Chem.*, **406**, 69 (1998).
94. N. M. Markovic and P. N. Ross, *J. Electrochem. Soc.*, **141**, 2590 (1995).
95. C. Z. Deng and M. J. Dignam, *J. Electrochem. Soc.*, **145**, 3513 (1998).
96. E. R. Vago and E. J. Calvo, *J. Chem. Soc. Faraday Trans.*, **91**, 2323 (1995).
97. J. O'M. Bockris and R. Aldu, *J. Electroanal. Chem.*, **448**, 189 (1998).
98. C. C. Cheng and T. C. Wen, *J. Appl. Electrochem.*, **27**, 355 (1997).
99. C. C. Cheng and T. C. Wen, *J. Electrochem. Soc.*, **143**, 1485 (1996).
100. D. T. Sawyer, 'Oxygen Chemistry', Oxford University Press, Oxford (1991).
101. C. A. Paliteiro, D.Phil. Thesis, Oxford University (1994).
102. K.-L. Hsueh, D.-T. Chin and S. Srinivasan, *J. Electroanal. Chem.*, **153**, 79 (1983).
103. S. Strbac, N. A. Anastesijevic and R. R. Adzic, *J. Electroanal. Chem.*, **323**, 179 (1992).
104. N. M. Markovic, I. M. Tidswell and P. N. Ross, *Langmuir*, **10**, 435 (1994).
105. J. O'M. Bockris and R. Aldu, *J. Electroanal. Chem.*, **448**, 189 (1998).
106. A. B. Anderson and T. V. Albu, *Electrochem. Comm.*, **1**, 203 (1999).
107. S. Mukerjee and S. Srinivasan, *J. Electroanal. Chem.*, **357**, 201 (1993).
108. S. Mukerjee, S. Srinivasan, M. P. Soriaga and J. McBreen, *J. Electrochem. Soc.*, **142**, 1409 (1995).
109. S. Gottesfeld, M. T. Paffett and A. Redondo, *J. Electroanal. Chem.*, **205**, 163 (1986).
110. B. C. Beard and P. N. Ross, *J. Electrochem. Soc.*, **137**, 3368 (1970).
111. K. T. Kim, J. T. Whang, Y. G. Kim and J. S. Chung, *J. Electrochem. Soc.*, **140**, 31 (1993).
112. M. Neergat, A. K. Shukla and K. S. Gandhi, *J. Appl. Electrochem.*, **31**, In press (2002).
113. J. Prakash and H. Joachim, *Electrochim. Acta*, **45**, 2289 (2000).

114. M. Hayashi, T. Hyodo, N. Miura and N. Yamazoe, *Electrochemistry*, **68**, 112 (2000).
115. T. Hyodo, M. Hayashi, S. Mitsutake, N. Miura and N. Yamazoe, *J. Ceram. Soc. Jpn.*, **105**, 412 (1997).
116. J. Ortiz and J. L. Gautier, *J. Electroanal. Chem.*, **391**, 111 (1995).
117. J. B. Goodenough, R. Manoharan and J. M. Zen, *J. Appl. Electrochem.*, **22**, 140 (1992).
118. A. K. Shukla, A. M. Kannan, M. S. Hedge and J. Gopalakrishnan, *J. Power Sources*, **35**, 163 (1991).
119. A. Widelov, N. M. Markovic and P. N. Ross, *J. Electrochem. Soc.*, **143**, 3504 (1996).
120. J. P. Collman, P. Denisevich, Y. Konai, M. Marroco, C. Koval and F. C. Anson, *J. Am. Chem. Soc.*, **102**, 6027 (1980).
121. K. Oyaizu, A. Haryono, J. Natori, H. Shinoda and E. Tuschida, *Bull. Chem. Soc. Jpn.*, **73**, 1153 (2000).
122. B. Schubert, E. Gocke, R. Schollhorn, N. Alonso-Vante and H. Tributsch, *Electrochim. Acta*, **41**, 1471 (1996).
123. C. Fischer, N. Alonso-Vante, S. Fiechter and H. Tributsch, *J. Appl. Chem.*, **25**, 1004 (1995).
124. O. Solorza-Feria, K. Ellmer, M. Giersig and N. Alonso-Vante, *Electrochim. Acta*, **39**, 1647 (1994).
125. N. Alonso-Vante, P. Borthieu, M. Fieber-Erdmann, H.-H. Streblow and E. Holub-Krappe, *Electrochim. Acta*, **45**, 4227 (2000).
126. O. Solorza-Feria, S. Citalem-Cigarroa, R. Rivera-Norriega and S. M. Fernandez-Valverde, *Electrochem. Commun.*, **1**, 585 (1999).
127. M. Pattati, R. H. Castellanos, P. J. Sebastian and X. Matthew, *Electrochem. Solid State Lett.*, **3**, 431 (2000).
128. H. Tributsch, 'ECS'99' Meeting, Portorož, Slovenia, September (1999).
129. A. S. Arico, S. Srinavasan and V. Antonucci, *Fuel Cells*, **1**, 1 (2001).
130. E. Peled, T. Duvdevani, A. Aharon and A. Melman, *Electrochem. Solid State Lett.*, **3**, 525 (2000).
131. S. Hietala, M. Koel, E. Skou, M. Elomaa and F. Sundholm, *J. Mater. Chem.*, **8**, 1127 (1998).
132. B. Bauer, D. J. Jones, J. Roziere, L. Tchicaya, G. Alberti, M. Casciola, L. Massinelli, A. Peraio, S. Besse and E. Ramunni, *J. New Mater. Electrochem. Syst.*, **3**, 93 (2000).
133. I. G. K. Andersen, E. K. Andersen, N. Knudsen and E. Skou, *Solid State Ionics*, **46**, 189 (1991).
134. A. S. Aricò, P. L. Antonucci, N. Giordano and V. Antonucci, *Mater. Lett.*, **24**, 399 (1995).
135. J. Wang, S. Wasmus and R. F. Savinell, *J. Electrochem. Soc.*, **142**, 4218 (1995).
136. P. Staiti, F. Lufrano, A. S. Arico, E. Passelacqua and V. Antonucci, *J. Membr. Sci.*, **188**, 71 (2001).
137. C. Yang, S. Srinavasan, A. S. Arico, P. Creti, V. Baglio and V. Antonucci, *Electrochem. Solid State Lett.*, **4**, A31 (2001).
138. A. S. Aricò, P. Creti, P. L. Antonucci and V. Antonucci, *Electrochem. Solid State Lett.*, **1**, 66 (1998).
139. P. Staiti, A. S. Arico, V. Baglio, F. Lufrano, E. Pasalacqua and V. Antonucci, *Solid State Ionics*, **145**, 201 (2001).
140. A. K. Shukla, P. A. Christensen, A. Hamnett and M. P. Hogarth, *J. Power Sources*, **55**, 87 (1995).
141. X. Ren, P. Zelenay, S. Thomas, J. Davey and S. Gottesfeld, *J. Power Sources*, **86**, 111 (2000).
142. A. S. Aricò, P. Creti, E. Modica, G. Monforte, V. Baglio and V. Antonucci, *Electrochim. Acta*, **45**, 4319 (2000).
143. M. K. Ravikumar and A. K. Shukla, *J. Electrochem. Soc.*, **143**, 2601 (1996).
144. K. Scott, W. M. Taama, P. Argyropoulos and K. Sundmacher, *J. Power Sources*, **83**, 204 (1999).
145. D. Buttin, M. Dupont, M. Straumann, R. Gille, J.-C. Dubois, R. Ornelas, G. P. Flebo, E. Ramunni, V. Antonucci, A. S. Arico, P. Creti, E. Modica, M. Pham-Thi and J. P. Ganne, *J. Appl. Electrochem.*, **31**, 275 (2001).
146. S. R. Narayanan, T. I. Valdez and F. Clara, 'Design and Development of Miniature Direct Methanol Fuel Cell Power Sources for Cellular Phone Applications', Fuel Cell Seminar Abstracts, Portland, OR, Oct. 30–Nov. 2, p. 795 (2000).
147. J. Bostaph, R. Koripella, A. Fisher, D. Zindel, J. Hallmark and J. Neutzler, '199th Meeting of the Electrochemical Society', Washington, DC, March 25–30, Abstract 97 (2001).

Chapter 19

Other direct-alcohol fuel cells

C. Lamy and E. M. Belgsir
Université de Poitiers, France

1 INTRODUCTION

Although research and development efforts have been undertaken to promote new clean sources of energy, for at least the next two decades, most of the fuel/energy technologies will continue to use the fossil fuel-based processes and the total emissions of greenhouse gases will continue to increase. For that reason, the potential development of fuel cells (FCs) will allow us to use renewable and environmentally friendly primary fuels, instead of fossil hydrocarbons. Hydrogen – the most abundant element in the nature – is the cleanest fuel, since its combustion only gives water, although its production from fossil fuels may also produce side-products, such as carbon dioxide. However, technological drawbacks concerning the storage and fuelling and the distribution infrastructure or the in situ production (e.g., by steam reforming; the production of hydrogen based on the steam reforming process has been proved successful, but it also produces carbon monoxide, which is a regulated pollutant that must be eliminated from the reformate gas, otherwise the performance and life of the fuel cell catalysts are drastically reduced) are driving research and development towards the direct liquid fuel cells (DLFCs), which are gaining ground as environmentally friendly technology.

During the 1980s, direct alcohol fuel cells (DAFCs) were overlooked because of their poor efficiency. However, important progress was made during the 1990s and as regards the predicted efficiencies alcohols may be promising candidates for alternative fuels to gaseous fuels, particularly for micro fuel cells (MFCs) able to power portable electronic devices. Among the several possible systems, we shall focus on the efforts involved in the direct combustion fuel cell (DCFC), using oxygenated organic compounds coming from biomass resources.

2 PRINCIPLE OF A DIRECT COMBUSTION FUEL CELL

The basic concepts of fuel cells have been discussed previously. A synoptic scheme of a proton exchange membrane fuel cell (PEMFC) operating on a $C_{x \geq 2}H_y O_z$ fuel type is presented in Figure 1.

The difference between such an ideal combustion process leading to CO_2 and the actual state-of-the-art depends fundamentally on the *electrocatalytic activation* of the anodic reaction. Far from the theory (cf. Section 4.1), it still seems virtually impossible to achieve, at reasonable overpotentials and low temperatures ($<100\,°C$), the total oxidation of fuels containing more than one carbon atom. The apparent reasons are related not only to the presence of the C–C bond, but also to the crucial activation of intermediate species, such as oxygenated species (the total oxidation of an alcohol fuel into CO_2 needs an extra-molecular oxygen atom. In aqueous medium, these oxygen atoms are provided by the activation of an H_2O molecule. However, most of the noble metal catalysts are not able to adsorb oxygenated species at low potentials and must be structurally modified by the addition of promoting elements to improve their catalytic properties). Thus, the development of a comprehensive mechanistic model would take into account the electrosorption and the electroactivation factors.

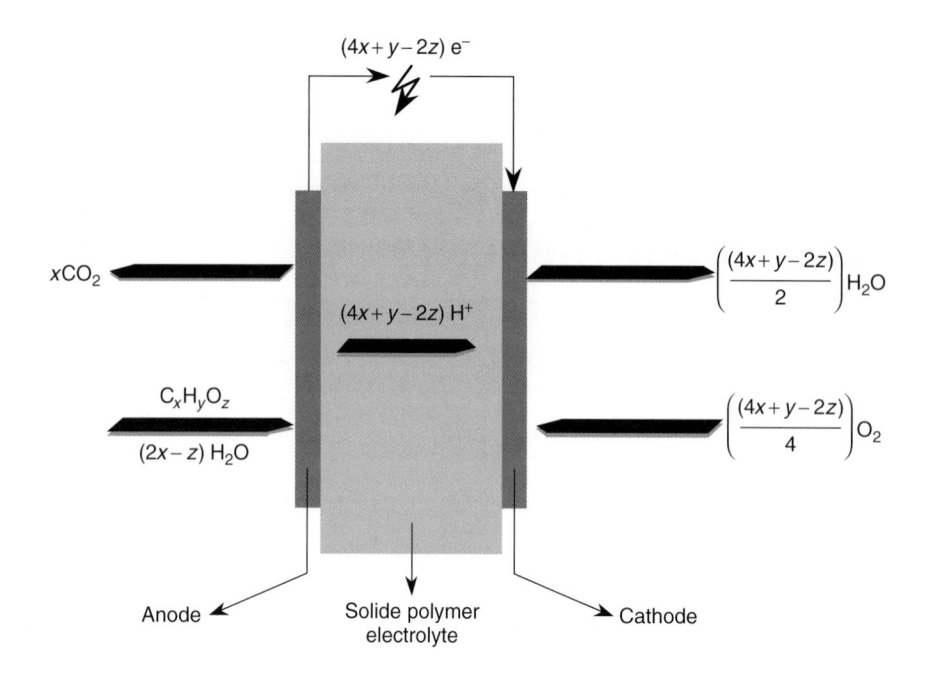

Figure 1. Sandwich configuration and reaction scheme of a PEMFC operating on a $C_{x\geq 2}H_yO_z$ fuel.

3 ARE DCFCs ENVIRONMENTALLY FRIENDLY?

The origin of the fuel is critical with regard to the environmental impact. The actual atmospheric carbon dioxide is out of balance because of the anthropogenic activity using abiotic resources and transferring the mineral carbon to the atmosphere (Figure 2).

However, the synergistic association of low emission technology and environmentally friendly production of fuels may promote a biomass fuel economy. The use of alcohols coming from biomass resources represents an opportunity to contribute to the clean air gains because they constitute a link to the natural carbon cycle (Figure 3). CO_2 is absorbed by plants during photosynthetic processes

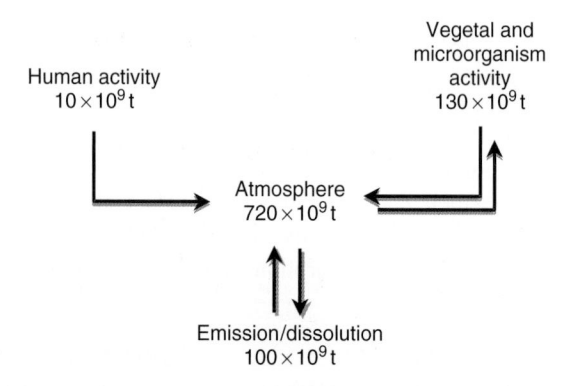

Figure 2. Environmental carbon mass balance.

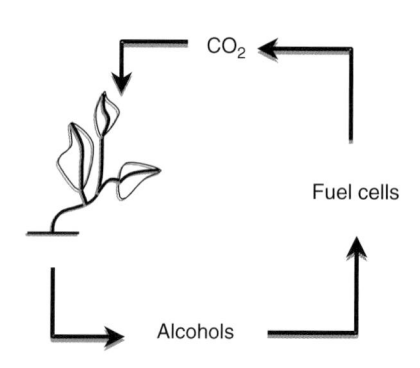

Figure 3. Carbon cycle.

to produce simple sugars which can be used or stored as starch, which is the raw material for alcohol production (methanol, ethanol, etc.)

Other aliphatic alcohols may be alternative fuels for DLFCs even if most of them are costly and have a limited supply. However, some alcohols (ethanol, ethylene glycol, glycerol, etc.) may be available and mass produced if a major scientific breakthrough is achieved in the development of new active and stable anode electrocatalysts.

4 ELECTROCATALYTIC OXIDATION OF OXYGENATED ORGANIC COMPOUNDS

In a DCFC the total electrooxidation to CO_2 of an aliphatic oxygenated compound C_xH_yO containing one oxygen atom (monoalcohols, aldehydes, ketones, ethers, etc.) involves the participation of water (H_2O) or of its adsorbed residue (OH) provided by the cathodic reaction (electroreduction of dioxygen).

The overall electro-oxidation reaction in acidic medium to reject the carbon dioxide produced can thus be written as follows:

$$C_xH_yO + (2x - 1)H_2O \longrightarrow xCO_2 + nH^+ + ne^- \quad (1)$$

with $n = 4x + y - 2$. Such an anodic reaction is very complicated from a kinetics point of view since it involves multielectron transfers and the presence of different adsorbed intermediates and several reaction products and by-products. However from thermodynamic data it is easy to calculate the reversible anode potential, the electromotive force under standard conditions, the theoretical efficiency and the energy density.

4.1 Thermodynamic data

According to reaction (1) the standard Gibbs energy change $-\Delta G_1^\circ$, allowing one to calculate the standard anode potential $E_1^\circ = -\Delta G_1^\circ/nF$, can be evaluated from the standard energy of formation ΔG_i^f of reactant i:

$$-\Delta G_1^\circ = x\Delta G_{CO_2}^f - \Delta G_{C_xH_yO}^f - (2x - 1)\Delta G_{H_2O}^f \quad (2)$$

In the cathodic compartment the electroreduction of oxygen occurs as follows:

$$\tfrac{1}{2}O_2 + 2H^+ + 2e^- \longrightarrow H_2O \quad (3)$$

with $\Delta G_2^\circ = \Delta G_{H_2O}^f = -237.1 \text{ kJ mol}^{-1}$, leading to a standard cathodic potential E_2°:

$$E_2^\circ = -\frac{\Delta G_2^\circ}{2F} = \frac{237.1 \times 10^3}{2 \times 96485}$$
$$= 1.229 \text{ V vs. SHE}$$

In the fuel cell the electrical balance corresponds to the complete combustion of the organic compound in presence of oxygen, as follows:

$$C_xH_yO + \left(x + \frac{y}{4} - \frac{1}{2}\right)O_2 \longrightarrow xCO_2 + \frac{y}{2}H_2O \quad (4)$$

with $\Delta G^\circ = (2x + y/2 - 1)\Delta G_2^\circ - \Delta G_1^\circ = x\Delta G_{CO_2}^f + y/2\Delta G_{H_2O}^f - \Delta G_{C_xH_yO}^f$, leading to the equilibrium standard e.m.f.:

$$E_{eq}^\circ = -\frac{\Delta G^\circ}{nF}$$
$$= -\frac{\Delta G_2^\circ}{2F} + \frac{\Delta G_1^\circ}{nF} = E_2^\circ - E_1^\circ$$

Then it is possible to evaluate the specific energy W_e in kWh kg^{-1}:

$$W_e = \frac{-\Delta G^\circ}{3600M}$$

with M the molecular mass of the compound, and knowing the enthalpy change ΔH° from thermodynamic data:

$$\Delta H^\circ = \left(2x + \frac{y}{2} - 1\right)\Delta H_2^\circ - \Delta H_1^\circ$$
$$= x\Delta H_{CO_2}^f + \frac{y}{2}\Delta H_{H_2O}^f - \Delta H_{C_xH_yO}^f$$

one may calculate the reversible energy efficiency under standard conditions:

$$\varepsilon_{rev} = \frac{\Delta G^\circ}{\Delta H^\circ}$$

As an example, the following calculations can be made in the case of the electro-oxidation of 1-propanol:

$$C_3H_7OH + 5H_2O \longrightarrow 3CO_2 + 18H^+ + 18e^- \quad (1a)$$
$$-\Delta G_1^\circ = 3\Delta G_{CO_2}^f - \Delta G_{C_3H_7OH}^f - 5\Delta G_{H_2O}^f \quad (2a)$$

$$-\Delta G_1^\circ = -3 \times 394.4 + 168.4$$
$$+ 5 \times 237.1 = 171 \text{ kJ mol}^{-1}$$
$$E_1^\circ = -\frac{\Delta G_1^\circ}{18F} = \frac{171 \times 10^3}{18 \times 96485}$$
$$= 0.098 \text{ V vs. SHE} \quad (2b)$$

$$C_3H_7OH + \frac{9}{2}O_2 \longrightarrow 3CO_2 + 4H_2O \quad (4a)$$
$$\Delta G^\circ = 9\Delta G_2^\circ - \Delta G_1^\circ = 3\Delta G_{CO_2}^f$$
$$+ 4\Delta G_{H_2O}^f - \Delta G_{C_3H_7OH}^f \quad (4b)$$

$$\Delta G^\circ = -3 \times 394.4 - 4 \times 237.1 + 168.4$$
$$= -1963 \text{ kJ mol}^{-1}$$

The standard electromotive force is thus

$$E^\circ_{eq} = -\frac{\Delta G^\circ}{18F} = \frac{1963 \times 10^3}{18 \times 96485} = \frac{237.1 \times 10^3}{2 \times 96485}$$

$$-\frac{171 \times 10^3}{18 \times 96485} = 1.229 - 0.098 = 1.131\,\text{V}$$

and the specific energy is

$$W_e = \frac{1963}{3600 \times 60} = 9.09\,\text{kWh kg}^{-1}$$

The enthalpy change of reaction (4) is

$$\Delta H^\circ = -3 \times 395.5 - 4 \times 285.8 + 302.6$$
$$= -2027\,\text{kJ mol}^{-1}$$

so that the reversible energy efficiency is

$$\varepsilon_{rev} = \frac{\Delta G^\circ}{\Delta H^\circ} = \frac{1963}{2027} = 0.968$$

Similar calculations can be made with other oxygenated fuels, including polyols (ethylene glycol), propargyl alcohol, ethers and polyethers [methoxymethane, CH_3OCH_3; methoxyethane, $CH_3OC_2H_5$; ethoxyethane, $C_2H_5OC_2H_5$; dimethoxymethane, $(CH_3O)_2CH_2$; trimethoxymethane, $(CH_3O)_3CH$; trioxane, $(CH_2O)_3$]. Table 1 summarizes the results obtained under standard conditions (25 °C, liquid phase).

For all oxygenated compounds listed in Table 1, the electromotive force varies from 1.2 to 1.0 V, which is very similar to that of an hydrogen/oxygen fuel cell (E°_{eq} = 1.23 V). The energy density varies between 50 and 100% of that of gasoline (10–11 kWh kg^{-1}) so that these compounds are good alternative fuels to hydrocarbons. Furthermore, the reversible energy efficiency ε_{rev} is close to 1, whereas that of the H_2/O_2 fuel cell is 0.83 at 25 °C.

4.2 Kinetic problems

The electrooxidation of aliphatic oxygenated compounds, even for the simplest one, methanol, involves the transfer of many electrons ($n = 6$ for methanol). The reaction mechanism is therefore complex, the oxidation reaction occurring through many successive and parallel paths and involving many adsorbed intermediates and by-products. The oxidation reaction needs a convenient electrocatalyst to increase the reaction rate and to modify the reaction pathway in order to reach the final step more rapidly, i.e. the production of carbon dioxide. The relative slowness of the reaction, and the difficulty of breaking the C–C bond at low temperatures (25–80 °C) lead to high anodic overvoltages η_a, which will greatly reduce the operating cell voltage (Figure 4).

In most cases η_a is at least 0.5 V for a reasonable current density (100 mA cm^{-2}) so that the cell voltage, including an overvoltage η_c = 0.3 V to 0.4 V for the cathodic reaction, will be of the order of 0.3 V, and the voltage efficiency will be $\varepsilon_E = 0.3/1.2 = 0.25$ under operating conditions. Such a drawback of the direct-alcohol fuel cell can only be removed by improving the kinetics of the electrooxidation of the fuel. This requires a relatively good knowledge of the reaction mechanisms, particularly of the rate-determining step, and a search for electrode materials (Pt/X binary and Pt/X/Y ternary electrocatalysts) with enhanced catalytic properties.

4.3 Reaction mechanisms

A detailed description of the reaction mechanism of complex multielectron transfer reactions, such as the oxidation of alcohols, needs a good knowledge of the adsorbed species and intermediate products, and a good description of the reaction paths and the rate-determining step. The elucidation of the reaction mechanisms therefore needs the simultaneous use of pure electrochemical methods (e.g. cyclic voltammetry, rotating disc electrodes, coulometry, etc.) and of physicochemical methods, such as in situ spectroscopic methods in the infrared[1] and UV/visible[2] range to identify the adsorbed intermediates, and on-line chromatographic[3] or mass spectrometric[4] techniques to analyze quantitatively the reaction products and by-products.

Table 1. Thermodynamic data associated with the electrochemical oxidation of some alcohols (under standard conditions).

Fuel	ΔG°_1 (kJ mol^{-1})	E°_1 (V vs. SHE)	ΔG° (kJ mol^{-1})	E°_{cell} (V)	W_e (kWh kg^{-1})	ΔH° (kJ mol^{-1})	ε_{rev}
CH_3OH	−9.3	0.016	−702	1.213	6.09	−726	0.967
C_2H_5OH	−97.3	0.084	−1325	1.145	8.00	−1367	0.969
C_3H_7OH	−171	0.098	−1963	1.131	9.09	−2027	0.968
$1-C_4H_9OH$	−409	0.177	−2436	1.052	9.14	−2676	0.910

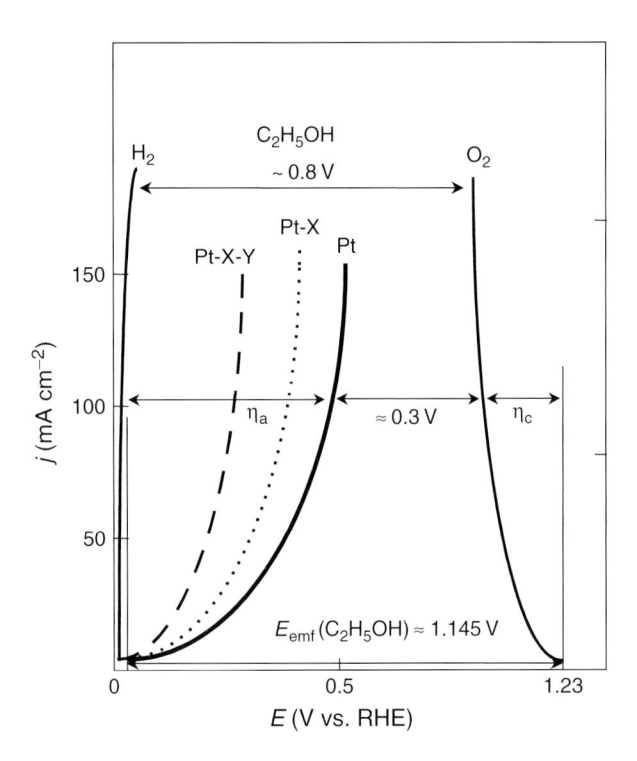

Figure 4. Current density j vs. potential E curves of an ethanol/oxygen fuel cell, compared with those of a hydrogen/oxygen fuel cell. Pt-X and Pt-X-Y stand for a binary and a ternary electrocatalyst, respectively.

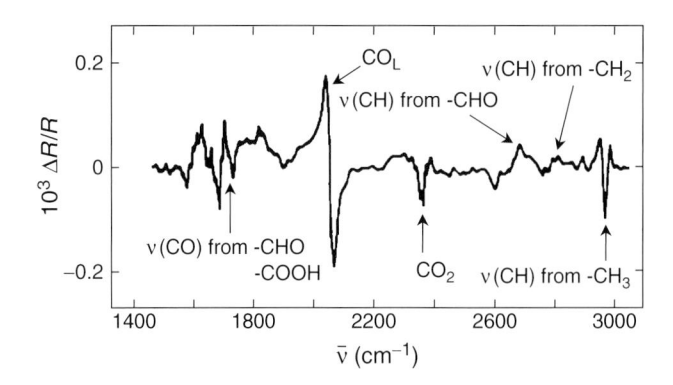

Figure 5. Infrared reflectance spectrum of the adsorbates resulting from the chemisorption of 0.1 M C_2H_5OH on a platinum polycrystalline electrode in acidic medium (0.5 M $HClO_4$, room temperature).

In the case of the electro-oxidation of the simplest alcohol, i.e. methanol, the reaction mechanism has been thoroughly investigated over the last four decades and it is now well established on platinum-based electrodes.[5] The main fact is the occurrence of adsorbed carbon monoxide, linearly and bridge-bonded to the catalytic surface (Pt-based electrodes), which blocks the active sites of the electrocatalysts. This poisoning intermediate, coming from the dissociative chemisorption of the alcohol, is also observed in the adsorption of several oxygenated aliphatic compounds, as is discussed below.

4.3.1 Aliphatic monoalcohols

The reaction mechanism for the electrooxidation of oxygenated aliphatic compounds with more than one carbon atom is much more complicated, since the number of electrons involved increases greatly, and the activation of the C–C bond is more difficult, particularly at low temperatures (20–120 °C).

The electrooxidation of ethanol has been investigated in detail over several years on different catalytic anodes (Pt, Rh, Ir, Au, Pt/X with X = Ru, Sn, Mo, etc., dispersed Pt-based electrodes). This reaction involves 12 electrons per molecule, as follows:

$$C_2H_5OH + 3H_2O \longrightarrow 2CO_2 + 12^+ + 12e^-$$

On a platinum electrode several adsorbed species and intermediate products were identified both by spectroscopic techniques in the infrared range[6] and by analytical techniques such as gas chromatography (GC) and liquid chromatography (HPLC)[7] or differential electrochemical mass spectrometry (DEMS).[4] The main adsorbed species resulting from the dissociative chemisorption of the molecule is linearly bonded CO_L, which behaves as a poisoning species (Figure 5).

Other adsorbed species and reaction intermediates such as acetaldehyde and acetic acid were identified by IR reflectance spectroscopy and liquid chromatography.[4,7,8] On most electrodes the main reaction products in acidic medium are acetaldehyde and acetic acid, which is a secondary product[9,10] (Schemes 1 and 2).

Further oxidation needs the breaking of the C–C bond, which is relatively difficult at room temperature with the usual electrocatalysts. However, on Pt electrodes together with the presence of adsorbed CO, some traces of CO_2 have been found by IR spectroscopy,[6] by DEMS[4] or by GC.[7] The dissociative electrosorption of ethanol in the so-called "hydrogen region" can lead to two different species via the adsorbed acetyl (Scheme 3).

Depending on the potential value, the methyl and CO moieties will desorb as methane and CO_2, respectively. Methane was identified by DEMS[4] at low potential (<0.1 V vs. RHE) and the linearly-bonded CO is known to be oxidized at potentials >0.6 V vs. RHE.

The electrooxidation of 1-propanol and 2-propanol on Pt and Pt/Ru electrodes has also been considered in some detail.[11–13] Using DEMS and Fourier transform infrared (FTIR) spectroscopy, Pastor et al.[12] were able to detect

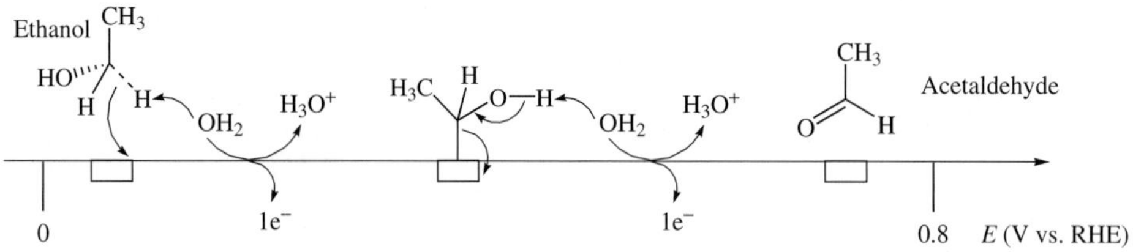

Scheme 1

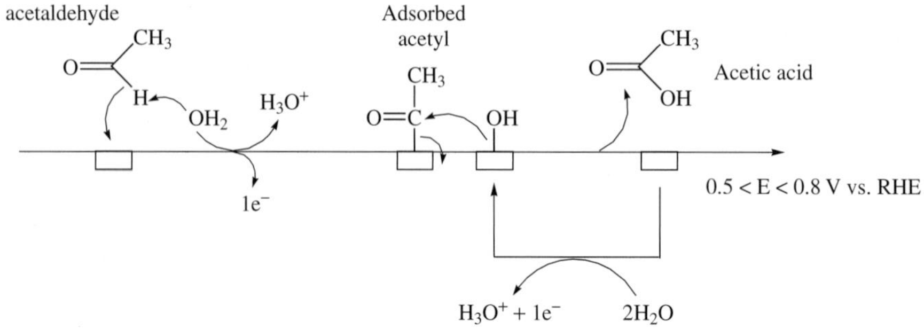

Scheme 2

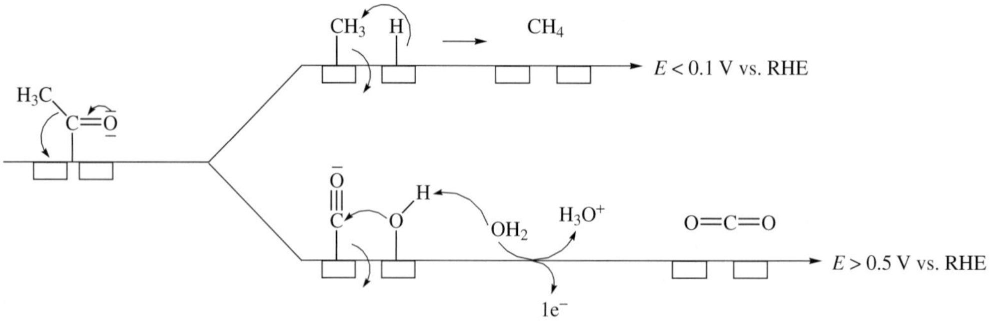

Scheme 3

the presence of propanal, propionic acid and CO_2 as the reaction products of the oxidation of 1-propanol on a Pt electrode. Some linearly-bonded CO was also detected at the electrode surface, in agreement with previous results.[11] On the other hand, using an electrodeposited $Pt_{0.89}Ru_{0.11}$ electrode, the same group showed by FTIR spectroscopy that the electrooxidation of 1- and 2-propanol is more complete, leading mainly to CO_2.[13] This proves again the possibility of cleaving the C–C bond at room temperature during the electrooxidation of alcohols.

1-Propanol and 2-propanol have also been investigated as fuels for direct-alcohol fuel cells under realistic fuel cell conditions.[14] The main products were propanal from 1-propanol and acetone from 2-propanol. Carbon dioxide was detected only in the case of the oxidation of 1-propanol.

The performances of the fuel cells operating on aliphatic alcohols show that ethanol is the most promising candidate among the aliphatic alcohols.

4.3.2 Aliphatic polyols

The oxidation of aliphatic diols, particularly of ethylene glycol, has also been studied in some detail.[15] In this case the overall oxidation reaction involves 10 electrons per molecule:

$$HOCH_2CH_2OH + 2H_2O \longrightarrow 2CO_2 + 10H^+ + 10e^-$$

However, partial oxidation usually occurs leading to several two-carbon intermediates such as glycolaldehyde, glyoxal,

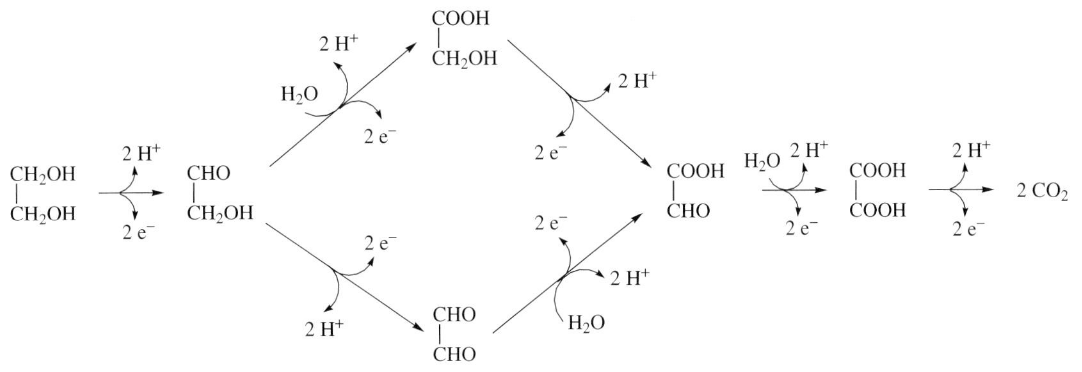

Scheme 4

glycolic acid, glyoxylic acid and oxalic acid, as shown by EMIR and FTIR spectroscopies[16] (Scheme 4).

In addition, some cleavage of the C–C bond occurs, leading to the formation of CO, which acts as a poisoning species of the platinum sites.[17] Both linearly and bridge-bonded CO were observed, the relative proportion of each adsorbed species depending strongly on the pH.[17]

Similarly, the electrocatalytic oxidation of 1,2-propanediol (1,2-PD) and 1,3-propanediol (1,3-PD) on a platinum electrode proceeds through several parallel steps involving many intermediate products, e.g. C_3 molecules, such as 3-hydroxypropanal, lactic acid, pyruvic acid and malonic acid in the case of 1,3-PD, and hydroxyacetone and lactic acid in the case of 1,2 PD.[18] As for ethylene glycol, IR reflectance spectroscopy shows clearly the presence of adsorbed CO, as a poisoning species resulting from the dissociation of the molecule, and CO_2 as a final reaction product. Some C_2 and C_1 molecules were also observed, such as acetic acid and formic acid.

The C_3 aliphatic triol glycerol has also been considered as a convenient fuel for fuel cells.[19] Very few studies have concerned the reaction mechanism of the electrooxidation of glycerol on noble metal electrodes (Pt and Au).[20–22] On a Pt electrode in acidic medium, the main reaction products are glyceraldehyde, glyceric acid, glycolic acid, oxalic acid and formic acid,[22] depending on the electrode potential. The reaction mechanisms are very complex and have still not yet been elucidated.

4.4 Catalytic role of the electrode material

The electrooxidation of complex molecules, such as aliphatic alcohols, needs the molecule to be activated by the electrode material. This is the field of electrocatalysis in which the nature and structure of the catalytic anode play a key role in orientating the reaction pathway in order to achieve complete oxidation of the alcohol to CO_2. This is the only way to recover all the energy from the molecule in terms of specific energy (from 6 to $9\,kWh\,kg^{-1}$).

Platinum is the basic electrode material in acidic medium, owing to its activity and stability, whereas gold is a good electrocatalyst only in alkaline medium. However, the oxidation reaction in alkaline medium stops at some intermediate steps, producing, e.g., carboxylates. Hence only acidic media are now considered in order to reject carbon dioxide from the fuel cell, although some years ago alkaline media were preferred for direct-glycol or direct-glycerol fuel cells, because the reactivity of these polyols is much higher in alkaline than in acidic media.

With all the aliphatic alcohols considered here, the chemisorption step always produces carbon monoxide, mainly CO_L, linearly-bonded to the electrode surface, leading to the poisoning of the catalytic sites.

Therefore, all efforts made during the last one to two decades have tended to focus on the development of bimetallic and trimetallic platinum-based electrocatalysts, in order to reduce the amount of adsorbed CO, or even to suppress its formation. Many metals have been considered for modifying the activity of platinum catalysts, but only a few of them (Ru, Sn, Mo, etc.) lead to improved performances.

The most studied bimetallic electrocatalyst is Pt/Ru, which enhances greatly the rate of oxidation of many alcohols (methanol, ethanol, propanol, etc.). In the case of the oxidation of methanol, a bifunctional mechanism was proposed by Watanabe and Motoo[23] to explain this synergistic effect. Dissociative chemisorption of the alcohol occurs on the platinum surface, whereas the water molecule is activated on ruthenium providing at low potentials the OH adsorbed species necessary to oxidize the alcohol functional group to CO_2. Electronic effects, i.e. the modification of the adsorption and catalytic properties of platinum by the presence of ruthenium, can also be mentioned.[24] The best atomic surface composition corresponds to $Pt_{0.80}Ru_{0.20}$, taking into account the number of surface sites necessary to accommodate the intermediate species arising from methanol adsorption for Pt sites and water adsorption for Ru sites.[25] Similar conclusions were obtained by Fujiwara et al., who observed that an optimum composition $Pt_{0.80}Ru_{0.20}$ gave an enhanced activity for the electrooxidation of ethanol.[26] However, using porous Pt, $Pt_{0.92}Ru_{0.08}$ and $Pt_{0.85}Ru_{0.15}$ electrodeposited electrodes, Ianiello et al. showed by cyclic voltammetry and DEMS that the addition of Ru to Pt decreases its electrocatalytic activity towards ethanol oxidation owing to the inactivity of Ru towards the adsorption of ethanol and the cleavage of the C–C bond.[27] However, for $Pt_{0.85}Ru_{0.15}$ higher activity is observed for the oxidation of the $-CH_2OH$ adsorbed residue from ethanol leading to CO_2, whereas for the oxidation of the $-CH_3$ residue pure Pt is a better electrocatalyst.

Other bimetallic and trimetallic electrocatalysts have also been investigated for the oxidation of aliphatic alcohols, particularly that of ethanol. In this case Pt/Sn and Pt/Ru/Sn are particularly efficient catalysts, much better than Pt/Ru. A Pt/Sn electrocatalyst directly dispersed on a proton exchange membrane (e.g. Nafion®) allows the oxidation overvoltage to be decreased by ca. 0.2 V and the oxidation current to be more than doubled[28] (Figure 6).

For practical applications in a direct-alcohol fuel cell, the catalytic material needs to be dispersed on a convenient substrate both to stabilize the catalyst nanoparticles and to reduce the amount of precious metals in order to reduce the electrode cost. High surface area materials (up to a few thousand $m^2 g^{-1}$), such as graphite, carbon blacks or carbon powders, are usually used to prepare platinum catalysts with a high dispersion corresponding to nanoparticles of the order of 1–2 nm.[29] Other supports also have be considered, such as the proton exchange membrane itself in a PEMFC, and electron-conducting polymers (polyaniline, polypyrrole, etc.). The electrodeposition of Pt-based catalysts on an ionic membrane can be realized by chemical reduction of a platinum salt, followed by the electroreduction of the second metallic salt.[30] For the oxidation of ethanol in a PEMFC this method allowed us to prepare very active Pt/Sn electrocatalysts.[28] Similarly, the dispersion of Pt/Sn electrocatalysts in a polyaniline film leads to efficient electrocatalysts for the oxidation of ethanol.[31] However, the electron-conducting polymers are not stable at temperatures >80 °C, so they cannot be used in a medium-temperature fuel cell.

5 STATE OF THE ART OF DIRECT COMBUSTION FUEL CELLS

5.1 Ethylene glycol fuel cell

Ethylene glycol has long been considered as a convenient fuel, particularly in alkaline media, where high current densities up to a few $A cm^{-2}$ can be obtained.[32–34] In the late 1970s, Siemens Research Laboratories (in Erlangen, Germany) developed a 225 W ethylene glycol/air fuel cell working in alkaline medium (6 M KOH) with electrolyte circulation.[35] The stack consisted of 52 single cells and was able to deliver 28 V at 4.5 A (125 W nominal power) and 16 V at 14 A (225 W peak power at 40 °C). The anode material consisted of a Pt/Pd/Bi trimetallic catalyst deposited on fibrous nickel as a conducting support with a total metal loading of $6.0 mg cm^{-2}$. The entire system is placed in a housing with the dimensions of a lead acid battery, leading to reasonable specific data: $55 Whr kg^{-1}$ and $5 W kg^{-1}$ (fuel cell + tank weight of 44 kg), $7.2 Wh l^{-1}$ and $6.6 W l^{-1}$ (fuel cell + tank volume of 34 l). However, the use of an alkaline medium prevents the oxidation reaction from proceeding to completion, with glycolate and oxalate as the main reaction products.

More recently, interesting results have been reported on the complete electrooxidation of ethylene glycol on the basis of a new nanoporous proton-conducting membrane.[36] These results show that ethylene glycol is less prone to crossing the membrane than methanol and shows up to 95% fuel utilization (9–10% higher than methanol). However, the energy conversion efficiency of methanol is still higher.

5.2 Ethanol fuel cell (and propanol fuel cell)

Ethanol appears as a very promising fuel for fuel cells, particularly for electric vehicles and portable devices. It can be easily produced from biomass by the fermentation

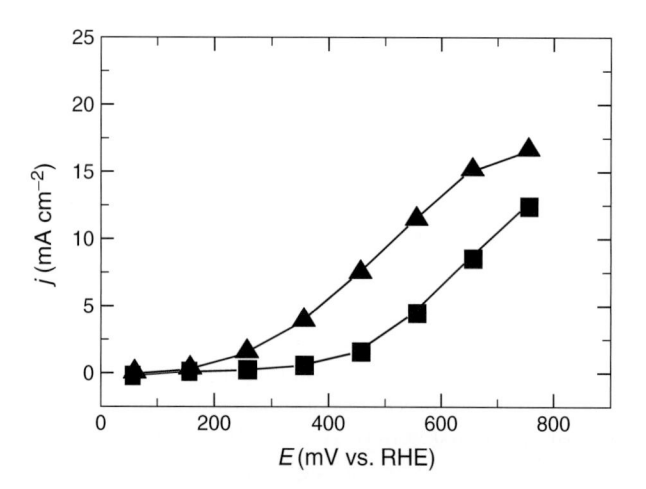

Figure 6. Stationary $j(E)$ curves for the oxidation of ethanol on Pt-Nafion and Pt-Sn-Nafion electrodes. (■) Pt/Nafion®; (▲) Pt/Sn/Nafion®.

of sugar-containing raw materials from agriculture. It has a good specific energy (8 kWh kg^{-1}), between those of methanol (6 kWh kg^{-1}) and gasoline (10–11 kWh kg^{-1}). It can be easily stored and distributed through the gas station network, as in Brazil. Finally, it is a relatively non-toxic fuel. Therefore, some recent investigations have attempted to demonstrate the feasibility of a direct ethanol fuel cell (DEFC).

Delime compared the electrocatalytic activity of different catalysts (Pt, Pt/Ru and Pt/Sn) directly deposited on a Nafion® 117 membrane towards the electrooxidation of ethanol.[37] The results obtained in a single 5 cm^2 PEM fuel cell fed with a liquid ethanol/water mixture at 90 °C are encouraging: at a current density of 60 mA cm^{-2} the cell voltage was 0.55 V with Pt/Sn (i.e. 33 mW cm^{-2} power density), 0.25 V with Pt/Ru and 0.05 V with pure Pt (Figure 7).

However, to improve the kinetics of the electrooxidation of ethanol, higher working temperatures are needed, of the order of 150 °C, which is far belong the stability of Nafion® membranes. Therefore, some attempts have been made using other kinds of protonic membranes.

Wang et al. used a phosphoric acid-doped polybenzimidazole (PBI) membrane, allowing them to work at temperatures as high as 170 °C.[14] They compared the behaviors of four aliphatic alcohols, methanol, ethanol, 1-propanol and 2-propanol (Figure 8).

The anode catalyst consisted of 4 mg cm^{-2} Pt/Ru alloy and the cathode of 4 mg cm^{-2} Pt black. A gaseous alcohol/water mixture was fed into the anode compartment, whereas humidified pure O$_2$ at atmospheric pressure was circulated in the cathode compartment. At a current density

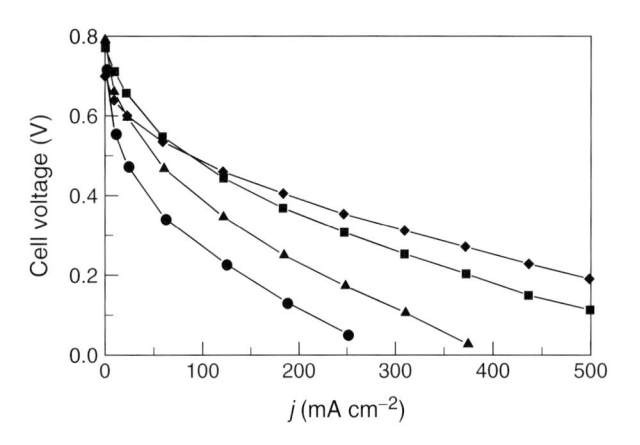

Figure 8. Comparison of the cell performances of a direct-alcohol fuel cell using a PBI membrane doped with H$_3$PO$_4$ (cell temperature 170 °C, 4 mg cm^{-2} Pt/Ru anode catalyst, 4 mg cm^{-2} Pt black cathode catalyst): (♦) methanol; (■) ethanol; (▲) 1-propanol; (●) 2-propanol. (Reproduced from Wang et al.[14] by permission of The Electrochemical Society, Inc.)

of 250 mA cm^{-2} the cell voltages were 0.35, 0.30, 0.17 and 0.05 V for methanol, ethanol, 1-propanol and 2-propanol, respectively. Analysis of the reaction products by on-line mass spectrometry showed that ethanal and CO$_2$ are the main reaction products of ethanol oxidation. An increase in the water to ethanol ratio increases the amount of CO$_2$ produced (from 20 to 32% when the ratio increases from 2 to 5). For the oxidation of 1-propanol the main products are propanal and CO$_2$, whereas for that of 2-propanol the main product is acetone, with small traces of CO$_2$. These results confirm that the primary alcohols are much more electroreactive than the secondary alcohols, in agreement with previous results.[16]

On the other hand, Arico et al. used an 80 μm thick composite membrane prepared by mixing and heat treating a Nafion® ionomer solution (5% w/w, Aldrich) with 3% w/w silica (Aerosil 200, Degussa).[38] The MEA was realized with a (1/1) Pt/Ru/C anode and a Pt/C cathode with a Pt loading of 2 mg cm^{-2} in both electrodes. The anodic compartment was fed with a 1 M ethanol aqueous solution under 4 bar (absolute) and the cathodic compartment with humidified oxygen under 5.5 bar (absolute), and the cell temperature was 145 °C. Under such experimental conditions they obtained a maximum power density of 110 mW cm^{-2} at 0.32 V and 0.35 A cm^{-2} (Figure 9).

A chromatographic analysis of the anode exhaust under stationary conditions (300 mA cm^{-2} at 0.35 V) allowed them to detect CO$_2$ as the main product (yield >90%) with a small amount of acetaldehyde (4%) and unreacted ethanol. These results are at variance with those obtained by Wang et al.,[14] maybe because of the low concentration of ethanol in the liquid feed.

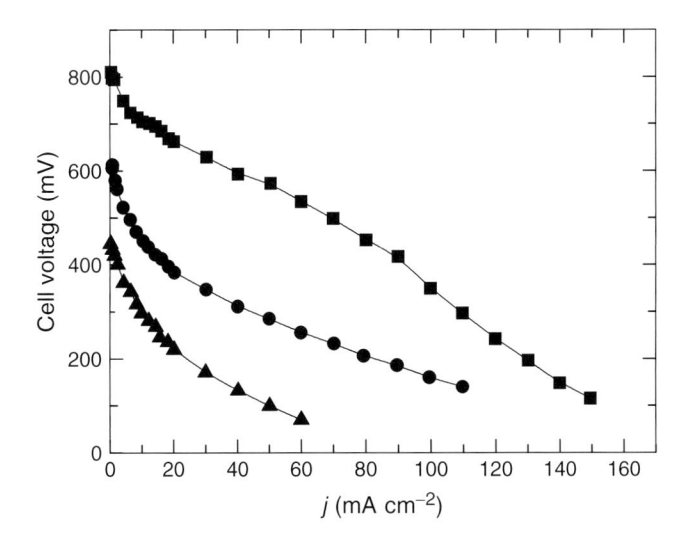

Figure 7. Comparison of the fuel cell characteristics of a direct ethanol fuel cell with Pt/X/Nafion® anodes (liquid ethanol/water mixture feed, cell temperature 90 °C, anode catalysts 1 mg cm^{-2}, oxygen catalyst 1 mg Pt cm^{-2}): (▲) Pt; (●) Pt/Ru; (■) Pt/Sn.

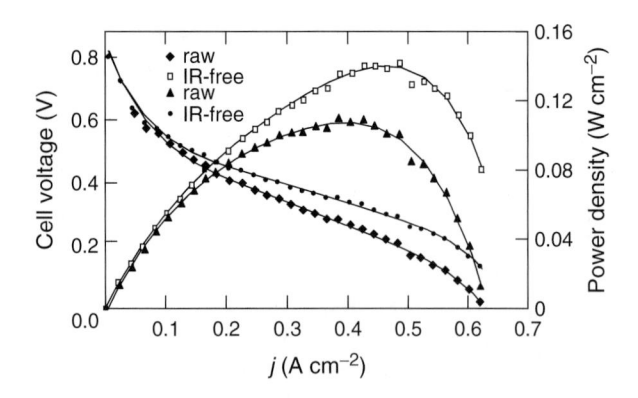

Figure 9. Polarization curves and power density of a direct ethanol fuel cell with a silica/Nafion® membrane (1 M ethanol, cell temperature 145 °C, p_{anode} 4 bar, $p_{cathode}$ 5.5 bar, anode catalyst 2 mg cm^{-2} (1/1) Pt/Ru/C, cathode catalyst 2 mg cm^{-2} Pt. (Reproduced from Arico et al.[38] by permission of The Electrochemical Society, Inc.)

6 PROSPECTS FOR OTHER OXYGENATED ORGANIC FUELS

6.1 Ether and aldehyde fuel cells

Some aliphatic ethers [dimethyl ether (DME)] and polyethers [dimethoxymethane (DMM), trimethoxymethane (TMM)] have also been considered interesting liquid or gaseous fuels, particularly because they can be easily produced from natural gas and they can advantageously replace methanol.[39–42] DMM and TMM, which have a lower vapor pressure than methanol, are oxidized with a high water/fuel ratio (4 and 5 H$_2$O per mole, respectively) and may be used at low concentrations limiting the transmembrane permeation.

Preliminary results with DMM and TMM, using a 5 cm^2 surface area single PEMFC were also obtained in our laboratory.[43] TMM and DMM oxidize relatively well at a Pt electrode, the oxidation potential onset being shifted towards more negative values by ca. 0.2 V compared with methanol (Figure 10).

In a liquid feed anode of a single PEMFC (with Pt/Ru and Pt/Sn electrocatalysts), the behavior of DMM and TMM is very close to that of methanol. Furthermore, the crossover of the Nafion® membrane is reduced compared with methanol, leading to higher cell voltages at low current densities.[39]

DME is also an interesting fuel, as shown by Müller et al.[41] using a PEMFC working at 130 °C. Polarization curves similar to those of methanol are obtained but with an improved Faradaic efficiency, resulting from the fact that DME is not oxidized at the cathode.

Otherwise, in connection with the previous mechanistic considerations, hydrated aldehydes may be convenient fuels for micro-DAFCs. The extra-molecular oxygen atom contribution is provided in solution during the hydration of the aldehyde (Scheme 5).

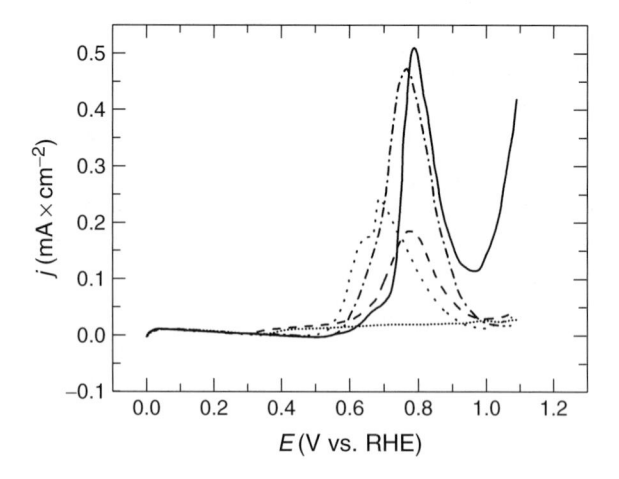

Figure 10. Cyclic voltammograms of a Pt electrode in the presence of several polyethers, formaldehyde and methanol (0.1 M organics in 0.1 M HClO$_4$, 22 °C, 5 mV s^{-1}): (solid line) DMM; (dashed line) TMM; (dotted line) trioxane; (dash dot dash line) methanol; (dash dot dot dash line) HCHO.

Scheme 5

Thus, total combustion will require electrocatalysts designed for the dehydrogenation process and eventually for the rupture of a C–C bond. However, the experimental results show that the adsorptive process involving hydrated aldehydes leads to the formation of linearly bonded CO, which acts as a poison. This means that the reactive species have lost their extra-molecular oxygen before or during the adsorption process (Scheme 6).

Scheme 6

This is particularly true for formaldehyde (HCHO) and glyoxal (OHC–CHO), the fundamental electrochemical properties of which have been extensively studied.[44] Both formaldehyde and glyoxal are oxidized on platinum-based electrodes, but at low potentials the activity of the electrode is drastically decreased owing to the presence of strongly adsorbed CO detected by electromodulated infrared spectroscopy.[44]

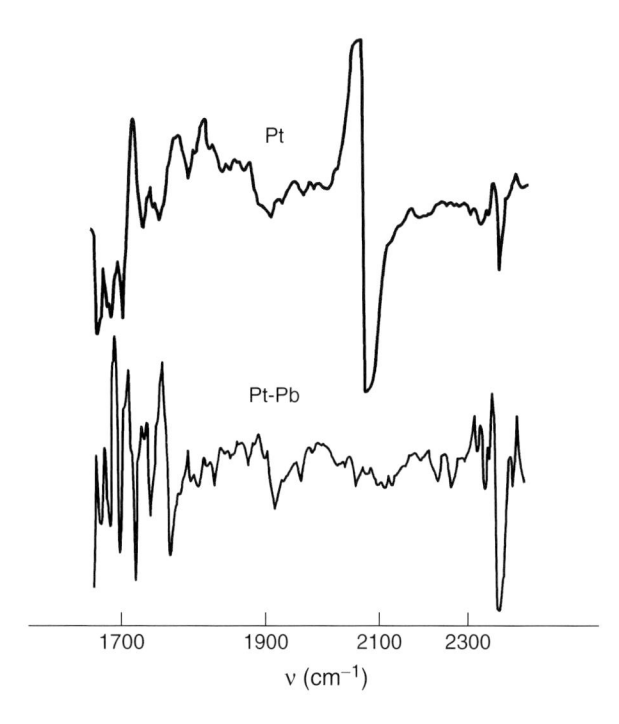

Figure 11. Electrochemically Modulated Infrared Reflectance spectra of the species resulting from the electrocatalytic oxidation of 0.1 M glyoxal on platinum (Pt) and platinum modified by lead adatoms (Pt/Pb) in 0.1 M $HClO_4$ at room temperature.

In the case of hydrated glyoxal, it was shown that the under potential deposition of lead adatoms strongly increased the activity of platinum at low potentials, avoiding the formation of the CO poisoning species (Figure 11).

Otherwise, under acidic conditions the hydrated aldehydes may undergo polymerization leading generally to the corresponding trimer (Scheme 7), which can be handled more safely because of its higher temperature of ebullition. However, its electrochemical reactivity is rather low.

Scheme 7

7 CONCLUSIONS

The development of direct liquid fuel cell technologies which can supply electrical power based on biomass economy will make its presence felt in several market applications. It is expected that these systems could replace most micro-power producing devices over the next 10–20 years. The challenge is not only technological, but above all scientific, because the high overpotential associated with alcohol oxidation (Figure 4) must be thoroughly reduced in order to obtain useful electromotive force and reasonable energy efficiency and power density. The poisoning species (CO) coming from the adsorption process, the permeability of the membrane and the sintering of the electrocatalysts are the main reasons for the high anodic overpotentials. Some topics discussed here contain forward-looking concepts which could contribute to the fundamental understanding of the activity and the development of adapted electrocatalysts.

REFERENCES

1. B. Beden and C. Lamy, 'Infrared Reflectance Spectroscopy', in "Spectroelectrochemistry – Theory and Practice", R. J. Gale (Ed), Plenum Press, New York, Chapter 5, pp. 189–261 (1988).

2. D. M. Kolb, 'UV–Visible Reflectance Spectroscopy', in "Spectroelectrochemistry – Theory and Practice", R. J. Gale (Ed), Plenum Press, New York, Chapter 4, pp. 87–188 (1988).

3. E. M. Belgsir, E. Bouhier, H. Essis-Yei, K. B. Kokoh, B. Beden, H. Huser, J.-M. Léger and C. Lamy, *Electrochim. Acta*, **36**, 1157 (1991).

4. T. Iwasita and E. Pastor, *Electrochim. Acta*, **39**, 531 (1994).

5. C. Lamy, J.-M. Léger and S. Srinivasan, 'Direct Methanol Fuel Cells – From a 20th Century Electrochemist's Dream to a 21st Century Emerging Technology', in "Modern Aspects of Electrochemistry", J. O'M. Bockris and B. E. Conway (Eds), Chapter 3, Plenum Press, New York, Vol. 34, pp. 53–117 (2000).

6. J. M. Pérez, B. Beden, F. Hahn, A. Aldaz and C. Lamy, *J. Electroanal. Chem.*, **262**, 251 (1989).

7. H. Hitmi, E. M. Belgsir, J.-M. Léger, C. Lamy and R. O. Lezna, *Electrochim. Acta*, **39**, 407 (1994).

8. S. C. Chang, L. W. Leung and M. J. Weaver, *J. Phys. Chem.*, **94**, 6013 (1990).

9. N. R. de Tacconi, R. O. Lezna, B. Beden, F. Hahn and C. Lamy, *J. Electroanal. Chem.*, **379**, 329 (1994).

10. G. Tremiliosi-Filho, E. R. Gonzalez, A. J. Motheo, E. M. Belgsir, J.-M. Léger and C. Lamy, *J. Electroanal. Chem.*, **444**, 31 (1998).

11. R. S. Gonçalves, J.-M. Léger and C. Lamy, *Electrochim. Acta*, **33**, 1581 (1988).

12. E. Pastor, S. Wasmus, T. Iwasita, M. C. Arevalo, S. Gonzalez and A. J. Arvia, *J. Electroanal. Chem.*, **353**, 97 (1993).

13. I. A. Rodrigues, J. P. I. De Souza, E. Pastor and F. C. Nart, *Langmuir*, **13**, 6829 (1997).

14. J. Wang, S. Wasmus and R. F. Savinell, *J. Electrochem. Soc.*, **142**, 5218 (1995).

15. C. Lamy, E. M. Belgsir and J.-M. Léger, *J. Appl. Electrochem.*, **31**, 799 (2001).
16. C. Lamy, *Electrochim. Acta*, **29**, 1581 (1984).
17. F. Hahn, B. Beden, F. Kadirgan and C. Lamy, *J. Electroanal. Chem.*, **216**, 169 (1987).
18. M. El Chbihi, D. Takky, F. Hahn, H. Huser, J.-M. Léger and C. Lamy, *J. Electroanal. Chem.*, **463**, 63 (1999).
19. H. Cnobloch and H. Kohlmüller, 27th ISE Meeting, Zurich, September 1976, Extended Abstract No. 282.
20. A. Kahyaoglu, B. Beden and C. Lamy, *Electrochim. Acta*, **29**, 1489 (1984).
21. R. S. Gonçalves, W. E. Triaca and T. Rabockai, *Anal. Lett.*, **18**, 957 (1986).
22. L. Roquet, E. M. Belgsir, J.-M. Léger and C. Lamy, *Electrochim. Acta*, **39**, 2387 (1994).
23. M. Watanabe and S. Motoo, *J. Electroanal. Chem.*, **60**, 267 (1975).
24. P. Waszczuk, A. Wieckowski, P. Zelenay, S. Gottesfeld, C. Coutanceau, J.-M. Léger and C. Lamy, *J. Electroanal. Chem.*, **511**, 55 (2001).
25. F. Vigier, F. Gloaguen, J.-M. Léger and C. Lamy, *Electrochim. Acta*, **46**, 4331 (2001).
26. N. Fujiwara, K. A. Friedrich and U. Stimming, *J. Electroanal. Chem.*, **472**, 120 (1999).
27. R. Ianniello, V. M. Schmidt, J. L. Rodriguez and E. Pastor, *J. Electroanal. Chem.*, **471**, 267 (1999).
28. F. Delime, J.-M. Léger and C. Lamy, *J. Appl. Electrochem.*, **29**, 1249 (1999).
29. M. Watanabe, S. Saegussa and P. Stonehart, *J. Electroanal. Chem.*, **271**, 213 (1989).
30. J. Wang, H. Nakajima and H. Kita, *Electrochim. Acta*, **35**, 323 (1990).
31. H. Laborde, A. Rezzouk, J.-M. Léger and C. Lamy, in 'Proceedings of the Symposium on Electrode Materials and Processes for Energy Conversion and Storage', S. Srinivasan, D. D MacDonald and A. C. Khandkar (Eds), The Electrochemical Society, Princeton, NJ, PV-94-23, pp. 275–293 (1994).
32. W. Hauffe and J. Heitbaum, *Electrochim. Acta*, **23**, 299 (1978).
33. B. Beden, F. Kadirgan, A. Kahyaoglu and C. Lamy, *J. Electroanal. Chem.*, **135**, 329 (1982).
34. J. D. E. McIntyre and W. F. Peck, in 'Proceedings of the 3rd Symposium on Electrode Processes', S. Bruckenstein, J. D. E. McIntyre, M. Miller and E. Yeager (Eds), The Electrochemical Society, Princeton, NJ, PV 80, p. 322 (1979).
35. H. Cnobloch, D. Gröppel, H. Kohlmüller, D. Kühl, H. Poppa and G. Siemsen, *Prog. Batteries Solar Cells*, **4**, 225 (1982).
36. E. Peled, T. Duvdevani, A. Aharon and A. Melman, *Electrochem. Solid-State Lett.*, **4**, A38 (2001); E. Peled, V. Livshits and T. Duvdevani, *J. Power Sources*, **106**, 245 (2002).
37. F. Delime, Ph.D. Thesis, University of Poitiers (1997).
38. A. S. Arico, P. Creti, P. L. Antonucci and V. Antonucci, *Electrochem. Solid-State Lett.*, **1**, 66 (1998).
39. S. R. Narayanan, E. Vamos, S. Surampudi, H. Franck, G. Halpert, G. K. Surya Prakash, M. C. Smart, R. Knieler, G. A. Olah, J. Kosek and C. Cropley, *J. Electrochem. Soc.*, **144**, 4195 (1997).
40. J. T. Wang, W. F. Lin, M. Weber, S. Wasmus and R. F. Savinell, *Electrochim. Acta*, **43**, 3821 (1998).
41. J. T. Müller, P. M. Urban, W. F. Hölderich, K. M. Colbow, J. Zhang and D. P. Wilkinson, *J. Electrochem. Soc.*, **147**, 4058 (2000).
42. A. J. Cisar, D. Weng and O. J. Murphy, US Patent, 6 054 228 (2000).
43. R. Colturi, DEA de Chimie Appliquées, University of Poitiers (1997).
44. E. M. Belgsir, Ph.D. Thesis, University of Poitiers (1990).

Chapter 20
Solid oxide fuel cells (SOFC)

P. Holtappels[1] and U. Stimming[2]

[1] *Swiss Federal Institute for Materials Testing and Research (EMPA), Duebendorf, Switzerland*
[2] *Technische Universität München, Garching, Germany*

1 INTRODUCTION

Solid oxide fuel cells are well known and increasing development has been undertaken since the 1960s. The characteristic features are an oxygen-ion conducting, solid electrolyte and the operating temperatures between 650 and 1000 °C. At these temperatures the electrode reactions are usually fast which allows for the use of nonnoble electro-catalysts as electrodes. Both the electrodes and the electrolyte form an all-solid-state-system, while the reacting species (fuel and oxygen) are in the gas phase.

Only two phases (the solid and the gas phase) have to be handled in the solid oxide fuel cell (SOFC). This allows for various designs and concepts. Since oxygen ions supplied by the electrolyte oxidize the fuel, both hydrogen and also carbon-containing species such as CO, CH_4 or higher hydrocarbons can be utilized directly. These fuels are provided from fossil energy sources, which are the main energy carriers to date.

During operation both electrical power and heat are produced in the SOFC. The high operating temperatures lead to high quality heat, which makes the SOFC particularly suited for combined heat and power production and combined cycle plants.

However, the high operating temperatures also pose specific problems to the SOFC: materials must be compatible and physical properties such as the thermal expansion and dimensional stability in reducing and oxidizing environments are crucial. Therefore, high requirements are on materials. This has been addressed in earlier reviews,[1] which also concentrate on hydrogen as fuel.

Particular problems are also imposed on the fuel cell periphery components, especially the high temperature stability and heat exchangers to control the heat management of the SOFC. Up to 1995, cells operating at temperatures between 900 and 1000 °C were developed for stationary applications in power plants. Since then significant progress has been made in lowering the operating temperature to between 700 and 850 °C by introducing the anode supported cell concept. Additionally systems in the size of a few kW up to 100 kW are becoming of interest not only for stationary co-generation of power and heat, but also for mobile applications.

The trends in both the designs and application of SOFC cells and systems to date will be addressed in this work.

2 PRINCIPLE SOFC

2.1 Thermodynamics

The principle of the SOFC is shown in Figure 1. The fuel cell consists of two separate chambers divided by the solid and gas tight oxygen-ion conducting electrolyte. In the air chamber, oxygen is reduced at the cathode to oxygen ions which are incorporated into the solid electrolyte. The fuels considered for SOFCs are reformats consisting of mostly hydrogen and carbon monoxide. In the fuel chamber, hydrogen and carbon monoxide are oxidized in

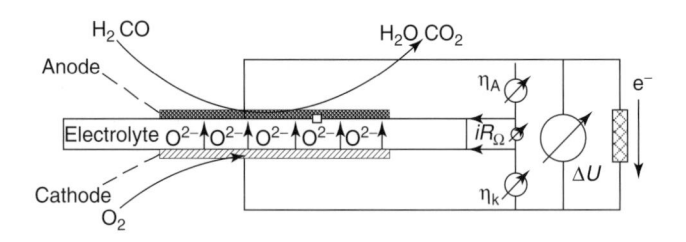

Figure 1. Principle of a Solid Oxide Fuel Cell.

presence of the oxygen ions from the electrolyte. Important is that, in contrast to the proton exchange membrane fuel cell (PEMFC), and similar to the molten carbonate fuel cell (MCFC) the water and carbon dioxide are formed at the anode.

The different electrochemical potentials in each chamber lead to a potential difference across the cell which corresponds to the Gibb's Free Enthalpy $\Delta G_{f,x,T}$ of the oxidation reaction of hydrogen and carbon monoxide, see equations (1), (2) and (3), (4), respectively.

$$H_2 + \tfrac{1}{2}O_2 \longrightarrow H_2O \tag{1}$$

$$U_0 = -\frac{\Delta G^0_{f,H_2O,T}}{2F} + \frac{RT}{2F}$$
$$\times \ln\left(\frac{\sqrt{p(O_2)}_{cathode}\, p(H_2)_{anode}}{p(H_2O)_{anode}}\right) \tag{2}$$

$$CO + \tfrac{1}{2}O_2 \longrightarrow CO_2 \tag{3}$$

$$U_0 = -\frac{\Delta G^0_{f,CO_2,T}}{2F} + \frac{RT}{2F}$$
$$\times \ln\left(\frac{\sqrt{p(O_2)}_{cathode}\, p(CO)_{anode}}{p(CO_2)_{anode}}\right) \tag{4}$$

The potential difference U_0 is the maximum available voltage to be obtained from one single cell and is determined by the kind of fuel, the partial pressure of the reacting species, hydrogen/carbon monoxide, and the operating temperature. R and F are the gas constant and the Faraday constant, respectively.

For the above reactions $\Delta G_{f,x}$ for the hydrogen and carbon monoxide reaction depend on temperature according to

$$\Delta G_{f,H_2O}(T) = -247.4 + 0.0541T \quad (kJ\,mol^{-1}) \tag{5}$$

$$\Delta G^0_{f,CO_2}(T) = -282.5 + 0.0866T \quad (kJ\,mol^{-1}) \tag{6}$$

The potential difference U_0 decreases with increasing temperature. The reason is the considerable contribution of the reaction entropy to the Gibbs Free Energy of the reactions. The entropy contribution to $\Delta G_{f,x}$ is also the reason that

the thermodynamic efficiency (7)

$$\varepsilon_{th} = \frac{\Delta G}{\Delta H} \tag{7}$$

is below 1 (0.7–0.9) and decreases further with increasing temperature.

There is no principle restriction that species other than hydrogen and carbon monoxide are being oxidized in presence of the oxygen ions from the electrolyte. Therefore, hydrocarbon species such as methane could also be considered to be oxidized and thus considered to be acting as fuels for SOFC's. The full electrochemical oxidation of methane to CO_2 and water (deep oxidation),

$$CH_4 + 4O^x_O \longrightarrow CO_2 + 2H_2O + 4V^{\cdot\cdot}_O + 8e^- \tag{8}$$

where O^x_O represents an oxygen ion in oxidation state-II on an oxygen site in the oxide electrolyte lattice, and $V^{\cdot\cdot}_O$ is a vacant oxygen site (charge is two times positive against the lattice). This would result in a thermodynamic efficiency close to 100% and almost independent of temperature. This high theoretical efficiency is advantageous in SOFCs and thus has attracted a significant interest among SOFC developers.

However, the direct electrochemical oxidation of methane requires an eight electron transfer which is hardly considered to occur simultaneously in one step, especially if competitive reaction pathways exist.

Alternative reaction pathways may involve the formation of CO (9)

$$CH_4 + 3O^x_O \longrightarrow CO + 2H_2O + 3V^{\cdot\cdot}_O + 6e^- \tag{9}$$

and followed by the water gas shift reaction (10)

$$CO + H_2O \longrightarrow CO_2 + H_2 \tag{10}$$

finally, the products H_2 and CO_2 from which hydrogen acts further as fuel.

In reaction (9) a total of 6 electrons are involved for the electrochemical CH_4 conversion which is still too high to occur in one single step.

A competitive pathway to the direct electrochemical oxidation of hydrocarbon species such as methane is reforming reactions. In presence of water and at the operating temperatures of above 600 °C, steam reforming (11) may take place yielding hydrogen and carbon monoxide as fuels.

$$CH_4 + H_2O \longrightarrow CO + 3H_2 \tag{11}$$

The steam reforming reaction is endothermic and thus offers the possibility to take up the heat released under operation. This will be discussed in particular in Section 4.2.

Another competitor to the direct oxidation is carbon formation due to cracking (12)

$$CH_4 \longrightarrow C + 2H_2 \quad (12)$$

The carbon formation should be avoided mainly because of safety reasons.

In multi-component fuels such as reformed methane, it is likely that equilibrium conditions are reached and both the hydrogen and carbon monoxide/oxygen equilibrium will control the respective oxygen partial pressure in the fuel compartment. The equilibrium cell potential will then be given by the difference in the oxygen partial pressures in both the air and the fuel compartment as described by equation (13)

$$\Delta U = \frac{RT}{4F} \ln \left(\frac{p_{O_2}}{P_{O_2}} \right) \quad (13)$$

The various processes such as electrochemical oxidation, reforming and cracking will strongly be influenced by the (electro-) catalytic properties of the fuel electrode (anode) material. This will be addressed in Section 2.2.

2.2 Cell components

2.2.1 Electrolyte

The name-giving component in the SOFC is the solid electrolyte. The function of the electrolyte is to separate the two gas atmospheres and to transport the oxygen ions without significant losses from the cathode to the anode. The electrolyte must therefore have sufficient oxygen-ion conductivity and also be chemically stable in a large oxygen partial pressure gradient from highly reducing to oxidizing conditions.

The transport of oxygen ions in the electrolyte occurs via oxygen vacancies in the oxygen sub lattice. The concentration of vacancies and their mobility determines the electrolyte conductivity. Materials developed until 1993 are summarized by Minh.[2] Among a number of suitable oxide ion conductors, zirconia stabilised in the conductive cubic phase with up to 10% yttria or scandia is the electrolyte most commonly used. Although higher conductive materials exists stabilised zirconia is the best compromise between high oxygen ion conductivity and matching secondary requirements with the other SOFC components such as chemical stability and durability. The conductivity is $>0.1\,\mathrm{S\,cm^{-1}}$ at 950–1000 °C and 0.03–0.003 S cm^{-1} at reduced temperatures of 600–800 °C (temperature dependence corresponds to an activation enthalpy of 0.8–1 eV).

In order to minimize the voltage losses in the electrolyte under operation of the fuel cell, the electrolyte layer should be as thin as possible. The thickness of the electrolyte is dependent on the cell design and the mechanical stability of the electrolyte. In the case of a flat electrolyte layer supporting the electrodes, a minimum thickness of 100–150 µm is required for zirconia stabilised with 8 mol % yttria. The mechanical stability can be increased by lowering the yttria content or by addition of alumina but both at the cost of ionic conductivity. Thinner electrolytes can only be obtained when the electrolyte layer is supported by another component (e.g., an electrode), or by a special shape.

Alternative electrolyte materials discussed to date which show higher conductivity are perovskites based on LaGaO$_3$[3] as well as doped ceria (fluorite). These materials are still under development and new results have revealed that the LaGaO$_3$ is chemically unstable under reducing atmospheres. The development is still ongoing.

2.2.2 Electrodes

General

In the SOFC electrode, the overall reactions mentioned above are locally distributed to the electrode/electrolyte interfaces in the air and fuel compartments. The electrode reaction between gaseous species, an electronic conductor and the oxygen ions provided from electrolyte take place at locations where all three phases are in close contact to each other. This region is called the three-phase boundary (TPB) region. High electrical power is obtained for high conversion rates (electrochemical reaction rates) leading to high current densities at low overpotentials. This requires highly active electrodes (high electro-catalytic activity) or high numbers of triple phase boundary contacts. The number of triple phase contact points is determined by the structure of the electrode. Small particles and pore sizes are advantageous for high length of the TPB.

Mixtures of the electronic conducting electrode material with an oxygen ion conductor (e.g., the electrolyte) leads to possible reaction sites also in the volume of the porous electrode. Therefore, the reaction zone is spread out into the volume of the porous electrode away from the electrode/electrolyte interface, which is advantageous for the electrochemical conversion rate. In cases of mixed oxygen ion and electron/hole conducting electrode materials, both oxygen ions and electrons are transported and therefore, the reaction site can be the whole mixed conductor surface exposed to the gas phase resulting in a similar extension of the reaction zone as observed for composites described above.

In the pores, gaseous reactants and products have to be transported to and from the TPB region. In order to diminish losses due to gas transport an open porous structure is required in the electrode.

The aspects mentioned above are determined by the (micro) structure of the electrode. In addition, the electrode material should also have a high electro-catalytic activity for the electrochemical reactions. At the operating temperature of the SOFC, the electrochemical kinetics can be considered to be fast and therefore, nonnoble metals and oxides can be used as electrode materials. However, the choice of materials is limited by so-called secondary requirements. The materials must be chemically compatible with the electrolyte and adjacent layers at operating conditions but also during fabrication. The material should match in the thermal expansion coefficients and – for practical use – be dimensionally and structurally stable under various reducing and oxidizing conditions. Suitable materials and structures for fuel and air electrode are addressed in the following.

Anode

At the anode, the fuel will be oxidized in the presence of oxygen ions from the electrolyte

$$\text{Red}_g + n\text{O}_O^x = \text{Ox}\text{O}_n + n\text{V}_O^{\cdot\cdot} + 2ne^- \qquad (14)$$

Equation (14) is written in the so called Kröger-Vinck notation, where O_O^x represents an oxygen ion (oxidation state-II) on an oxygen site in the oxide electrolyte lattice, $V_O^{\cdot\cdot}$ is a vacant oxygen site in the electrolyte lattice (charge 2 times positive against the lattice). Red and Ox refer to the reduced and oxidized fuel species and g to the gas phase. n is the number of electrons involved in the electrochemical reaction (see Table 1).

At the reducing conditions in the fuel electrode, metals are stable under a wide range of operation conditions. Several materials and alloys have been investigated and Ni was chosen because of its high electrochemical activity for the hydrogen oxidation reaction, costs and chemical compatibility reasons. In order to adjust the thermal expansion coefficient of a Ni containing anode to that of the electrolyte, the Ni is mixed to a composite with the electrolyte material, the so-called Ni-yttria stabilized zirconia (YSZ) cermet. In the Ni-YSZ cermet, the ceramic electrolyte forms a matrix that prevents Ni from agglomeration and thus, maintains the porous and highly dispersed microstructure of the anode. At Ni contents of around 40% the cermet are electronic conductors (500–1800 S cm^{-1}[2]) and show a thermal expansion coefficient close to that of YSZ. A typical micrograph is shown in Figure 2.

Ni-YSZ cermet has been the state-of-the-art anode material developed up to high performance for hydrogen oxidation.[4] The electrochemical activity of Ni-YSZ electrodes reaches values of $R_F \approx 0.15\,\Omega\,\text{cm}^2$ at 850 °C and $R_F < 0.1\,\Omega\,\text{cm}^2$ (950–1000 °C).

The processes of the hydrogen oxidation reaction in Ni-YSZ cermet electrodes have been investigated for both technical electrodes and model systems. The results reveal that the reaction is indeed limited by the length of the TPB region.[5, 6] Possible steps in the hydrogen oxidation reaction at the TPB region are adsorption, surface diffusion, and charge transfer on both the Ni and the YSZ electrolyte surface, as shown in Figure 3. Different models postulating different limiting steps are reported and the results are, however, not conclusive. The differences may arise from variations in the microstructures and the impurity elements introduced from different precursor materials and fabrication techniques, which are likely to affect the electrode performance. In high performing electrodes with area specific resistances below 0.1 $\Omega\,\text{cm}^2$, additional limitations due to gas phase transport are observed.[7]

The Ni-YSZ cermet electrodes show good performance only for hydrogen but severe problems occur when operating the cell in pure or reformed hydrocarbon fuels. Ni has a high activity for coke formation, which deteriorates the electro-catalytic activity and the microstructure. In order to prevent coking, high water contents are required in the fuel, which reduce the equilibrium potential and thus the maximum voltage of the cell, leading to a reduced power output.

Table 1. Reacting species in SOFCs.

Red	Ox	Number of electrons transferred in the overall electrochemical reaction
H_2	H_2O	2
CO	CO_2	2
CH_4	$CO_2 + 2H_2O$	8
	$CO + 2H_2O$	6

Figure 2. Micrograph of a state-of-the art Ni-YSZ cermet electrode. (Reproduced from Holtappels[40] (2000) by permission of Gesellschaft Deutscher Chemiker.)

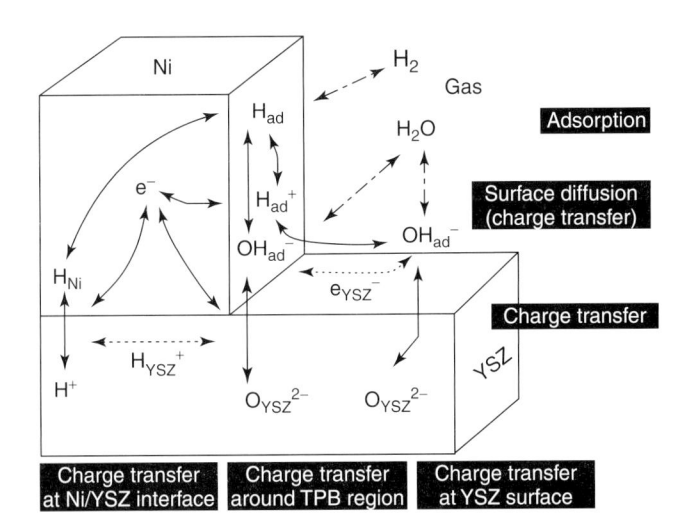

Figure 3. Possible electrochemical processes in the hydrogen/water reaction at the TPB region around the Ni/YSZ contact including solid state diffusion in both the Nickel and the solid electrolyte as well as surface diffusion and adsorption at the electrode/electrolyte surfaces. Indexes 'ad' denote adsorbed species, 'YSZ', and 'Ni' refer to species in the YSZ and Ni phase, respectively.

For the oxidation of carbon monoxide, unstable reaction rates have been observed and related to carbon deposition on Ni even at conditions where carbon formation is not thermodynamically stable.[8]

Fossil fuels such as pipeline natural gas, gasoline and diesel contain sulfur either naturally or added for leak detection. Traces of sulfur above 50 ppm may lead to a severe decrease in the anode performance due to catalyst deactivation.

Another limitation in the technical application of Ni-YSZ cermets is the re-oxidation of Ni, which involves phase and dimensional changes. Only a few developers have reported redox stable Ni-YSZ cermet structures and also for these Ni-YSZ cermets the number of redox cycles is limited. Both the carbon deposition and re-oxidation form the boundary conditions for the operating range of the Ni-YSZ cermet electrode.

As an alternative to Ni-YSZ cermet anodes, electronic and mixed ionic-electronic conducting ceramics have been investigated. Development of alternative fuel electrodes is concentrating on ceramic materials such as Gd-doped ceria, Ti- and Y-doped zirconia and lanthanum chromite as fuel electrodes. The ceramics, in general, show a very low, even negligible, activity for coking and also stability against redox-cycles (fuel air).

From these alternative materials only ceria based anodes such as $Ce_{0.6}Gd_{0.4}O_{1.8}$ have been developed up to a performance competitive to the Ni-YSZ cermet.[9] By addition of a dispersed Ni catalyst into the porous ceria structure, performance of $0.06\,\Omega\,cm^2$ at $1000\,°C$ in hydrogen and similar to Ni-YSZ have been reported. No carbon deposition has been observed at high methane contents and redox stability and long-term stability proven. The ceria electrode does not fully match the thermal expansion coefficient of the YSZ electrolyte and expands upon reduction. The resulting problems of thermo-mechanical stresses at the ceria/YSZ interface have been overcome by introducing an anchoring construction at the interface to the YSZ electrolyte. A disadvantage is the poor electronic conductivity in ceria that requires an additional current collector layer or the fabrication of metal-ceria cermets.

Ceria based composite anodes Cu/YSZ/ceria[10] and Ni/ceria[11] are being developed for direct application of hydrocarbon fuels. Direct feeding of methane, propane and higher hydrocarbons up to decane has been reported, without carbon deposition. For these electrodes direct oxidation of methane and higher hydrocarbons is postulated (see Section 3.2.2). In contrast, for ceria electrodes investigated extensively in hydrogen and methane, cracking instead of direct oxidation of methane is indicated.

Cathode
At the cathode, gaseous oxygen is reduced and the resulting oxygen ions are incorporated into the electrolyte

$$O_{2,g} + 4e^- + 2V_O^{\cdot\cdot} = 2O_O^x \qquad (15)$$

The cathode environment is highly oxidizing which excludes the use of metals as electrode materials. The state-of-the-art SOFC cathode material is an electronically conducting oxide ceramic based on the perovskite lanthanum manganite, $LaMnO_3$ in which La is partly replaced by Sr (strontium doped lanthanum manganite $(La, Sr)MnO_3$ (LSM)). These perovskites are electronic p-type conductors, for which the electrical properties are determined by the La/Sr ratio.

The exact composition of the perovskite also affects the chemical properties of the cathode. The chemical compatibility with the YSZ electrolyte is especially important. Formation of $La_2Zr_2O_7$ and SrO due to solid-state reactions deactivates the cathode performance and thus, has to be avoided. This undesired reaction does take place significantly with an excess of Mn (1–10%) at temperatures below $1300\,°C$.[12]

Similar to the Ni-YSZ cermet, the LSM can also be fabricated as a composite with the YSZ electrolyte as the second solid phase. In those composites the reaction zone may be extended into the porous cathode.[13] At $950\,°C$ area specific resistance of $0.04\,\Omega\,cm^2$ and $0.23\,\Omega\,cm^2$ at $750\,°C$ are state-of-the-art.

The electronic conductivity of LSM is at around $200\,S\,cm^{-1}$ for Sr contents of 15%. This conductivity is relatively low compared to f.i. Ni-YSZ cermets. In order to ensure optimized current collection in the cathode, multiple layer electrodes are advantageous. Two layer cathodes are frequently prepared in which the layer closer to the solid electrolyte is optimized for electrochemical performance while the top layer is optimized for electronic conduction and porosity. As current collectors, other perovskites such as La(Sr)CoO$_3$ are better suited because of their higher electronic conductivity, but are incompatible to YSZ. Multiple layer cathodes with gradually changing structure and composition from YSZ to LSM to LSCo have been developed (functionally graded cathodes). The principle and the microstructure of a nine-layer cathode are shown in Figure 4. These advanced multiple-layer cathodes show high electrochemical performance corresponding to 0.07–0.12 Ω cm^2 at 850 °C and 0.2–0.4 Ω cm^2 at 750 °C. The temperature dependence of the cathodes corresponds to an activation enthalpy of around 140 kJ mol^{-1} that is significantly less than that observed for pure LSM cathodes (200 kJ mol^{-1}).[14]

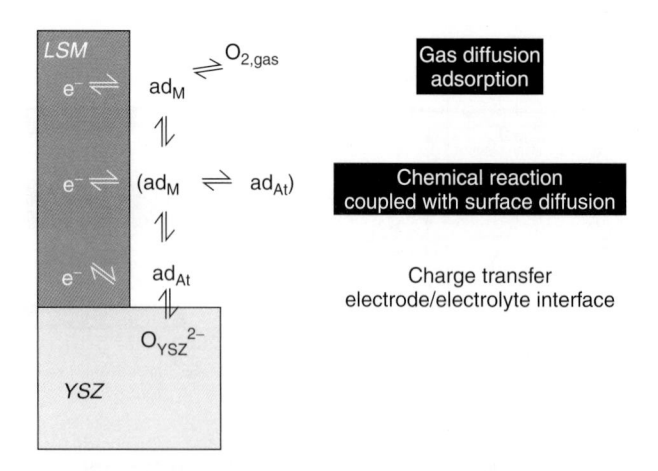

Figure 5. Electrochemical processes around the TPB region at the LSM/YSZ contact in an SOFC cathode including adsorption and surface diffusion processes. ad$_M$ and ad$_{At}$ refer to molecularly and atomically adsorbed oxygen species. O$_{YSZ}^{2-}$ represents an oxygen ion (oxidation state-II) on an oxygen site in the oxide electrolyte lattice.

Significant attention has been paid to the understanding of the electrochemical processes in SOFC LSM cathodes.[15] The electrochemically active area is assumed to be around the three phase boundary where the electronic conducting perovskite, the YSZ electrolyte and the gas phase are in contact with each other. It is generally assumed that the oxygen reduction is a multiple step reaction involving adsorption and surface diffusion in a region around the TPB (see Figure 5). For high performance electrodes, the mass transport via the gas phase inside the porous structure adds to the losses for the electrodes. An improvement of the electro-catalytic activity of LSM is reported by the addition of noble metal catalysts.[16] This is an indirect indication that bond breaking (O–O bond) is one of the important and rate determining steps in the oxygen reduction reaction.

The materials costs are an important aspect for the cathode. The costs of lanthanum manganite are strongly dependent on the purity of the rare earth element La, since the separation of the rare earth elements is very costly. The materials costs for the cathode can be significantly reduced if mixtures of lanthanides instead of pure lanthanum could be used.

3 DESIGN OF SOFC CELLS AND STACKS

3.1 General aspects

The cell components described above (electrodes and electrolyte) are all in the solid phase. As an all-solid state

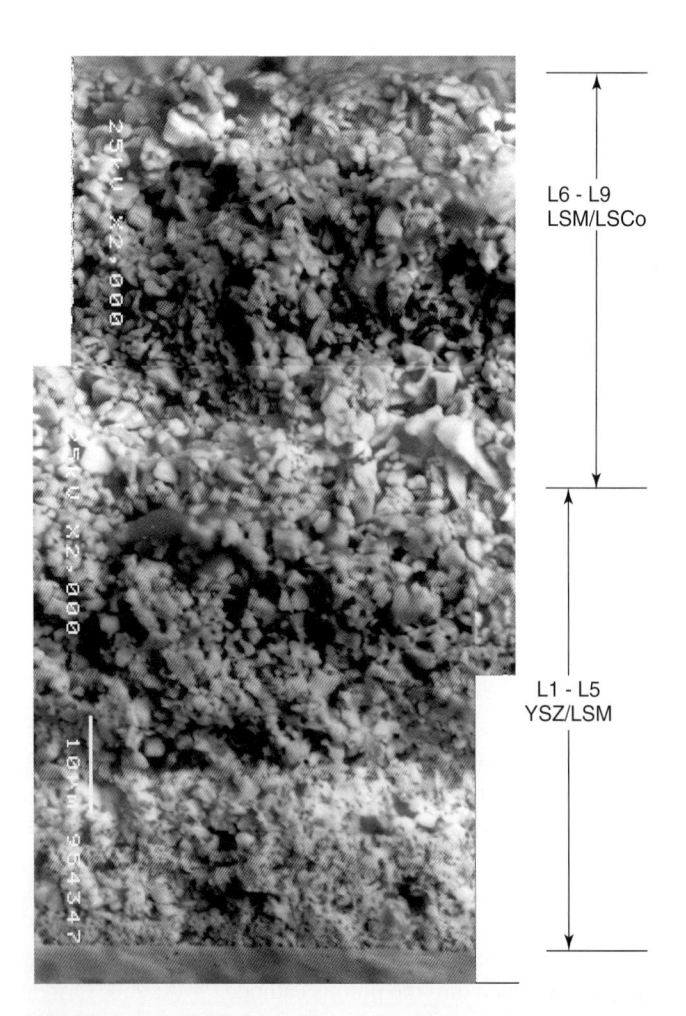

Figure 4. Microstructure of a functionally graded cathode: Unpolished cross section of a nine layer YSZ/LSM/LSCo cathode sintered 3 times at 1100 °C in air. (Reproduced from Holtappels and Bagger (2002)[14] with permission from Elsevier Science.)

device, various cell designs can be applied to SOFCs. The power output of one cell is usually less than 1 V (0.7 V is a 'standard' operation voltage) and assuming a cell power output of $0.5-1\,W\,cm^2$ $50-100\,W$ may be obtained from one cell with an active area of $100\,cm^2$. Higher power requires an assembly of cells, which has to be considered in order to choose a particular cell design.

The criteria which a cell design has to meet are:

1. Separation of the two gas compartments (air/fuel);
2. Optimized power output of the cells: this includes active electrodes, optimal fuel utilization, and optimal electrical contact;
3. Thermal management of the cells and stack (uptake of heat released during operation);
4. Cell and stack geometries allowing fabrication techniques suitable for mass production.

For certain applications the power/mass and power/volume ratio is important.

3.1.1 Assembly

Interconnection
In order to connect multiple cells, an electronic conductor is needed, which generally is exposed to both highly reducing and highly oxidizing conditions at the anode and cathode, respectively. An additional requirement is the dimensional stability in a gradient of $p(O_2)$ and matching thermal expansion coefficient with the electrodes and the electrolyte. For some cell designs the interconnection should be able to be mechanically machined in order to incorporate the gas channels.

The choice of the interconnection material for a gas tight electrical contact between the adjacent cells is strongly dependent upon the operating temperature of the cells. At temperatures above $850\,^\circ C$ high temperature-resistant Cr-alloys or ceramic interconnects based on $LaCrO_3$ must be applied. These high temperature-resistant materials show their particular disadvantages of cost and practical use. The Cr-alloy releases CrO_3, which poisons the air electrode (cathode) and deteriorates performance. In order to prevent the Cr-release the interconnection is coated at the air side with a protection layer made from $La(Sr)MnO_3$ or $La(Sr)CoO_3$.[17] The ceramic interconnect $La(Sr)CrO_3$ undergoes expansion upon reduction which may lead to bending and loss of electrical contact to the electrodes, especially for assemblies of flat plates.[18]

At temperatures up to $800\,^\circ C$ ferritic steel (Ni–Cr–Fe alloys) can be used. Compared to the high temperature resistant Cr-alloys and the ceramic $LaCrO_3$, materials costs are significantly reduced and machining is easier. However, the formation of oxide scales and the release of chromium under oxidizing conditions makes it advantageous to apply a protection layer at the cathode side.

Sealing
Sealing is a very important issue for SOFC developers but little information is publicly available. Sealing materials and composition depend on the operating temperature and the area that has to be sealed. In general, the sealing must be gas tight, an electrical (both ionic and electronic) insulator and thermally and chemically compatible with adjacent materials.

Development is concentrating on glasses with compositions that are adjusted to a desired viscosity and without any negative influence on the other components (e.g., the electrochemical activity of the electrodes by evaporation of Si, P). In the glass phase these materials show good sealing properties.

The glass phase is not thermodynamically stable (meta-stable) and may undergo re-crystallization. This leads to embrittlement and most probably crack formation, often below the operating temperature of an SOFC. Also traces of elements Na, K, Ca and Ba which are always present in technically produced materials increase the potential of re-crystallization.

Fabrication of SOFC components
The fabrication of SOFC cell components uses ceramic fabrication techniques.[19] Two general methods can be applied (i) the wet ceramic processing or (ii) the gas phase preparation route. In the wet ceramic processing route, the ceramic powders are mixed with solvent and binder to form a slurry. The slurry can be deposited as a layer or formed in a die to the desired shape. For the application of flat layers with large areas, deposition methods such as screen printing, tape casting, spraying, pressing, calendaring, slip (vacuum) or dip coating are used. The techniques are suitable for mass production but generally limited to simple geometries. Complex structures from single materials can be fabricated by extrusion of the slurry. In a firing step, binder and solvent will be removed and connections between the powder particles formed to build the final structure.

In contrast to the wet ceramic routes, ceramics can also be deposited from the gas phase. This application route is more complex but allows the deposition of thin layers below $1\,\mu m$. The final structure is formed during deposition of the material and, therefore, no additional thermal treatment is needed. Techniques are 'thermal spraying' (also plasma spraying) ($15\,000\,^\circ C$) which yield gas tight layers when performed in a vacuum, and 'chemical/electrochemical vapor deposition' where either a chemical or electrochemical reaction is performed simultaneously with the deposition of precursors.

3.2 Detailed description of the designs

3.2.1 Tubular designs

The best design is based on tubular cells and was developed at Siemens-Westinghouse (Figure 6), referred to as Westinghouse design.[20] The cell is built onto a porous cathode support of 1–2 cm in diameter made of LSM fabricated by extrusion. By subsequent layers the cathode (LSM), electrolyte (YSZ) and anode layer Ni-YSZ, are deposited by thermal spraying, chemical vapor deposition (CVD)/electrochemical vapor deposition (EVD), respectively. The thickness of the cathode, electrolyte and anode layer are around 2 mm, 40 µm and 100 µm respectively. Cells with a length of 150 cm with an electrochemically active cell area of 1036 cm^2 can be mass produced.

In this cell concept, the air is supplied from the inner side of the tube while the fuel chamber surrounds the tubular cell. The cells are connected in series by using a ceramic LaCrO$_3$-interconnection, and in parallel by using Ni-felts. The current flows along the porous cathode and anode layer as indicated in Figure 6 leading to relatively long current pathways through the poor conducting LSM.

The arrangement of the cells in a bundle is shown in Figure 7. The bundle is divided in two compartments; one surrounding the electrochemically active cell area, while in a separate zone the combustion of fuel and air takes place. Sealing is only required in the colder areas and not considered a problem in this set-up.

The cells and bundles have been demonstrated for more than 25 000 h of operation. The power density achieved is around 0.2 W cm^{-2} at a fuel utilization of ~90% and operating temperature of 1000 °C under atmospheric pressure. The performance is well analyzed and its dependence on temperature, partial pressure is described in Ref. [21]. The major disadvantage in the cell design that limits the performance is the ohmic loss due to the long current pathways in the thin and porous layers. In order to improve the current pathway through the cells and also to optimize the volumetric power density, the so-called high power density design is developed. In the somewhat flattened tube, bridges lead to a reduction of the internal losses due to improved current flow (Figure 6). Theoretical performance of 0.45 W cm^{-2} are expected. The costs are targeted to 1000 \$ kW^{-1} by industrial mass production.

The current pathway through the cell can be shortened by decreasing the diameter of the tubular cell. A development towards this direction are tubular electrolytes with a diameter of 2–3 mm (so called micro tubes), which are coated with the fuel and air electrodes on the inner and outer side, respectively.[22] Using open ended tubes, the combustion of unused fuel provides the heat to operate even small stacks without an external heater for remote applications. Manifolding and electrical contact can be made in the cold parts of such a bundle. The tubular electrolytes are fabricated by extrusion and show a very high thermo-mechanical

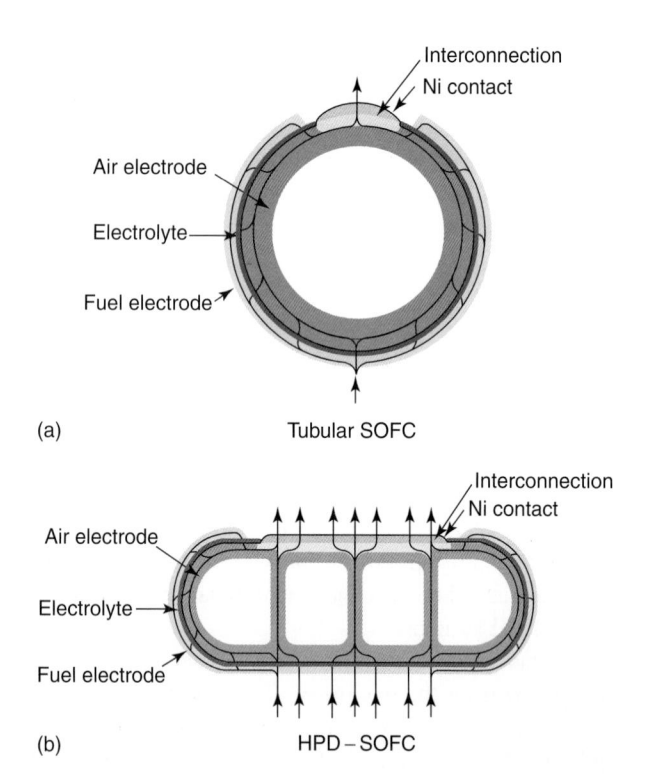

Figure 6. SOFC cell designs developed at Siemens-Westinghouse. (a) Conventional tubular design with long current pathways in the circumference of the tubes; (b) future 'High Power Density (HPD)' design with improved current pathway through the cell. Incorporated ribs in the 'flattend tube' act as bridges for the current pathway through the cell. (Reproduced from Singhal (1999)[20] with permission from The Electrochemical Society, Inc.)

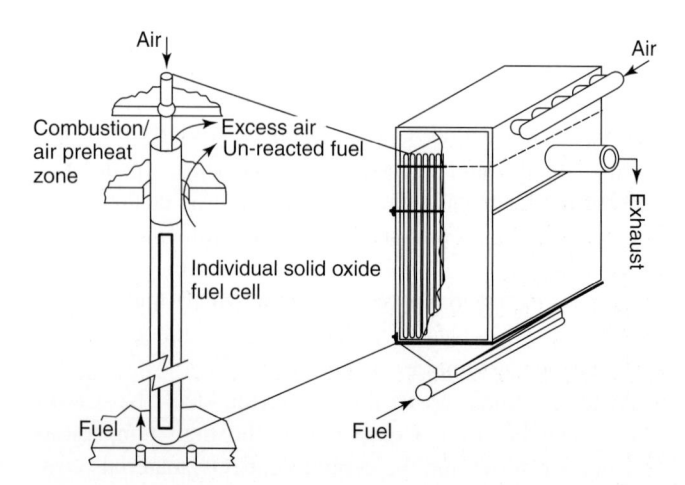

Figure 7. Module of tubular SOFC cells at Siemens Westinghouse (With permission of Siemens Westinghouse.)

stability, which allows for high thermal gradients over the cells.

In the tubular designs described above, one cell is fabricated on one tube. An alternative tubular cell concept is the arrangement of multiple cells on one tube and connected in series. In this concept, the current will be determined by only one segment of the tube, while the voltage is summed up over all cells on one tube. Such a design is pursued by Mitsubishi Heavy Industries (Figure 8).[23] The cells are fabricated in subsequent layers onto a porous substrate and connected by a thin gas tight interconnection layer. Tubes of 720 mm length have been prepared and stacks up to 10 kW (nominal) and 21 kW maximum power output tested. The electrical efficiency was 35% at 75% fuel utilization. Degradation rates are 1–2% per 1000 h.

3.2.2 Flat plate designs

Since the 1980s, cells in the shape of flat plates have been developed. The schematic is shown in Figure 9. The cell consisting of anode/electrolyte and cathode layer is sandwiched between the interconnection plates. The SOFC stack is built up by multiple cells stapled on top of each

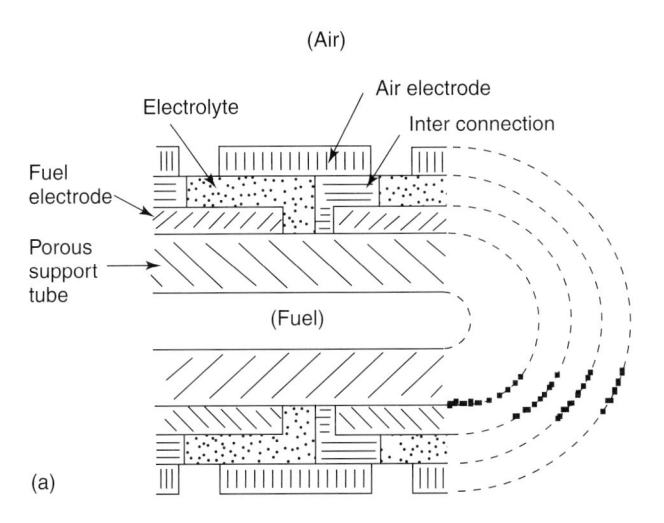

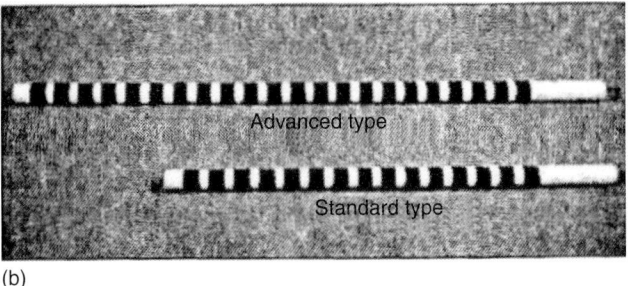

Figure 8. Tubular SOFC design developed at Mitsubishi Heavy Industries. (a) Cross section, (b) photograph of the tubes up to 72 cm length.[23] (Reproduced from Mori *et al*.[23] (2001) with permission of The Electrochemical Society.)

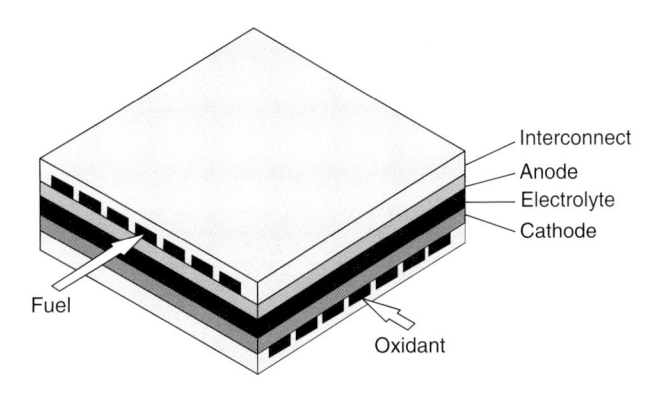

Figure 9. Concept of the flat plate design for planar cells.

other in which the components in Figure 9 form the 'repeat unit'. In this concept, the current flow is perpendicular to the cell surface resulting in short current pathways in a stack.

In the flat plate concept the interconnection must fulfill two functions: (a) it forms the electrical contact between the repeating elements and (b) supplies fuel gas and air to the anode and the cathode. Depending on the geometry of the cell, the fuel and airflow can be in either the same direction (co-flow), opposite directions (counter flow), or perpendicular to each other (cross-flow). Since fuel is being consumed and heat is released during operation of the SOFC, the flow design affects the current-potential distribution as well as the temperature distribution over the entire cell area.

The gas supply to the interconnection channels is made by the manifold (Figure 10). This can be realized in two different ways: (i) in an internal manifold the gas supply to the channels is included into the interconnection plates, or (ii) in an external manifold. All flow designs can be realized by machining the desired structures into the interconnection plate. This requires more complex structures in the interconnection plate and thus, increase the effort for machining. Simpler interconnection structures (only channels) require a separate (external) manifold. This is particularly advantageous for high temperature resistant alloys or ceramics that are difficult to machine, although it limits the flow design to cross flow.

In the flat plane design, at least one of the cell components must form the supporting part. Similar to the tubular design, this can either be the electrolyte [self supported electrolyte (SSE)] or one of the porous electrodes, e.g., the anode [anode supported electrolyte (ASE)]. The difference between the two different flat plate designs is shown in Figure 11. The particular advantage of the anode supported electrolyte cells is the potential to lower the operating temperatures of the SOFC cell. While decreasing the electrolyte thickness to below 10 μm, the ohmic losses in the YSZ electrolytes become acceptable even

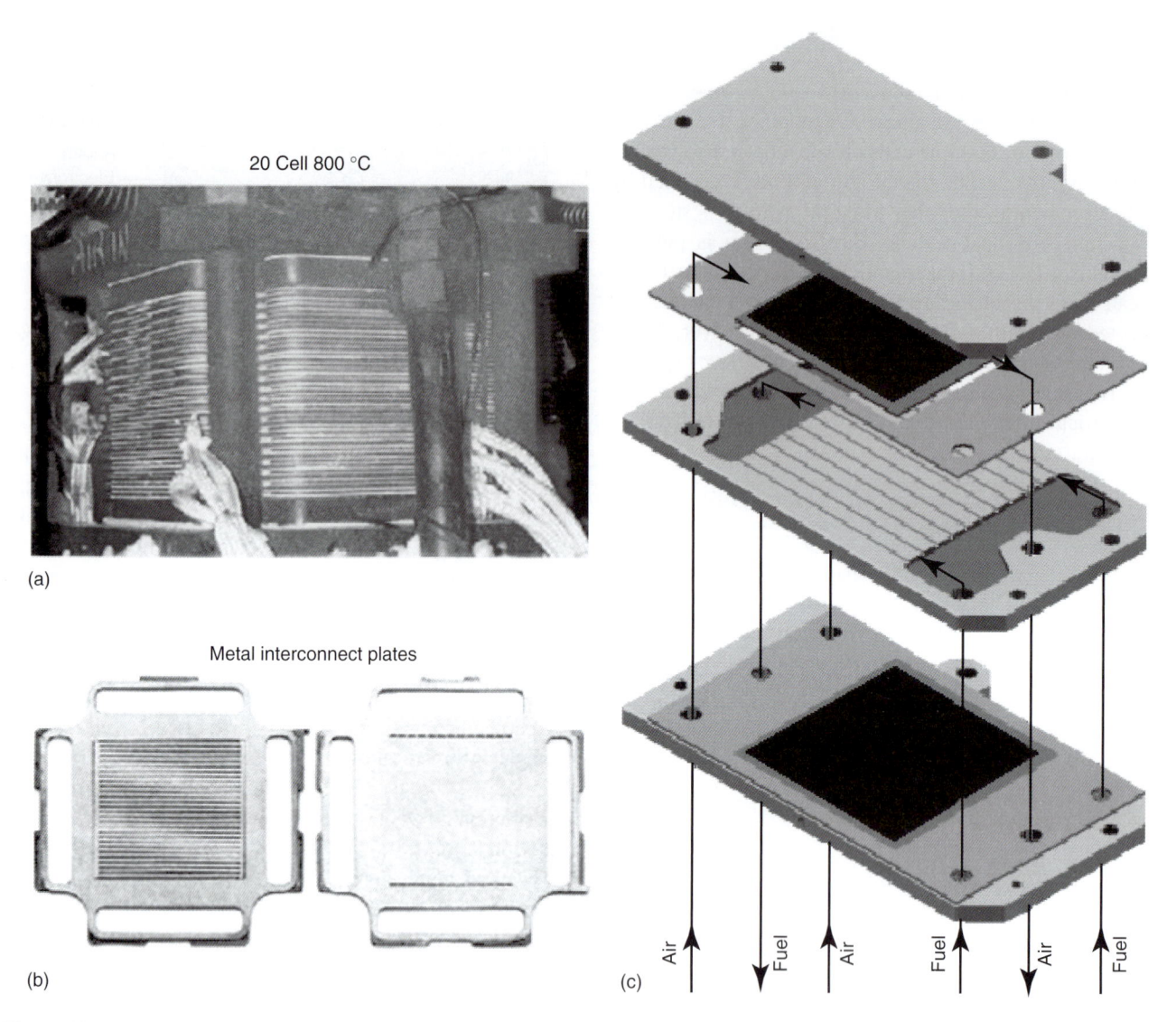

Figure 10. Interconnection plates for internal manifolding: (a) assembled 20-cell SOFC-stack made from anode supported cells, (b) cross flow. (c) co- and counter-flow. (Reproduced from de Haart et al. (2001)[28] with permission of Forschungszentrum Jülich.)

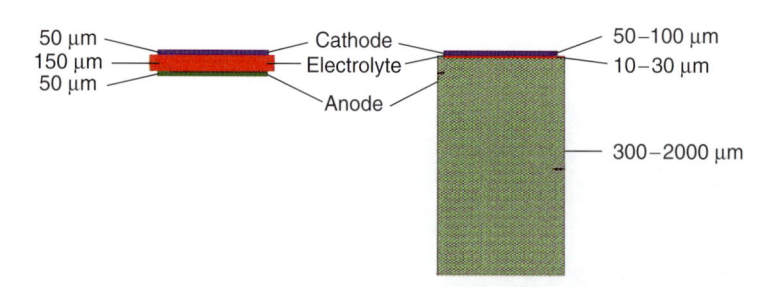

Figure 11. Comparison of self supported electrolyte cells (left) and anode supported electrolyte cells (right) in the flat plate design.

at temperatures of around 700 °C. At present, both cell concepts are being pursued for flat plate cells. The recent status for both designs will be discussed in the following.

Self supported electrolyte cells (SSE)
Planar flat electrolyte supported cells and stacks operating at around 930–950 °C have been developed worldwide, e.g., by Siemens,[1] Ceramic Fuel Cell Limited,[24] but

also research centers such as Risø DK and ECN/InDec, NL. Interconnection materials are Cr-alloys (the Plansee alloy) or ceramic (doped lanthanum chromite). Stacks with both internal and external manifolding have been demonstrated. The cell performances reported by various developers ranges from 0.18 to 0.26 W cm^{-2} at a fuel utilization of 50–80% and operating temperatures between 920 and 1000 °C. Stacks up to more than 5 kW have been demonstrated operating on both hydrogen and methane.

Sealing and costs, especially the materials costs, of the interconnection plates are realized to be one of the major problems relating to the high operating temperatures above 850 °C. Furthermore, the stack is sensitive to thermal gradients which cannot be avoided under operation of the SOFC. Although the distinct temperature distribution in the cell is different for the various flow designs (cross-, co- and counter-flow), the thermal management of the planar electrolyte supported cell stack is a severe problem. At present, systems based on electrolyte supported cells are under development at Ceramic Fuel Cell Limited (CFCL). Other developers such as Siemens have stopped their activities on the SSE cells and are pursuing either the tubular or the anode-supported cell designs (see Section 3.2.2).

A slightly modified cell concept based on a flat plate design is the Sulzer HEXIS-cell.[25] A schematic is presented in Figure 12. The cell is ring shaped (outer diameter of 120 mm) and equipped with an inner hole forming a channel for fuel supply in the stack. Sealing is only required at the inner rim of the fuel channel. At the outer rim, the cells are not sealed, allowing for combustion of unused fuel with ambient air. The interconnection is specially designed to transfer the heat released in this combustion process back to the cell in order to maintain a homogeneous temperature distribution over the cell area. The name HEXIS stands for Heat Exchange Integrated System, which refers to the interconnection as integrated heat exchanger. The

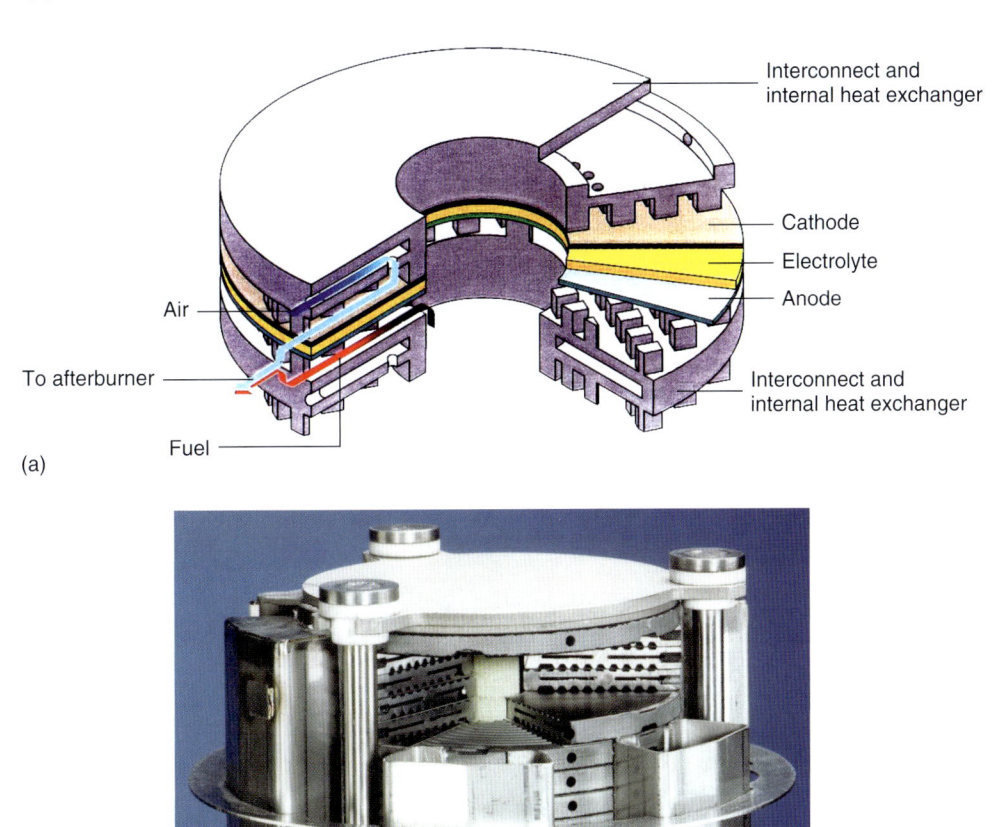

Figure 12. Cell design of the HEXIS cells.

interconnection plate is made from Cr-alloy (Plansee) and equipped with channels for fuel and air supply.[26]

The cells made from 10%Sc-doped zirconia as have been tested under hydrogen and reformed hydrocarbon fuels at temperatures around 930 °C. The cell voltage in the unloaded state is approximately 10% lower than the theoretical equilibrium potential difference due to the unsealed rim of the cells. Power densities of 0.18 W cm^{-2} have been achieved. The cells showed stability against thermal and re-oxidation cycles.

Supported electrolyte cells
The anode-supported electrolyte cells have attracted high interest in recent years. After Jülich demonstrated the feasibility of the anode supported electrolyte design,[27] most SOFC developers world wide have started their own development line of anode supported electrolyte cells. Considerable progress has been made in improving the performance of the anode-supported electrolyte cells. State-of-the-art cells have performances up to 1 W cm^{-1} at 750–850 °C (0.5 W cm^{-2} at 0.7 V) in hydrogen fuels. Typical current–voltage characteristics are given in Figure 13.[28] Stacks of 1–2 kW have been successfully tested in hydrogen but also in pre-reformed natural gas.

At operating temperatures below 850 °C, ferritic steel is used as the interconnection. Experience from stack tests indicates that a protection layer at the air side of the interconnection is needed in order to avoid oxidation of the interconnection which would decrease the electronic conductivity at the interconnection/cathode contact.

At Global Thermoelectric a compression gasket is used as sealing.[29] A glassy layer will only be used to close the pores at the edge of the anode substrate. Twenty to thirty successful thermal cycles have been reported. The sealing should also be tolerant to vibrations. The increased thermal cycling and mechanical stability of the stack assembly are significant improvements and highly advantageous for the practical use of SOFCs especially in nonstationary applications.

Because of the disadvantages of the Ni with respect to the re-oxidation stability, the C-formation activity and the high reforming activity, alternative support materials are of growing interest. At operating temperatures below 850 °C various metals may be applicable. However, during fabrication with conventional sintering procedures, higher temperatures are applied that exceed the stability region of metals such as ferritic steel. With the application of techniques that keep the fabrication temperature below 850 °C (such as thermal spraying) felts (ferritic steel) can be used as support for the anode, electrolyte and cathode layer. Plasma sprayed cells on steel supports have been prepared and power densities of 0.6 W cm^{-2} in hydrogen at 900 °C reported.[30]

A different concept of supported cells is being pursued at Rolls Royce Aerospace, UK.[31] Resembling the supported tubular cells (Figure 8), multiple cells per surface have also been arranged on porous flat plate supports and connected in series by interconnection layers (Integrated Planar (IP)-SOFC design). The set-up is shown in Figure 14. Similar to the Mitsubishi Heavy Industries (MHI) design, in this arrangement the current is determined by a one cell

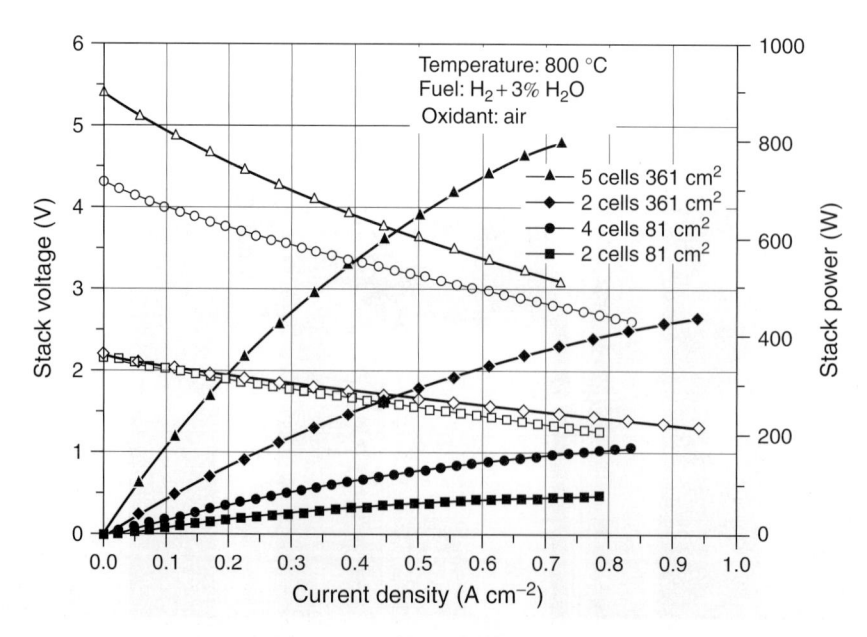

Figure 13. Performance data for anode supported cell stacks. Open symbols refer to the I/U characteristics, solid symbols to the stack power. (Reproduced from de Haart et al. (2001)[28] with permission of The Electrochemical Society, Inc.)

(a)

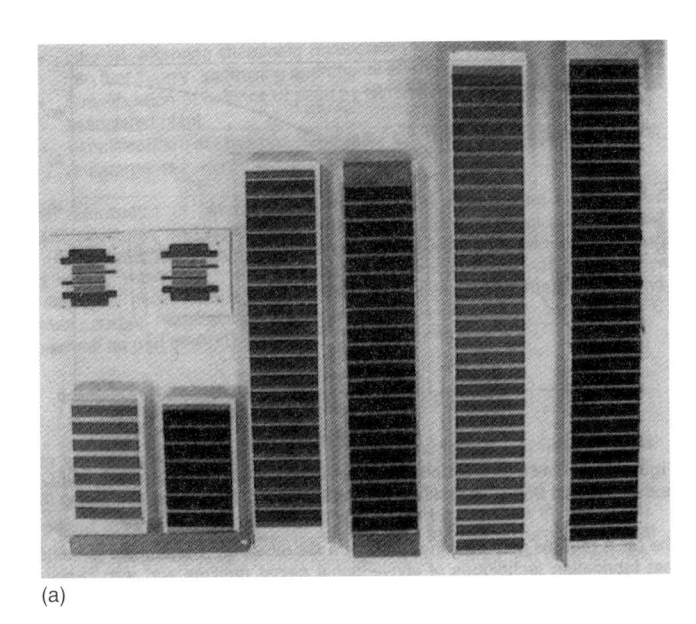

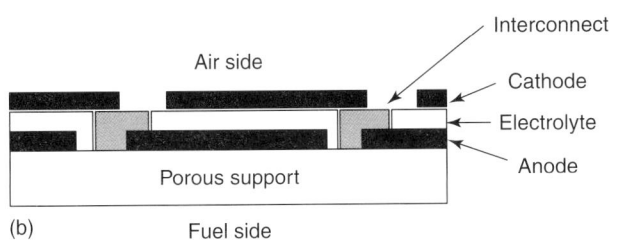

(b)
Air side — Interconnect, Cathode, Electrolyte, Anode
Porous support
Fuel side

Figure 14. IP-SOFC design. (Reproduced from Day (2000)[31] with permission of the European Fuel Cell Forum.)

segment, whereas the voltage is the sum of all individual cells on one support. Power out of 0.54 W cm^{-2} at 950 °C is reported for hydrogen as fuel.

3.2.3 Mono-layer-block SOFC

In the above described tubular and flat plate designs, one cell component, either the electrolyte or a porous layer, act as support for the cell. Alternatively, the mechanical stability of the ceramic cells may be increased by a more complex structure, which is self-supporting. As an example, the honeycomb structured cell from MHI is shown in Figure 15.[32] In this design, the cell forms a block which introduced the name 'mono layer block'-design. The design is related to an earlier monolithic SOFC design developed at Allied signal and described in Ref. [1]. The advantage here is that no additional structure for gas supply is required and simple (unstructured plates) can be used as interconnections. Mono-layer block (MOLB) cells and stacks are pursued at Mitsubishi Heavy Industries and reach performance of 0.22 W cm^{-2} in hydrogen. Stacks up to 5 kW have been tested.

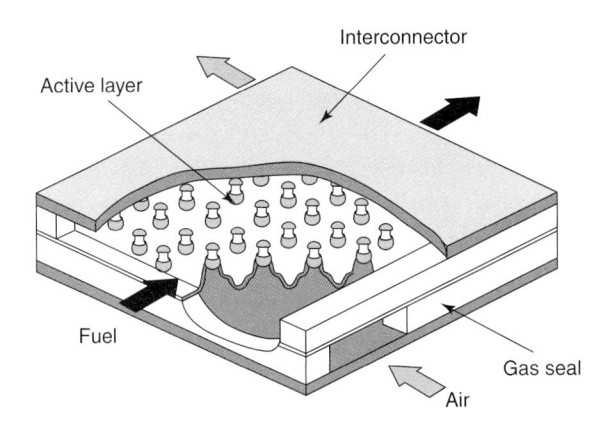

Figure 15. MOLB SOFC-cells. (Reproduced from Sakaki *et al.* (2001)[32] with permission of The Electrochemical Society, Inc.)

3.2.4 Single chamber SOFCs

When operating the SOFC at reduced temperatures, the electro-catalytic activity of the materials becomes more and more important. The different catalytic properties of metals and oxides for the fuel oxidation reaction and the oxygen reduction form the basis for the so-called 'single chamber SOFC'. At temperatures below the combustion temperature of a hydrocarbon fuel and air mixture, the oxygen reduction reaction is more prominent at oxides such as LSM, whereas the hydrocarbon oxidation is activated at metallic catalysts, e.g., Ni. By exposing a gas mixture containing hydrocarbons and air to a cell with those different electrode materials, electrochemical potential differences can be established across the cell. If the respective electrode reactions can proceed sufficiently fast, current can be drawn from the cell. With ethane as fuel at 500 °C, cell voltages of 0.9 V have been achieved for unloaded cells. The cells produced a maximum power of 0.4 W cm^{-2}. Since the concept allows for simple cell and stack designs, the performance is very promising for future SOFC systems.

3.3 Summary

To date the basic differences between the various SOFC systems are in the designs. Although a number of different materials have been investigated, yttria or scandia doped zirconia, Ni-YSZ cermets and Sr-doped lanthanum manganite are the materials used as electrolyte, anode and cathode, respectively. These materials are applied in the various designs and for operating temperatures from 750 to 1000 °C. The best design developed sofar is the tubular cell concept from Siemens-Westinghouse that is considered for operating temperatures above 950 °C. Although the cost estimates are optimistic to reach the targeted $1000 per kW, the design requires relatively expensive fabrication methods and shows low volumetric power densities (even in the high power density cell design).

Microtubes several millimeters in diameter have an increased volumetric power density (because the surface to volume ratio increases with decreasing diameter), but however, mass production of high performing cells at tolerable costs has to be demonstrated.

The flat plate design shows promising perspectives for mass production using cheap fabrication techniques and high volumetric power densities. However, due to problems related to the high operating temperatures, electrolyte supported cells are getting out of focus. There is a clear trend to date, that anode-supported flat plate cells will become dominant.

The anode-supported cells offer the possibility to reduce the operating temperature so that cheap and easy to handle alloys such as ferritic steels can be used to construct stacks. The lower operating temperatures are also advantageous for the design of periphery components in the SOFC system, which is addressed in the following sections.

With respect to the direct conversion of hydrocarbons, only cells with ceria as the anode material are at a development stage interesting for cell developers. However, in order to benefit from both a reduced operating temperature and direct application of the hydrocarbon fuel, new anode concepts are required for anode supported cells. The ceria is too poor an electronic conductor to allow it to be applied as anode supported cells.

4 SOFC SYSTEMS

4.1 General aspects and periphery components

The advantage of SOFCs, with respect to liquid and fossil fuels, is the capability to electrochemically oxidize both hydrogen and carbon monoxide. These reactants are easily obtained by reforming hydrocarbons containing fuels such as natural gas, gasoline, diesel/heating oil and coal. Fossil fuels are, to date, the most important energy carriers for electrical power, traction and heat for industrial and also domestic use. So far, the largest progress has been made in the development of stationary SOFC for combined heat and power applications in the range from 1 to 100 kW. Concepts exist for larger power plants, including gas and steam turbines in combined cycles. In these areas the SOFC competes in particular with the molten carbonate fuel cell (MCFC). Recently, the SOFC has also been discussed for mobile and even transportable systems for which liquefied natural gas/gasoline and diesel are the fuel of choice. Here SOFCs are becoming competitors to the low temperature PEMFC systems.

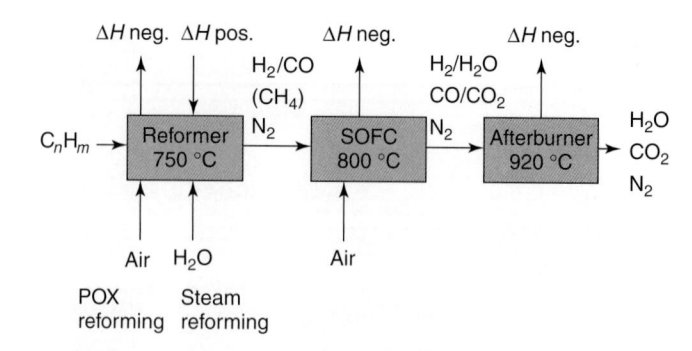

Figure 16. Principle of an SOFC system: fuel processing, air conditioning stack, off-gas treatment. ΔH neg., ΔH pos. correspond to heat evolved or consumed, respectively.

The general principle/set-up of an SOFC system using hydrocarbon fuels is shown in Figure 16. The key parts, in addition to the SOFC stack, are the fuel processing, the air conditioning and the off-gas treatment. The specific details of the various parts depend on the system size, the kind of fuel and considered applications (combined heat and power, combined cycle, power only) as well as on the operating conditions and design of the SOFC.

4.1.1 Fuel processing

The fuel processing includes gas cleaning, reforming, and if liquid or solid fuels are used, the gasification/evaporation additionally.

In the reforming process hydrocarbons are converted into a gas containing mostly hydrogen, carbon monoxide with little amounts of water and carbon dioxide depending upon the nature of the reforming process.

In the steam reforming process (already described in Section 2.1) hydrogen and carbon monoxide (syn gas) are obtained that can directly be fed to the SOFC. The steam reforming reaction is mostly considered for natural gas but is also possible for syn gas production from coal. The reaction generally proceeds quickly at Ni-containing catalysts. The process is endothermic, which means that process heat has to be supplied to the reformer unit.

Gasoline and diesel are mostly converted into hydrogen and carbon monoxide by partial oxidation with air. Within the POX reformer the fuel vapor and air (at air numbers of about 0.4) are mixed and converted within a hot zone into a gas mixture of mainly CO, H_2 and nitrogen. The partial oxidation reaction is exothermic and heat has to be removed from the POX reformer during operation of the system.

$$C_nH_{2n+2} + \frac{n}{2}O_2 \longrightarrow nCO + n+1H_2 \quad (16)$$

The POX reformer can be built as thermal POX reformer with temperatures between 1200 and 1300 °C or as a

Table 2. Typical product distribution in a POX reformer at 750 °C and air number of 0.4.

Species	H_2	CO	CH_4	N_2	CO_2	O_2	H_2O	Rest
Fraction (mol%)	23	25	0.35	50	0.6	0	0.3	<0.1

catalytic POX reformer at temperatures between 700 and 900 °C. Due to the catalyst a good conversion is reached already at the reduced temperature. The typical product distribution of a catalytic POX reformer for a temperature of 750 °C is given in Table 2.

A combination of both the steam reforming and the partial oxidation reaction is the auto thermal reforming. The process is more difficult to control but can be tailored to be either slightly exothermic or endothermic.

In fossil fuels such as oil, diesel and gasoline, sulfur is naturally present as mercaptanes or disulfides. In the form of thiophene it is added to pipeline natural gas for leak detection. These odorants can cause a deactivation of both the SOFC cell and the reformer catalysts and, thus, have to be removed prior to entering the reformer unit. For mercaptanes and disulfides ZnO can be used to trap the sulfur. Thiophenes require a hydrodesulfurization step in which the thiophene is converted to H_2S, which can be removed by ZnO.

4.1.2 Air conditioning

The assembly of ceramic layers in the SOFC is limited with respect to tolerable thermo-mechanical stresses. As a consequence precaution has to be taken that both the air and fuel gas are at an acceptable temperature prior to entering the SOFC stack. As pointed out in Section 3 this is particularly important for planar, flat plate designs.

Both fuel and air are supplied to the SOFC system at its operation temperature between 800 and 1000 °C depending on the cell design. After a reforming process the temperature of the pre-treated fuel gas is generally close to the operating temperature. Air is usually supplied from the surroundings at ambient temperature and, thus, requires a separate pre-heating step. Additionally filters are needed to remove dust particles. The transport of air through the system requires additional blowers, fans or compressors.

4.1.3 Off gas treatment

In the SOFC stack the fuel is not fully utilized. The stack off-gas still contains residual energy, which can be used as heat or electricity. In small SOFC systems the residual energy will usually be thermally used to provide additional heat. In a combustion reaction the fuel is completely oxidized into water and CO_2 which can be released into the environment. The combustion reaction can be adiabatic in a conventional burner or catalytic (catalytic burner). The lower temperature in the catalytic burner is advantageous with respect to NO_x-formation.

At a certain size of the SOFC system, the installation of gas and steam turbines is considered, which adds to the electrical power produced in the plant.

4.1.4 Thermal management of SOFC systems

Under operation of the SOFC, heat is released in addition to electrical power (at a ratio of 1/1 as a rough estimate). Therefore, almost similar amounts of electrical and thermal energy have to be handled during the operation of the SOFC system. Additional heat sources in an SOFC system can be the (POX)-reformer and the off-gas burner. The steam reformer and air pre-heater require heat and, thus, form heat sinks. In order to make efficient use of the fuel energy content, the coupling of heat sources and heat sinks is required. The design of the flow sheet and the thermal management of the system is an important issue for SOFC systems, particularly for the planar flat plate cell designs.

The transport of heat out of the SOFC-stack is usually performed by the air stream. The transferable amount of heat depends upon the airflow rate and the temperature difference between cell/stack inlet and outlet. The smaller the temperature gradients allowed over the cell, the higher the air flow rates needed in order to take up a certain amount of heat. The required airflow rates determine the design of periphery components such as blowers, compressors and filters.

The amount of heat released from the SOFC stack can, in principle, be reduced by using direct or indirect internal steam reforming as an additional heat sink. This would reduce the airflow and thus the effort for air transport through the system.

As seen from the considerations made, the properties of the ceramic fuel cell imposes specific aspects to the design of the overall SOFC system. This distinguishes the SOFC system from MCFC and PEMFCs. The impact of the SOFC operation upon the overall system performance (expressed in the efficiency) and the resulting economic aspects (here expressed as the cost of electricity) is described in the following section.

4.2 System analysis

4.2.1 Operation of the SOFC system (efficiency/costs)

As seen from the considerations made above the SOFC system is a complex coupling of various components, each with its own characteristics. An analysis of the system with respect to efficiency and economics has been presented in Ref. [33] for an SOFC co-generation system operating at 950 °C and using de-sulfurized pipeline natural gas as fuel. In this analysis a flat plate planar stack design is considered which restricts the tolerable temperature difference over the cell to a maximum of 100 K. Several operation parameters, e.g., the cell operating potential, degree of internal reforming, temperature span over the cell, fuel utilization, and pressure drop have been analyzed.

The impact of the degree of internal reforming is shown in Figure 17. The higher the degree of internal reforming, the higher is the electrical efficiency of the system and the less the costs for electricity production. A similar positive influence was observed for higher heat spans allowed over the cell. The main reason for these results is that both parameters allow the reduction of the amount of air fed into the stack. Internal reforming serves as a heat sink for the heat produced under operation, while a higher heating span allows for lower entrance temperatures of the gases.

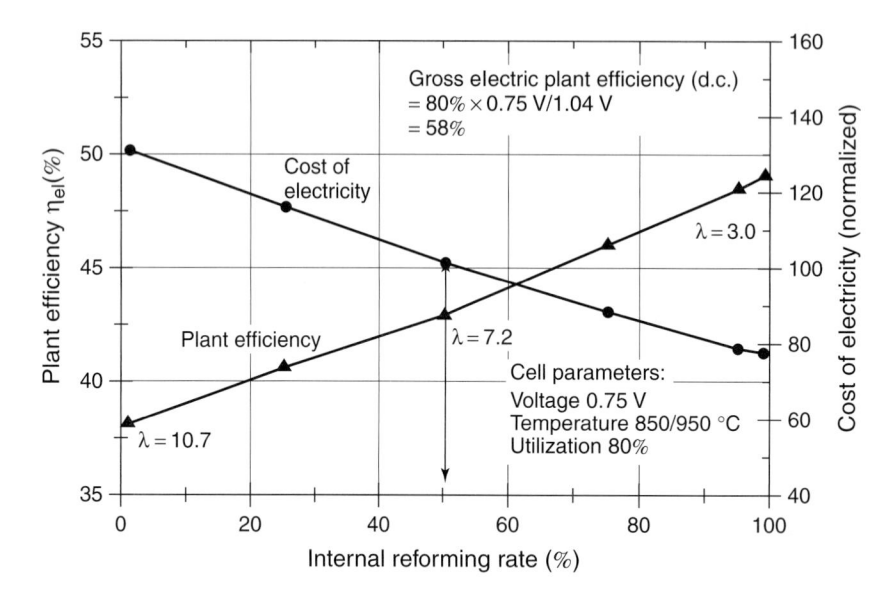

Figure 17. Effect of internal reforming on the efficiency and economics of an 200 kW$_{el}$-SOFC system for combined heat and power supply. (Reproduced from Riensche (1998)[33] with permission from Elsevier Science.)

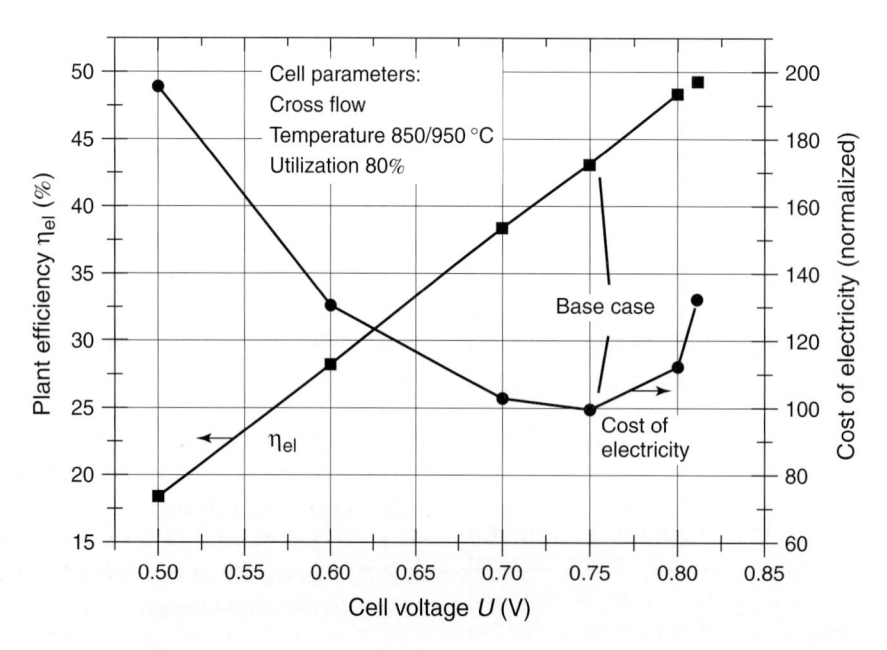

Figure 18. Effect of cell voltage on the efficiency and economics of an 200 kW$_{el}$-SOFC system for combined heat and power supply. (Reproduced from Riensche (1998)[33] with permission from Elsevier Science.)

This reduces significantly the effort for air supply and air preheating and, thus, is advantageous for both system efficiency and costs.

Like every fuel cell type, the SOFC can be operated at various cell voltages. The effect of the cell potential is shown in Figure 18. Increasing the cell voltage leads to a continuous increase in the efficiency as expected for increasing ratios of the actual cell voltage and the equilibrium potential difference. However, at the same time the cell current and, thus, the cell and stack power output decreases. As a consequence, larger active areas mean bigger stacks are needed in order to install a certain targeted power (here 200 kW). This leads to an increase in stack and materials costs, which overcompensate the back pay for the high electrical efficiency.

4.3 Stationary SOFC systems

At present SOFC systems up to 100 kW, and based on the tubular Westinghouse cells, have been demonstrated for co-generation of heat and power. Siemens-Westinghouse Power Corporation installed a test plant of 100 kW in Arnhem, the Netherlands which operated for more than 16 000 h in total. The system design is shown in Figure 19. The hot off-gas from the stack is subsequently used to preheat the air and produce steam prior to being supplied to the local district heating system.

With Pipeline natural gas as fuel the electrical efficiency was 43% with a fuel utilization of 90% without

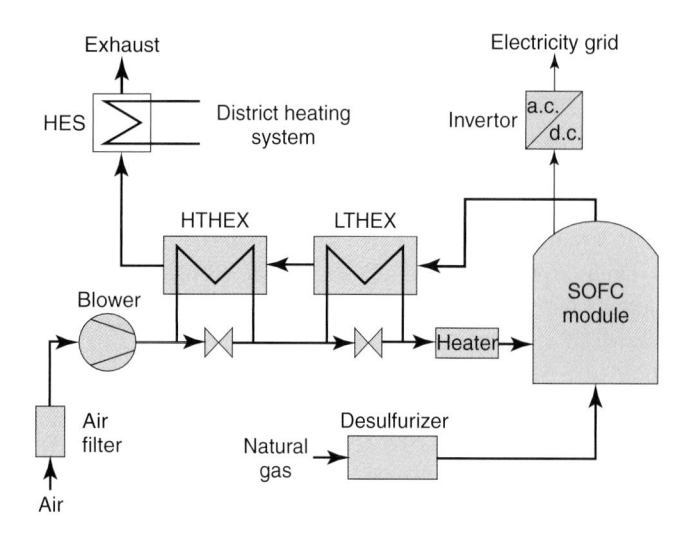

Figure 19. Flow sheet of a 100 kW$_{el}$ SOFC system for combined heat and power production from Siemens Westinghouse Power Corporation. The plant has been installed in Arnhem, the Netherlands and operated for more than 5000 h. (With permission of Siemens Westinghouse.)

degradation.[20] The plant has been transferred to a second customer and re-operated successfully.

The particular advantage of stationary SOFC systems is that the hot SOFC off-gas can be used in a gas or steam turbine in a combined cycle. For this application operation of the SOFC-stack at elevated pressure is advantageous for the turbine operation. A pressurized SOFC system of 250 kW coupled with a microturbine has been installed by Siemens-Westinghouse in California.[20] From this plant

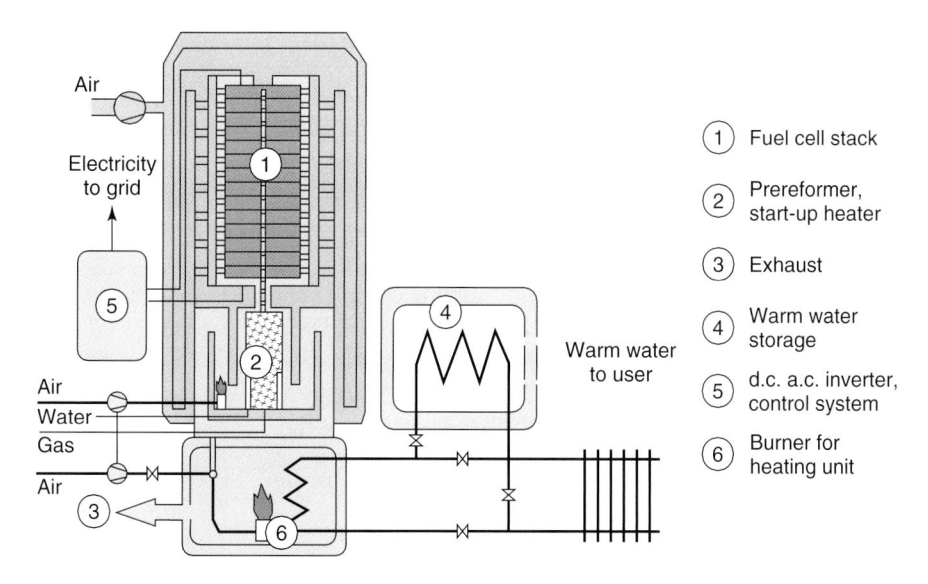

Figure 20. System concept of the Sulzer Hexis SOFC System of 1 kW$_{el}$ considered for domestic heat and power supply. (Sulzer Innotec). The system consists of fuel cell stack (1) pre-reformer serving as start-up heater, (2) exhaust, (3) warm water storage, (4) direct current (d.c.)/alternating current (a.c.) inverter control system, (5) and burner for heating unit. (6) The system is considered for integration in the infrastructure of a building. (Reproduced from Schmidt[41] (1998) with permission from Elsevier Science.)

experiences for the operation of co-generation systems will be achieved. Design studies and analysis have been performed for SOFC systems up to 300 MW$_{el}$ at Siemens-Westinghouse.[34]

Recently, SOFC systems for domestic/small business application have been attracting considerable interest. The first system currently in the field test phase is the Hexis system developed by Sulzer for combined heat and power supply with a power output of 1 kW$_{el}$. Figure 20 shows the design of the system which is intended for installation in a single household.[35] Because of the integrated heat exchanger and burner in the stack concept (see Section 3.2.2), the overall system design can be kept simple. A reformer is placed at the entrance to the stack and coupled with a heat exchanger for hot water supply. The electrical efficiency of the Hexis system is 35% while the total efficiency is 90% for pipeline natural gas as fuel.

Development of systems in the size of a few kW for domestic or small business applications have started at Global Thermoelectric, based on the anode substrate SOFC cell concept[29] and at Ceramic Fuel Cells Limited (AU)[36] based on electrolyte supported cells. Recently Acumentrics, USA, announced it would develop SOFC systems of this size-based on electrolyte, supported tubular cells several millimeters in diameter as described in Section 3.2.1.

4.4 Mobile applications

Since 1998, SOFC systems for mobile (nonstationary) applications have been under development. Here the emphasis is paid to SOFC's operating on gasoline and diesel. Targeted applications in the automotive area are auxiliary power units for passenger cars or trucks.[37]

The principle set-up for the auxiliary power unit (APU) developed at BMW/Delphi is shown in Figure 21. The system consists of a POX-reformer, SOFC stack and the waste energy recovery unit, that is an afterburner with integrated air pre-heater. In this system, an anode-supported SOFC stack fabricated from Global thermoelectric is used, enabling stack operating temperatures at around 750 °C (see Section 3.2.2). The hot components are placed into two thermally isolated compartments, allowing for individual thermal management of SOFC-stack and waste energy recovery unit in one and the POX reformer in the second compartment. The system has been operated with a stack power density of 0.37 W cm^{-2} at 750 °C and using a POX-reformate.

Although the feasibility of the concept has been proven, further optimization of thermal cycling stability and, especially, mass and volume are clearly needed. In contrast

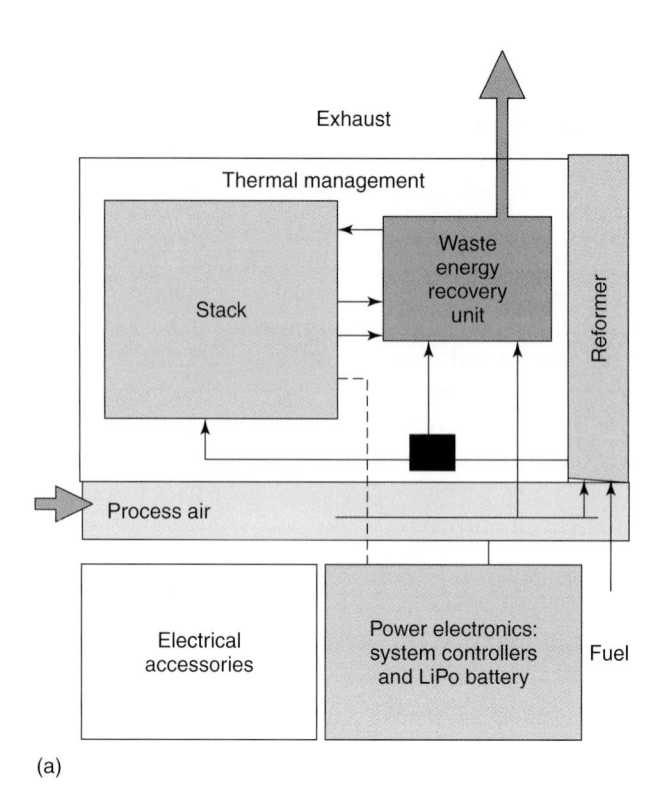

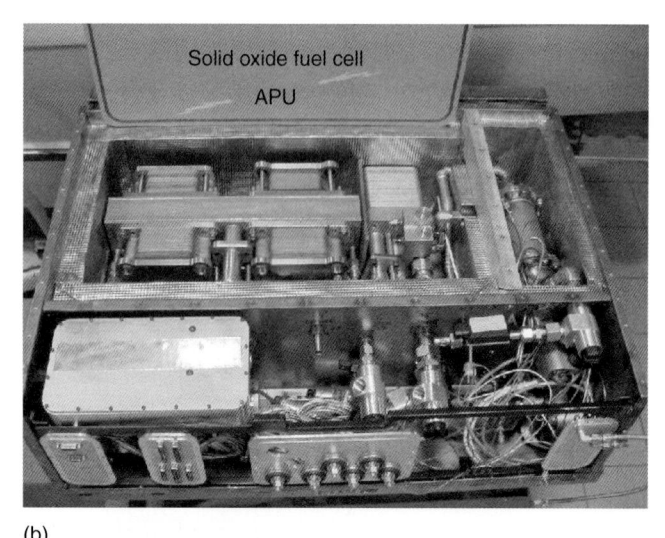

Figure 21. Flow sheet and photograph of the Delphi/BMW-auxiliary power unit considered for mobile application in passenger cars. (Reproduced by permission of Delphi Corporation.)

to larger and stationary SOFC systems, space restrictions rule out the use of volume-consuming heat exchangers in most mobile applications. Thus, alternative ways of handling the air and heat flow have to be developed. For APUs an active cooling of the hot box has been proposed.[38] The principle are an insulation built from multiple layers with space in between them allowing a process air stream inside the insulation. By variation of the air stream, the heat transfer through the insulation and, thus, the heat flow out of the hot compartments can be controlled.

Portable mobile SOFC systems based on tubular cells are currently under development at Honeywell (former Allied Signal) for even smaller systems of 0.5 kW.[39]

5 CONCLUDING REMARKS

In recent years, a paradigm shift has been observed in SOFC development. In addition to the originally considered application of SOFCs in large and stationary power plants and combined with a gas or steam turbine in a topping cycle, significant progress has been made in the development of smaller systems in the kW size for both stationary and mobile applications.

The driving force for these trends appears to be the particular advantage of SOFCs for operation on fossil fuels such as natural gas, oil/diesel and gasoline. The trends are supported by re-orientation of utilities to a more decentralized energy market, which will still be strongly dependent upon those fuels. Instead of large power plants, smaller units and virtual power plants are considered. Virtual power plants are an assembly of, e.g., domestic CHP systems that are centrally controlled by a utility. This offers a high flexibility and dynamics for supplying electrical power with high efficiency at alternating demands. The increasing demand of electricity in cars promotes the development of SOFCs, especially in the 1–20 kW size. It could be expected that synergetic effects in the development of both small systems for domestic and automotive application will accelerate the overall progress in the SOFC development. However, the future outcome will strongly depend upon cost reduction and reliability of SOFC technology.

REFERENCES

1. A. Hamou and J. Guindet, 'The CRC Handbook of Solid State Electrochemistry', H. J. Bouwmeester and P. Gellings (Eds), CRC Press, Inc, Boca Raton, FL (1997).
2. N. Q. Minh, *J. Am. Ceram. Soc.*, **76**, 563 (1993).
3. (a) I. Ishihara, M. Honda, T. Shibayama, H. Minami, H. Nishiguchi and Y. Takita, *J. Electrochem. Soc.*, **145**, 3177 (1998); (b) A. Matrazek, D. Kobertz, L. Singheiser and K. Hilpert, 'SOFC VII', "The Electrochemical Society Proceedings", Pennington, NJ, PV01-16, pp. 319 (2001).
4. S. Linderoth, 'Proceedings of the 4th European Solid Oxide Fuel Cell Forum, Luzern', A. J. McEvoy (Ed), European Fuel Cell Forum, Oberrohrbach, pp. 19 (2000).
5. P. Holtappels, I. C. Vinke, L. G. J. de Haart and U. Stimming, *J. Electrochem. Soc.*, **146**, 2976 (1999).
6. M. Brown, S. Primdahl and M. Mogensen, *J. Electrochem. Soc.*, **146**, 475 (1999).
7. S. Primdahl and M. Mogensen, *J. Electrochem. Soc.*, **146**, 2827 (1999).
8. P. Holtappels, L. G. J. de Haart, M. Mogensen, U. Stimming and I. C. Vinke, *J. Appl. Electrochem.*, **29**(5), 561 (1999).
9. M. Mogensen, *J. Electroceram.*, **5**, 141 (2000).
10. H. Kim, S. Park, J. M. Vohs and R. Gorte, *J. Electrochem. Soc.*, **147**, A693 (2001).
11. E. P. Murray, T. Tsai and S. A. Barnet, *Nature*, 400 (1999).
12. G. Stochniol, E. Syskakis and A. Naoumidis, *J. Am. Ceram. Soc.*, **78**, 929 (1995).
13. (a) M. Juhl, S. Primdahl, C. Manon and M. Mogensen, *J. Power Sources*, **61**, 173 (1996); (b) K. Sasaki and L. J. Gauckler, *Proc. Int. Symp. Struc. Func. Mater.*, **3**, 651 (1995).
14. P. Holtappels and C. Bagger, *J. Eur. Ceram. Soc.*, **22**, 41 (2002).
15. M. J. Joergensen and M. Mogensen, *J. Electrochem. Soc.*, **148**, A433 (2001).
16. J. W. Erning, Th. Hauber, U. Stimming and K. Wippermann, *J. Power Sources*, **61**, 205 (1996).
17. (a) E. Batawi, A. Plas, W. Straub, K. Honnegger and R. Diethelm, 'Solid Oxide Fuel Cells (SOFC6)', "The Electrochemical Society Proceedings", S. C. Singhal and M. Dokiya (Eds), The Electrochemical Society, Pennington, NJ, PV99-19, 767 (1999); (b) C. Gindorf, L. Singheiser, K. Hilpert, M. Schroeder, M. Martin, H. Greiner and F. Richter, 'SOFC6', "The Electrochemical Society Proceedings", S. C. Singhal and M. Dokiya (Eds), The Electrochemical Society, Pennington, NJ, PV99-19, 774 (1999).
18. P. H. Larsen, P. V. Hendriksen and M. Mogensen, 'Proceedings of the 3rd European Solid Oxide Fuel Cell Forum, Nantes', P. Stevensen (Ed), Solid Oxide Fuel Cell Forum, Luzern, 181 (1998).
19. T. A. Ring, 'Fundamentals of Ceramic Powder Processing and Synthesis', Academic Press, San Diego (1996).
20. S. C. Singhal, 'Solid Oxide Fuel Cells (SOFC6)', "Electrochemical Society Proceedings Series", S. C. Singhal and M. Dokiya (Eds), PV99-19, 39 (1999).
21. 'Fuel Cell Handbook' 5th edition, SOFC, EG&G Services, Parson National Technical Information Service, US Department of Commerce, VA, Chapter 8, pp. 8.1–8.24 (2000).
22. J. Van herle, J. Sfeir, R. Ihringer, N. M. Sammes, G. Tompsett, K. Kendall, K. Yamada, C. Wen, M. Ihara, T. Kawada and J. Mizusaki, '4th European Solid Oxide Fuel Cell Forum, Lucerne', A. J. McEvoy (Ed), European Fuel Cell Forum, Oberrohrdorf, Vol. 1, pp. 251 (2000).
23. H. Mori, H. Omura, N. Hisatome, K. Ikeda and K. Tomida, 'Solid Oxide Fuel Cells (SOFC VI)', "Electrochemical Society Proceedings", S. C. Singhal and M. Dokiya (Eds), The Electrochemical Society, Pennington, NJ, PV99-19, 53 (1999).

24. S. P. S. Badwal and K. Foger, 'Proceedings of the 3rd European Solid Oxide Fuel Cell Forum, Nantes', P. Stevensen (Ed), Solid Oxide Fuel Cell Forum, Luzern, 95 (1998).
25. R. Diethelm, M. Schmidt, K. Honnegger and E. Batawi, 'Solid Oxide Fuel Cells (SOFC6)', "Electrochemical Society Proceedings", S. C. Singhal and M. Dokiya (Eds), The Electrochemical Society, Pennington, NJ, PV99-19, 60 (1999).
26. W. Glatz, E. Batawi, M. Janousek, W. Kraussler, R. Zach and G. Zobl, 'Solid Oxide Fuel Cells (SOFC6)', "Electrochemical Society Proceedings", S. C. Singhal and M. Dokiya (Eds), The Electrochemical Society, Pennington, NJ, PV99-19, 783 (1999).
27. H. P. Buchkremer, U. Diekmann, L. G. J. de Haart, H. Kabs, U. Stimming and D. Stöver, 'Solid Oxide Fuel Cells, (SOFC V)', "Electrochemical Society Proceedings", U. Stimming, S. C. Singhal, H. Tagawa and W. Lehnert (Eds), The Electrochemical Society, Pennington, NJ, PV97-40, 160 (1997).
28. L. G. J. de Haart, I. C. Vinke, A. Janke, H. Ringel and F. Tietz, 'Solid Oxide Fuel Cells (SOFC VII)', "Electrochemical Society Proceedings", H. Yokokawa and S. C. Singhal (Eds), The Electrochemical Society, Pennington, NJ, PV01-16, 111 (2001).
29. D. Gosh, E. Tang, D. Prediger, M. Pastula and R. Boersma, 'Solid Oxide Fuel Cells (SOFC VII)', "Electrochemical Society Proceedings", H. Yokokawa and S. C. Singhal (Eds), The Electrochemical Society, Pennington, NJ, PV01-16, 100 (2001).
30. G. Schiller, T. Franco, R. Henne, M. Lang and R. Ruckdäschl, 'Solid Oxide Fuel Cells (SOFC VII)', "The Electrochemical Society Proceedings", H. Yokokawa and S. C. Singhal (Eds), The Electrochemical Society, Pennington, NJ, PV01-16, 885 (2001).
31. M. J. Day, 'Proceedings of the 4th European Solid Oxide Fuel Cell Forum, Lucerne', A. J. McEvoy (Ed), European Fuel Cell Forum, Oberrohrdorf, Vol. 1, 133 (2000).
32. Y. Sakaki, A. Nakanishi, M. Hattori, H. Miyamoto, H. Aiki and K. Takenobu, 'Solid Oxide Fuel Cells (SOFC VII)', "The Electrochemical Society Proceedings", H. Yokokawa and S. C. Singhal (Eds), The Electrochemical Society, Pennington, NJ, PV01-16, 72 (2001).
33. E. Riensche, U. Stimming and G. Unverzagt, *J. Power Sources*, **73**, 251 (1998).
34. W. L. Lundberg and S. E. Veyo, 'Solid Oxide Fuel Cells (SOFC VII)', "The Electrochemical Society Proceedings", H. Yokokawa and S. C. Singhal (Eds), The Electrochemical Society, Pennington, NJ, PV01-16, 78 (2001).
35. R. Diethelm, M. Schmidt, K. Honnegger and E. Batawi, 'Solid Oxide Fuel Cells (SOFC6)', "The Electrochemical Society Proceedings", S. C. Singhal and M. Dokiya (Eds), The Electrochemical Society, Pennington, NJ, PV99-19, 783 (1999).
36. B. Godfrey and K. Föger, 'Solid Oxide Fuel Cells (SOFC VII)', "The Electrochemical Society Proceedings", H. Yokokawa and S. C. Singhal (Eds), The Electrochemical Society, Pennington, NJ, PV01-16, 120 (2001).
37. S. Mukerjee, M. J. Grieve, K. Hattiner, M. Faville, J. Noetzel, K. Keegan, D. Schumann, D. Armstrong, D. England, J. Haller and C. DeMinco, 'Solid Oxide Fuel Cells (SOFC VII)', "The Electrochemical Society Proceedings", H. Yokokawa and S. C. Singhal (Eds), The Electrochemical Society, Pennington, NJ, PV01-16, 173 (2001).
38. D. Arthur, 'Final Report', Little Inc., Ref. 71316 Jan (2001).
39. N. Minh, A. Anumakonda, R. Doshi, J. Guan, S. Huss, G. Lear, K. Montgomery, E. Ong and J. Yamanis, 'Solid Oxide Fuel Cells (SOFC VII)', "The Electrochemical Society Proceedings", H. Yokokawa and S. C. Singhal (Eds), The Electrochemical Society, Pennington, NJ, PV01-16, 190 (2001).
40. P. Hoptappels, in 'Metalle in der Elektrochemie-Gewinmung, Abscheidung, Korrosion,' W. Plieth and J. Russow (Eds), GDCh-monographien Bd. 18, Gesellscharft Deutscher Chemiker, Frankfurt, p. 18 (2000).
41. M. Schmidt, *Fuel Cell Bull.*, **1**, 9 (1998).

Chapter 21
Biochemical fuel cells

E. Katz, A. N. Shipway and I. Willner
Institute of Chemistry and The Farkas Center for Light-Induced Processes, The Hebrew University of Jerusalem, Jerusalem, Israel

1 INTRODUCTION

During the 20th century, energy consumption increased dramatically and an unbalanced energy management exists. While there is no sign that this growth in demand will abate (particularly amongst the developing nations), there is now an awareness of the transience of nonrenewable resources and the irreversible damage caused to the environment. In addition, there is a trend towards the miniaturization and portability of computing and communications devices. These energy-demanding applications require small, light power sources that are able to sustain operation over long periods of time, particularly in remote locations such as space and exploration. Furthermore, advances in the medical sciences are leading to an increasing number of implantable electrically operated devices (e.g., pacemakers). These items need power supplies that will operate for extremely long durations as maintenance would necessitate surgery. Ideally, implanted devices would take advantage of the natural fuel substances found in the body, and thus would continue to draw power for as long as the subject lives. Biofuel cells potentially offer solutions to all these problems, by taking nature's solutions to energy generation and tailoring them to our own needs. They take readily available substrates from renewable sources and convert them into benign by-products with the generation of electricity. Since they use concentrated sources of chemical energy, they can be small and light, and the fuel can even be taken from a living organism (e.g., glucose from the blood stream).

Biofuel cells use biocatalysts for the conversion of chemical energy to electrical energy.[1-8] As most organic substrates undergo combustion with the evolution of energy, the biocatalyzed oxidation of organic substances by oxygen or other oxidizers at two-electrode interfaces provides a means for the conversion of chemical to electrical energy. Abundant organic raw materials such as methanol, organic acids, or glucose can be used as substrates for the oxidation process, and molecular oxygen or H_2O_2 can act as the substrate being reduced. The extractable power of a fuel cell (P_{cell}) is the product of the cell voltage (V_{cell}) and the cell current (I_{cell}) (equation (1)). Although the ideal cell voltage is affected by the difference in the formal potentials of the oxidizer and fuel compounds ($E_{ox}^{o\prime} - E_{fuel}^{o\prime}$), irreversible losses in the voltage (η) as a result of kinetic limitations of the electron transfer processes at the electrode interfaces, ohmic resistances and concentration gradients, lead to decreased values (equation (2)).

$$P_{cell} = V_{cell} \times I_{cell} \quad (1)$$
$$V_{cell} = E_{ox}^{o\prime} - E_{fuel}^{o\prime} - \eta \quad (2)$$

Similarly, the cell current is controlled by the electrode sizes, the ion permeability and transport rates across the membrane separating the catholyte and anolyte compartments of the biofuel cell (specifically, the rate of electron transfer at the respective electrode surfaces). These different parameters collectively influence the biofuel cell power, and for improved efficiencies, the V_{cell} and I_{cell} values should be optimized.

Biofuel cells can use biocatalysts, enzymes or even whole cell organisms in one of two ways.[1–8] Either (i) the biocatalysts can generate the fuel substrates for the cell by biocatalytic transformations or metabolic processes, or (ii) the biocatalysts may participate in the electron transfer chain between the fuel substrates and the electrode surfaces. Unfortunately, most redox enzymes do not take part in direct electron transfer with conductive supports, and therefore a variety of electron mediators (electron relays) are used for the electrical contacting of the biocatalyst and the electrode.[9] Recently, novel approaches have been developed for the functionalization of electrode surfaces with monolayers and multilayers consisting of redox enzymes, electrocatalysts and bioelectrocatalysts that stimulate electrochemical transformations at the electrode interfaces.[10] The assembly of electrically contacted bioactive monolayer electrodes could be advantageous for biofuel cell applications as the biocatalyst and electrode support are integrated. This chapter summarizes recent advances in the tailoring of conventional microbial-based biofuel cells and describes novel biofuel cell configurations based on biocatalytic interface structures integrated with the cathodes and anodes of biofuel cells.

2 MICROBIAL-BASED BIOFUEL CELLS

The use of entire microorganisms as microreactors in fuel cells eliminates the need for the isolation of individual enzymes, and allows the active biomaterials to work under conditions close to their natural environment, thus at a high efficiency. Whole microorganisms can be difficult to handle, however, requiring particular conditions to remain alive, and their direct electrochemical contact with an electrode support is virtually impossible.

Microorganisms have the ability to produce electrochemically active substances that may be metabolic intermediaries or final products of anaerobic respiration. For the purpose of energy generation, these fuel substances can be produced in one place and transported to a biofuel cell to be used as a fuel. In this case the biocatalytic microbial reactor produces the biofuel and the biological part of the device is not directly integrated with the electrochemical part (Figure 1a). This scheme allows the electrochemical part to operate under conditions that are not compatible with the biological part of the device. The two parts can even be separated in time, operating completely individually. The most widely used fuel in this scheme is hydrogen

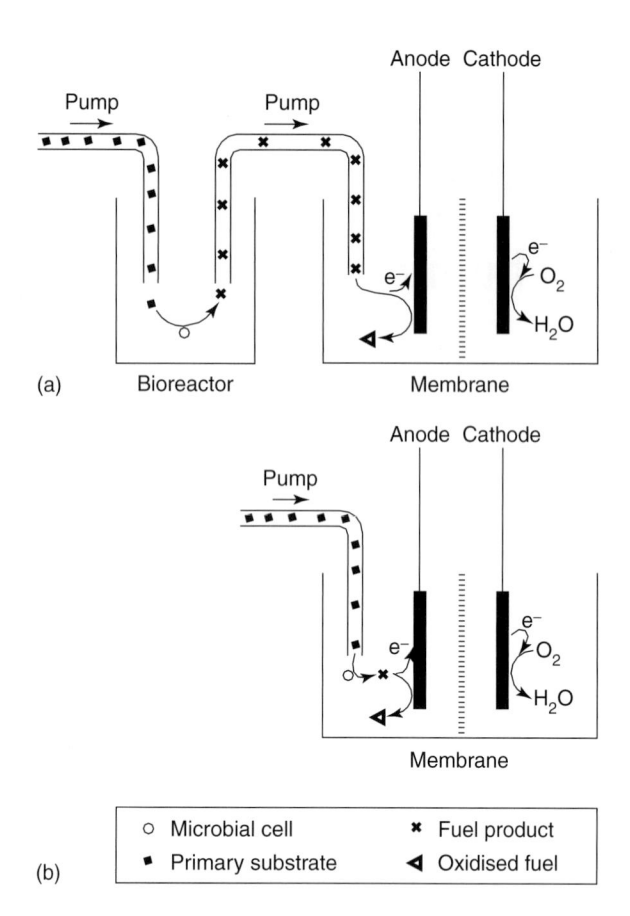

Figure 1. Schematic configuration of a microbial biofuel cell: (a) with a microbial bioreactor providing fuel separated from the anodic compartment of the electrochemical cell; (b) with a microbial bioreactor providing fuel directly in the anodic compartment of the electrochemical cell.

gas, allowing well-developed and highly efficient H_2/O_2 fuel cells to be conjugated with a bioreactor.

According to another approach, the microbiological fermentation process proceeds directly in the anodic compartment of a fuel cell, supplying the anode with the in situ produced fermentation products (Figure 1b). In this case the operational conditions in the anodic compartment are dictated by the biological system, so they are significantly different from those in conventional fuel cells. At this point we have a real biofuel cell and not a simple combination of a bioreactor with a conventional fuel cell. This configuration is also often based on the biological production of hydrogen gas, but the electrochemical oxidation of H_2 is performed in the presence of the biological components under mild conditions. Other metabolic products (e.g., formate, H_2S) have also been used as fuels in this kind of system.

A third approach involves the application of artificial electron transfer relays that can shuttle electrons between the microbial biocatalytic system and the electrode. The mediator molecules take electrons from the biological electron transport chain of the microorganisms and transport them to the anode of the biofuel cell. In this case, the biocatalytic process performed in the microorganisms becomes different from the natural one since the electron flow goes to the anode instead of to a natural electron acceptor. Since the natural electron acceptor is usually more efficient, it can compete with the desired scheme, so it is usually removed from the system. In most cases, the microbiological system operates under anaerobic conditions (when O_2 is removed from the system), allowing electron transport to the artificial electron relays and, finally, to the anode.

2.1 Microbial bioreactors producing H_2 for conventional fuel cells

Various bacteria and algae, for example *Escherichia coli*, *Enterobacter aerogenes*, *Clostridium butyricum*, *Clostridium acetobutylicum*, and *Clostridium perfringens* have been found to be active in hydrogen production under anaerobic conditions.[11–16] The most effective H_2 production is observed upon fermentation of glucose in the presence of *Clostridium butyricum* (strain IFO 3847, 35 µmol h^{-1} H_2 evolution by 1 g of the microorganism at 37 °C).[17] This conversion of carbohydrate to hydrogen is achieved by a multienzyme system. In bacteria the route is believed to involve glucose conversion to 2 mol of pyruvate and 2 mol of NADH by the Embden–Meyerhof pathway. The pyruvate is then oxidized through a pyruvate–ferredoxin oxidoreductase producing acetyl-CoA, CO_2, and reduced ferredoxin. NADH–ferredoxin oxidoreductase oxidizes NADH and reduces ferredoxin. The reduced ferredoxin is reoxidized by the hydrogenase to form hydrogen. As a result, 4 mol of hydrogen are produced from 1 mol of glucose under ideal conditions (equations (3)–(6)). However, only ca. 1 mol of H_2 per 1 mol of glucose was obtained under optimal conditions in a real system. Since the H_2 yield is only ca. 25% of the theoretical yield,[18] the improvement of hydrogen production by genetic engineering techniques and screening of new hydrogen-producing bacteria is possible for enhanced energy conversion. Glucose is an expensive substrate, and industrial wastewater containing nutritional substrates for H_2-producing bacteria have been successfully applied to produce hydrogen later used in a fuel cell.[17]

$$\text{Glucose} + 2\text{NAD}^+ \xrightarrow{\text{Multienzyme Embden–Meyerhof pathway}} 2\text{Pyruvate} + 2\text{NADH} \quad (3)$$

$$\text{Pyruvate} + \text{Ferredoxin}_\text{ox} \xrightarrow{\text{Pyruvate–ferredoxin oxidoreductase}} \text{Acetyl-CoA} + CO_2 + \text{Ferredoxin}_\text{red} \quad (4)$$

$$\text{NADH} + \text{Ferredoxin}_\text{ox} \xrightarrow{\text{NADH-ferredoxin oxidoreductase}} \text{NAD}^+ + \text{Ferredoxin}_\text{red} \quad (5)$$

$$\text{Ferredoxin}_\text{red} + 2\text{H}^+ \xrightarrow{\text{Hydrogenase}} \text{Ferredoxin}_\text{ox} + H_2 \quad (6)$$

The immobilization of hydrogen-producing bacteria, *Clostridium butyricum*, has great value because this stabilizes the relatively unstable hydrogenase system. In order to stabilize the biocatalytic performance, the bacteria were introduced into polymeric matrices, e.g., polyacrylamide,[17] agar gel,[19, 20] and filter paper.[18] The immobilized microbial cells continuously produced H_2 under anaerobic conditions for a period of weeks, whereas nonimmobilized bacteria cells were fully deactivated in less than 2 days.[19]

A H_2/O_2 fuel cell (Pt-black/nickel mesh anode and Pd-black/nickel mesh cathode separated by a nylon filter and operated at room temperature) was connected to a bioreactor (Jar-fermentor) producing H_2.[19, 20] The H_2 gas produced was collected and transported to the anodic compartment of the fuel cell, where the gas was used as a fuel (equation (7)). The current and voltage output were dependent on the rate of hydrogen production in the fermentor. For example, an open-circuit voltage (V_oc) of 0.95 V and short-circuit current density (i_sc) of 40 mA cm^{-2} were obtained at the H_2 flow of 40 ml min^{-1}. The biofuel cell operating at steady-state conditions for 7 days reveled a continuous current of between 500 and 550 mA.[20]

$$H_2 \longrightarrow 2H^+ + 2e^- \text{ (to anode)} \quad (7)$$

2.2 Integrated microbial-based biofuel cells producing electrochemically active metabolites in the anodic compartment of biofuel cells

Microbial cells producing H_2 gas during fermentation have been immobilized directly in the anodic compartment of a H_2/O_2 fuel cell.[21, 22] A rolled Pt-electrode was introduced into a suspension of *Clostridium butyricum* microorganisms, then the suspension was polymerized with acrylamide to form a gel.[21] The fermentation was conducted directly at the electrode surface, supplying the anode with the H_2 fuel. In this case some additional by-products of the fermentation process (hydrogen, 0.60 mol; formic acid, 0.20 mol; acetic acid, 0.60 mol; lactic acid,

0.15 mol)[21] could also be utilized as additional fuel components. For example, pyruvate produced according to equation (3) can be alternatively oxidized to formate through a pyruvate–formate lyase (equation (8)).[17, 21] The metabolically produced formate is directly oxidized at the anode when the fermentation solution passes the anode compartment (equation (9)). The biofuel cell that included ca. 0.4 g of wet microbial cells (ca. 0.1 g of dry material) yielded upon optimal operating conditions the outputs $V_{cell} = 0.4$ V and $I_{cell} = 0.6$ mA.[21]

$$\text{Pyruvate} \xrightarrow{\text{Pyruvate–formate lyase}} \text{Formate} \quad (8)$$

$$HCOO^- \longrightarrow CO_2 + H^+ + 2e^- \text{(to anode)} \quad (9)$$

It should be noted that in the case that a Pt-black electrode is used as an anode, oxidation of the original substrate utilized by the microorganisms in the fermentation process (e.g., glucose) can contribute to the anodic current. Thus, the H_2 provided by the microorganisms is the main, but not the only source of the anodic current.[23]

Other fuels have also been produced by microorganisms in the anodic compartments of biofuel cells. There are many microorganisms producing metabolically reduced sulfur-containing compounds (e.g., sulfides, S^{2-}, HS^-, sulfites, SO_3^{2-}). Sulfate-reducing bacteria (e.g., *Desulfovibrio desulfuricans*) form a specialized group of anaerobic microbes that use sulfate (SO_4^{2-}) as a terminal electron acceptor for respiration. These microorganisms yield S^{2-} while using a substrate (e.g., lactate) as a source of electrons (equation (10)). This microbiological oxidation of lactate with the formation of sulfide has been used to drive an anodic process in biofuel cells.[24, 25] The metabolically produced sulfide was oxidized directly at an electrode, providing an anodic reaction that produces sulfate or thiosulfate (equations (11) and (12)).

$$\text{Lactate} + SO_4^{2-} + 8H^+ \xrightarrow{\text{Bacteria}} S^{2-} + 4H_2O$$
$$+ \text{Pyruvate} \quad (10)$$

$$S^{2-} + 4H_2O \longrightarrow SO_4^{2-} + 8H^+$$
$$+ 8e^- \text{(to anode)} \quad (11)$$

$$2S^{2-} + 3H_2O \longrightarrow S_2O_3^{2-} + 6H^+$$
$$+ 8e^- \text{(to anode)} \quad (12)$$

The fermentation solution was composed of a microbe suspension (ca. 10^8 nonimmobilized cells per mL), with the nutritional substrates (mainly lactate) under anaerobic conditions. Accumulation of sulfides in the medium results in the inhibition of the metabolic bacteria process because of their interaction with iron containing proteins (e.g., cytochromes), causing the electron transport systems to be blocked. To prevent the toxic effect of H_2S, the anode should effectively oxidize it. However, many metallic electrodes are poisoned by sulfide because of its strong and irreversible adsorption. Thus, porous graphite electrodes were used (100 cm^2, impregnated with 10% (w/w) cobalt hydroxide, which in the presence of S^{2-} undergoes a transition into a catalytically highly active cobalt oxide/cobalt sulfide mixture).[24, 25] The biocatalytic anode was combined with an oxygen cathode (porous graphite electrode, 100 cm^2 geometrical area, activated with iron(II) phthalocyanine and vanadium(V) compounds) separated with a cation-exchange membrane in order to maintain anaerobic conditions in the anodic compartment. In a test study,[25] the electrical output of the biofuel element composed of three cells connected in series was $V_{oc} = 2.8$ V and $I_{sc} = 2.5$–4.0 A ($i_{sc} = $ ca. 30 mA cm^{-2}). The element was loaded discontinuously for a period of 18 months, about 6 A being drawn from the cell for 40–60 min daily.

Microbiological fermentation under aerobic conditions utilizes O_2 as a terminal electron acceptor. It has been shown that aerobic fermentation of *Saccharomyces cerevisiae* or *Micrococcus cerificans* bacteria in the presence of glucose as the nutritional substrate in an anodic compartment of a biofuel cell results in an anodic current.[26–29] A biofuel cell in such a system works as an O_2-concentration cell utilizing the potential difference produced at the cathode and anode due to the oxygen consumption in the anodic compartment.

Table 1 summarizes the electrical output obtained in biofuel cells operating without electron transfer mediators and using the natural products of microbial fermentation (e.g., H_2, H_2S) as the current providing species.

2.3 Microbial-based biofuel cells operating in the presence of artificial electron relays

Reductive species generated by metabolic processes inside microbial cells are isolated from the external world by a microbial membrane. Thus, contact of the microbial cells with an electrode usually results in a very minute electron transfer across the membrane of the microbes.[30] In some specific cases, however, direct electron transfer from the microbial cells to an anode surface is still possible. The metal-reducing bacterium *Shewanella putrefaciens* MR-1 has been reported to have cytochromes in its outer membrane.[31] These electron carriers (i.e., cytochromes) are able to generate anodic current in the absence of terminal electron acceptors (under anaerobic conditions).[32, 33] However, this is a rather exceptional example.

Table 1. Examples of microbial-based biofuel cells utilizing fermentation products for their oxidation at anodes.[a]

Microorganism	Nutritional substrate	Fermentation product	Biofuel cell voltage	Biofuel cell current or current density	Anode[c]	Ref.
Clostridium butyricum	Waste water	H_2	0.62 V (at 1 Ω)	0.8 A (at 2.2 V)	Pt-blackened Ni, 165 cm^2 (5 anodes in series)	[19]
Clostridium butyricum	Molasses	H_2	0.66 V (at 1 Ω)	40 mA cm^{-2} (at 1 Ω)[b]	Pt-blackened Ni, 85 cm^2	[20]
Clostridium butyricum	Lactate	H_2	0.6 V (oc)[d]	120 μA cm^{-2} (sc)[e]	Pt-black, 50 cm^2	[21]
Enterobacter aerogenes	Glucose	H_2	1.04 V (oc)	60 μA cm^{-2} (sc)	Pt-blackened stainless steel, 25 cm^2	[22]
Desulfovibrio desulfuricans	Dextrose	H_2S	2.8 V (oc)	1 A	Graphite, Co(OH)$_2$ impregnated (3 anodes in series)	[25]

[a] In most studies the biofuel anode was conjugated with an O_2-cathode.
[b] The value calculated from other data using Ohm's law.
[c] The anode surface area is given as a geometrical area.
[d] Open-circuit measurements.
[e] Short-circuit measurements.

Low molecular weight redox species may assist the shuttling of electrons between the intracellular bacterial space and an electrode. However, there are many important requirements that such a mediator should satisfy in order to provide an efficient electron transport from the bacterial metabolites to the anode: (a) The oxidized state of the mediator should easily penetrate the bacterial membrane to reach the reductive species inside the bacterium. (b) The redox potential of the mediator should fit the potential of the reductive metabolite (the mediator potential should be positive enough to provide fast electron transfer from the metabolite, but it should not be so positive as to prevent significant loss of potential). (c) Neither oxidation state of the mediator should interfere with other metabolic processes (should not inhibit them or be decomposed by them). (d) The reduced state of the mediator should easily escape from the cell through the bacterial membrane. (e) Both oxidation states of the mediator should be chemically stable in the electrolyte solution, they should be well soluble, and they should not adsorb on the bacterial cells or electrode surface. (f) The electrochemical kinetics of the oxidation process of the mediator-reduced state at the electrode should be fast (electrochemically reversible).[8]

Many different organic and organometallic compounds have been tested in combination with bacteria to test the efficiency of mediated electron transport from the internal bacterial metabolites to the anode of a biofuel cell. Thionine has been used extensively as a mediator of electron transport from *Proteus vulgaris*[34–38] and from *Escherichia coli*.[38,39] Other organic dyes that have been tested include benzylviologen, 2,6-dichlorophenolindophenol, 2-hydroxy-1,4-naphthoquinone, phenazines (phenazine ethosulfate, safranine), phenothiazines (alizarine brilliant blue, *N,N*-dimethyl-disulfonated thionine, methylene blue, phenothiazine, toluidine blue) and phenoxazines (brilliant cresyl blue, gallocyanine, resorufin).[36,38–44] These organic dyes were tested with *Alcaligenes eutrophus, Anacystis nidulans, Azotobacter chroococcum, Bacillus subtilis, Clostridium butyricum, Escherichia coli, Proteus vulgaris, Pseudomonas aeruginosa, Pseudomonas putida* and *Staphylococcus aureus* bacteria, usually using glucose and succinate as substrates. Among the dyes tested, phenoxazine, phenothiazine, phenazine, indophenol, bipyridilium derivatives, thionine and 2-hydroxy-1,4-naphthoquinone were found to be very efficient in maintaining relatively high cell voltage output when current was drawn from the biofuel cell.[36] Some other dyes did not function as effective mediators because they are not rapidly reduced by microorganisms, or they lacked sufficiently negative potentials. Ferric chelate complexes (e.g., Fe(III)EDTA) were successfully used with *Lactobacillus plantarum, Streptococcus lactis* and *Erwinia dissolvens*, oxidizing glucose.[45]

Since thionine has frequently been used as a mediator in microbial biofuel-based cells, mono and disulfonated derivatives of thionine have been applied to determine the effect of hydrophilic substituents on the mediation of electron transfer from *Escherichia coli* to an anode.[46] Changing from thionine to 2-sulfonated thionine and 2,6-disulfonated thionine results in an increase in the efficiency of the mediated electron transport. This increase is reflected by changes in the biofuel cell current under a 560 Ω load – 0.35, 0.45 and 0.6 mA for thionine, monosulfo- and

Table 2. Redox potentials of electron relays used in microbial-based biofuel cells and the kinetics of their reduction by microbial cells.[a]

Redox relay	Structural formula	Redox potential (V vs. NHE)[b]	Rate of reduction (μmol (g dry wt)$^{-1}$ s^{-1})[c]
2,6-Dichlorophenol-indophenol		0.217	0.41
Phenazine ethosulphate		0.065	8.57
Safranine-O		−0.289	0.07
N,N-Dimethyl-disulphonated thionine		+0.220	0.33
New Methylene Blue		−0.021	0.20
Phenothiazinone		+0.130	1.43
Thionine		+0.064	7.10
Toluidine Blue-O		+0.034	1.47
Gallocyanine		+0.021	0.53

Table 2. (*continued*)

Redox relay	Structural formula	Redox potential (V vs. NHE)[b]	Rate of reduction (μmol (g dry wt)$^{-1}$ s^{-1})[c]
Resorufin		−0.051	0.61

[a]The data are taken from Ref. [38].
[b]$E^{\circ\prime}$ at pH 7.0. NHE, normal hydrogen electrode.
[c]The dye reduction by *Proteus vulgaris* at 30 °C, with 50 μM dye and 0.10–0.15 mg (dry wt) mL^{-1} of microbial cells. The oxidizable substrate is glucose.

disulfo-derivatives, respectively. The low efficiencies of the biofuel cells operating with thionine and 2-sulfonated thionine were attributed to interference to electron transfer by adsorption of the mediator on the microbial membrane. Table 2 summarizes the structures and redox potentials of electron transfer mediators as well as the rate constants of their reduction by microorganisms. It should be noted that the overall efficiency of the electron transfer mediators also depends on many other parameters, and in particular on the electrochemical rate constant of mediator re-oxidation, which depends on the electrode material.

Since an electron transfer mediator needs to meet many requirements, some of which are mutually exclusive, it is not possible to reach perfect conditions for electron transport from a bacterial cell to an electrode. A mixture of two mediators can be useful in optimizing the efficiency. A solution containing thionine and Fe(III)EDTA was applied to mediate electron transport from *Escherichia coli*, oxidizing glucose as a primary substrate to an anode.[47] Although both mediators can be reduced by the *Escherichia coli*, thionine is reduced over 100 times faster than Fe(III)EDTA. The electrochemical oxidation of the reduced thionine is much slower than oxidation of Fe(II)EDTA, however. Therefore, electrons obtained from the oxidation of glucose in the presence of *Escherichia coli* are transferred mainly to thionine under the operational conditions of the cell. The reduced thionine is rapidly re-oxidized by Fe(III)EDTA, the rate of which has been shown to be very fast, $k_{et} = 4.8 \times 10^4$ M^{-1} s^{-1}. Finally, the reduced chelate complex, Fe(II)EDTA, transfers electrons to the anode by the electrode reaction of a Fe(III)EDTA/Fe(II)EDTA couple with a sufficiently large rate constant, $k_{el} = 1.5 \times 10^{-2}$ cm s^{-1}. One more example of the enhanced electron transport in the presence of a mixture of mediators has been shown for *Bacillus* strains oxidizing glucose as a primary substrate. The biofuel cell was operating in the presence of methylviologen (MV^{2+}) and 2-hydroxy-1,4-naphthoquinone or Fe(III)EDTA.[16] Methylviologen can efficiently accept electrons from the bacterial cells, but its reduced state (MV$^{\bullet+}$) is highly toxic for the bacteria and immediately inhibits the fermentation process. In the presence of a secondary mediator that has a more positive potential, MV$^{\bullet+}$ is efficiently re-oxidized to MV^{2+}. The reduced secondary mediator (quinone or Fe(II)EDTA) then transports the electrons to the anode.

Engineering of the electrochemical cell provides a means of enhancing the electrical contact between a biocatalytic system and an anode and to improve the cell output. The interfacial contact has been increased using a three-dimensional packed bed anode.[48] The anode compartment was filled with graphite particles mixed with *Escherichia coli* bacteria, glucose as a primary electron source, and 2-hydroxy-1,4-naphthoquinone as a diffusional mediator. The bed electrode provided an active surface area 27.4-fold larger than the cross-sectional area of the anodic compartment, thus, the anodic current extracted was increased by an order of magnitude over a plate electrode.

In order to organize an integrated biocatalytic assembly in the anode compartment of a biofuel cell, the microbial cells and the electron transfer mediator should be co-immobilized at the anode surface. This goal is, however, very difficult to achieve since the electron transfer mediator needs to have some freedom to reach the intracellular bacterial space to interact with the metabolic species. In fact, the co-immobilization of bacterial cells with electron transfer mediators is much more difficult than the co-immobilization of redox enzymes with their respective mediators because in the case of microbial cells, the active species are isolated within a cellular membrane. This problem has been addressed using several methods. Neutral red (**1**), an organic dye known to be an active diffusional mediator for electron transport from *Escherichia coli*,[40] has been covalently linked to a graphite electrode by making an amide bond between a carboxylic group on the electrode surface and an amino group of the dye[49] (Figure 2a). The mediator-modified electrode was used as an anode in the presence of *Escherichia coli*, and the surface-bound dye provided electron transfer

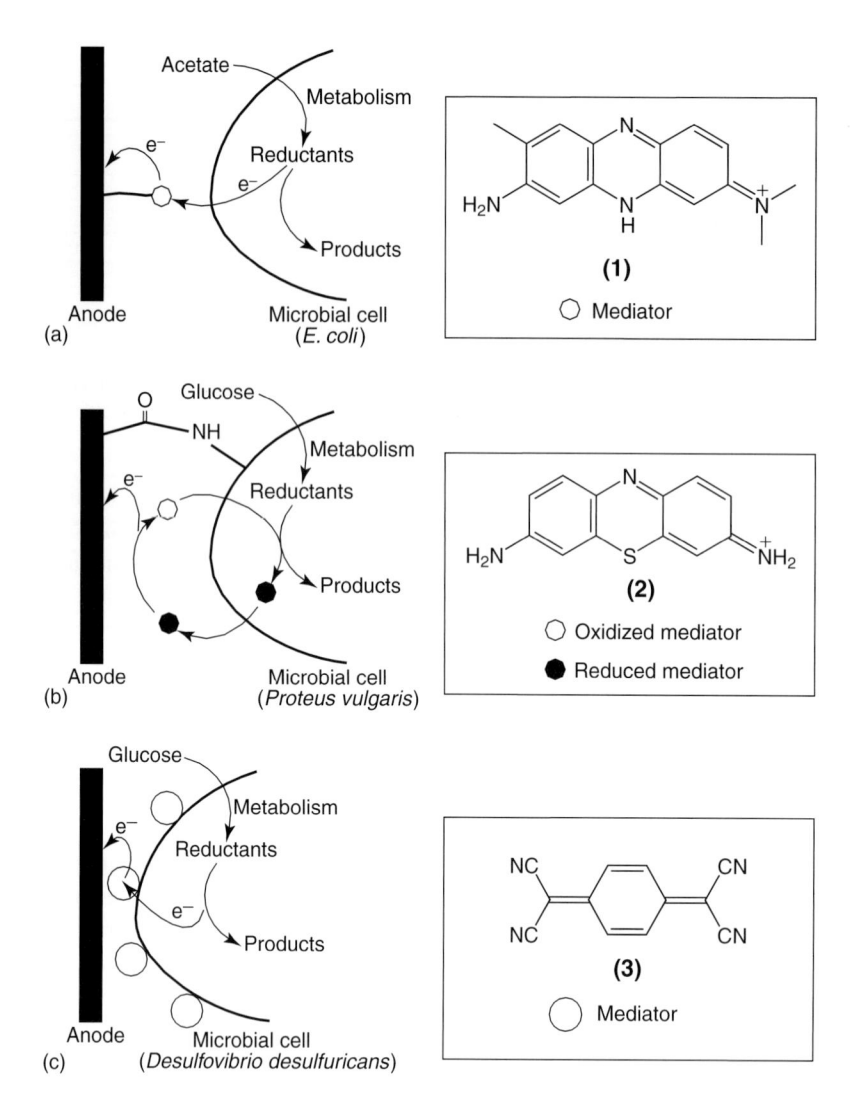

Figure 2. Electrical wiring of microbial cells to the anode of the electrochemical cell using electron transfer mediators: (a) a diffusional mediator shuttling between the microbial suspension and the anode surface; (b) a diffusional mediator shuttling between the anode and microbial cells covalently linked to the electrode; (c) a mediator adsorbed on the microbial cells providing the electron transport from the cells to the anode.

from the microbial cells to the conductive support under anaerobic conditions. In this case, only those bacteria that reached the modified electrode surface were electrically contacted. *Proteus vulgaris* microbial cells have been covalently bound to an oxidized carbon electrode surface by making amide bonds between carboxylic groups on the electrode surface and amino groups of the microbial membrane[50] (Figure 2b). An electrode modified with the attached microorganisms was applied as an anode in a biofuel cell in the presence of glucose as a primary reductive substrate and thionine (**2**) as a diffusional electron mediator. The microbe-modified anode showed an enhanced current output and better stability in comparison with a system composed of the same components but with nonimmobilized cells. *Desulfovibrio desulfuricans* microbial cells were covered by a polymeric derivative of viologen or modified by tetracyanoquinodimethane-2,2-(cyclohexa-2,5-diene-1,4-diylidene)bis(propane-1,3-dinitrile) (TCNQ) (**3**) adsorbed on the surface of the microbial cell[51] (Figure 2c). The microbial cells functionalized with the electron transfer mediators were applied in a biofuel cell providing a current to an anode in the absence of a diffusional mediator.

Microbial cells have also been grown in the presence of various nutritional substrates. For example, *Proteus vulgaris* bacteria were grown using glucose, galactose, maltose, trehalose and sucrose as primary electron donors and used in a biofuel cell with thionine as a diffusional electron transfer mediator.[34, 35, 37] Hydrocarbons such as *n*-hexadecane (using *Micrococcus cerificans*)[52] and methane (using *Pseudomonas methanica*)[53] have also been used as fuels to maintain anodic current in the anodic compartment of a biofuel cell. It has been shown that biofuel cell

Table 3. Examples of microbial-based biofuel cells utilizing electron relays for coupling of the intracellular electron transfer processes with electrochemical reactions at anodes.[a]

Microorganism	Nutritional substrate	Mediator	Biofuel cell voltage	Current or current density	Anode[c]	Ref.
Pseudomonas methanica	CH_4	1-Naphthol-2-sulfonate indo-2,6-dichlorophenol	5–6 V (oc)[d]	2.8 µA cm^{-2} (at 0.35 V)	Pt-black, 12.6 cm^2	[54]
Escherichia coli	Glucose	Methylene blue (**15**)	0.625 V (oc)	–	Pt, 390 cm^2	[42]
Proteus vulgaris, Bacillus subtilis, Escherichia coli	Glucose	Thionine	0.64 V (oc)	0.8 mA (at 560 Ω)	Reticulated vitreous carbon, 800 cm^2	[36]
Proteus vulgaris	Glucose	Thionine	350 mV (at 100 Ω)[b]	3.5 mA (at 100 Ω)	Reticulated vitreous carbon, 800 cm^2	[35]
Proteus vulgaris	Sucrose	Thionine	350 mV (at 100 Ω)[b]	3.5 mA (at 100 Ω)	Carbon	[34]
Escherichia coli	Glucose	Thionine	390 mV (at 560 Ω)[b]	0.7 mA (at 560 Ω)	–	[46]
Lactobacillus plantarum, Streptococcus lactis	Glucose	Fe(III)EDTA	0.2 V (oc)	90 µA (at 560 Ω)[b]	–	[45]
Erwinia dissolvens	Glucose	Fe(III)EDTA	0.5 V (oc)	0.7 mA (at 560 Ω)[b]	–	[45]
Proteus vulgaris	Glucose	2-Hydroxy-1,4-naphthoquinone	0.75 V (oc)	0.45 mA (at 1 kΩ)	Graphite felt, 1 g (0.47 m^2 g^{-1})	[51]
Escherichia coli	Acetate	Neutral Red (**1**)	0.25 V (oc)	1.4 µA cm^{-2} (sc)[e]	Graphite, 100 cm^2	
Escherichia coli	Glucose	Neutral Red (**1**)	0.85 V (oc)	17.7 mA (sc)	Graphite felt, 12 g (0.47 m^2 g^{-1})	[40]
Escherichia coli	Glucose	2-Hydroxy-1,4-naphthoquinone	0.53 V (at 10 kΩ)	0.18 mA cm^{-2} (sc)	Glassy carbon, 12.5 cm^2	[49]

[a] In most studies the biofuel anode was conjugated with an O_2 cathode.
[b] The value calculated from other data using Ohm's law.
[c] The anode surface is given as a geometrical surface.
[d] Open-circuit measurements.
[e] Short-circuit measurements.

performance depends heavily on the primary substrate used in the fermentation process. The metabolic process in the bacteria is very complex, involves many enzymes, and may proceed by many different routes. It has been shown that a mixture of nutritional substrates can result even in higher extractable current than any single component alone.[54] It is possible to achieve maximum fuel cell efficiency just by changing the carbon source and thus inducing various metabolic states inside the microorganism. Table 3 summarizes the electrical output of microbial biofuel cells operating with different electron transfer mediators as species providing anodic current, and using different nutrients.

3 ENZYMATIC BIOFUEL CELLS

Microbial biofuel cells require the continuous fermentation of whole living cells performing numerous physiological processes, and thus dictate stringent working conditions. In order to overcome this constraint, the redox enzymes responsible for desired processes may be separated and purified from living organisms and applied as biocatalysts in biofuel cells rather than using the whole microbial cells.[1–6] That is, rather than utilizing the entire microbial cell apparatus for the generation of electrical energy, the specific enzyme(s) that oxidize(s) the target fuel-substrate may be electrically contacted with the electrode of the biofuel cell element. Enzymes are still sensitive and expensive chemicals, and thus special ways for their stabilization and utilization must be established.

Upon utilizing enzymes as catalytically active ingredients in biofuel, one may apply oxidative biocatalysts in the anodic compartments for the oxidation of the fuel-substrate and transfer of electrons to the anode, whereas reductive biocatalysts may participate in the reduction of the oxidizer in the cathodic compartment of the biofuel cell. Redox enzymes lack, however, direct electrical communication with electrodes due to the insulation of the redox center from the conductive support by the protein matrices.

Several methods have been applied to electrically contact redox enzymes and electrode supports.[9, 10, 55]

In the following sections, the engineering of biocatalytic electrodes for the oxidation of potential fuel substrates (biocatalytic anodes) and for the reduction of oxidizers (biocatalytic cathodes) is described. These electrodes are then integrated into biofuel cell elements and the output efficiencies of the bioelectronic devices are addressed.

3.1 Anodes for biofuel cells based on enzyme-catalyzed oxidative reactions

The electrochemical oxidation of fuels can be biocatalyzed by enzymes communicating electrically with electrodes. Different classes of oxidative enzymes (e.g., oxidases, dehydrogenases) require the application of different molecular tools to establish this electrical communication.[9, 10] Electron transfer mediators shuttling electrons between the enzyme active centers and electrodes are usually needed for the efficient electrical communication of flavin adenine dinucleotide FAD-containing oxidases (e.g., glucose oxidase (GOx)). NAD(P)$^+$-dependent dehydrogenases (e.g., lactate dehydrogenase) require NAD(P)$^+$-co-factor and an electrode catalytically active for the oxidation of NAD(P)H and regeneration of NAD(P)$^+$ to establish an electrical contact with the electrode.

3.1.1 Anodes based on the bioelectrocatalyzed oxidation of NAD(P)H

The nicotinamide redox co-factors (NAD$^+$ and NADP$^+$) play important roles in biological electron transport, acting as carriers of electrons and activating the biocatalytic functions of dehydrogenases, the vast majority of redox enzymes. The application of NAD(P)$^+$-dependent enzymes (e.g., lactate dehydrogenase, EC 1.1.1.27; alcohol dehydrogenase, EC 1.1.1.71; glucose dehydrogenase, EC 1.1.1.118) in biofuel cells allows the use of many organic materials such as lactate, glucose and alcohols as fuels. The biocatalytic oxidation of these substrates requires the efficient electrochemical regeneration of NAD(P)$^+$-co-factors in the anodic compartment of the cells. The biocatalytically produced NAD(P)H co-factors participating in the anodic process transports electrons from the enzymes to the anode, and the subsequent electrochemical oxidation of the reduced co-factors regenerates the biocatalytic functions of the system.

In aqueous solution at pH 7.0, the thermodynamic redox potential ($E^{\circ\prime}$) for NAD(P)$^+$/NAD(P)H is ca. -0.56 V (vs. the saturated calomel electrode (SCE)) – sufficiently negative for anode operation. Electrochemistry of NAD(P)H has been studied extensively, and it has been demonstrated that the electrochemical oxidation process is highly irreversible and proceeds with large overpotentials (η) (ca. 0.4, 0.7 and 1 V vs. SCE at carbon, Pt and Au electrodes, respectively).[9, 56] Strong adsorption of NAD(P)H and NAD(P)$^+$ (e.g., on Pt, Au, glassy carbon and pyrolytic graphite) generally poisons the electrode surface and inhibits the oxidation process. Furthermore, NAD(P)$^+$ acts as an inhibitor for the direct oxidation of NAD(P)H, and adsorbed NAD(P)H can be oxidized to undesired products that lead to the degradation of the co-factor (e.g., to NAD$^+$-dimers). Thus, the noncatalyzed electrochemical oxidation of NAD(P)H is not appropriate for use in workable biofuel cells. For the efficient electrooxidation of NAD(P)H, mediated electrocatalysis is necessary.[9, 56] Several immobilization techniques have been applied for the preparation of mediator-modified electrodes: The mediator molecules can be adsorbed directly onto electrodes, incorporated into polymer layers, or covalently linked to functional groups on electrode surfaces.[56]

A biofuel cell based on the electrocatalytic regeneration of NAD$^+$ at a modified anode has been developed.[57] Glucose dehydrogenase (EC 1.1.1.47) was immobilized in a porous glass located in the anode compartment of the biofuel cell. The enzyme oxidized the substrate (glucose) and produced the reduced state of the co-factor (NADH). The reduced co-factor reached the anode surface diffusionally (Meldola Blue was adsorbed at the graphite electrode, ca. 1×10^{-9} mol cm^{-2}), where it was oxidized to NAD$^+$. The biocatalytic anode was coupled to a Pt-cathode that reduced water to hydrogen and the biofuel cell provided $V_{oc} = 300$ mV and $i_{sc} = 220\,\mu\text{A cm}^{-2}$ over a period of several hours.

The covalent coupling of redox mediators to self-assembled monolayers on Au-electrode surfaces has an important advantage for the preparation of multi-component organized systems.[58] Pyrroloquinoline quinone (PQQ, (4)) can be covalently attached to amino groups of a cystamine monolayer assembled on a Au surface (Figure 3a). The resulting electrode demonstrates good electrocatalytic activity for NAD(P)H oxidation, particularly in the presence of Ca^{2+}-cations as promoters (Figure 3b).[59] A quasi-reversible redox wave at the formal potential, $E^{\circ\prime} = -0.155$ V (vs. SCE at pH 8.0) is observed, corresponding to the two-electron redox process of the quinone units (Figure 3b, curve 1). Coulometric analysis of the quinone redox wave indicates that the PQQ surface coverage on the electrode is 1.2×10^{-10} mol cm^{-2}, a value that is typical for monolayer coverage. The electron transfer rate constant was found to be $k_{et} = 8\,\text{s}^{-1}$. Figure 3b, curve 2, shows a cyclic voltammogram of a PQQ-functionalized electrode upon the addition of NADH (10 mM) in the presence of Ca^{2+}-ions. An electrocatalytic anodic current is observed in the presence

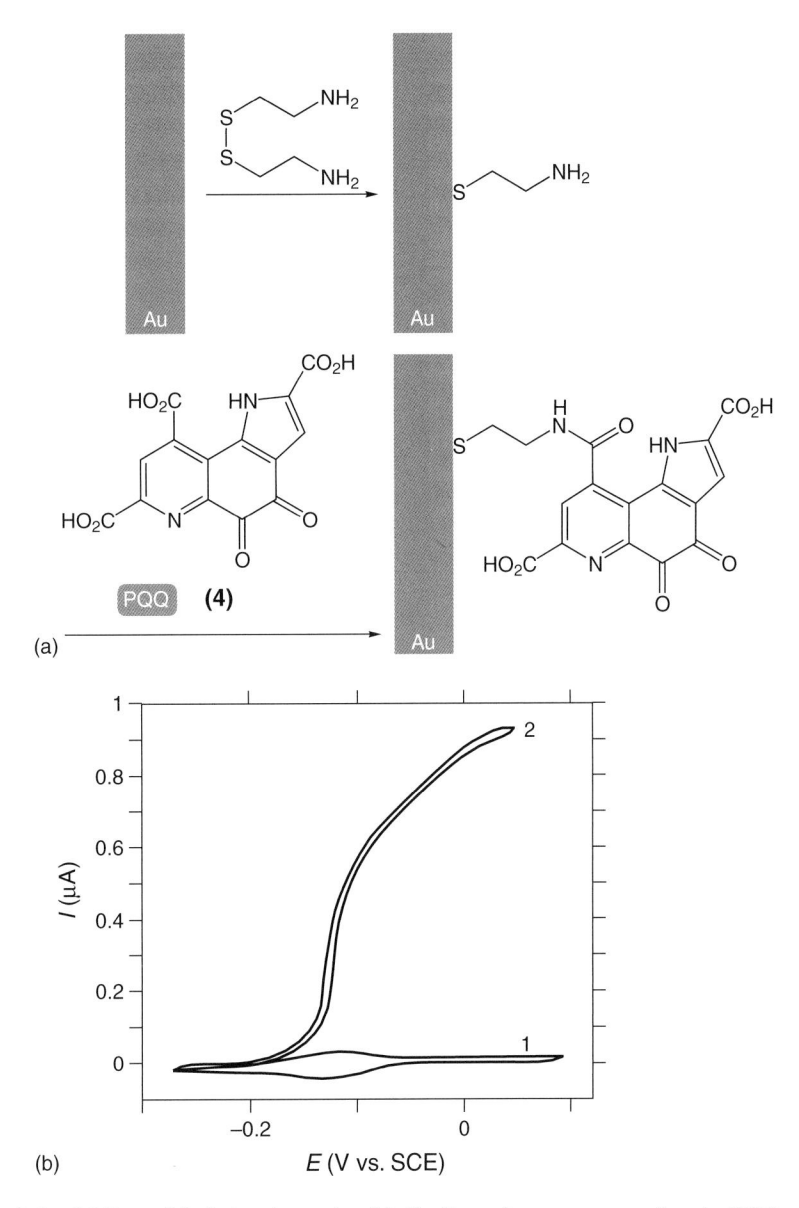

Figure 3. (a) Assembly of the PQQ-modified Au-electrode. (b) Cyclic voltammograms of a Au-PQQ electrode (geometrical area 0.2 cm^2, roughness factor ca. 1.5) in the presence of: (1) 0.1 M Tris–buffer, pH 8.0; (2) 10 mM NADH and 20 mM Ca^{2+}. Recorded at a scan rate of 1 mV s^{-1}.

of NADH, implying the effective electrocatalyzed oxidation of the co-factor, equations (13) and (14).

$$\text{NADH} + \text{PQQ} + \text{H}^+ \longrightarrow \text{NAD}^+ + \text{PQQH}_2 \quad (13)$$

$$\text{PQQH}_2 \longrightarrow \text{PQQ} + 2\text{H}^+ + 2\text{e}^- \text{(to anode)} \quad (14)$$

NAD(P)$^+$-dependent enzymes electrically contacted with electrode surfaces can provide efficient bioelectrocatalysis for NAD(P)H oxidation. For example, diaphorase (EC 1.6.4.3) was applied to oxidize NADH using a variety of quinone compounds, several kinds of flavins and viologens as mediators.[60] The bimolecular reaction rate constants between the enzyme and mediators whose redox potentials are more positive than -0.28 V (vs. SCE) at pH 8.5 can be as high as 10^8 M^{-1} s^{-1}, suggesting that the reactions are diffusionally controlled. A biofuel cell was developed based on enzymes (producing NADH upon the biocatalytic oxidation of primary substrate) and diaphorase (electrically contacted via an electron relay and providing bioelectrocatalytic oxidation of the NADH to NAD$^+$, (**5**)).[60] A number of NAD$^+$-dependent enzymes (alcohol dehydrogenase, EC 1.1.1.1; aldehyde dehydrogenase, EC 1.2.1.5; formate dehydrogenase, EC 1.2.1.2) provided a sequence of biocatalytic reactions resulting in the oxidation of methanol to formaldehyde and finally CO$_2$ (Figure 4). The reduced co-factor (NADH) produced in all the steps was biocatalytically

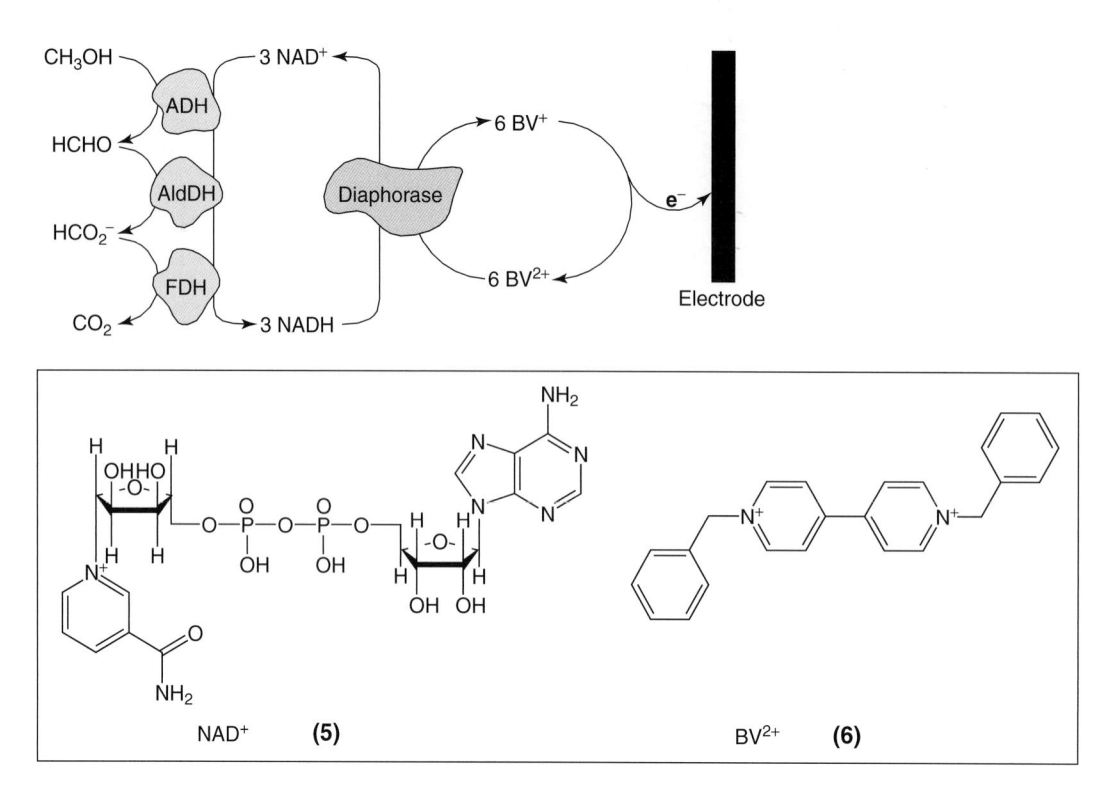

Figure 4. Schematic configuration of methanol/dioxygen biofuel cell. NAD$^+$-dependent dehydrogenases oxidize CH$_3$OH to CO$_2$; diaphorase (D) catalyzes the oxidation of NADH to NAD$^+$ using N,N'-dibenzyl-4,4-bipyridinium (benzylviologen, BV^{2+}) as the electron acceptor. BV$^+$ is oxidized to BV^{2+} at a graphite anode and thus, releases electrons for the reduction of dioxygen at a platinum cathode. ADH: alcohol dehydrogenase; AldDH: aldehyde dehydrogenase; FDH: formate dehydrogenase.

oxidized by diaphorase. This diaphorase was electrochemically contacted by a diffusional electron relay (benzylviologen, BV^{2+}, (**6**)) that provided enzyme regeneration and anodic current. The biocatalytic anode was conjugated with an O$_2$-cathode to complete the biofuel cell. The total reaction in the biofuel cell is methanol oxidation by O$_2$. The biofuel cell provided $V_{oc} = 0.8$ V and a maximum power density of ca. 0.68 mW cm^{-2} at 0.49 V. It should be noted, however, that this multienzyme system was utilized in a nonorganized configuration, where all biocatalysts exist as diffusional components in the cell.

In view of the high cost of NAD(P)$^+$/NAD(P)H co-factors, practical applications require their immobilization together with enzymes. Nevertheless, the covalent coupling of natural NAD(P)$^+$ co-factors to an organic support results in a substantial decrease of their functional activity. Mobility is vital for its efficient interaction with enzymes, and thus attention has been paid to the synthesis of artificial analogs of the NAD(P)$^+$ co-factors carrying functional groups separated from the bioactive site of the co-factor by spacers.[61, 62] The spacer is usually linked to N-6 position of the NAD(P)$^+$ molecule, and provides some flexibility for the bioactive part of the co-factors, allowing them to be associated with the enzyme molecules. Structure–activity relationships of the artificial functionalized NAD(P)$^+$-derivatives have been studied with different enzymes and the possibility to substitute natural NAD(P)$^+$ with these artificial analogs has been demonstrated.[61, 63] An efficient electrode that acts as an anode in the presence of an NAD(P)$^+$-dependent enzyme should include three integrated, electrically contacted components: The NAD(P)$^+$-co-factor that is associated with the respective enzyme and a catalyst that allows the efficient regeneration of the co-factor.

Electrodes functionalized with co-factor monolayers can form stable affinity complexes with their respective enzymes.[64–66] These interfacial complexes can be further crosslinked to produce integrated bioelectrocatalytic matrices consisting of the relay-units, the co-factor and the enzyme molecules. Electrically contacted biocatalytic electrodes of NAD$^+$-dependent enzymes have been organized by the generation of affinity complexes between a catalyst/NAD$^+$-monolayer and the respective enzymes.[65, 66] A PQQ monolayer covalently linked to an amino-functionalized NAD (N^6-(2-aminoethyl)-NAD$^+$, (**7**)) was assembled onto a Au-electrode (Figure 5a). The resulting monolayer-functionalized electrode binds NAD$^+$-dependent enzymes (e.g., L-lactate dehydrogenase

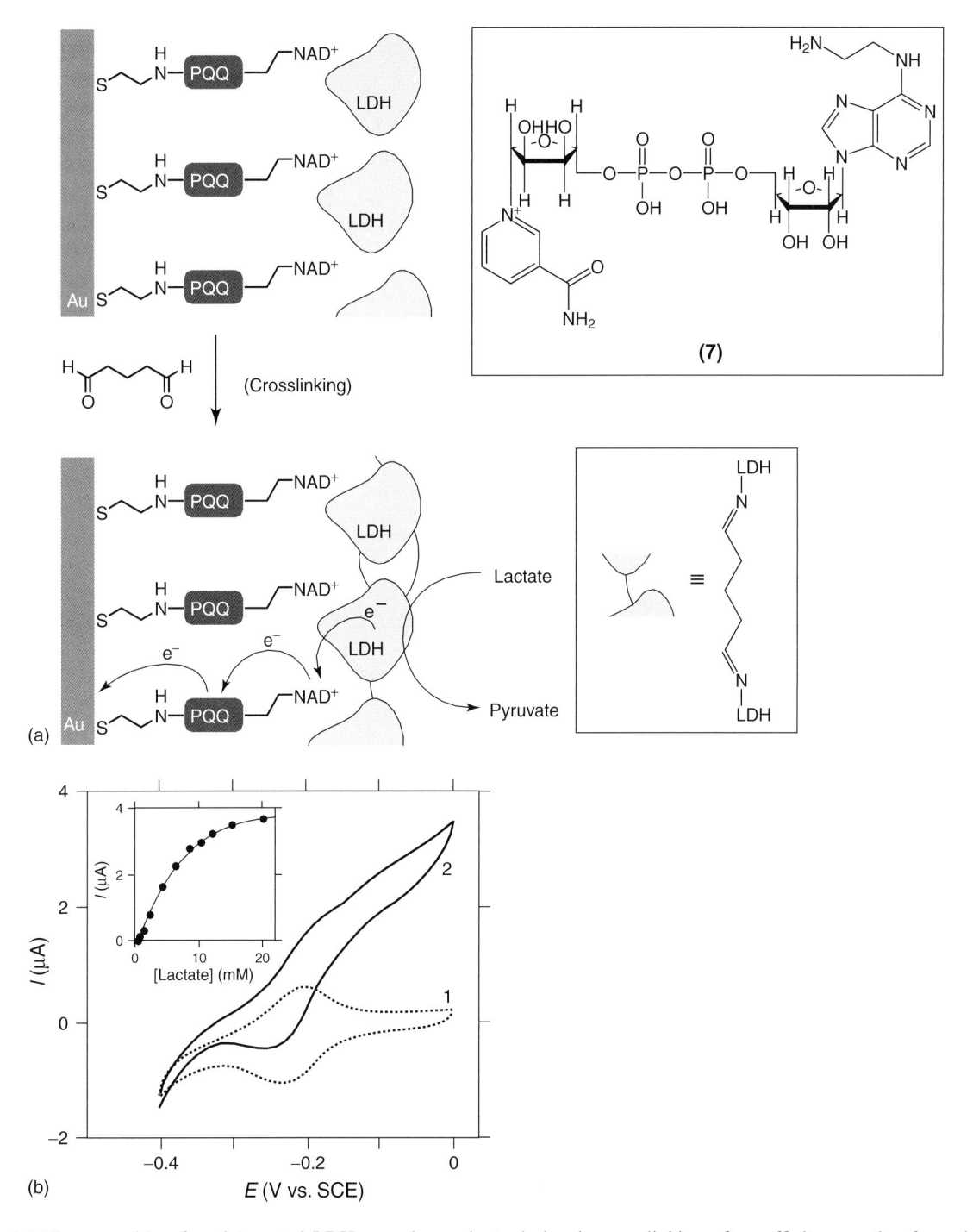

Figure 5. (a) The assembly of an integrated LDH monolayer-electrode by the crosslinking of an affinity complex formed between the LDH and a PQQ/NAD$^+$ monolayer-functionalized Au-electrode. (b) Cyclic voltammograms of the integrated crosslinked PQQ/NAD$^+$/LDH electrode (geometrical area ca. 0.2 cm^2, roughness factor ca. 15): (1) in the absence of lactate; (2) with lactate, 20 mM. Data recorded in 0.1 M Tris–buffer, pH 8.0, in the presence of 10 mM CaCl$_2$, under Ar, scan rate, 2 mV s^{-1}. Inset: Amperometric responses of the integrated electrode at different concentrations of lactate upon the application of a constant potential corresponding to 0.1 V vs. SCE. (Adapted from Scheme 4 and Figure 2 in Ref. [66]. Reprinted with permission. Copyright 1997 American Chemical Society.)

(LDH) EC 1.1.1.27) by affinity interactions between the co-factor and the biocatalyst. These enzyme electrodes electrocatalyze the oxidation of their respective substrates (e.g., lactate). The crosslinking of the enzyme layer, using glutaric dialdehyde, generates a stable, electrically contacted electrode. Figure 5(b) shows the electrical responses of a crosslinked layered PQQ/NAD$^+$/LDH electrode in the absence (curve 1) and the presence (curve 2) of lactate, and the inset shows the respective calibration curve corresponding to the amperometric output

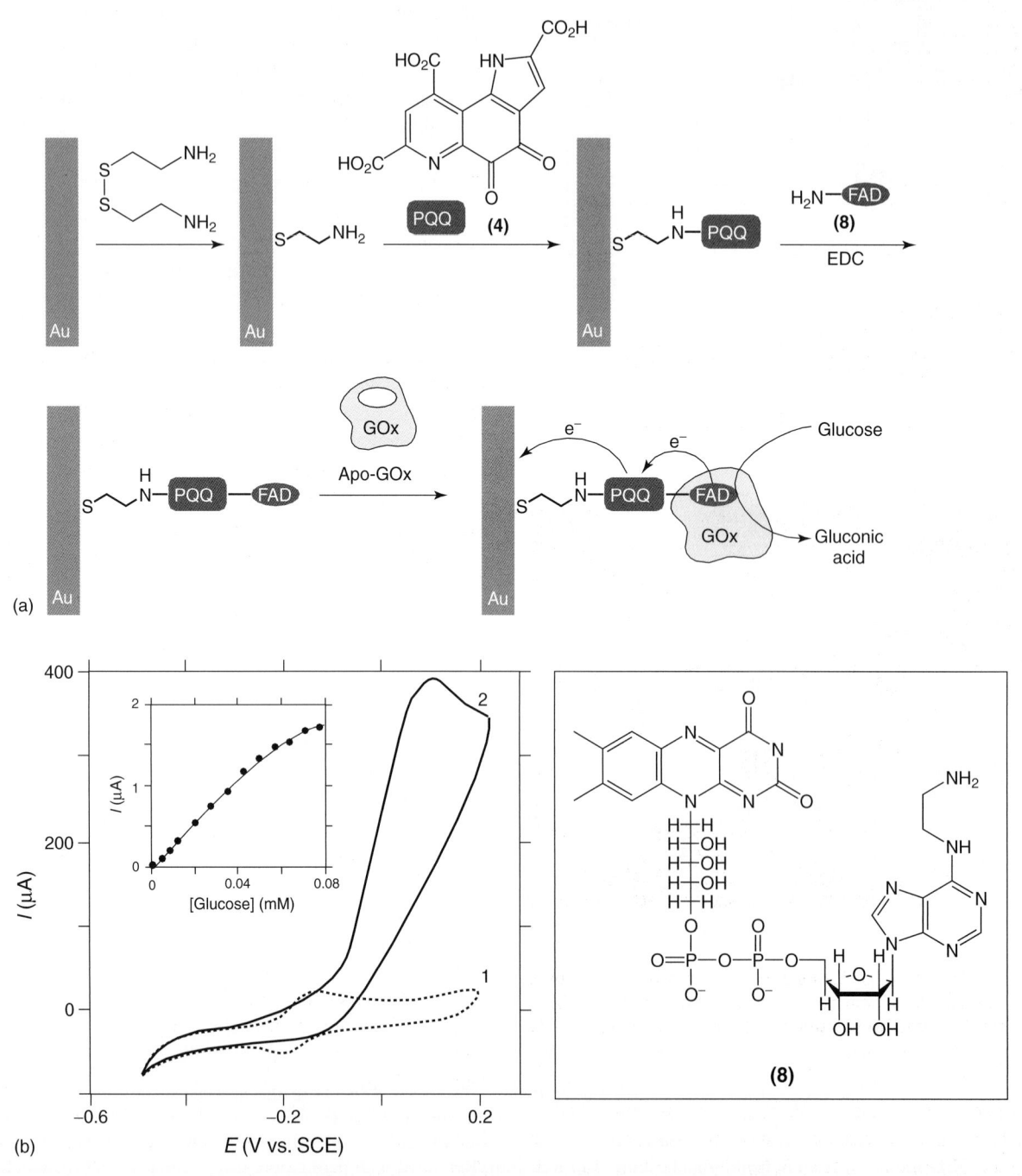

Figure 6. (a) The surface-reconstitution of apo-GOx on a PQQ-FAD monolayer assembled on a Au-electrode (geometrical area ca. 0.4 cm^2, roughness factor ca. 20). (b) Cyclic voltammograms of the GOx-reconstituted PQQ-FAD-functionalized Au-electrode: (1) in the absence of glucose; (2) with glucose, 80 mM. Recorded in 0.1 M phosphate buffer, pH 7.0, under Ar, at 35 °C, scan rate, 5 mV s^{-1}. Inset: Calibration curve corresponding to the current output (measured by chronoamperometry, $E = 0.2$ V vs. SCE) of the PQQ-FAD-reconstituted glucose oxidase enzyme-electrode at different concentrations of glucose. (Adapted from Scheme 1 and Figure 1 in Ref. [71]. Reprinted with permission. Copyright 1996 American Chemical Society.)

of the integrated LDH layered electrode at different lactate concentrations. This system exemplifies a fully integrated rigid biocatalytic matrix composed of the enzyme, co-factor and catalyst. The complex between the NAD^+-co-factor and LDH aligns the enzyme on the electrode support, thereby enabling effective electrical communication between the enzyme and the electrode, while the PQQ sites provide the regeneration of NAD^+.

3.1.2 Flavoenzyme-functionalized electrodes as anode-elements: oxidation of glucose by GOx reconstituted on an FAD/PQQ-monolayer-functionalized electrode

The electrical contacting of redox enzymes that defy direct electrical communication with electrodes can be established by using synthetic or biologically active charge carriers as intermediates between the redox center and the electrode.[10] The overall electrical efficiency of an enzyme-modified electrode depends not only on the electron transport properties of the mediator, but also on the transfer steps occurring in the assembly. Diffusional electron relays have been utilized to shuttle electrons between oxidative enzymes and anodes of biofuel cells, providing the bioelectrocatalyzed oxidation of organic fuels (e.g., methanol).[67–69] A sequence of biocatalytic reactions was applied to achieve the stepwise oxidation of methanol to CO_2. In order to accomplish superior electron contacting, the mediator may be selectively placed in an optimum position between the redox center and the enzyme periphery. In the case of surface-confined enzymes, the orientation of the enzyme-mediator assembly with respect to the electrode can also be optimized. A novel means for the establishment of electrical contact between the redox center of flavoenzymes and their environment based on a reconstitution approach has recently been demonstrated.[70]

The organization of a reconstituted enzyme aligned on an electron relay-FAD monolayer was recently realized by the reconstitution of apo-glucose oxidase (apo-GOx) on a surface functionalized with a relay-FAD monolayer (Figure 6a).[71, 72] PQQ (**4**) was covalently linked to a base cystamine monolayer at a Au-electrode, and N^6-(2-aminoethyl)-FAD (**8**) was then attached to the PQQ relay units. Apo-GOx (obtained by the extraction of the native FAD-co-factor from GOx (EC 1.1.3.4)) was then reconstituted onto the FAD units of the PQQ-FAD-monolayer architecture to yield a structurally aligned, immobilized, biocatalyst on the electrode with a surface coverage of 1.7×10^{-12} mol cm^{-2}. The resulting enzyme-reconstituted PQQ-FAD-functionalized electrode revealed bioelectrocatalytic properties. Figure 6(b) shows cyclic voltammograms of the enzyme electrode in the absence and the presence of glucose (curves a and b, respectively). When the glucose-substrate is present, an electrocatalytic anodic current is observed, implying electrical contact between the reconstituted enzyme and the electrode surface. The electrode constantly oxidizes the PQQ site located at the protein periphery, and the PQQ-mediated oxidation of the FAD-center activates the bioelectrocatalytic oxidation of glucose (equations (15)–(17)). The resulting electrical current is controlled by the recycling rate of the reduced FAD by the substrate. Figure 6(b), inset, shows the derived calibration curve corresponding to the amperometric output of the enzyme-reconstituted electrode at different concentrations of glucose. The resulting current densities are unprecedentedly high (300 mA cm^{-2} at 80 mM of glucose).

$$FAD + glucose + 2H^+ \longrightarrow FADH_2$$
$$+ \text{gluconic acid} \quad (15)$$
$$FADH_2 + PQQ \longrightarrow FAD + PQQH_2 \quad (16)$$
$$PQQH_2 \longrightarrow PQQ + 2H^+$$
$$+ 2e^- (\text{to anode}) \quad (17)$$

Control experiments reveal that without the PQQ component, the system does not exhibit electron-transfer communication with the electrode surface, demonstrating that the PQQ relay unit is, indeed, a key component in the electro-oxidation of glucose.[71, 72] The electron-transfer turnover rate of GOx with molecular oxygen as the electron acceptor is around 600 s^{-1} at 25 °C. Using an activation energy of 7.2 kcal mol^{-1}, the electron-transfer turnover rate of GOx at 35 °C is estimated to be ca. 900 s^{-1}.[71, 72] A densely packed monolayer of GOx (ca. 1.7×10^{-12} mol cm^{-2}) that exhibits the theoretical electron-transfer turnover rate is expected to yield an amperometric response of ca. 300 mA cm^{-2}. This indicates that the reconstituted GOx on the PQQ-FAD monolayer exhibits an electron-transfer turnover with the electrode of similar effectiveness to that observed for the enzyme with oxygen as a natural electron acceptor. Indeed, the high current output of the resulting enzyme-electrode is preserved in the presence of O_2 in the solution.

3.2 Cathodes for biofuel cells based on enzyme-catalyzed reductive reactions

The biocatalytic reduction of oxidizers (e.g., dioxygen, hydrogen peroxide) has attracted much less attention than the biocatalytic oxidation of fuels. Nonetheless, in order to construct a biofuel cell element, it is essential to design a functional cathode for the reduction of the oxidizer that is coupled to the anode and allows the electrically balanced current flow. Conventional O_2-reducing cathodes used in fuel cells are usually not compatible with biocatalytic anodes since high temperatures and pressures are applied for their operation. Thus, biocatalytic reductive processes at the cathode should be considered as a strategy to design all biomaterial-based functional fuel cells.

3.2.1 Bioelectrocatalytic cathodes for the reduction of peroxides

Hydrogen peroxide is a strong oxidizer ($E^{\circ\prime} = 1.535$ V vs. SCE), yet its electrochemical reduction proceeds with a very high overpotential. The bioelectrocatalyzed reduction of H_2O_2 has been accomplished in the presence of various peroxidases (e.g., horseradish peroxidase, EC 1.11.1.7).[73] Microperoxidase-11 (MP-11, (**9**)) is an oligopeptide consisting of 11 amino acids and a covalently linked Fe(III)-protoporphyrin IX heme site.[74] The oligopeptide is obtained by the controlled hydrolytic digestion of cytochrome c and it corresponds to the active-site microenvironment of the cytochrome. MP-11 reveals several advantages over usual peroxidases: It has a much smaller size, high stability and exhibits direct electrical communication with electrodes since its heme is exposed to the solution.

MP-11 was covalently linked to a cystamine monolayer self-assembled on a Au-electrode.[75] The MP-11 (**9**) structure suggests two different modes of coupling of the oligopeptide to the primary cystamine monolayer: (i) linkage of the carboxylic functions associated with the protoporphyrin IX ligand to the monolayer interface; (ii) coupling of carboxylic acid residues of the oligopeptide to the cystamine residues. These two modes of binding reveal similar formal potentials – $E^{\circ\prime} = -0.40$ V vs. SCE (Figure 7a). The electron transfer rates of the two binding modes of MP-11 were kinetically resolved using chronoamperometry,[76] and appear in an approximately 1 : 1 ratio. The interfacial electron transfer rates to the heme sites linked to the electrode by the two binding modes are 8.5 and 16 s^{-1}. Coulometric analysis of the MP-11 redox wave, corresponding to the reversible reduction–oxidation of the heme (equation (18)), indicates a surface coverage of 2×10^{-10} mol cm^{-2}.

$$[\text{heme}-\text{Fe(III)}] + e^- \longrightarrow [\text{heme}-\text{Fe(II)}] \qquad (18)$$

Figure 7(b) shows cyclic voltammograms of the MP-11-functionalized electrode recorded at positive potentials in the absence of H_2O_2 (curve 1) and in the presence of added H_2O_2 (curve 2). The observed electrocatalytic cathodic current indicates the effective electrobiocatalyzed reduction of H_2O_2 by the functionalized electrode. It should be noted that the electrocatalytic current for the reduction of H_2O_2 in aqueous solutions is observed at much more positive potentials than the MP-11 redox potential registered in the absence of H_2O_2 (cf. Figure 7a and b). The reason for this potential shift is the result of the formation of the Fe(IV) intermediate species in the presence of H_2O_2 (equations (19)–(21)). Control experiments reveal that no electroreduction of H_2O_2 occurs at the bare Au electrode within this potential window.

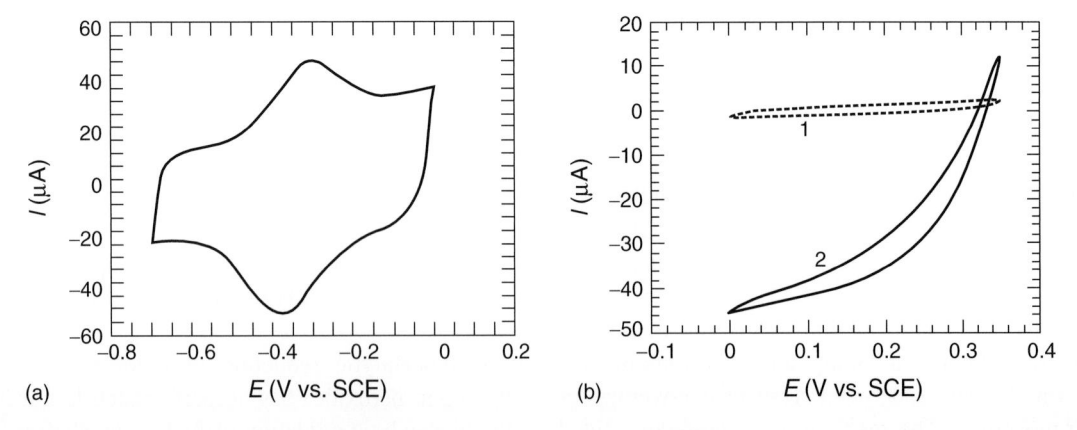

$$[\text{heme}-\text{Fe(III)}] + H_2O_2 \longrightarrow [\text{heme}-\text{Fe(IV)}{=}\text{O}]^{\bullet+} + H_2O \qquad (19)$$

$$[\text{heme}-\text{Fe(IV)}{=}\text{O}]^{\bullet+} + e^-(\text{from cathode}) + H^+$$
$$\longrightarrow [\text{heme}-\text{Fe(IV)}-\text{OH}] \qquad (20)$$

$$[\text{heme}-\text{Fe(IV)}-\text{OH}] + H^+ + e^-(\text{from cathode})$$
$$\longrightarrow [\text{heme}-\text{Fe(III)}] + H_2O \qquad (21)$$

Figure 7. (a) Cyclic voltammogram of the MP-11-modified Au-electrode (geometrical area ca. 0.2 cm^2, roughness factor ca. 15) in 0.1 M phosphate buffer, pH 7.0, under Ar atmosphere, scan rate 50 mV s^{-1}. (b) Cyclic voltammograms of the MP-11 modified electrodes recorded at positive potentials in 0.1 M phosphate buffer, pH 7.0, scan rate 10 mV·s^{-1}, (1) without H_2O_2, (2) in the presence of 5 mM H_2O_2. (Adapted from Figures 3 and 4 in Ref. [87]. Reproduced by permission of The Royal Society of Chemistry.)

The biocatalytic reduction of oxidizers in nonaqueous solutions immiscible with water is important since it can be coupled to biocatalytic oxidative processes through liquid/liquid interfaces. Some enzymes,[77] particularly peroxidases,[73] can function in nonaqueous solutions. A horseradish peroxidase (HRP)-modified electrode has been applied for the biocatalytic reduction of organic peroxides in nonaqueous solvents.[78] The biocatalytic activity of enzymes, particularly of HRP,[79] however, is usually lower (sometimes by an order of magnitude) in organic solvents than in water. MP-11 monolayer-modified electrodes have demonstrated high activity and stability for the electrocatalytic reduction of organic hydroperoxides in acetonitrile and ethanol.[80]

In order to perform a biocatalytic cathodic reaction in a medium immiscible with an aqueous solution, an MP-11-modified electrode was studied in a dichloromethane electrolyte (Figure 8, inset).[81] A quasi-reversible redox wave was observed for the heme center of MP-11 at $E^{\circ\prime} = -0.30\,\text{V}$ (vs. aqueous SCE, in dichloromethane that includes tetrafluoroborate). Coulometric assay of the redox wave indicated a surface coverage of ca. $3 \times 10^{-10}\,\text{mol}\,\text{cm}^{-2}$. Figure 8 shows cyclic voltammograms of the MP-11-functionalized electrode in the absence of an organic peroxide (curve 1), and in the presence of cumene peroxide (**10**) (curve 2). The observed electrocatalytic cathodic current indicates the effective bioelectrocatalyzed reduction of (**10**) by the functionalized electrode. The sequence of electron transfers leading to the reduction of the peroxide is summarized in equations (12) and (13). It should be noted that the MP-11 bioelectrocatalyzed reduction of organic peroxides in nonaqueous solutions does not include the intermediate formation of the Fe(IV) species, and proceeds at the MP-11 potential corresponding to the Fe(III)–Fe(II) redox transformation (equation (18)).

$$[\text{heme–Fe(II)}] + \text{Cumene peroxide (10)} \longrightarrow$$
$$[\text{heme–Fe(III)}] + \text{Cumene alcohol (11)} \quad (22)$$
$$[\text{heme–Fe(III)}] + e^-(\text{from cathode}) \longrightarrow$$
$$[\text{heme–Fe(II)}] \quad (23)$$

3.2.2 Bioelectrocatalytic cathodes for the reduction of dioxygen

The direct electrochemical reduction of dioxygen proceeds with very large overpotentials (e.g., at ca. $-0.3\,\text{V}$ vs. SCE at a bare Au electrode, pH 7). Thus, catalysts are required in order to utilize oxygen reduction in fuel cells. The four-electron transfer reduction of O_2 to water, without the

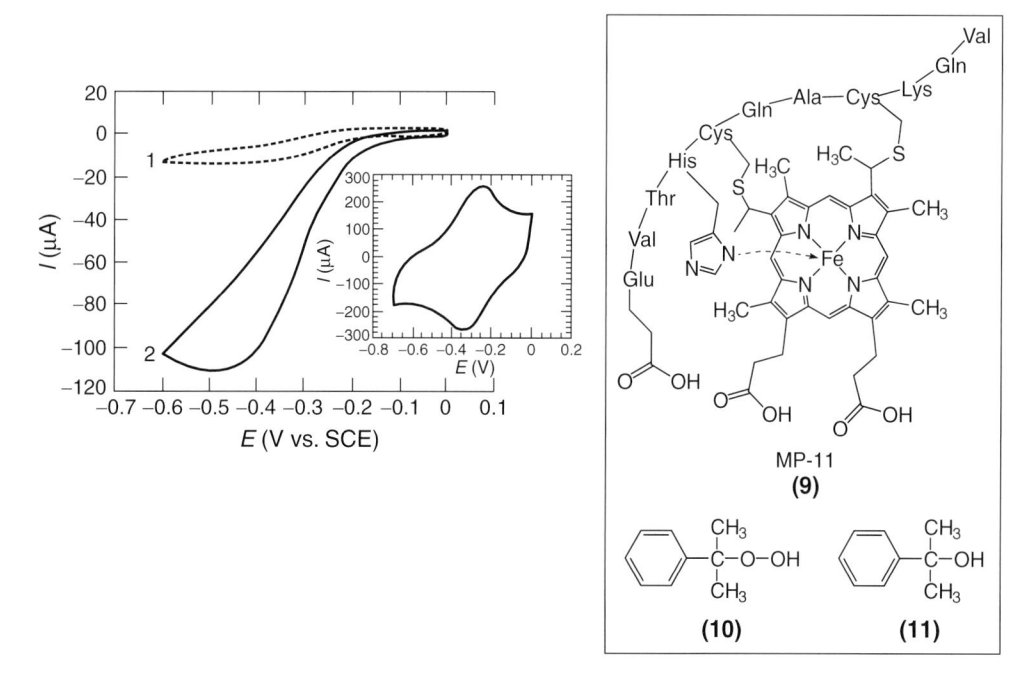

Figure 8. Cyclic voltammograms of the MP-11-functionalized Au-electrode (geometrical area ca. $0.4\,\text{cm}^2$, roughness factor ca. 20): (a) in the absence of an organic peroxide; (b) in the presence of cumene peroxide, $5 \times 10^{-3}\,\text{M}$. Potential scan rate, $5\,\text{mV}\,\text{s}^{-1}$. Inset: Cyclic voltammogram of the MP-11-monolayer-modified Au electrode in the absence of cumene peroxide. Potential scan rate, $50\,\text{mV}\,\text{s}^{-1}$; Ar atmosphere; electrolyte composed of a dichloromethane solution with $0.05\,\text{M}$ tetra n-butylammonium tetraphenylborate (TBATFB). Structures of MP-11 (**9**), cumene peroxide (**10**) and cumene alcohol (**11**) are shown. (Adapted from Figure 1(B) in Ref. [81]. Reproduced by permission of The Royal Society of Chemistry.)

formation of peroxide or superoxide, is a major challenge for the future development of biofuel cell elements, since such reactive intermediates would degrade the biocatalysts in the system.

Biocatalytic systems composed of enzymes and their respective electron transfer mediators (e.g., bilirubin oxidase, EC 1.3.3.5,[82] or fungal laccase, EC 1.10.3.2,[83] with 2,2′-azino-bis-(3-ethylbenzothiazoline-6-sulfonate) as a mediator) are able to biocatalyze the electroreduction of O_2 to H_2O effectively at ca. 0.4 V (vs. SCE), significantly decreasing the overpotential. These systems, however, are composed of dissolved enzymes and mediators operating via a diffusional path that is unacceptable for technological applications. Organized layered enzyme systems are much more promising for their use in biocatalytic cathodes.

Cytochrome c that includes a single thiol group at the 102-cysteine-residue (yeast iso-2-cytochrome c from *Saccharomyces cerevisiae*) was assembled as a monolayer on a Au-electrode by the covalent linkage of the thiol functionality to a maleimide monolayer-modified electrode as outlined in Figure 9. The quasi-reversible cyclic voltammogram of the electrode ($E^{o\prime} = 0.03$ V vs. SCE) (Figure 10a) indicated that the resulting heme–protein exhibits direct electrical contact with the electrode, probably the result of the structural alignment of the heme–protein on the electrode. Coulometric assay of the redox wave indicated a protein coverage of 8×10^{-12} mol cm^{-2}. Taking into account the cytochrome c (Cyt c) diameter (ca. 4.5 nm), this surface coverage corresponds to a densely packed monolayer. In a variety of biological transformations, Cyt c acts as

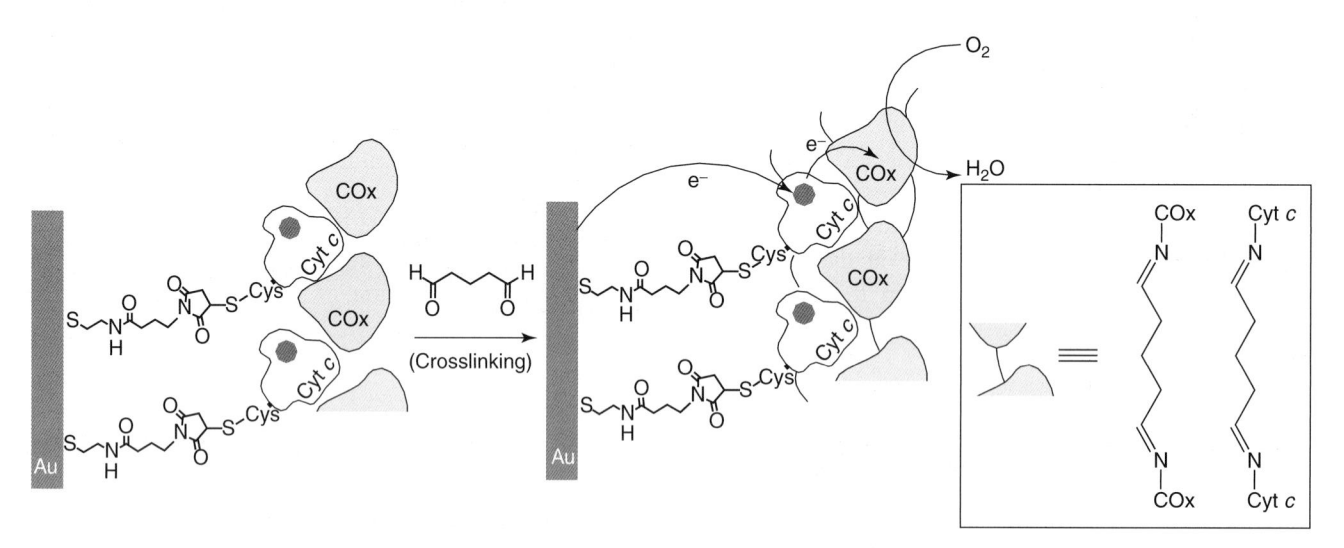

Figure 9. Assembly of the integrated bioelectrocatalytic Cyt c/COx-electrode. (Adapted from Scheme 4, Ref. [85]. Reproduced by permission of The Royal Society of Chemistry.)

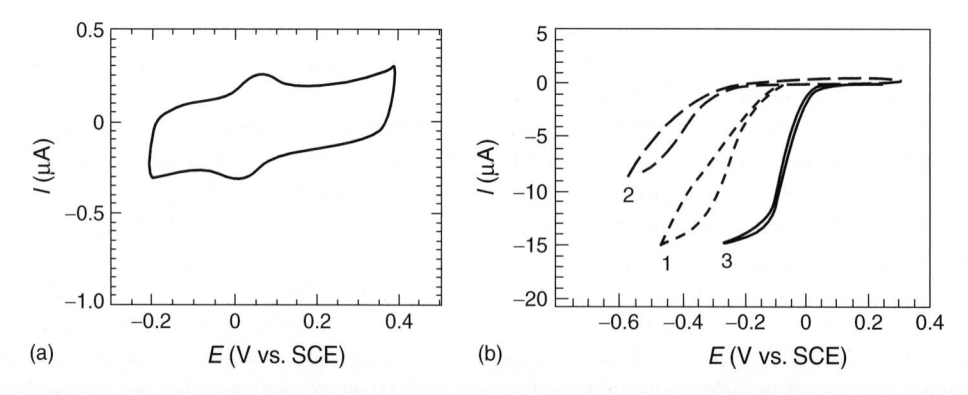

Figure 10. (a) Cyclic voltammogram of the Cyt c monolayer-modified Au-electrode (geometrical area ca. 0.2 cm^2, roughness factor ca. 1.5) measured under argon, potential scan rate 100 mV s^{-1}. (b) Cyclic voltammograms obtained in an O_2-saturated electrolyte solution at: (1) a bare Au-electrode; (2) the Cyt c monolayer-modified Au-electrode (3) the Cyt c/Cox assembly modified Au-electrode; potential scan rate 10 mV s^{-1}. The experiments were performed in 0.1 M phosphate buffer, pH 7.0. (Adapted from Figures 1(A) and 5 in Ref. [85]. Reproduced by permission of The Royal Society of Chemistry.)

an electron relay through the formation of an inter-protein complex. The association constant between the Cyt c monolayer and cytochrome oxidase (COx) was determined[64] to be $K_a = 1.2 \times 10^7 \, M^{-1}$, and an integrated Cyt c/COx layered electrode was prepared as outlined in Figure 9.[84, 85] The Cyt c monolayer electrode was interacted with COx to generate the affinity complex on the surface, which was then crosslinked with glutaric dialdehyde. A similar Cyt c/COx assembly was organized on a Au-quartz crystal microbalance surface. Microgravimetric analyses indicates that the surface coverage of COx on the base Cyt c monolayer is ca. $2 \times 10^{-12} \, mol \, cm^{-2}$. This surface density corresponds to an almost densely packed monolayer of COx.

Figure 10(b), curve 1, shows the cyclic voltammogram of a bare Au-electrode in the presence of O_2 (the background electrolyte equilibrated with air). The cathodic wave of the O_2-electroreduction is observed at ca. -0.3 V vs. SCE. This reduction wave is negatively shifted in the presence of the Cyt c monolayer electrode (Figure 10b, curve 2), implying that the heme/protein layer is inactive as a biocatalyst for the reduction of O_2. In fact, the Cyt c monolayer enhances the overpotential for the reduction of dioxygen due to hydrophobic blocking of the electrode surface. Figure 10(b), curve 3, shows a cyclic voltammogram of the layered Cyt c/COx crosslinked electrode in the presence of O_2. An electrocatalytic wave is observed at ca. -0.07 V (vs. SCE), indicating that the Cyt c/COx layer does act as a biocatalytic interface for the reduction of dioxygen. In a control experiment, a COx monolayer was assembled on the Au-electrode without the base Cyt c layer. No bioelectrocatalytic activity towards the reduction of O_2 was observed. Thus, the effective bioelectrocatalyzed reduction of O_2 by the Cyt c/COx interface originates from the direct electrical communication between the Cyt c and the electrode and the electrical contact in the crosslinked Cyt c/COx assembly. The electron transfer to Cyt c is followed by electron transfer to COx, which acts as an electron storage biocatalyst for the concerted four-electron reduction of O_2, (equations (24)–(26)) (the two-electron reduction of O_2 yields H_2O_2, while a concerted four-electron reduction of O_2 generates H_2O).

$$\text{Cyt } c_{ox} + e^- \text{(from cathode)} \longrightarrow \text{Cyt } c_{red} \quad (24)$$

$$4\text{Cyt } c_{red} + \text{COx}_{ox} \longrightarrow 4\text{Cyt } c_{ox} + \text{COx}_{red} \quad (25)$$

$$\text{COx}_{red} + O_2 + 4H^+ \longrightarrow \text{COx}_{ox} + 2H_2O \quad (26)$$

Rotating disk electrode (RDE) experiments were performed to estimate the electron transfer rate constant for the overall bioelectrocatalytic process corresponding to the reduction of O_2.[85] The calculated number of electrons involved in the reduction of O_2 ($n = 3.9 \pm 0.2$) and the electrochemical rate constant ($k_{el} = 5.3 \times 10^{-4} \, cm \, s^{-1}$) were found from the Koutecky–Levich plot. The overall electron transfer rate constant ($k_{overall} = k_{el}/\Gamma_{COx} = 6.6 \times 10^5 \, M^{-1} \, s^{-1}$) was calculated taking into account the surface density of the bio-electrocatalyst ($\Gamma_{COx} = 2 \times 10^{-12} \, mol \, cm^{-2}$). To determine the limiting step in the bioelectrocatalytic current formation, the experimental diffusion-limited current density ($11.6 \, \mu A \, cm^{-2}$) was compared to the calculated current density assuming primary electron transfer to Cyt c to be the rate-limiting step. Taking into account the electron transfer rate constant to Cyt c ($20 \, s^{-1}$) and the surface density of Cyt c ($\Gamma_{Cyt \, c} = 8 \times 10^{-12} \, mol \, cm^{-2}$), the calculated current density of the system is $15.5 \, \mu A \, cm^{-2}$. Since the calculated current density is only slightly higher than the experimental value, it was assumed that the primary electron transfer process from the electrode to the Cyt c monolayer is the limiting step in the overall bioelectrocatalytic reduction of O_2. The detailed analysis of the interfacial electron-transfer rate-constants, and the use of the various electrochemical techniques to identify the kinetic parameters of the bioelectrocatalytic transformations occurring at the electrode, are described here in order to emphasize that the physical characterization of the systems is essential to optimize the electrode performance. That is, for each of the functional electrodes the complex sequence of reactions must be resolved kinetically in order to determine the rate-limiting step. Once the rate-limiting step is identified, biomaterial engineering on the respective redox protein may be undertaken in order to optimize its electron-transfer functionality, e.g., electrical communication with the electrode, inter-protein electron transfer, mediated electron transfer, etc.

3.3 Biofuel cells based on layered enzyme-electrodes

The previous sections have addressed the engineering of electrodes and the assembly of separate biocatalytic anodes and cathodes. For the design of complete biofuel cells, it is essential to couple the cathode and anode units into integrated devices. The integration of the units is not free of limitations. The oxidizer must not react with the biocatalyst relay, nor co-factor units at the anode interface, as this would decrease or prohibit the biocatalyzed oxidation of the fuel substrate. Furthermore, for synchronous operation of the biofuel cell, charge compensation between the two electrodes must be attained, and the flow of electrons in the external circuit must be compensated by cation-transport in the electrolyte solution. To overcome these limitations, the catholyte and anolyte solutions may be compartmentalized.

Alternatively, the bioelectrocatalytic transformations at the electrodes may be driven efficiently enough that interfering components do not perturb the cell operation. In any biofuel cell, either the bioelectrocatalytic transformations or the transport process is a rate-limiting step controlling the cell efficiency. The mechanistic characterization and understanding of the biofuel cell performance is therefore important as it provides a means for the further optimization of the cell efficiency.

3.3.1 A biofuel cell based on PQQ and MP-11 monolayer-functionalized electrodes

The bioelectrocatalyzed reduction of H_2O_2 by MP-11 and oxidation of NADH by PQQ has been used to design a biofuel cell using H_2O_2 and NADH as the cathodic and anodic substrates (Figure 11).[86] For the optimization of the biofuel cell element, the potentials of the functionalized electrodes were determined (vs. the reference

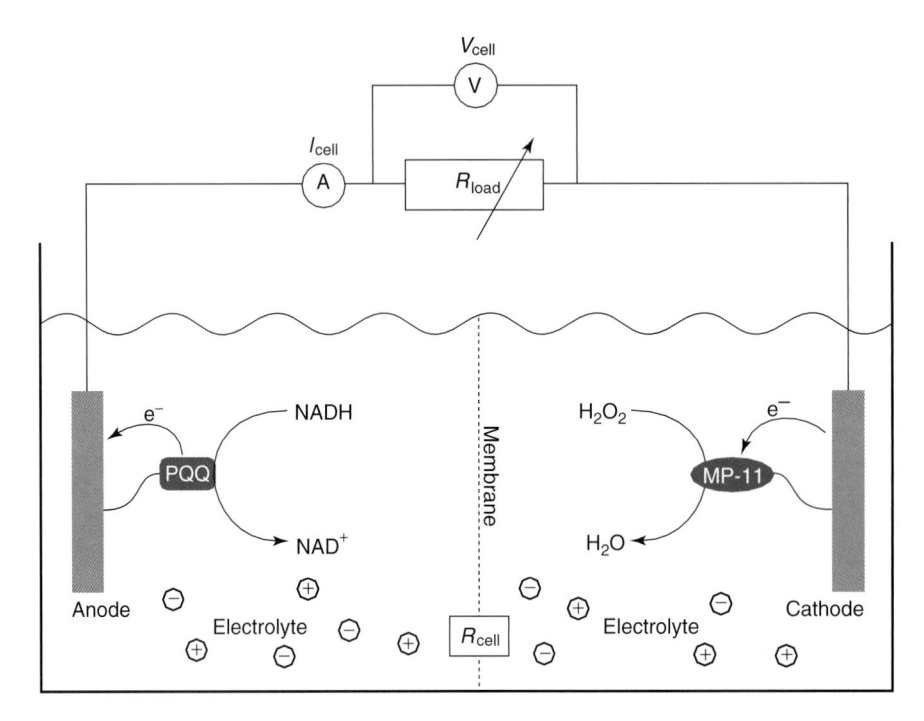

Figure 11. Schematic configuration of a biofuel cell employing NADH and H_2O_2 as fuel and oxidizer substrates and PQQ- and MP-11-functionalized-electrodes as catalytic anode and cathode, respectively. (Adapted from Scheme 1 in Ref. [86]. Reproduced by permission of Elsevier Science.)

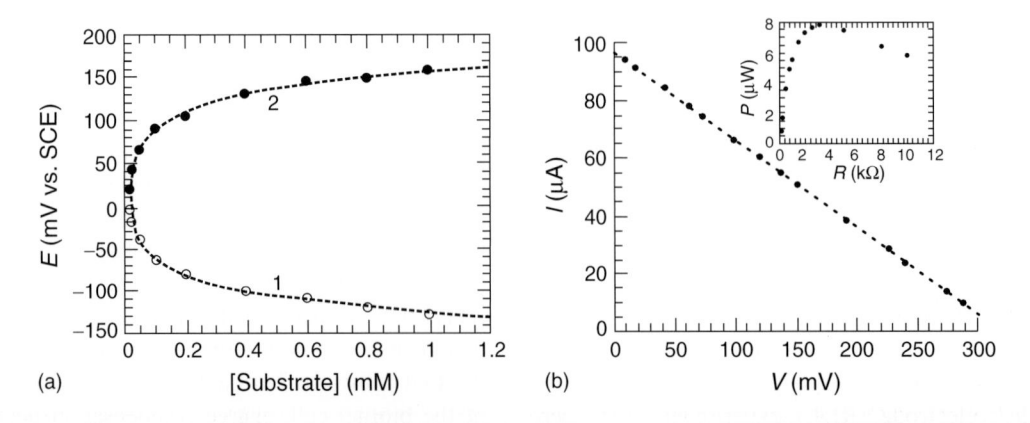

Figure 12. (a) Potentials of: (1) the PQQ-functionalized Au-electrode as a function of NADH concentration; (2) the MP-11-functionalized Au-electrode as a function of H_2O_2 concentration. The potentials of the modified electrodes were measured vs. SCE. (b) Current–voltage behavior of the PQQ-anode/MP-11-cathode biofuel cell measured at different loading resistances. Inset: electrical power extracted from the biofuel cell at different external loads. (Adapted from Figures 3 and 5 in Ref. [86]. Reproduced by permission of Elsevier Science.)

electrode, SCE) as a function of the cathodic and anodic substrate concentrations. Figure 12(a) shows the potential of the PQQ-electrode at different concentrations of NADH (curve 1) and the potential of the MP-11-electrode at different H_2O_2 concentrations (curve 2). The potentials of the PQQ monolayer-electrode and the MP-11-functionalized electrode are negatively shifted and positively shifted as the concentrations of NADH and H_2O_2 are elevated, respectively. The potentials of the electrodes reveal Nernstian-type behavior reaching saturation at high substrate concentrations (ca. 1×10^{-3} M). From the saturation potential values of the PQQ- and MP-11-functionalized electrodes, an open-circuit voltage of the cell of ca. 0.3 V was estimated. Taking into account the surface density of the catalysts (1.2×10^{-10} and 2×10^{-10} mol cm^{-2} for PQQ and MP-11, respectively), their interfacial electron transfer rate constants (ca. 8 and 14 s^{-1} for PQQ and MP-11, respectively) and the number of electrons participating in a single electron transfer event (2 and 1 for PQQ and MP-11, respectively), one may derive the theoretical limit of the current densities that can be extracted by the catalytically active electrodes (ca. 185 and 270 µA cm^{-2} for the PQQ and MP-11 electrodes, respectively).

The biofuel cell performance was examined at 1×10^{-3} M of each of the fuel and oxidizer. The cell voltage rose upon increasing the external load resistance and levels off to a constant value of ca. 310 mV at ca. 50 kΩ. Upon an increase of the load resistance, the cell current dropped and reached almost zero at a resistance of ca. 50 kΩ. Figure 12(b) shows the current–voltage behavior of the biofuel cell at different external loads. The cell yields a short-circuit current (I_{sc}) and open-circuit voltage (V_{oc}) of ca. 100 µA and 310 mV, respectively. The short-circuit current density was ca. 30 µA cm^{-2}, which is almost one order of magnitude less than the theoretical limits for the catalyst-modified electrodes. Thus, the interfacial kinetics of the biocatalyzed transformations at the electrodes is probably not the current-limiting step. The power extracted from the biofuel cell ($P_{cell} = V_{cell}I_{cell}$) is shown in Figure 12(b), inset, for different external loads, and reaches a maximum of 8 µW at an external load of 3 kΩ. The ideal voltage–current relationship for an electrochemical generator of electricity is rectangular. The linear dependence observed for this biofuel cell has a significant deviation from the ideal behavior and yields a fill factor of the biofuel cell of $f \approx 0.25$ (equation (27)). This deviation from the ideal rectangular V_{cell}–I_{cell} relationship results from mass transport losses reducing the cell voltage below its reversible thermodynamic value. It should also be noted that in this study NADH is used as the fuel. In a real biofuel cell, NADH should be generated in situ from an abundant substrate and the corresponding NAD$^+$-dependent dehydrogenase (e.g., alcohol or lactate acid in the presence of alcohol dehydrogenase or LDH, respectively).

$$f = P_{cell} \times I_{sc}^{-1} \times V_{oc}^{-1} \qquad (27)$$

3.3.2 Biofuel cells based on GOx and MP-11 monolayer-functionalized electrodes

The bioelectrocatalyzed reduction of H_2O_2 by the MP-11 monolayer electrode, and the oxidation of glucose by the reconstituted GOx-monolayer electrode allow us to design biofuel cells using H_2O_2 and glucose as the cathodic and anodic substrates (Figure 13).[87] Figure 14(a) shows the potentials of the GOx monolayer electrode at different concentrations of glucose (curve 1) and the potentials of the MP-11 monolayer electrode at different concentrations of H_2O_2 (curve 2). The potentials of the GOx monolayer electrode and of the MP-11 monolayer electrode are negatively shifted and positively shifted as the concentrations of the glucose and H_2O_2 are elevated, respectively. The potentials of the electrodes reveal Nernstian-type behavior, reaching saturation at high substrate concentrations of ca. 1×10^{-3} M. From the saturated potential values of the GOx and MP-11 monolayer electrodes, the theoretical limit of the open-circuit voltage of the cell is estimated to be ca. 320 mV. The short-circuit current (I_{sc}) generated by the cell is 340 µA. Taking into account the geometrical electrode area (0.2 cm^2) and the electrode roughness factor (ca. 15), the current generated by the cell can be translated into a current density of ca. 114 µA cm^{-2}. The theoretical limit of the current density extractable from the MP-11 monolayer electrode is ca. 270 µA cm^{-2} (surface coverage × interfacial electron transfer rate × Faraday constant). For the GOx monolayer electrode the maximum extractable current density was estimated to be ca. 200 µA cm^{-2} based on the surface coverage of the reconstituted GOx (1.7×10^{-12} mol cm^{-2}) and the turnover rate of the enzyme (ca. 600 s^{-1}). Thus, the observed short-circuit current density of the cell is probably controlled and limited by the bioelectrocatalyzed oxidation of glucose. This suggests that increasing the GOx content associated with the electrode could enhance the current density and the extractable power from the cell.

The biofuel cell performance was examined at concentrations of 1×10^{-3} M of each substrate. The cell voltage increases as the external load resistance is elevated, and at an external load of ca. 50 kΩ, it levels off to a constant value of ca. 310 mV. Upon increasing the external load, the current drops and is almost zero at an external load of 100 kΩ. Figure 14(b) shows the current–voltage behavior of the biofuel cell at different external loads. The linear dependence observed for the biofuel cell has

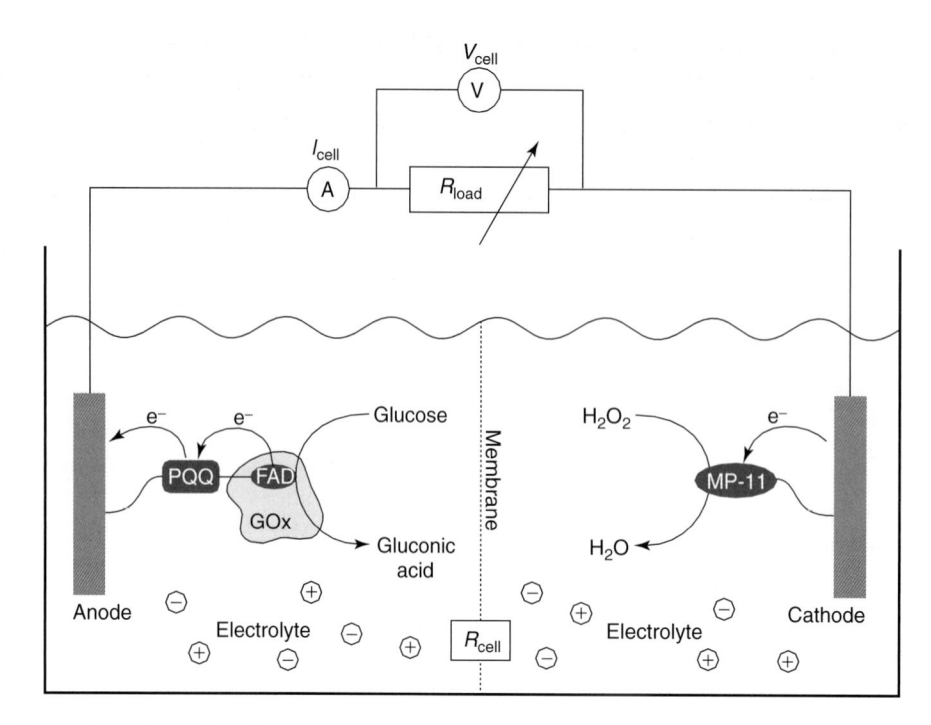

Figure 13. Schematic configuration of a biofuel cell employing glucose and H_2O_2 as a fuel and an oxidizer, respectively. GOx reconstituted onto a PQQ-FAD-monolayer and MP-11-functionalized Au-electrodes act as the biocatalytic anode and cathode, respectively. (Adapted from Scheme 3 in Ref. [87]. Reproduced by permission of The Royal Society of Chemistry.)

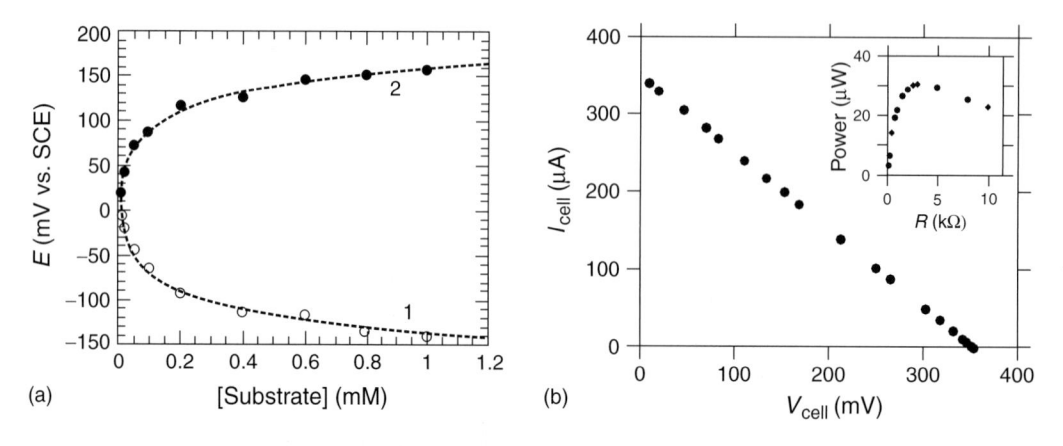

Figure 14. x(a) Potential of the PQQ-FAD/GOx-modified Au-electrode as a function of glucose concentrations (1) and potential of the MP-11-functionalized Au-electrode as a function of H_2O_2 concentrations (2), potentials were measured vs. SCE. (b) Current–voltage behavior of the GOx-anode/MP-11-cathode biofuel cell at different external loads. Inset: electrical power extracted from the biofuel cell at different external loads. (Adapted from Figures 5 and 7 in Ref. [87]. Reproduced by permission of The Royal Society of Chemistry.)

significant deviation from the ideal rectangular behavior and a fill factor of ca. 0.25. This deviation results from mass transport losses reducing the cell voltage below its reversible thermodynamic value. The power extracted from the biofuel element is shown in Figure 14, inset, for different external loads. The maximum power obtained is 32 µW at an external load of 3 kΩ. The biofuel cell voltage and current outputs are identical under Ar and air. This oxygen-insensitivity of the bioelectrocatalytic process at the anode originates from the effective electrical contact of the surface-reconstituted GOx with the electrode support, as a result of its alignment.[71, 72]

The stability of the biofuel cell was examined at the optimal loading resistance of 3 kΩ as a function of time.[87] The power decreases by about 50% after ca. 3 h of cell operation. This loss could originate from the depletion of the fuel substrate, leakage of the fuel or oxidizer into the wrong compartment or the degradation of the biocatalysts. Since the cell voltage appears to be stable, the current also decreases by the same factor. Integration of the current output yields the charge that passes through the cell, and it was thus calculated that ca. 25% of the fuel was consumed upon operation of the cell for ca. 3 h. Thus, 25% of the total decrease in the current output can be attributed to the loss of the fuel concentration. Recharging the cell with the fuel substrate and oxidizer could compensate for this component of the decrease in the current output.

Charge transfer processes across the interface between two immiscible electrolyte solutions can provide an additional potential difference between cathodic and anodic reactions due to the potential difference at the liquid/liquid interface. Many different interfacial liquid/liquid systems have been studied using numerous experimental approaches.[88] The application of two immiscible solvents that exhibit perspectives for enhancing the biofuel cell output has not been used previously. The reduction of cumene peroxide in dichloromethane, electrocatalyzed by the MP-11 monolayer electrode, and the oxidation of glucose in aqueous solution, bioelectrocatalyzed by the reconstituted GOx-monolayer-electrode, enables us to design a liquid/liquid interface biofuel cell using cumene peroxide and glucose as the cathodic and anodic substrates.[81] Figure 15(a) shows the potential of the GOx-monolayer electrode at different concentrations of glucose (curve 1) and the potential of the MP-11-monolayer electrode at different concentrations of cumene peroxide in dichloromethane (curve 2). The potentials of the GOx-electrode and the MP-11-electrode are negatively and positively shifted, respectively, as the substrate concentrations are elevated. The potentials of the electrodes reveal Nernstian-type behavior, showing a logarithmic increase and reaching saturation at high concentrations of the substrates. The saturated potential values of the anode and cathode are reached at ca. 1×10^{-3} M of glucose and 1×10^{-3} M of cumene peroxide, respectively, and from the saturated potential values of the GOx- and MP-11-monolayer electrodes, the theoretical limit of the open-circuit voltage of the cell is estimated to be ca. 1.0 V. It should be noted that the potentials extrapolated to zero concentrations of the substrates show a large difference (ca. 700 mV) which results from the potential jump at the liquid/liquid interface. The phase separation of the fuel and oxidizer is the origin for the enhanced efficiency of the cell. The cell reveals an open-circuit voltage of ca. 1.0 V and a short-circuit current density of ca. 830 μA cm^{-2}. The maximum power output of the cell is 520 μW at an optimal loading resistance of 0.4 kΩ (Figure 15b).

3.3.3 A noncompartmentalized biofuel cell based on GOx and Cyt c/Cox monolayer-electrodes

The next generation of biofuel cells could utilize complex, ordered enzyme or multi-enzyme systems immobilized on both electrodes, that may permit the elimination of the need for compartmentalization of the anode and the cathode. The tailoring of efficient electron transfer at the enzyme-modified electrodes could enable specific biocatalytic transformations that compete kinetically with any chemical reaction of the electrode or of the biocatalysts with interfering substrates (e.g., substrate transport from

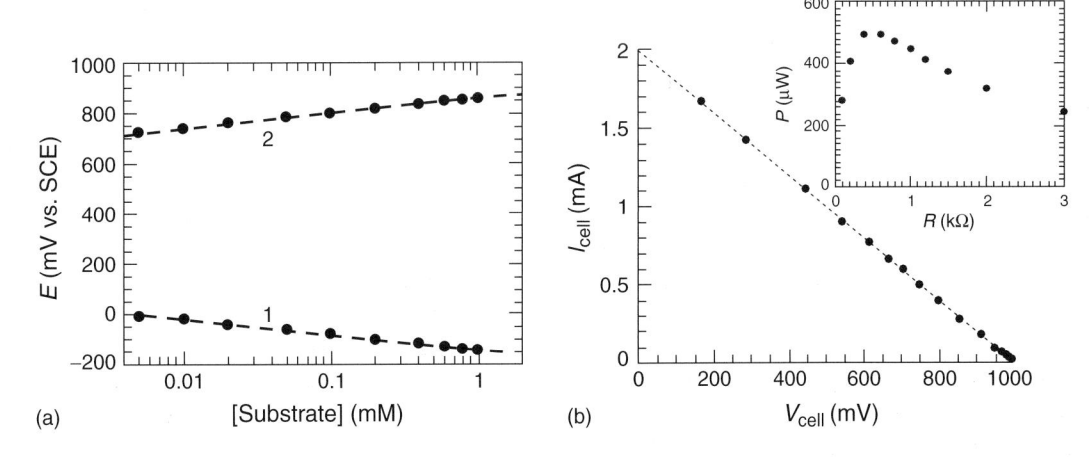

Figure 15. (a) Potential of: (1) the PQQ-AD/GOx-modified Au-electrode as a function of glucose concentration in 0.01 M phosphate buffer, pH 7.0, and 0.05 M TBATFB; (2) the MP-11-functionalized Au-electrode as a function of cumene peroxide concentration in a dichloromethane solution, 0.05 M TBATFB. Potentials were measured vs. aqueous SCE. (b) Current–voltage behavior of the biofuel cell at different external loads. Inset: electrical power extracted from the biofuel cell at different external loads. The biocatalytic cathode and anode (Au discs of ca. 0.6 cm^2) were immersed in a dychloromethane solution (lower phase) and an aquous solution (upper layer), respectively. (Adapted from Figures 2 and 4 in Ref. [81]. Reproduced by permission of The Royal Society of Chemistry.)

the counter compartment, oxygen, etc.). This would enable the design of noncompartmentalized biofuel cells where the biocatalytic anode and cathode are immersed in the same phase with no separating membrane. In a working example, an anode consisting of GOx reconstituted onto a PQQ-FAD monolayer (cf. Section 3.1.2) for the biocatalyzed oxidation of glucose was coupled to a cathode composed of an aligned Cyt c/COx couple that catalyzes the reduction of O_2 to water (cf. Section 3.2.2) (Figure 16a).[84] Since the reconstituted GOx provides extremely efficient biocatalyzed oxidation of glucose that is unaffected by oxygen, the anode can operate in the presence of oxygen.

Thus, the biofuel cell uses O_2 as an oxidizer and glucose as a fuel without the need for compartmentalization. The cell operation was studied at different external loads (Figure 16b), and achieved a fill factor of ca. 40% with a maximum power output of $4\,\mu W$ at an external load of $0.9\,k\Omega$. The relatively low power extracted from the cell originates mainly from the small potential difference between the anode and cathode. The bioelectrocatalyzed oxidation of glucose occurs at the redox potential of the PQQ-electron mediator, $E^{\circ\prime} = -0.125\,V$ (vs. SCE at pH 7.0), whereas the redox potential of Cyt c is $E^{\circ\prime} = 0.03\,V$. This yields a potential difference of only 155 mV

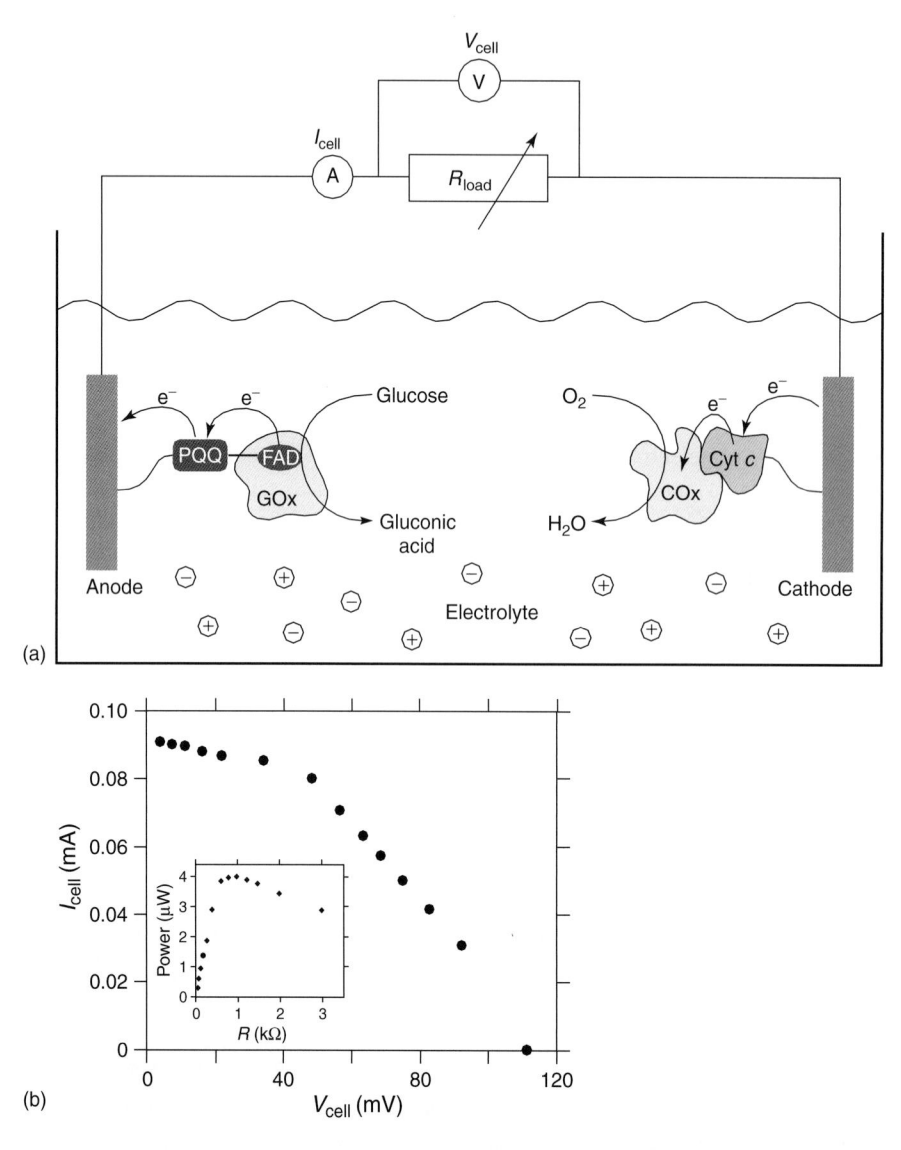

Figure 16. (a) Schematic configuration of a noncompartmentalized biofuel cell employing glucose and O_2 as fuel and oxidizer, and using PQQ-FAD/GOx and Cyt c/COx-functionalized Au-electrodes as biocatalytic anode and cathode, respectively. (b) Current–voltage behavior of the biofuel cell at different external loads. Inset: electrical power extracted from the biofuel cell at different external loads. The biocatalytic cathode and anode (ca. $0.8\,cm^2$ geometrical area, roughness factor ca. 1.3) were assembled in a thin-layer electrochemical cell with a distance of $5\,\mu m$ between the electrodes. (Adapted from Scheme 1 and Figure 1 in Ref. [84]. Reproduced by permission of Elsevier Science.)

between the anode and cathode. By the application of electron mediators that exhibit more negative potentials, the extractable power from the cell could be enhanced. The major advance of this system is its operation in a non-compartmentalized biofuel cell configuration. This suggests that the electrodes may be used as an in vivo electrical energy generation device utilizing as fuel and oxidizer glucose and O_2 from the bloodstream. Such in vivo electrical energy generation devices may be power sources for implantable machinery devices, e.g., pacemakers or insulin pumps.

4 CONCLUSIONS

Biofuel cells for the generation of electrical energy from abundant organic substrates can be organized by various approaches. One approach involves the use of microorganisms as biological reactors for the fermentation of raw materials to fuel products, e.g., hydrogen, that are delivered into a conventional fuel cell. The second approach to utilize microorganisms in the assembly of biofuel cells includes the in situ electrical coupling of metabolites generated in the microbial cells with the electrode support using diffusional electron mediators. A further methodology to develop biofuel cells involves the application of redox enzymes for the targeted oxidation and reduction of specific fuel and oxidizer substrates at the electrode supports and the generation of the electrical power output.

Towards this goal, it is essential to tailor integrated enzyme-electrodes that exhibit electrical contact and communication with the conductive supports. The detailed characterization of the interfacial electron transfer rates, biocatalytic rate-constants and cell resistances is essential upon the construction of the biofuel cells. Identification of the rate-limiting steps allows then the development of strategies to improve and enhance the cell output. Chemical modification of redox enzymes with synthetic units that improve the electrical contact with the electrodes provides a general means to enhance the electrical output of biofuel cells. The site-specific modification of redox enzymes and the surface-reconstitution of enzymes represent novel and attractive means to align and orient biocatalysts on electrode surfaces. The effective electrical contacting of aligned proteins with electrodes suggests that future efforts might be directed towards the development of structural mutants of redox proteins to enhance their electrical communication with electrodes. The stepwise nanoengineering of electrode surfaces with relay-co-factor-biocatalyst units by organic synthesis allows us to control the electron transfer cascades in the assemblies. By tuning the redox potentials of the synthetic relays or biocatalytic mutants, enhanced power outputs from the biofuel cells may be envisaged.

The configurations of the biofuel cells discussed in this paper can theoretically be extended to other redox enzymes and fuel substrates, allowing numerous technological applications. The production of electrical energy from biomass substrates using biofuels could complement energy sources from chemical fuel cells. An important potential use of biofuel cells is their in situ assembly in human body fluids, e.g., blood. The extractable electrical power could then be used to activate implanted devices such as pacemakers, pumps (e.g., insulin pumps), sensors and prosthetic units.

ACKNOWLEDGEMENTS

The research was supported by the Enriqus Berman Foundation.

REFERENCES

1. C. Van Dijik, C. Laane and C. Veeger, *Recl. Trav. Chim. Pays-Bas.*, **104**, 245 (1985).
2. W. J. Aston and A. P. F. Turner, *Biotechnol. Gen. Eng. Rev.*, **1**, 89 (1984).
3. L. B. Wingard Jr, C. H. Shaw and J. F. Castner, *Enzyme Microb. Technol.*, **4**, 137 (1982).
4. A. P. F. Turner, W. J. Aston, I. J. Higgins, G. Davis and H. A. O. Hill, 'Applied Aspects of Bioelectrochemistry: Fuel Cells, Sensors, and Bioorganic Synthesis', Presented at the Fourth Symposium on Biotechnology in Energy Production and Conservation, C. D. Scott (Ed), Interscience, New York, **12**, 401 (1982).
5. G. T. R. Palmore and G. M. Whitesides, 'Microbial and Enzymatic Biofuel Cells', in "Enzymatic Conversion of Biomass for Fuels Production", M. E. Himmel, J. O. Baker and R. P. Overend (Eds), ACS Symposium Series No. 566, American Chemical Society, Washington, DC, pp. 271–290 (1994).
6. A. T. Yahiro, S. M. Lee and D. O. Kimble, *Biochim. Biophys. Acta*, **88**, 375 (1964).
7. F. D. Sisler, 'Biochemical Fuel Cells', in "Progress in Industrial Microbiology", D. J. D. Hockenhull (Ed), J. & A. Churchill, London, Vol. 9, pp. 1–11 (1971).
8. S. Wilkinson, *Autonomous Robots*, **9**, 99 (2000).
9. P. N. Bartlett, P. Tebbutt and R. C. Whitaker, *Prog. Reaction Kinetics*, **16**, 55 (1991).
10. I. Willner and E. Katz, *Angew. Chem. Int. Ed.*, **39**, 1180 (2000).
11. K. Lewis, *Bacteriol. Rev.*, **30**, 101 (1966).
12. R. K. Thauer, F. H. Kirchniawy and K. A. Jungermann, *Eur. J. Biochem.*, **27**, 282 (1972).

13. S. Raeburn and J. C. Rabinowitz, *Arch. Biochem. Biophys.*, **146**, 9 (1971).

14. K. A. Jungermann, R. K. Thauer, G. Leimenstoll and K. Deker, *Biochim. Biophys. Acta*, **305**, 268 (1973).

15. R. K. Thauer, K. A. Jungermann and K. Deker, *Bacteriol. Rev.*, **41**, 100 (1977).

16. T. Akiba, H. P. Bennetto, J. L. Stirling and K. Tanaka, *Biotechnol. Lett.*, **9**, 611 (1987).

17. S. Suzuki and I. Karube, *Appl. Biochem. Bioeng.*, **4**, 281 (1983).

18. S. Suzuki, I. Karube, H. Matsuoka, S. Ueyama, H. Kawakubo, S. Isoda and T. Murahashi, *Ann. N.Y. Acad. Sci.*, **413**, 133 (1983).

19. I. Karube, S. Suzuki, T. Matunaga and S. Kuriyama, *Ann. N.Y. Acad. Sci.*, **369**, 91 (1981).

20. S. Suzuki, I. Karube, T. Matsunaga, S. Kuriyama, N. Suzuki, T. Shirogami and T. Takamura, *Biochimie*, **62**, 353 (1980).

21. I. Karube, T. Matsunaga, S. Tsuru and S. Suzuki, *Biotechnol. Bioeng.*, **19**, 1727 (1977).

22. S. Tanisho, N. Kamiya and N. Wakao, *Bioelectrochem. Bioeng.*, **21**, 25 (1989).

23. C. C. Liu, N. A. Carpenter and J. G. Schiller, *Biotechnol. Bioeng.*, **20**, 1687 (1978).

24. M. J. Cooney, E. Roschi, I. W. Marison, C. Comniellis and U. von Stockar, *Enzyme Microbiol. Technol.*, **18**, 358 (1996).

25. W. Habermann and E. H. Pommer, *Appl. Microbiol. Biotechnol.*, **35**, 128 (1991).

26. H. A. Videla and A. J. Arvia, *Biotechnol. Bioeng.*, **17**, 1529 (1975).

27. E. A. Disalvo and H. A. Videla, *Biotechnol. Bioeng.*, **23**, 1159 (1981).

28. S. J. Yao, A. J. Appleby, A. Geisel, H. R. Cash and S. K. Wolfson, Jr, *Nature*, **224**, 921 (1969).

29. S. K. Wolfson, Jr, L. B. Wingard, Jr, C. C. Liu and S. J. Yao, 'Biofuel Cells', in "Biomedical Applications of Immobilized Enzymes and Proteins", T. M. S. Chang (Ed), Plenum Press, Vol. 1, pp. 377–389 (1977).

30. M. J. Allen, 'Biofuel Cells', in "Methods in Microbiology", J. R. Norris and D. W. Ribbon (Eds), Academic Press, New York, pp. 247–283 (1972).

31. C. R. Myers and J. M. Myers, *J. Bacteriol.*, **174**, 3429 (1992).

32. H. J. Kim, M. S. Hyun, I. S. Chang and B. H. Kim, *J. Microbiol. Biotechnol.*, **9**, 365 (1999).

33. B. H. Kim, H. J. Kim, M. S. Hyun and D. H. Park, *J. Microbiol. Biotechnol.*, **9**, 127 (1999).

34. H. P. Bennetto, G. M. Delaney, J. R. Mason, S. D. Roller, J. L. Stirling and C. F. Thurston, *Biotechnol. Lett.*, **7**, 699 (1985).

35. C. F. Thurston, H. P. Bennetto, G. M. Delaney, J. R. Mason, S. D. Roller and J. L. Stirling, *J. Gen. Microbiol.*, **131**, 1393 (1985).

36. G. M. Delaney, H. P. Bennetto, J. R. Mason, S. D. Roller, J. L. Stirling and C. F. Thurston, *J. Chem. Technol. Biotechnol.*, **34B**, 13 (1984).

37. N. Kim, Y. Choi, S. Jung and S. Kim, *Biotechnol. Bioeng.*, **70**, 109 (2000).

38. S. D. Roller, H. P. Bennetto, G. M. Delaney, J. R. Mason, S. L. Stirling and C. F. Thurston, *J. Chem. Technol. Biotechnol.*, **34B**, 3 (1984).

39. H. P. Bennetto, J. L. Stirling, K. Tanaka and C. A. Vega, *Biotechnol. Bioeng.*, **25**, 559 (1983).

40. D. H. Park and J. G. Zeikus, *Appl. Environ. Microbiol.*, **66**, 1292 (2000).

41. I. Ardeleanu, D. G. Margineanu and H. Vais, *Bioelectrochem. Bioenerg.*, **11**, 273 (1983).

42. J. B. Davis and H. F. Yarbrough, Jr, *Science*, **137**, 615 (1962).

43. R. A. Patchett, A. F. Kelly and R. G. Kroll, *Appl. Microbiol. Biotechnol.*, **28**, 26 (1988).

44. G. Kreysa, D. Sell and P. Krämer, *Ber. Bunsen-ges. Phys. Chem.*, **94**, 1042 (1990).

45. C. A. Vega and I. Fernández, *Bioelectrochem. Bioeng.*, **17**, 217 (1987).

46. A. M. Lithgow, L. Romero, I. C. Sanchez, F. A. Souto and C. A. Vega, *J. Chem. Res., Synop.*, **5**, 178 (1986).

47. K. Tanaka, C. A. Vega and R. Tamamushi, *Bioelectrochem. Bioeng.*, **11**, 289 (1983).

48. D. Sell, P. Krämer and G. Kreysa, *Appl. Microbiol Biotechnol.*, **31**, 211 (1989).

49. D. H. Park, S. K. Kim, I. H. Shin and Y. J. Jeong, *Biotechnol. Lett.*, **22**, 1301 (2000).

50. R. M. Allen and H. P. Bennetto, *Appl. Biochem. Biotechnol.*, **39**, 27 (1993).

51. D. H. Park, B. H. Kim, B. Moore, H. A. O. Hill, M. K. Song and H. W. Rhee, *Biotechnol. Techniques*, **11**, 145 (1997).

52. H. A. Videla and A. J. Arvia, *Experientia*, **18**, 667 (1971).

53. W. van Hees, *J. Electrochem. Soc.*, **112**, 258 (1965).

54. M. J. Allen, *Electrochim. Acta*, **11**, 1503 (1966).

55. I. Willner and B. Willner, *Trends Biotechnol.*, **19**, 222 (2001).

56. I. Katakis and E. Dominguez, *Mikrochim. Acta*, **126**, 11 (1997).

57. B. Person, L. Gorton, G. Johansson and A. Torstensson, *Enzyme Microb. Technol.*, **7**, 549 (1985).

58. I. Willner and A. Riklin, *Anal. Chem.*, **66**, 1535 (1994).

59. E. Katz, T. Lötzbeyer, D. D. Schlereth, W. Schuhmann and H.-L. Schmidt, *J. Electroanal. Chem.*, **373**, 189 (1994).

60. G. T. R. Palmore, H. Bertschy, S. H. Bergens and G. M. Whitesides, *J. Electroanal. Chem.*, **443**, 155 (1998).

61. M. Maurice and J. Souppe, *New J. Chem.*, **14**, 301 (1990).

62. A. F. Bückmann and V. Wray, *Biotechnol. Biochem.*, **15**, 303 (1992).

63. J. Hendle, A. F. Bückmann, W. Aehle, D. Schomburg and R. D. Schmid, *Eur. J. Biochem.*, **213**, 947 (1993).

64. A. B. Kharitonov, L. Alfonta, E. Katz and I. Willner, *J. Electroanal. Chem.*, **487**, 133 (2000).

65. E. Katz, V. Heleg-Shabtai, A. Bardea, I. Willner, H. K. Rau and W. Haehnel, *Biosens. Bioelectron.*, **13**, 741 (1998).

66. A. Bardea, E. Katz, A. F. Bückmann and I. Willner, *J. Am. Chem. Soc.*, **119**, 9114 (1997).
67. E. V. Plotkin, I. J. Higgins and H. A. O. Hill, *Biotechnol. Lett.*, **3**, 187 (1981).
68. G. Davis, H. A. O. Hill, W. J. Aston, I. J. Higgins and A. P. F. Turner, *Enzyme Microb. Technol.*, **5**, 383 (1983).
69. P. L. Yue and K. Lowther, *Chem. Eng. J.*, **33**, B69 (1986).
70. A. Riklin, E. Katz, I. Willner, A. Stocker and A. F. Bückmann, *Nature*, **376**, 672 (1995).
71. I. Willner, V. Heleg-Shabtai, R. Blonder, E. Katz, G. Tao, A. F. Bückmann and A. Heller, *J. Am. Chem. Soc.*, **118**, 10 321 (1996).
72. E. Katz, A. Riklin, V. Heleg-Shabtai, I. Willner and A. F. Bückmann, *Anal. Chim. Acta*, **385**, 45 (1999).
73. T. Ruzgas, E. Csöregi, J. Emneus, L. Gorton and G. Marko-Varga, *Anal. Chim. Acta*, **330**, 123 (1996).
74. P. A. Adams, 'Microperoxidases and Iron Porphyrins', in "Peroxidases in Chemistry and Biology", J. Everse, K. E. Everse and M. B. Grisham (Eds), CRC Press, Boca Raton, FL, Vol. 2, Chapter 7, pp. 171–200 (1991).
75. T. Lötzbeyer, W. Schuhmann, E. Katz, J. Falter and H.-L. Schmidt, *J. Electroanal. Chem.*, **377**, 291 (1994).
76. E. Katz and I. Willner, *Langmuir*, **13**, 3364 (1997).
77. A. M. Klibanov, *Trends Biochem. Sci.*, **14**, 141 (1989).
78. J. Li, S. N. Tan and J. T. Oh, *J. Electroanal. Chem.*, **448**, 69 (1998).
79. L. Yang and R. W. Murray, *Anal. Chem.*, **66**, 2710 (1994).
80. A. N. J. Moore, E. Katz and I. Willner, *J. Electroanal. Chem.*, **417**, 189 (1996).
81. E. Katz, B. Filanovsky and I. Willner, *New J. Chem.*, **23**, 481 (1999).
82. S. Tsujimura, H. Tatsumi, J. Ogawa, S. Shimizu, K. Kano and T. Ikeda, *J. Electroanal. Chem.*, **496**, 69 (2001).
83. G. Tayhas, R. Palmore and H.-H. Kim, *J. Electroanal. Chem.*, **464**, 110 (1999).
84. E. Katz, I. Willner and A. B. Kotlyar, *J. Electroanal. Chem.*, **479**, 64 (1999).
85. V. Pardo-Yissar, E. Katz, I. Willner, A. B. Kotlyar, C. Sanders and H. Lill, *Faraday Discussions*, **116**, 119 (2000).
86. I. Willner, G. Arad and E. Katz, *Bioelectrochem. Bioeng.*, **44**, 209 (1998).
87. I. Willner, E. Katz, F. Patolsky and A. F. Bückmann, *J. Chem. Soc., Perkin Trans.*, **2**, 1817 (1998).
88. A. G. Volkov and D. W. Deamer, 'Liquid/Liquid Interface Theory and Methods', CRC, Boca Raton, FL (1996).

Chapter 22
Metal/air batteries: The zinc/air case

O. Haas[1], F. Holzer[1], K. Müller[1] and S. Müller[2]

[1] Paul Scherrer Institute, Villigen, Switzerland
[2] Euresearch, Bern, Switzerland

1 INTRODUCTION

1.1 Basic aspects of metal/air batteries

It is the special advantage of metal/air batteries that only the active metallic material has to be stored in the battery. The reactive component for the positive electrode, oxygen, is absorbed from the ambient air during discharge and is returned there during charging. For this basic reason, such systems can attain higher values of specific charge and energy density than other electrochemical energy storage systems. For many applications and in particular for traction, this advantage may be decisive. As a corollary, however, there is a slight change in weight of the battery during charge and discharge. Since the specific charge of the oxygen which passes through the oxygen electrode is very high (3350 Ah kg^{-1}), this weight change is normally less than 5% of the total battery weight.

Because of the distinctly better performance of the oxygen electrode and lower corrosion of the metal electrode, strongly alkaline electrolytes are preferred for metal/air batteries. In alkaline electrolytes the anion (OH$^-$) produced at the oxygen electrode is used to form metal hydroxide as a cell reaction product. In principle, acidic electrolytes could also be used. In such a case a metal salt is formed as the cell reaction product and the anion of this salt has to be provided by the acidic electrolyte solution. The OH$^-$ produced at the oxygen electrode leads to a pH increase at this electrode, which may be rather substantial when acidic electrolytes are used. This will shift the oxygen electrode potential to lower working potentials and lower cell potentials are the result when the solution next to the Zn electrode remains acidic.

The fact that in Zn/air cells, air oxygen is used implies that the electrolyte must be in contact with the environment via a porous oxygen electrode, hence a suitable water management must be provided in order to maintain a constant electrolyte level and carbonation of the electrolyte caused by carbon dioxide from the air must be avoided to the largest possible extent, otherwise the life of the battery would be shortened significantly (see Section 2.2).

Some metals other than zinc can also be used in metal/air batteries. Their electrochemical parameters, as well as those of a typical metal hydride, are given in Table 1.

The metals cadmium[1] and iron[2, 3] have already been used in alkaline rechargeable batteries employing a metal oxide counterelectrode. They can basically be used as well in rechargeable metal/air batteries. The hydrogen overpotential is high enough at these base metals to allow their cathodic redeposition after discharge in aqueous solutions. By thermodynamic arguments it cannot be excluded that some self-discharge will occur. This effect is minor in the case of cadmium and zinc but quite important in the case of iron.

Aluminum and magnesium have low equivalent masses and hence a high theoretical value of specific charge. They also have a rather negative redox potential, so that a metal/air battery containing these metals as the negative electrode will have a very high theoretical specific energy. However, the highly negative potential of the metal electrode in these batteries leads to a high rate of self-discharge in aqueous electrolytes. In a number of electrolytes though, these metals tend to passivate, which is a practical advantage for the use of these electrodes

Table 1. Theoretical electrochemical parameters for electrode materials of interest in connection with metal/air batteries.

Electrode material	Equivalent mass[b] (g)	Specific. charge[b] (Ah kg^{-1})	Standard electrode potential (V)	Electrode potential[c] at pH 14 (V)	Theoretical metal/O$_2$ cell voltage[c] (V)
Al	25.99	1032	−1.66	−2.38	2.78
Cd	73.20	366	−0.40	−0.80	1.20
Fe	44.92	597	−0.44	−0.87	1.27
Li	23.94	1120	−3.04	−3.04	3.44
MeH$_x$[a]	70.00	370	0.00	−0.83	1.23
Mg	29.16	919	−2.38	−2.70	3.10
Na	39.98	671	−2.71	−2.71	3.11
Zn	40.69	659	−0.49(pH 0)	−1.26	1.67
O$_2$	8.00	3350	1.23	0.401	–

[a] Approximate values for the equivalent mass and theoretical specific energy in AB$_5$ alloys, such as LaNi$_5$.
[b] Equivalent mass is here calculated with respect to the product of discharge (e.g., Al(OH)$_3$, rather than Al (MW$_{Al(OH)_3}$: 77.98 g mol^{-1}, EM: 77.98/3 = 25.99 g).
[c] Referring to the stable form of the discharge product reported in the Pourbaix diagram.

in batteries, but at the same time the passivation leads to considerable discharge overpotentials and thus to a lower practical energy density and lower energy yield.

Direct recharging of aluminum and magnesium electrodes in batteries with aqueous electrolytes is not feasible; the electrode material can only be regenerated by electrolysis in molten salts or organic electrolytes outside the battery.

Occasionally even the strongly basic metals lithium[4] and sodium have been proposed for metal/air batteries, but on account of their vigorous reactions with water, aqueous electrolytes are problematic.

In a molten-salt electrolyte, hydrogen evolution at these negative electrodes could be avoided, but few electrolytes exist in which an oxygen counterelectrode would work. Alkaline oxide melts could be used in principle, but they have very high melting points and tend to form peroxides at the oxygen electrode. Solid electrolytes, such as ZrO$_2$/Y$_2$O$_3$ could also be used, but then the metal of the metal/air battery would have to be in its molten state and should be able to absorb the oxide formed in order not to passivate at the electrolyte/metal interface. Up to now, not much work has been done on such systems.

Table 1 also provides some typical figures for a metal hydride electrode, which basically could be used as the negative electrode in a battery with an air electrode as the positive electrode.[5–7]

Rather than undergoing the metal/metal ion redox reaction, these alloys reversibly absorb and desorb hydrogen. The discharge process yields protons (which recombine to water with hydroxyl ions of the electrolyte), while water electrolysis (which yields hydrogen at the metal hydride electrode and oxygen at the air electrode) is used to recharge the metal hydride with atomic hydrogen diffusing into the metal lattice, where it is stored. In view of the extremely low equivalent mass of hydrogen and its high energy storage density, the metal hydride electrodes have a relatively high charge density. The theoretical value for the specific charge of LaNi$_5$H$_6$ is given in Table 1. The practical specific charge of alloys used in nickel/metal hydride batteries approaches values of 250–400 Ah kg^{-1}. In nickel/metal hydride storage batteries, more than thousand cycles and up to 90 Wh kg^{-1} have been achieved with such metal hydride electrodes.

The metal lattices must be optimized by suitable alloy formation to build electrodes with a high hydrogen storage density, high rate of hydrogen diffusion and high reversibility (long cycle life). Titanium-nickel[8–11] and lanthanum/nickel[12, 13] alloys, which form multiphase mixtures, are of the highest interest today. Amorphous structures appear to be superior to crystalline structures.[11]

However, when these alloys are used in metal hydride/air batteries, the cell voltage is rather low (about 0.6–0.8 V, as in fuel cells), which is a disadvantage. In addition, high rates of self-discharge and some safety problems must be expected because of hydrogen pressure at the hydride electrode and the contact with air at the porous air electrode, where gaseous hydrogen could escape. Technical solutions exist but tend to raise the cost of practical batteries. Metal hydride/air cells have the great advantage, however, that they could also be chemically recharged with hydrogen to regenerate the metal hydride electrodes.

1.2 Zinc electrodes

Metallic zinc has excellent electrochemical properties with fast electrode kinetics and a high overpotential towards hydrogen evolution, thus aqueous electrolytes can be used to deposit Zn electrochemically even though it has a rather negative standard potential ($U_o = -0.76$ V vs. normal hydrogen electrode (NHE)). Zinc is also a good energy storage material. The transformable free energy (ΔG) of the oxidation reaction with oxygen leads to $1.37\,\text{kWh}\,\text{kg}^{-1}$ for Zn, based on the molecular weight of Zn and as a comparison, to $33\,\text{kWh}\,\text{kg}^{-1}$ for hydrogen, based on the molecular weight of hydrogen[1]. Thus, the specific energy stored in Zn that can be oxidized to ZnO is 24 times lower than that of hydrogen being oxidized to water, but due to the high density of the metallic zinc compared to hydrogen, its energy density ($\text{kWh}\,\text{l}^{-1}$) is about four times higher than that of liquid hydrogen or 11 times higher than that of compressed hydrogen at 300 bar. The special cryogenic or pressure vessels necessary for hydrogen were disregarded when calculating these factors. (Note: density of zinc $= 7.14\,\text{g}\,\text{cm}^{-3}$, density of liquid hydrogen $= 0.07\,\text{g}\,\text{cm}^{-3}$ and density of gaseous hydrogen at 300 bar $= 0.027\,\text{g}\,\text{cm}^{-3}$.)

The practical cell potentials of Zn/air batteries are about 1.5 times higher than those of H_2/O_2 fuel cells, which is another practical advantage of zinc over hydrogen. In fact, zinc is very largely used in combination with alkaline aqueous electrolytes and different counterelectrodes, such as MnO_2 and NiOOH. In combination with oxygen electrodes, it is widely used in primary cells for hearing aids and other special applications. Rechargeable Zn/air cells or Zn/air fuel cells have not yet been commercialized to a comparable extent, but in many aspects their state of development is already rather advanced.

1.3 Basic aspects of zinc/air batteries

For practical applications, the most promising system available up to now and in the foreseeable future is the zinc/air system: Energy density, price, availability and environmental compatibility of all components are strong arguments for this battery.

The overall electrode reactions occurring in zinc/air batteries with alkaline electrolyte (pH 14) are as follows:

At the negative electrode:

$$Zn + 4OH^- = Zn(OH)_4^{2-} + 2e^-$$
$$E_o = -1.266\,\text{V vs. NHE}$$

Zinc oxide precipitates from supersaturated zincate solutions according to the following reaction:

$$Zn(OH)_4^{2-} = ZnO + H_2O + 2OH^-$$

At the positive electrode:

$$\tfrac{1}{2}O_2 + H_2O + 2e^- = 2OH^- \quad E_o = +0.401\,\text{V vs. NHE}$$

For the overall battery reaction:

$$Zn + \tfrac{1}{2}O_2 + H_2O = ZnO \quad E_o = +1.667\,\text{V}$$

In practical zinc/air batteries, the theoretical cell voltage cannot be attained because of high overpotentials of the reactions at the oxygen electrode. An electrode reaction yielding H_2O_2 is superimposed during discharge on the electrode reaction reported above for the oxygen electrode:

$$O_2 + H_2O + 2e^- = HO_2^- + OH^- \quad E_o = -0.076\,\text{V}$$

This reaction leads to lower cell discharge voltages. Catalysts for hydrogen peroxide decomposition when incorporated into the air electrode serve to improve the cell voltage by decreasing the concentration of HO_2^-, at the same time they protect the air electrode surface from chemical attack by the hydrogen peroxide species.

In practice, the open-circuit voltage of zinc/air cells is about 1.35 V, their working voltage is between 1.0 and 1.2 V depending on the current drain.

1.4 Problem areas

It was seen above that from a thermodynamic and kinetic point of view, zinc/air batteries are attractive. However, an important problem for rechargeable batteries with zinc as the negative electrode arises from the changes in structure and shape of the zinc electrode occurring during repeated charge/discharge cycles (zinc dissolution and redeposition) due to dissimilar current distributions during charge and discharge.

A major problem in rechargeable batteries with an air electrode as the positive electrode is that of realizing a highly active, long-lived bifunctional air electrode. High energy efficiencies can only be achieved when electrochemical reduction of oxygen at discharge and oxygen evolution during charging occur with low overpotentials. In addition, this electrode must resist oxidative attack. These problems are current research topics. Satisfactory results have been achieved using $La_{0.6}Ca_{0.4}CoO_3$ perovskite or $LaNiO_3$ spinel or other catalytically active metal oxides as electrocatalysts[14, 15] (see Section 2.2.2).

In a mechanically rechargeable zinc/air battery, the zinc of the negative electrode is regenerated outside the cell from zinc discharge products withdrawn from the battery. For the air (oxygen) electrode, this implies that only its oxygen reduction function is needed during consecutive discharge periods; long electrode life is more readily achieved in this system when care is taken that the oxygen electrode is not damaged while the zinc electrode is exchanged.

Basically, one could continuously feed zinc and thus use the cell in a fuel cell mode where the consumable electrode would be replaced by a zinc band, zinc rod, or zinc slurry passing through the cell. Such systems are under investigation and will be discussed in later sections.

2 ELECTRICALLY RECHARGEABLE ZINC/AIR BATTERIES

2.1 Electrically rechargeable zinc electrodes

2.1.1 Zinc electrode structure

One of the most important tasks in the development of electrically rechargeable zinc batteries is that of realizing a zinc electrode structure which is stable over many hundred charge/discharge cycles.

These cycles involve dissolution and redeposition of zinc, leading to morphology changes of the zinc electrode, more precisely to the formation of dendrites and to changes in the outer geometry ("shape change") and inner geometry ("densification") of the electrode.

The dendrites lead to short circuits between the electrodes in the battery cells, while the morphology changes as a whole are accompanied by passivation phenomena and by progressive loss of electrical contact between active material and the electrode structure, hence to even lower capacities of the zinc electrode in consecutive cycles.

All three phenomena: dendrites, shape change and densification, are consequences of the high solubility of zinc electrode discharge products. Another important factor are the local differences in electrolyte composition developing within the cells in consecutive charge/discharge cycles.

According to McBreen and Gannon[16] and Cairns et al.,[17, 18] the shape changes of the zinc electrode are caused by differences in the current density distribution over the different zones of the zinc electrode during discharge and charging. The higher density of the electrode mass observed by Cairns et al. in the center of the zinc electrode relative to peripheral regions is said to arise from mutual effects of current density distribution and concentration gradients in the electrolyte.

Choi and Bennion[19] attribute the morphological changes of the zinc electrode during cycling (shape change, densification) to convective electrolyte flows along the surface of the zinc electrode during discharge and charging. In these flows, gravity is said to be of minor importance. An essential driving force for the convective flows is electroosmosis. The flows lead to differences in the concentrations of zincate and hydroxide ions in the electrolyte and to non-uniform current density distributions across the electrodes. Shape changes of the zinc electrode during cycling are supposed to be prevented to a large extent when these convective flows are stopped, for instance by separators.

Einerhand et al.[20, 21] showed with in situ radiotracer (^{65}Zn) experiments that due to density gradients developing during repeated cycling, zinc material gradually moves from the top to the bottom of the electrode. During battery cycling, an electrolyte flow arises as a result of density gradients in the solution layer adjacent to the zinc electrode and of the volume variations of the battery electrolyte.

2.1.2 Zinc deposition

The zinc electrode's morphology depends on the deposition conditions defined by such factors as the zincate concentration, the electrolyte concentration gradients and the current density.

Strong concentration gradients of zincate ions that are detrimental to deposit morphology, can be prevented from developing in electrolytes when the zincate ion concentration is kept relatively low. However, apart from the fact that this is not practical in high-drain batteries, low zincate concentrations give rise to mossy morphology of the zinc deposited electrochemically from the solution. Undesirable mossy deposits also arise when using high KOH concentrations, high electrolyte temperature and poor zinc ion transport in the electrolyte.

The ideal deposit form is difficult to define; the optimum depends on the system design as a whole. However, it is possible to generalize literature observations[22–27] concerning the morphology and other properties of the zinc deposits and the conditions of electrochemical (cathodic) deposition from alkaline solutions, as follows.

To a first approximation, low current densities ($\sim 20\,\text{mA}\,\text{cm}^{-2}$) initially lead to smooth deposits, but later the deposit morphology changes to mossy forms. Higher current densities ($\sim 50\,\text{mA}\,\text{cm}^{-2}$) will produce powdery or granular deposits. Very high current densities which lead to diffusion-controlled deposition (~ 100–$150\,\text{mA}\,\text{cm}^{-2}$) produce dendritic deposits.

According to Frenay et al.,[22] in particular, an optimum zinc deposit in the form of dendritic powder suitable for batteries with alkaline electrolyte can be produced at high cathodic current densities (100–$150\,\text{mA}\,\text{cm}^{-2}$), temperatures not exceeding $50\,°\text{C}$, a zinc ion concentration of $25\,\text{g}\,\text{l}^{-1}$ and a NaOH concentration of $240\,\text{g}\,\text{l}^{-1}$. Mossy zinc deposits can be prevented by strong electrolyte stirring, which decreases the diffusion layer thickness at the electrode surface.

Cachet, Ströder and Wiart[28] studied zinc electrode kinetics in strongly alkaline solutions in the presence of certain additives. They found that tetrabutylammonium bromide suppresses or, rather, limits dendrite formation in repeated charge/discharge cycles. Fluorinated surfactants, for instance Atochem's F1110 ($C_6F_{13}C_2H_4[OC_2H_4]_{12}OH$), are supposed to stabilize the grain structure of zinc in repeated charge/discharge cycles and suppress mossy zinc deposition in the charging process.

The fluorinated organic agents (F1110), also certain brighteners and lead ions added in concentrations of 10^{-5}–10^{-3} M to the electrolyte promote compact deposits; their effect increases in the order cited.

2.1.3 Zinc dissolution

The conditions of zinc dissolution are of considerable importance for the electrolyte composition after discharge and hence during subsequent charging and have for this reason been studied by a number of workers.

Cachet, Saidani and Wiart[29, 30] used impedance measurements to study anodic zinc dissolution in zincate-saturated, strongly alkaline electrolytes. They deduced a two-step mechanism for electrochemical zinc dissolution. After the first electron transfer step, the univalent zinc ($Zn(OH)_{ad}$) is assumed to form a thin, reactive intermediate phase on the zinc electrode surface. This phase constitutes a separation between the zinc metal and the electrolyte. Its conductivity depends on the potential. When during anodic dissolution the potential becomes a little more positive the $Zn(OH)_{ad}$ phase grows thicker, but also more porous. This porous layer has a high ionic conductivity and in the strongly alkaline electrolyte, it now becomes thinner by dissolution. The analogous (inverse) mechanism is formulated by the above authors for cathodic zinc deposition.[29]

Prentice, Chang, and Shan[31] performed rotating-disk studies from which they concluded in a similar way that anodic zinc dissolution and anodic passivation of the zinc electrode involve an intermediate ($ZnOH_{ad}$) species with univalent zinc. According to

$$Zn + OH^- = Zn(OH)_{ad} + e^-$$
$$Zn(OH)_{ad} + 2OH^- = Zn(OH)_3^- + e^-$$
$$Zn(OH)_3^- + OH^- = Zn(OH)_4^{2-}$$

this intermediate leads to a soluble oxidation product that will hardly block the electrode. On the other hand, at slightly more positive potentials the intermediate is further oxidized according to

$$Zn(OH)_{ad} + OH^- = Zn(OH)_2 + e^-$$

and forms a hydroxide deposit which then blocks the electrode. This deposit dissolves in strongly alkaline electrolytes, forming zincate.

Sunu and Bennion[32] concluded from experimental work and model considerations that three effects are important for the charge/discharge behavior of the zinc electrode in zincate-saturated alkaline electrolytes:

- electrolyte depletion at the phase boundary (electrode surface) by consumption of OH^- ions during discharge,
- blocking and clogging of pores by reaction products,
- passivation of the electrode (surface) by insoluble zinc oxide deposits on the electrode.

2.1.4 Zinc corrosion reactions

It must be remembered when using zinc electrodes in rechargeable batteries that zinc is a metal which by thermodynamic principles can undergo corrosion reactions in aqueous electrolytes, not only at open circuit, but also under the conditions of discharge and of charging. These reactions are attended by hydrogen evolution and imply reduced current yields and/or discharge capacity. Hydrogen evolution should therefore be suppressed by inhibitors and by a high hydrogen overpotential at the electrode materials.

To do so, different additives are used in the zinc electrode and/or electrolyte, which have the purpose of suppressing the corrosion reactions via their blocking effect on cathodic hydrogen evolution without having a detrimental effect on anodic zinc ion formation during discharge.

Mercury has long been used as an excellent additive suppressing corrosion of the zinc electrode, but its adverse health and environmental effects have led to the decision to replace it entirely by other agents. The effect of additives on the amount of hydrogen evolved on zinc in alkaline solution were summarized in a paper by Keily and Sinclair.[33] As an example, quaternary ammonium compounds when added in a concentration of 0.1 M to the alkaline 0.5 M zincate electrolyte bring about a significant decrease in the corrosion rate. A distinct decrease in zinc corrosion is seen as well when adding 0.01 M silicic acid to the electrolyte, but this additive lacks long-term efficacy.[33, 34]

These results are limited, in that they have been derived from observations of gas evolution without polarization or parallel tests under the conditions of charging or discharge. The corresponding studies were concerned with corrosion in general, rather than zinc used in rechargeable batteries.

Mansfeld and Gilman[35] found that Zn/Pb alloys (with 0.4 wt% of Pb) and Zn/Al alloys (with 0.4 wt% of Al), which had been prepared from highly pure metals, exhibit a distinctly lower rate of corrosion than highly pure Zn (9 ml H_2 $cm^{-2} \cdot day^{-1}$ for the alloys, 65 ml H_2 $cm^{-2} \cdot day^{-1}$ for 99.999% Zn). On the other hand, trace impurities of Cu, Ni and Fe give rise to a distinct increase in corrosion rate, as expected, because they reduce the hydrogen overpotential. However, it was also found by Mansfeld and Gilman[35] that lead ions when present in the electrolyte induce cathodic zinc deposition in the form of finer crystals; both the charging and discharge process exhibit a 100 mV higher overpotential for the zinc reaction.

Yoshizawa, Miura and Ohta[36] found that a liquid In/Ga alloy is a perfect, even superior substitute for mercury. Zinc corrosion and hydrogen evolution at the zinc are completely suppressed when the zinc surface is wetted with this alloy. Also, alloy formation with indium, lead and bismuth (individually or in combination) is very effective.[36] When these alloys are employed in combination with the above perfluoroalkyl-polyethyleneoxide as an additive to the electrolyte, which forms a surface film on the zinc, a further improvement of the zinc electrode reaction is obtained, viz., higher current yields because of suppression of the hydrogen evolution reaction.[36] The same observation was made with cadmium and thallium alloys,[37] which in combination with organic inhibitors (3-[4-nitrophenyl]-5-amino-1,2,4-triazole) inhibit the corrosion reactions without having a detrimental effect on zinc ionization during discharge.

It is as yet an open question whether the approaches discussed in this section are feasible in battery technology. They may be successful technically but risk failing economically.

In patents of Electric Fuel Limited (EFL), Israel,[38, 39] oxides of Sb, Bi, Cd, Ga, In, Pb, Tl and Sn in concentrations between 0.05 and 4 wt% are claimed as additives to a zinc paste to be used as the electrode material in zinc/air batteries. Zinc powder as the starting material should have a grain size of 100–500 μm, a density between 0.3 and 1.4 $g\,cm^{-3}$ and a specific surface area of 0.5–6 $m^2\,g^{-1}$. It is made into an active mass with KOH solution and the above oxides. Polyacrylic acid, graphite, carbon fibers and $Ca(OH)_2$ are mentioned as further additives.

2.1.5 Reversible zinc electrodes

Zn electrodes retaining their discharge products

Zinc electrodes which retain their discharge products, rather than forming soluble discharge products, can be employed both in electrically rechargeable zinc/air batteries and in zinc/air batteries which are mechanically "recharged" (or "refueled" after regeneration of the zinc electrode outside the system, see Section 3.3). The requirements that must be met for the two system types differ considerably.

For a zinc electrode in electrically rechargeable batteries, the most important requirement is the absence of significant changes in structure during many consecutive charge/discharge cycles (see Section 2.1), since such changes adversely affect the electrical round-trip efficiency. In addition, any zinc oxide and hydroxide forming on the electrode must not cause mechanical blocking of the electrode nor interfere with the conduction path. Electrodes meeting this requirement are prepared as pasted electrodes where a mixture of zinc oxide powder, cellulose, polytetrafluoroethylene (PTFE) binder and water and optionally zinc powder is stirred to a paste that is slurried or pressed into a current collector consisting of lead-plated copper[40, 41] and then dried at temperatures slightly above 100 °C (see Figure 1). These electrodes have a porosity close to 50%.[42]

As mentioned in Section 2.1.1, hydrogen evolution and uneven current distribution must be avoided. This aspect was investigated by several groups.[16, 41, 43–47]

By experience it has become good practice to use a sizeable excess of zinc in the negative electrode. Thus, one third of the zinc present in the electrode should still be in the oxide form when the electrode is "fully" charged; in this way overcharging and the formation of hydrogen gas can be avoided. Also, one third of the zinc present in the electrode should still be in the metallic form when the electrode is "fully" discharged in order to keep the active material in a good condition, thus extending the cycle life of the battery. These two zinc reserves together lead to the recommendation of a three-fold excess of zinc in the electrode.[41]

However, Müller, Holzer and Haas have recently reported cycle life experiments and discharge performance with Zn/O_2 or Zn/air cells in which two thirds of the nominal

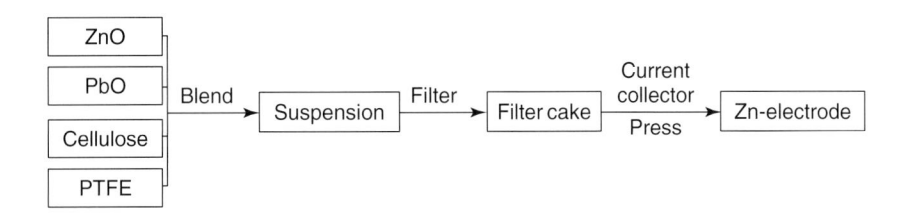

Figure 1. Preparation of pasted zinc electrodes. (Reproduced from Müller et al. (1995).[43])

capacity of the zinc electrode was used. They proposed a zinc electrode with one third of zinc oxide present at the "fully" charged state and almost no metallic zinc at the "fully" discharged state.[42, 48] They thus increased the cycled capacity of the zinc electrode very considerably.

The available capacity of zinc electrodes can also be raised by adding conductive fibers (chemically metallized short-staple polymer fibers)[49] or acetylene black[50] to the electrode.

Even cellulose fibers lacking electronic conduction led to an improved discharge characteristic of the zinc electrode when added to the active mass.[40, 42] This can be attributed on one hand to hydrophilicity of the fiber material, which leads to improved wetting of the electrodes and enhanced OH^- ion transport in the porous electrode. On the other hand the structure of the electrode is improved so as to prolong the cycle life. The best results were found with fibers more than 5 mm long, which were present in the electrode at a concentration of 10 wt%.[40]

Zn electrodes with soluble discharge products

The zinc electrodes can also be operated as electrodes with soluble discharge products, by adding enough KOH to render the zinc oxide soluble. Ross[51, 52] incorporated an inert porous, reticulated flow-through electrode for Zn deposition and dissolution. A porous metal foam, usually Cu, thus served as current collector/substrate for a zinc electrode that was also porous. In order to cycle this porous Zn electrode, the electrolyte had to be pumped through the system. By optimizing the electrolyte, the Zn was deposited as mossy Zn on the metal foam. The reported cycle life in Zn/air cells (1.5 Ah, 25 cm^2) was 150 cycles at a charge rate of 3 h.

Müller *et al*.[43, 53] introduced a zinc electrode which combined the features of electrodes with soluble and insoluble discharge products. During the first charge the Zn-containing electrolyte was pumped through the cell so that metallic zinc would be formed homogeneously throughout the foam. After the first charge, the electrolyte was changed to a ZnO-saturated electrolyte so that discharge and charge occurred as in electrodes retaining their discharge products. However, after a few such cycles, due to zinc redistribution and dissolution, the battery had to be recharged once more in the same way as during first charge.

Bronoel *et al*.[54] constructed cells with a large excess of electrolyte so that, in the oxidized state, the greater part of the zinc dissolved as zincate ions. Cells with flowing electrolyte and cells with stationary electrolyte were investigated as cell types. The second type required a combination of additives (e.g., Pb^{2+} ions and fluorinated compounds) in solution and pulsed current during charge in order to obtain a good life cycle.

2.2 Air electrodes for electrically rechargeable zinc/air batteries

2.2.1 Electrode structure

The key task in any technical realization of electrically rechargeable air-based batteries is the development of an efficient, long-lived bifunctional oxygen electrode. At bifunctional ("two-way") oxygen electrodes, O_2 is consumed during discharge and generated during charge of the battery. Only optimized "one way" O_2-consuming electrodes are commercially available and are used in fuel cells and primary metal/air batteries (see also Section 3.2).

Cathodic oxygen reduction occurs in a series of consecutive steps:

- diffusion of oxygen to the three-phase boundary, which is the reaction zone of oxygen electrodes;
- adsorption of the oxygen at the three-phase boundary;
- electron transfer to the adsorbed oxygen (electrochemical step);
- elimination of the reaction products (OH^-, H_2O_2, H_2O) via the electrolyte or gas phase.

Anodic oxygen evolution roughly consists of the inverse sequence of steps.

From these steps, it is evident that high-performance oxygen electrodes, both bifunctional ones and electrodes merely working as cathodes, should contain an optimum pore system for ion and oxygen transport and a large, catalytically active phase boundary for charge transfer in oxygen reduction and evolution. These requirements are taken care of by porosity engineering, which must maximize the extent of the liquid-pore/gas-pore contact ("three-phase boundary") in an electrode comprising solids (catalysts, electronic conductor, structural components), liquid (the electrolyte as ionic conductor) and gas phase. Stable hydrophobic zones for gas transport as well as stable hydrophilic zones with a high degree of wetting by the electrolyte are needed in order to stabilize the three-phase boundary and the transport processes involved in oxygen reduction. The stability of the hydrophobic/hydrophilic boundary is achieved by engineering a relatively steep boundary between a hydrophilic inner layer (containing wettable catalyst and conductor particles) and a hydrophobic outer layer (containing hydrophobic binder and/or filler). This engineering also includes the use of fibrous elements, which could be hydrophobic (to favor gas inflow) or hydrophilic (to favor liquid transport). The hydrophobic air-diffusion layer and the hydrophilic catalyst layer should not be thicker than a few tenths of a millimetre so that transport hindrance in the electrode will be minimal.

Active carbon or carbon black having a large surface area is most often used as basic constituent of these electrodes.

Some graphite is added to the carbon black to improve electrical conductivity, which is important at higher current drains. Activation of the carbon material and deposition of oxygen reduction catalysts on the active material are additional steps serving to accelerate oxygen reduction. Proven methods of activation include treatments with CO_2 or steam at temperatures of 800–1000 °C, which gives rise to the formation of basic functional groups on the carbon surface.

Fine-mesh metal screens or expanded metal (advantageously made of nickel) and carbon fiber mats were found to be satisfactory as a current collector. They work best when incorporated into or pressed onto the electrode's gas-diffusion layer.

For a long life cycle, diffusion of carbon dioxide and water loss or gain through the porous electrode should be avoided. Protection against water loss or flooding and against excessive CO_2 inflow can be achieved by means external to the actual electrode, such as lengthened transport paths (diffusion/convection management) and active transport (fans only operated during current drain). The inflow of CO_2 can be prevented by the use of scrubbed air; the substances used for scrubbing may have a scrubbing capacity strongly depending on the humidity of the scrubbed air.[105]

Water management may also be required, e.g., electrolyte recirculation may have to be integrated into the system. This is an action often found necessary for extended operability of the zinc electrode.

2.2.2 Catalysts and substrates

The selection of an appropriate bifunctional catalyst is an important prerequisite for an efficient electrode.

Oxygen evolution and, more particularly, oxygen reduction in aqueous solutions occur at a high overpotential, even at moderately high temperatures $[E(O_{2,\text{gen.}}) > E(O_{2,\text{red.}})]$. The reaction rates are higher in alkaline than in acidic electrolytes. This irreversibility of the reactions constitutes the major problem in the development of oxygen or air electrodes that will function in both the anodic and cathodic mode. Efficiency losses and serious lifetime problems of the catalyst and its support material have to be overcome by more active catalysts and stable support materials.

Many kinds of electrocatalysts have been investigated as the key components for these electrodes. Proven catalysts which will accelerate oxygen reduction are silver and modified transition metal complexes with nitrogen-containing macrocycles; and for oxygen evolution, metal oxides and spinels, but most of them are not simultaneously highly stable and active in both of these processes. In addition, some aspects, such as mechanical erosion by the evolving gas must be taken into account when developing bifunctional oxygen electrodes.

Noble metals, such as Pt and Pt/Ru alloys,[55–60] are electroactive in both reactions but are expensive. In a zinc/air battery, moreover, catalysts containing platinum metals are not particularly favored, inasmuch as mere traces of these metals dissolving on account of inevitable corrosion reactions when reaching the zinc electrode will drastically lower its hydrogen overvoltage and thus lead to undesirable hydrogen production.

So far, mixed oxides,[61–63] oxides with a perovskite structure,[64–66] silver,[67–69] graphitic materials catalyzed with metal oxides,[70, 71] and pyrochlores[62, 63, 72] have been tried and used as bifunctional electrocatalysts.

The state of the art has been reviewed by Swette et al.,[73] Goodenough et al.,[74] and Kinoshita.[75] Studies concerning the mechanism of the oxygen reduction/evolution reaction at perovskite catalysts were performed by Bockris et al.[76, 77] Useful guiding principles for optimized materials design have been formulated by Goodenough et al.[74]

Current research is particularly active in the area of perovskites. These are ionic compounds having a cubic or orthorhombic crystal structure and the general formula ABX_3, where A and B are cations and X is an anion (for the catalysts discussed here, this is always the oxygen anion). A good example of a compound having this structure is $LaNiO_3$. Perovskites are interesting oxygen electrode catalysts because of their high oxide ion mobility. By partial substitution of A by a cation A' or of B by a cation B' with a different valency, ionic defects or a change in valence state of the catalytically active metal B can be induced in the perovskite in order to influence its catalytic activity and conductivity.

Bockris and Otagawa[77] reached the following conclusions when studying perovskites: The catalytic activity is due primarily to the transition metal (component B) in the perovskite. It increases with increasing number of d-electrons in the transition metal, in the series of V → Cr → Mn → Fe → Co → Ni. The number of electrons in the antibonding M^Z–OH orbitals also increases in this order. The electrocatalytic function of the perovskite is governed by the interaction between the reactive center of the catalyst (the transition metal) and the OH intermediates. Rate determining for the electrode process is the breaking of the M^Z–OH bond, that is, detachment of the intermediate M^Z–OH; formation of this bond is less critical for the reaction rate on perovskites. It was found that the more active perovskites are those which are uncharged in a strongly alkaline solution, that is, exhibit a weaker binding of the OH^- groups. Unfortunately, the same perovskites are more highly susceptible to corrosion.

For the preparation of perovskites, suitable methods reported include pure solid-state synthesis and liquid/solid synthesis.

Solid-state synthesis relies on reactions in mixtures of powdered metal salts. Metal salts, such as hydroxides, nitrates, carbonates, acetates and oxalates from which the anion is readily eliminated are preferred as the precursors. High temperatures are needed (more than 1000 °C) in order to reach sufficiently high reaction rates of perovskite formation. However, these temperatures lead to a drastic loss in surface area due to sintering processes. Therefore, the solid-state reaction products are less suitable for catalyst applications.

Liquid/solid synthesis involves the thermal reaction (dehydration, oxidation) of a gel-like metal salt mixture obtained when evaporating metal salt solutions with appropriate stoichiometries. This leads to more homogeneous products with a higher surface area. Vacuum decomposition of the metal salt mixtures at 300 °C and subsequent calcining in an oxygen flow at 500 °C produces specific surface areas of 30 m^2 g^{-1}. Other variants of liquid/solid synthesis are spray pyrolysis, freeze-drying and coprecipitation.

Kannan, Shukla and co-workers[15, 61] have studied the following compounds with perovskite structure with respect to their success as bifunctional catalysts: $NiCo_2O_4$, $LaNiO_3$, $LaCoO_3$, $La_{0.1}Sr_{0.9}FeO_3$ and $La_{0.7}Sr_{0.3}MnO_3$. The $LaNiO_3$ catalyst proved to be best in its performance. The potential difference between battery charge (oxygen evolution, $i = 100$ mA cm^{-2}) and discharge (oxygen reduction, $i = 30$ mA cm^{-2}), which is a good measure of activity of a bifunctional catalyst, was 0.6 V.

The above authors believe, however, that a Ni^{2+}/Ni^{3+} redox reaction in the catalyst material is superimposed upon (occurs concomitantly with) the oxygen electrode reactions, which implies a permanent change in state of the catalyst and should not be favorable for its long-term stability.

Shimizu et al.[78] conducted studies at bifunctional oxygen electrodes with La/Ca/Co perovskite catalysts. With a $La_{0.6}Ca_{0.4}CoO_3$ catalyst, a potential difference of about 1 V was observed between oxygen evolution and oxygen reduction in zinc-ion-containing, alkaline (7 M KOH) electrolyte when both the anodic and cathodic current density was 100 mA cm^{-2}. This performance remained constant without any failures over up to 100 cycles run in 10 min intervals. Over short periods of time current densities of up to 1 A cm^{-2} could be drawn from the electrodes.

Striebel et al.[79] used pulsed laser deposition to produce thin films of $La_{0.6}Ca_{0.4}CoO_3$, $La_{0.6}Ca_{0.4}MnO_3$ and $La_{0.5}Sr_{0.5}FeO_3$ on stainless steel substrates and studied the kinetics of oxygen reduction and evolution in 0.1 M KOH with the rotating ring-disk technique. The activities of these three metal oxide films in oxygen reduction were found to decrease in the order of $La_{0.6}Ca_{0.4}MnO_3 > La_{0.5}Sr_{0.5}FeO_3 > La_{0.6}Ca_{0.4}CoO_3$ while the Tafel slope was 90 mV dec^{-1}. The activities in oxygen evolution decreased in the order of $La_{0.6}Ca_{0.4}MnO_3 > La_{0.6}Ca_{0.4}CoO_3 > La_{0.5}Sr_{0.5}FeO_3$ while the associated Tafel slope was 60 mV dec^{-1}.

Müller, Striebel and Haas,[66] who extended the work of Shimizu et al. mentioned above, were able to substantially improve the lifetime of these bifunctional oxygen electrodes. This was possible when using graphitized active carbon XC 72 (2700 °C) as a corrosion-resistant, highly conducting support for the $La_{0.6}Ca_{0.4}CoO_3$ catalysts. They developed a new technology to fabricate gas diffusion electrodes. The active (catalyst) layer and the diffusion layer were manufactured separately by a rolling process and then combined to an electrode with a metal screen as the current collector, by a combined pressing and sintering process (see Figure 2).

Recently the activity and stability of bifunctional air electrodes was investigated as a function of the carbon black and graphite material used as catalyst support.[80] The activity for oxygen reduction was strongly influenced by the surface area and wettability of the carbon material. With different carbon materials in the electrodes, no significant differences of overpotential were detected for the oxygen evolution reaction, which takes place at the perovskite. Both reactions have been investigated by applying ac impedance spectroscopy.[81] It was concluded from an analysis of these spectra that the electrode's active surface area decreases during oxygen evolution. A large catalyst loading (perovskite) improved the electrode stability. Among the carbon materials tested (HSAG 100 and 300, Vulcan as received and graphitized and Ketjenblack), the graphitized Vulcan XC 72 showed superior long-term stability under conditions of continuous oxygen reduction and evolution. However, a heat treatment in oxidizing atmosphere was found to be necessary for the graphitized Vulcan XC 72 in order to develop its potential as a catalyst support. A lifetime of more than 2000 h was achieved with electrodes containing graphitized Vulcan which had been thermally activated.[80] Oxygen reduction and generation was performed at ±6 mA cm^{-2} and the resulting potentials were 1.2 and 2.0 V vs. Zn, respectively.

It is known that cobalt and nickel oxides and their mixed oxides (which have spinel structure) are suitable as catalysts for H_2O_2 decomposition in alkaline solutions. For this reason a number of authors have recommended these metal oxides as electrode catalysts for oxygen evolution[82–87] and also for oxygen reduction[71] in alkaline solutions. However, in substance it must be concluded from these publications that the stability of cobalt oxides with spinel structure very probably is not sufficiently high for satisfactory lifetimes in zinc/air batteries. Moreover, a bifunctional application of cobalt oxides in a practical oxygen electrode is not described in the papers cited.

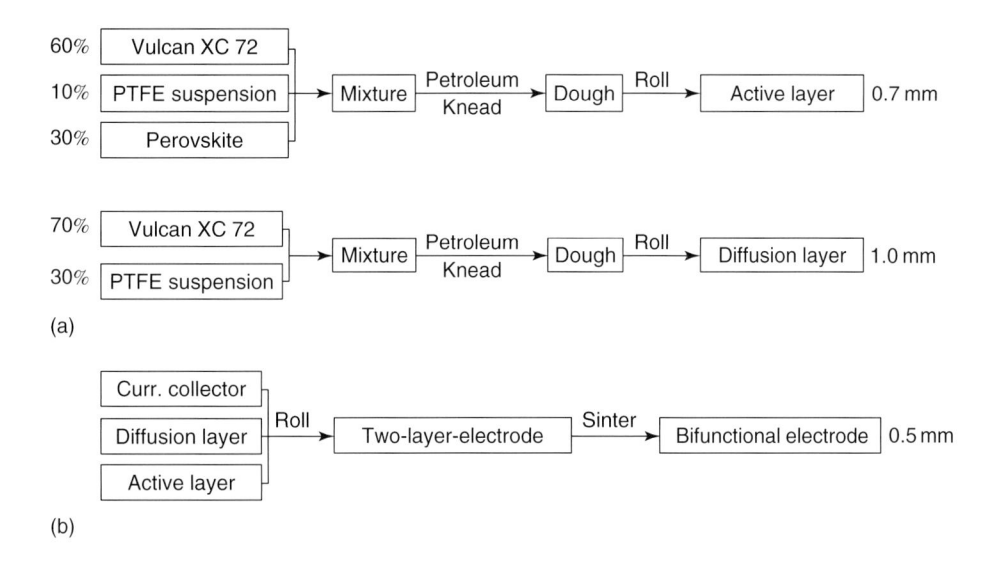

Figure 2. Preparation process of bifunctional oxygen electrodes for alkaline electrolytes. (Reproduced from Holzer et al. (1998).[48])

In studies involving pyrochlores, ($Pb_2Ir_2O_{7-y}$, $PbBiRu_2O_{7-y}$) as bifunctional catalysts for oxygen electrodes,[62, 88] cathodic current densities of up to $10\,mA\,cm^{-2}$ and anodic current densities of up to $7.5\,mA\,cm^{-2}$ were drawn from the electrodes, the potential difference between oxygen evolution and reduction being about 0.6 V under these conditions. There was no drop in performance over 100 cycles. Unfortunately, the costs of these catalyst materials are such that a practical application cannot be contemplated. Moreover, as pointed out above, noble metals always constitute a risk with respect to satisfactory hydrogen overpotential at the zinc electrodes. (Some additional comments on catalyst and its supports are given in Sections 3.2.2 and 3.2.3.)

2.3 Separators and electrolyte

Separators of suitable shape and design are generally useful in suppressing or diminishing the formation of dendrites and in maintaining dimensional stability of the zinc electrode over many charge/discharge cycles. Separators are a necessity in rechargeable zinc batteries; they must meet the following requirements:

- Chemical stability toward the oxygen evolved during the charging process, toward the concentrated KOH and other electrolyte components, even at elevated temperatures (up to 80 °C).
- Electrochemical stability at the potentials of the electrodes with which they are in contact.
- Good wettability by the electrolyte and abundant electrolyte uptake.
- Low resistance to ionic (electrolytic) conduction.
- Homogeneous pore structure and high mechanical stability so as to block gas permeation and dendrite growth.
- Management of ionic transport in the electrolyte.

Fleece-type polymer materials on the basis of polyacrylamide, polyacrylic acid and poly(vinylpyrrolidone) copolymerized with divinylbenzene[89] have been tested, but have not by themselves been able to completely suppress structural changes of the zinc electrode. Still, with these separators 350 charge/discharge cycles of the batteries have been demonstrated.

A separator to a large extent meeting the requirements formulated above is the Celgard product marketed by the Hoechst Celanese Corporation of North Carolina (USA). The material of these separators is both homoporous and microporous and the separators themselves are laminated from a number of layers of the material.

2.4 Electrically rechargeable zinc/air battery systems

Each of the two types of zinc electrodes described in Section 2.1, viz., those with the discharge products dissolved in the electrolyte and those with the discharge products retained in the electrode, can be used in the mechanically and electrically rechargeable batteries which will be discussed in this section. Figure 3 shows the potentials of the zinc and air electrode during four charge and discharge cycles of an electrically rechargeable Zn/air battery with two electrodes in series.

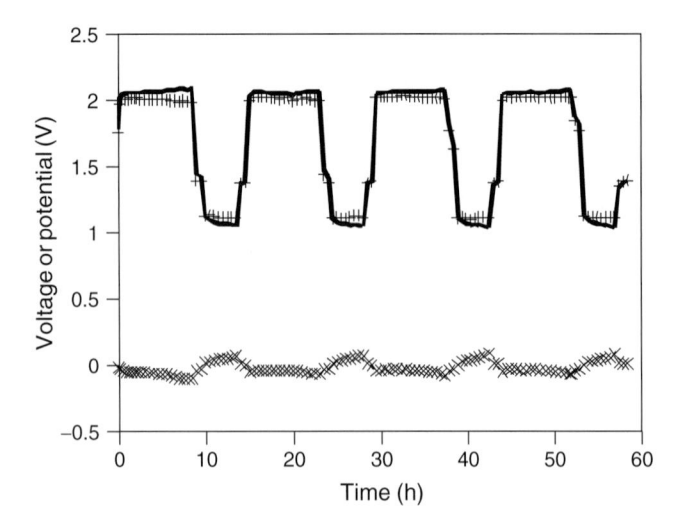

Figure 3. Charge/discharge cycles for an electrically rechargeable Zn/air battery with two electrodes in series. Total cell voltage (−), potential of the zinc electrode (×), potential of the bifunctional air electrode(+). All potentials are measured against a zinc reference electrode.

2.4.1 Batteries with auxiliary electrodes for their recharge

Due to the fact that it is difficult to obtain a sufficient service life, electrocatalytic activity and corrosion resistance of the battery's air electrode, the possibility has been considered repeatedly to develop a three-electrode cell, where an auxiliary electrode is used for oxygen evolution during battery charging. In a parallel-plate cell, this oxygen-evolving electrode should be placed between the zinc electrode and the oxygen diffusion electrode and be electrically insulated from the latter, for instance by arranging an additional separator between the two oxygen electrodes.[90] The auxiliary electrode could consist of a nickel screen (a structure as open as possible to the electrolyte is required in order to have unhindered ion transport during battery discharge) and should be activated with oxygen evolution catalysts (heavy metal oxides) for oxygen evolution.

During charging, the (anodic) current must be supported by the auxiliary electrode, which then evolves oxygen, while the gas diffusion electrode serving as the oxygen-reducing electrode during discharge, which could be damaged by superoxide and peroxide, must be cut off from this anodic process. During discharge the auxiliary electrode remains idle.

This design solution has been discussed,[91, 92] but not in great depth, the reason being that it raises new problems, such as the risk of short-circuiting by additional electronic conductors, the need to equip each individual cell in the battery with a logic circuit switch with high current-carrying capability, the high cost associated with the corresponding solid-state devices and the need for a special electrolyte circuit through which the oxygen evolved during charging is eliminated from the system.[90]

Lindström et al.[5] offered an interesting design solution for arranging a similar auxiliary electrode in a metal hydride/air battery. Here, the oxygen-evolving auxiliary anode was arranged between two negative (hydride) electrodes which in turn are surrounded by two air electrodes, thus the two oxygen electrodes (the "anode" for charging and the "cathode" for discharge) cannot interfere, though it has been reported that the inverted distribution of electrode reaction products obtained during charging and discharge is a disadvantage.

2.4.2 Batteries with bifunctional oxygen electrodes having separate active layers for charge and discharge

The design of these bifunctional electrodes is such that in a multilayer electrode, two active layers with different wetting behavior (graded hydrophobicity or, rather, graded pore radius structure) exist and that a first active layer with the higher wettability supporting oxygen evolution during charging is directly adjacent to the free electrolyte phase. A second active layer with somewhat lower capillary pressure, where oxygen is reduced during discharge, is situated between the first active layer and the hydrophobic backing, while the free air space is on the other side of the hydrophobic backing. Each of the two active layers has the appropriate catalysts, either for oxygen evolution or for oxygen reduction.

During charging, gas is evolved which generates an excess pressure forcing the electrolyte from the second active layer that is sensitive against peroxide. During discharge the air pressure should be such that the electrolyte penetrates into both active layers and oxygen reduction can take place in the layer which contains the sensitive oxygen-reduction catalysts.[89]

It remains difficult to manage the exact position of the electrolyte layers in the air electrode during the different operating phases of the battery. Traces of electrolyte remaining in the second active layer during the charging phase induce corrosion phenomena by H_2O_2 in the catalyst system of this layer.

Development efforts toward zinc/air batteries employing this design have been reported by Duperray.[89] In the air electrode, the active layer for oxygen reduction was provided on the side of the electrolyte with a porous nickel or silver layer for oxygen evolution. A PTFE membrane on the side of the gas phase served as the barrier layer. Expanded nickel screen was used for the current collector. The negative electrode was produced in situ from a ZnO electrode, that is, a zinc electrode mounted in the battery in the discharged condition. In 5 h charge/discharge cycles at a

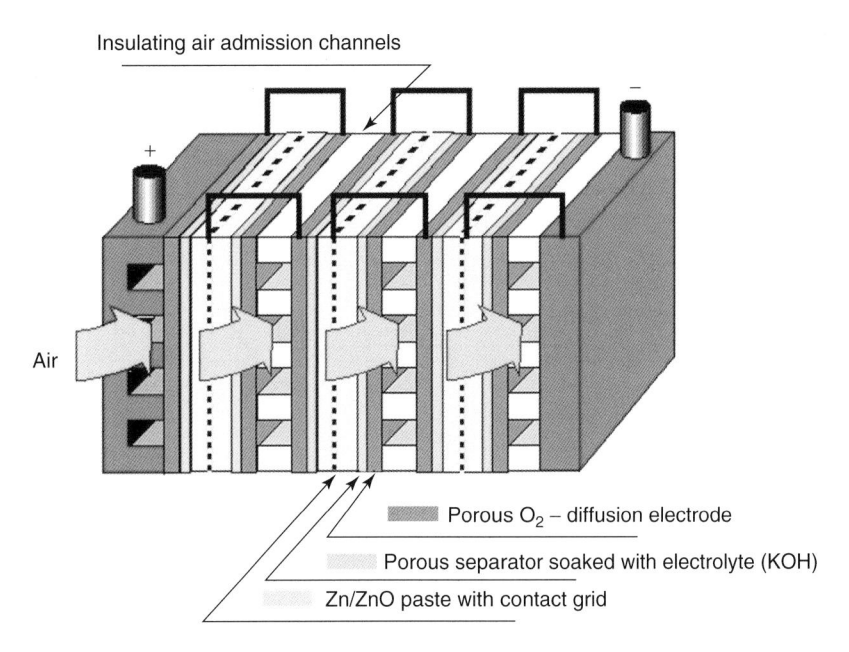

Figure 4. Schematic view of a monopolar rechargeable zinc/air-battery arrangement with zinc electrodes sandwiched between two oxygen electrodes and external series connection. (From Paul Scherrer Institute.)

current density of $10\,\text{mA}\,\text{cm}^{-2}$ about 50% of the theoretical capacity could be drawn from the cell. After 60 cycles of this type the capacity dropped off. The data yield a figure of $28\,\text{W}\,\text{kg}^{-1}$ for the specific power and of $77\,\text{Wh}\,\text{kg}^{-1}$ for the specific energy at peak load. The diagnosed failure mode of this cell was shape change of the zinc electrode, inasmuch as nothing had been done in the experiments to counteract it. It was found that the bifunctional oxygen electrode was still without performance loss after the above 60 cycles.

2.4.3 Batteries with bifunctional oxygen electrodes having a common active layer for charge and discharge

In these batteries, the catalysts of the oxygen electrode support both the reduction and evolution of oxygen, or are so selected that one catalyst type is insensitive to the reaction products (for instance peroxide, superoxide) arising at the other type. In principle two design versions are possible, with a monopolar or a bipolar cell arrangement, respectively. Figures 4 and 5 provide a schematic view of these two battery systems.

Electrically rechargeable zinc/air batteries using bifunctional air electrodes have been developed and studied over the last 10 years by several Research Institutes and Companies, such as Lawrence Berkeley Laboratories (LBL; USA), Eltech Research Corporation together with Matsi (USA), and Paul Scherrer Institute (CH). LBL and Matsi have focused their efforts on zinc/air batteries with soluble zinc electrodes. The service life of the LBL test cell is limited

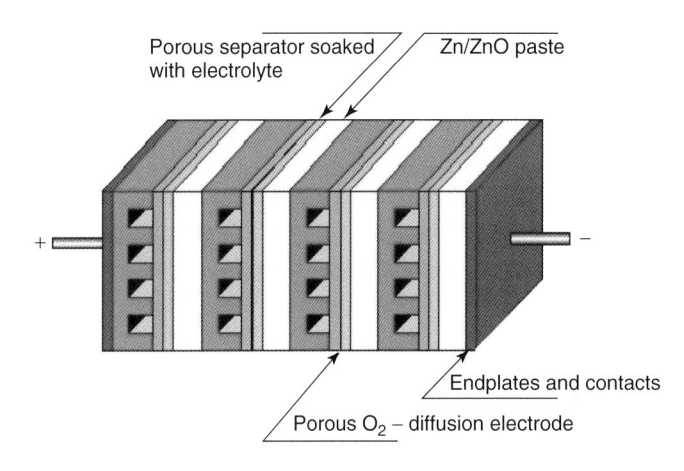

Figure 5. Schematic view of a bipolar zinc/air-battery arrangement with internal series connection.

by the bifunctional air electrodes (Electromedia Corporation, USA). A cycle life of below 50 cycles was obtained at 6 h charge and 3 h discharge rates for a smaller laboratory test cell.

The Paul Scherrer Institute employed a pasted insoluble zinc electrode in its battery system. Specific energies in the range of $70-100\,\text{Wh}\,\text{kg}^{-1}$ were reported for their electrically rechargeable zinc air system. They were able to achieve an important improvement of these electrodes particularly with respect to their lifetime. With their bifunctional air electrodes and pasted Zn electrodes, a maximum cycle life of 450 cycles (3000 h) was obtained at 6 h charge and 3 h discharge rates in 2.4 Ah test cells. Considering the weight of the battery components and the peak power

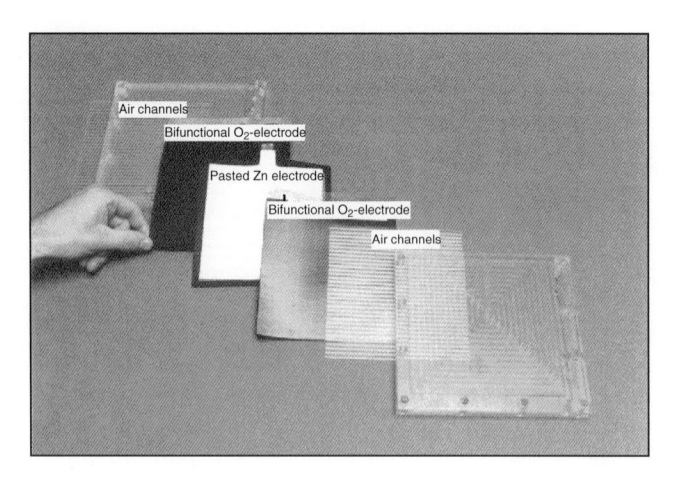

Figure 6. Components for a single 200 cm^2 Zn/air cell with a nominal capacity of 20–30 Ah. (Reproduced from Holzer *et al.* (1998).[48])

available from the test cells operated with air and with oxygen, one can calculate specific peak powers in excess of the medium-term power targets of 150 W kg^{-1} set by USABC for batteries to be used in electric vehicles. Up-scaling of the battery components demonstrated the feasibility of a 20 Ah/12 V electrically rechargeable zinc/air battery with a service life of ca. 2000 h (ca. 200 cycles). The state of the art of this technology represents a realistic starting point for developing a commercial product.

The 20 Ah/12 V battery developed at Paul Scherrer Institute is shown in Figure 7. The components necessary for a single cell are shown in Figure 6. Typical current/voltage and current/power curves of a two-cell zinc/air battery with a nominal capacity of 30 Ah is shown in Figure 8.

This technology has been transferred to ChemTEK/Zoxy AG, Karlsruhe and Freiberg (Germany), who wish to produce similar rechargeable zinc/air batteries. If the industrial

Figure 7. A 20 Ah/12 V zinc/air battery (without battery case), 10 cells connected in series. (Reproduced from Müller *et al.* (1997).[40])

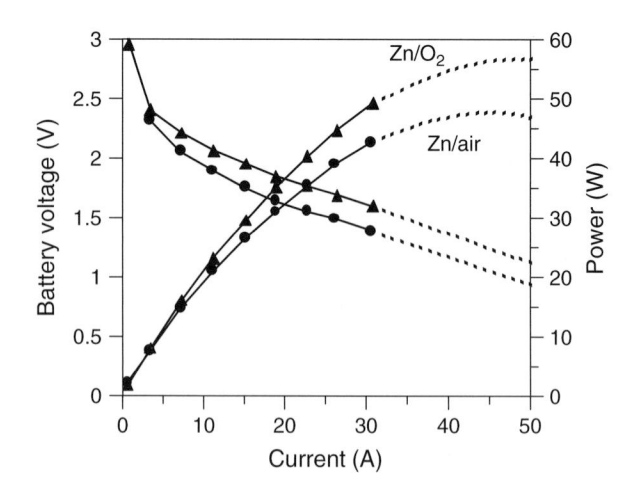

Figure 8. Current/voltage and current/power curves of a (▲) Zn/O$_2$ and a (●) Zn/air battery. Each battery had two cells connected in series and a nominal capacity of 30 Ah. The Zn/O$_2$ battery was charged to 17.7 Ah (59%), the Zn/air battery to 20 Ah (67%). (Reproduced from Holzer *et al.* (1998).[48])

effort for a production line is successful, this system may be introduced in the market in the near future.

3 MECHANICALLY RECHARGEABLE ZINC/AIR BATTERIES

3.1 Zinc electrodes for mechanically rechargeable zinc/air batteries

Morphology and shape changes of the zinc electrode can be avoided quite generally with batteries not recharged electrically. In fact, regeneration of the zinc electrode material can occur outside the battery in a separate electrolysis plant. From this zinc a new electrode is produced and inserted into the battery case.

Two basically different procedures are used:

(i) The zinc discharge products are made to completely transfer into the liquid electrolyte phase where they form a super-saturated, colloidal solution stabilized by additives.
(ii) The zinc discharge products are made to remain as solids within the electrode (which forms a pocket or cartridge) or are made to precipitate from the electrolyte, according to the following overall reaction:

$$Zn + 4OH^- - 2e^- = Zn(OH)_4^{2-}$$
$$= ZnO + 2OH^- + H_2O$$

The ZnO is then dissolved as zincate in an excess of alkali and active zinc is regenerated electrolytically from this solution.

3.1.1 Zinc discharge products dissolved in the electrolyte

Aspects of regeneration

Different models of zinc/air batteries have been developed where zincate as the discharge product is made to form a genuine or colloidal solution in the electrolyte during discharge. Three different versions of the process by which the corresponding zinc electrodes can be regenerated outside the battery are practiced:

(i) Strongly adhering, large surface area zinc is deposited cathodically on a collector electrode prepared from a highly porous support (porosity up to 97%) of foam metal (e.g., foam copper) or metallized plastic foam coated with a thin, continuous layer of a metal with high hydrogen overpotential (such as lead or compact zinc).[93]

(ii) Dendritic, yet well-adhering zinc, is deposited cathodically at a rough electrode surface from zincate-saturated electrolyte. The good adhesion of the porous, large surface area zinc deposit on the substrate is secured by electrolyte additives, such as lead ions (about 1% of the metal ion concentration), which have the effect of producing a more ductile deposit, raise the cathodic overpotential of the process and inhibit hydrogen evolution. Sometimes the electrode is additionally compacted (by compression, possibly adding some binder) to offer higher strength and better ease of handling when being inserted into the battery.[94]

(iii) Zinc is deposited as powder using high current densities in a fluid-bed electrolysis cell where mass transport coefficients are high.[95] For discharge, this zinc powder is operated as a fixed-bed electrode or pressed into holders in the form of a cartridge. Magnesium has been proposed as the cathode in this version, since the zinc powder will not adhere to this substrate, while nucleation and crystal growth at this substrate are such that the resulting zinc powder has optimum grain structure and grain size.[96]

Zinc electrodes producing discharge products dissolved in a liquid electrolyte have stirred practical interest only for mechanically rechargeable batteries where the zinc electrode is regenerated in an external electrolysis process. However, the first of the above three deposition process versions could be employed in a directly rechargeable zinc battery.

A special kind of zinc powder electrode is the zinc suspension or slurry electrode, where both discharge and recharging occur under continued stirring of a slurry of the active material in contact with a collector electrode. The slurry can be more or less viscous. It is normal in this kind of fluid-bed reactor that both in the battery discharge mode and in the regeneration mode all mechanical components of the system in contact with the slurry: the separators, the electrodes and others, are subject to enhanced abrasion.[96, 97]

It is important to remember when electrolytically generating active zinc for batteries that the high current densities required to deposit zinc powder of optimum grain size or metallic zinc with a highly open, porous structure imply a higher overpotential at the oxygen electrode, hence higher energy consumption for the process at the counterelectrode.

Aspects of the electrolyte

For a soluble zinc electrode where the discharge products dissolve in the electrolyte during discharge the electrolyte must have a large capacity for uptake of the zinc ions.

It is a well-known observation that anodic oxidation of a zinc electrode in concentrated KOH solution produces a solution with much higher zincate concentration than that which corresponds to the ZnO/KOH reaction equilibrium and is obtained when dissolving ZnO in KOH. This implies that the solution generated by anodic oxidation of zinc in KOH is highly supersaturated with zincate.

Vedel and Debiemme-Chauvy[98] concluded on the basis of ample electrochemical and physical studies of zincate-supersaturated solutions that the enhanced solubility of zincate is due to the formation of oxodihydroxyzincate ($[ZnO(OH)_2]^{2-}$) and dioxozincate ($[ZnO_2]^{2-}$) ions which upon supersaturation are produced from tetrahydroxyzincate ions ($[Zn(OH)_4]^{2-}$). The latter are the predominant species in strongly alkaline solutions not saturated with zincate. The reaction occurs via the following reactions liberating H_2O as the solvent:

$$[Zn(OH)_4]^{2-} = [ZnO(OH)_2]^{2-} + H_2O,$$
$$[ZnO(OH)_2]^{2-} = [ZnO_2]^{2-} + H_2O$$

It could be demonstrated by light scattering measurements in freshly prepared, supersaturated solutions that colloid formation is not involved.

Attempts have been made, on the other hand, to use additives to the electrolyte, such as silicate, to further enhance the ability of the electrolyte to take up zinc ions in a colloidal form:

$$xZnO + H_2O + zSiO_3^{2-} = (ZnO)_x(H_2O)_y(SiO_3)_z^{2-}$$

This colloidal solution is directly subjected to electrolysis.

In this manner, the threshold concentration for a precipitation of passivating zinc hydroxide from the electrolyte is raised to a value of 300 g Zn l^{-1} of 12 M KOH.[99, 100]

Partial substitution of KOH by LiOH in the electrolyte also contributes to higher solubility of the zinc ions. Adding

$25\,\text{g}\,\text{l}^{-1}$ of LiOH to 12 M KOH is supposed to suppress blocking of the oxygen electrode by precipitating ZnO.[100]

3.1.2 Zinc electrodes retaining their discharge products as a solid

Electrode design
Zinc electrodes which retain their discharge products, rather than forming soluble discharge products, can be employed both in directly rechargeable zinc/air batteries and in zinc/air batteries that are mechanically "recharged" ("refueled" by regeneration of the zinc electrode outside the system). The requirements that must be met for the two system types differ considerably.

In a cartridge-type electrode for mechanically rechargeable batteries, all that matters is to avoid as much as possible any loss of zinc ions from the electrode and to let electrolyte access the zinc surface to the extent required for highly complete anodic oxidation of the zinc (some 80–90%). The electrode does not need a more complete stabilization, inasmuch as it is regenerated in a separate process after discharge. The zinc electrode is produced by pressing zinc powder moistened with electrolyte into the electrode cartridge. It is desirable to have a relatively low density of the zinc,[101] of about 0.3–$0.4\,\text{g}\,\text{cm}^{-3}$, in order to obtain a high porosity and a large two-phase boundary between the zinc (BET surface area: about $1\,\text{m}^2\,\text{g}^{-1}$) and the electrolyte. This facilitates discharge, but leads to lower values of specific charge of the zinc electrode.

Electrolyte design
To avoid dissolution of the discharge product, the electrolyte should be zinc ion-saturated if prevention of zinc dissolution during discharge is desired. The solubility of zinc ions can be lowered by a factor of five relative to that in 7 M KOH when the KOH concentration is lowered to about 3 M and the electrolyte is saturated with zincate at the outset. However, this implies a loss in electrolyte conductivity by about 25%.

There is a practical limit to lowering the OH^- concentration, not only in order to maintain sufficient electrolyte conductivity but also in order to avoid a loss of activity of the oxygen electrode. Also, at higher current drains, which may be desirable, a rise in temperature occurs in the battery, which leads to an undesirable rise in zinc ion solubility.

To lower the zinc ion solubility even more, anion substitution was tested with fluoride, phosphate, borate, oxalate and carbonate.[18] It was seen that the anions have a much lower effect on zinc ion solubility than OH^- concentration decrease, even in view of the low solubility of ZnF_2. Borate and carbonate have a buffering effect,[47] which lowers the activity of the hydroxyl ions, thus leads to lower zinc ion solubility but also to some lowering of zinc electrode activity.

Borate and fluoride ions also suppress dendrite formation, that is, they have an effect on metal overpotential and thus on electrocrystallization.

Carbonate additions to the electrolyte have been studied and recommended by Cairns *et al.*[17] for nickel/zinc batteries but are not suitable for zinc/air batteries, since the pH changes occurring in the pores of the electrode during charging and discharge lead to carbonate precipitation and thus blocking of the pores.[102]

The solubility of zincate can be lowered when adding $Ca(OH)_2$ to the electrode or to the electrolyte, since the solubility of the resulting calcium zincate $[Ca(OH)_2\cdot 2Zn(OH)_2\cdot 2H_2O]$ is substantially lower than that of potassium zincate.[41] The capacity loss per cycle can be reduced from 0.47 to 0.13% in 31% KOH. The desired process is favored by the electrode reactions, since at the negative electrode during discharge, OH^- ions are consumed while at the same time the zinc ion concentration close to the zinc electrode increases, hence the zinc ions produced by discharge tend to undergo direct precipitation as zinc hydroxide in the electrode. During charging the concentration of OH^- ions in the porous zinc electrode increases, which facilitates dissolution of the zinc hydroxide present there and the discharge of the zinc ions to metallic zinc.

For a complete precipitation of all zinc ions in the form of calcium zincate, one calcium ion would be needed for every two zinc ions produced by discharge, which would considerably lower the charge density of the battery. In practice, enough $Ca(OH)_2$ is added to have one calcium ion for every four to five zinc ions.[103]

Charging mode
As pointed out earlier, a constant charging current density of at least $100\,\text{mA}\,\text{cm}^{-2}$ is the optimum for the deposition of electrochemically active zinc. However, this current density may have adverse effects on the counterelectrode, the battery could heat up even at this current density on account of internal resistances or concentration polarization could develop. It would then be advantageous to use pulsed currents with pulses of $100\,\text{mA}\,\text{cm}^{-2}$.[54] Bronoel *et al.* recommended charging pulses of about 2.5 ms, with a pause of 7.5 ms.

Prevention of electrode shape change
Gunther and Bendert[104] summarized the working conditions under which a "shape change" can be prevented or at least reduced, as follows:

1. lower the electrolyte mobility near the electrode, for instance by using gel electrolytes;

2. lower the zinc solubility in the electrolyte (e.g., by low KOH concentration in the electrolyte, addition of $Ca(OH)_2$ to the electrodes, in order to form insoluble calcium zincate during discharge);
3. design the cell so as to avoid regions with strong electrolyte flow along the zinc electrode (e.g., no electrolyte reservoirs close to the corners of the zinc electrode);
4. generally reduce electrolyte movement in the cell (e.g., by minimizing the amount of free electrolyte in the cell, by proper selection of the separator system which will suppress electroosmotic flow or by avoiding gassing and thus convection induced by gas bubbles during the charging process with electrodes having excess charging capacity).

3.2 Air electrodes for mechanically rechargeable zinc/air batteries

3.2.1 Principles

The requirements of an air electrode for mechanically rechargeable ("refuelable") zinc/air batteries are similar to what has been discussed in Section 2.2.1 but it needs no bifunctionality. Its function is oxygen reduction, which should be a four-electron step (to hydroxyl ions) or a two-electron step (to hydrogen peroxide) coupled with rapid chemical peroxide decomposition at a catalyst:

$$2H_2O_2 \longrightarrow O_2 + 2H_2O$$

regenerating one half of the oxygen previously reduced to peroxide, thus allowing the reaction, in a final analysis, to go to completion as well.

The catalyst promoting chemical decomposition of the peroxide species need not be the same as that promoting reduction of the oxygen molecules. It should be efficient, since peroxide and other intermediate species generated prior to molecular oxygen might attack carbon materials in the electrode, which would impair conductivity, catalytic activity for oxygen reduction, structural integrity and would produce carbon dioxide, which in turn would form carbonate with the alkaline electrolyte, leading to undesired changes in the electrode's hydrophobic/hydrophilic balance, clogging of the electrode and electrolyte consumption.

The structural requirements of an air electrode have been discussed in Section 2.2.1.

3.2.2 Catalysts

Most of the oxygen reduction catalysts for air electrodes of mechanically rechargeable zinc/air batteries are also used in reversible air electrodes or alkaline fuel cells (see discussion in Section 2.2.2). Platinum-based catalysts used by Fielder and Singer,[106] Gao et al.,[107] Kera and Nakagishi[108] and Kuwaba and Matsuoka.[109] Ioroi et al.[110] added V, Ce or Zr oxides to the platinum. Additional work including special electrode engineering has been done by Kordesch et al.[111] and Celiker.[112] Nickel alloys were used by Sofronkov et al.,[113] Co/Ag by Kublanovskii et al.,[114] amorphous Ag by Teng et al.,[115, 116] and Ag doped with Fe, Pt, Bi, Ni and Ti by Cho et al.[117]

Kivisaari et al.[118] used cobalt tetramethoxyphenylporphyrin (CoTMPP) macrocyclic systems as a catalyst for oxygen reduction, and Lamminen et al.[119] used MnO_2.

While it is not possible to draw a conclusion as to the "champion metal combination" for applications in a zinc/air fuel cell or battery, it will be safe to say that platinum is overly expensive and in a rechargeable battery it might easily spoil the zinc anode's efficiency. Silver-based catalysts appear an optimum choice for high-performance zinc/air applications but their high positive potential at open circuit leads to oxidation of the carbon support. Cobalt and nickel are preferably used in the form of (mixed) oxides and are especially useful for bifunctional oxygen electrodes.

3.2.3 Catalyst supports

In addition to the discussion in Section 2.2.2, a few research papers are mentioned here to show the relevance of carbon materials for successful electrode catalysis. Thus, Ohms prepared carbons by pyrolysis of polyacrylonitrile.[120] This is a common route to active carbon materials as well as to such relatively inert materials as glassy carbon. However, by conducting the preparation in the presence of transition metals, Ohms obtained catalytically active materials for oxygen reduction reaction (ORR) in alkaline and acidic solutions. Porosity in the catalyst layer can be engineered using ammonium bicarbonate.[121]

Horita et al.[122] used oxygen plasma-treated acetylene black for the ORR in alkaline solutions. They stressed the importance of wettability (which is obtained by said plasma treatment) and fine pores. Goodenough and Manoharan claimed in a review that oxygen reduction actually requires more hydrophobic catalysts.[74] We may conclude that it is probably the combination of a hydrophilic carbon with a more hydrophobic oxide catalyst environment (including a hydrophobic binder) which enables access of gaseous oxygen and elimination of ionic products to occur in close proximity to each other.

An interesting observation was the performance gain attained when in an oxygen electrode for zinc/air fuel cells, polythiophene was used as support for the catalyst instead of a more inert carbon support.[123]

3.3 Batteries with external regeneration of the zinc electrode

This battery type has the feature that after discharge of the battery and consumption of the active zinc, the zinc electrode is withdrawn and regenerated to active zinc by cathodic reduction outside the battery. "Recharging" or "charging" of the battery occurs when instead of the component being consumed, a fresh porous zinc electrode or zinc powder cartridge or a moderately viscous zinc powder slurry or paste are put mechanically into the battery. The electrodes needed for this battery have been discussed in Section 3.1.

An electrode exchange always implies some electrolyte exchange or at least electrolyte refreshing. Since all rechargeable zinc/air batteries use strongly alkaline electrolyte solutions, care must be taken to prevent or counteract undesirable carbonate formation in the electrolyte during recharge.

3.3.1 Development of batteries with dissolved discharge products

In this battery type, discharge of the zinc electrode, which can be a porous zinc electrode or a zinc powder electrode, leads to the formation of discharge products dissolving in the electrolyte phase. Directly from the electrolyte or from the precipitates dissolved in excess alkali, new porous zinc electrodes or powder zinc electrodes are then produced electrolytically outside the battery and inserted as a "refill" of new active material into the battery.

The first battery of this kind was tested by Gulf General Atomic Company at the beginning of the 1960s.[124] In it the electrolyte was vigorously recirculated in order to continuously eliminate the zincate and zinc oxide as discharge products from the zinc electrode. The 19 kWh test battery was operated at a temperature of 65 °C. It had a specific energy of 60 Wh kg^{-1} and a specific power of 24 W kg^{-1}, its life was less than 100 h.

Evans and Savaskan[125, 95] of the LBL have resumed development work on this battery type more recently using fixed-bed, zinc powder electrodes. Their laboratory cells had oxygen electrodes with surface areas of 80 and 400 cm^2 (see Figure 9a).

The fixed bed consisted of zinc grains with a particle size between 0.3 and 0.6 mm or of zinc-coated copper powder with a particle size between 0.5 and 0.6 mm, was separated from the space of the positive electrode by a diaphragm and had a graphite or copper rod as the current collector. The electrolyte for both the anode and cathode space was KOH solution having an initial concentration of 45% (12 M); natural convection made the anolyte flow through

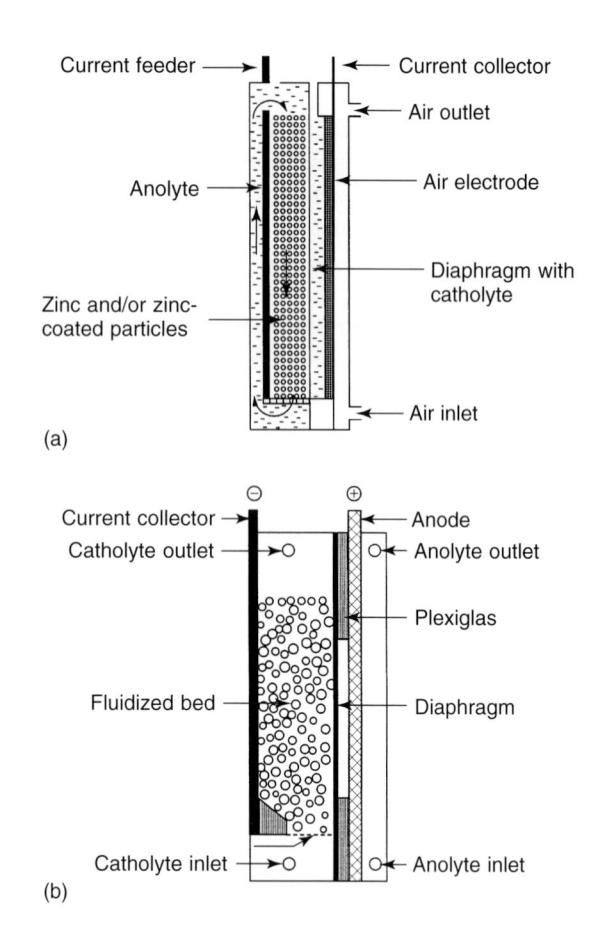

Figure 9. (a) Schematic of a laboratory Zn/air cell employing a packed-bed anode, (b) side view of a laboratory fluidized-bed cell for ZnO regeneration. (Reproduced from Haas *et al.* (1996)[145] with permission from John Wiley & Sons, Ltd.)

the fixed bed. The oxygen electrode was separated from the diaphragm by 3 mm of electrolyte. The positive electrodes tested in this cell were multilayer oxygen electrodes AE-20 of Electromedia Corporation and AC65-1072 of Alupower Corporation.

Using their test results for a laboratory cell, the Berkeley authors extrapolated the performance figures to a battery with a peak power of 55 kW in a vehicle subjected to the Simplified Federal Urban Driving Schedule test for electric traction batteries. Depending on the load, which was assumed to be between 150 and 97 W kg^{-1}, specific energies between 100 and 228 Wh kg^{-1} were quoted. For the three situations envisaged, a battery refill should provide driving ranges of 240, 498 and 221 km for said vehicle. It is difficult to check this extrapolation, inasmuch as the authors quoted 128.5 kWh for the amount of energy drawn from 63 kg of zinc in the 55 kW battery in their second example. Complete use of this amount of zinc corresponds to a theoretical amount of energy of only 61 kWh.

Regeneration of the battery is supposed to occur by draining or sucking the electrolyte from the battery together with the residual metal powder bed. The discharge should occur in such a way that all the products are found in a true or colloidal solution in the electrolyte phase while the particles of the fixed-bed electrode have merely decreased in size but not in number. Then redeposition of the zinc directly on these particles should be possible in a subsequent fluidized bed electrolysis (see Figure 9b).

The cell to be used for regeneration of the zinc electrode is quite similar in its design to the zinc/air battery. The current collector is made of graphite, copper or lead. A diaphragm separates the two electrode compartments. It is mechanically reinforced and protected in view of the strong agitation of the electrode particles. The oxygen-evolving electrode is of the known type of dimensionally stable anodes. The fluidization is produced through catholyte flow and is most favorable when the volume expansion is 25% relative to the initial bed. The catholyte is zincate-saturated 45% KOH. The anolyte is 25% KOH, which has a higher conductivity. The electrolysis current densities were between 100 and 1000 mA cm^{-2}, the associated electrolysis voltages were between 2.2 and 6.0 V.

An optimized process with optimum selections for electrodes and separators is said to consume 1.92 kWh per kg of zinc when working at a current density of 100 mA cm^{-2} or 2.08 kWh per kg of zinc when working at a current density of 200 mA cm^{-2} (the corresponding electrolysis voltages are 2.22 and 2.40 V). On this basis the authors calculate an overall efficiency for electric energy storage of 46–50%, though evidently without considering the energy consumed for handling and maintenance.

It is an advantage of the system that regeneration of the battery by draining of exhausted and refilling of fresh slurry is a manipulation coming quite close to the ordinary refueling at a gasoline station. Thus, the battery system works like a fuel cell. Similar studies have been reported from the Lawrence Livermore National Laboratory by Noring et al.[126]

The refueling principle has been discussed in connection with the zinc/air batteries being developed in collaboration with Kummerow Corporation by BAT International (Burbank, CA, USA).[127] The batteries are expected to offer 160 Wh of energy storage per kg and an associated driving range of 400–480 km. Their minivan was reported as having been driven a record distance of 478 miles on a single charge. In view of protection of the air electrodes during refill, apparently it is preferable to refuel by exchanging the entire battery pack.

Slightly different working principles have been followed by Pauling[128] of ChemTEK (Karlsruhe, Germany) who instead of a fixed bed of particles employed porous zinc on a conducting support. These electrodes were produced directly by electrolysis. During discharge, they dissolved forming zincate up to the point where the electrolyte was supersaturated with zincate. Composition and amount of this electrolyte were selected so that practically all of the discharge product remained in the electrolyte as a true or colloidal solution, so that this solution could be employed without any prior processing in electrolytic regeneration of the zinc electrode.

The cell voltage at open circuit was reported to be 1.45 V, the working voltage was 1.1 V and the specific energy 110 Wh kg^{-1}. For a battery of 210 Ah a weight of 2.4 kg was reported, which at a working voltage of 1.1 V would actually imply a specific energy of only 96 Wh kg^{-1}.

The latest developments reported by ChemTEK use yet another electrode arrangement: here the discharge products remain in an electrode cartridge, largely undissolved. This type of zinc electrode is the subject of Section 3.1.3.

3.3.2 Development of batteries with a zinc slurry or powder paste electrode

In these batteries, a fluid paste prepared from zinc powder and electrolyte or a zinc slurry in the electrolyte is moved continuously past a collector electrode during discharge, so that the zinc can be oxidized to zincate (Figures 10 and 11).

Regeneration of the slurry or paste occurs analogously in a separate electrolysis process.

Development efforts to realize this working principle were already reported in the 1970s from Japan and particularly from France (Compagnie Générale d'Electricité).[97] A concentric space was created between an inner, tubular

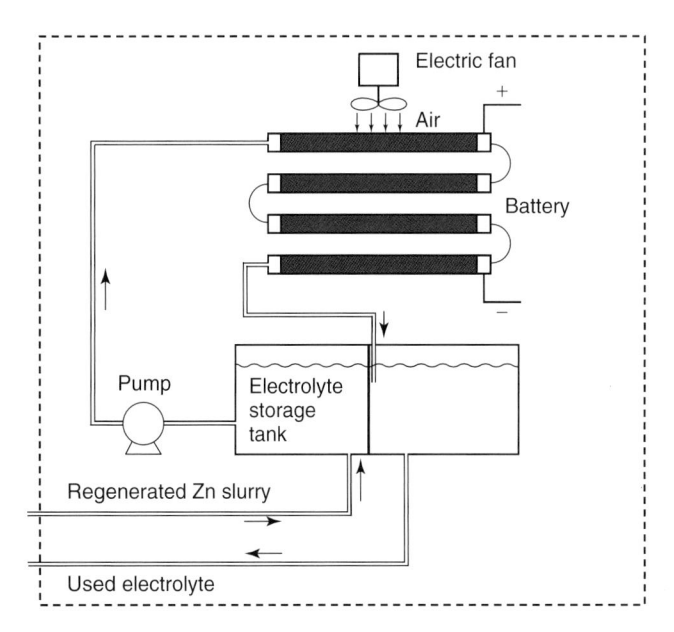

Figure 10. Schematic diagram of the Compagnie Générale d'Electricité Zn/air system with battery and electrolyte storage tank. (Reproduced from Haas et al. (1996)[145] with permission from John Wiley & Sons, Ltd.)

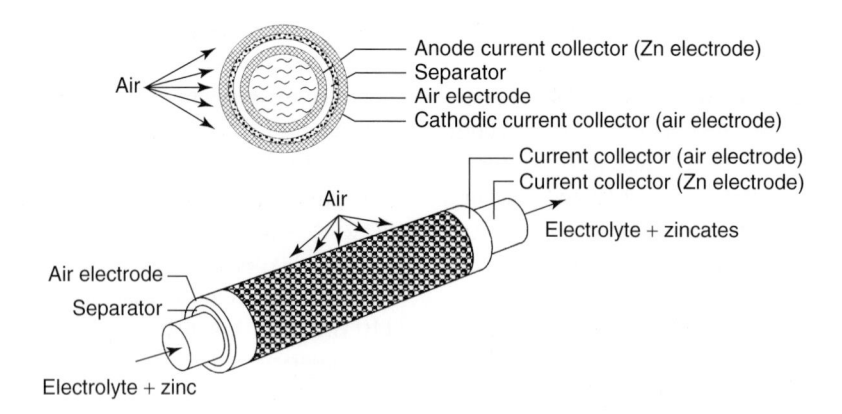

Figure 11. Schematic of the Compagnie Générale d'Electricité tubular Zn/air cell design. (Reproduced from Haas et al. (1996)[145] with permission from John Wiley & Sons, Ltd.)

collector electrode (working as the negative electrode) and an outer, tubular oxygen electrode. This space contained a separator. A zinc powder/KOH slurry was passed along the collector electrode with a flow velocity of $0.8\,m\,s^{-1}$. Discharge produced a solution supersaturated with zinc. This battery was designed for a specific energy of $100\,Wh\,kg^{-1}$ and a specific power of $80\,W\,kg^{-1}$ and was supposed to support 500 3 h discharge cycles. In the experiments a peak power of 75 W per tube was obtained at 50 °C with air electrodes having a surface area of $330\,cm^2$ and a cell voltage of 0.9 V. For the zinc electrode, a polarization of merely 20 mV was reported for a current density of $300\,mA\,cm^{-2}$ of the collector electrode surface. A complete battery weighing 300 kg was reported to have reached a specific energy of $84\,Wh\,kg^{-1}$ and at the same time a specific power of $82\,W\,kg^{-1}$.[96, 97, 99] It has also been claimed that 600 battery discharge cycles have been achieved, each cycle followed by renewal of the zinc slurry but without replacing the oxygen electrodes.[96] (It would be more appropriate here to speak of a zinc/air fuel cell, rather than zinc/air battery!)

Tubular cells were used, too, to regenerate the zincate/zinc hydroxide slurry. The electrolysis voltage for regeneration was 2.4 V (this is more than in other electrolysis processes for zinc powder regeneration, for which values of 2.2–2.3 V have been reported). Under the assumption of a working voltage of 1 V for this battery, the energy efficiency would then be a mere 40% when this is calculated from the voltage difference between battery discharge and zinc regeneration.

Figure 10 shows the overall design concept of this Compagnie Générale d'Electricité zinc/air battery. Units of up to 15 kW power were tested between 1970 and 1980 by Compagnie Générale d'Electricité in France. Under the assumption of a production volume of 10 000 batteries, the authors predicted a manufacturing price in 1977 of $52\,\$\,kg^{-1}$.[96] As traction batteries these zinc/air batteries should then have lower investment costs than the traditional lead/acid batteries and the operating cost of the vehicle in Europe should be lower than that of vehicles with carburetor fuels.[97] These promising development efforts were shelved late in the 1970s.

Starting with the work of Foller in the early 1980s at the Continental Group's Energy Systems Laboratory (Cupertino, CA, USA),[99] the French work was then taken up at the Pinnacle Research Institute (Cupertino, CA, USA). Foller worked with small laboratory cells using electrodes of a few square centimeters. He assumed that improved oxygen electrodes should by now be available in view of recent large development efforts and that improved collector materials for the slurry electrode and improved separator materials should also be available, thus greatly improving the chances for technical success of the slurry design. He even proposed to use a hydrogen diffusion anode as the counterelectrode when cathodically regenerating the zinc electrode (or zinc slurry), which would raise the energy efficiency of the system as a whole, because a hydrogen electrode has lower overpotentials than an oxygen electrode.

By extrapolating the performance of laboratory cells to a 300 kg traction battery, Foller[99] arrived at figures of 27 kWh for the stored energy and 37 kW for continuous power, which is equivalent to a specific energy of $90\,Wh\,kg^{-1}$ and a specific power of $123\,W\,kg^{-1}$.

From a maximum feasible zinc content of $350\,g\,l^{-1}$ in silicate-containing 12 M KOH (as colloidal zinc oxide after discharge, with an equivalent amount of zinc powder in the charged state), the theoretical capacity is 287 Ah, which when combined with a discharge voltage of 1.1 V leads to a figure for the energy density of 315 Wh per liter of the energy vector. Assuming a discharge current density of $80\,mA\,cm^{-2}$, an average discharge voltage of 1.18 V, a charging current density of $150\,mA\,cm^{-2}$, an average electrolysis voltage of 2.4 V and a parasitic energy consumption of about 16% for pumps, self-discharge, shunt currents and losses due to less than 100% current yields, Foller calculated an energy efficiency of 41.8%.

Alcazar et al. of the Pinnacle Research Institute[100] reported development efforts toward a 50 kW battery consisting of 370 bicells (a bicell implies a zinc electrode sandwiched between two air electrodes; this concept unfortunately does not admit bipolar battery design), each with an air electrode surface area of 200 cm^2, which had an empty weight of 197.5 kg and a tank filling of 516 l of zinc powder slurry weighing 557 kg. The battery, to be used in a 2 ton vehicle, was supposed to have an energy of 136 kWh and the rather low power of 5.5 kW, from which one can calculate a specific energy of 180 Wh kg^{-1} and a specific power of 66 W kg^{-1}. This high specific energy content of the zinc slurry could be attained by adding LiOH (25 g l^{-1}) to the electrolyte (12 M KOH), which acts to raise the zinc ion solubility from 87 to 279 g l^{-1}.

The activity loss of the oxygen electrode observed in life tests was attributed to a precipitation of ZnO in the transport pores of this electrode. The voltage loss appearing during battery operation is essentially the result of ohmic losses in the zinc/electrolyte slurry.

3.3.3 Development of batteries with undissolved discharge products in exchange cartridges

In this section, we report on a variety of systems that have the common denominator that during discharge, a paste consisting of zinc powder and electrolyte undergoes oxidation in an electrode cartridge. The composition of the electrolyte, the amount of electrolyte and the arrangement of the electrodes is such that the discharge products remain in the cartridge. After discharge, only this cartridge needs to be replaced. Some freshening up of the electrolyte is also required because of carbonate formation in the electrolyte and of additives lowering the solubility of the zinc ions.

Inserting a new cartridge largely takes care of the electrolyte renewal, since the amount of free electrolyte existing outside the cartridge and separators is very small in batteries of this type (Figure 12).

Regeneration occurs by dissolution of the consumed active mass of the zinc electrode in alkali and subsequent electrolytic production of zinc powder, which serves as the starting material for fresh paste in new cartridges.

The first development efforts toward batteries of this type were by General Motors,[129, 130] again in the 1960s. In the final battery, each zinc electrode was sandwiched between two oxygen diffusion electrodes. A module consisted of 49 such cells and six modules were combined to a battery of 35 kWh. The specific energy of this battery was 120 kWh kg^{-1}, at the same time it had a very low specific power of only about 20 W kg^{-1}, which was due to low load capability of the oxygen electrodes and, since the electrolyte was not kept in circulation, discharge products blocked the electrode. The battery was operated for less than 100 h.[131]

In the late 1980s, the General Motors design of zinc electrode operation in a zinc/air battery was adopted and advanced by EFL.[132, 133] A paste of zinc powder, electrolyte and additives was pressed into a copper screen with a protecting layer to form electrode cartridges, these were mechanically introduced into the battery container for discharge. The separators enveloping the electrode were moistened with an electrolyte in which the zinc ions are poorly soluble. After discharge, the cartridges were extracted from the battery, the zinc oxide and hydroxide was dissolved in KOH to form zincate and the solution was subjected to electrolysis in order to regenerate the zinc powder.

Concerning regeneration of the zinc powder, EFL provided the following data:[101] Electrolysis occurs between nickel anodes and magnesium cathodes at a current density of 100–200 mA cm^{-2} and an electrolyte temperature of 40–70 °C in 7–8 M KOH while the zincate concentration corresponds to 30–40 g Zn^{2+} per liter. The resulting zinc powder should have a low density (0.3 g cm^{-3}) and the largest possible specific surface area (1 m^2 g^{-1}), so that the paste electrode made from it has minimal polarization during discharge. The corrosion rate of this zinc was reported to be 1% per week at 30 °C.

In the above electrolysis process, nickel electrodes with fiber structure obtained from Zipper Technik (Germany) were said to have exhibited the lowest oxygen overpotential.

The design changes made by EFL in the General Motors battery type were not very important, but the performance figures were found to be greatly improved. This was attributed to significantly improved performance of the oxygen electrodes, higher utilization of the active mass and a more perfect electrode arrangement. Cells have an average working voltage of 1.16 V, a module of 12 cells in series was reported to have a capacity of 272 Ah and an energy content of 3.7 kWh. With a total mass of 14 kg and a volume of 10.7 l, the specific energy is calculated to be 191 Wh kg^{-1} (referring to 5 h discharge and 71% utilization of the zinc). For peak power, a value of 113 W kg^{-1} was reported. More energy can be drawn from the battery at low load, so that a specific energy of over 200 Wh kg^{-1} might be reached.

EFL made a cost forecast of 100 \$ kWh^{-1} of storage capacity, which is far below the cost for conventional batteries (200 \$ kWh^{-1} for lead/acid batteries). For better comparison, the battery life should be included into these estimates, as the specific cost for all the kilowatt hours of

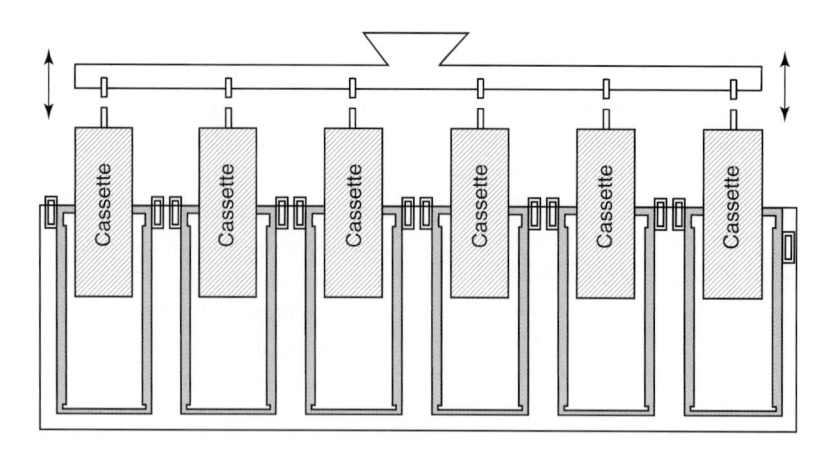

Figure 12. EFL Zn/air battery concept with mechanically rechargeable Zn electrode cartridges. (Reproduced from Haas et al. (1996)[145] with permission from John Wiley & Sons, Ltd.)

electrical energy stored during the battery's operating life would be an interesting figure.

A large-scale test series involving German companies where this battery, amongst others, was used for vehicle traction, was undertaken in 1997. It was planned to equip a fleet of 66 vehicles with batteries and to operate in an area around the city of Bremen for the German Postal Service. The question was to be answered whether a fleet of electric vehicles operated with mechanically rechargeable zinc/air batteries is economically and ecologically advantageous.[134] For the battery used in the Bremen tests, EFL indicated a specific energy of over 200 Wh kg^{-1} and a specific power of 100 W kg^{-1} to be maintained down to a depth of discharge of 80%. The batteries were assembled from the required number of 6.25 kWh modules, each consisting of 22 cells in series. An MB410 Mercedes-Benz 4.6-ton transporter was fitted with a battery with 24 such modules, which represented a stored energy of 150 kWh. This battery, which weighed 800 kg, was equipped with heat and air-flow management. The transporter was to have an autonomy of about 300 km with one battery charge.[135] On an average, this autonomy should be sufficient for one week of operation of such a postal truck.

In 1996, ChemTEK of Karlsruhe started reporting on their battery system "ZOXY", which resembles the EFL system insofar as its electrolyte composition and zinc electrode design are such that the products of discharge will for the most part remain as a precipitate within the negative electrode. However, in contrast to what is done in EFL technology, regeneration of the negative electrode in ZOXY batteries occurs without dissolution of the reaction products in KOH, that is, directly by electrolysis with an oxygen counterelectrode in KOH. Data of ChemTEK indicate that gradually, with increasing number of charge/discharge cycles, the zinc electrodes suffer a capacity loss, hence after any ten discharges an anodic dissolution of the discharge products from the depleted negative electrode must occur prior to cathodic redeposition of the zinc.

The following performance figures have been given by ChemTEK for the system:[136]

- Energy density: 160–210 Wh l^{-1}
- Specific energy: 220–300 Wh kg^{-1}
- Specific power: 60–80 W kg^{-1}
- Operating temperatures: -20 to $+40\,°C$
- Air electrode life: 4 years in continuous operation

This new battery of ChemTEK was successful in setting a world record in driving distance with a single battery charge at below-freezing ambient temperatures (a minibus was driven 760 km in Utah in 1996).

With two dedicated service stations (in Karlsruhe and Oberderdingen) a fleet test had been planned for the second half of 1997 in cooperation with the Municipal Enterprises of Karlsruhe and the Electric Utility Company Badenwerk in the Karlsruhe area. The Energy Research Foundation of the Federated state of Baden-Württemberg also provided support for development of the ZOXY system.

The operating cost for a minitransporter powered by a 400 kg battery set with an energy content of 60 kWh (that is, with a specific energy of 150 Wh kg^{-1}) has been calculated under the assumption that on an average the vehicle will require 150 Wh km^{-1}. This figure is 0.36 DEM km^{-1} (roughly 0.2 US\$ km^{-1}). The daily driving distance of the vehicle is assumed to be 100 km, the autonomy for one battery charge is 300 km. The batteries are recharged overnight in a service station with a capacity sufficient for 14 vehicles. Electrical energy is made available at the reduced night-time rate.

Very little has been reported up to now by ChemTEK concerning electrode design and electrolyte management. It appears that the design principles resemble those known from earlier work by the Central Laboratory for Chemical Power Sources (Sofia) and by Dr. I. Iliev, one of the principal investigators of this laboratory, who for more than 20 years have been developing and producing primary Zn/air batteries in Sofia.[137, 138]

4 INDUSTRIAL ZINC/AIR SYSTEMS

4.1 AER Energy Resources

The accent of the AER (Atlanta, GA, USA) developments in zinc/air power sources has been on the side of the air electrode. In a 1994 patent,[139] a slanted distribution of the oxygen reduction and evolution catalysts in the bifunctional air electrode is described, which is supposed to be such that O_2 evolution will not occur at the reduction catalyst, while gas evolution would not occur inside the cell during discharge (thus, the formation of gas pockets would be prevented). Reduction catalysts include CoTMPP, cobalt oxide, lanthanum nickel cobaltate and NiS, while evolution catalysts include WC, $FeWO_4$ and WS_2. Preferably plural catalysts are incorporated into the electrode structure. It is not quite clear, though, to what extent the nickel current collector also participates as a catalytically active component.

A later patent[140] describes a coated oxygen evolution catalyst. The coating consists of bonded graphite particles and is intended to trap the carbon particles present within the electrode structure and protect them against corrosion. The coating is also intended to trap reactive species (oxygen radicals) produced at the oxygen evolution catalyst, long enough so that they will react with each other and become inoffensive, rather than diffuse and attack carbon black (present as catalyst support in the electrode). The patent also describes the provisions made to manage air-flow to the air electrodes, electrolyte leakage from the air electrodes, excessive humidity exchange at the air electrodes and dendrite growth at the zinc electrode. The electrolyte is soaked up into porous separator layers.

A related patent[141] claims that an evaporated thin-film carbon layer on the O_2 evolution catalyst particles will lead to higher power output during discharge. It may be speculated that this is a conductivity effect, since the O_2 evolution catalyst does not participate in cell discharge.

Mechanical means have been introduced[142] by AER to provide oxygen to the air electrode when needed while humidity exchange and CO_2 access are largely inhibited by an extended tubular approach structure (diffusion air manager technology), which serves to preserve cell capacity under dry and highly wet conditions. This technology was developed for throw-away zinc/air batteries (for cellular phones, portable computers) but is applicable to rechargeable or refuelable zinc/air batteries. The fans or other mechanical devices needed for operation of air diffusion management are said to absorb about 10% of the cell's output.

An overview of the AER technology spectrum was given by Sieminski.[143]

4.2 EFL

EFL (Jerusalem, Israel) is well known for their 3.3 Ah zinc/air cells for mobile phones (15.5 g, 5 cm^3, specific energy 240 Wh kg^{-1}, energy density 700 Wh l^{-1}), a commercial product introduced in 1999. Further developments in this field have focused on batteries able to support digital modes requiring pulses of up to 2 A.

Electric Fuel's electric vehicle branch (Electric Fuel Transportation Corporation) has introduced electric vehicle power in the form of zinc anode cassettes. The high-energy zinc/air battery is coupled with a high-power auxiliary battery to provide range as well as acceleration (joint development of this all-electric hybrid system with GE). Tests performed in the Summer of 2001 indicate that a transit bus based on this system is feasible for the operating conditions applicable in the New York business district (the bus was driven 97 miles, while 90 miles is a typical daily work load). A German industry consortium on the other hand is collaborating in the development of delivery vehicles. These applications employ modules with 47 cells having 40–57 V operating voltage, a capacity of 325 Ah, an energy content of 17.4 kWh and a peak power (at 80% DOD) of 8 kW. Such a module weighs 88 kg and occupies a volume of 79 l, thus its energy density is 200 Wh kg^{-1}. The cassettes contain an electrochemically (re)generated zinc particle slurry. Regeneration occurs at centralized facilities.

4.3 Metallic Power

The technology of Metallic Power (Carlsbad, CA, USA) is described by Colborn.[144] Metallic Power concentrates on fuel cells where 0.5–0.8 mm diameter zinc pellets are the fuel. The zincate produced by fuel cell operation is flushed from the fuel cell continuously by a flow of KOH electrolyte pumped between the power-generating fuel cell and a reprocessing unit. The zincate is transformed to zinc oxide and reprocessed to zinc pellets in this unit; the round-trip efficiency was reported to be 50–60%. The zinc pellets are introduced into each of the cells of a battery from hoppers. Another option for refueling is by replaceable refueling cartridges. The air electrode is boosted by an air blower. A golf cart was powered by Metallic Power in late 1998.

In contrast to Electric Fuel Ltd., Metallic Power favors delocalized reprocessing units. Technical specifications for such a unit are available (5 kg ZnO equivalent to 4 kWh reprocessed with home power within 24 h). In the short run, not having to build up an infrastructure for zinc reprocessing is an advantage. In the long run, some doubt may be raised with respect to Metallic Power's claim that the amount of zinc first installed in the cell "can be regenerated over and over, forever": this would imply that the combination of fuel cell and reprocessing unit operates as a perfectly rechargeable battery. Apart from inevitable electrochemical inefficiencies, Metallic Power's claim ignores the presence of CO_2 causing electrolyte problems in the long run.

The first 1.5 kW units have been shipped for field tests in May 2001. Briggs & Stratton of Milwaukee, makers of gasoline engines for outdoor power equipment, are one of the companies participating in this testing.

4.4 Powercell Corporation

Powercell's (Burlington/Cambridge, MA, USA) Power-Block® is a large, power-engineered unit weighing 2.7 tons and has an energy capacity of 100 kWh. The power rating is said to be 100 kW, while recharge should occur at 20 kWh h^{-1}. The units are designed to supply three-phase, 60 Hz, 480 V AC power, so that grid power can be replaced or peak shaving performed. The electrochemical "heart" of the PowerBlock is a "Zinc-Flow®" zinc/air fuel cell.

4.5 Zinc Air Power Corporation

Zinc Air Power (Cleveland, OH, USA) specializes in electrically rechargeable zinc/air batteries. The critical cycle life improvement needed for direct electrical recharging has been achieved with what Zinc Air Power calls a "Tricell" three-electrode cell design, which is used in combination with inductively coupled AC charging of the cells. The initial target market are electroscooters (using 3600 W, 15 l units). Owing to the Tricell design, monofunctional air electrodes can be used; their newest types are said to achieve 590 mW cm^{-2}. The third electrode was described in WO 00/63985 as a support structure coated with a layer of lanthanum nickelate and silver oxide.

4.6 EVonyx

EVonyx (EV, Hawthorne, NY, USA) is another company aiming at the electroscooter market. This company offers RPC (Reveo power cell) disposable zinc/air batteries as well as rechargeable and refuelable zinc/air batteries. Their air electrodes are capable of over 500 mA cm^{-2} in air (over 2 A cm^{-2} in oxygen); their zinc electrodes also operate at 500 mA cm^{-2} in a zinc/air fuel cell. An RPC-powered electric car ran 217 miles without recharge in October 2000. The company recently leased production facilities in Taiwan.

4.7 ChemTEK/Zoxy AG

ChemTEK/Zoxy AG (Karlsruhe and Freiberg, Germany) are manufacturers of prototype "Zoxy" zinc/air fuel cells. With their technology, record distances have been driven by electric vehicles, both in freezing and in hot temperatures during 1997 in the USA. The system is based on the idea that for refueling, the zinc anodes are replaced together with the electrolyte. Applications of the fuel cell pack are caravans, worldwide satellite container tracking and construction site illumination. Their PZ 100 fuel battery delivers 120 Ah from a module weighing 450 g (nominal current: 1 A, peak current: 15 A, electrolyte: 140 ml of KOH solution). A prototype scooter battery weighs 1.5 kg and has a nominal capacity of 250 Ah (250 Wh at 20 A).

Together with Daimler-Chrysler, ChemTEK/Zoxy is developing all-electric hybrid vehicles where the high-energy fuel battery is supplemented by high-power supercapacitors.

When deep discharge is prevented, a Zoxy can then work as an electrically rechargeable storage battery.

5 CONCLUSIONS

The zinc/air battery has long been recognized as an attractive power source for electric vehicle applications. For this application, major development efforts have focused on mechanically rechargeable batteries. These have the advantage of fast recharge. The first batteries of this kind were developed and tested by Gulf General Atomic Company and General Motors at the beginning of the 1960s. In the late 1980s, the General Motors design of zinc electrode operation in a zinc/air battery was adopted and advanced by EFL using exchangeable Zn cartridges. The EFL cells were successfully tested in electric vehicle tests in Germany. 150 kWh batteries (800 kg) were fitted into several Mercedes-Benz transporters, giving them an autonomy of about 300 km with one battery charge. EFL made a cost forecast of 100 $ kWh^{-1} of storage capacity, which would be far below the cost for conventional batteries. Most recently Kummerow Corporation reported a record driven distance of 478 miles on single charge for their minivan equipped with a mechanically rechargeable zinc/air battery. However, very little has been reported about the service life of mechanically rechargeable zinc/air batteries and there is no production line available to produce these batteries in large numbers, which would be necessary to achieve the price goal mentioned above.

Electrically rechargeable zinc/air batteries using bifunctional air electrodes have been developed and studied over the last 10 years by several Research Institutes and Companies, such as LBL (USA), Eltech Research Corporation together with Matsi (USA) and the Paul Scherrer Institute (CH). While LBL and Matsi have focused their efforts on zinc/air batteries with soluble zinc electrodes, the Paul Scherrer Institute employed a pasted insoluble zinc electrode in its battery system. Specific energies in the range of 70–100 Wh kg^{-1} were reported for the electrically rechargeable system, even though only part of the active mass was cycled. The service life of the LBL test cell was limited by the bifunctional air electrodes. A cycle life below 50 cycles was obtained at 6 h charge and 3 h discharge rates for a small laboratory test cell. Paul Scherrer Institute was able to achieve an important improvement of these electrodes particularly with respect to their cycle life. With their bifunctional air electrodes and pasted Zn electrodes a maximum cycle life of 450 cycles (3000 h) could be reached in 2.4 Ah test cells. Scale-up of the battery components will lead to a 20 Ah/12 V electrically rechargeable zinc/air battery offering a service life of ca. 2000 h (ca. 200 cycles).

Over the last decade, the interest of battery manufacturers was absorbed by efforts made in developing primary zinc/air button cells capable of substituting the environmentally harmful HgO/Zn batteries. Electrically rechargeable zinc/air systems, however, are not far from a paced market introduction if industrial efforts to realize a production line are successful.

NOTES

^{1}The comparison of H$_2$/air fuel cells with zinc/air cells is not quite correct since the discharge reaction product has to be stored in the cell in the case of Zn/air cells but not in that of H$_2$/air fuel cells. The specific energy of Zn as a fuel should therefore rather be based on the molecular weight of ZnO, which gives 1.09 kWh kg^{-1} rather than the above 1.37 kWh kg^{-1}.

REFERENCES

1. O. C. Wagner, *J. Electrochem. Soc.*, **116**, 693 (1969).
2. L. Öjefors and L. Carlsson, *J. Power Sources*, **2**, 287 (1977).
3. H. Cnobloch, D. Gröppel, W. Nippe and F. von Sturm, *Chem. Ing. Techn.*, **45**, 203 (1973).
4. E. L. Littauer and K. C. Tsai, in "Proceedings 26th Power Sources Conf., Atlantic City", NJ, pp. 57 (1974).
5. T. Sakai, T. Iwaki, Z. Ye, D. Noréus and O. Lindström, Presented at the Electrochemical Society Fall Meeting, Miami, FL, pp. 108 (1994).
6. S. Gamburzev, O. A. Velev, R. Danin, S. Srinivasan and A. J. Appleby, *Electrochem. Soc. Proc.*, **97–18**, 726 (1997).
7. N. Vassal, E. Salomon and J. F. Fauvarque, *Electrochem. Soc. Proc.*, **97–18**, 869 (1997).
8. K. Beccu, US Patent No. 3,824,131 (1974).
9. K. D. Beccu and M. A. Gutjahr, Presented at Power Sources 5, the 9th Int. Symp., Brighton, UK, Sept. (1974).
10. K. Sapru, B. Reichmann, A. Reger and S. Ovshinsky, US Patent No. 4,623,597 (1986).
11. S. Ovshinsky, *et al.*, *Science*, **4**, 176 (1993).
12. H. A. C. M. Bruning, J. H. N. van Vocht, F. F. Westendorp and H. Zijlstra, US Patent No. 4,216,274 (1980).
13. B. G. A. Percheron, J. C. Achard, J. Loriers and M. Bonnemay, *et al.*, US Patent No. 4,107,405 (1978).
14. A. M. Kannan, A. K. Shukla and S. Sathyanarayana, *J. Electroanal. Chem.*, **281**, 339 (1990).
15. R. Manoharan and A. K. Shukla, *Electrochim. Acta*, **30**, 205 (1985).
16. J. McBreen and E. Gannon, *Electrochim. Acta*, **26**, 1439 (1981).
17. T. C. Adler, F. R. McLarnon and E. J. Cairns, *J. Electrochem. Soc.*, **140**, 289 (1993).
18. J. T. Nichols, F. R. McLarnon and E. J. Cairns, *Chem. Eng. Commun.*, **37**, 355 (1985).
19. K. W. Choi and D. N. Bennion, *J. Electrochem. Soc.*, **123**, 1628 (1976).
20. R. E. F. Einerhand, W. Visscher, J. J. M. de Goeij and E. Barendrecht, *J. Electrochem. Soc.*, **138**, 1 (1991).
21. R. E. F. Einerhand, W. Visscher, J. J. M. de Goeij and E. Barendrecht, *J. Electrochem. Soc.*, **138**, 7 (1991).
22. J. Frenay, M. Elboudjani and E. Ghali, in 'Proc. Int. Symp. Electrometallurgical Plant Practice', Montreal, Canada, pp. 85, Oct. (1990).
23. L. McVay, LBL Report, No. 30843 (1991).
24. J. St-Pierre and D. L. Piron, *J. Electrochem. Soc.*, **139**, 10 (1992).
25. J. St-Pierre and D. L. Piron, *J. Appl. Electrochem.*, **20**, 163 (1990).
26. M. G. Chu, J. McBreen and G. Adzic, *J. Electrochem. Soc.*, **128**, 228 (1981).
27. M. Y. Abyaneh, J. Hendrix, W. Visscher and E. Bahrendrecht, *J. Electrochem. Soc.*, **129**, 2654 (1982).
28. C. Cachet, U. Ströder and R. Wiart, *Electrochim. Acta*, **27**, 903 (1982).
29. C. Cachet, B. Saidani and R. Wiart, *J. Electrochem. Soc.*, **138**, 678 (1991).

30. C. Cachet, B. Saidani and R. Wiart, *J. Electrochem. Soc.*, **139**, 644 (1992).
31. G. Prentice, Y. C. Chang and X. Shan, *J. Electrochem. Soc.*, **138**, 890 (1991).
32. W. G. Sunu and D. N. Bennion, *J. Electrochem. Soc.*, **127**, 2007 (1980).
33. T. Keily and T. J. Sinclair, *J. Power Sources*, **6**, 47 (1981).
34. M. Cenek and O. Kouril, *NASA Tech. Transl.*, **F-14**, 693 (1973).
35. F. Mansfeld and S. Gilman, *J. Electrochem. Soc.*, **117**, 1328 (1970).
36. H. Yoshizawa, A. Miura and A. Ohta, *Electrochem. Soc. Proc.*, **93–18**, 241 (1993).
37. V. S. Shaldaev, N. N. Tomashova and A. D. Davydov, *Prot. Met.*, **28**, 400 (1993) [English Translation].
38. J. Goldstein, A. Meitav and M. Kravitz, US Patent No. 5,232,798 (1993).
39. J. Goldstein and A. Meitav, US Patent No. 5,209,096 (1993).
40. S. Müller, F. Holzer and O. Haas, *Electrochem. Soc. Proc.*, **97–18**, 859 (1997).
41. R. Jain, T. C. Adler, F. R. McLarnon and E. J. Cairns, *J. Appl. Electrochem.*, **22**, 1039 (1992).
42. S. Müller, F. Holzer and O. Haas, *J. Appl. Electrochem.*, **28**, 895 (1998).
43. S. Müller, F. Holzer, O. Haas, C. Schlatter and C. Comninellis, *Chimia*, **49**, 27 (1995).
44. J. Bressan and R. Wiart, *J. Appl. Electrochem.*, **9**, 43 (1979).
45. E. G. Gagnon and Y. M. Wang, *J. Electrochem. Soc.*, **134**, 2091 (1987).
46. R. A. Sharma, *J. Electrochem. Soc.*, **135**, 1875 (1988).
47. K. Bass, P. J. Mitchell, G. D. Wilox and J. Smith, *J. Power Sources*, **39**, 273 (1992).
48. F. Holzer, S. Müller and O. Haas, in "Proceedings of the 38th Power Source Conference", Cherry Hill, NJ, USA, pp. 354–357, June 8–11 (1998).
49. W. Taucher, L. Binder and K. Kordesch, *J. Appl. Electrochem.*, **22**, 95 (1992).
50. E. Frackowiak, 'Power Sources 13', J. Thomson (Ed), Academic Press, London, pp. 225 (1991).
51. P. N. Ross, in "Proceedings of the 21st Intersoc. Energy Convers. Eng. Conf.", San Diego, USA, pp. 1066–1072, Aug. 25–29 (1986).
52. P. N. Ross, US Patent No. 4,842,963 (1989).
53. S. Müller, O. Haas, C. Schlatter and C. Comninellis, *J. Appl. Electrochem.*, **28**, 305 (1998).
54. G. Bronoel, A. Millot and N. Tassin, *J. Power Sources*, **34**, 243 (1991).
55. S. Motoo, M. Watanabe and N. Furuya, *J. Electroanal. Chem.*, **160**, 351 (1984).
56. M. Watanabe, M. Tomikawa and S. Motoo, *J. Electroanal. Chem.*, **195**, 81 (1985).
57. M. Watanabe, M. Uchida and S. Motoo, *J. Electroanal. Chem.*, **199**, 311 (1986).
58. H. A. Gasteiger, N. Markovic, P. N. Ross, Jr and E. J. Cairns, *J. Phys. Chem.*, **97**, 12 020 (1993).
59. H. A. Gasteiger, N. Markovic, P. N. Ross, Jr and E. J. Cairns, *J. Electrochem. Soc.*, **141**, 1795 (1994).
60. H. A. Gasteiger, N. Markovic, P. N. Ross, Jr and E. J. Cairns, *J. Phys. Chem.*, **98**, 617 (1994).
61. A. M. Kannan, A. K. Shukla and S. Sathyanarayana, *J. Power Sources*, **25**, 141 (1989).
62. L. Swette and J. Giner, *J. Power Sources*, **22**, 399 (1988).
63. R. Larsson and L. Y. Johan, *J. Power Sources*, **32**, 253 (1990).
64. E. Budevski, I. Iliev, A. Kaisheva, A. Despic and K. Krsmanovic, *J. Appl. Electrochem.*, **19**, 323 (1989).
65. D. O. Ham, G. Monitz and E. J. Tailor, *J. Power Sources*, **22**, 409 (1988).
66. S. Müller, K. Striebel and O. Haas, *Electrochim. Acta*, **39**, 1661 (1994).
67. L. Carlson and L. Öjeford, *J. Electrochem. Soc.*, **127**, 525 (1980).
68. H. Cnobloch, D. Groeppel, D. Kuehl, W. Nippe and G. Simsen, 'Power Sources 5', D. H. Collins (Ed), Academic Press, London, pp. 261 (1975).
69. A. Gibney and D. Zuckerbrod, 'Power Sources 9', J. Thomson (Ed), Academic Press, London, pp. 143 (1982).
70. B. C. Beard and P. N. Ross, Jr, *J. Electrochem. Soc.*, **137**, 3368 (1990).
71. S. P. Jiang and A. C. C. Tseung, *J. Electrochem. Soc.*, **137**, 3442 (1990).
72. J. Prakash, D. A. Tryk and E. B. Yeager, *J. Electrochem. Soc*, **146**, 4145 (1999).
73. L. Swette and N. Kackley, *J. Power Sources*, **29**, 423 (1990).
74. J. B. Goodenough and R. Manoharan, *Electrochem. Soc. Proc.*, **92–11**, 523 (1992).
75. K. Kinoshita, 'Electrochemical Oxygen Technology', Wiley, New York (1992).
76. J. O'M. Bockris and T. Otagawa, *J. Phys. Chem.*, **87**, 2960 (1983).
77. J. O'M. Bockris and T. Otagawa, *J. Electrochem. Soc.*, **131**, 290 (1984).
78. Y. Shimizu, K. Uemura, H. Matsuda, N. Miura and N. Yamazoe, *J. Electrochem. Soc.*, **137**, 3430 (1990).
79. K. A. Striebel, C. D. Deng and E. J. Cairns, *Electrochem. Soc. Proc.*, **95–26**, 112 (1995).
80. S. Müller, F. Holzer, H. Arai and O. Haas, *J. New Mater. Electrochem. Systems*, **2**, 227 (1999).
81. H. Arai, S. Müller and O. Haas, *J. Electrochem. Soc*, **147**, 3584 (2000).
82. R. N. Singh, M. Hamdani, J. F. Koenig, G. Poillerat, J. L. Gautier and P. Chartier, *J. Appl. Electrochem.*, **20**, 442 (1990).
83. K. Lian, S. J. Thorpe and D. W. Kirk, *Electrochim. Acta*, **37**, 169 (1992).
84. L. Brossard and C. Messier, *J. Appl. Electrochem.*, **23**, 379 (1993).

85. S. P. Jiang and A. C. C. Tseung, *J. Electrochem. Soc.*, **138**, 1216 (1991).
86. M. R. G. de Chialvo and A. C. Chialvo, *Electrochim. Acta*, **38**, 2247 (1993).
87. M. R. G. de Chialvo and A. C. Chialvo, *Electrochim. Acta*, **36**, 1963 (1991).
88. A. M. Kannan and A. K. Shukla, *J. Power Sources*, **35**, 113 (1991).
89. G. Duperray, G. Marcellin and B. Pichon, 'Power Sources 8', J. Thompson (Ed), Academic Press, London, pp. 489 (1981).
90. R. A. Putt and G. W. Merry, LBL-Report No. 31184 (1991).
91. S. Hattori, M. Yamura, C. Kawamura and S. Yoshida, 'Power Sources 4', D. H. Collins (Ed), Oriel Press, London, pp. 361 (1973).
92. S. Takahashi, *J. Power Sources*, **7**, 331 (1982).
93. K. A. Striebel, F. R. McLarnon and E. J. Cairns, *J. Power Sources*, **47**, 1 (1994).
94. H. J. Pauling, Fa. Chemtek GmbH, Karlsruhe, private communication (1996).
95. G. Savaskan, T. Huh and J. W. Evans, *J. Appl. Electrochem.*, **22**, 909 (1992).
96. A. J. Appleby, J. Jaquelin and J. P. Pompon, in "Proc. Int. Automotive Eng. Congr.", Detroit (1977).
97. A. J. Appleby and M. Jacquier, *J. Power Sources*, **1**, 1730 (1976).
98. C. Debiemme-Chouvy and J. Vedel, *J. Electrochem. Soc.*, **138**, 2538 (1991).
99. P. C. Foller, *J. Appl. Electrochem.*, **16**, 529 (1986).
100. H. B. S. Alcazar, P. D. Nguyen and A. P. Pinnacle, in "Proc. Power Sources Conf.", Atlantic City, **33**, 434 (1988).
101. J. R. Goldstein, N. Lapidot, M. Aguf, I. Gektin and M. Givon, *Electrochem. Soc. Proc.*, **97–18**, 881 (1997).
102. S. Müller, F. Holzer, G. Masanz, S. Boss and O. Haas, PSI Bericht No. 96–19, Nov. (1996).
103. J. S. Chen and L. F. Wang, *J. Appl. Electrochem.*, **26**, 227 (1996).
104. R. G. Gunther and R. M. Bendert, *J. Electrochem. Soc*, **134**, 782 (1987).
105. J.-F. Drillet, F. Holzer, T. Kallis, S. Müller and V. M. Schmidt, *Phys. Chem. Chem. Phys.*, **3**, 368 (2001).
106. W. L. Fielder and J. Singer, NASA Tech. Memo, No. 102,580 (1990).
107. Y. Gao, F. Noguchi, T. Mitamura and H. Kita, *Electrochim. Acta*, **37**, 1327 (1992).
108. Y. Kera and T. Nakagishi, *Konki Daigaku Rikogakubu Kenkyu Hokoku*, **23**, 115 (1987).
109. K. Kuwaba and A. Matsuoka, *Jpn. Kokai Tokkyo Koko*, 55,807 (1998).
110. T. Ioroi, K. Yasuda and H. Takenaka, *Jpn. Kokai Tokkyo Koko*, 342,965 (2000).
111. H. Kordesch, K.-H. Steininger and K. Tomantschger, in "Proc. 23rd Intersoc. Energy Convers. Eng. Conf.", Denver, CO, Vol. 2, pp. 283, July 31–Aug. 5 (1988).
112. H. Celiker, M. A. Al-Saleh, S. Gultekin and A. S. Al-Zakri, *J. Electrochem. Soc.*, **136**, 1671 (1991).
113. A. Sofronkov, L. Korolenko and V. Petrosyan, *Pr. Nauk. Akad. Ekon. Im. Oskara Langego Wroclawiu*, **526**, 168 (1990).
114. V. S. Kublanovskii, M. O. Danilov and S. P. Antonov, *Ukr. Khim. Zh. Russ. Ed.*, **56**, 780 (1990).
115. J. Teng, L. Jin and L. Tang, *Dianhuaxue*, **3**, 428 (1997).
116. J. Teng, L. Jin and L. Tang, *Dianyuan Jishu*, **22**, 21 (1998).
117. J. Y. Cho, Y. W. Kim, J. Kim and J. S. Lee, *Hwahak Konghak*, **31**, 475 (1993).
118. J. Kivisaari, J. Lamminen, M. J. Lampinen and M. Viitanen, Technical Report, PB90-210956 (1990).
119. J. Lamminen, J. Kivisaari, M. J. Lampinen, M. Viitanen and J. Vuorisalo, *J. Electrochem. Soc*, **138**, 905 (1991).
120. D. Ohms, S. Herzog, R. Franke, V. Neumann, *et al.*, *J. Power Sources*, **38**, 327 (1992).
121. J. Y. Cho, J. H. Kim, Y. W. Kim, H. J. Kim and J. S. Lee, *Hwahak Konghak*, **32**, 86 (1994).
122. K. Horita, Y. Nakaosa and H. Oki, *Tanso*, **154**, 220 (1992).
123. G. Ciric-Marjanovic and S. Mentus, in "Proceedings of the International Conference Fundamental Aspects Physical Chemistry", Belgrade, pp. 279–281, Sept. 23–25 (1998).
124. R. R. Shipps, in "Proc. 20th Power Sources Conf.", Atlantic City, NJ, USA, pp. 86 (1966).
125. J. W. Evans and G. Savaskan, *J. Appl. Electrochem.*, **21**, 105 (1991).
126. J. Noring, S. Gordon, A. Maimoni, M. Spragge and J. F. Cooper, "Abstracts of the Electrochemistry Society Meeting", Honolulu, USA, pp. 24, May (1993).
127. S. L. Wilkinson, C&EN, p. 18, October 13 (1997).
128. H. J. Pauling, Technical Information, ChemTek, Karlsruhe, Germany (1994).
129. S. M. Chodosh, B. Jagid and E. Katsoulis, 'Power Sources 2', D. H. Collins (Ed), Pergamon Press, London, pp. 423 (1970).
130. R. R. Witherspoon, E. L. Zeitner and H. A. Schulte, in "Proc. Intersoc. Energy Convers. Eng. Conf.", Boston, Vol. 6, pp. 96, Aug. 3–5 (1971).
131. B. D. McNicol and D. A. J. Rand, 'Power Sources for Electric Vehicles', Elsevier (1984).
132. B. Koretz, J. R. Goldstein, Y. Harats and M. Y. Korall, in "ISATA Symp. Proc., Florence, Italy", **25**, 582 (1992).
133. J. Whartman and B. Koretz, in "Proc. Annual Int. Zinc Conf., Scottdale, AR", **2** (1993).
134. N. Lossau, *VDI-Nachrichten Magazin*, Sept. (1993).
135. B. Böndel, *Wirtschaftswoche*, p. 68, July 2 (1993).
136. H. J. Pauling, G. Weissmüller and E. Hoffmann, in "Jahrestagung GDCh", Fachgruppe Angew. Elektrochemie, Wien, Abstract pp. 35 (1997).
137. E. Budevski, I. Iliev and R. Varbov, *Z. Phys. Chem. Leipzig*, **261**, 716 (1980).
138. S. Gamburtzev, A. Kaisheva, I. Iliev and E. Budevski, *Isv. Otd. Khim. Nauki Bulg. Akad. Nauk*, **8**, 443 (1975).

139. V. R. Shepard, Y. G. Smalley and D. R. Bentz, US Patent No. 5,306,579 (1994).
140. J. A. Read, US Patent No. 6,127,060 (2000).
141. M. N. Golovin and I. Kuznetsov, US Patent No. 6,069,107 (2000).
142. AER, Press release: "White paper", August 21 (1998).
143. D. Sieminski, Presented at the Knowledge Foundation Conference, Washington, April (2001).
144. Colborn, 'Battery Power Products and Technology', (1999).
145. O. Haas, S. Müller and K. Wiesener, *Chemie Ingenieur Technik*, **68**, 524 (1996).

Chapter 23
Seawater aluminum/air cells

J. P. Iudice de Souza[1] and W. Vielstich[2]
[1] *Universidade Federale do Pará, Belém, Brazil*
[2] *IQSC, São Carlos, Universidade de São Paulo, Brazil*

1 ALUMINUM AS AN ANODE FOR METAL/AIR CELLS

Aluminum has long attracted attention as a potential battery anode because of its negative thermodynamic potential of -1.662 V in neutral (saline) electrolytes and -2.31 V in alkaline electrolytes (vs. the standard hydrogen electrode (SHE), and its high Ah capacity of $2.980\,\text{Ah}\,\text{kg}^{-1}$. From a volume standpoint, aluminum should yield $8.04\,\text{Ah}\,\text{cm}^{-3}$, compared with 2.06 for lithium and 5.85 for zinc. In addition, aluminum is an abundant and relatively inexpensive metal.

However, in an aqueous media, the aluminum surface is covered with a protective oxide layer, which causes a decrease in the open circuit potential and also a "delayed action", that is, the time lag before the cell reaches its maximum operating voltage when the circuit is closed.[1] The oxide film can be removed by dissolution in concentrated alkali solutions, or the presence of aggressive ions like chloride ions creates an extensive localized attack. However, in addition to the "activation" of the aluminum anode, wasteful corrosion takes place, resulting in less than 100% utilization of the metal and an evolution of hydrogen.

The aluminum/oxygen system was first demonstrated in the early 1960s by Zaromb[2] and Trevethan *et al.*[3] who found that the addition of zinc oxide or certain organic inhibitors, e.g., alkyldimethylbenzyl ammonium salts, to the electrolyte significantly decreases the corrosion of amalgamated aluminum anodes in 10 M sodium or potassium hydroxide solutions. The reaction at the cathode is the electroreduction of oxygen, which can be sustained at practical rates using gas diffusion electrodes.

1.1 Oxidation of aluminum in alkaline, neutral and acid electrolytes

Due to different electrode reactions at the aluminum anode, depending on the pH of the solution, the aluminum/air cell voltages also vary with the pH. In strongly *alkaline* solutions, all potentials at standard conditions and versus SHE are

Anode
$$2\text{Al} + 6\text{OH}^- \longrightarrow 2\text{Al(OH)}_3 + 6\text{e}^- \quad \varphi^{00} = -2.31\,\text{V} \tag{1}$$

Cathode
$$\frac{3}{2}\text{O}_2 + 3\text{H}_2\text{O} + 6\text{e}^- \longrightarrow 6\text{OH}^- \quad \varphi^{00} = 0.401\,\text{V} \tag{2}$$

Brutto[1]
$$2\text{Al} + \frac{3}{2}\text{O}_2 + 3\text{H}_2\text{O} \longrightarrow 2\text{Al(OH)}_3 \quad E^{00} = 2.711\,\text{V} \tag{3}$$

Al(OH)_3 is a crystalline form of aluminum hydroxide. With sufficient OH^- concentration in the electrolyte, we have:

$$2\text{Al(OH)}_3 + 2\text{OH}^- \longrightarrow 2\text{Al(OH)}_4^-$$

and

Brutto² $2Al + \frac{3}{2}O_2 + 3H_2O + 2OH^- \longrightarrow 2Al(OH)_4^-$

$$E^{00} = 2.751 \text{ V} \qquad (4)$$

In *neutral* electrolytes (saline or seawater), at the anode H^+ ions are formed, and at the cathode OH^- ions:

Anode

$2Al + 6H_2O \longrightarrow 2Al(OH)_3 + 6H^+ + 6e^-$

$$\varphi^{00} = -1.662 \text{ V} \qquad (5)$$

Cathode

$\frac{3}{2}O_2 + 3H_2O + 6e^- \longrightarrow 6OH^- \quad \varphi^{00} = 0.401 \text{ V} \qquad (2)$

Brutto

$2Al + \frac{3}{2}O_2 + 3H_2O \longrightarrow 2Al(OH)_3 \quad E^{00} = 2.063 \text{ V} \qquad (3)$

And in *acid* electrolytes, we have:

Anode

$2Al \longrightarrow 2Al^{3+} + 6e^- \quad \varphi^{00} = -1.662 \text{ V} \qquad (6)$

Cathode

$\frac{3}{2}O_2 + 6H^+ + 6e^- \longrightarrow 3H_2O \quad \varphi^{00} = 1.229 \text{ V} \qquad (7)$

Brutto

$2Al + \frac{3}{2}O_2 + 6H^+ \longrightarrow 2Al^{3+} + 3H_2O$

$$E^{00} = 2.891 \text{ V} \qquad (8)$$

In aqueous media, Al is prone to *parasitic reactions*:

$Al + 3OH^- \longrightarrow Al(OH)_3 + 3e^- \qquad (9)$

$2H_2O + 2e^- \longrightarrow H_2 + 2OH^- \qquad (10)$

The competing hydrogen evolution can lead to low anodic potentials, described as a "negative difference effect".[4] Parasitic reactions may also lower the coulombic efficiency of Al oxidation, although under several conditions these effects may be minimized, as will be shown in Section 3 of this chapter.

1.2 Effect of Cl⁻ ions on the aluminum oxide layer

The study of the electrochemical behavior of aluminum in different aqueous solutions, particularly regarding the role played by the oxide layer or pitting corrosion due to the presence of chloride ions, leads to various mechanisms proposed to explain the breakdown of oxide films.[5] One of them takes into account the migration of chloride ions through the film. Breakdown occurs when chloride reaches the metal/film interface. An investigation using radioactively labeled chloride has shown that chloride does not enter the oxide film but that it is chemisorbed onto the oxide surface and acts as a reaction partner, aiding dissolution via the formation of oxide–chloride complexes.[6]

When a freshly polished aluminum electrode is dipped into the 0.5 M NaCl solution, the corrosion potential moves rapidly towards anodic values on formation of the oxide layer. Then it shifts in the cathodic direction during the first hour of immersion. This evolution is attributed to the adsorption of chloride ions.[7] For increasing immersion times, the corrosion potential moves towards a more positive value and keeps a constant value after ca. 3 h of immersion due to the formation of a porous layer of alumina. Bockris[8] has shown that passive layers on aluminum involve a porous pre-layer of $Al(OH)_3$ and Al_2O_3. Part of the Al_2O_3 is $AlO(OH)$, which is a fibril structure. Breakdown involves penetration of these fibrils by Cl^-, which displaces OH^- and diffuses out in the form of $AlCl_n^{m+}$ complexes.

A number of organic compounds have been described as aluminum corrosion inhibitors in neutral and acidic media, the majority being nitrogen-containing compounds.[5] The protection effect is explained by the competitive adsorption of the inhibitor, which prevents the adsorption of chloride ions. Vielstich *et al.*[9] have tested Al/air cells with several chloride ions containing electrolytes, e.g., NaCl, KCl, sea water, and various additions were made to reduce corrosion of the aluminum anodes, e.g., sulfates, chromates, citric and malic acids.

2 THERMODYNAMIC AND EFFECTIVE ENERGY DENSITIES OF ALUMINUM/AIR CELLS IN COMPARISON TO OTHER SYSTEMS

Metal/air cells contain metal (negative) anodes and oxygen (positive) cathodes, with oxygen usually obtained from air. These batteries can be classified as semi-fuel cells, as oxygen is continuously supplied at the cathode. Some important electrochemical features are summarized in Table 1.[1]

In comparison to other metals, aluminum is characterized by the high values of 2.980 Ah kg^{-1} and 8.100 Wh kg^{-1}. Only lithium has a Ah g^{-1} equivalent higher than aluminum. The anode potentials are given for the case of alkaline electrolytes. In relation to the components of the

Table 1. Characteristics of metal/air cells with alkaline electrolytes. (Reproduced from Li and Bjerrum (2002).[1])

Cell	Anode reaction	Anode potential (V) vs. SHE	Metal equiv. (Ah g^{-1})	Cell voltage (oxygen cathode, V)		Specific energy (kWh kg^{-1})	
				Theoretical	Operating	of metal	of cell reactants
Li/air	$Li + OH^- = LiOH + e^-$	−3.05	3.86	3.45	2.4	13.3	3.9
Al/air	$Al + 3OH^- = Al(OH)_3 + 3e^-$	−2.35	2.98	2.70	1.2–1.6	8.1	2.8
Mg/air	$Mg + 2OH^- = Mg(OH)_2 + 2e^-$	−2.69	2.20	3.09	1.2–1.4	8	2.8
Ca/air	$Ca + 2OH^- = Ca(OH)_2 + 2e^-$	−3.01	1.34	3.42	2.0	4.6	2.5
Fe/air	$Fe + 2OH^- = Fe(OH)_2 + 2e^-$	−0.88	0.96	1.28	1.0	1.2	0.8
Zn/air	$Zn + 2OH^- = Zn(OH)_2 + 2e^-$	−1.25	0.82	1.65	1.0–1.2	1.3	0.9

cell reaction, the specific energy of the Al/air cell again is second on the list.

Finally, in Table 2 equivalent data of the Al/air system are compared with some secondary batteries. The conventional aqueous solution systems, such as lead/acid and nickel/cadmium batteries, suffer from low energy density and environmental pollution. As can be noted in Table 2, the electrochemical characteristics of the Al/air system looks superior.

Up to now, however, most of the attractive theoretical data of aluminum/air cells could not be obtained in practice. A commercial application is still missing. The reasons are: (1) the handling of the final products, especially in neutral solution (the gel type $Al(OH)_3$ material formed results in problems); (2) the strong corrosion in acid and alkaline electrolytes; and (3) the formation of passive oxide layers.

3 INFLUENCE OF SMALL ADDITIONS OF OTHER METALS TO ALUMINUM

One way to improve the anode performance is the addition of inhibitors to make the electrolyte less corrosive. Another way is to add small amounts of some elements to the aluminum metal. This modification of aluminum behavior by addition of alloying agents is known as

Table 2. Comparison of conventional batteries and Al/air cell.

Batteries	Capacity (Ah kg^{-1})	Open circuit voltage (V)		Theoretical specific energy (Wh kg^{-1})
		Theoretical	measured	
Lead/acid	83	2.0	2.0	170
Ni/Cd	181	1.4	1.2	240
Zn/AgO	199	1.6	1.4	310
Li/SOCl$_2$	120	3.6	3.4	430
Al/air	2980	2.70	1.2–1.6	8100

"activation". The mechanism of the aluminum activation is related to the effects: (i) controlling the thickness of the oxide layers; (ii) reducing the rate of hydrogen evolution; and (iii) controlling the dissolution morphology. Aluminum alloys based on high purity grade metal doped with elements such as Ga, In, Sn, Zn, Mg, Ca, Pb, Hg, Mn and Tl have been investigated.[10–13]

3.1 Problems related to the grade of aluminum anodes

The presence of certain impurities, such as iron and copper, in the aluminum can markedly affect the electrochemical behavior. For example, the corrosion rate is found to be influenced by the iron content in the metal.[14] Therefore, in order to achieve a desirable anode performance, it is necessary to use super-purity metals (99.999%). However, the high cost in energy of such a grade of aluminum (40 kWh kg^{-1}) reduces the overall thermodynamic efficiency of the battery (taking into account the Faradaic efficiency and the overpotential of both the anodic and cathodic reactions) to about 10%. Moreover, its production price, 10–20 times greater than industrial 2N7 aluminum, does not satisfy the requirements for commercial application. To reduce significantly the battery cost, it is desirable to develop aluminum anodes based on high-purity smelter-grade metal (99.8%). Quaternary alloys based on 99.8% pure aluminum containing Pb, Ga and In have been tested in NaOH based solutions.[15] The properties of aluminum alloy electrodes for use in seawater or other neutral electrolytes have been explored, including additions of Ga, In, Tl and small quantities of P and Mn.[10]

3.2 Electrochemical behavior of aluminum alloys in Cl$^-$ containing electrolytes

To be employed as anode material in batteries using saline solutions, the passive film formed on the aluminum surface has to be modified. The passivity of aluminum can be overcome by adding suitable alloying elements such as Zn, In, Sn and Ga. Also, it has been demonstrated that it is possible to activate aluminum by the addition of Hg^{2+}, In^{3+} and Sn^{4+} ions in neutral solutions. El Shayeb et al.[16] have studied the effect of Ga^{3+} ions on the electrochemical behavior of Al, Al/Sn, Al/Zn and Al/Zn/Sn alloys in chloride solutions. It has been found that the deposition of Ga on the electrode surface after a cathodic polarization (-2.0 V vs. the saturated calomel electrode (SCE)) causes a pronounced activation of the Al 99.99% electrode. On the other hand, Ga^{3+} ions have no activating influence on the dissolution of Al 99.61% and other Al alloy electrodes in 0.6 M NaCl, and there was no Ga deposition on the surface. Kliskic et al.[17] found that a content of Sn up to 0.4 wt% in the alloy leads to an accelerated attack of the alloy in 2.0 M NaCl solution. Cathodic polarization accelerates these phenomena, causing a thinner oxide film, both at Al and Al/Sn alloys. It is believed that superactivity occurs as a consequence of a localized de-filming process brought about by the retention and agglomeration of high mobile metallic species on the electrode surface.

4 ALUMINUM/AIR CELLS USING SALINE ELECTROLYTES

Despic et al.[10] were the first who explored the Al/air batteries with a saline electrolyte. A suitable electrolyte is a 12% solution of sodium chloride, which is near the maximum conductivity. In marine applications, sea water (c.a. 3% NaCl) can be used as the electrolyte. The batteries are characterized by high specific energies (Table 2) and, if constructed in such a way that the aluminum plates can be refilled, they have to be considered as (solid fuel) fuel cells or mechanically rechargeable secondary batteries. These kind of saline cells are harmless for the ordinary user and are ecologically acceptable without restrictions. Another major advantage is the infinite shelf-life in the inactive state with easy activation by the addition of water.

4.1 Model of a mechanically rechargeable battery

Budevski et al.[18] have investigated the performance of an experimental model of a battery run on sea water, satisfying power and energy requirements of a medium-size yacht designed for long periods at sea. The battery (Figure 1) was designed for a continuous supply of 24 W (12 V, 2 A) of power with periodic replacements of the electrolyte. It was composed of 20 modules. A flat plate aluminum alloy electrode was placed between two air electrodes of 200 cm^2 area each. The interelectrode distance at the start was about 2 mm. Significant space for excess electrolyte was left at the bottom of the cell. The aluminum electrodes were in the form of flat plates ($20 \times 10 \times 0.8$ cm in size), made of an Al/Sn/Ga/Mg alloy. The polarization characteristics of the alloy in 2 M NaCl and the corrosion rate were

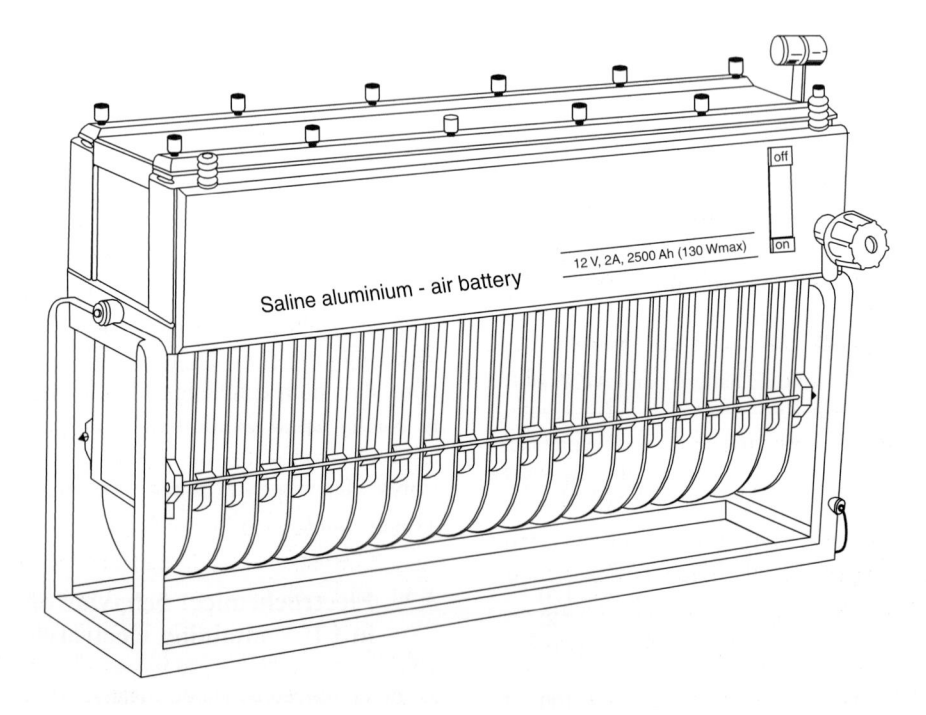

Figure 1. A 24 W (12 V) saline aluminum/air battery. (Reproduced from Budevski et al. (1989).[18])

measured, and it was observed that the alloy exhibits very low corrosion at open circuit (of only 0.3 mA cm^{-2}) and a relatively low so-called negative difference effect, i.e., the increase in the rate of hydrogen evolution with increasing in anodic current density, promising a material efficiency in an anodic discharge of 89%. The flat polarization plateau of the anode up to 200 mA cm^{-2} indicates that the limitations in the power density are not on the side of the aluminum electrode. Discharge curves for a load of 2 A at room temperature over 8 h are shown in Figure 2. Considering the total battery weight (including electrolyte) of 40 kg, a specific energy value of 577 Wh kg^{-1} is arrived at, which is a very attractive figure.

Summarizing, the authors have concluded that saline aluminum/air batteries supplied with appropriately active alloy anodes and running on sea water can serve as a reliable large-capacity medium-power sources for sea-going applications.

The saline battery is essentially an energy battery suitable for use in low-power requirements. It was shown that changes in specific energy and specific power with a quantity of aluminum in the cell (plate thickness) oppose each other.[19] Hence, the optimization of power capacity and energy capacity can be achieved. For high-power applications, however, they should be used in a hybrid combination with some typical power battery (e.g., lead/acid or Ni/Cd).

4.2 Battery with a continuous exchange of saline electrolyte

Unlike other batteries the capacities of which are limited by the amount of electrochemically active substances, in the Al/air battery, and especially the saline type, it is the acceptance of the reaction product by the electrolyte that limits its use. Hence, while using one and the same set of anodes, the electrolyte has to be replaced several times, depending on the ratio of aluminum capacity to electrolyte acceptance capacity. However, in marine applications, where the seawater can easily be replenished, or even caused to flow through the battery, it is only the aluminum-based specific energy that counts. From Figure 3, it is possible to visualize the concept of an Al/air battery operating in a seawater electrolyte.[9]

Ritschel and Vielstich[9] developed Al/air cells without a diaphragm or membrane, applying improved carbon-based air cathodes and additions to the neutral electrolyte such as citric acid (Figure 4). With these cells open to sea water even with voltages of about 1.0 V and Ah efficiencies for the aluminum of only 60–70%, more than 2000 Wh kg^{-1} could be obtained.

Due to the alkaline pH at the cathode, silver catalyzed carbon electrodes could be used.

A long-term test in 3% NaCl solution and in seawater showed stationary behavior over 400 h. Using 99.9% pure Al as the anode and a constant load of 10 mA cm^{-2}, a cell voltage of 0.4 V has been obtained. With Al/In (0.1%), an increase in the cell voltage to 0.8–1.0 V at room temperature was observed (see Figure 4).

4.3 Battery for sub-sea applications

Shen et al.[20] developed an Al/oxygen (oxygen dissolved in the sea) seawater battery for sub-sea applications.

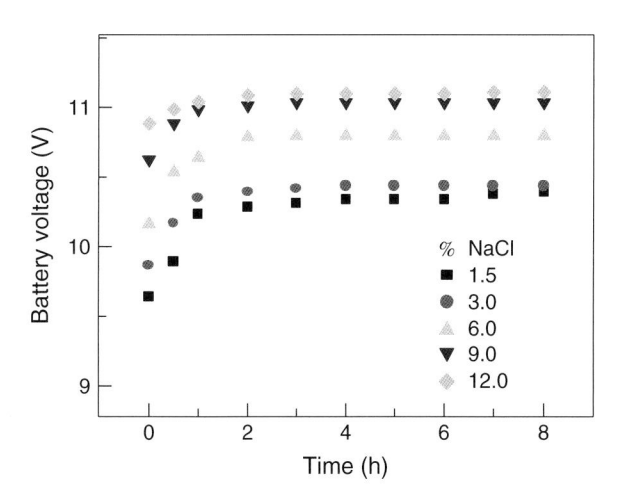

Figure 2. Cell voltage of the saline aluminum/air battery of Figure 1 for a load of 2 A and different NaCl concentrations (room temperature). (Reproduced from Budevski et al. (1989).[18])

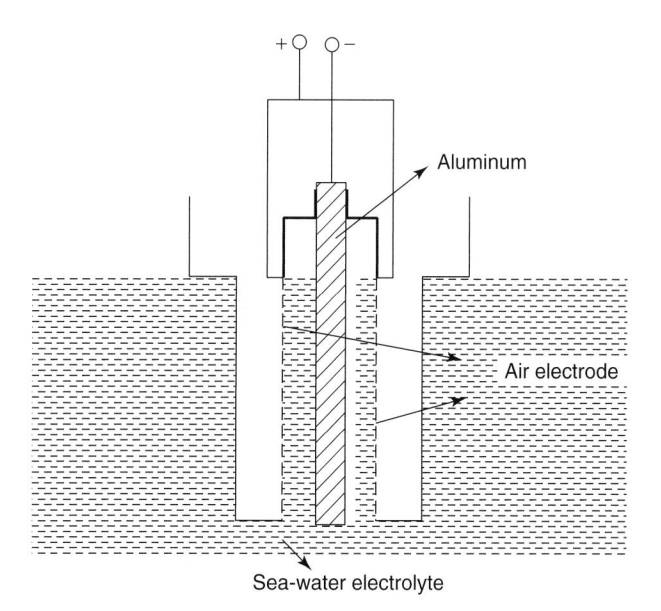

Figure 3. Model of a seawater/air battery with continuous exchange of saline electrolyte. (Reproduced from Ritschel and Vielstich (1979)[9] with permission from Elsevier Science.)

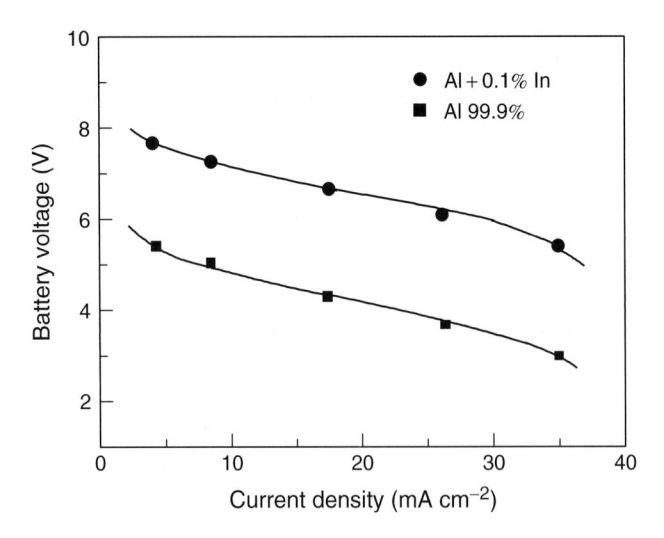

Figure 4. Cell voltage/current density plots for different aluminum anodes (closed six-cell battery, 3 M NH_4Cl + citric acid, room temperature). (Reproduced from Ritschel and Vielstich (1979)[9] with permission from Elsevier Science.)

Long-term tests using Teflon-bonded Co_3O_4/C oxygen reduction cathodes have shown that the performance of the constructed battery was stable over about 80 days. The aluminum alloy used as the anode contained 0.63% Mg, 0.14% Sn, 0.04% Ga and 0.0018% Fe. The operating voltage was about 1.4 V at current densities of $0.1\,mA\,cm^{-2}$, using an anode/cathode gap of 0.5 cm. A conceptual design showed that such battery has an energy density of $1\,kWh\,kg^{-1}$ over an operating period of 1 year, significantly higher than conventional primary batteries considered for use in sub-sea environments.

5 DEVELOPMENT PROBLEMS AND POSSIBLE APPLICATIONS

The theoretical energy density for an Al/air cell is $8.1\,kWh\,kg^{-1}$. As was shown above, practical values between 500 and $2000\,Wh\,kg^{-1}$ have been achieved. This means that there is still significant scope for anode development. In the 1980s, the US Department of Energy set the Aluminum/Air Traction Battery Program an energy yield target of $4.3\,kWh\,kg^{-1}$ at $200\,mA\,cm^{-2}$ with a power density of $0.6\,W\,cm^{-2}$ at $600\,mA\,cm^{-2}$.[21] Improved alloys had been studied for both saline and alkaline electrolytes. The alkaline system provides both high specific energy and power density, whereas the saline system possesses a reasonably high specific energy density with a much lower power.

The particular advantages of the saline system are its infinite shelf life (if stored without electrolyte and activated simply by salt water or ocean water), its low corrosion rates and its stable discharge characteristics.

Issues for research and development are:

- aluminum-based anodes with improved polarization and corrosion characteristics,
- provision of air electrodes matched to each application both in terms of performance and cost, and
- for closed cells: optimization of electrolyte management to maximize cell life or to maintain a specific reaction product concentration.

The saline batteries are suitable for special applications, such as emergency lighting, reserve power, long-lasting silent power for communication equipment and lighting on yachts and other marine objects, lighting for camping and so on, where the high energy content of such batteries is a clear advantage.

REFERENCES

1. Q. Li and N. J. Bjerrum, *J. Power Sources*, **110**, 1 (2002).
2. S. Zaromb, *J. Electrochem. Soc.*, **109**, 1125 (1962).
3. L. Trevethan, D. Bockstie and S. Zaromb, *J. Electrochem. Soc.*, **110**, 267 (1963).
4. S. Licht, G. Levitin, R. Tel-Vered and C. Yarnitzky, *Electrochem. Commun.*, **2**, 329 (2000).
5. L. Garrigues, N. Pebere and F. Dabosi., *Electrochem. Acta*, **41**, 1209 (1996).
6. F. D. Bogar and R. T. Foley, *J. Electrochem. Soc.*, **119**, 462 (1972).
7. R. Ambat and E. S. Dwarakadasa, *J. Appl. Electrochem.*, **24**, 911 (1994).
8. J. O'M Bockris and Lj. V. Minevski, *J. Electroanal. Chem.*, **349**, 375 (1993).
9. M. Ritschel and W. Vielstich, *Electrochem. Acta*, **24**, 885 (1979); M. Ritschel and W. Vielstich, German Patent DE 2819 117 C2, Oct 31 (1979).
10. A. R. Despic, D. M. Drozic, M. M. Purenavic and N. Cjkovic, *J. Appl. Electrochem.*, **6**, 527 (1976).
11. N. Fitzpatrick and G. Scamans, *New Scientist*, **111**, 34 (1986).
12. D. D. MacDonald, K. H. Lee, A. Moccari and D. Harrington, *Corrosion Sci.*, **44**, 652 (1988).
13. A. Mance, D. Cerovic and A. Mihajlovic, *J. Appl. Electrochem.*, **14**, 459 (1984).
14. M. L. Doche, F. Novel-Catin, R. Durand and J. J. Rameau, *J. Power Sources*, **65**, 197 (1997).
15. V. Kapali, S. Venkatakrishma, V. Balaramachandran, K. B. Sarangapani, M. Ganesan, M. A. Kulandeinathan and A. S. Medeen, *J. Power Sources*, **39**, 263 (1992).
16. H. A. El Shayeb, F. M. Abd El Wahab and S. Zein El Abedin, *Corrosion Sci.*, **43**, 643 (2001).

17. M. Kliskie, J. Radosevic, S. Gudic and M. Smith, *Electrochem. Acta*, **43**, 3241 (1998).
18. E. Budevski, I. Iliev, A. Kaisheva, A. R. Despic and K. Krsmanovic, *J. Appl. Electrochem.*, **19**, 323 (1989).
19. A. R. Despic, *J. Appl. Electrochem.*, **15**, 191 (1985).
20. P. K. Shen, A. C. C. Tseung and C. Kuo, *J. Power Sources*, **47**, 119 (1994).
21. G. Scamans, 'Development of the Aluminum/Air Battery', Presented at a meeting of the Electrochemical Technology Group of the SCI, London, Oct. 8 (1985).

Chapter 24
Energy storage via electrolysis/fuel cells

J. Divisek and B. Emonts
Institute for Materials and Processes in Energy Systems, Forschungszentrum Jülich GmbH, Jülich, Germany

1 INTRODUCTION

Energy storage via electrolysis/fuel cells is coupled with the hydrogen storage medium as the product of energy conversion in an electrolyzer and as the fuel gas of electricity generation in a fuel cell. In connection with the use of renewable energy sources, emission-free conversion and storage can be described and thus a contribution made to the still unclarified questions of future energy supply:

- How can energy supply be secured in view of increasingly scarce fossil fuel reserves?
- What energy supply options will exist if anthropogenic CO_2 emissions have to be significantly reduced in the near future?
- How can a time- and performance-related decoupling of energy supply and energy demand be achieved?

These challenges are not the only motivation for the development of energy storage systems with electrolyzer and fuel cell, because even today there is a demand for grid- and fuel-independent as well as emission-free energy supply systems.

1.1 Operation area for storage systems

Storage systems with hydrogen as the storage medium may be applied wherever energy is to be provided autonomously from renewable sources in a decentralized manner or for lack of grid connection over prolonged periods of time. An established technology with which these requirements can only be partially fulfilled is a system composed of a renewable energy source including battery and a diesel/generator unit. The main difference between the energy system with hydrogen storage and that with a diesel/generator unit is the provision of diesel fuel for securing the supply of a consumer with electric energy at any time. Furthermore, a system with temporary motor generator drive in diesel operation is not emission-free. The possibility of autonomous, all-year energy supply makes the system with hydrogen storage interesting for applications in which grid connection and fuel provision involve extremely high costs. In general, such applications are located in grid- and access-distant regions with high seasonal differences in the yield of renewable energy sources. This means that the location of application is characterized by the necessity of providing large fuel quantities at high specific transportation cost for the fuel. Irrespective of the cost situation, an electrolyzer/hydrogen/fuel cell system offers the possibility of also supplying a consumer who is subject to the specific boundary conditions of air pollution control with electric energy without harmful exhaust gas release. This can be of decisive significance for regions with permanent smog and in applications for measuring climatic changes.

Another field of application for a hydrogen-based energy storage system, which results from the increasing use of wind power and its feeding into the supply grid, is grid-coupled energy buffering. In this case, it is important that, upon the occurrence of an energy oversupply at grid nodes with temporary feeding of high wind powers, this energy is efficiently stored without any time lag and fed into the grid again at times of potential supply deficiency. This makes

Table 1. Projects for energy storage systems.

Project	Description	Performance characterization
PHOEBUS Jülich[1–4]	Autonomous, all-year electricity supply of an office building with PV	PV unit: 42 kW$_P$
		Electrolyzer: 26 kW; 7 bar Fuel cell: 5.6 kW Energy production: 15 MWh$_{el}$/a Peak load: 15 kW
Autonomous solar-energy house, Freiburg[7]	Complete supply of an energy-optimized residential building with PV	PV unit: 4.2 kW$_P$
		Electrolyzer: 2 kW Fuel cell: 0.5 kW Energy production: 1.8 MWh$_{el}$/a; 0.9 MWh$_{th}$/a
Solar–Wasserstoff–Bayern (SWB); Neunburg vorm Wald[5]	Demonstration plant for the production and application of solar hydrogen	PV unit: 365 kW$_P$
		Electrolyzer: 111 kW; 80 bar; 100 kW; 1.5 bar; 100 kW; 32 bar
SAPHYS, Rome, Italy[8, 9]	Demonstration, test and research plant for solar hydrogen technologies	PV unit: 5.6 kW$_P$
		Electrolyzer: 5 kW; 20 bar Fuel cell: 3 kW
HYSOLAR, Germany, Saudi Arabia[6]	R&D as well as demonstration of the production and use of solar hydrogen	Two systems of coupled PV-electrolyzer unit: 350 kW, 10 kW
Schatz Solar Hydrogen Project, Arcata, USA[10]	All-year aquarium ventilation by means of PV and hydrogen storage	PV unit: 9.2 kW$_P$
		Electrolyzer: 5.8 kW Fuel cell: 0.6 kW
Compact, seasonal energy storage, Helsinki, Finland[11]	Component tests for autonomous electricity supply of electrical systems	Electrolyzer: 30 W
		Fuel cell: 100 W

it possible to operate a stable grid while maintaining maximum base-load supply.

Energy storage systems with solar-produced hydrogen have been explored and operated worldwide with different objectives. Table 1 provides an overview of the projects with their special features and characteristic performance data. On the one hand, the aim is to test and demonstrate the operation of alternative energy supply systems on a large technical scale (SWB, HYSOLAR), on the other hand, complete systems have been built up for special applications (PHOEBUS, solar energy house, Schatz, Helsinki). The designs for autonomous energy systems with hydrogen storage realized within the framework of theoretical studies (see also Refs. [12–19]) assume that no surplus energy is lost and the hydrogen storage tank is equally filled at the end and beginning of a scenario year. However, different system configurations can only be evaluated realistically with a design approach which considers plant operation over many years and includes costs as a quality function.[20]

1.2 Energy storage technology

The base case of a system configuration with energy storage via electrolyzer, gas tank and fuel cell consists of electricity production, short-term storage and electric current collection. An autonomous energy supply system is schematically represented in Figure 1.

The configuration is determined by a direct current (d.c.) busbar to which all components are connected directly or via adapted power electronics. The generator, which consists of a photovoltaics unit (PV) or a wind energy converter (WEC) or a hybrid system (PV-WEC), produces energy for both direct consumption and storage. Short-term storage in the range from hours to a few days is effected via a battery which also assumes the function of efficient and spontaneous decoupling of energy supply and energy demand. Long-term storage compensates for seasonal differences between electricity production and consumption or serves as interim storage for large amounts of electric surplus energy. It consists of an electrolyzer,

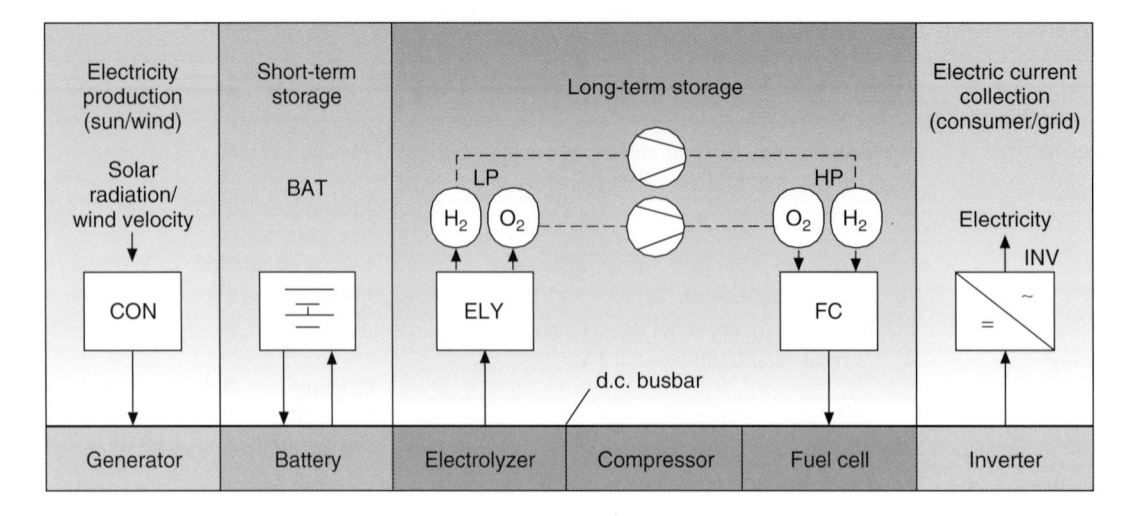

Figure 1. Schematic set-up of a system with energy storage.

which feeds hydrogen and oxygen into buffer tanks after drying and catalytic gas cleaning, two compressors and high-pressure tanks for storage at elevated energy density as well as a fuel cell for reconversion into electricity. Current collection is generally effected by a consumer or by the grid via an inverter connected to the d.c. busbar. A central energy management system[21] serves to monitor all energy flows and states of charge of the storage systems. It predefines the power requirements for fuel cell and electrolyzer. The main control parameter is the state of battery charge. If the state of battery charge drops below a lower limit in periods of low yield, the fuel cell is operated to supply the consumer. If the state of battery charge exceeds an upper limit in periods of high yield, the electric surplus energy is used to produce hydrogen and oxygen with the electrolyzer. Hydrogen production thus not only depends on the yield conditions, but also on consumer behavior and on the adjustment parameters of the energy management of the complete system. The energy management system is optimized so that high system efficiency and a long service life are achieved for the electrochemical components.

2 ELECTROLYSIS/FUEL CELL SYSTEM TECHNOLOGY

2.1 Introductory remarks

According to the classification of different forms of energy storage technologies,[22] the storage route via electrolysis/fuel cells belongs to the group of storage technologies exclusively based on (electro)chemical methods. The process of electrolyzing water to hydrogen and oxygen and totally recombining the gases in a fuel cell is an important example of the combination of storage as chemicals followed by electrochemical conversion. Water electrolysis is a commercially proven method for continuous production of hydrogen by converting the electrical energy into chemical energy. In contrast to this, a fuel cell is an electrochemical cell which can continuously convert the chemical energy of a fuel and an oxidant into electrical energy. The idea of connecting both systems into one operational unit is very old. Grove electrolyzed water and then used the evolved gases to produce electricity. Water electrolysis has since then developed into an industrial process, and is in fact the oldest known electrochemical technology. It has been introduced as a useful industrial way of producing hydrogen and oxygen. The fuel cells, on the contrary, despite their long history and development during the past 40 years, have experienced very few applications up to now. The most important have been their use as an electrical energy source for the manned spacecraft of the NASA Space Shuttle programs. At the present time, they are aimed at applications for electric vehicle propulsion, remote power plants, cogeneration, industrial waste utilization, emergency power supply or for military purposes. There are two possible ways of fuel cell operation: firstly, using carbon-containing fuels, and secondly, pure hydrogen. The major difference between a fuel cell operating on a fuel such as natural gas and one operating on hydrogen is that with the latter no fuel processing is required. In a fuel cell system operating on hydrogen such as in the reverse case of an electrolysis/fuel cell, the fuel cell unit is likely to be inexpensive compared with one based on natural gas. Only the stack itself and a feedback loop for cooling and water removal are necessary. This is also the fuel cell type which can be used for the purpose of water electrolysis/fuel cell hybrid systems.

2.2 Water electrolysis

In general, the water electrolysis cell is composed of two electronic conductive electrodes in contact with the ionic conductive electrolyte. When an electric current is passed through the cell containing water, the water molecules are split into hydrogen and oxygen according to the general equation

$$H_2O + \text{electric energy} \longrightarrow H_2 + \tfrac{1}{2}O_2 \quad (1)$$

Water is a poor ionic conductor and for this reason a conductive electrolyte must be present either as an addition to the water or as a solid-state electrolyte so that the water splitting reaction can proceed at technically acceptable cell voltage and current densities. The electrolyte itself does not undergo any changes during the electrolysis. Both alkaline and acidic electrolytes can be used in water electrolysis. The water splitting reaction (1) is a sum of two partial electrochemical reactions at the anode (oxygen evolution) and cathode (hydrogen evolution), which can take place in an alkaline or acidic medium. In each case the net sum reaction (1) is the same.

Alkaline medium:

$$\text{cathode:} \quad 2H_2O + 2e^- \longrightarrow H_2 + 2OH^-$$
$$\text{anode:} \quad 2OH^- \longrightarrow \tfrac{1}{2}O_2 + H_2O + 2e^- \quad (2)$$

Acidic medium:

$$\text{cathode:} \quad 2H_3O^+ + 2e^- \longrightarrow 2H_2O + H_2$$
$$\text{anode:} \quad 3H_2O \longrightarrow 2H_3O^+ + \tfrac{1}{2}O_2 + 2e^- \quad (3)$$

Equation (2) gives the basic electrode reactions in industrial alkaline water electrolysis, while equation (3) describes the electrode reactions in an acidic medium. In the alkaline cell, the ionic current is given by the flux of hydroxyl ions from the cathode to the anode, and in an acidic cell, by the flux of protons from the anode to the cathode. Both cases are shown in Figure 2.

As shown in Figure 2, the main components of the water electrolysis cell are the anode, cathode and separator. Both electrodes must be corrosion-resistant in the electrolyte at the corresponding potentials, have a good electronic conductivity and structural integrity. The electrodes should catalyze the electrochemical evolution of gases (hydrogen and oxygen) as effectively as possible. In the acidic medium, the basic reactions (equation 3) suggest one water molecule per proton as the water transport process in the cell. The complete hydration shell of the proton consists of still more associated water molecules, however. If, for example, the Nafion® membrane is used as an electrolyte, the measurements of the electroosmotic drag coefficient n_d show that the water transport is a strong function of the temperature.[23] This temperature dependency can be

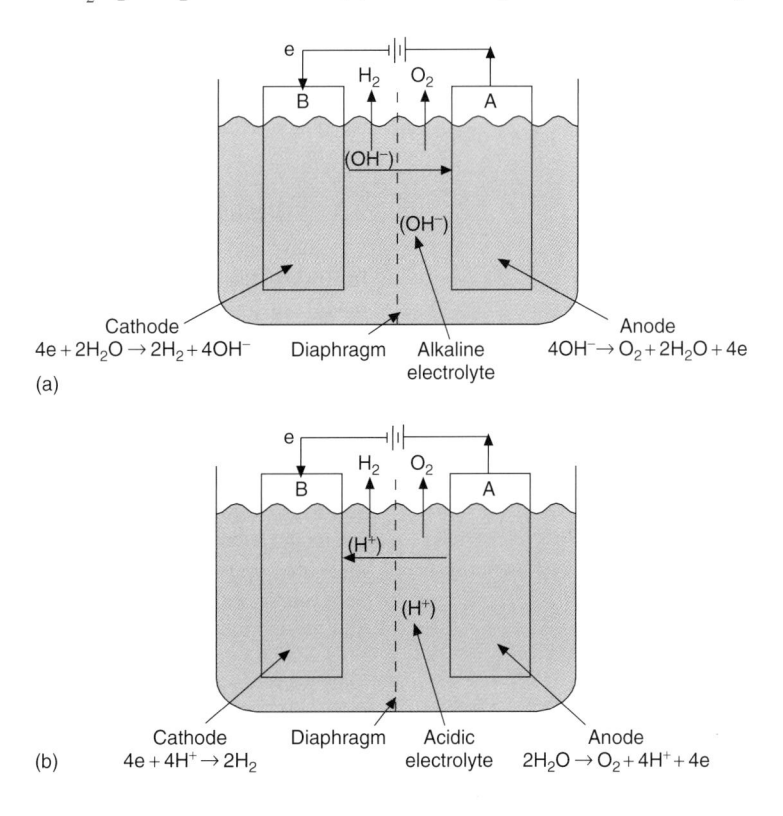

Figure 2. Schematic of water electrolysis. (a) Alkaline electrolyte; (b) acidic electrolyte.

expressed as

$$n_d = 0.1345 \times \exp(0.0092T) \quad (4)$$

The energy consumption of the water electrolysis based on the thermodynamic treatment usually starts with the estimation of the Gibbs–Helmholtz equation of the water splitting reaction (1). According to LeRoy et al.,[24] a more realistic dependency takes into account the properties of the educt (water) and the products (hydrogen, oxygen, water vapor) under the effective working conditions of the water electrolyzer. In this case the corresponding reversible voltage can be expressed as

$$U_{\text{rev},T,p} = -\frac{\Delta G_f^0(\text{H}_2\text{O}_{(l)})_T}{2F} + \frac{RT}{2F} \ln \frac{(p - p_w)^{3/2} p_w^0}{p_w} \quad (5)$$

in which p_w is the partial water vapor pressure and p_w^0 is the water vapor pressure under standard conditions. The voltage ($U_{\text{rev},T,p}$ in V) corresponding to the standard molar Gibbs energy $\Delta G_{f,T}^0$ involved in equation (5) is given by the formula[24]

$$U_{\text{rev},T}^0 = 1.5184 - 1.5421 \times 10^{-3}T$$
$$+ 9.523 \times 10^{-5}T \cdot \ln T + 9.84 \times 10^{-8}T^2 \quad (6)$$

For the estimation of the energy consumption and the heat evolution during water electrolysis, in addition to the reversible voltage, knowledge of the thermoneutral voltage U_{tn} (enthalpic voltage $U_{\Delta H}$) is also necessary as well as of the higher heating value voltage (U_{HHV}). These voltages are defined by equation (7)

$$U_{\Delta H} = -\frac{\Delta H_f^0(\text{H}_2\text{O}_{(l)})_T}{2F}$$
$$U_{\text{HHV},T,p} = U_{\Delta H} + \frac{(H_{T,p} - H_{298}^0)_{w(l)}}{2F} \quad (7)$$

The temperature dependence of the corresponding standard voltages is given by equation (8)[24]

$$U_{\Delta H}^0 = 1.5187 - 9.763 \times 10^{-5}T$$
$$- 9.50 \times 10^{-8}T^2 \quad [\text{V}]$$
$$U_{\text{HHV},T}^0 = 1.4146 + 2.205 \times 10^{-4}T$$
$$+ 1.0 \times 10^{-8}T^2 \quad [\text{V}] \quad (8)$$

According to the definitions shown, the higher heating value voltage represents the enthalpy content of the product gases at the electrolysis temperature relative to water at standard conditions (298 K). The voltage difference $U_{\text{HHV},T} - U_{\Delta H}$ therefore means the energetic voltage equivalent necessary to heat 1 mol of feedwater from 298 K to the operating temperature. Now, with U_{obs} being the observed cell voltage, the energetic efficiency η_{EL} of the water electrolyzer may be defined by equation (9)

$$\eta_{\text{EL}} = \frac{U_{\text{HHV}}}{U_{\text{obs}}} \quad (9)$$

The observed cell voltage written now as U_{cell} is noticeably higher than that given by the thermodynamic values only. The actual cell voltage has to overcome the electrical resistance in the electrodes, in the electrolyte between the electrodes and in the separator in the case of alkaline electrolysis. In addition there are also the different kinds of overvoltages, which can be kinetically or mass-transport determined. The actual cell voltage is then given by the relation

$$U_{\text{cell}} = U_{\text{rev}} + \eta_a + |\eta_c| + j \cdot R_\Sigma \quad (10)$$

where η_a, η_b are the overvoltages at the anode and the cathode, j is the current density and R_Σ the sum of the surface-specific electrical resistances in the electrolyzer. The electrical resistance is determined by the electrode and separator material, cell design, as well as the physical conditions of electrolysis (wetting, bubble formation, electrolyte flux, temperature, etc.). A very important condition is also the electrode activation, which strongly determines the overvoltages. As an example, the real resistivity values of a Norsk Hydro electrolyzer[25] are given in Table 2 (electrolyte 25% KOH, 80 °C, 1 bar).

2.3 Alkaline water electrolysis

In industrial plants an alkaline medium has been preferred up to now, mainly because of the lower investment costs and minor corrosion problems. Alkaline electrolyzers are composed of many cells or electrode pairs. The unit cell consists of the cell frame, anolyte and catholyte chamber, usually being encased by bipolar cell walls, and

Table 2. Ohmic resistances and cell voltages in a water electrolysis cell. (After Andreassen.[25])

Sum ohmic resistances	1.11 Ω cm^2
Sum ohmic losses ($j = 0.2\,\text{A}\,\text{cm}^{-2}$)	0.22 V
Sum overvoltages (anodic + cathodic) ($j = 0.2\,\text{A}\,\text{cm}^{-2}$)	0.32 V
Reversible cell voltage, U_{rev}	1.19 V
Total cell voltage, U_{cell} ($j = 0.2\,\text{A}\,\text{cm}^{-2}$)	1.73 V

the diaphragm separating the production gases hydrogen and oxygen from each other. As a separator a diaphragm made of a hydroxide-resistive polymer composite material Zirfon® has recently been suggested.[26] In order to reduce the internal cell resistance, the zero-gap configuration is adopted today by the majority of electrolyzer manufacturers. In this configuration the electrode materials (porous metals, extended metal sheets, perforated plates) are pressed on either side of the diaphragm, so that the product gases are forced to leave the electrodes as bubbles at the reverse side, away from the separator. Then they do not interfere with the current passage through the electrolyte between the electrodes. In principle, the gas/electrolyte volume ratio should be low. The cell frame and the electrodes are made of nickel-plated steel. The electrodes are activated by a suitable coating to lower the overvoltages. The basic elements of a typical alkaline electrolyzer are shown in Figure 3.

Generally, a KOH solution is the most frequently used alkaline component because of the good conductivity. The hydroxide concentration may be between 25 and 30 wt%. For corrosion reasons, the working temperature has to be lower than 100 °C, usually about 80 °C. The operating pressure in the industrial plants can vary from 1 to 30 bar, for some special purposes, such as the water electrolysis/fuel cell hybrid systems, the operating pressure should be preferably 100 bar to avoid additional compression work.

A photo of such a high-pressure system developed by Forschungszentrum Jülich is shown in Figure 4.

2.4 Polymer electrolyte membrane (PEM) water electrolysis

In practice acid solutions are not used in water electrolysis. Instead, acid membranes have been introduced as electrolytes, so that the acidic water electrolysis is implemented in the membrane cells. The original SPE electrolyzer technology is derived from General Electric fuel cell work. In these cells, a polymer membrane electrolyte, mainly Nafion®, is the key component of the electrolyzer. Catalysts and current collectors in contact with the membrane must be acid-resistant. Platinum is used as the catalyst on the cathode side and iridium or its alloys on the anode side. Because of anodic gas evolution, the oxygen electrode must be more resistant than the oxygen electrode in the polymer electrolyte fuel cell (PEFC), i.e., carbon cannot be used as the electrode material and current collectors are made from titanium, tantalum or niobium. The choice depends on temperature, pressure and current density. Accordingly, only a perfluorinated polymer can be used as the separator and its replacement by other polymers is unlikely, since stability problems are predominant and restricting when cheaper

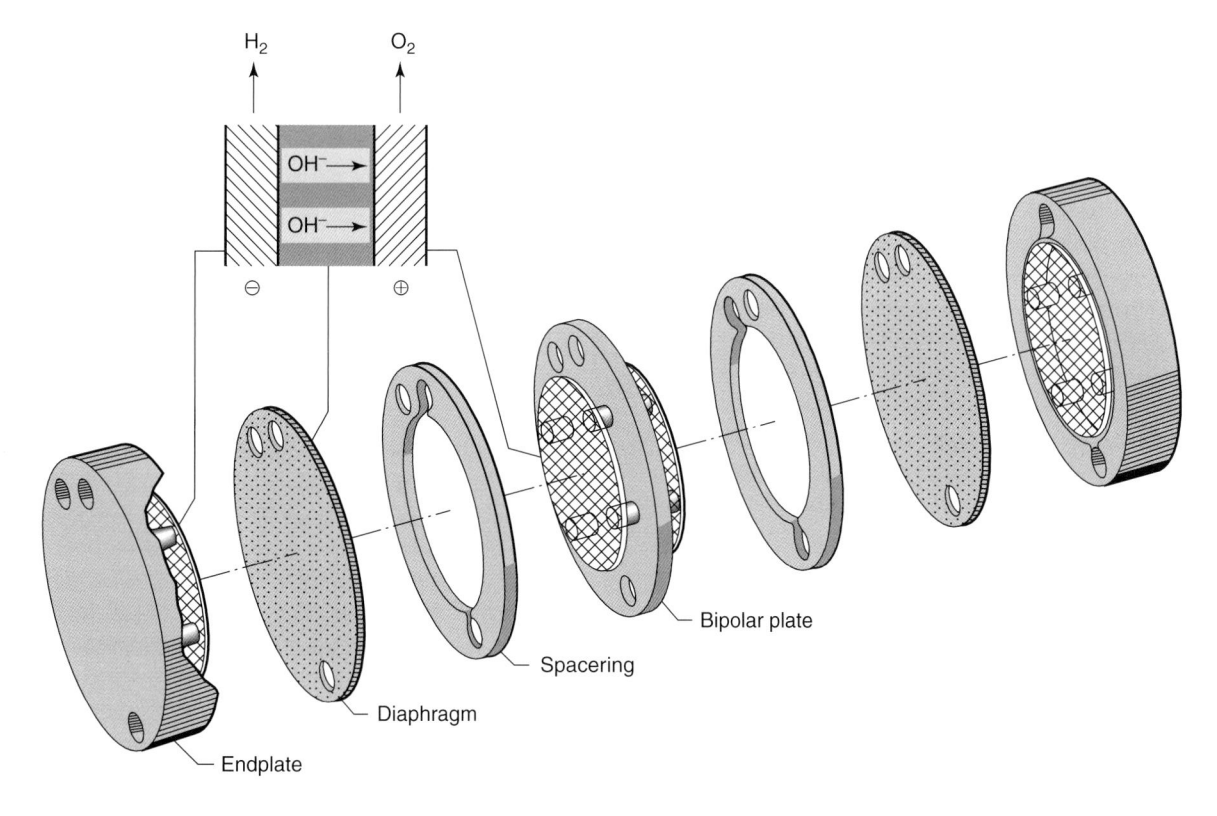

Figure 3. Exploded view of one electrolyzer cell.

Figure 4. High-pressure alkaline electrolyzer developed in Forschungszentrum Jülich.

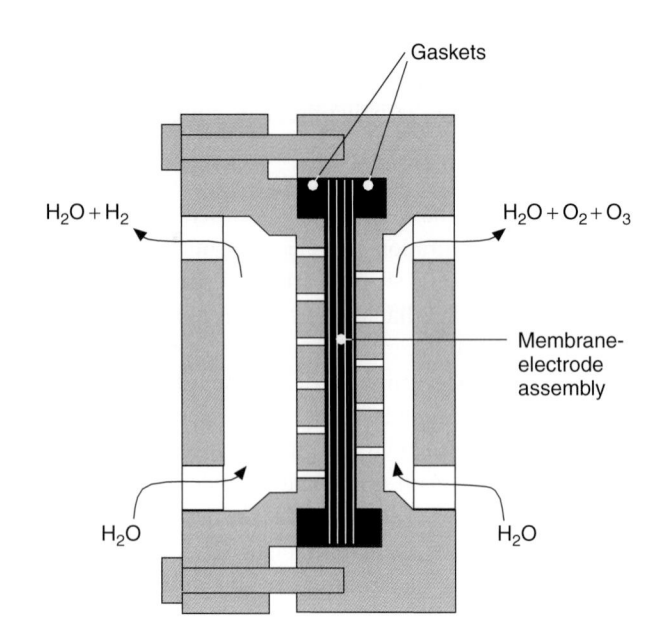

Figure 5. Scheme of an acidic membrane water electrolyzer. (Adapted from Stucki et al. (1985).[28].)

alternatives to Nafion® are evaluated. Results of SPE experiments with newly developed membranes (PBI, polyether sulfones, etc.) must be awaited.[27]

Because of the solid electrolyte, only pure water is circulated through cells. In this respect, the SPE electrolyzer technology is simpler than the corresponding alkaline variant. The SPE cell can operate over a wide range of temperatures up to 150 °C and high pressures up to 200 atm. The cells have the ability to operate at considerably higher current densities than do alkaline cells. They have, however, severe materials and cost problems which are still prohibitive with respect to the practical application of SPE water electrolysis. The SPE technology principle has also been used for development of the MEMBREL water electrolysis cell by Brown Boveri Company. Lately, the MEMBREL cell has been adapted for in situ anodic generation of ozone with PbO_2 anodes in electrolyte free-water.[28] The basic scheme of the electrolyzer is given in Figure 5.

2.5 Fuel cells

A fuel cell is an electrochemical device which can continuously convert a fuel's chemical energy, especially of hydrogen, back into electrical energy. If we consider equation (1) as a water electrolysis reaction converting the electrical energy into the chemical energy of hydrogen gas, the reverse reaction which takes place in a fuel cell could be written as

$$H_2 + \tfrac{1}{2}O_2 \longrightarrow H_2O + \text{electric energy} \quad (11)$$

The electrochemical reactions (2) and (3) and the thermodynamic reactions (5)–(8) are valid correspondingly. We can also define the energetic efficiency of the fuel cell η_{FC} as

$$\eta_{FC} = \frac{U_{obs}}{U_{HHV}} \quad (12)$$

Analogously, the actual cell voltage can also be given in principle by the relation

$$U_{cell} = U_{rev} - \eta_a - |\eta_c| - j \cdot R_\Sigma \quad (13)$$

In contrast to water electrolysis, however, the cell voltage of a fuel cell is more sensitive to the irreversibility of the oxygen reaction at low current densities and greatly restricted by mass transport at high current densities, so that the cell voltage curve is usually expressed as a fitting curve,[29, 30] in the form of

$$U_{cell} = U_{rev} + b \cdot \log j_0 - \eta_a - b \cdot \log j - j \cdot R_\Sigma \\ - \alpha j^k \ln(1 - \beta j) \quad (14)$$

Equation (14) is considerably more complex than equation (13). The parameters in equation (14) are partially determined by the Tafel terms of the oxygen reaction (exchange current density j_0, Tafel slope b), and partially derived from mathematical and statistical considerations of the current–voltage curve (α, k, β).

In the case of fuel cell/electrolysis system technology, which requires H_2/O_2 fuel cells, two cell types for low temperature applications can be taken into consideration: the alkaline fuel cell (AFC) and polymer electrolyte fuel cell (H_2-PEFC). In the AFC the electrolyte is concentrated (35–50 wt%) potassium hydroxide, which is a good OH^- ion conductor. Since oxygen reduction kinetics are more rapid in alkaline electrolytes than in acid ones, this cell can achieve higher cell voltage and consequently higher energy efficiency than the acidic cell. In addition, the use of non-noble metal electrocatalysts is feasible in AFCs. A major disadvantage of this cell type is that generally AFCs are restricted to specialized applications where CO_2 is absent and pure H_2 and O_2 are utilized. In the case of energy storage technology based on the combination of fuel cells and electrolysis, this restriction does not apply and the AFC can be used for this purpose. In the PEFC, the electrolyte is an ion exchange membrane (fluorinated sulfonic acid polymer or, more recently, less expensive polymers) which is a good proton conductor. The only liquid in this cell is water, thus electrolyte corrosion problems are negligible, but due to the danger of the membrane drying out, water management of the cell is critical. The sum of advantages versus disadvantages of the two cells are:

- the AFC has a KOH solution as the electrolyte; there is no free corrosive liquid in the PEFC;
- the PEFC needs noble metal catalysts; the AFC can operate with cheap common metals;
- the fluorinated polymer electrolyte in the PEFC is expensive; the AFC separator is much cheaper;
- the AFC cell has to handle liquid electrolyte; the PEFC cell is simple to fabricate and operate;
- water and heat management is critical for efficient PEFC operation; the AFC is not so sensitive in this respect;
- immobilization of liquid electrolyte is necessary in the AFC for efficient operation; solid electrolyte in the PEFC possesses a high natural stability.

2.6 Water electrolysis/fuel cell hybrid system

Since a fuel cell reaction is the reverse of water electrolysis (cf. equation (11)), it seems very simple and natural to integrate the two cells in one device. These reverse or regenerative fuel cells can work in principle with the alkaline electrolyte (combination alkaline water electrolysis and AFC)[31] as well as with the PEM/PEFC cell, which operates with a solid polymer electrolyte.[32] The requirement that the hydrogen and oxygen electrodes operate in the dual function creates considerable material stability problems, particularly at the oxygen electrode. Viable candidate materials must meet the following demands: high resistance to chemical corrosion and electrochemical oxidation, high bifunctional electrocatalytic activity and good electrical conductivity. For this purpose, mixed metal oxides mainly of the ABO_x form were investigated for both oxygen reduction and evolution in alkaline media. As one example, the perovskite material $LaCoO_3$ was developed.[33] Also an alloy catalyst of the composition $Na_xPt_3O_4$ was identified as a very promising material.[34]

However, most attempts were directed towards the acidic membrane fuel cell/electrolysis systems. The energy out to energy in ratio for this type of regenerative fuel cell is between 55 and 65%.[35] To solve the stability problem connected with the oxygen electrode catalyst, a new concept of bifunctional electrode has been developed.[36] In contrast to the classical electrodes described so far, in this concept oxidation and reduction reactions are assigned to the electrodes and not the catalysts. Thus both the oxidation reactions of oxygen evolution (electrolysis mode) and hydrogen oxidation to protons (fuel cell mode) proceed alternatively at the same electrode, and the two remaining reduction reactions, such as hydrogen evolution and oxygen reduction, at the other electrode. Using this concept it is less difficult to produce suitable electrode materials for both electrodes. For both reduction reactions platinum is the best catalyst, whereas the catalyst for the oxidation electrode can be selected on the basis of the electrochemical activity for both reactions that take place, i.e., oxygen evolution and hydrogen oxidation. In the cell itself, the catalytic activity for oxygen evolution is a more prevailing property than for the hydrogen oxidation, which is the more reversible reaction. For this reason, the Ir/Ru catalyst should be used, as this alloy has been identified as the best catalyst for the oxygen evolution reaction.[36]

3 GAS STORAGE SYSTEM TECHNOLOGY

3.1 Introductory remarks

The storage of regeneratively produced electric energy in the form of large quantities of hydrogen over a prolonged time is a prerequisite for the all-year operation of an autonomous energy supply system. The economically most efficient form of gas storage is in tanks at high pressure. In this case, the demands made on a suitable compression process are to bring hydrogen and oxygen as the electrolysis product gases efficiently to a high pressure (>100 bar) at low throughput. In the following, several small-scale hydrogen compression processes will be presented. This includes mechanical compressors, which are commercially available, as well as electrochemical and thermal compressors which, at present, are still at the research and development stage. In considering the individual techniques, priority is given to the expenditure of energy. For the individual processes, the specific electric driving power, P_{spec}, and the contribution of compressor power, A_V, to the lower heating value of the hydrogen compressed are evaluated as equation (15),

$$P_{spec} = \frac{P_{el}}{V_{H_2}} \qquad (15)$$

where P_{el} represents the electric power fed to the compressor and V_{H_2} is the compressor throughput. With $\Delta H^0_{H_2}$ being the lower heating value of hydrogen (3.54 kWh m$_n^{-3}$) the compressor power A_V is defined as the dependency equation (16).

$$A_V = \frac{P_{spec}}{\Delta H^0_{H_2}} \qquad (16)$$

3.2 Mechanical compression processes

Mechanical compressors for low throughputs are available in two designs: piston compressor and metal diaphragm compressor. As a rule, both are multistage designs. Whereas the diaphragm compressors are always driven electrically, the piston compressors can be driven either pneumatically or electrically.

3.2.1 Pneumatically driven piston compressors

In pneumatically driven piston compressors the compression energy is obtained from the expansion energy of a working gas. The working gas, generally compressed air from a supply grid, must be available at a working pressure corresponding to the compressor rating. In order to determine the energy consumption of the pneumatic compressor, it is necessary to determine the electrical energy required for providing the working gas. This value is specified as 0.12 kWh m$_n^{-3}$ for air as the working medium in the product information of relevant compressors. Measurements of air consumption, compressed hydrogen amount as well as final pressure at an existing hydrogen storage tank show that the specific driving power increases steeply with rising pressure ratio. In the case of compression with a pressure ratio between 10 and 35, the total energy demand is about 80–97% of the energy content of the hydrogen compressed. For a pressure ratio of 20 the specific driving power is thus 3.05 kWh m$_n^{-3}$ and its contribution to the higher heating value of the hydrogen compressed is 85.5%. A purely thermodynamic consideration of the compression process shows that the theoretical values of A_V should be in the above range of the pressure ratio between 4 and 6%. However, actual consumption is far above that theoretically possible due to the occurrence of friction losses, flow losses on the working gas side and dead volumes in the cylinders.

3.2.2 Electrically driven mechanical compressors

A number of losses can be avoided by directly driving the pistons or the metal diaphragm of a mechanical compressor by an electric motor instead of using the expansion energy of a working medium. Electrically driven compressors therefore show a clearly lower energy demand than those driven pneumatically. According to manufacturer specifications the P_{spec} and A_V values for compression with a pressure ratio of 20 are 0.32 kWh m$_n^{-3}$ and 9% for a four-stage piston compressor. The values slightly decrease to 0.3 kWh m$_n^{-3}$ and 8.6% for a two-stage metal diaphragm compressor. A comparison with the theoretical values for polytropic, multistage compression with a pressure ratio of 20 shows that the losses due to friction and dead volume are also considerable here. The associated values for P_{spec} and A_V are about 0.13 kWh m$_n^{-3}$ and 3.7%.

3.2.3 High-pressure electrolyzers

In high-pressure electrolysis the hydrogen and oxygen product gases are directly produced at high pressure. The actual pressure increase takes place prior to electrolysis by feeding liquid water at the operating pressure of electrolysis. The energy demand for this step is negligible. In the electrolyzer stack itself the water is decomposed at high pressure in the further course of the process. When the electrolysis temperature is low enough, the approximate pressure dependence of the reversible voltage as the isothermal theoretical decomposition voltage, $U_{0,T}$, can be derived from equation (5) as isothermal change of state in the Nernst equation:

$$\begin{aligned} U_{0,T,p} &= U^0_{rev,T} + \frac{RT}{2F} \ln\left(\frac{p}{p_0}\right)^{3/2} \\ &= U^0_{rev,T} + \frac{3RT}{4F} \ln\frac{p}{p_0} \end{aligned} \qquad (17)$$

In equation (17) $U^0_{rev,T}$ is defined by equation (6), p is the pressure of the electrolysis process and p_0 is the standard pressure (1.01325 bar).

According to equation (17) the value of the theoretical decomposition voltage increases with rising pressure. At an operating pressure of 120 bar and the electrolysis temperature of 80 °C $U_{0,T}$ increases by just over 9% to 1.29 V compared to $U^0_{\text{rev},T}$. In the real electrolysis process, this additional energy demand can be nearly compensated by the reduction in electrolyte resistance due to decreasing gas bubble volumes with increasing pressure. The additional energy demand of electrolysis at 120 bar decreases to 0.024 kW m_n^{-3} according to simulation calculations, which corresponds to a contribution of about 0.7% to the higher heating value of the hydrogen produced. It has also been confirmed by operational investigations at a 5 kW high-pressure electrolyzer for pressures of up to 120 bar that the real decomposition voltage is nearly pressure-independent and only influenced by load-dependent inhibitions due to charge carrier transfer, gas diffusion and reaction.[37]

The advantage of high-pressure electrolysis in terms of energy and apparatus is possibly conflicting with safety-engineering considerations at elevated operating pressures. The use of 30% potassium hydroxide (KOH) in advanced electrolyzers as the circulated electrolyte leads to a foreign gas fraction of hydrogen in oxygen and vice versa due to the solubility of the gases. Since the solubility of a gas varies in proportion to pressure, the foreign gas fraction increases with rising operating pressure. Foreign gas fractions are additionally increased by parasitic currents and by the permeation especially of hydrogen through the membrane. The product mixtures established in an optimized system must be checked for their safety-engineering acceptability after operationally relevant determination of the explosion limits.

3.3 Electrochemical compression by polymer electrolyte membrane cells

The setup of an electrochemical compression cell is nearly identical to that of a PEFC. Since the same components and materials are used, the mechanisms are also largely comparable. The application of a voltage leads to proton transport through the polymer membrane, which can be used to compress hydrogen. The reaction equation for this process is

$$H_2(p_1) \longrightarrow 2H^+ + 2e^- \longrightarrow H_2(p_2) \quad (18)$$

Since the membrane is only proton-conducting in the moist state, water must be added to the hydrogen on the anode side. Molecular hydrogen is ionized into protons at the anode while electrons are removed. Due to the potential difference applied, the hydrated protons permeate through the membrane and molecular hydrogen is formed again at the cathode while electrons are supplied. This leads to a net hydrogen flow from the anode to the cathode towards an even higher pressure level.

The cell voltage to be applied for compressing the hydrogen is composed of the thermodynamic potential difference and the effective electrical internal resistance of the cell. The pressure-dependent thermodynamic potential difference follows from the Nernst equation. The effective resistance contains the ohmic losses in the membrane as well as the electrochemical overvoltages at the phase boundaries. Due to the differential-pressure-dependent hydrogen permeation from the cathode back to the anode, further losses occur for which, however, a negligible fraction of specific compression energy is to be applied at technically relevant current densities of 0.5 A cm^{-2}. For this current density the specific throughputs to be expected are in the range of 0.2 l_n (h cm^{-2}). At a pressure ratio of 20 and an effective electrical internal resistance of the cell of 0.36 $\Omega\,cm^2$ the electric demand for electrochemical hydrogen compression (0.55 kWh m_n^{-3}) is about 16% of the higher heating value of the hydrogen compressed.

Since the two loss fractions of permeate backflow and electrical resistance behave complementarily as a function of current density, there exists a current density ($j < 0.2$ A cm^{-2}) at which the electric energy taken up is minimal. Since the hydrogen throughput of electrochemical compression varies in proportion to the current density, it is low in this operating range amounting to $j < 0.2$ A cm^{-2}. Consequently, the cell area must be enlarged. This involves high material expenditure and the introduction of a suitable support structure. However, the specific throughput remains comparatively low so that the process is suitable for applications with low hydrogen throughput.

3.4 Thermal hydrogen compression by metal hydrides

All the processes of hydrogen storage presented above have in common that they need electrical driving energy. Due to their sorption characteristics, metal hydrides are capable of compressing hydrogen using only thermal energy. In this process, the hydride is contained in a reactor which enables good hydrogen supply and removal as well as the most efficient heat exchange possible (cf. Figure 6).

If heat is supplied to a supercharged reactor in equilibrium (see point 1 in Figure 6) in the closed state, hydrogen desorbs and the pressure increases ($1 \rightarrow 2$). When the desired desorption pressure, p_{des}, is reached, the reactor is opened and hydrogen can be given off at constant pressure and further heat supply at temperature T_{des} into a high-pressure storage tank until the state of equilibrium in point 3 is reached ($2 \rightarrow 3$). The reactor is then closed again. Heat supply is interrupted and the reactor cooled. Hydrogen is absorbed due to heat removal, which leads to a drop in pressure ($3 \rightarrow 4$). When the absorption pressure, p_{abs}, is reached, the reactor is opened and supplied from a low-pressure storage tank. The metal hydride takes up

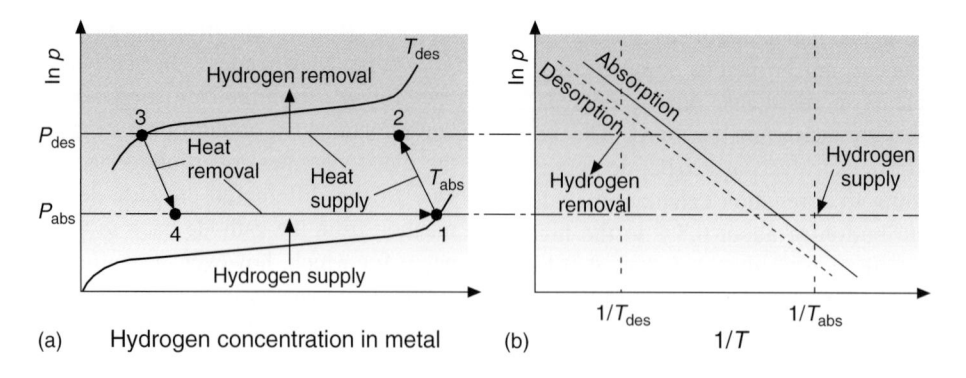

Figure 6. Concentration–pressure isotherm diagram (a) and Van't Hoff diagram (b) of single-stage thermal hydrogen compression.

the hydrogen supplied while continuing to remove heat at temperature T_{abs} until the equilibrium state in point 1 is reached (4 → 1). The cyclic process is then repeated.

With optimized reactors and using titanium- and manganese-based metal hydrides with alloying additions of zirconium, vanadium, iron and nickel, it is possible to achieve a specific maximum throughput of $130\,l_n$ $(h\,kg_{MH})^{-1}$ in the single-stage process at a desorption pressure of 20 bar. At a storage pressure of 120 bar the two-stage process achieves a maximum specific throughput of $23\,l_n\,(h\,kg_{MH})^{-1}$. The desorption times associated with the maximum values of hydrogen throughput are 20 and 50 min. The respective specific energy demand and its contribution to the higher heating value of the hydrogen compressed is $1.5\,kWh\,m_n^{-3}$ and 44% for the operating point with maximum throughput at 20 bar and single-stage process flow as well as $4\,kWh\,m_n^{-3}$ and 117% at 120 bar and two-stage process flow. These operating results determined in[38] show the very high energy demand for thermal compression. However, this is low-temperature heat at about 100 °C, which can be provided by thermal solar collectors or waste heat from industrial processes. Furthermore, the high costs of the complete system are a negative factor. Thermal hydrogen compressors based on metal hydrides offer themselves above all where low throughputs at high final pressures are required and favorable conditions prevail for the provision of low-temperature heat.

4 PROCESS ENGINEERING OF STORAGE SYSTEMS VIA ELECTROLYSIS/FUEL CELLS

In addition to the components described above for energy conversion and energy storage, trouble-free operation as a seasonal energy storage system or a grid-connected energy buffer requires the integration of adapted power electronics and of an optimized energy management system. The power electronics ensures the operation-oriented connection of the energy conversion components to electricity supply and electricity consumption. The energy management system realizes the demand-oriented distribution of energy flows to the storage and consumer paths. A system built up in this way with an energy storage unit composed of electrolyzer, hydrogen tank and fuel cell can only be designed energetically and economically with the aid of a sophisticated design procedure based on dynamic system simulation with validated component models.

4.1 Electrical power preparation system

The electrolyzer and fuel cell are electrically connected to the d.c. busbar via adapted d.c. power controllers. If the voltage level of the d.c. busbar and battery is higher than that of the fuel cell and electrolyzer, the electrolyzer is connected via a down-converter and the fuel cell via an up-converter. Both converters should be of modular design so that the individual modules can be connected and disconnected depending on requirements. This provides an almost homogeneous efficiency over the entire power range of the converter. Taking internal consumptions into account, the efficiency of a down-converter is in the range of 88–89% and that of an up-converter 90–91%.

A low-cost and simple alternative to coupling the electrolyzer and fuel cell to the system with d.c. busbar is the use of circuit breakers with semiconductor elements. The prerequisite for trouble-free operation is a mutual adaptation of the components connected to each other by selecting the respective optimum number of cells. The optimization criterion is the currents between the components which should not exceed predefined limit values. A switch unit replacing the d.c. power controller consists of one or several high-current field-effect transistors. These are intelligent semiconductor switches, so-called smart power switches, equipped with integrated protection and driving circuitry.

Non-steady-state operating regimes of the electrolyzer or fuel cell are realized by a control of the current average values by means of pulse width modulation. The associated frequent switching operations are harmless for semiconductor switches in contrast to mechanical switches.

Alternating current coupling combines high system efficiency with high flexibility in dimensioning the components. Coupling with the alternating current (a.c.) busbar is effected via a rectifier for the electrolyzer and via an inverter for the fuel cell. Both electric converters are characterized by a high efficiency of 90–98% and a wide range of commercially available units.

4.2 Energy management system

The highest level of controlling a system with hydrogen energy storage is called energy management. It monitors all energy flows and states of storage charge and predefines the power requirements for fuel cell and electrolyzer. This ensures that energy deficits are covered by the fuel cell at any time. This means that the fuel cell is constantly in operation and only on irradiance-rich days in the stand-by mode without power output. The electrolyzer is also in operation throughout the year.[20] If in addition to hydrogen storage a battery is also integrated into the system for short-term storage, the main control parameter used is the state of battery charge (SOC). The energy management system can thus be used to achieve high system efficiency and a long service life of the electrochemical components.[21]

An energy management system optimized in this way predefines power levels and states of components so that the complete system is always in a reliable operating state. The SOC is kept within predefined limits by the energy management system predefining the operating states of electrolyzer and fuel cell. The effect of energy management on battery, electrolyzer, fuel cell and hydrogen tank is demonstrated in Figure 7.

The battery charge profile in Figure 7(a) shows that the energy management system only allows values between 30 and 80%. Higher states of charge significantly deteriorate the efficiency of the battery due to the occurrence of gassing losses. Exhaustive discharges reduce the service life of the electricity storage unit. Whenever minimum and maximum values of battery charge are reached, the energy management system connects the fuel cell (30%) or the electrolyzer (80%). The disconnecting points are selected here as 5% hysteresis at 35% for the fuel cell and at 75% for the electrolyzer. Depending on the yield profile, the switching points should be adapted aiming at a minimum number of switching operations of the electrolyzer and fuel cell. The switching thresholds kept constant here over the year can be varied to optimize operation as a function of time. The electrolysis is operated either at constant load (see Figure 7b) or in its permissible power range so that the battery current becomes zero. The power thus "floats" according to the actual surplus, one also speaks of "float" operation. The power of the fuel cell is generally set to a constant value (compare Figure 7b).

The essential condition which must be fulfilled for achieving system autonomy is an at least equal filling level of the hydrogen tank at the end of the period under consideration in comparison to the starting point. In addition, the tank must never be empty. Figure 7(c) shows that the tank as well as the fuel cell and electrolyzer have been optimally dimensioned. The tank is half filled at the beginning of the year and this amount is sufficient for the remaining fuel cell season. In the electrolyzer season just the correct amount of hydrogen was produced which ensured that the tank reached its maximum pressure only

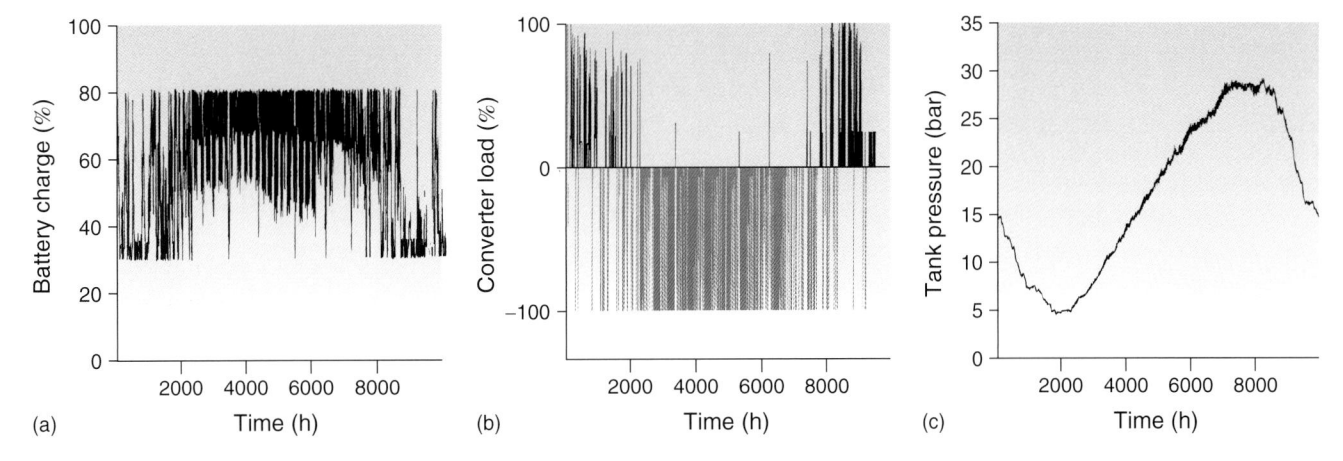

Figure 7. (a) Annual profiles of the state of battery charge. (b) Operating states of electrolyzer (negative converter load) and (c) fuel cell (positive converter load) H_2 tank pressure.

temporarily. At the end of the year the initial tank status is reached again.

In order to protect the battery against excessively high states of charge and the associated losses due to gassing, two additional switching thresholds are provided above the described switching threshold for regular electrolysis operation. In the case of an excessively high SOC, the electrolyzer is run at full load until the adjusted charge hysteresis is passed through. If for reasons of cost the electrolyzer is not designed for the peak load of the generators, this surplus is fed into the possibly existing electricity grid or the generators are regulated down to protect the battery against full charge. For the fuel cell there are also two modes of operation: normally, it is operated at high efficiency in the partial load range. If the SOC is markedly too low, the battery is protected against prolonged exhaustive discharge by full load operation of the fuel cell.

The energy management system also regulates the pressure of gases in the buffer storage tanks by connecting or disconnecting the mechanical compressor. In addition, it performs monitoring and control functions.

4.3 Process and system analysis

With the methods of process and system analysis it is possible to simulate and evaluate a complex system under realistic conditions. As a simulation result, on the one hand, the energy balance over the period under consideration is obtained and, on the other hand, information about the dimensions and thus costs of the integrated components is provided.

4.3.1 Energy balance and system efficiency

The base case shown in Figure 8 is assumed in order to evaluate the energy balance of the energy storage unit with electrolysis and fuel cell. This concept is based on the following features:

1. d.c. busbar permits coupling to electricity production, short-term storage and current collection;
2. electrolyzer and fuel cell are coupled to d.c. busbar via d.c. power controllers;
3. low-pressure electrolyzer with maximum operating pressure of 7 bar;
4. process gas compressors are supplied with electric energy from the system;
5. fuel cell for cathodic oxygen supply;
6. waste heat utilization of the fuel cell is taken into account.

Figure 8 shows the configuration of a hydrogen energy storage system for the base case with the efficiencies of the individual components as well as the incoming and outgoing energy flows in units of energy (UE). A total of 90.7 of 100 UE are passed to the down-converter in the form of electric d.c. current for hydrogen and oxygen production. The compressors for hydrogen and oxygen need 9.3 UE to compress the gases. This corresponds to a contribution of 13.5% to the heating value of the hydrogen compressed. The efficiency data for the tanks describe the losses over one year caused by leakages. The comparatively high efficiency of the fuel cell is a consequence of H_2/O_2 operation. The use of waste heat from the fuel cell delivers 26.7 UE to a low-temperature application. Ultimately, 32.7 of 100 UE fed in on an annual average can be used in the form of direct current.

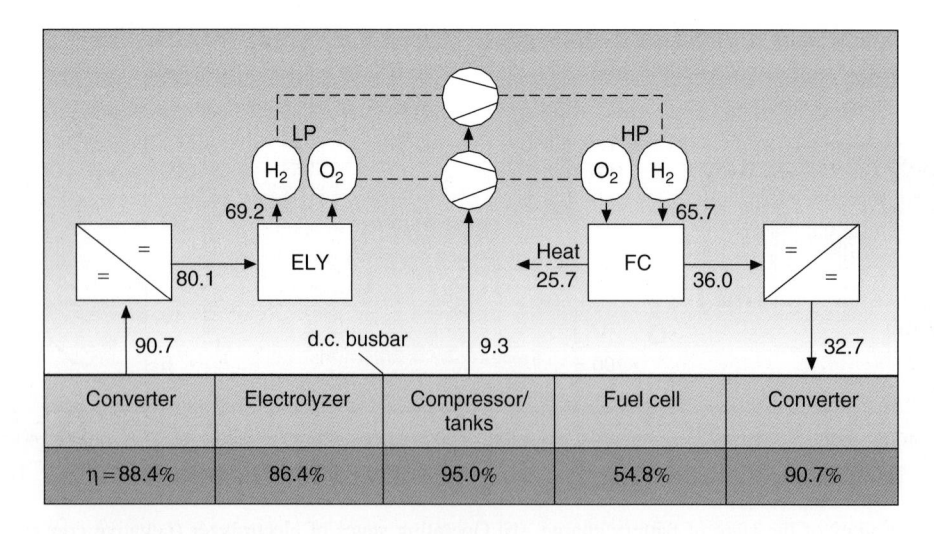

Figure 8. Energy balance of a hydrogen energy storage system.

Alternative concepts to the base case discussed above are considered to be the integration of a high-pressure electrolyzer, the omission of storage and the use of oxygen as well as the replacement of the d.c. power controllers by semiconductor circuit breakers.

In comparison to the base configuration, the system with high-pressure electrolyzer differs from the energy and control perspective by the different efficiency of the electrolyzer and the omission of the buffer tanks and compressors. With an annual working efficiency of 87.7% the annual yield increases to 36.6 UE in the case of feeding 100 UE into the energy storage system with high-pressure electrolyzer. If air is used instead of pure oxygen in the fuel cell, this leads to a drastic increase of the amounts of gas to be passed through the fuel cell. Due to the operating overpressure on the air side and the pressure loss in the gas diffusers, this means a considerable energy demand. Moreover, the achievable efficiency of the fuel cell is lower. Depending on the mode of operation and the fuel cell concept, an efficiency decrease by 30% must be expected. This gives an efficiency of 38.4% of net electricity production for the fuel cell. Ultimately, the savings in investment costs by omitting an oxygen path are faced with a reduction in energy yield of the energy storage system with H_2/air fuel cell. If 100 UE are annually transferred to the storage unit, the system only provides 22.9 UE of electricity. Whether the higher fraction of heat produced can be used will depend on both the apparatus expenditure and the mode of operation. As an optimization approach under development it is proposed to use novel cyclically clocked electronic circuit breakers – smart power switches – instead of the d.c. power controllers employed for coupling the electrolyzer and fuel cell to the d.c. busbar. Apart from low investment costs due to mass production, wear-free switching and high reliability, the semiconductor switches composed of high-current field-effect transistors are characterized by a high efficiency of about 99%. An energy storage system with electrolyzer designed for the voltage level of the battery and correspondingly adapted fuel cell coupled to a d.c. busbar via these switches achieves an annual energy yield of 39.5 UE for feeding 100 UE.

4.3.2 Cost analysis

According to the general cost models, the total cost, C, of a system is determined by adding the investment costs, C_{inv}, to the operating costs, C_{op}, over 20 years of operation (cf. also equation (19)).

$$C = C_{inv} + C_{op} \quad (19)$$

For a simple analysis, the costs may be regarded as invariable over the period under consideration, i.e., all costs are assumed for a fixed point in time without taking interest and price variations over the operating period into consideration. In order to determine the investment costs of an energy storage system with electrolyzer, gas tank and fuel cell, first, the costs of these components are summed up as shown in equation (20)

$$C_{comp} = C_{ELY} + C_{TANK} + C_{FC} \quad (20)$$

The individual component costs are calculated from the design data and the respective specific costs. For the electrolyzer and fuel cell the design variable is the installed electrical output and the specific costs are related to this output. In contrast, the design variable for the tank is the hydrogen capacity, which is proportional to a specific energy content. The specific costs of the tanks for hydrogen and oxygen are correspondingly related to the energy content of the storable hydrogen. The component costs thus determined are provided with an allowance for investment costs for additional mounting expenditures, $f_{mount} = 1.2$, and an allowance for engineering, documentation, mounting, maintenance, disassembly and disposal, $f_{other} = 2.12$.

$$C_{inv} = C_{comp} f_{mount} f_{other} \quad (21)$$

The investment costs (cf. equation 21) cover all costs for the system over the entire service life inasmuch as they can be directly apportioned among the design data for the most important components. Added to these costs are the operating costs, C_{op}, which result from the energy balance of system operation over many years. For a hydrogen storage system this means consideration of the costs C_{ADD} for the purchase of additional hydrogen and oxygen in the case of empty tanks. These costs can be calculated according to equation (22) from the costs of hydrogen and oxygen relative to the H_2 energy content and from the energy content of additionally purchased hydrogen:

$$C_{op} = C_{ADD} = k_{H_2} E_{ADD} \quad (22)$$

For the base case of an energy storage system described in Section 4.3.1 and on the assumption of the design variables described in Ref. [20] and specified in Table 3 with the associated specific costs for the components of a seasonal hydrogen storage system, the cost allocation shown in Figure 9 is obtained.

More than three-quarters of the component costs (39%) are spent on the electrolyzer (15%) and the fuel cell (15%). The other investment costs represent 59% of the total cost. Since these costs are proportional to the component costs, the electrolyzer and fuel cell have by far the greatest influence on the total cost. One of the most important tasks in optimizing energy storage systems with electrolyzer and

Table 3. Component costs for a seasonal hydrogen storage system.

Description	Design variables	Specific costs	Component costs
H_2 and O_2 gas storage including compressors	$C_{TANK} = 9040\, m_n^3 H_2$	$k_{TANK} = 11\, €\, kW^{-1}$	$K_{TANK} = 350\,000\, €$
Electrolyzer including peripherals and down-converter	$P_{ELY} = 50\, kW$	$k_{ELY} = 12\,500\, €\, kW^{-1}$	$K_{ELY} = 625\,000\, €$
Fuel cell including peripherals and up-converter	$P_{BZ} = 30\, kW$	$k_{BZ} = 20\,000\, €\, kW^{-1}$	$K_{BZ} = 600\,000\, €$

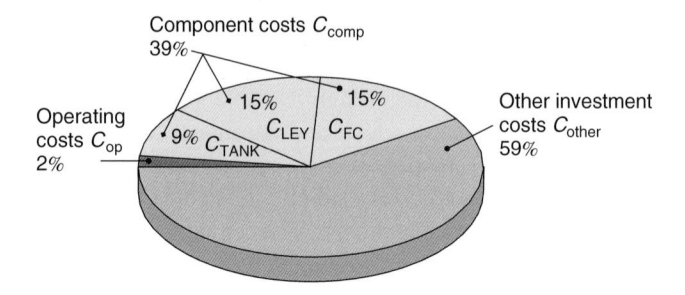

Figure 9. Cost allocation for a seasonal hydrogen storage system.

fuel cell is therefore to further reduce the costs of these two components. Specific target costs of 100 €/kW for fuel cell stacks with polymer membrane have to be obtained, assuming 500–750 €/kW for the total system. This would reult in a total cost of 20 000 €/kW, including service costs over a period of 20 years.

REFERENCES

1. H. Barthels, W. A. Brocke, H.-G. Groehn and P. Ritzenhoff, in 'Proceedings of ISES Solar World Congress', Budapest, Hungary, August 23–27, Vol. 2, pp. 421–430 (1993).
2. H. Barthels, W. A. Brocke, H.-G. Groehn, H. Mai, J. Mergel and P. Ritzenhoff, in 'Proceedings of 12th European Photovoltaic Solar Energy Conference', Amsterdam, The Netherlands, April, 11–15, pp. 439–443 (1994).
3. H. Barthels, W. A. Brocke, K. Bonhoff, H.-G. Groehn, G. Heuts, M. Lennartz, H. Mai, L. Schmid, J. Mergel and P. Ritzenhoff, *Int. J. Hydrogen Energy*, **23**, 295 (1998).
4. C. Meurer, H. Barthels, W. A. Brocke, B. Emonts and H.-G. Groehn, *Int. J. Hydrogen Energy*, **67**, 131 (1999).
5. A. Szyska, *Int. J. Hydrogen Energy*, **23**, 849 (1998).
6. A. Abaoud and H. Steeb, *Int. J. Hydrogen Energy*, **23**, 445 (1998).
7. W. Stahl, A. Goetzberger and K. Voss, 'Das energieautarke Solarhaus', C. F. Müller (Ed), Verlag, Heidelberg, pp. 66–80 (1997).
8. S. Galli and M. Stefanoni, *Int. J. Hydrogen Energy*, **22**, 453 (1997).
9. S. Galli, M. Stefanoni, K. Havre, P. Borg, W. A. Brocke and J. Mergel, in 'Proceedings of 11th World Hydrogen Energy Conference', Stuttgart, Germany, June 23–28 (1996).
10. P. A. Lehman, C. E. Chamberlin, G. Pauletto and M. A. Rocheleau, *Int. J. Hydrogen Energy*, **22**, 465 (1997).
11. J. P. Vanhanen, P. D. Lund and J. S. Tolonen, *Int. J. Hydrogen Energy*, **23**, 267 (1998).
12. J. Winter, 'Wasserstoff als Energieträger: Technik, Systeme, Wirtschaft', Springer-Verlag, Berlin, pp. 205–243 (1989).
13. R. Pötter, *Fotschrittsberichte VDI*, **6**, 376 (1997).
14. C. Müller, 'Interaktives Modell für den Betrieb von Photovoltaikanlagen mit Energiespeicherpfaden', PhD Thesis, GH Duisburg, Germany.
15. P. Beckhaus, G. Krost and C. Müller, 'Simulation Tool for Planning and Verification of Autonomous Photovoltaic Systems', Presented at 33rd Universities Power Engineering Conference, Edinburgh (1998).
16. D. Heinemann, 'Zur Steuerung regenerativer Energieversorgungssysteme unter Verwendung von Energiewettervorhersagen', PhD Thesis, University of Oldenburg, Germany (1990).
17. H. G. Beyer, 'Zur Bestimmung des energetischen Verhaltens regenerativer Elektrizitätsversorgungssysteme unter besonderer Berücksichtigung statistischer Charakteristiken des Windes und der Solarstrahlung', PhD Thesis, University of Oldenburg, Germany (1988).
18. P. Ritzenhoff, *Ber. Forschungs. Jülich*, 3150 (1995).
19. O. Ulleberg, 'Stand-Alone Power Systems for the Future', ITEV Rapport 1998, p. 11 (1998).
20. H. C. Meurer, *Ber. Forschungs. Jülich*, 3793 (2000).
21. W. A. Brocke and H. Barthels, in 'Proceedings of the ISES Solar World Congress', Budapest, Hungary, August 23–27, Vol. 2, p. 431 (1993).
22. K. Kordesch, 'Electrochemical Energy Storage', in "Comprehensive Treatise of Electrochemistry", Vol. 3, J. O'M. Bockris, B. E. Conway, E. Yeager and R. E. White (Eds), Plenum Press, New York, pp. 124 (1981).
23. X. Ren, W. Henderson and S. Gottesfeld, *J. Electrochem. Soc.*, **144**, L267 (1997).
24. R. L. LeRoy, Ch. T. Bowen and D. J. LeRoy, *J. Electrochem. Soc.*, **127**, 1954 (1980).
25. K. Andreassen, 'Hydrogen Production by Electrolysis', in "Hydrogen Power: Theoretical and Engineering Solutions", T. O. Saetre, (Ed), Kluwer, Dordrecht, p. 91 (1998).

26. Ph. Vermeiren, W. Adriansens, J. P. Moreels and R. Leyesn, *Int. J. Hydrogen Energy*, **23**, 321 (1998).
27. C. A. Linkous, H. R. Anderson, R. W. Kopitzke and G. L. Nelson, *Int. J. Hydrogen Energy*, **23**, 525 (1998).
28. S. Stucki, G. Theis, R. Kötz, H. Devantay and H. J. Christen, *J. Electrochem. Soc.*, **132**, 367 (1985).
29. J. Kim, S.-M. Lee, S. Srinivasan and C. E. Chamberlin, *J. Electrochem. Soc.*, **142**, 2670 (1995).
30. G. Squadrito, G. Maggio, E. Passalacqua, F. Lufrano and A. Patti, *J. Appl. Electrochem.*, **29**, 1449 (1999).
31. A. Winsel, in 'Elektrochemische Energieumwandlung einschließlich Speicherung', DECHEMA-Monographien, Vol. 92, Verlag Chemie, Weinheim, p. 21 (1982).
32. K. Ledjeff, A. Heinzel, V. Peinecke and F. Mahlendorf, *Int. J. Hydrogen Energy*, **19**, 453 (1994).
33. D. O. Ham, G. Moniz and E. J. Taylor, *J. Power Sources*, **22**, 409 (1988).
34. L. Sweete, N. Kackley and S. A. McCatty, *J. Power Sources*, **36**, 323 (1991).
35. R. Baldwin, M. Pham, A. Leonida, J. McElroy and T. Nalette, *J. Power Sources*, **29**, 399 (1990).
36. J. Ahn and R. Holze, *J. Appl. Electrochem.*, **22**, 1167 (1992).
37. H. Janßen, B. Emonts, H.-G. Groehn, H. Mai, R. Reichel and D. Stolten, 'High Pressure Electrolysis – The Key Technology for Efficient H_2 Production', Presented at HYPOTHESIS IV, Stralsund, Germany, September 9–14 (2001).
38. K. Bonhoff, *Ber. Forschungs. Jülich*, 3571 (1998).

Contents for Volumes 2, 3 and 4

VOLUME 2: Electrocatalysis

Contributors to Volume 2	vii
Foreword	xi
Preface	xv
Abbreviations and Acronyms	xvii

Part 1: Introduction — 1

1. What is electrocatalysis? — 3
   W. Vielstich
2. The role of adsorption — 4
   C. Sánchez/E. Leiva
3. Understanding electrocatalysis: From reaction steps to first-principles calculations — 14
   C. Sánchez/E. Leiva
4. Electrode potential as parameter — 36
   C. Sánchez/E. Leiva
5. Catalysis by UPD metals — 47
   C. Sánchez/E. Leiva
6. Outer sphere reactions — 62
   C. Sánchez/E. Leiva
7. The NEMCA effect — 65
   C. Sánchez/E. Leiva

Part 2: Theory of electrocatalysis — 69

8. Electrode potential and double layer — 71
   S. Trasatti
9. Reaction mechanism and rate determining steps — 79
   S. Trasatti
10. Adsorption – Volcano curve — 88
    S. Trasatti
11. Theoretical aspects of some prototypical fuel cell reactions — 93
    E. Leiva/C. Sánchez
12. Theory of electrochemical outer sphere reactions — 132
    E. Leiva/C. Sánchez
13. Theory of the NEMCA effect — 145
    E. Leiva/C. Sánchez

Part 3: Methods in electrocatalysis — 151

14. Cyclic voltammetry — 153
    W. Vielstich
15. Product analysis — 163
    R. M. Torresi/S. Wasmus
16. Vibrational spectroscopy — 191
    A. Rodes/J. M. Pérez/A. Aldaz
17. Electrochemical impedance spectroscopy — 220
    E. Ivers-Tiffée/A. Weber/H. Schichlein
18. Ex-situ surface preparation and analysis: Transfer between UHV and electrochemical cell — 236
    H. E. Hoster/H. A. Gasteiger
19. Structure sensitive methods: AFM/STM — 266
    L. A. Kibler/D. M. Kolb
20. EXAFS, XANES, SXS — 279
    R. R. Adžić/J. X. Wang/B. M. Ocko/J. McBreen
21. Normalization of porous active surfaces — 302
    F. C. Nart/W. Vielstich
22. Rotating thin-film method for supported catalysts — 316
    T. J. Schmidt/H. A. Gasteiger
23. Combinatorial catalyst development methods — 334
    T. E. Mallouk/E. S. Smotkin
24. Numerical simulations of electrocatalytic processes — 348
    M. T. M. Koper

Part 4: The hydrogen oxidation/evolution reaction — 359

25. Reaction mechanisms of the H_2 oxidation/evolution reaction — 361
    M. W. Breiter
26. The hydrogen electrode reaction and the electrooxidation of CO and H_2/CO mixtures on well-characterized Pt and Pt-bimetallic surfaces — 368
    N. M. Marković
27. Oxidation reactions in high-temperature fuel cells — 394
    A. J. McEvoy
28. New CO-tolerant catalyst concepts — 408
    M. Watanabe
29. Hydrogen evolution reaction — 416
    A. Lasia

Part 5: The oxygen reduction/evolution reaction — 441

30. Reaction mechanisms of the O_2 reduction/evolution reaction — 443
    M. Gattrell/B. MacDougall

434   Contents for Volumes 2, 3 and 4

31 Oxygen reduction reaction on smooth single crystal electrodes   465
    P. N. Ross Jr.
	43 Methanol and formic acid oxidation on ad-metal modified electrodes 635
	    P. Waszczuk/A. Crown/S. Mitrovski/A. Wieckowski

32 O_2 reduction on the Pt/polymerelectrolyte interface   481
    K. Ota/S. Mitsushima
	44 Methanol effects on the O_2 reduction reaction 652
	    S. Gilman/D. Chu

33 Fundamental kinetics/transport processes in MEAs   490
    E. A. Ticianelli/E. R. Gonzalez
	45 Oxidation of C_2 molecules 662
	    O. A. Petrii

34 O_2 reduction and structure-related parameters for supported catalysts   502
    S. Mukerjee/S. Srinivasan
	46 Oscillations and other dynamic instabilities 679
	    K. Krischer/H. Varela

35 Oxide-based ORR catalysts 520 *J. B. Goodenough/B. L. Cushing*	
36 Chevrel phases and chalcogenides 534 *N. Alonso-Vante*	

Part 7: Other energy conversion related topics   703

47 Hydrogenation reactions   705
    E. Pastor/J. L. Rodríguez
37 Macrocycles   544
    J. H. Zagal
48 CO_2-reduction, catalyzed by metal electrodes   720
    Y. Hori
38 Poisons for the O_2 reduction reaction   555
    U. A. Paulus/T. J. Schmidt/H. A. Gasteiger
49 Electrochemical supercapacitors and their complementarity to fuel cells and batteries   734
    B. E. Conway
39 O_2-reduction at high temperature: MCFC   570
    K. Hemmes/J. R. Selman
40 O_2-reduction at high temperatures: SOFC   587
    E. Ivers-Tiffée/A. Weber/H. Schichlein
50 Technical characteristics of PEM electrochemical capacitors   747
    J. A. Kosek/B. M. Dweik/A. B. LaConti

Part 6: Oxidation of small organic molecules   601

41 Methanol and CO electrooxidation   603
    T. Iwasita

Contents for Volumes 1, 3 and 4   761

Subject Index   767

42 Formic acid oxidation   625
    J. M. Feliu/E. Herrero

VOLUME 3: Fuel Cell Technology and Applications, Part 1

Contributors to Volumes 3 and 4	**ix**
Foreword	**xv**
Preface	**xix**
Abbreviations and Acronyms	**xxi**

Part 1: Sustainable energy supply **1**	**Part 2: Hydrogen storage and hydrogen generation** **79**
1 Alternative fuels and prospects – Overview 3 *J. M. Ogden*	***Development prospects for hydrogen storage*** ***81***

2 Natural gas for power generation and the automotive market   25
    T. Theisen
	6 High pressure storage 83
	    R. Funck

3 Methanol from fossil and renewable resources   39
    R. Edinger/G. Isenberg/B. Höhlein
	7 Liquid hydrogen technology for vehicles 89
	    J. Wolf

4 Synthetic hydrocarbons as long-term fuel option   49
    J. G. Price/B. Jager/L. Dancuart/M. J. Keyser/J. van der Walt
	8 Hydride storage 101
	    G. Sandrock

5 Solar and wind energy coupled with electrolysis and fuel cells 62 *R. Wurster/J. Schindler*	

Chemical hydrogen storage devices **113**

9 Aqueous borohydride solutions 115
    S. Suda
10 Ammonia crackers 121
    V. Hacker/K. Kordesch

Reforming of methanol and fuel processor development **129**

11 Catalyst development and kinetics for methanol fuel processing 131
    B. A. Peppley/J. C. Amphlett/R. F. Mann
12 Methanol reformer design considerations 141
    J. B. Hansen
13 Mixed POX/steam-reforming reactor design considerations 149
    J. Reinkingh/M. Petch

Fuel processing from hydrocarbons to hydrogen **157**

14 Steam reforming, ATR, partial oxidation: catalysts and reaction engineering 159
    J. R. Rostrup-Nielsen/K. Aasberg-Petersen
15 Sulfur removal methods 177
    A. Hidalgo-Vivas/B. H. Cooper
16 Catalyst development for water-gas shift 190
    J. R. Ladebeck/J. P. Wagner
17 Membrane reactor concepts 202
    S. Wieland/T. Melin
18 PROX catalysts 211
    L. Shore/R. J. Farrauto
19 Autothermal reforming 219
    F. Baumann/S. Wieland/M. Himmen/D. W. Agar
20 Alternative design possibilities for integrated fuel processors 229
    G. E. Voecks

Well-to-wheel efficiencies **243**

21 Well-to-wheel efficiencies of different fuel choices 245
    B. Höhlein/G. Isenberg/R. Edinger/T. Grube

Hydrogen safety, codes and standards **255**

22 Hydrogen safety, codes and standards for vehicles and stationary applications 257
    G. Scheffler/R. Wurster/J. Schindler

Part 3: Polymer electrolyte membrane fuel cells and systems (PEMFC) 269

Bipolar plate materials and flow field design **271**

23 Basic materials corrosion issues 273
    D. A. Shores/G. A. Deluga
24 Performance and durability of bipolar plate materials 286
    G. O. Mepsted/J. M. Moore
25 Metal bipolar plates and coatings 294
    J. Wind/A. LaCroix/S. Braeuninger/P. Hedrich/C. Heller/M. Schudy
26 Graphite-based bipolar plates 308
    K. Roßberg/V. Trapp
27 Serpentine flow field design 315
    D. P. Wilkinson/O. Vanderleeden

28 Interdigitated flow field design 325
    T. V. Nguyen/W. He
29 Two-phase flow and transport 337
    C.-Y. Wang

Membrane materials **349**

30 Perfluorinated membranes 351
    M. Doyle/G. Rajendran
31 First principles modeling of sulfonic acid based ionomer membranes 396
    S. J. Paddison
32 Composite perfluorinate membranes 412
    M. Nakao/M. Yoshitake
33 Hydrocarbon membranes 420
    K. D. Kreuer
34 High-temperature membranes 436
    J. S. Wainright/M. H. Litt/R. F. Savinell
35 Inorganic/organic composite membranes 447
    D. J. Jones/J. Rozière
36 Membrane/electrode additives for low-humidification operation 456
    J.-C. Lin/H. R. Kunz/J. M. Fenton

Electro-catalysts **465**

37 Pt alloys as oxygen reduction catalysts 467
    D. Thompsett
38 High dispersion catalysts including novel carbon supports 481
    T. Tada
39 Development of CO-tolerant catalysts 489
    K. Ruth/M. Vogt/R. Zuber
40 Manufacture of electrocatalyst powders by a spray-based production platform 497
    M. Hampden-Smith/P. Atanassova/P. Atanassov/T. Kodas
41 Precious metal supply requirements 509
    C. Jaffray/G. Hards

Membrane-electrode-assembly (MEA) **515**

42 Diffusion media materials and characterisation 517
    M. F. Mathias/J. Roth/J. Fleming/W. Lehnert
43 Principles of MEA preparation 538
    S. S. Kocha
44 Catalyst coated composite membranes 566
    S. Cleghorn/J. Kolde/W. Liu
45 Novel catalysts, catalysts support and catalysts coated membrane methods 576
    M. K. Debe

State-of-the-art performance and durability **591**

46 Beginning-of-life MEA performance – Efficiency loss contributions 593
    H. A. Gasteiger/W. Gu/R. Makharia/M. F. Mathias/B. Sompalli
47 Durability 611
    D. P. Wilkinson/J. St-Pierre
48 Effect of ionic contaminants 627
    T. Okada
49 Mechanisms of membrane degradation 647
    A. B. LaConti/M. Hamdan/R. C. McDonald
50 Reliability issues and voltage degradation 663
    M. Fowler/R. F. Mann/J. C. Amphlett/B. A. Peppley/P. R. Roberge

VOLUME 4: Fuel Cell Technology and Applications, Part 2

Contributors to Volumes 3 and 4	ix
Foreword	xv
Preface	xix
Abbreviations and Acronyms	xxi

Part 3: Polymer electrolyte membrane fuel cells and systems (PEMFC) — 679

(Continued from previous volume)

System design and system-specific aspects — 681

51 System design for stationary power generation — 683
    F. Barbir
52 System design for vehicle applications: DaimlerChrysler — 693
    G. Konrad/M. Sommer/B. Loschko/A. Schell/A. Docter
53 System design for vehicle applications: GM/Opel — 714
    D. A. Masten/A. D. Bosco

Air-supply components — 725

54 Air-supply components — 727
    S. Pischinger/O. Lang

Applications based on PEM-technology — 743

55 Special applications using PEM-technology — 745
    A. B. LaConti/L. Swette

Part 4: Alkaline fuel cells and systems (AFC) — 763

56 Stack materials and design — 765
    K. Kordesch/V. Hacker
57 System design and applications — 774
    K. Strasser
58 A comparison between the alkaline fuel cell (AFC) and the polymer electrolyte membrane (PEM) fuel cell — 789
    K. Kordesch/M. Cifrain

Part 5: Phosphoric acid fuel cells and systems (PAFC) — 795

59 Stack materials and stack design — 797
    R. D. Breault
60 Catalyst studies and coating technologies — 811
    D. A. Landsman/F. J. Luczak
61 Experience with 200 kW PC25 fuel cell power plant — 832
    J. M. King/B. McDonald

Part 6: Direct methanol fuel cells and systems (DMFC) — 845

62 Transport/kinetic limitations and efficiency losses — 847
    J. Müller/G. Frank/K. Colbow/D. Wilkinson
63 New materials for DMFC MEAs — 856
    M. Neergat/K. A. Friedrich/U. Stimming
64 System design for transport applications — 878
    A. Lamm/J. Müller
65 DMFC system design for portable applications — 894
    S. R. Narayanan/T. I. Valdez/N. Rohatgi

Part 7: Molten carbonate fuel cells and systems (MCFC) — 905

66 Stack material and stack design — 907
    Y. Mugikura
67 Electrolyte and material challenges — 921
    J. Hoffmann/C.-Y. Yuh/A. Godula Jopek
68 System design — 942
    M. Farooque/H. Ghezel-Ayagh
69 Durability — 969
    Y. Fujita

Part 8: Solid oxide fuel cells and systems (SOFC) — 983

Materials — 985

70 Current electrolytes and catalysts — 987
    T. Kawada/J. Mizusaki
71 Low temperature electrolytes and catalysts — 1002
    O. Yamamoto
72 MEA/cell preparation methods: Europe/USA — 1015
    D. Stöver/H. P. Buchkremer/J. P. P. Huijsmans
73 MEA/cell preparation methods: Japan/Asia — 1032
    M. Suzuki
74 Interconnects — 1037
    K. Hilpert/W. J. Quadakkers/L. Singheiser

Stack and system design — 1055

75 Internal reforming — 1057
    K. Eguchi
76 System design — 1070
    J. J. Hartvigsen/S. Elangovan/A. C. Khandkar

New concepts — 1087

77 New microtube concepts — 1089
    K. Kendall
78 Direct hydrocarbon SOFCs — 1098
    S. A. Barnett
79 Novel electrolytes operating at 400–600 C — 1109
    T. Ishihara

Part 9: Primary and secondary metal/air cells   1123

80 Alkaline methanol/air power devices   1125
    G. A. Koscher/K. Kordesch

Part 10: Portable fuel cell systems   1131

81 Portable direct methanol fuel cell systems   1133
    S. R. Narayanan/T. I. Valdez
82 Portable PEM systems   1142
    A. Heinzel/C. Hebling
83 Small-size PEM systems for special applications   1152
    F. N. Büchi

Part 11: Current fuel cell propulsion systems   1163

PEM fuel cell systems for cars/buses   *1165*

84 DaimlerChrysler fuel cell activities   1167
    H.-P. Schmid/J. Ebner
85 General Motors/OPEL fuel cell activities – Driving towards a successful future   1172
    A. Rodrigues/M. Fronk/B. McCormick
86 Honda fuel cell activities   1180
    S. Matsuo
87 Hy.Power – A technology platform combining a fuel cell system and a supercapacitor   1184
    P. Dietrich/F. Büchi/A. Tsukada/M. Bärtschi/R. Kötz/
    G. G. Scherer/P. Rodatz/O. Garcia/M. Ruge/M. Wollenberg/
    P. Lück/A. Wiartalla/C. Schönfelder/A. Schneuwly/P. Barrade

PEM fuel cell systems for submarines   *1199*

88 H_2/O_2-PEM-fuel cell module for an air independent propulsion system in a submarine   1201
    K. Strasser

AFC fuel cell systems   *1215*

89 Automotive development   1217
    V. Hacker/K. Kordesch
90 Space-shuttle fuel cell   1224
    H. A. Wagner

Part 12: Electric utility fuel cell systems   1231

91 PEMFC fuel cell systems   1233
    J. Garche/L. Jörissen
92 MCFC fuel cell systems   1260
    M. Bischoff/M. Farooque/S. Satou/A. Torazza
93 SOFC fuel cell systems   1276
    S. E. Veyo/S. Fukuda/L. A. Shockling/W. L. Lundberg

Part 13: Future prospects of fuel cell systems   1291

94 Life-cycle analysis of fuel cell system components   1293
    M. Pehnt
95 Market concepts, competing technologies and cost challenges for automotive and stationary applications   1318
    T. Lipman/D. Sperling
96 Potential economic impact of fuel cell technologies   1329
    E. Jochem/E. Schirrmeister

Contents for Volumes 1 and 2   1337

Subject Index   1341

Subject Index

1880–1950's German R&D 181–2
1930's research 168–73
1940's research 168–73
1950's research 173–94
1960's research 179–94
1970's Japanese R&D 188
1991 Japanese R&D 189
acid fuel cells (ACFC) 47–51, 195–7
acid reactions 231–4
acidic electrolytes 419–20, 409–10
activation 411–2
adsorption 17–18
AEG–TELEFUNKEN 186
AER Energy Resources 403
AESI *see* Apollo Energy Systems Inc.
AFC *see* alkaline fuel cells
air...
   conditioning 349
   cooled cells 134–42
   electrodes 164–7, 388–91, 397
   feed 74–5
   fuel cells 267–79
   sintering processes 253–6
alcohol fuel cells 323–33
aldehydes 332–3
aliphatic...
   ethers 332–3
   monoalcohols 327–8
   oxygenated compounds 324–30
   polyols 328–9
alkaline electrolytes
   aluminum oxidation 409–10
   carbonates 205, 232–3
   energy storage 419–21
   hydrazine fuel cells 283–5
   hydrogen/oxygen (air) fuel cells 267–79
   research discoveries 163
alkaline fuel cells (AFC)
   energy storage 423
   kinetics 37
   principles 267–79
   R&D 180, 194–5
   thermodynamics 37
alkaline water electrolysis 420–1
Allis-Chalmers Company 270, 281, 283
Alsthom 283
aluminum...
   air cells 409–14
   alloys 411–2
   anodes 409–10, 411
   hydroxide 410
   inhibition 411–2
   metal/air batteries 382

   oxidation 409–10
   oxide layers 410
   oxygen systems 409–10
America *see* United States of America
ammonia 202–3
anaerobic digestion gas 291
ancillary systems evolution 296–7
anodes
   Baur, E. 164
   catalysts 299, 283
   enzymatic biofuel cells 364–9
   exhaust 331–2
   flavoenzyme-functional electrodes 369
   gas diffusion 77
   materials 237–8, 244, 250, 260
   microbial biofuel cells 357–8, *359*
   performance improvements 308–14
   reactions 302
   solid oxide fuel cells 338–9
   supported electrolyte cells 346–7
apoenzymes 369
Apollo Energy Systems Inc. (AESI) *196, 206,* 274–6, 279
APU *see* auxiliary power units
aqueous carbonate electrolytes 301–4
Argonne National Laboratory 253
assemblies, membrane-electrode 105–11, 207–8, 305–7
assembly, solid oxide fuel cells 341
atmospheric power generation 129
Austin A-40 271–2, *273*
Austria 189, 278–9
automobiles
   air-cooled cells 138–40
   Germany 187
   solid oxide fuel cells 352–3
   see also vehicles
auxiliary electrodes 392
auxiliary power units (APU) 352–3
aviation 187

Bacon, F. T. 168–71, 179–81, 194–5, 269–70
Ballard Power Systems 141
balloons 151–2
banned solid oxide fuel cells 225–7, 252
base reactions 231–4
Battelle Institut 183
batteries 148, 154–5, 382–405, 412–4
Baur, E. 163–4, 167
BBC/ABB 185
Becquerel, E. 160
Belgium 189–90
Bell and Spigot 225–7, 252
Berl, W. G. 167

Berlzelius, J. J. 155–6
bicarbonate electrolytes 205, 302–4
bifunctional oxygen electrodes 388–94
bimetallic electrocatalyst 329–30
binary diffusion coefficient 87
biochemical fuel cells 355–79
bioelectrocatalytic cathodes 370–3
bioelectrocatalyzed oxidation 364–9
bipolar plates 70–2, 88–93, 244–5
bonded electrodes 298–9
Bonn University 182
Bosch, R. 185
Boyle, R. 150–1
Braunschweig Technical University 182
Brazil 192
Bremen tests 402
Britain 179–80
Bulgaria 190
Butler–Volmer equation 34–5, 80

cadmium 382
Canada 192–3
carbon dioxide separation 262
carbon electrodes 167–8
carbon monoxide 295, 307–14, 335–7
carbon plate electrodes 270–1
carbon-containing fuel performance 302–4
carbonate electrolytes 301–4
Carnot cycles 26–30
Carnot factor 29
Carnot, S. 146
cars *see* automobiles; vehicles
catalysis discovery 155–6
catalysts
    air electrodes 397
    direct methanol fuel cells 314–7
    electrically rechargeable zinc/air batteries 389–91
    hydrazine fuel cells 282–5
    oxygen reduction 314–7
cathodes
    Baur, E. 164–7
    catalysts 298–9
    cathodic potential 325–6
    diffusion 77–8
    dissolution 244
    enzymatic biofuel cells 369–73
    exhaust 317
    high temperature fuel cells 250
    mass transfer simulations 62–6
    materials 237–8, 244, 250
    reactions 303
    solid oxide fuel cells 339–40
    thermodynamic stability 237–8
cations 8
Cavendish, H. 150–1
cell...
    components 337–40
    contamination 299
    decay 299–300
    designs 305–7, 340–8
    mass transfer 241
    performance 297–8
    potential 48, 50
    reaction efficiencies 26–30
    stacks 111–2, 293–5
    thermodynamics 21–5
    voltage 23–4, 27–8, 409–10
ceramic solid oxide fuel cells 117–8
CFD *see* computational fluid dynamic
charge/discharge layers 392–4
charging mode 396
Charles, J. A. C. 152
chemical...
    diffusion 240
    energy conversions 28
    potential gradients 240
    reactions 3–5, 26
    reactivity 249
ChemTek 402–3, 404
Chevrel phases 316–7
China 193
Chloride Electric Storage Company 179–80, 281, 283
chloride ions 410
chlorine containing electrolytes 411–2
chlorine/hydrogen cells 5–6
chromium poisoning 258–9
circulating electrolytes 269
coal 29–30, 146, 147, 160
coefficients
    binary diffusion 87
    drag 61, 87, 109–10
    effective mass transfer 91–2
    electroosmotic drag 61, 87, 109–10
    heat transfer 103–5, 121, 127
    mass transfer 91–2
    net water transfer 87
    water transfer 87
cogeneration 127–31, 347
combined heat/power applications 287–8
combustion processes 323
components
    electrochemical cells 3–12
    phosphoric acid electrolyte fuel cells 292–6
compression 424–6
computational fluid dynamic (CFD) software 83–4
concentrated solution theory 60
conductivity
    air-cooled cells 136–7
    conduction 100–1, 115
    ionic 3, 8–12
    ionomeric polymers 10–11
    low temperature fuel cells 101–2
    superionic conductors 11
configurations, direct methanol fuel cells 305–7
conservation laws 55, 85, 146
continuous exchange batteries 413
convection
    high temperature fuel cells 115, 121
    low temperature fuel cells 100, 101–3
    mass transfer 240
conversion efficiency 220–1, 248–9, 261–2
corrosion
    aluminum/air cells 411
    electrically rechargeable zinc/air batteries 386–7

Subject Index **441**

high temperature fuel cells   223, 244–6
phosphoric acid electrolyte fuel cells   294, 299–300
cost analysis   256–7, 349–51, 429–30
COx *see* Cytochrome oxidase
creep   223
current
 aqueous carbonate electrolytes   303
 collectors   244–5
 density
  electrochemical kinetics   33–5
  high temperature fuel cells   117–8, 124–7
  mass transfer   50–1, 65–6, 87
  porous electrode theory   209
 electrochemical kinetics   31
 mass transfer   86–7, 91
 per area   8
 potential characteristics   375–9
 voltage characteristics   4–5, 7, 375–9
cyclic voltammograms   369–73
Cytochrome c (Cyt c)   377–9
Cytochrome oxidase (COx)   377–9

Daimler-Benz   187
Daniell, J. F.   154–5
Darcy's law   56, 57, 84
Darmstadt Technical University   186
Davtyan, O. K.   172–3
Davy, Sir H.   156
DECHMA *see* The Deutsche Gellschaft Für Chemische Technik and Biotechnologie
decomposers   281–3
degradation stability   221
Denmark   190
deposition   241, 312–3, 385–6
description conventions, electrochemical cells   24–5
The Deutsche Gellschaft Für Chemische Technik and Biotechnologie (Society for Chemical and Biotechnology DECHMA)   186
The Deutsches Zentrum fürLuft- und Raumfahrt (DLR)   187
development
 history   219–63
 research, 1950's and 1960's   179–94
Dewar, J.   152
diffuse electrode layers   15–18
diffusion
 binary coefficient   87
 electrodes   392
 gas phase transport   55–6
 layers   300
 low temperature fuel cells   101
 mass flux   87
 mass transfer   73–88
digester gas   291
dilute solution theory   58
dioxygen reduction   371–3
direct combustion fuel cells (DCFC)   323, 330–2
direct methanol fuel cells (DMFC)
 aqueous carbonate electrolytes   302–4
 cell designs   305–7
 configurations   305–7
 heat transfer   101, 106–13
 kinetics   42–3

mass transfer   72, 88–93
 principles   305–19
 R&D   199–201
 thermodynamics   42–3
direct-alcohol fuel cells (DAFC)   323–33
discharge layers   392–4
discharge products   387–8, 395–403
dissolution   241, 386
dissolved discharge products   398–9
dissolved fuels   201–2
distributed generators   289–91
DLE *see* The Deutsches Zentrum fürLuft- und Raumfahrt
DMFC *see* direct methanol fuel cells
Döberiener, J. W.   149–50, 155–6
Doppel Skelett Katalysator Electrodes *see* double skeleton catalyst
Dornier systems   187, 277–8
double layer electrode–electrolyte interfaces   13–14
double skeleton catalyst (DSK) electrodes   173–4, 177–8
drag coefficient   61, 87, 109–10
Dresden Technical University   185
DSK *see* double skeleton catalyst

EAC *see* Electric Auto Corporation
early paper review   145–56
effective efficiency   26–30
effective energy density   410–1
effective mass transfer coefficients   91–2
efficiency
 Carnot cycles   26–30
 cell reactions   26–30
 factor   209
 high temperature fuel cells   248–9
 solid oxide fuel cell systems   349–51
electric...
 energy conversions   28
 vehicles   147–8
 work   26
Electric Auto Corporation (EAC)   279
Electric Fuel Limited (EFL)   387, 403
Electric Power Research Institute (EPRI)   222
electrical...
 compressors   424
 current   31
 energy   148
 potential gradient   240
 potentials   31
 power   426–7
electrically rechargeable batteries   385–94
electrified interfaces   13–14, 21–5
electrocatalysis
 activity   323, 331
 biofuel cells   355–79
 introduction   36–43
 oxidation   324–30
 oxygenated organic compound electrooxidation   329–30
electrochemical...
 active metabolites   357–8, *359*
 cells   3–12, 24–5
 compressors   425
 oxidation   336
 potentials   21–5, 336
 reactions

electrochemical... (continued)
    current density   33–5
    electrode kinetics   53–4
    high temperature fuel cells   119–20
    kinetics   31–5, 53–4
electrochemistry
  historical overview   152–5
  methanol electrooxidation   311
electrodes
  batteries   392, 399–401
  Baur, E.   164
  double skeleton   173–4, 177–8
  electrolyte interfaces   13–20
  exchange current density   33–5
  kinetics   53–4, 326
  materials   329–30
  Nernst equation   22
  pasted zinc   393–4
  phosphoric acid electrolyte fuel cells   298–9
  potentials   13–15, 23–4
  rechargeable batteries   392
  redox   22, 358–63
  research discoveries   168
  reversible zinc   387–8
  shape change prevention   396–7
  solid oxide fuel cells   337
  solution interfaces   6–8
  thermodynamics   21–5
  zinc   382–405
electrokinetic phenomena   59–60
electrolysis   3–4, 416–30
electrolytes
  alkaline fuel cells   268
  aluminum oxidation   409–10
  aluminum/air cells   412–4
  circulating   269
  electrode interfaces   13–20
  high temperature fuel cells   243, 248
  losses   223–4, 246
  low temperature fuel cells   205
  materials   249–50, 259–60
  melts   12
  migration   299
  redistribution   223
  seawater aluminum/air cells   409–14
  solid oxide fuel cells   337, 344–7
  zinc/air batteries   391, 395–6
electrolytic phase transport   57–62
electromagnetic radiation   115
electromotive force (EMF)   15–20, 23–4
electroneutrality   55, 58
electrons
  carriers   358–63
  relays   358–63
  transfer   3–5, 32–3, 356–79
  tunneling   32–3
electroorganic chemistry   355–79
electroosmotic drag coefficient   61, 87, 109–10
electrooxidation   308–14, 324–30
electrostatics   146
Electrovan   271–2
Elenco fuel cell system   276

energy
  balances   99–107, 121, 125, 428–9
  conservation   145–6
  conversions   28, 145–6, 220–1
  density   410–1
  efficiency diagrams   29–30
  management systems   427–8
  sources   146–7
  storage systems   416–30
  supply   29–30
  traction source effective efficiencies   29–30
enthalpy
  fuel cell reactions   27–8
  high temperature fuel cells   116–7, 121, 232
  low temperature fuel cells   99, 106
  solvation   8
entropy   27–8, 106–7
environment, direct combustion fuel cells   324
enzymatic biofuel cells   363–79
enzyme co-factors   364
enzyme-catalyzed reactions   364–73
enzyme-electrodes   373–9
equations of momentum transfer   83
equilibria   230–42
ethanol   327–8, 330–2
ethers   332–3
ethylene glycol   330
Euler, K.-J   158, 160
EUREKA Bus Project   276
European development   189–92, 224–5, 228
European Space Agency   277–8
European space shuttle HERMES   276–8
evaporation   241, 294
EVonyx   404
exchange cartridges   401–3
exchange current density   33–5
external regeneration   398–403

fabrication
  deposition   241, 312–3, 385–6
  mass transfer   241
  solid oxide fuel cell components   341
FAD see flavin adenine dinucleotide
Faraday, M.   153
Fick's law   50, 55–6
flat plate designs   343–7
flavin adenine dinucleotide (FAD)   364, 369
flavoenzyme-functional electrodes   369
flow
  fields   70–94
  pressure driven   56–7, 60–1
flowsheet variation   129
fluid dynamics   72–88
fluid temperatures   103–5, 116, 127
fluid velocity   101–13
formaldehyde   332–3
Forschungszentrum Jülich (FZJ)   187
France   190–1
Franck–Condon principle   32–3
Fraunhofer Gesellschaft   187
Fraunhofer Initiative   141
friction   78–81

fuel cell history
    early papers 145–56
    R&D 157–210
fuel flexibility 262
fuel processing 230–42, 295, 348–9
functional biochemical designs 355–79

Galvani, A. 152–3
Galvani potential 15–20, 21–5
galvanic cells 5–6, 36–43
gas...
    crossover 223
    diffusion 74, 77, 208–9
    diffusion electrodes 208–9
    electrodes 22–3, 208–9
    evolution 92–3
    flow 70–94
    phase transport 55–7
    storage systems 423–6
    streams 76–7
gaseous fuels 202–5
gaseous voltaic batteries 158–60
Gautherot, N. 156–7
GE see General Electric Co
Gemini Program 206, 274–5
General Electric Co (GE) 198–9, 204, 207–8
German Center for Aviation and Space Exploration 187
German R&D groups 181–7
Gibbs energy change 325–6
Gibbs Free Enthalpy 336
Gläßner, A. 163
Global Thermoelectric 346
glucose oxidase (GOx) 369, 375–9
glyoxal 332–3
von Goethe, J. W. 152, 155–6
Goüy–Chapman theory 16–17
GOx see glucose oxidase
Greek symbols
    heat transfer 132
    mass transfer 68, 94
Greifswald University 185
Grove, Sir W. R.
    air electrodes 165
    batteries 155
    discoveries 157–62
    low temperature fuel cells 146
    R&D 195–6, 203–5
Grubb, W. T. 198–9, 207
Grünegerg, G. 283

Haber, F. 163
heat...
    capacity 99–113
    conduction 100–1
    conductivity 136–7
    convection 100, 101–3
    energy conversions 28
    engines 28–9
    flux 136–7
    management 108–13
    power applications 287–8

    removal 127–31
    sources 116–7
    transfer 99–113, 115–32, 134–42
    transport 70–94
Heise, G. W. 167
Helmholtz planes 13, 16–20
HERMES 276–8
high operating temperature electrolysis (HOT EllY) 228
high temperature fuel cells (HTFC)
    development history 219–63
    heat transfer 115–32
    mass transfer 88
    materials science 229–30
    superionic conductivity 11
high-pressure electrolysis 424–5
Hindenburg 152
historical overview
    early paper review 145–56
    high temperature fuel cells 219–63
    R&D 157–210
Hitachi 284
HOECHST, BASF AG 186
Hoechst falling film fuel cell system 278
HOR see hydrogen oxidation reactions
HTFC see high temperature fuel cells
humidified gas streams 76–7
hybrid energy storage systems 423
hydrated aldehydes 333
hydrated glyoxal 333
hydrated metal cations 8
hydrazine fuel cells 281–5
hydrocarbons 203–5, 336–7
hydrodynamics 70–94
hydrogen...
    alkaline fuel cells 37
    based energy storage systems 416–7
    compression 425–6
    diffusion electrodes 165–7
    discovery 150–2
    energy storage systems 416–7
    German R&D 187
    microbial biofuel cells 357
    oxidation reactions (HOR) 50, 53–4
    oxygen cells 5–6, 267–79
    oxygen chains 156–7
    solid oxide fuel cells 335–7
    transport 85
hydrogen peroxide 283–4
hydrophilic diffusion electrodes 167

ideal efficiencies 26–30
IEM see ion exchange membranes
IGBT see insulated gate bipolar transistors
India 193–4
industrial zinc/air batteries 403–4
infrared (IR) spectroscopy 311–2
inner electrode layers 17–18
inner potential see Galvani potential
inorganic substances 202
Institute of Gas Technology (IGT) 221–2
insulated gate bipolar transistors (IGBT) 296
integrated heat exchangers 123–4

interconnects
    high temperature fuel cells 228–9, 257–9
    materials 238–40, 250–1, 260
    solid oxide fuel cells 341
interdigitated flow fields 71–2
interfaces
    double layer 13–14
    electrified 13–14, 21–5
    electrode-electrolyte 13–20
    mass transfer 47–68
    phase 47–68
    solution 6–8
interfacial statistical thermodynamics 15–18
invariance 301–2
inverters 296
ions
    chloride 410
    conductivity 3, 8–12
    exchange membranes (IEM) 207–8
    ionomeric polymer conductivity 10–11
    melts 11–12
    migration 12
    solutions 6–8, 9–10
IR see infrared
iron 382
Italy 191

Jacques, W. W. 162–3, 180
Janlochkoff, P. 160
Japan 187–9, 224, 228–9, 290–1
Jet Propulsion Laboratory (JPL) 319
Joule loss constraints 259
JPL see Jet Propulsion Laboratory
Junger, E. W. 165
Justi, E. W. L. 173–4, 194

kinetics
    Carnot cycles 26–30
    cell reaction efficiencies 26–30
    electrochemical cells 3–12
    electrochemical reactions 31–5
    electrode–electrolyte interfaces 13–20
    electrooxidation oxygenated organic compounds 326
    fuel cell types 36–43
    zinc/air batteries 384
Knudsen diffusion 56, 57
Kordesch, K. V.
    alkaline fuel cells 271–2
    hydrazine fuel cell motorcycle 282
    R&D 194, 202–3
    research discoveries 174–6

lactate dehydrogenase 364
landfills 291
Langer, C. 161, 179, 196
lateral heat conductivity 136–7
Lavoisier, A-L. 149, 150–2
layered enzyme-electrodes 373–9
LBST see Ludwig Bölkow Systemtechnik
ligand–metal separation 33

lighting research history 148
Linde Group 186–7
line-commutated inverters 296
liquid feed direct methanol fuel cells 91–3
liquid phase transport 57
liquid-cooled cells 134–5
load efficiencies 28
local current density 87
local equilibrium approximation 240
local membrane conductivity 87
local overpotentials 87
localized structures 8
Lomonossow, M. V. 149
low power air-cooled cells 140–2
low temperature fuel cells (LTFC) 99–113, 145–210
Ludwig Bölkow Systemtechnik (LBST) 186
LURGI 186–7

McMillan–Mayer theory 7, 16
magnesium 382
magnetism 146
Mannesmann 186–7
marine applications 412–4
mass...
    conservation 85
    flux 87
    transfer 47–68, 70–94, 230–42, 260
materials
    balances 54–5, 79
    cathodes 237–8, 244, 250
    corrosion 294, 299–300
    electrodes 329–30
    electrolytes 249–50, 259–60
    future trends 261–3
    high temperature fuel cells 219, 229–45, 249–51
    interconnects 238–40, 250–1, 260
    transfer 230–42
mathematical mass transfer modeling 47–68
matrix materials 243–4
Matsushita electric industrial company 284
Matteucci, C. 157
maximum obtainable work 26, 28–9
Mayer, J. R. 146
de Mayerne, T. 150–1
MCFC see molten carbonate fuel cells
MEA see membrane and electrode assemblies
measurements
    cell voltages 23–4
    electrode potentials 23–4
mechanical compressors 424–5
mechanically rechargeable batteries 384–5, 394–403, 412–3
membrane conductivity 87
membrane and electrode assemblies (MEA) 105–11, 207–8, 305–7
membrane phase transport 57–62
Mercury Program 274–5
Messer-Griesheim 186–7
metal...
    air cells 382–405, 409–11
    hydrides 425–6
    interconnects 228–9, 239–40, 250–1, 257–9
metal–vacuum boundaries 14

metallic electrodes   168
metallic oxide catalysts   314–7
Metallic Power   403–4
methane   336
methanol
  aqueous carbonate electrolytes   302–4
  cathode exhaust   317
  electrooxidation   308–14, 326
  kinetics   42–3, 326
  oxidation mechanism   308–14
  thermodynamics   42–3
  tolerant oxygen reduction   316–7
  vapor-feed fuel cells   307–8
  see also direct methanol...
MHI *see* Mitsubishi Heavy Industries
Michigan Technological University (MTU)   187
micro fuel cells (MFC)   323
micro-gas turbine generator (MTG)   227
micro-turbine generators   131–2
microbial biofuel cells   356–63
microperoxidase-11 (MP-11)   374–7
military applications   205, 270, 291–2
Mitsubishi Heavy Industries (MHI)   346–7
mixed alkali carbonate melts   237
mobile air-cooled cells   138–40
mobile solid oxide fuel cells   352–3
mobilities   9–10
modeling   47–68, 81–8
molar conductivity   9–10
molar free energy   26
MOLB *see* mono-layer-block
molecular dynamic method   7, 18–20
molecule interactions, electrode–electrolyte interfaces   17–18
molten carbonate electrolytes   234–6
molten carbonate fuel cells (MCFC)
  development history   219–63
  heat transfer   117–8
  Japan   189
  kinetics   38–40
  mass transfer   71
  R&D   180, 189
  thermodynamics   38–40
molten salts   11–12
momentum transfer   82–8
Mond, L.   161, 179, 196
mono-layer-block (MOLB) cells   347
monoalcohols   327–8
monolayer-functional electrodes   369, 374–9
monolithic stacks   253
monopolar alkaline fuel cells   276
Monte Carlo method   7, 18–20
Moonlight Program   188–9
motorcycles   282
Motorola   319
MP-11 *see* microperoxidase-11
MTU *see* Michigan Technological University
Muchot, A.   152
multielectron transfer reactions   326–9

Nafion   10–11, 73, 306, 314–8
NASA Apollo fuel cell system   274–6
natural fuel cells types   36–7

Nernst equation   22–3
Nernst, W.   163, 196
net overall current density   209
net water transfer coefficient   87
The Netherlands   138, 191–2
neutral electrolytes   409–10
NHE *see* normal hydrogen electrodes
nickel catalysts   282–5
nickel oxide   223–4, 232–3
nicotinamide redox cofactors   364–9, 374–5
Niederreither, J.   168
noble metals   389
normal hydrogen electrodes (NHE)   54
NovArs, Germany   *141*
numerical mass transfer simulations   62–6
Nyberg,   167

off gas treatments   349
ohmic limitations   63–6
Ohm's law   53
oil   147
on-site power plants cells   288–9
operation, biochemical fuel cells   355–79
organic compound electrooxidation   324–30
organic dissolved fuels   201–2
organic fuels   301–2
ORR *see* oxygen reduction reactions
Ostwald, W.   161–2
overpotentials   31, 87, 240
oxidation   308–14, 324–30
oxides
  aluminum   410
  catalysts   314–7
  electrically rechargeable zinc/air batteries   389
  interconnect materials   239, 250
  nickel   223–4, 232–3
  solubility   243–5
  see also solid...
oxygen...
  diffusion electrodes   392
  discovery   150–2
  electrodes   178, 382, 388–92
  feed   75–6
  hydrogen (air) fuel cells   267–79
  hydrogen chain investigation   156–7
  mass transfer limitations   63–6
  potential gradient, mass transfer   240
  reduction reactions (ORR)
    aqueous carbonate electrolytes   303
    Butler–Volmer equation   80
    electrocatalysts   314–7
    mass transfer   49, 53–4
  solid-oxide fuel cells   40–1
  transport   85
oxygenated organic compounds   324–30
oxygenated organic fuels   332–3

PAFC *see* phosphoric acid fuel cells
parallel channel ribbed flow field plates   70–1
parasitic reactions   246–7
partial stabilized zirconia   259–60

partially humidified gas streams 76–7
particle size 240
pasted electrodes 393–4, 399–401
Peclet numbers 106
PEFC *see* polymer electrolyte fuel cells
PEM *see* polymer electrolyte membranes; proton exchange membranes
PEMFC *see* polymer electrode membrane fuel cells
perfluorosulfonic acid membranes 318
performance
    anodes 308–14
    carbon-containing fuels 302–4
    cell 297–8
    direct methanol fuel cells 318–9
periphery components 348–9
perovskites 339–40, 389–90
peroxide reduction 370–1
phase diagrams 233
phase flow 86–8
phase interfaces 47–68
phosphoric acid fuel cells (PAFC)
    kinetics 37–8
    mass transfer 70–1
    principles 287–300
    R&D 197–8
    thermodynamics 37–8
Photon fuel cell system 277
physical chemistry roots 148–50
Pipeline 351
piston compressors 424
planar flat electrolyte supported cells 344–6
planar solid oxide fuel cells 120–3, 129
planar stacks 257–9
Planté, G. 155
plates
    bipolar 70–2, 88–93, 244–5
    carbon electrodes 270–1
    flat designs 343–7
platinum
    alloys 298–9
    anodes 308–14
    catalysts 298–9, 284–5
    cathodes 298–9
    methanol electrooxidation 308–14
    oxygen reduction electrocatalysts 315
    oxygenated organic compound electrooxidation 329
    ruthenium anodes 299, 313–4
    underpotential deposition 312–3
pneumatically driven piston compressors 424
polarization
    curves 49–51, 63–4, 271–3
    mass transfer 49–51, 63–4, 80–1
polymer electrode membrane fuel cells (PEMFC) 106–13
polymer electrolyte fuel cells (PEFC) 198–9, 423, 425
polymer electrolyte membranes (PEM) 47–68, 71–88, 304, 421–2
polymeric perfluorosulfonic electrolytes 10–11
polyols 328–9
polytetrafluoroethylene (PTFE) 207–8, 306
porous baked carbon plate electrodes 270–1
porous electrode theory 51–5, 208–9
porous media diffusion 74, 78–80

portable air-cooled cell applications 140–2
potential difference 336
potential energy 14, 33
potentials, electron relays 358–63
powder paste electrodes 399–401
power conditioning 295–6
power density 331
power plants 288–91
Powercell Corporation 404
PQQ *see* pyrroloquiniline quinone
pressure driven flow 56–7, 60–1
pressure drops 139–40
pressure electrolysis 424–5
Priestley, J. 150
process analysis 428–30
process engineering 426–30
process parameter variations 129
propanol 325–6, 328, 330–2
proton exchange membrane fuel cell (PEMFC), R&D 180
proton exchange membranes (PEM) 47–68, 134–42
PTFE *see* polytetrafluoroethylene
pyrroloquiniline quinone (PQQ) 364–9, 374–5

radiation 100, 115–6
Raney nickel 173–4, 282–5
reactant distribution 294
reaction enthalpy 27–8
reaction entropy 27–8, 106–7
reaction rate distribution 303
rechargeable batteries 384–403, 412–4
redox
    cells 163
    cofactors 364–9, 374–5
    electrodes 22
    ligand–metal separation 33
    potential energy 33
    potentials 358–63
    thermodynamics 231–4
reformers 241–2, 295, 350–1
reorganization energy 33
research and development (R&D)
    direct methanol fuel cells 199–201
    dissolved fuels 201–2
    Europe 189–92
    fuel cell history 157–210
    gaseous fuels 202–5
    Germany 181–2
    Japan 187–9
    low temperature fuel cells 194–205
    phosphoric acid fuel cells 197–8
    polymer electrolyte fuel cells 198–9
researchers 156–62
reversible zinc electrodes 387–8
Ritter, J. W. 153, 156, 157–60
de la Rive, A. A. 157
Rolls Royce Aerospace, UK. 346
Roman symbols, mass transfer 67–8
Ruhrchemei 182
Russia 180–1, 277

saline electrolytes 412–4
salts, ionic conductivity 11–12
Schaeffer, G. J. 179, 196–7
Scharf, P. 161
Scheele, C. W. 150
Schmid, A. 165–6
Schoenbein, C. F. 156–60
Schumacher, E. A. 167
SCR see silicon controlled rectifiers
seal-less tubular stacks 225–7, 251–3
sealing 246
seawater cells 409–14
secondary batteries 148
segmented-cell-in-series solid oxide fuel cells 225–7, 257
self supported electrolyte cells (SSE) 344–6
self-commutated inverters 296
separators 391
shape factor cells 115–6
Shell 179
shorting 246–7
Siemens
   air-cooled cells 138–40
   alkaline fuel cells 276–7
   direct methanol fuel cells 318–9
   hydrazine fuel cells 284–5
   R&D 183–5
   stationary solid oxide fuel cells 351
silicon controlled rectifiers (SCR) 296
simulation techniques 18–20
single chamber solid oxide fuel cells 347
single phase energy balances 105–7
sintering 223, 253–6
slurry electrodes 395
SOC see state of battery charge
Society for Chemical and Biotechnology 186
von Soemmerring, S. T. 152
SOFC see solid oxide fuel cells
software 83–4
solar energy 187
solar-produced hydrogen 417
solid...
   electrolytes 236–7
   ionic conductors 11
   low temperature fuel cells 105
   oxide fuel cells (SOFC)
      development history 219–63
      heat transfer 115–32
      Japan 189
      kinetics 40–1
      mass transfer 71
      principles 335–53
      segmented-cell-in-series 225–7, 257
      thermodynamics 40–1
      wet/sintering processes 253–6
      see also tubular
   phase conductivity 53
   polymer electrolytes (SPE) 10–11, 41–2, 305–8
   temperature distribution 121–3
solubility 243–5, 302–4
soluble discharge products 388
soluble zinc electrodes 395–6
solute-solvent interactions 7–8

solution interfaces 6–8
space
   Germany 187
   low temperature fuel cells 205
   shuttle orbiter 275–6
SPE see solid polymer electrolytes
species diffusion mass flux 87
species transport 73–4
specific adsorption 17–18
SSE see self supported electrolyte cells
stability 230–42
stabilization energy 231–2
stabilized zirconia 259–60
stacks
   air-cooled cells 135–8
   efficiency 130–1
   high temperature fuel cells 260–1
   phosphoric acid electrolyte fuel cells 293–5
   seal-less tubular 225–7, 251–3
   solid oxide fuel cells 251
standard Gibbs energy change 325–6
state of battery charge (SOC) 425–8
static electrolytes 268
stationary air-cooled cell applications 140–2
stationary phosphoric acid electrolyte fuel cells 287–91, 295–6
stationary solid oxide fuel cells 351–2
steam-driven vehicles 147
Stefan–Maxwell equations 55–6, 57, 60
structures, hydrated metal cations 8
sub-sea batteries 413–4
subscript symbols, mass transfer 68, 93–4
substrates 389–91
Südchemei 186–7
sulfide catalysts 316–7
sulfonic acid 197
Sulzer HEXIS-cells 345–6, 351
superionic conductivity 11
superscript symbols, mass transfer 68, 93–4
supported electrolyte cells 344–7
Sweden 191, 278
Switzerland 138
symbols
   heat transfer 131–2
   mass transfer 67–8, 93–4
systems
   efficiency 428–9
   electrolysis/fuel cell analysis 428–30
   high temperature fuel cells 261–3
   phosphoric acid electrolyte fuels 292–3, 296–7
   solid oxide fuel cells 348–53

tactical military applications 291–2
Tafel slopes 298–9
TBP see three-phase boundary
technical universities 182, 185, 186
technology revolutions 297–300
Teflon bonded electrodes 298–9
temperature...
   air-cooled cell profiles 136
   distributions 117–8, 124–8
   gradients 240

temperature... (continued)
　　mass transfer　78–81
　　see also high...; low...
terminal velocity　8
ternary oxide catalysts　316–7
ternary sulfide catalysts　316–7
TFMSA see trifluoromethane sulfonic acid
thermal...
　　conductivity　101–2
　　cycling　223
　　diffusivity　101
　　hydrogen compression　425–6
　　management　296–7, 349
thermodynamics
　　acid reactions　231–4
　　aluminum/air cells　410–1
　　Carnot cycles　26–30
　　cell reaction efficiencies　26–30
　　cells　3–12, 21–5, 237–8
　　efficiencies　26–30
　　electrified interfaces　21–5
　　electrochemical cells　3–12
　　electrodes　13–20, 21–5, 237–8
　　electrolyte interfaces　13–20
　　electrooxidation　325–6
　　fuel cell types　36–43
　　metal/air cells　410–1
　　molten carbonate electrolytes　234–6
　　redox reactions　231–4
　　solid electrolytes　236–7
　　solid oxide fuel cells　335–7
　　stability　237–8
　　zinc/air batteries　384
three phase interfaces　47–68
three-dimensional mass transfer models　82–8
three-phase boundary (TPB) regions　337–8
thyristors　296
Tobler, J.　167
traction source effective efficiencies　29–30
transfer
　　coefficients　87
　　heat　99–113, 115–32, 134–42
　　mass　47–68, 70–94, 230–42, 260
　　materials　230–42
　　momentum　82–8
　　multielectron reactions　326–9
　　water　87
transport
　　electrolytic phase　57–62
　　gas phase　55–7
　　heat　70–94
　　hydrogen　85
　　liquid phase　57
　　mass transfer　52–3, 61
　　membrane phase　57–62
　　modes　47–68
　　oxygen　85
　　polymer electrolyte membrane cells　73–4
　　species　73–4
　　uncharged species　62
　　water　59, 70–94

transportation see automobiles; vehicles
treatment, waste-water plants　291
trifluoromethane sulfonic acid (TFMSA)　197
trimetallic electrocatalyst　329–30
tubular solid oxide fuel cells
　　heat transfer　115, 123–7, 129
　　high temperature fuel cell development　229
　　new challenges　256–7
　　principles　342–3
　　seal-less　225–7, 251–3
　　Westing Power Corporation　251–3
tubular zinc/air battery cells　400
two phase flow　86–8
two phase interfaces　47–68
two-dimensional mass transfer models　81–2

UCC see Union Carbide Corporation
UK see United Kingdom
uncharged species transport　62
underpotential deposition (UPD)　312–3
undissolved discharge products　401–3
Union Carbide Corporation (UCC)　194–5, 270–4, 283
United Kingdom (UK)　179–80
United States of America (USA)
　　Department of Energy　222
　　R&D groups　180
　　US Air Force　270
　　US Army　291–2
United Technology Corporation　222
universities　182, 185, 186
UPD see underpotential deposition
USA see United States of America
USSR　180–1, 277

valence stability　231–2
vapor feed direct methanol fuel cells　89–91
VARTA　163, 182–3
Varta Eloflux system　275
vehicles
　　electric　147–8
　　fuel research history　147–8
　　phosphoric acid electrolyte fuel cells　292
　　see also automobiles
velocity, terminal　8
Vielstich, W.　176–8
Volta, A.　153, 157–60
Volta potential　15–20
voltage...
　　cells　23–4, 27–8, 409–10
　　current characteristics　4–5, 7, 375–9
　　electromotive force　15–20, 23–4
voltaic batteries　158–60
Volta's pile　153–5
volume expansion mismatching　249

Warltire, J.　151
waste-water treatment plants　291
water
　　activity　87
　　balance　66–7

coefficients 87
dipoles 18–20
electrolysis 419–22, 423
fuel research history 146
management 108–13, 296–7, 304
molecule interactions 17–18
transfer coefficient 87
transport 59, 70–94
treatment plants 291
Westing Power Corporation (WHPC) 219–20, 225–9, 251–3, 342–3, 351
Westphal, C. 161, 179
wet-seals 223
wet/sintering processes 253–6
wind 146
Winsel, A. 173–4

wood 146
work 26, 28–9

yttria-stabilized zirconia (YSZ) 237–8, 337–53
Yuasa 285

Zentrum für Sonnenenergie und Wasserstoffforschung (ZSW) 187
Zinc Air Power Corporation 404
zinc electrodes 382–405
zinc/air systems 209–10, 382–405
zirconia
    electrolytes 225
    yttria-stabilized 237–8, 337–53
ZSW see Zentrum für Sonnenenergie und Wasserstoffforschung